Lecture Notes in Mathematics 2254

More information about this subseries at http://www.springer.com/series/3114

Fondazione C.I.M.E., Firenze

C.I.M.E. stands for *Centro Internazionale Matematico Estivo*, that is, International Mathematical Summer Centre. Conceived in the early fifties, it was born in 1954 in Florence, Italy, and welcomed by the world mathematical community: it continues successfully, year for year, to this day.

Many mathematicians from all over the world have been involved in a way or another in C.I.M.E.'s activities over the years. The main purpose and mode of functioning of the Centre may be summarised as follows: every year, during the summer, sessions on different themes from pure and applied mathematics are offered by application to mathematicians from all countries. A Session is generally based on three or four main courses given by specialists of international renown, plus a certain number of seminars, and is held in an attractive rural location in Italy.

The aim of a C.I.M.E. session is to bring to the attention of younger researchers the origins, development, and perspectives of some very active branch of mathematical research. The topics of the courses are generally of international resonance. The full immersion atmosphere of the courses and the daily exchange among participants are thus an initiation to international collaboration in mathematical research.

C.I.M.E. Director (2002 – 2014)
Pietro Zecca
Dipartimento di Energetica "S. Stecco"
Università di Firenze
Via S. Marta, 3
50139 Florence
Italy
e-mail: zecca@unifi.it

C.I.M.E. Director (2015 –)
Elvira Mascolo
Dipartimento di Matematica "U. Dini"
Università di Firenze
viale G.B. Morgagni 67/A
50134 Florence
Italy
e-mail: mascolo@math.unifi.it

C.I.M.E. Secretary
Paolo Salani
Dipartimento di Matematica "U. Dini"
Università di Firenze
viale G.B. Morgagni 67/A
50134 Florence
Italy
e-mail: salani@math.unifi.it

CIME activity is carried out with the collaboration and financial support of INdAM (Istituto Nazionale di Alta Matematica)

For more information see CIME's homepage: http://www.cime.unifi.it

Matthias Hieber • James C. Robinson •
Yoshihiro Shibata

Mathematical Analysis of the Navier-Stokes Equations

Cetraro, Italy 2017

Giovanni P. Galdi • Yoshihiro Shibata

Editors

 Springer

FONDAZIONE
CIME
ROBERTO CONTI
CENTRO INTERNAZIONALE MATEMATICO ESTIVO
INTERNATIONAL MATHEMATICAL SUMMER CENTER

Authors
Matthias Hieber
Department (FB) of Mathematics
Technische Universität Darmstadt
Darmstadt, Germany

James C. Robinson
Mathematics Institute
University of Warwick
Coventry, UK

Yoshihiro Shibata
Department of Mathematics
Waseda University
Tokyo, Japan

Editors
Giovanni P. Galdi
MEMS Department
University of Pittsburgh
Pittsburgh, PA, USA

Yoshihiro Shibata
Department of Mathematics
Waseda University
Tokyo, Japan

ISSN 0075-8434   ISSN 1617-9692 (electronic)
Lecture Notes in Mathematics
C.I.M.E. Foundation Subseries
ISBN 978-3-030-36225-6   ISBN 978-3-030-36226-3   (eBook)
https://doi.org/10.1007/978-3-030-36226-3

Mathematics Subject Classification (2010): Primary: 35Q30, 76D05, 35Q35; Secondary: 65Mxx, 65Nxx

This Springer imprint is published by the registered company Springer Nature Switzerland AG.
The registered company address is: Gewerbestrasse 11, 6330 Cham, Switzerland

Preface

As is well known, the Navier–Stokes equations constitute one of the most active and attractive areas of research, from both theoretical and applied viewpoints. In particular, especially over the last two decades, they have been the focus of a number of fundamental mathematical contributions from different perspectives. Probably, this rapid growth may be also due to the circumstance that, since the year 2000, the question of existence of global, regular solutions corresponding to initial data of unrestricted size has been declared as one of the main open Millennium Problems by the Clay Mathematical Institute. Yet, in spite of all efforts, to date, the involved mathematicians unanimously agree that very little is still known about these equations and that their deep secrets are still far from being uncovered.

The principal objective of the CIME school on "Mathematical Analysis of the Navier–Stokes Equations: Foundations and Overview of Basic Open Problems," in Cetraro, September 4–8 2017, was to provide series of lectures devoted to several fundamental and diverse aspects of the Navier–Stokes equations. The present volume collects some of them, including boundary layers, fluid–solid interactions, free surface and complex fluid problems (Professor Matthias Hieber, TU Darmstadt, Germany); questions of existence, uniqueness, and regularity (Professor James C. Robinson, University of Warwick, UK); and local and global well-posedness and asymptotic behavior for free boundary problems (Professor Yoshihiro Shibata, Waseda University, Japan).

It is our distinct pleasure to thank all lecturers and participants for their enthusiastic—scientific and social—contribution to the success of the school, the Fondazione CIME and its scientific committee for giving us the opportunity to organize this event, and all the staff for their invaluable help.

Pittsburgh, PA, USA   Giovanni P. Galdi
Tokyo, Japan   Yoshihiro Shibata

Contents

Contents

Chapter 1
Analysis of Viscous Fluid Flows: An Approach by Evolution Equations

Matthias Hieber

Preface

This course of lectures discusses various aspects of viscous fluid flows ranging from boundary layers and fluid structure interaction problems over free boundary value problems and liquid crystal flow to the primitive equations of geophysical flows. We will be mainly interested in strong solutions to the underlying equations and choose as mathematical tool for our investigations the theory of evolution equations. The models considered are mainly represented by semi- or quasilinear parabolic equations and from a modern point of view it is hence natural to investigate the underlying equations by means of the maximal L^p-regularity approach.

For this reason, we start these lectures by an introduction to Cauchy problems and sectorial operators. The latter are the starting point for the functional calculus of bounded, holomorphic functions, which will be then extended to the operator-valued H^∞-calculus. The extension leads us to the notion of $\mathcal{R}$-bounded families of operators, which will be the key for boundedness results of the H^∞-calculus in this setting. It implies the Kalton–Weis theorem on the closedness of the sum of two commuting, sectorial operators and via the extension of Mikhlin's theorem to Banach spaces having the UMD-property, also the characterization theorem of maximal L^p-regularity for parabolic evolution equations in terms of $\mathcal{R}$-boundedness of its resolvent. We then proceed with quasilinear parabolic evolution equations and present three important results for these equations: local well-posedness, the generalized principle of linearized stability implying under suitable assumptions the global existence of strong solutions for data close to an equilibrium point and on the existence of global strong solutions in the presence of compact embeddings and strict Lyapunov functionals.

M. Hieber (✉)
Department (FB) of Mathematics, TU Darmstadt, Darmstadt, Germany
e-mail: hieber@mathematik.tu-darmstadt.de

© Springer Nature Switzerland AG 2020

G. P. Galdi, Y. Shibata (eds.), *Mathematical Analysis of the Navier-Stokes Equations*, Lecture Notes in Mathematics 2254,
https://doi.org/10.1007/978-3-030-36226-3_1

Coming back to the main aims to these notes, well-posedness results for viscous fluid flows, we start the second part with a discussion of balance laws for general heat conducting fluids and deduce from there the fundamental equations of viscous fluid flows, the incompressible and compressible Navier–Stokes equations. We continue with the analysis of the Stokes equation in a half space $\mathbb{R}^n_+$ within the L^p-setting. The associated operator, the negative Stokes operator, is shown to be a sectorial operator on $L^p(\mathbb{R}^n_+)$. As a consequence of the results in Part A we obtain the property of maximal L^p–L^q-regularity for the Stokes equation on $\mathbb{R}^n_+$. A localization procedure yields then the corresponding regularity results for the Stokes equations on standard domains.

We then consider stability questions for Ekman boundary layers. The latter are explicit stationary solutions of the Navier–Stokes equations in the rotational framework. We then show that the Ekman layer is asymptotic stable provided the Reynolds number involved is small enough. We continue with moving and free boundary value problems. In fact, a fluid-rigid body interaction problem will be discussed for the situation of compressible fluids. Maximal regularity of the linearized equation in Lagragian coordinates allows us to prove a local well-posedness result for strong solutions. Strong solutions for the two-phase problem for generalized Newtonian fluids are again obtained by a fixed point argument in the associated space of maximal regularity. We finally study also the primitive equations, which are a model for oceanic and atmospheric flows and are derived from the Navier–Stokes equations by assuming a hydrostatic balance for the pressure term. We show that these equations are globally strongly well-posedness for arbitrary large initial data lying in critical spaces. Finally, we show that the primitive equations may be obtained as the limit of anisotropically scaled Navier–Stokes equations. The approach to the results concerning the primitive equations are based again on maximal L^p-regularity estimates, this time for the hydrostatic Stokes operator.

My sincere thanks go to the Fondazione CIME for their very kind support and hospitality during a Summer Course on 'Mathematical Analysis of the Navier–Stokes Equations: Foundations and Overview of Basic Open Problems' held at Cetraro in September 2017, in which these lectures notes were developed.

Part A: Parabolic Evolution Equations

This part of these notes mainly concerns the theory of evolution equations. Aiming for applications to viscous fluid flows we are mainly interested in parabolic evolution equations.

We start by introducing the basic concept of distributions, the Schwartz space $\mathcal{S}$ and its dual space $\mathcal{S}'$, the Fourier transform on $\mathcal{S}$ and $\mathcal{S}'$ and discuss also briefly various types of function spaces and their basic properties. Furthermore, we define Fourier multipliers for $L^p(\mathbb{R}^n)$. The celebrated theorem due to Mikhlin on L^p-

boundedness of translations invariant operators on $L^p(\mathbb{R}^n)$ for $1 < p < \infty$ as well as its analogue in the period setting will be used later on in many occasions. This theorem allows us further to introduce the Hilbert transform as well as the Riesz transforms as bounded operators on $L^p(\mathbb{R})$ and $L^p(\mathbb{R}^n)$, respectively.

In Sect. 1.2 we generalize this concept to the vector-valued setting, i.e. given a Banach space X, we discuss X-valued distributions, the Bochner integral and basic theorems for singular integral operators with operator-valued kernels. Banach spaces having the UMD-property will be definded as spaces for which the Hilbert transform acts as a bounded operator on $L^p(\mathbb{R}; X)$ for some $p \in (1, \infty)$.

In the sequel, we consider in Sect. 1.3 semigroups of operators and their generators. The classical theorems due to Hille-Yosida and Lumer-Philipps will be proven as well as the characterization theorem and smoothing properties for holomorphic semigroups.

In Sect. 1.4 we introduce sectorial operators. They are the starting point for the functional calculus of bounded, holomorphic functions, which, including important examples, will be investigated in this section. An extendend H^∞-calculus allows us to define fractional powers of sectorial operators and their properties in an elegant way.

Section 1.6 deals with the operator-valued H^∞-calculus. The extension of the scalar-valued H^∞-calculus to the X-valued functions leads us to the notion of $\mathcal{R}$-bounded families of operators. This notion will be the key for the boundedness result of the H^∞-calculus in this setting. It implies the Kalton–Weis theorem on the closedness of the sum of two commuting, sectorial operators and via the extension of Mikhlin's theorem to Banach spaces having the UMD-property, also the characterization theorem of maximal L^p-regularity for parabolic evolution equations in terms of $\mathcal{R}$-boundedness of its resolvent.

The final section of this first part discusses quasilinear parabolic evolution equations. Here we present the approach based on the theory of maximal L^p-regularity. This property will be the key property, when presenting three important results for these equations: local well-posedness, the generalized principle of linearized stability implying under suitable assuptions the global existence of strong solutions for data close to an equilibrium point and on the existence of global strong solutions in the presence of compact embeddings and strict Lyapunov functionals.

1.1 Basics on Distributions, Fourier Transforms and Sobolev Spaces

In this section we collect basic facts on distributions, the Fourier transforms and function spaces and also introduce the notation being used later on.

Distributions and Fourier Transforms

Let us begin with the notion of a multiindex. A *multiindex* $\alpha = (\alpha_1, \dots, \alpha_n) \in \mathbb{N}_0^n$ is an ordered n-tuple of nonnegative integers and we denote by $|\alpha| = \alpha_1 + \dots + \alpha_n$ the order of α. For $x = (x_1, \ldots, x_n) \in \mathbb{R}^n$ and a multiindex α, we set $x^\alpha = x_1^{\alpha_1} \cdots x_n^{\alpha_n}$. For a multiindex α, $\partial^\alpha f$ denotes the derivative $\partial_1^{\alpha_1} \cdots \partial_n^{\alpha_n} f$ of some function f on $\mathbb{R}^n$.

We denote by $\mathcal{D}(\mathbb{R}^n)$ (or by $C_c^\infty(\mathbb{R}^n)$) the space of all complex-valued C^∞-functions on $\mathbb{R}^n$ with compact support and by $\mathcal{S}(\mathbb{R}^n)$ the *Schwartz space* of all smooth, rapidly decreasing functions on $\mathbb{R}^n$, i.e.

$$\mathcal{S}(\mathbb{R}^n) := \{\varphi \in C^\infty(\mathbb{R}^n) : \|\varphi\|_{m,\alpha} < \infty \text{ for all } m \in \mathbb{N}_0, \alpha \in \mathbb{N}_0^n\},$$

where

$$\|\varphi\|_{m,\alpha} := \sup_{x \in \mathbb{R}^n} (1 + |x|^m)|\partial^\alpha \varphi(x)|).$$

The family of all seminorms $\| \cdot \|_{m,\alpha}$ defines a topology and $\mathcal{S}(\mathbb{R}^n)$ equipped with this topology becomes a Fréchet space. The space $C_c^\infty(\mathbb{R}^n)$ is a dense subspace of $\mathcal{S}(\mathbb{R}^n)$.

We call $\mathcal{D}(\mathbb{R}^n)'$ the space of all *distributions*, i.e. linear maps $f : \varphi \mapsto\, <\varphi, f>$ of $\mathcal{D}(\mathbb{R}^n)$ into $\mathbb{C}$ such that for each compact set $K \subset \mathbb{R}^n$ there exist $m \in \mathbb{N}$ and a constant $C > 0$ such that

$$| <\varphi, f> | \leq C \sup_{|\alpha| \leq m} \sup_{x \in \mathbb{R}^n} |\partial^\alpha \varphi(x)|$$

for all $\varphi \in \mathcal{D}(\mathbb{R}^n)$ with supp $\varphi \subset K$.

We denote by $\mathcal{S}(\mathbb{R}^n)'$ the space of all *temperate distributions*, i.e. continuous linear maps from $\mathcal{S}(\mathbb{R}^n)$ into $\mathbb{C}$. Observe that $\mathcal{S}(\mathbb{R}^n)'$ is then embedded in a natural way into $\mathcal{D}(\mathbb{R}^n)$. We equipp $\mathcal{D}(\mathbb{R}^n)'$ with the topology arising from the duality with $\mathcal{D}(\mathbb{R}^n)$, i.e. a net (f_j) of distributions converges to 0 in $\mathcal{D}(\mathbb{R}^n)'$ if and only if $<\varphi, f_j> \to 0$ for all $\varphi \in \mathcal{D}(\mathbb{R}^n)$.

It is well known that any locally integrable function $f : \mathbb{R}^n \to \mathbb{C}$ may be identified with a distribution and that any function f belonging to $L^p(\mathbb{R}^n)$ for $1 \leq p \leq \infty$ or the constant function 1 are temperate distributions.

In the following, we describe in which way differentiation, Fourier transform and convolution can be extended from functions to distributions. To this end, let $g : \mathbb{R}^n \to \mathbb{C}$ be a C^∞-function. Then, obviously, $\varphi \cdot g \in \mathcal{D}(\mathbb{R}^n)$ for $\varphi \in \mathcal{D}(\mathbb{R}^n)$.

The derivatives $\partial_j f$ $(j = 1, \ldots, n)$ of a distribution $f \in \mathcal{D}(\mathbb{R}^n)'$ are defined in $\mathcal{D}(\mathbb{R}^n)'$ by

$$< \varphi, \partial_j f >:= -< \partial_j \varphi, f >, \quad \varphi \in \mathcal{D}(\mathbb{R}^n).$$

Note also that ∂_j maps $\mathcal{S}(\mathbb{R}^n)'$ into itself and that this notation is consistent when differentiable functions are identified with distributions. Integration by parts shows for higher order derivatives we have

$$< \varphi, \partial^\alpha f >:= (-1)^{|\alpha|} < \partial^\alpha \varphi, f >, \quad \varphi \in \mathcal{D}(\mathbb{R}^n).$$

Given functions f and g, the *convolution* $f * g$ of f with g is defined by

$$(f * g)(x) := \int_{\mathbb{R}^n} f(x - y)g(y)dy$$

whenever the integral exists. Note that $\psi * \varphi \in \mathcal{S}(\mathbb{R}^n)$ and that the map $\psi \mapsto \psi * \varphi$ is continuous provided $\varphi, \psi \in \mathcal{S}(\mathbb{R}^n)$. Hence, the convolution $\varphi * f$ of $\varphi \in \mathcal{S}(\mathbb{R}^n)$ with a tempered distribution $f \in \mathcal{S}(\mathbb{R}^n)'$ can be defined as

$$< \psi, \varphi * f >:=< \psi * \check{\varphi}, f >, \quad \psi \in \mathcal{S}(\mathbb{R}^n),$$

where $\check{\varphi}(x) := \varphi(-x)$. In this case, $\varphi * f \in \mathcal{S}(\mathbb{R}^n)'$. Note that

$$\partial^\alpha (\varphi * f) = (\partial^\alpha \varphi) * f, \quad \varphi \in \mathcal{S}(\mathbb{R}^n), f \in \mathcal{S}(\mathbb{R}^n)'$$

for all α.

We now consider the Fourier transform and begin with its classical definition in $L^1(\mathbb{R}^n)$.

Definition 1.1.1 For $f \in L^1(\mathbb{R}^n)$, the *Fourier transform* $\mathcal{F}f$ of f is defined by

$$\hat{f}(\xi) := (\mathcal{F}f)(\xi) := \int_{\mathbb{R}^n} e^{-ix \cdot \xi} f(x)dx, \quad \xi \in \mathbb{R}^n,$$

where $x \cdot \xi := \sum_{j=1}^n x_j \xi_j$.

The Fourier inversion theorem, see e.g. [69], says that the Fourier transform $\mathcal{F}$ is a linear and topological isomorphism of $\mathcal{S}(\mathbb{R}^n)$ and that $\mathcal{F}^{-1}$ is given by

$$(\mathcal{F}^{-1}\varphi)(\xi) = (2\pi)^{-n}(\mathcal{F}\varphi)(-\xi), \quad \varphi \in \mathcal{S}(\mathbb{R}^n), \xi \in \mathbb{R}^n. \tag{1.1.1}$$

The Fourier transform hence induces an isomorphism of $\mathcal{S}(\mathbb{R}^n)'$ by

$$< \varphi, \mathcal{F}f >:=< \mathcal{F}\varphi, f >, \quad \varphi \in \mathcal{S}(\mathbb{R}^n), f \in \mathcal{S}(\mathbb{R}^n)',$$

which is also denoted by $\mathcal{F}$.

The following properties of the Fourier transform, which are elementary for functions, extend due to the above setting to distributions.

Lemma 1.1.2 *For $\varphi \in \mathcal{S}(\mathbb{R}^n)$ and $f \in \mathcal{S}(\mathbb{R}^n)'$ the following assertions hold.*

$$\mathcal{F}^{-1} f = (2\pi)^{-n} (\mathcal{F} f)^{\vee} = (2\pi)^{-n} \mathcal{F} \check{f}, \quad \text{where } < \varphi, \check{f} >:=< \check{\varphi}, f >$$

$$\mathcal{F} \partial^\alpha f = (i\xi)^\alpha \mathcal{F} f,$$

$$\mathcal{F}(\varphi * f) = (\mathcal{F}\varphi) \cdot (\mathcal{F} f).$$

The following Plancherel's theorem is a very classical result on Fourier transforms.

Proposition 1.1.3 (Plancherel) *Let $\varphi, \psi \in \mathcal{S}(\mathbb{R}^n)$. Then*

$$< \mathcal{F}\varphi, \overline{\mathcal{F}\psi} >= (2\pi)^n < \varphi, \overline{\psi} >$$

and the Fourier transform $\mathcal{F}$ extends to a bounded linear operator on $L^2(\mathbb{R}^n)$ such that $(2\pi)^{-n/2} \mathcal{F}$ is unitary.

Note that $\overline{\psi}$ denotes the complex conjugate of $\psi \in \mathcal{S}(\mathbb{R}^n)$.

Fourier Multipliers

We are now considering the concept of so-called Fourier multipliers. Assume that $m \in L^\infty(\mathbb{R}^n)$ takes values in $\mathbb{C}$. For $\varphi \in \mathcal{S}(\mathbb{R}^n)$, define $m\widehat{\varphi} \in \mathcal{S}(\mathbb{R}^n)'$ by $< \psi, m\widehat{\varphi} >:=< m, \widehat{\varphi} \cdot \psi >$. We then look for conditions on the function m such that the mapping

$$\varphi \mapsto (m\widehat{\varphi})^{\vee}, \quad \varphi \in \mathcal{S}(\mathbb{R}^n)$$

becomes continuous on $L^p(\mathbb{R}^n)$ for $1 \le p \le \infty$.

Definition 1.1.4 Let $1 \le p \le \infty$. A function $m \in L^\infty(\mathbb{R}^n)$ is called a *Fourier multiplier* for $L^p(\mathbb{R}^n)$ if $\mathcal{F}^{-1}(m\widehat{\varphi}) \in L^p(\mathbb{R}^n)$ for $\varphi \in \mathcal{S}(\mathbb{R}^n)$ and there exists a constant $C > 0$ such that

$$\|\mathcal{F}^{-1}(m\widehat{\varphi})\|_p \le C \|\varphi\|_p, \quad \varphi \in \mathcal{S}(\mathbb{R}^n).$$

Given $p \in [1, \infty)$ and a Fourier multiplier m for $L^p(\mathbb{R}^n)$, the map $\varphi \mapsto \mathcal{F}^{-1}(m\widehat{\varphi})$ extends to a bounded, linear operator

$$T_m : f \mapsto \mathcal{F}^{-1}(m\widehat{f})$$

on $L^p(\mathbb{R}^n)$. If $p = \infty$, the extension is weak*-continuous. The space consisting of all Fourier multipliers for $L^p(\mathbb{R}^n)$ is denoted by $\mathcal{M}_p(\mathbb{R}^n)$. Equipped with the norm

$$\|m\|_{\mathcal{M}_p(\mathbb{R}^n)} := \|T_m\|_{\mathcal{L}(L^p(\mathbb{R}^n))},$$

$\mathcal{M}_p(\mathbb{R}^n)$ is a Banach space. It follows from Plancherel's theorem that $\mathcal{M}_2(\mathbb{R}^n) = L^\infty(\mathbb{R}^n)$.

A very useful sufficient conditions for a given function m to belong to $\mathcal{M}_p(\mathbb{R}^n)$ for $1 < p < \infty$ is given by the *Mikhlin multiplier theorem*. In order to formulate this result, let $j = \min\{k \in \mathbb{N}, k > \frac{n}{2}\}$ and consider the Banach space

$$\mathcal{M}_M := \{m : \mathbb{R}^n \to \mathbb{C}; m \in C^j(\mathbb{R}^n \setminus \{0\}), |m|_M < \infty\},$$

where the norm $|\cdot|_M$ is defined by

$$|m|_M := \max_{|\alpha| \le j} \sup_{\xi \in \mathbb{R}^n \setminus \{0\}} |\xi|^\alpha |\partial^\alpha m(\xi)|.$$

Then the following holds.

Theorem 1.1.5 (Mikhlin) *Let* $1 < p < \infty$. *Then* $\mathcal{M}_M \hookrightarrow \mathcal{M}_p(\mathbb{R}^n)$.

For proofs and generalizations of the results stated above and for more information on this topic we refer to [11, 69, 139, 140, 148].

Consider one instance of Mikhlin's theorem for $n = 1$. For $\varepsilon > 0$, we define $(1/t)_\varepsilon \in L^1_{loc}(\mathbb{R})$ by

$$(1/t)_\varepsilon(\tau) := \tau^{-1} \chi_{[|\tau| \ge \varepsilon]}(\tau), \quad \tau \in \mathbb{R},$$

so that

$$< (1/t)_\varepsilon, \varphi > = \int_{|\tau| \ge \varepsilon} \frac{\varphi(\tau)}{\tau} d\tau \quad \text{for } \varphi \in \mathcal{S}(\mathbb{R}).$$

Then there exists a unique temperate distribution $\mathrm{pv}(1/t)$, called the *principal value* of $1/t$, such that

$$< \mathrm{pv}(1/t), \varphi > := \lim_{\varepsilon \to 0} \int_{|\tau| \ge \varepsilon} \frac{\varphi(\tau)}{\tau} d\tau, \quad \varphi \in \mathcal{S}(\mathbb{R}).$$

For $u \in \mathcal{S}(\mathbb{R})$ we define the *Hilbert transform* Hu of u by

$$Hu := \frac{1}{\pi} \mathrm{pv}(1/t) * u, \quad u \in \mathcal{S}(\mathbb{R}).$$

The symbol m of the translation invariant operator H is given by

$$(\mathrm{pv}(1/t))\hat{} = -i\pi \operatorname{sign}.$$

Then $\operatorname{sign} \in \mathcal{M}_M$ and by Mikhlin's theorem, $\operatorname{sign} \in \mathcal{M}_p(\mathbb{R})$ for $1 < p < \infty$. We thus have the following result.

Proposition 1.1.6 *Let* $1 < p < \infty$. *Then the Hilbert transform is a bounded linear operator on* $L^p(\mathbb{R})$.

The *Riesz transforms* are n-dimensional analogues of the Hilbert transform, with properties analogous to those of the Hilbert transform on $\mathbb{R}$. More precisely, for $j = 1, \ldots, n$ consider the functions

$$m_j(\xi) := -i\frac{\xi_j}{|\xi|}, \quad \xi \in \mathbb{R}^n \setminus \{0\},$$

and define the j-th Riesz transform R_j by

$$R_j\varphi := \mathcal{F}^{-1}(m_j\widehat{\varphi}), \quad \varphi \in \mathcal{S}(\mathbb{R}^n).$$

Since $m_j \in \mathcal{M}_M$ for all $j = 1, \ldots, n$, we have the following result.

Proposition 1.1.7 *Let* $1 < p < \infty$ *and* $j = 1, \ldots, n$. *Then the Riesz transforms* R_j *are bounded, linear operators on* $L^p(\mathbb{R}^n)$.

It is useful to note that the Riesz transforms satisfy

$$\sum_{j=1}^{n} R_j^2 = -Id.$$

We will also consider Fourier multipliers for the space $L^p((-\pi, \pi)^n)$ for $1 < p < \infty$. We say that a sequence $(a_k)_{k \in \mathbb{Z}^n} \subset \mathbb{C}$ is said to be a Fourier multiplier on $L^p((-\pi, \pi)^n)$ if there exists a constant $C > 0$ such that

$$\left\| \sum_{k \in \mathbb{Z}^n} a_k c_k e^{i<k,\cdot>} \right\|_{L^p((-\pi,\pi)^n)} \leq C \left\| \sum_{k \in \mathbb{Z}^n} c_k e^{i<k,\cdot>} \right\|_{L^p((-\pi,\pi)^n)} \quad (1.1.2)$$

for any sequence $(c_k)_{k \in \mathbb{Z}^n}$ with $c_k \neq 0$ for only finitely many $k \in \mathbb{Z}^n$. Given a Fourier multiplier $(a_k)_{k \in \mathbb{Z}^n}$ for $L^p((-\pi, \pi)^n)$, the mapping

$$\sum_{k \in \mathbb{Z}^n} c_k e^{i<k,\cdot>} \mapsto \sum_{k \in \mathbb{Z}^n} a_k c_k e^{i<k,\cdot>}$$

extends uniquely to a bounded operator T_a on $L^p((-\pi, \pi)^n)$ with norm $\|T_a\|$ being the greatest lower bound of the set of all constants such that (1.1.2) holds for any finite sequence $(c_k)_{k \in \mathbb{Z}^n} \subset \mathbb{C}$.

A classical result due to Marcinkiewicz gives a sufficient criteria for a sequence to be a Fourier multiplier for $L^p((-\pi, \pi)^n)$ for $1 < p < \infty$. In order to formulate this result, let $D_\nu = I_{\nu_1} \times \ldots \times I_{\nu_n}$ for $\nu \in \mathbb{N}_0^n$ and denote by $(D_\nu)_{\nu \in \mathbb{N}_0^n}$ be a dyadic decomposition of $\mathbb{Z}^n$, where $I_0 = \{0\}$ and $I_j = \{m \in \mathbb{Z} : 2^{j-1} \leq |m| \leq 2^j\}$ for $j \in \mathbb{N}$.

Theorem 1.1.8 (Marcinkiewicz) *Let* $1 < p < \infty$ *and* $(a_k)_{k \in \mathbb{Z}^n} \subset \mathbb{C}$ *be a sequence satisfying*

$$\sup_{\nu \in \mathbb{N}_0^n} var_{D_\nu} a < \infty.$$

Then the sequence $(a_k)_{k \in \mathbb{Z}^n}$ *is a Fourier multiplier for* $L^p((-\pi, \pi)^n)$ *and there exists a constant* $C = C(n, p) > 0$ *such that*

$$\|T_a\|_{\mathcal{L}(L^p((-\pi, \pi)^n))} \leq C \sup_{\nu \in \mathbb{N}_0^n} var_{D_\nu} a.$$

The following Corollary of Theorem 1.1.8 will be used in Sect. 1.16.

Corollary 1.1.9 *Let* $1 < p < \infty$, $n \in \mathbb{N}$ *and* $(a_k)_{k \in \mathbb{Z}^n} \subset \mathbb{C}$ *be a sequence such that* $a_k = m(k)$ *for all* $k \in \mathbb{Z}^n \setminus \{0\}$ *and some function* $m \in C^n(\mathbb{R}^n \setminus \{0\})$. *Suppose that*

$$[m] := \sup_{\gamma \in \{0,1\}^n} \sup_{\xi \neq 0} |\xi^\gamma D^\gamma m(\xi)| < \infty.$$

Then the sequence $(a_k)_{k \in \mathbb{Z}^n}$ *is a Fourier multiplier for* $L^p((-\pi, \pi)^n)$ *and there exists a constant* $C = C(n, p) > 0$ *such that*

$$\|T_a\|_{\mathcal{L}(L^p((-\pi, \pi)^n))} \leq C \max\{[m], |a_0|\}.$$

Function Spaces on Domains

The concept of distributions described above can be extended to the case of distributions on an open subset Ω of $\mathbb{R}^n$. We denote by $\mathcal{D}(\Omega) := C_c^\infty(\Omega)$ the space of *test functions* on Ω, i.e. C^∞-functions of compact support in Ω. The space $\mathcal{D}(\Omega)'$ of *distributions* on Ω is defined to be the space of all linear functionals f on $\mathcal{D}(\Omega)$ such that for each compact set $K \subset \Omega$ there exists $m \in \mathbb{N}$ and a constant $C > 0$ such that

$$| < \varphi, f > | \leq C \sup_{|\alpha| \leq m} \sup_{x \in \Omega} |\partial^\alpha \varphi(x)|$$

for all $\varphi \in \mathcal{D}(\Omega)$ with $\text{supp}\varphi \subset K$. Again, locally integrable functions on Ω can be identified with distributions and the derivatives ∂_j of a distribution f are defined by

$$< \varphi, \partial_j f >:= - < \partial_j \varphi, f >, \qquad \varphi \in \mathcal{D}(\Omega).$$

The *Sobolev spaces* $W^{m,p}(\Omega)$ are defined for $m \in \mathbb{N}$ and $1 \leq p \leq \infty$ by

$$W^{m,p}(\Omega) := \{f \in L^p(\Omega) : \partial^\alpha f \in L^p(\Omega) \text{ for all } \alpha \in \mathbb{N}_0^n \text{ with } |\alpha| \leq m\},$$

where $\partial^\alpha f$ is understood in the sense of distributions. Equipped with the norm

$$\|f\|_{W^{m,p}(\Omega)} := \left(\sum_{|\alpha| \leq m} \|\partial^\alpha f\|_p^p \right)^{1/p},$$

the space $W^{m,p}(\Omega)$ becomes a Banach space. Moreover, the closure of $\mathcal{D}(\Omega)$ in $W^{m,p}(\Omega)$ is denoted by $W_0^{m,p}(\Omega)$, i.e.

$$W_0^{m,p}(\Omega) := \overline{C_c^\infty(\Omega)}^{\|\cdot\|_{W^{m,p}}}.$$

For $p = 2$, one often uses the notation $H^m(\Omega) := W^{m,2}(\Omega)$ and $H_0^m(\Omega) := W_0^{m,2}(\Omega)$. Then, equipped with the equivalent norm

$$\|f\|_{H^m(\Omega)} := \left(\sum_{|\alpha| \leq m} \|\partial^\alpha f\|_2^2 \right)^{1/2},$$

$H^m(\Omega)$ is a Hilbert space with the inner product

$$(f|g)_{H^m(\Omega)} = \sum_{|\alpha| \leq m} \int_\Omega \partial^\alpha f \, \overline{\partial^\alpha g} dx.$$

By Plancherel's Theorem 1.1.3 and Lemma 1.1.2,

$$H^m(\mathbb{R}^n) = \{ f \in L^2(\mathbb{R}^n) : \xi^\alpha \mathcal{F}f \in L^2(\mathbb{R}^n) \text{ for all multiindices } \alpha \text{ with } |\alpha| \leq m \},$$

and $H^m(\Omega)$ coincides also with the space consisting of all functions in $L^2(\mathbb{R}^n)$ such that

$$\xi \mapsto (1 + |\xi|^2)^{m/2} \mathcal{F}f(\xi) \in L^2(\mathbb{R}^n).$$

The analysis of viscous fluid flow described in Part B relies on various types of function spaces. In particular, we mention here

(a) the homogenous Sobolev spaces $\widehat{W}^{m,p}(\Omega)$ for $m \in \mathbb{N}_0$ and $1 \leq p \leq \infty$,
(b) the Bessel-Potential spaces $H^{s,p}(\Omega)$ for $s \in \mathbb{R}$ and $1 < p < \infty$,
(c) the Besov spaces $B_{p,q}^s(\Omega)$ for $s \in \mathbb{R}$ and $1 \leq p, q \leq \infty$,
(d) the Triebel-Lizorkin spaces $F_{p,q}^s(\Omega)$ for $s \in \mathbb{R}$ and $1 \leq p, q \leq \infty$,
(e) the real interpolation spaces

$$W_p^s(\Omega) = \left(W^{m,p}(\Omega), W^{m+1,p}(\Omega) \right)_{\theta,p}$$

for $m \in \mathbb{N}_0$, $s \in (m, m+1)$, $\theta = s - [s]$ and $1 < p < \infty$, often called Sobolev-Slobodeckij spaces,

(f) and the complex interpolation spaces

$$H^{s,p}(\Omega) = \left[W^{m,p}(\Omega), W^{m+1,p}(\Omega) \right]_\theta$$

for $1 \leq p \leq \infty$, $0 < \theta < 1$, $s = (1-\theta)m + \theta(m+1) = m + \theta$, defined on open sets $\Omega \subset \mathbb{R}^n$. We do not give precise definitions of these spaces here and refer the reader e.g. to the a classical monograph by Triebel [149] or the recent one by Amann [12].

1.2 The Vector Valued Setting

As before, we start this section by introducing basic facts on distributions and Fourier transforms, now however in the *vector-valued setting*.

Banach Space Valued Distributions
We note that the proofs of the assertions listed below on distributions are only straightforward modifications of the ones for the scalar case described in Sect. 1.1.

Let X be a Banach space. We denote by $\mathcal{S}(\mathbb{R}^n; X)$ the Schwartz space of smooth rapidly decreasing X-valued functions on $\mathbb{R}^n$. Then

$$\mathcal{D}(\mathbb{R}^n; X) \overset{d}{\hookrightarrow} \mathcal{S}(\mathbb{R}^n; X) \overset{d}{\hookrightarrow} \mathcal{E}(\mathbb{R}^n; X),$$

where $\mathcal{D}(\mathbb{R}^n; X)$ is the space of all X-valued C^∞-functions on $\mathbb{R}^n$ with compact supports, as usual equipped with the inductive limit topology, and $\mathcal{E}(\mathbb{R}^n; X) := C^\infty(\mathbb{R}^n; X)$ equipped with the bounded convergence topology.

The space $\mathcal{S}'(\mathbb{R}^n; X)$ of X-valued temperate distributions is defined by $\mathcal{S}'(X) := \mathcal{L}(\mathcal{S}(\mathbb{R}^n); X)$; the spaces $\mathcal{E}'(\mathbb{R}^n; X)$ and $\mathcal{D}'(\mathbb{R}^n; X)$ are defined analogously. Then

$$\mathcal{E}'(\mathbb{R}^n; X) \hookrightarrow \mathcal{S}'(\mathbb{R}^n; X) \hookrightarrow \mathcal{D}'(\mathbb{R}^n; X).$$

Finally, the space $\mathcal{O}_M(\mathbb{R}^n; X)$ of all X-valued slowly increasing smooth functions on $\mathbb{R}^n$ consists of all $u \in \mathcal{E}(\mathbb{R}^n; X)$ such that, given $\alpha \in \mathbb{N}_0^n$, there exists $m_\alpha \in \mathbb{N}$ and $C_\alpha > 0$ such that

$$\|\partial^\alpha u(x)\| \leq C_\alpha (1 + |x|^2)^{m_\alpha}, \quad x \in \mathbb{R}^n.$$

Observe that $\mathcal{S}(\mathbb{R}^n; X) \hookrightarrow \mathcal{O}_M(\mathbb{R}^n; X) \hookrightarrow \mathcal{S}'(\mathbb{R}^n; X)$. We also write $\mathcal{S}(\mathbb{R}^n) := \mathcal{S}(\mathbb{R}^n; \mathbb{K})$, $\mathcal{S}'(\mathbb{R}^n) := \mathcal{S}'(\mathbb{R}^n; \mathbb{K})$ and $\mathcal{O}_M(\mathbb{R}^n) := \mathcal{O}_M(\mathbb{R}^n; \mathbb{K})$, where $\mathbb{K} \in \{\mathbb{C}, \mathbb{R}\}$.

We next turn our attention to the Bochner integral. To this end, given an interval I in $\mathbb{R}$, bounded or unbounded, or a rectangle I in $\mathbb{R}^2$, we call a function $g : I \to X$ a *simple* function if it is of the form $g(t) = \sum_{j=1}^n x_j \chi_{S_j}(t)$ for some $n \in \mathbb{N}$, $x_j \in X$

and Lebesgue measurable sets $S_j \subset I$ with finite Lebesgue measure $\mu(S_j)$; g is called a *step function* if S_j can be chosen to be an interval. A function $f : I \to X$ is called *measurable* if there exists a sequence of simple functions g_n such that $f(t) = \lim_{n\to\infty} g_n(t)$ for almost all $t \in I$. For a simple function $g : I \to X$ of the form $g = \sum_{j=1}^{n} x_j \chi_{S_j}$ we define

$$\int_I g(t)dt := \sum_{j=1}^{n} x_j \mu(S_j).$$

A function $f : I \to X$ is then called *Bochner integrable* if there exist simple functions g_n such that $g_n \to f$ pointwise almost everywhere and $\lim_{n\to\infty} \int_I \|f(t) - g_n(t)\|dt = 0$. In this case, the *Bochner integral of f on I* is given by

$$\int_I f(t)dt := \lim_{n\to\infty} \int_I g_n(t)dt.$$

The class of Bochner integrable functions can be characterized as follows.

Proposition 1.2.1 (Bochner) *A function $f : I \to X$ is Bochner integrable if and only if f is measurable and $\|f\|$ is integrable. If f is Bochner integrable, then*

$$\left\| \int_I f(t)dt \right\| \le \int_I \|f(t)\|dt.$$

For $1 \le p < \infty$, we denote by $L^p(I; X)$ the space of all measurable functions $f : I \to X$ such that

$$\|f\|_p := \left(\int_I \|f(t)\|^p dt \right)^{1/p} < \infty.$$

Furthermore, we set $L^\infty(I; X)$ to be the space of all measurable functions $f : I \to X$ such that

$$\|f\|_\infty := \operatorname{ess\,sup}_{t\in I} \|f(t)\| < \infty.$$

With the usual identifications, $L^p(I; X)$ becomes a Banach space for all $1 \le p \le \infty$. We note that the above theory of integration works in the same way, when the interval I is replaced by a measurable set in $\mathbb{R}^n$. It follows from Fubini's theorem that for $1 \le p < \infty$ there is an isometric isomorphism between $L^p(I \times \Omega; X)$ and $L^p(I; L^p(\Omega; X))$ for any measurable set $\Omega \subset \mathbb{R}^n$ given by $f \mapsto g$, where

$$(g(s))(t) := f(s, t).$$

We further note that for $1 \leq p < \infty$ the norm of a function $f \in L^p(\mathbb{R}^n; X)$ can be calculated by

$$\|f\|_{L^p(\mathbb{R}^n;X)}$$
$$= \sup \left\{ \left| \int_{\mathbb{R}^n} < g(x), f(x) > dx \right| : g \in L^{p'}(\mathbb{R}^n; X'), \|g\|_{L^{p'}(\mathbb{R}^n;X')} \leq 1 \right\}.$$
(1.2.1)

Given $u \in L^1(\mathbb{R}^n; X)$ we define the *Fourier transform* of u by

$$\mathcal{F}u(\xi) := \hat{u}(\xi) := \int_{\mathbb{R}^n} e^{-i<x,\xi>} u(x) dx, \quad \xi \in \mathbb{R}^n,$$

where as above $< x, \xi > = \sum_{j=1}^n x_j \xi_j$. The *Fourier inversion theorem* guarantees that the Fourier transform $\mathcal{F}$ is an isomorphism on $\mathcal{S}(\mathbb{R}^n; X)$ and that

$$\mathcal{F}^{-1}u = (2\pi)^{-n}\check{\hat{u}} = (2\pi)^{-n}\hat{\check{u}}, \quad u \in \mathcal{S}(\mathbb{R}^n; X)$$
(1.2.2)

where $\check{u}(x) := u(-x)$ for $x \in \mathbb{R}^n$ denotes the reflection of u. The Fourier transform $\hat{u} := \mathcal{F}u$ of $u \in \mathcal{S}'(\mathbb{R}^n; X)$ is defined by

$$\hat{u}(\varphi) := u(\hat{\varphi}), \quad \varphi \in \mathcal{S}(\mathbb{R}^n).$$

Defining $\check{u}(\varphi) := u(\check{\varphi})$ for $u \in \mathcal{D}'(\mathbb{R}^n; X)$ and $\varphi \in \mathcal{D}(\mathbb{R}^n)$ it follows that $\mathcal{F}$ is an isomorphism on $\mathcal{S}'(\mathbb{R}^n; X)$ and that (1.2.2) holds for $u \in \mathcal{S}'(\mathbb{R}^n; X)$.

Let $\varphi \in \mathcal{S}(\mathbb{R}^n)$ and $u \in \mathcal{S}'(\mathbb{R}^n; X)$. Then the convolution $u * \varphi$ of u and φ is defined by

$$(u * \varphi)(x) := u(\tau_x \check{\varphi}), \quad x \in \mathbb{R}^n,$$

where $\tau_a \varphi(x) := \varphi(x - a)$ for $x \in \mathbb{R}^n$ and $a \in \mathbb{R}^n$. Moreover, the convolution theorem states that

$$(u * \varphi\hat{)} = \hat{u}\hat{\varphi}.$$

Finally, let H be a Hilbert space. Then $L^2(\mathbb{R}^n; H)$ is a Hilbert space with respect to the inner product

$$(u|v)_2 := (u|v)_{L^2(\mathbb{R}^n;H)} := \int_{\mathbb{R}^n} u \cdot \overline{v} \, dx.$$

We note that Plancherel's theorem carries over to the situation of Hilbert space valued functions. More precisely, the following holds.

Proposition 1.2.2 (Plancherel's Theorem in Hilbert Spaces) *Let H be a Hilbert space and $u, v \in \mathcal{S}(\mathbb{R}^n; H)$. Then*

$$(\hat{u}|\hat{v})_2 = (2\pi)^n (u|v)_2$$

Moreover, $(2\pi)^{-n/2}\mathcal{F}$ is a unitary operator on $L^2(\mathbb{R}^n; H)$.

Singular Integrals: Classical Theory

Singular integral operators will be an important tool in the analysis of the Stokes equation in the following sections. We now state two results on the L^p-boundedness of singular integral operators in the vector valued setting.

Consider a kernel operator of the form

$$(Tf)(x) = \int_{\mathbb{R}^n} K(x, y) f(y) d\mu(y), \tag{1.2.3}$$

where the kernel K is singular near $x = y$, and (1.2.3) is meaningful only in some limiting sense. More precisely, we assume that for $f \in L^2(\mathbb{R}^n; X)$ with compact support, the integral in (1.2.3) converges absolutely for a. a. $x \in (\mathrm{supp} f)^C$ and that (1.2.3) holds for these x, where the kernel K is assumed to belong to $L^1_{loc}(\mathbb{R}^n \times \mathbb{R}^n, \mathcal{L}(X, Y))$.

Theorem 1.2.3 *Let X and Y be Banach spaces and let T be defined as in (1.2.3). Assume that there exists a constant $C > 0$ such that*

(i) $\|Tf\|_{L^2(\mathbb{R}^n;Y)} \leq C\|f\|_{L^2(\mathbb{R}^n;X)}$ for all $f \in L^2(\mathbb{R}^n; X)$
(ii) There exists a constant $r > 1$ such that for all $y_1 \in \mathbb{R}^n, \delta > 0$

$$\int_{B(y_1, r\delta)^C} \|K(x, y_1) - K(x, y_2)\|_{\mathcal{L}(X,Y)} d\mu(x) \leq C \text{ if } y_2 \in B(y_1, \delta)$$

Then, for all $1 < p < 2$, there exists a constant C_p (depending only on C and p) such that

$$\|Tf\|_{L^p(\mathbb{R}^n;Y)} \leq C_p\|f\|_{L^p(\mathbb{R}^n;X)}, \text{ for } f \in L^p(\mathbb{R}^n; X).$$

We note that the proof of the scalar-valued version of Theorem 1.2.3 given e.g. in [140] extends without any difficulties to the situation considered above.

Remark 1.2.4 It is worthwhile to point out the following variant of Theorem 1.2.3. Assume that T defined as in (1.2.3) satisfies the condition

$$\|Tf\|_{L^r(\mathbb{R}^n;Y)} \leq C\|f\|_{L^r(\mathbb{R}^n;X)}, \qquad f \in L^r(\mathbb{R}^n; X)$$

for some $r \in (1, \infty)$ and condition (ii) of Theorem 1.2.3. Then T admits a bounded extensions to L^p for $1 < p < r$. The proof follows the lines of the proof given in [140] and is left to the reader.

Consider now the special case of convolution operators. More specifically, let X and Y be Banach spaces and suppose that $K \in L^1_{loc}(\mathbb{R}^n \setminus \{0\}, \mathcal{L}(X, Y))$. Then the integral

$$Tf(x) := \int_{\mathbb{R}^n} K(x - y)f(y)dy \qquad (1.2.4)$$

is well defined for $f \in L^\infty(\mathbb{R}^n; X)$ with compact support and $x \in (suppf)^C$. For operators of the form (1.2.4), the condition (ii) of Theorem 1.2.3 is equivalent to the *Hörmanders's condition*

$$\int_{|x|>2|y|} \|K(x - y) - K(x)\| dx \leq C < \infty, \quad y \in \mathbb{R}^n \setminus \{0\}. \qquad (1.2.5)$$

The following result due to Benedek, Calderón and Panzone says that operators of the form (1.2.4) are bounded operators on $L^p(\mathbb{R}^n; X)$, for $1 < p < \infty$, provided this holds for some $r \in (1, \infty)$ and condition (1.2.5) is satisfied.

Theorem 1.2.5 (Benedek, Calderón, Panzone) *[16] Suppose that $T \in \mathcal{L}(L^r(\mathbb{R}^n; X) L^r(\mathbb{R}^n; Y))$ for some $r \in (1, \infty)$. Assume that T may be represented by (1.2.4) for $f \in L^\infty(\mathbb{R}^n; X)$ with compact support and $x \in (supp f)^C$, and that (1.2.5) is satisfied. Then T admits a bounded extension to $L^p(\mathbb{R}^n; X)$ for $p \in (1, \infty)$. Moreover, there exists a constant C such that*

$$\|Tf\|_{L^p(\mathbb{R}^n;Y)} \leq C\|f\|_{L^p(\mathbb{R}^n;X)}, \quad 1 < p < \infty.$$

Proof We note that for $p \in (1, r)$ the assertion follows from Remark 1.2.4 since for translation invariant operators of the form (1.2.4), the condition ii) of Theorem 1.2.3 is equivalent to (1.2.5). The remaining case where $p \in (r, \infty)$ follows by a duality argument. □

Hilbert Transform and UMD-Spaces
An interesting application of Theorem 1.2.5 concerns the Hilbert transform on $L^p(\mathbb{R}; X)$. We already saw in the previous Sect. 1.1 that the Hilbert transform acts boundedly on $L^p(\mathbb{R})$. As in the scalar case, we define for $u \in \mathcal{S}(\mathbb{R}^n; X)$ the *Hilbert transform Hu* of u by

$$Hu := \frac{1}{\pi} pv(1/t) * u, \quad u \in \mathcal{S}(\mathbb{R}^n; X).$$

For the time being, assume that for some $p \in (1, \infty)$ we have $\|Hu\|_{L^p(\mathbb{R};X)} \leq C\|u\|_{L_p(\mathbb{R};X)}$ for $u \in \mathcal{S}(\mathbb{R}; X)$. Then there exists a unique extension of H to a bounded operator on $L^p(\mathbb{R}; X)$. This extension, also denoted by H, is called the *Hilbert transform* on $L^p(\mathbb{R}; X)$. Since $\widehat{Hu} = -i\,sign(\cdot)\hat{u}$, it follows from Plancherel's Theorem 1.2.2 that

$$(Hu|Hv)_2 = \frac{1}{2\pi}(\widehat{Hu}|\widehat{Hv})_2 = \frac{1}{2\pi}(\hat{u}|\hat{v})_2 = (u|v)_2$$

for $u, v \in \mathcal{S}(\mathbb{R}; X)$. By density, H is an isomorphism on $L^2(\mathbb{R}; X)$ satisfying $H^{-1}v = -Hv$ for $v \in L^2(\mathbb{R}; X)$. We summarize these observations in part (a) of the following proposition.

Proposition 1.2.6 *Let X be a Banach space.*

(a) *The Hilbert transform is a unitary operator on $L^2(\mathbb{R}; X)$ provided X is a Hilbert space.*

(b) *Suppose that the Hilbert transform is bounded on $L^p(\mathbb{R}; X)$ for some $p \in (1, \infty)$. Then it is bounded on $L^q(\mathbb{R}; X)$ for all $q \in (1, \infty)$.*

Proof Note that assertion (b) is an immediate consequence of Theorem 1.2.5 since

$$\int_{|x|>2|y|} \|K(x-y) - K(x)\| dx = \int_{|x|>2|y|} \left| \frac{1}{x-y} - \frac{1}{x} \right| dx \leq C.$$

$\square$

Let us remark that the Hilbert transform is not bounded on $L^p(\mathbb{R}; X)$, in general. Banach spaces for which the Hilbert transform is, however, bounded for some (and then all) $p \in (1, \infty)$ play an important role in the regularity theory of parabolic evolution equations. We finish this section with the definition of such spaces.

Definition 1.2.7 A Banach space X is called a *UMD-space* if the Hilbert transform is bounded on $L^p(\mathbb{R}; X)$ for some (and then all) $p \in (1, \infty)$.

The word *UMD* stands for the property of *unconditional martingale differences* and goes back to the work of Burkholder [24] and Bourgain [20]. For a comprehensive treatment of these spaces we refer to the monographs [87] and [8]. The following remark collects some basic facts and examples of *UMD*-spaces.

Remarks 1.2.8

(a) Let us remark here that every finite dimensional Banach space or any Hilbert space is a *UMD*-space.

(b) *UMD* spaces are reflexive. The converse is not true.

(c) If X is a *UMD* space and (Ω, μ) a σ-finite measure space, then $L^p(\Omega, \mu; X)$ has the *UMD*-property provided $1 < p < \infty$.

(d) All closed linear subspaces of *UMD*-spaces are *UMD* spaces.

(e) $L^1(\Omega, \mu)$ or spaces of continuous functions $C(K)$ do not have the *UMD* property.

1.3   Cauchy Problems and Semigroups

We start this section by investigating the well-posedness of the abstract Cauchy problem

$$(ACP) \qquad \begin{cases} u'(t) & = Au(t), \quad t \geq 0, \\ u(0) & = x, \end{cases} \qquad (1.3.1)$$

where A is a densely defined, closed operator on a Banach space X. Note that A is called *closed* whenever for a sequence $(x_j) \in D(A)$ with $x_j \to x$ and $Ax_j \to y$ it follows that $x \in D(A)$ and $Ax = y$.

Cauchy Problems and Semigroups

The following definition of a C_0-semigroup and its generator is fundamental for our approach.

Definition 1.3.1 A C_0-*semigroup* on a Banach space X is a strongly continuous function $T : \mathbb{R}_+ \to \mathcal{L}(X)$ satisfying

$$T(0) = I \text{ and } T(t + s) = T(t)T(s), \quad s, t \geq 0.$$

Its *generator* A is defined by

$$Ax := \lim_{t \to 0} \frac{T(t)x - x}{t}$$

with domain $D(A) := \{x \in X : \lim_{t \to 0} \frac{T(t)x - x}{t}$ exists $\}$.

Then $D(A)$ is dense in X and A is a closed, linear operator. The operator A is called the infinitesimal generator of T since A coincides with the strong derivative of T in 0. Furthermore, we call a C_0-semigroup a *bounded semigroup*, if there exists a constant $M > 0$ such that $\|T(t)\| \leq M$ for all $t > 0$; it is called a *contraction semigroup* provided

$$\|T(t)\| \leq 1 \text{ for all } t > 0.$$

In the following lemma we collect basic properties of semigroups and their generators.

Lemma 1.3.2 *Let T be a C_0-semigroup on X with generator A and $x \in D(A)$. Then*

(a) $T(t)x \in D(A)$ for all $t \geq 0$,
(b) $AT(t)x = T(t)Ax$ for all $t \geq 0$,

(c) *the mapping* $t \mapsto T(t)x$ *is differentiable for all* $t > 0$ *and*

$$\frac{d}{dt}T(t)x = AT(t)x, \quad t > 0.$$

For the proof of these basic properties, we refer e.g. to [13], Chapter 3.1.

Before proceeding further, some comments about resolvents and the resolvent set of A are in order. Given $\lambda \in \mathbb{C}$ and an operator A on X, λ is said to belong to the *resolvent set* $\varrho(A)$ of A if $\lambda - A$ is invertible. In this case we write $R(\lambda, A) = (\lambda - A)^{-1}$. Note that if $\varrho(A)$ is non-empty, then A is closed. The function $R(\cdot, A) : \varrho(A) \to \mathcal{L}(X)$ is the *resolvent* of A. The *spectrum* of A is defined as

$$\sigma(A) := \mathbb{C}\backslash\varrho(A);$$

the *spectral bound* is given by $s(A) := \sup\{Re\lambda : \lambda \in \sigma(A)\}$. The *point spectrum* $\sigma_p(A)$ of A consists of all the eigenvalues of A, i.e.,

$$\sigma_p(A) := \{\lambda \in \mathbb{C} : \ker(\lambda - A) \neq \{0\}\}.$$

We now collect basic properties of resolvents in the following lemma.

Lemma 1.3.3 *Let A be an operator on X. Then*

(a) $\varrho(A)$ *is open and* $\sigma(A)$ *is closed in* $\mathbb{C}$.
(b) *If* $\mu \in \varrho(A)$ *and* $\lambda \in \mathbb{C}$ *with* $|\lambda - \mu| < \|R(\mu, A)\|^{-1}$, *then* $\lambda \in \varrho(A)$ *and*

$$R(\lambda, A) = \sum_{n=0}^{\infty} (\mu - \lambda)^n R(\mu, A)^{n+1}.$$

(c) $R(\cdot, A)$ *is holomorphic on* $\varrho(A)$ *and*

$$R(\mu, A)^{(n)} = (-1)^n n! R(\mu, A)^{n+1}, \quad n \in \mathbb{N}.$$

(d) *It* $\lambda, \mu \in \varrho(A)$, *then*

$$R(\lambda, A) - R(\mu, A) = (\mu - \lambda)R(\lambda, A)R(\mu, A). \tag{1.3.2}$$

The above Eq. (1.3.2) is called the *resolvent equation*.

Consider now an arbitrary C_0-semigroup T with generator A. Then it is not difficult to prove that there exist constants $M, \omega > 0$ such that $\|T(t)\| \leq Me^{\omega t}$ for all $t > 0$ and that

$$R(\lambda, A)x = \int_0^{\infty} e^{-\lambda t} T(t)x dt, \quad x \in X, Re\lambda > \omega,$$

where $R(\lambda, A) = (\lambda - A)^{-1}$. This means in particular that $\lambda \in \varrho(A)$ for all $\lambda \in (\omega, \infty)$.

One can prove the stronger assertion that generators of C_0-semigroups are precisely those operators whose resolvent is a Laplace transform. More precisely, we have the following result.

Proposition 1.3.4 ([13, Thm. 3.1.7]) *Let* $T : [0, \infty) \rightarrow \mathcal{L}(X)$ *be a strongly continuous function,* $\omega \in \mathbb{R}$ *and* A *be an operator on* X *such that* $(\omega, \infty) \subset \varrho(A)$ *and*

$$R(\lambda, A)x = \int_0^\infty e^{-\lambda t} T(t)x\, dt, \quad x \in X, \, Re\, \lambda > \omega.$$

Then T *is a* C_0-*semigroup on* X *with* A *its generator.*

After the more algebraic definition of a C_0-semigroup in Definition 1.3.1 and the approach based on the Laplace transform in Proposition 1.3.4 we now show that the well-posedness of (ACP_0) is very closely related to the fact that A is the generator of a C_0-semigroup. By a *classical solution* of (1.3.1) we understand a function $u \in C^1(\mathbb{R}_+; X)$ such that $u(t) \in D(A)$ for all $t \geq 0$ and (1.3.1) holds for all $t \geq 0$. If a classical solution exists, then $u(0) = x \in D(A)$.

Proposition 1.3.5 *Let* A *be a closed operator on* X. *Then* A *generates a* C_0-*semigroup if and only if* $\rho(A) \neq \emptyset$ *and for all* $x \in D(A)$ *there exists unique, classical solution of* (ACP).

It is thus interesting to characterize generators of C_0-semigroups by properties of the operators A or their resolvents. In the following, we first prove the Hille-Yosida theorem, which characterizes generators of C_0-semigroups in terms of a resolvent estimate for real λ.

Theorem 1.3.6 (Hille-Yosida)   *Let* A *be a densely defined operator on* X. *Then* A *generates a* C_0-*semigroup* T *on* X *satisfying* $\|T(t)\| \leq 1$ *for all* $t \geq 0$ *if and only if* $(0, \infty) \subset \rho(A)$ *and*

$$\|\lambda R(\lambda, A)\| \leq 1, \quad \lambda > 0. \tag{1.3.3}$$

Proof Observe first that if A is the generator of a contraction semigroup T, then the assertion follows immediately from Proposition 1.3.4.

Conversely, assume that A satisfies (1.3.3). For $\lambda > 0$, we define the *Yosida approximation* of A by

$$A_\lambda := \lambda^2 R(\lambda, A) - \lambda Id = \lambda A R(\lambda, A).$$

Then, for $x \in D(A)$,

$$A_\lambda x \rightarrow Ax \text{ as } \lambda \rightarrow \infty.$$

Since A_λ is a bounded operator on X, we may define the semigroup T_λ generated by A_λ by

$$T_\lambda(t) := e^{tA_\lambda} = e^{-\lambda t}e^{\lambda^2 t R(\lambda, A)} = e^{-\lambda t}\sum_{j=0}^{\infty}\frac{(\lambda^2 t)^j}{j!}R(\lambda, A)^j.$$

The assumption $\|\lambda R(\lambda, A)\| \le 1$ for all $\lambda > 0$ implies that

$$\|T_\lambda(t)\| \le 1.$$

Thus T_λ is a contraction semigroup on X having A_λ as its generator.

Next, due to the resolvent Eq. (1.3.2),

$$A_\mu T_\lambda(t) = T_\lambda(t)A_\mu, \quad \lambda, \mu > 0, t > 0.$$

Hence, for $x \in D(A)$ we thus obtain by Lemma 1.3.2(c),

$$T_\lambda(t)x - T_\mu(t)x = \int_0^t \frac{d}{ds}[T_\mu(t-s)T_\lambda(s)x]ds = \int_0^t T_\mu(t-s)T_\lambda(s)(A_\lambda x - A_\mu x)ds.$$

Thus $\|T_\lambda(t)x - T_\mu(t)x\| \le t\|A_\lambda x - A_\mu x\| \to 0$ as $\lambda, \mu \to \infty$. Therefore,

$$T(t)x := \lim_{\lambda\to\infty} T_\lambda(t)x$$

exists for all $t \ge 0$ and all $x \in D(A)$. Since $\|T_\lambda(t)\| \le 1$ for all $\lambda > 0$ and all $t > 0$, the above limit exists for all $x \in X$ and thus T is a contraction semigroup on X. The proof of the fact that A is the generator of T is left to the reader as an exercise. □

There is a second characterization of contraction semigroups which turns out to be quite useful when dealing with differential operators. To this end, we call an operator A in X *dissipative* if

$$\|(\lambda - A)x\| \ge \lambda\|x\| \text{ for all } x \in D(A), \lambda > 0. \tag{1.3.4}$$

We then have the following so called Lumer–Phillips theorem.

Theorem 1.3.7 (Lumer–Phillips) *Let A be a densely defined operator on X. Then A generates a C_0-semigroup of contractions on X if and only if A is dissipative and $(\lambda - A)D(A) = X$ for some (or all) $\lambda > 0$.*

Proof Assume that A is the generator of a contraction semigroup. Then the assertion follows from the Hille-Yosida Theorem 1.3.6.

In order to prove the converse implication, note that by assumption $\lambda_0 - A$ is invertible and that $\|R(\lambda_0, A)\| \le \lambda_0^{-1}$ for some $\lambda_0 > 0$. Hence, $\Lambda := \varrho(A)\cap(0, \infty)$

is non empty, and thus A is closed. Proving furthermore that Λ is open and closed in $(0, \infty)$ it follows that $\Lambda = (0, \infty)$ and thus $(0, \infty) \subset \varrho(A)$. The assertion follows then from (1.3.4) and the Hille-Yosida Theorem. $\qquad\square$

Remark 1.3.8 Dissipative operators acting on Hilbert spaces H may be characterized as follows. Denote by $(\cdot|\cdot)$ the inner product in H and let A be an operator on H. Then A is dissipative if and only if $Re(Ax|x) \leq 0$ for all $x \in D(A)$.

Holomorphic Semigroups

We now turn our attention to bounded holomorphic semigroups. To this end, for $\phi \in (0, \pi)$, we define the sector Σ_θ in the complex plane by

$$\Sigma_\theta := \{\lambda \in \mathbb{C}\backslash\{0\} : |\arg\lambda| < \theta\}.$$

The definition of such semigroups reads as follows.

Definition 1.3.9 Let $\theta \in (0, \frac{\pi}{2}]$. A C_0-semigroup T is called a bounded holomorphic semigroup of angle θ if T has a bounded holomorphic extension to $\Sigma_{\theta'}$ for each $\theta' \in (0, \theta)$.

If we do not want to specify the angle, we call T a *bounded holomorphic semigroup* if T is a bounded holomorphic semigroup of angle θ for some $\theta \in (0, \frac{\pi}{2}]$.

The following characterization theorem for holomorphic semigroups is of fundamental importance for many parabolic problems. Generators of bounded analytic semigroups are closely related to so-called sectorial operators, which we will study in more detail in the following section. For this reason, we will also postpone the proof of the following Theorem 1.3.10 to this section; see Remark 1.4.8(a).

Theorem 1.3.10 *Let A be an operator in X and $\theta \in (0, \frac{\pi}{2}]$. Then the following assertions are equivalent.*

(i) A generates a bounded holomorphic semigroup of angle θ.
(ii) $\Sigma_{\theta+\frac{\pi}{2}} \subset \rho(A)$ and

$$\sup_{\lambda \in \Sigma_{\theta+\frac{\pi}{2}-\varepsilon}} \|\lambda R(\lambda, A)\| < \infty \text{ for all } \varepsilon > 0.$$

We remark that if one is not interested in the angle of holomorphy in the above theorem, it suffices to verify condition (ii) above in a right half plane, only.

Corollary 1.3.11 *An operator A on X generates a bounded holomorphic semigroup on X if and only if $\{z \in \mathbb{C} : Rez > 0\} \subset \rho(A)$ and*

$$M := \sup_{Re\lambda>0} \|\lambda R(\lambda, A)\| < \infty. \tag{1.3.5}$$

In order to prove the assertion of this corollary it suffices by the above Theorem 1.3.10 to show that condition (1.3.5) implies assertion (ii) of Theorem 1.3.10. To this end, set $c := 1/2M$ and for $s \in \mathbb{R}\backslash\{0\}$ and $-c|s| < r \leq 0$ let $\lambda := c|s|+r+is$.

Then $|\lambda - (r + is)| = c|s| \leq 1/2 \|R(\lambda, A)\|^{-1}$. It follows that $r + is \in \rho(A)$ and

$$\|(r + is)R(r + is, A)\| \leq 2M(c + 1).$$

Thus (ii) is satisfied with $\theta = \arctan c$. $\qquad\square$

Next, we define holomorphic semigroups as follows.

Definition 1.3.12 An operator A is said to generate a *holomorphic semigroup* on X if there exists $\omega \geq 0$ such that $A - \omega$ generates a bounded holomorphic semigroup.

The following examples of holomorphic semigroups are of special interest in the following sections.

Examples 1.3.13

(a) (*Selfadjoint Operators*).

Every selfadjoint operator in a Hilbert space H which is bounded above by ω is the generator of a bounded holomorphic semigroup of angle $\pi/2$ on H satisfying

$$\|T(z)\| \leq e^{\omega Rez}, \qquad Rez > 0.$$

In fact, by the spectral theorem we may assume that $H = L^2(\Omega, \mu)$ and that A is given by $Af = m \cdot f$ for $f \in D(A) = \{f \in H : mf \in H\}$ and a measurable function $m : \Omega \to (-\infty, \omega]$. Thus $T(z)f(x) := e^{zm(x)}f(x)$ for $Rez > 0$ and $x \in \Omega$ defines a bounded, holomorphic semigroup on H with generator A.

(b) (*The Laplacian and the Gaussian semigroup*).

The solution of the heat equation $u_t - \Delta u = 0, t > 0$ on $\mathbb{R}^n$ with initial data $u(0) = f \in X$ is governed by the Gaussian semigroup T, which for X being one of the spaces $L^p(\mathbb{R}^n)$ for $1 \leq p < \infty$, $C_0(\mathbb{R}^n)$ or $BUC(\mathbb{R}^n)$ can be represented as

$$T(t)f(x) := (4\pi t)^{-n/2} \int_{\mathbb{R}^n} f(x - y)e^{-|y|^2/4t}dy, \quad t > 0, f \in X, x \in \mathbb{R}^n.$$

This semigroup is a bounded holomorphic semigroup on X of angle $\pi/2$. Its generator is the Laplacian Δ_X on X equipped with maximal domain, i.e.

$$D(\Delta_X) = \{f \in X : \Delta f \in X\} \text{ and } \Delta_X f = \Delta f.$$

If $X = L^p(\mathbb{R}^n)$ and $1 < p < \infty$, then by Mikhlin's Theorem 1.1.5, $D(\Delta_X)$ coincides with $W^{2,p}(\mathbb{R}^n)$. Mikhlin's theorem implies also that in this case $\sigma(\Delta) = (-\infty, 0]$.

(c) (*The Dirichlet Laplacian Δ_D in $L^p(\mathbb{R}^n_+)$ for $1 < p < \infty$*).

Consider the Dirichlet Laplacian Δ_D in $L^p(\mathbb{R}^n_+)$ for $1 < p < \infty$ defined by

$$Au = \Delta u, \quad D(A) = W^{2,p}(\mathbb{R}^n_+) \cap W_0^{1,p}(\mathbb{R}^n_+). \tag{1.3.6}$$

Then A is densely defined, injective and has dense range. Let $E : L^p(\mathbb{R}_+^n) \to L^p(\mathbb{R}^n)$ be the extension operator by 0 and $R : L^p(\mathbb{R}^n) \to L^p(\mathbb{R}_+^n)$ be the restriction of a function $u \in L^p(\mathbb{R}^n)$ to $L^p(\mathbb{R}_+^n)$. Then, for $\lambda \in \mathbb{C}$ with $|\arg(\lambda)| < \pi$,

$$R(\lambda, A) = RR(\lambda, \Delta)E - RSR(\lambda, \Delta)E, \tag{1.3.7}$$

where S denotes the reflection of a function in the normal coordinate, i.e., S is given by $Su(x_1, \dots, x_{n-1}, x_n) = u(x_1, \dots, x_{n-1}, -x_n)$. Hence, Δ_D generates a bounded holomorphic semigroup on $L^p(\mathbb{R}_+^n)$.

d) (*The Dirichlet Laplacian* Δ_D *in spaces of bounded functions*).
Using (1.3.7) and Example (b), it follows that the Dirichlet Laplacian generates a bounded holomorphic semigroup also on $BUC(\mathbb{R}_+^n)$, $C_0(\mathbb{R}_+^n)$ and $L^\infty(\mathbb{R}_+^n)$, which in the latter case is, however, not strongly continuous. Note that its domain does *not* coincide with the spaces given in (1.3.6).

When compared with arbitrary C_0-semigroups, holomorphic semigroups show many particular properties. The so-called smoothing effect of holomorphic semigroups can be shown elegantly by means of the functional calculus developed in the following Sect. 1.4. We hence postpone the proof of the following Theorem 1.3.14 again to Sect. 1.4; see Remark 1.4.8(b).

Theorem 1.3.14 *Let A be the generator of a bounded holomorphic semigroup on X, $x \in X$ and $n \in \mathbb{N}$. Then*

$$\sup_{t>0} \|t^n A^n e^{tA}\| < \infty,$$

and there exists a unique function

$$u \in C^\infty((0, \infty); X) \cap C([0, \infty); X) \cap C^1((0, \infty), D(A))$$

satisfying the Cauchy problem $u'(t) = Au(t)$ for $t > 0$ and $u(0) = x$.

The following perturbation result for generators of holomorphic semigroups is in particular useful when treating lower order perturbations of differential operators.

Theorem 1.3.15 (Relatively Bounded Perturbations) *Let A be the generator of a holomorphic semigroup on X. Assume that $B : D(A) \to X$ is an operator such that for every $\varepsilon > 0$ there exists a constant $b \geq 0$ such that*

$$\|Bx\| \leq \varepsilon \|Ax\| + b\|x\|, \quad x \in D(A).$$

Then $A + B$ generates a holomorphic semigroup.

Proof Assume first that A generates a bounded analytic semigroup. By Theorem 1.3.10 there exists $\theta \in (0, \pi/2]$ such that $\Sigma_{\theta+\frac{\pi}{2}} \subset \rho(A)$ and

$\sup_{\lambda \in \Sigma_{\theta + \frac{\pi}{2}}} \|\lambda R(\lambda, A)\| =: M < \infty$. It follows from the assumption that, given $\varepsilon > 0$, there exists $b \geq 0$ such that for $x \in X$

$$\|BR(\lambda, A)x\| \leq \varepsilon(M + 1)\|x\| + \frac{bM}{\lambda}\|x\|, \quad \lambda \in \Sigma_{\theta + \frac{\pi}{2}}.$$

Choosing ε small enough, it follows that $\|BR(\lambda, A)\| \leq q < 1$ for $|\lambda|$ sufficiently large. Hence $Id - BR(\lambda, A)$ is invertible and since

$$\lambda - (A + B) = (Id - BR(\lambda, A))(\lambda - A), \quad \lambda \in \Sigma_{\theta + \frac{\pi}{2}},$$

it follows that $\lambda - (A + B)$ is invertible for all $\lambda \in \Sigma_{\theta + \frac{\pi}{2}}$ with λ sufficiently large and that $\|R(\lambda, A + B)\| \leq \frac{C}{|\lambda|}$ for all $\lambda \in \Sigma_{\theta + \frac{\pi}{2}}$ with λ sufficiently large. The general case follows by applying the above proof to $A - \omega$. □

Real Interpolation Spaces

We conclude this section with some remarks concerning the real interpolation space $(X, D(A))_{\theta, p}$ which is defined for generators A of analytic semigroups T as follows. Let us first recall from Theorem 1.3.14 that there exists a constant $C > 0$ such that

$$\|e^{tA}x\| + \|tAe^{tA}x\| \leq C\|x\|$$

for $x \in X$ and $t \in [0, 1]$. Further, for $\theta \in (0, 1)$ and $p \in (1, \infty)$ we set

$$\phi_{\theta, p}(x, t) := t^{1 - \theta - 1/p}\|Ae^{tA}x\|_X.$$

Definition 1.3.16 For $\theta \in (0, 1)$ and $p \in (1, \infty)$ the *real interpolation space* $(X, D(A))_{\theta, p}$ is defined as

$$\left(X, D(A)\right)_{\theta, p} := \{x \in X : \phi_{\theta, p}(x, \cdot) \in L^p(0, 1)\}$$

equipped with the norm

$$\|x\|_{\theta, p} := \|x\| + \|\phi_{\theta, p}(x, \cdot)\|_{L^p(0, 1)}.$$

Remarks 1.3.17

(a) We remark first that the space $(X, D(A))_{\theta, p}$ is obviously an intermediate space between X and $D(A)$ in the sense that $D(A) \subset \left(X, D(A)\right)_{\theta, p} \subset X$ with continuous embeddings.

(b) The real interpolation space $(X, D(A))_{\theta, p}$ equipped with the above norm is a Banach space. This can be seen by standard methods.

(c) Replacing the interval $(0, 1)$ in the above definition by the interval $(0, T)$ for some $T > 0$ leads to the same space $(X, D(A))_{\theta, p}$ as above, however with an equivalent norm.

(d) It is sometimes more convenient to denote the above real interpolation space by $D_A(\theta, p)$, i.e. we set $D_A(\theta, p) := (X, D(A))_{\theta, p}$ with θ and p as above.

1.4 Sectorial Operators and Bounded H^∞-Functional Calculus

In this section we introduce and investigate the class of sectorial operators, a class of operators which play an important role in the analysis of the Stokes equation. We also introduce the concept of a bounded H^∞-functional calculus for sectorial operators.

Sectorial Operators and Bounded H^∞-Calculus
Throughout this section, X always denotes a complex Banach space. For $\theta \in (0, \pi)$ we recall that the sector Σ_θ in the complex plane is given by

$$\Sigma_\theta := \{\lambda \in \mathbb{C}\backslash\{0\} : |\arg \lambda| < \theta\}.$$

Definition 1.4.1 (Sectorial Operators) Let X be a Banach space and A be a closed, densely defined linear operator on X. Then A is called *sectorial of angle* $\phi \in (0, \pi)$ if

(S1) A is injective and has dense range,
(S2) $\sigma(A) \subset \overline{\Sigma_\phi}$ and

$$\sup\{\|\lambda R(\lambda, A)\|_{\mathcal{L}(X)} : \lambda \in \mathbb{C}\backslash\overline{\Sigma_{\phi'}}\} < \infty \text{ for all } \phi' \in (\phi, \pi).$$

The class of sectorial operators on X of angle ϕ will be denoted by $\mathcal{S}_\phi(X)$, and $\mathcal{S}(X)$ denotes the class of sectorial operators of some angle ϕ. Moreover, we call the infimum of all ϕ for which (S2) is satisfied, the *spectral angle* ϕ_A of A.

The following elementary properties of sectorial operators will be useful later on.

Lemma 1.4.2 *Let A be a sectorial operator on X. Then, for $x \in X$*

$$\lim_{t \to \infty} t(t + A)^{-1}x = x, \quad \lim_{t \to 0} t(t + A)^{-1}x = 0,$$
$$\lim_{t \to \infty} A(t + A)^{-1}x = 0, \quad \lim_{t \to 0} A(t + A)^{-1}x = x.$$

In particular, $D(A^k) \cap R(A^k)$ is dense in X for all $k \in \mathbb{N}$.

The proof of the above assertions is based on the identity

$$Id - t(t + A)^{-1} = A(t + A)^{-1}$$

and on the fact that the sets $\{t(t+A)^{-1} : t > 0\}$ and $\{A(t+A)^{-1} : t > 0\}$ are bounded in $\mathcal{L}(X)$. Given $x \in X$ and $n, k \in \mathbb{N}$, define

$$x_{n,k} := (1 + n^{-1}A)^{-k}A^k(n^{-1} + A)^{-k}x.$$

Then $x_{n,k} \in D(A^k) \cap R(A^k)$ and $x_{n,k} \to x$, as $n \to \infty$, by the above assertion.

We now describe the construction of a functional calculus for sectorial operators which is being inspired by the classical Dunford calculus. To this end, for $\theta \in (0, \pi]$, we introduce the space $H^\infty(\Sigma_\theta)$ of holomorphic functions as

$$H^\infty(\Sigma_\theta) = \{f : \Sigma_\theta \to \mathbb{C}, f \text{ holomorphic and bounded}\}. \tag{1.4.1}$$

Equipped with the norm

$$|f|_\infty^\theta = \sup\{|f(\lambda)| : |\arg \lambda| < \theta\}, \tag{1.4.2}$$

$H^\infty(\Sigma_\theta)$ is a Banach algebra. For the time being, assume that $A \in \mathcal{S}(X)$ is bounded and invertible. Fix $\theta > \phi_A$. Then the classical Dunford calculus for bounded linear operators applies. In this situation the spectrum $\sigma(A)$ is a compact subset of Σ_θ and choosing a closed path Γ_A in Σ_θ surrounding $\sigma(A)$ counterclockwise, we define

$$f(A) := \frac{1}{2\pi i} \int_{\Gamma_A} f(\lambda)R(\lambda, A)d\lambda, \quad f \in H^\infty(\Sigma_\theta). \tag{1.4.3}$$

Noting that Γ_A is compact, the above integral is well defined. It is well known that this formula defines an algebra homomorphism from $H^\infty(\Sigma_\theta)$ to $\mathcal{L}(X)$.

Relation (1.4.3) will be the starting point of our definition of a boundeded holomorphic functional calculus for sectorial operators A on X. First we consider functions $f \in H^\infty(\Sigma_\theta)$ having some decay at 0 and infinity. More precisely, for $\lambda \in \Sigma_\theta$ define $\varrho(\lambda) := \frac{\lambda}{(1+\lambda)^2}$ and assume that

$$|f(\lambda)| \leq C|\varrho(\lambda)|^\varepsilon, \quad \lambda \in \Sigma_\theta \tag{1.4.4}$$

for some $C > 0$ and $\varepsilon > 0$. For $\theta \in (0, \pi)$ we then set

$$H_0^\infty(\Sigma_\theta) := \{f \in H^\infty(\Sigma_\theta) : \text{ there exists } C, \varepsilon > 0 \text{ such that (1.4.4) holds}\}. \tag{1.4.5}$$

Given a sectorial operator $A \in \mathcal{S}(X)$ and $f \in H_0^\infty(\Sigma_\theta)$ where $\theta > \phi_A$, we define

$$f(A) := \frac{1}{2\pi i} \int_{\partial \Sigma_{\theta'}} f(\lambda)R(\lambda, A)d\lambda,$$

where $\phi_A < \theta' < \theta$. It then follows that this integral is absolutely convergent in $\mathcal{L}(X)$. This definition of $f(A)$ allows further to define a functional calculus for A on $H_0^\infty(\Sigma_\theta)$ having the following properties.

Proposition 1.4.3 *Let X be a Banach space and $A \in \mathcal{S}_\phi(X)$. For $\phi < \theta' < \theta$ and $f \in H_0^\infty(\Sigma_\theta)$ set*

$$\Phi_A(f) := f(A) := \frac{1}{2\pi i} \int_{\partial \Sigma_{\theta'}} f(\lambda) R(\lambda, A) d\lambda. \tag{1.4.6}$$

Then $\Phi_A : H_0^\infty(\Sigma_\theta) \to \mathcal{L}(X)$ is a linear mapping with the following properties:

(a) the above integral is independent of the particular choice of $\theta' \in (\phi, \theta)$.
(b) $\Phi_A(f \cdot g) = \Phi_A(f)\Phi_A(g)$, $f, g \in H_0^\infty(\Sigma_\theta)$.
(c) Let (f_n), $f \in H_0^\infty(\Sigma_\theta)$ be uniformly bounded and assume that $f_n(\lambda) \to f(\lambda)$ for $\lambda \in \Sigma_\theta$. Then

$$\lim_{n \to \infty} \Phi_A(f_n \cdot g) = \Phi_A(f \cdot g) \text{ in } \mathcal{L}(X)$$

for all $g \in H_0^\infty(\Sigma_\theta)$.
(d) If $f(\lambda) = \lambda(\mu_1 - \lambda)^{-1}(\mu_2 - \lambda)^{-1}$ with $\mu_1, \mu_2 \notin \overline{\Sigma_\theta}$, then

$$f(A) = AR(\mu_1, \Lambda)R(\mu_2, A).$$

Note that the functional calculus developed in Proposition 1.4.3 does not yield a satisfying calculus so far, since it does not apply to standard functions such as $\lambda \mapsto (\mu - \lambda)^{-1}$. However, we will improve the above construction by the following approximation procedure. To this end, we introduce the following approximate identity.

Example 1.4.4 (Approximate Identity) In the situation of Proposition 1.4.3, define for $n \geq 2$ the functions $\varrho_n : \Sigma_\theta \to \mathbb{C}$ by $\varrho_n(\lambda) := n(n + \lambda)^{-1} - (1 + n\lambda)^{-1}$. Then

$$\varrho_n(A) = n(n + A)^{-1} - n^{-1}(n^{-1} + A)^{-1}.$$

In fact, $\varrho_n \in H_0^\infty(\Sigma_\theta)$ for all $n \geq 2$ and the above equality follows from Proposition 1.4.3(d). Moreover, the range of $\varrho_n(A)^k$ equals $D(A^k) \cap Rg(A^k)$ for all $k \in \mathbb{N}$ and $\lim_{n \to \infty} \varrho_n(A)^k x = x$ for all $x \in X$. The latter assertion follows from Lemma 1.4.2 since $\lim_{n \to \infty} \varrho_n(A)x = x$ and hence $\lim_{n \to \infty} \varrho_n(A)^k x = x$.

We will now construct a closed extension of Φ_A defined in Proposition 1.4.3 whose domain $H_A^\infty(\Sigma_\theta)$ will consist of all $f \in H^\infty(\Sigma_\theta)$ for which $f(A)$ can be defined as a bounded operator on X. More precisely, for $A \in \mathcal{S}(X)$, $\theta > \phi_A$ and $f \in H_0^\infty(\Sigma_\theta)$ set

$$|||f|||_A := \|f\|_{H^\infty(\Sigma_\theta)} + \|f(A)\|_{\mathcal{L}(X)}$$

and define $H_A^\infty(\Sigma_\theta)$ to be the class of functions $f \in H^\infty(\Sigma_\theta)$ for which there exists a sequence $(f_n) \subset H_0^\infty(\Sigma_\theta)$ with $f_n(\lambda) \to f(\lambda)$ for all $\lambda \in \Sigma_\theta$ and $\sup_{n \in \mathbb{N}} |||f_n(A)|||_A < \infty$.

Then $H_A^\infty(\Sigma_\theta)$ is a subalgebra of $H^\infty(\Sigma_\theta)$. We will now show that for $f \in H_A^\infty(\Sigma_\theta)$ the limit

$$\overline{\Phi_A}(f)(x) := \lim_{n \to \infty} \Phi_A(f_n)(x), \quad x \in X$$

exists and defines a functional calculus on $H_A^\infty(\Sigma_\theta)$.

Proposition 1.4.5 (McIntosh's Convergence Lemma) *Let $A \in \mathcal{S}(X)$ and $\theta > \phi_A$. Then there exists an extension $\overline{\Phi_A} : H_A^\infty(\Sigma_\theta) \to \mathcal{L}(X)$ of Φ_A defined in Proposition 1.4.3 satisfying the following properties:*

(a) *$\overline{\Phi_A}$ is linear and multiplicative.*
(b) *For $\mu \notin \overline{\Sigma_\theta}$, $r_\mu(\lambda) = (\mu - \lambda)^{-1} \in H_A^\infty(\Sigma_\theta)$ and $\overline{\Phi_A}(r_\mu) = R(\mu, A)$.*
(c) *Let $f \in H^\infty(\Sigma_\theta)$ and $f_n \in H_A^\infty(\Sigma_\theta)$ such that $f_n(\lambda) \to f(\lambda)$ for all $\lambda \in \Sigma_\theta$ and $|||f_n|||_A < C$ for all $n \in \mathbb{N}$ and some $C > 0$. Then $f \in H_A^\infty(\Sigma_\theta)$, $\lim_{n \to \infty} \overline{\Phi_A}(f_n)(x) = \overline{\Phi_A}(f)(x)$ for all $x \in X$ and $\|\overline{\Phi_A}\| \leq C$.*

For a proof of this lemma we refer to [113] or [95].

We are now in the position to introduce the concept of operators having a bounded H^∞-calculus on X.

Definition 1.4.6 Let $A \in \mathcal{S}(X)$ and $\theta \in (\phi_A, \pi)$. Then A is said to admit a *bounded $H^\infty(\Sigma_\theta)$-calculus* if there exists a constant $C > 0$ such that

$$\|f(A)\|_{\mathcal{L}(X)} \leq C \|f\|_{H^\infty(\Sigma_\theta)} \quad \text{for all } f \in H_0^\infty(\Sigma_\theta). \tag{1.4.7}$$

The infimum of such θ is called the H^∞-*angle* of A and is denoted by ϕ_A^∞. The class of sectorial operators on X admitting a bounded H^∞-calculus for some angle $\theta \in (0, \pi)$ is denoted by $\mathcal{H}^\infty(X)$.

Remarks 1.4.7

(a) Note that due to Proposition 1.4.5, for $\theta > \phi_A$, $A \in \mathcal{S}(X)$ admits a bounded $H^\infty(\Sigma_\theta)$-calculus if and only if there exists a bounded algebra homomorphism $H^\infty(\Sigma_\theta) \to \mathcal{L}(X)$, $f \mapsto f(A)$, satisfying the convergence properties (c) of Proposition 1.4.5 and such that

$$f(A) = \frac{1}{2\pi i} \int_{\partial \Sigma_{\theta'}} f(\lambda) R(\lambda, A) d\lambda$$

for $f \in H_0^\infty(\Sigma_\theta)$ and $\theta > \theta' > \phi_A$.
(b) We also remark that $A \in \mathcal{S}(X)$ admits a bounded H^∞-calculus of angle $\theta > \phi_A$ if and only if $H^\infty(\Sigma_\theta) = H_A^\infty(\Sigma_\theta)$.

(c) Let $A \in \mathcal{S}(X)$ and $\theta > \phi_A$. Assume that $f \in H_0^\infty(\Sigma_\theta)$ is in addition analytic in a neighborhood of 0. Then $f \in H_A^\infty(\Sigma_\theta)$ and

$$\overline{\Phi_A}(f) = \frac{1}{2\pi i} \int_{\partial \Sigma(\delta)} f(\lambda) R(\lambda, A) d\lambda,$$

where $\Sigma(\delta) := \Sigma_{\theta'} \cup B(0, \delta)$ for $\theta' \in (\phi_A, \theta)$ and δ being small enough so that f is analytic on $\Sigma(\delta)$. Note that in this situation

$$\|\overline{\Phi_A}(f)\| \leq \frac{1}{2\pi} \sup_{\lambda \in \partial \Sigma(\delta)} \|\lambda R(\lambda, A)\| \int_{\partial \Sigma(\delta)} |f(\lambda)||\lambda|^{-1} d\lambda.$$

(d) Since $\varrho_n(\lambda) \to 1$ and $\varrho_n(A)x \to x$ for all $x \in X$ by Example 1.4.4 we have $1 \in H_A^\infty(\Sigma_\theta)$ and $\Phi_A(1) = Id$.
(e) Let A be a sectorial operator in a Banach space X such that $D(A')$ is dense in X'. Then $A \in \mathcal{H}^\infty(X)$ if and only if $A' \in \mathcal{H}^\infty(X')$ and $\phi_A^\infty = \phi_{A'}^\infty$.

Note that for a given sectorial operator A the class $H_A^\infty(\Sigma_\theta)$ contains many interesting functions; in particular it contains the functions $\lambda \mapsto \lambda^n e^{-\lambda z}$. They allow us to derive elegant proofs of the basic theorems for bounded holomorphic semigroups by the functional calculus described above as follows.

Remarks 1.4.8 Let $A \in \mathcal{S}(X)$ with $\phi_A < \pi/2$.

(a) For $z \in \mathbb{C}$ with $|\arg(z)| < \pi/2 - \phi_A$ set $e_z(\lambda) := e^{-\lambda z}$. Then $e_z \in H_A^\infty(\Sigma_\theta)$ for $\theta \in (\phi_A, \frac{\pi}{2} - |\arg(z)|)$ by Remark 1.4.7(c). Setting $T(z) := e_z(A)$, we obtain the assertion of Theorem 1.3.10 thus by our functional calculus, this remark and by noting that the semigroup property $T(z_1)T(z_2) = T(z_1 + z_2)$ follows from $e_{z_1} e_{z_2} = e_{z_1+z_2}$.
(b) Similarly, we obtain the assertion of Theorem 1.3.14 again from Remark 1.4.7 (c) by noting that $\{t^n A^n e^{tA} : t > 0\} = \{\varphi(tA) : t > 0\}$ with $\varphi(\lambda) = \lambda^n e^{-\lambda}$ and since $\int |\varphi(t\lambda)| \frac{d\lambda}{|\lambda|} < \infty$ and independent of $t > 0$.

Examples
We now present two basic examples of operators, the derivative operator $\frac{d}{dt}$ and the Laplacian Δ, respectively, both admitting a bounded H^∞-calculus on L^p-spaces.

Example 1.4.9 (The Time Derivative on $\mathbb{R}$, $\mathbb{R}_+$ and $(0, T)$)

(a) *The derivative operator on $L^p(\mathbb{R})$.*

For $1 < p < \infty$ define $A_\mathbb{R}$ on $L^p(\mathbb{R})$ by

$$A_\mathbb{R} u = \frac{d}{dt} u, \quad D(A_\mathbb{R}) = W^{1,p}(\mathbb{R}).$$

It is easy to see that $A_\mathbb{R}$ is densely defined, injective and has dense range. In order to show that $A_\mathbb{R}$ is sectorial, consider for $\lambda \in \mathbb{C}$ with $|\arg \lambda| > \theta > \frac{\pi}{2}$ and

$g \in \mathcal{S}(\mathbb{R})$ the equation $\lambda u - A_{\mathbb{R}}u = g$. Applying the Fourier transform yields $\lambda \widehat{u}(\tau) - i\tau \widehat{u}(\tau) = \widehat{g}(\tau)$ for all $\tau \in \mathbb{R}$. This equation admits the unique solution $\widehat{u} = m_\lambda \widehat{g}$ with $m_\lambda(\tau) = (\lambda - i\tau)^{-1}$. By Mikhlin's Theorem 1.1.5, the operator T_{m_λ} given by $T_{m_\lambda}f = \mathcal{F}^{-1}m_\lambda\mathcal{F}f$ for $f \in \mathcal{S}(\mathbb{R})$ has a continuous extension to $L^p(\mathbb{R})$ satisfying $\|T_{m_\lambda}\| \leq \frac{C}{|\lambda|}$ for some $C > 0$ independent of λ. Moreover, we have $R(\lambda, A_{\mathbb{R}}) = T_{m_\lambda}$ for all such λ. Hence $A_{\mathbb{R}}$ is a sectorial operator in $L^p(\mathbb{R})$ of angle $\frac{\pi}{2}$.

Furthermore, let $f \in H_0^\infty(\Sigma_\theta)$ for $\theta > \frac{\pi}{2}$ and $u \in \mathcal{S}(\mathbb{R})$. Then

$$f(A_{\mathbb{R}}) = \frac{1}{2\pi i} \int_{\partial \Sigma_{\theta'}} f(\lambda) R(\lambda, A_{\mathbb{R}}) d\lambda$$

for $\theta' \in (\frac{\pi}{2}, \theta)$ and applying the Fourier transform yields

$$\mathcal{F}(f(A_{\mathbb{R}})u)(\tau) = \frac{1}{2\pi i} \int_{\partial \Sigma_{\theta'}} f(\lambda)(\lambda - i\tau)^{-1}\widehat{u}(\tau) d\lambda = f(i\tau)\widehat{u}(\tau), \quad \tau \in \mathbb{R}.$$

Cauchy's theorem and Mikhlin's multiplier theorem imply that $\tau \mapsto f(i\tau)$ defines a Fourier multiplier for $L^p(\mathbb{R})$ and we verify that

$$\|f(A_{\mathbb{R}})\|_{\mathcal{L}(L^p(\mathbb{R}))} \leq C\|f\|_{H^\infty(\Sigma_\theta)}, \quad f \in H_0^\infty(\Sigma_\theta),$$

for some $C > 0$. We hence proved that the derivative operator $A_{\mathbb{R}} = \frac{d}{dt}$ in $L^p(\mathbb{R})$ equipped with the domain $W^{1,p}(\mathbb{R})$ admits a bounded H^∞-calculus on $L^p(\mathbb{R})$ with $\phi_{A_{\mathbb{R}}}^\infty = \frac{\pi}{2}$.

(b) *The time derivative on $L^p(\mathbb{R}_+)$*

Consider next the derivative operator $A_{\mathbb{R}_+}$ defined on $L^p(\mathbb{R}_+)$ with domain $W_0^{1,p}(\mathbb{R}_+)$. As before, we see that $A_{\mathbb{R}_+}$ is densely defined, injective and has dense range. Given $u \in L^p(\mathbb{R}_+)$, we call $Eu \in L^p(\mathbb{R})$ the extension of u to $\mathbb{R}$ by 0. Obviously, $E : L^p(\mathbb{R}_+) \to L^p(\mathbb{R})$ is a continuous operator. The restriction of E to $W_0^{1,p}$ is obviously again continuous from $W_0^{1,p}(\mathbb{R}_+)$ to $W^{1,p}(\mathbb{R})$. Further, the restriction operator $R : L^p(\mathbb{R}) \to L^p(\mathbb{R}_+)$ defined by $Ru := u_{|\mathbb{R}_+}$ is also continuous. Hence, for $\lambda \in \mathbb{C}$ with $|\arg \lambda| > \frac{\pi}{2}$, the resolvent $R(\lambda, A_{\mathbb{R}_+})$ of $A_{\mathbb{R}_+}$ is given by

$$R(\lambda, A_{\mathbb{R}_+}) = RR(\lambda, A_{\mathbb{R}})E.$$

Thus $A_{\mathbb{R}_+}$ is sectorial of angle $\frac{\pi}{2}$. Furthermore, for $f \in H_0^\infty(\Sigma_\theta)$ with $\theta > \frac{\pi}{2}$ we obtain

$$f(A_{\mathbb{R}_+})v = Rf(A_{\mathbb{R}})Ev, \quad v \in L^p(\mathbb{R}_+).$$

It thus follows that the derivative operator $A_{\mathbb{R}_+}$ in $L^p(\mathbb{R}_+)$ equipped with the domain $W_0^{1,p}(\mathbb{R}_+)$ admits a bounded H^∞-calculus on $L^p(\mathbb{R}_+)$ with $\phi_{A_{\mathbb{R}_+}}^\infty = \frac{\pi}{2}$.

(c) *The Time Derivative on $L^p(0, T)$.*

For $T > 0$ the derivative operator $A_{(0,T)}$ on $L^p(0, T)$ is defined by

$$Au := u', \qquad D(A_{(0,T)}) = \{u \in W^{1,p}(0, T) : u(0) = 0\}$$

In this case $A_{(0,T)}$ is even surjective. Denoting by $E_{(0,T)}$ and $R_{(0,T)}$ the extension and restriction operators, from $L^p(0, T) \to L^p(\mathbb{R}_+)$ and $L^p(\mathbb{R}_+) \to L^p(0, T)$, respectively, we see that for $\lambda \in \mathbb{C}$ with $|\arg \lambda| > \frac{\pi}{2}$, the resolvent $R(\lambda, A_{(0,T)})$ of $A_{(0,T)}$ is given by

$$R(\lambda, A_{(0,T)}) = R_{(0,T)} R(\lambda, A_{\mathbb{R}_+}) E_{(0,T)}.$$

As above

$$f(A_{(0,T)})v = R_{(0,T)} f(A_{\mathbb{R}_+}) E_{(0,T)} v, \quad v \in L^p(0, T), f \in H_0^\infty(\Sigma_\theta),$$

and thus the derivative operator $A_{(0,T)}$ in $L^p(0, T)$ is invertible and admits a bounded H^∞-calculus on $L^p(0, T)$ with $\phi_{A_{(0,T)}}^\infty = \frac{\pi}{2}$.

Example 1.4.10 (The Negative Laplacian on $\mathbb{R}^n$ and $\mathbb{R}_+^n$)

(a) *The negative Laplacian $-\Delta$ in $L^p(\mathbb{R}^n)$.*
   As a second example, we consider the negative Laplacian $-\Delta$ in $L^p(\mathbb{R}^n)$ for $1 < p < \infty$ defined as

$$Au = -\Delta u, \quad D(A) = W^{2,p}(\mathbb{R}^n).$$

Then $-\Delta$ is densely defined and injective. Weyl's lemma implies that A has dense range in $L^p(\mathbb{R}^n)$.
   As above we see that for $\lambda \in \mathbb{C}$ with $|\arg \lambda| > \theta > 0$ and $g \in \mathcal{S}(\mathbb{R}^n)$ the equation $\lambda u - Au = g$ has a unique solution u given by $T_{m_\lambda} g = \mathcal{F}^{-1} m_\lambda \mathcal{F} g$, where m_λ is now given by $m_\lambda(\xi) = (\lambda + |\xi|^2)^{-1}$ for $\xi \in \mathbb{R}^n$. Mikhlin's theorem implies that $m_\lambda \in \mathcal{M}_p(\mathbb{R}^n)$ and that $\|\lambda m_\lambda\|_{\mathcal{M}_p} \leq C$ for some C independent of θ. Hence $-\Delta$ is sectorial with angle $\phi_A = 0$. Similarly as above we see that for $f \in H_0^\infty(\Sigma_\theta)$ with $\theta > 0$ and $u \in \mathcal{S}(\mathbb{R}^n)$

$$\mathcal{F}(f(A)u)(\xi) = \frac{1}{2\pi i} \int_{\partial \Sigma_{\theta'}} f(\lambda)(\lambda + |\xi|^2)^{-1} \widehat{u}(\xi) d\lambda = f(-|\xi|^2) \widehat{u}(\xi), \, \xi \in \mathbb{R}^n.$$

Again, Cauchy's theorem and Mikhlin's multiplier theorem imply that $\xi \mapsto f(-|\xi|^2)$ defines a Fourier multiplier for $L^p(\mathbb{R}^n)$ and we verify that

$$\|f(A)\|_{\mathcal{L}(\mathbb{R}^n)} \leq C\|f\|_{H^\infty(\Sigma_\theta)}, \quad f \in H_0^\infty(\Sigma_\theta),$$

for some $C > 0$. Hence, the negative Laplacian $A = -\Delta$ in $L^p(\mathbb{R}^n)$ with domain $W^{2,p}(\mathbb{R}^n)$ admits a bounded H^∞-calculus on $L^p(\mathbb{R}^n)$ with $\phi_A^\infty = 0$.

(b) *The negative Dirichlet Laplacian* $-\Delta_D$ *in* $L^p(\mathbb{R}^n_+)$.

Consider the negative Dirichlet Laplacian $-\Delta_D$ in $L^p(\mathbb{R}^n_+)$ for $1 < p < \infty$ defined as in Example 1.3.13(c) by

$$Au = -\Delta u, \quad D(A) = W^{2,p}(\mathbb{R}^n_+) \cap W_0^{1,p}(\mathbb{R}^n_+).$$

Formula (1.3.7) implies that

$$f(A) = Rf(-\Delta)E - RSf(-\Delta)E,$$

where $-\Delta$ is defined as in (a). Hence, $-\Delta_D$ admits a bounded H^∞-calculus on $L^p(\mathbb{R}^n_+)$ with $\phi_A^\infty = 0$.

Further Results

We conclude this section with stating several classes of operators allowing for a bounded H^∞-calculus. For a proof of these assertions, we refer e.g. to [75, 95] or [88].

Proposition 1.4.11

(a) *Let H be a Hilbert space, $A \in \mathcal{S}(H)$, and assume that $-A$ generates a contraction semigroup on H. Then A has a bounded H^∞-calculus on H and $\phi_A^\infty = \phi_A$.*

(b) *Let $-A$ be the generator of a positive contraction semigroup or a bounded group on $L^p(\Omega)$, where $1 < p < \infty$ and $\Omega \subset \mathbb{R}^n$ is open. Then A has a bounded $H^\infty(\Sigma_\theta)$-calculus on $L^p(\Omega)$ for all $\theta \in (\pi/2, \pi]$.*

Furthermore, let $-A$ be an elliptic differential operators of order $2m$ on $L^p(\Omega)$ subject to general boundary conditions, where $\Omega \subset \mathbb{R}^n$ is a domain of regularity-class C^{2m}. Then, roughly speaking, A admits a bounded H^∞-calculus on $L^p(\Omega)$ provided the top-order coefficients of A are Hölder continuous. For details, see [37]. We mention here explicitly only the examples of the Dirichlet and Neumann Laplacian. To this end, let $\Omega \subset \mathbb{R}^n$, $n \in \mathbb{N}$, be a bounded domain with boundary of class C^2, $1 < p < \infty$ and let $-\Delta_D$ be the negative Dirichlet Laplacian on $L^p(\Omega)$ defined by

$$\Delta_D u = \Delta u \quad \text{for} \quad u \in D(\Delta_D) = W^{2,p}(\Omega) \cap W_0^{1,p}(\Omega).$$

We also define the negative Laplacian subject to Neumann boundary conditions $-\Delta_N$ as

$$\Delta_N u = \Delta u \quad \text{for} \quad u \in D(\Delta_N) = \{u \in W^{2,p}(\Omega) : \partial_\nu u = 0 \text{ on } \partial\Omega\}.$$

Proposition 1.4.12 *The negative Dirichlet Laplacian* $-\Delta_D$ *as well as the negative Neumann Laplacian* $-\Delta_N$ *admit a bounded* H^∞*-calculus on* $L^p(\Omega)$ *for* $1 < p < \infty$ *with* $\phi^\infty_{-\Delta_D} = \phi^\infty_{-\Delta_N} = 0$.

Remark 1.4.13 In Sect. 1.14 on nematic liquid crystal flow we consider the Neumann Laplacian also in the space $H^{1,p}(\Omega)$, where $\Omega \subset \mathbb{R}^n$ is above. We then set

$$\Delta_N^1 u = \Delta u \quad \text{for} \quad u \in D(\Delta_N) = \{u \in W^{3,p}(\Omega) : \partial_\nu u = 0 \text{ on } \partial\Omega\}.$$

It then can be shown that $-\Delta_N^1$ admits also a bounded H^∞-calculus on $H^{1,p}(\Omega)$ for $1 < p < \infty$ with $\phi^\infty_{-\Delta_N^1} = 0$.

1.5   Fractional Powers

In this section we introduce fractional powers of sectorial operators. The approach we are using is based on an extended H^∞-calculus. Note that so far we are unable to define operators A^n, $A^{1/2}$ or more generally A^α for $\alpha \in \mathbb{C}$ by the methods developed in Sect. 1.4, since these operators as well as the functions $\lambda \mapsto \lambda^\alpha$ are unbounded on suitable sectors Σ_θ. We hence extend the functional calculus described in Sect. 1.4 to polynomially bounded functions. Before we start with this extended functional calculus, let us remark that in the sequel we do not assume that $0 \in \rho(A)$.

In the following, let $\theta \in (0, \pi)$ and let $\alpha \geq 0$. Then $H_\alpha(\Sigma_\theta)$ is defined to be the space of all holomorphic functions $f : \Sigma_\theta \to \mathbb{C}$ for which

$$\|f\|_{H_\alpha(\Sigma_\theta)} := \sup\{|\varrho(\lambda)|^\alpha |f(\lambda)| : \lambda \in \Sigma_\theta\} < \infty,$$

where ϱ is defined as in Sect. 1.4 as $\varrho(\lambda) = \lambda(1+\lambda)^{-2}$. For $f \in H_\alpha(\Sigma_\theta)$ and $k \in \mathbb{N}$ with $k > \alpha$ we obtain

$$f(\lambda) = \varrho(\lambda)^{-k}(\varrho^k f)(\lambda).$$

Since the second factor above belongs to $H_0^\infty(\Sigma_\theta)$, it follows from Proposition 1.4.3 that $(\varrho^k f)(A)$ is a bounded operator on X. The following definition is thus natural.

Definition 1.5.1 Let $A \in \mathcal{S}(X)$, $\theta \in (\phi_A, \pi)$ and $\alpha \geq 0$. For $f \in H_\alpha^\infty(\Sigma_\theta)$ and $k \in \mathbb{N}$ with $k > \alpha$ set

$$f(A) := \varrho(A)^{-k}(\varrho^k f)(A)$$
$$D(f(A)) := \{x \in X : (\varrho^k f)(A)x \in D(\varrho(A)^{-k})\}.$$

The multiplicativity of the functional calculus implies that $f(A)$ is well defined.

It is now natural to define the fractional powers A^z for $z \in \mathbb{C}$ by applying the extended functional calculus to the function $f_z : \Sigma_\theta \to \mathbb{C}, \lambda \mapsto \lambda^z = e^{z \log \lambda}$, where we use the branch of the logarithm which is holomorphic in $\mathbb{C} \setminus \{\mathbb{R}_-\}$. In fact, the estimate

$$|f_z(\lambda)| = |r^z||e^{i\varphi z}| \leq |\lambda|^{Rez} e^{\theta |Imz|}, \quad \lambda = re^{i\varphi} \in \Sigma_\theta,$$

implies that $f_z \in H_{Rez}(\Sigma_\theta)$ for all $\theta < \pi$.

Definition 1.5.2 For $A \in \mathcal{S}(X)$, $\theta \in (\phi_A, \pi)$ and $z \in \mathbb{C}$, the operator A^z is defined by

$$A^z := f_z(A), \quad z \in \mathbb{C}.$$

It is then clear that A^z is a well defined, closed and densely defined operator on X. In the following we state the basic properties of the fractional power operators A^z.

Theorem 1.5.3 For $A \in \mathcal{S}(X)$, the fractional powers A^z, $z \in \mathbb{C}$, are closed and injective operators satisfying $\overline{D(A)} = \overline{R(A)} = X$. Moreover,

(a) if $Rez_1 < Rez_2 < 0 < Rez_3 < Rez_4$, then

$$D(A^{z_1}) \subset D(A^{z_2}), \ D(A^{z_3}) \supset D(A^{z_4}),$$
$$R(A^{z_1}) \subset R(A^{z_2}), \ R(A^{z_3}) \supset R(A^{z_4}).$$

(b) $A^{z_1+z_2} = A^{z_1} A^{z_2}$ for all $z_1, z_2 \in \mathbb{C}$ with $Rez_1 \cdot Rez_2 > 0$.
(c) $(A^z)^{-1} = A^{-z}, \quad z \in \mathbb{C}.$

The proof of the above properties of A^z is quite technical and we refrain from giving it here. For a detailed proofs of the above result, we refer to [95], Theorem 15.15.

For exponents $\alpha \in \mathbb{R}$ we have the following multiplicative rule.

Proposition 1.5.4 Let $A \in \mathcal{S}(X)$ and $\alpha \in \mathbb{R}$ with $\phi_A |\alpha| < \pi$. Then $A^\alpha \in \mathcal{S}(X)$ and $\phi_{A^\alpha} = |\alpha| \phi_A$. Furthermore,

$$(A^\alpha)^z = A^{\alpha z}, \quad z \in \mathbb{C}.$$

Moreover, for any $\alpha, \beta, \gamma \in \mathbb{R}$ satisfying $\alpha < \beta < \gamma$ with $\gamma - \alpha < \pi/\phi_A$, the momentum inequality

$$\|A^\alpha u\| \leq C \|A^\alpha u\|^{\frac{\gamma-\beta}{\gamma-\alpha}} \|A^\gamma u\|^{\frac{\beta-\alpha}{\gamma-\alpha}}, \quad u \in D(A^\alpha) \cap D(A^\gamma) \tag{1.5.1}$$

is valid for some constant M independent of u.

We remark that in the above situation $D(A + \eta)^\alpha = D(A^\alpha)$ for any $\eta > 0$ and that there exists a constant $C > 0$ such that

$$\|(A + \eta)^\alpha - A^\alpha u\| \leq C\eta^\alpha \|u\|, \qquad u \in D(A^\alpha). \tag{1.5.2}$$

If in the situation of Proposition 1.5.4 the operator A admits a bounded H^∞-calculus on X and $\alpha > 0$, then A^α also admits a bounded H^∞-calculus provided $\alpha \phi_A^\infty < \pi$. More precisely, the following result holds.

Corollary 1.5.5 *Let $A \in \mathcal{H}^\infty(X)$, $\alpha > 0$ such that $\alpha \phi_A^\infty < \pi$. Then*

$$A^\alpha \in \mathcal{H}^\infty(X) \quad \text{and} \quad \phi_{A^\alpha}^\infty \leq \alpha \phi_A^\infty.$$

In order to prove this assertion, let $f \in H_0^\infty(\Sigma_\theta)$ for $\theta \in (\alpha \phi_A^\infty, \pi)$ and ϱ_n as in Example 1.4.4. Then,

$$R(\mu, A^\alpha)\varrho_n(A)x = \frac{1}{2\pi i} \int_{\partial\Sigma_{\theta/\alpha}} (\mu - \lambda^\alpha)^{-1} \varrho_n(\lambda) R(\lambda, A)x \, d\lambda$$

for $x \in X$, $\mu \in \partial\Sigma_\theta$ and Fubini's theorem implies that

$$f(A^\alpha)\varrho_n(A)x = \frac{1}{2\pi i} \int_{\partial\Sigma_\theta} f(\lambda)\varrho_n(\lambda) R(\lambda, A^\alpha)x \, d\lambda$$

$$= (\frac{1}{2\pi i})^2 \int_{\partial\Sigma_{\theta/\alpha}} \int_{\partial\Sigma_\theta} f(\mu)(\mu - \lambda^\alpha)^{-1} \varrho_n(\lambda) R(\mu, A)x \, d\lambda \, d\mu$$

$$= \frac{1}{2\pi i} \int_{\partial\Sigma_{\theta/\alpha}} f(\lambda^\alpha)\varrho_n(\lambda) R(\lambda, A)x \, d\lambda = \varrho_n(A) f(\cdot^\alpha)(A)x.$$

Letting $n \to \infty$, we see that $f(A^\alpha)x$ and $f(\cdot^\alpha)(A)x$ coincide and that

$$\|f(A^\alpha)\| = \|f(\cdot^\alpha)(A)\| \leq C\|(\cdot^\alpha)\|_{H^\infty(\Sigma_{\theta/\alpha})} \leq \|f\|_{H^\infty(\Sigma_\theta)}, \quad f \in H_0^\infty(\Sigma_\theta).$$

The assertion thus follows.

We finish this section by considering the fractional power spaces

$$X_\alpha := (D(A^\alpha), \|\cdot\|_\alpha), \qquad \|x\|_\alpha := \|x\| + \|A^\alpha x\|, \qquad 0 < \alpha < 1,$$

for sectorial operators A on X. If, in addition $A \in \mathcal{H}^\infty(X)$, then it is possible to identify X_α in many cases as a concrete function space by the following theorem.

Theorem 1.5.6 *Let $A \in H^\infty(X)$ for some Banach space X. Then*

$$X_\alpha \simeq [X, D(A)]_\alpha, \qquad 0 < \alpha < 1,$$

where $[X, D(A)]_\alpha$ denotes the complex interpolation space of order α.

For a proof of this fact, we refer e.g. to [148].

1.6 Operator-Valued H^∞-Calculus, $\mathcal{R}$-Boundedness, Fourier Multipliers and Maximal L^p-Regularity

In this section we extend the bounded H^∞-calculus developed in the previous section to the operator-valued setting.

To this end, denote by $\mathcal{A}$ the subalgebra of $\mathcal{L}(X)$ of all bounded operators that commute with resolvents of a given sectorial operator A on a Banach space X. We then consider bounded holomorphic functions $F : \Sigma_\theta \to \mathcal{A}$. Roughly speaking, we show that the functional calculus for scalar valued, bounded holomorphic functions described in Sect. 1.4 can be extended to operator-valued functions $F : \Sigma_\theta \to \mathcal{A}$ provided the range $\{F(\lambda) : \lambda \in \Sigma_\theta\}$ of F is not only uniformly bounded, but $\mathcal{R}$-bounded. This notion will be of fundamental importance in the operator-valued setting.

Starting from this extension we may then deduce fairly easily a result on the closedness of the sum of two sectorial operators which allows further to characterize the property of maximal L^p-regularity for parabolic evolution equations.

We begin with the notion of $\mathcal{R}$-bounded families of operators.

$\mathcal{R}$-Bounded Families of Operators

Definition 1.6.1 Let X and Y be Banach spaces. A family of operators $\mathcal{T} \subset \mathcal{L}(X, Y)$ is called $\mathcal{R}$-*bounded*, if there is a constant $C > 0$ and $p \in [1, \infty)$ such that for each $N \in \mathbb{N}$, $T_j \in \mathcal{T}$, $x_j \in X$ and for all independent, symmetric, $\{-1, 1\}$-valued random variables ε_j on a probability space $(\Omega, \mathcal{M}, \mu)$, the inequality

$$\| \sum_{j=1}^N \varepsilon_j T_j x_j \|_{L^p(\Omega; Y)} \leq C \| \sum_{j=1}^N \varepsilon_j x_j \|_{L^p(\Omega; X)} \qquad (1.6.1)$$

is valid. The smallest such C is called the $\mathcal{R}$-*bound of* $\mathcal{T}$, we denote it by $\mathcal{R}(\mathcal{T})$.

Some remarks about properties of $\mathcal{R}$-bounded families of operators are in order.

Remarks 1.6.2

(a) If $\mathcal{T} \subset \mathcal{L}(X, Y)$ is $\mathcal{R}$-bounded then it is also uniformly bounded, with

$$\sup\{\|T\| : T \in \mathcal{T}\} \leq \mathcal{R}(\mathcal{T}).$$

This follows from the definition of $\mathcal{R}$-boundedness with $N = 1$, since $\|\varepsilon_1\|_{L^p(\Omega)} = 1$.

(b) The definition of $\mathcal{R}$-boundedness is independent of $p \in [1, \infty)$. This observation follows directly from *Kahane's inequality*: For any Banach space X and $1 \leq p, q < \infty$ there is a constant $C(p, q, X)$ such that

$$\| \sum_{j=1}^N \varepsilon_j x_j \|_{L^p(\Omega; X)} \leq C(p, q, X) \| \sum_{j=1}^N \varepsilon_j x_j \|_{L_q(\Omega; X)}, \qquad (1.6.2)$$

for each $N \in \mathbb{N}$, $x_j \in X$, and for all independent, symmetric, $\{-1, 1\}$-valued random variables ε_j on a probability space $(\Omega, \mathcal{M}, \mu)$.

(c) Assume that X and Y are Hilbert spaces. Then $\mathcal{T} \subset \mathcal{L}(X, Y)$ is $\mathcal{R}$-bounded if and only if $\mathcal{T}$ is uniformly bounded.

In fact, let $\mathcal{T}$ be uniformly bounded by $C > 0$. Then, choosing $p = 2$ it follows that

$$\| \sum_{j=1}^{N} \varepsilon_j T_j x_j \|_{L_2(\Omega; Y)}^2 \leq C^2 \| \sum_{j=1}^{N} \varepsilon_j x_j \|_{L_2(\Omega; X)}^2$$

since the ε_j are independent, hence orthogonal in $L_2(\Omega)$.

(d) Let $X = Y = L^p(G)$ for some open $G \subset \mathbb{R}^n$. Then $\mathcal{T} \subset \mathcal{L}(X, Y)$ is $\mathcal{R}$-bounded if and only if there is a constant $M > 0$ such that the following *square function estimate* holds:

$$\| (\sum_{j=1}^{N} |T_j f_j|^2)^{1/2} \|_{L^p(G)}$$

$$\leq M \| (\sum_{j=1}^{N} \|f_j\|^2)^{1/2} \|_{L^p(G)}, \quad N \in \mathbb{N}, f_j \in L^p(G), T_j \in \mathcal{T}. \quad (1.6.3)$$

This is a consequence of *Khintchine's inequality*: For each $p \in [1, \infty)$ there is a constant $K_p > 0$ such that

$$K_p^{-1} \| \sum_{j=1}^{N} \varepsilon_j a_j \|_{L^p(\Omega)} \leq (\sum_{j=1}^{N} |a_j|^2)^{1/2} \leq K_p \| \sum_{j=1}^{N} \varepsilon_j a_j \|_{L^p(\Omega)}, \quad (1.6.4)$$

for all $N \in \mathbb{N}$, $a_j \in \mathbb{C}$, and for all independent, symmetric, $\{-1, 1\}$-valued random variables ε_j on a probability space $(\Omega, \mathcal{M}, \mu)$.

(e) The following assertion is known as *Kahane's contraction principle*. Let $N \in \mathbb{N}$, $x_j \in X$, ε_j independent, symmetric, $\{-1, 1\}$-valued random variables on a probability space $(\Omega, \mathcal{M}, \mu)$, and $\alpha_j, \beta_j \in \mathbb{C}$ such that $|\alpha_j| \leq |\beta_j|$, for each $j = 1, \ldots, N$. Then

$$\| \sum_{j=1}^{N} \alpha_j \varepsilon_j x_j \|_{L^p(\Omega; X)} \leq 2 \| \sum_{j=1}^{N} \beta_j \varepsilon_j x_j \|_{L^p(\Omega; X)}.$$

The above Remark 1.6.2(d) gives a very useful sufficient condition for the $\mathcal{R}$-boundedness of kernel operators in $L^p(G)$. More precisely, we have the following lemma.

Lemma 1.6.3 *Let $G \subset \mathbb{R}^n$ be open and $\mathcal{T} = \{T_\mu : \mu \in M\} \subset \mathcal{L}(L^p(G; \mathbb{C}^m))$ a family of kernel operators of the form*

$$[T_\mu f](x) = \int_G k_\mu(x, y) f(y) dy, \quad x \in G, \ f \in L^p(G; \mathbb{C}^m),$$

which are dominated by a kernel k_0, i.e.

$$|k_\mu(x, y)| \leq k_0(x, y), \quad \text{for a.a. } x, y \in G, \text{ and all } \mu \in M.$$

Then $\mathcal{T} \subset \mathcal{L}(L^p(G; \mathbb{R}^m))$ is $\mathcal{R}$-bounded, provided T_0 is bounded in $L^p(G)$.

Proof By Remark 1.6.2(d) we only have to verify the square function estimate (1.6.3). Due to L^p-boundedness of the dominating operator T_0, we have

$$\|(\sum_{j=1}^N |T_j f_j|^2)^{1/2}\|_{L^p(G)} \leq \|(\sum_{j=1}^N (T_0|f_j|)^2)^{1/2}\|_{L^p(G)}$$

$$\leq \|T_0(\sum_{j=1}^N |f_j|^2)^{1/2}\|_{L^p(G)} \leq \|T_0\|_{\mathcal{L}(L^p(G))}\|(\sum_{j=1}^N |f_j|^2)^{1/2}\|_{L^p(G)}.$$

$\square$

We remark further that $\mathcal{R}$-bounds behave like norms. More precisely, let X, Y, Z be Banach spaces and $\mathcal{T}, \mathcal{S} \subset \mathcal{L}(X, Y)$ be $\mathcal{R}$-bounded. Then

$$\mathcal{T} + \mathcal{S} = \{T + S : T \in \mathcal{T}, \ S \in \mathcal{S}\}$$

is $\mathcal{R}$-bounded as well, and $\mathcal{R}(\mathcal{T} + \mathcal{S}) \leq \mathcal{R}(\mathcal{T}) + \mathcal{R}(\mathcal{S})$. Moreover, let $\mathcal{T} \subset \mathcal{L}(X, Y)$ and $\mathcal{S} \subset \mathcal{L}(Y, Z)$ be $\mathcal{R}$-bounded. Then $\mathcal{S}\mathcal{T} = \{ST : T \in \mathcal{T}, \ S \in \mathcal{S}\}$ is $\mathcal{R}$-bounded, and $\mathcal{R}(\mathcal{S}\mathcal{T}) \leq \mathcal{R}(\mathcal{S})\mathcal{R}(\mathcal{T})$.

Another useful criteria for $\mathcal{R}$-boundedness of a given family of operators is given in the following lemma.

Lemma 1.6.4 *Let $G \subset \mathbb{C}$ be open, $K \subset G$ compact, and suppose that $H : G \to \mathcal{L}(X, Y)$ is a holomorphic function. Then $H(K) \subset \mathcal{L}(X, Y)$ is $\mathcal{R}$-bounded.*

In order to prove this assertion, fix $z_0 \in K$. Since H is holomorphic in G there exists a ball $B_r(z_0)$ such that the power series representation

$$H(z) = \sum_{k=0}^\infty H^{(k)}(z_0) \frac{(z - z_0)^k}{k!}, \quad |z - z_0| \leq r,$$

of H is absolutely convergent and $\rho_0 := \sum_{k=0}^\infty \|H^{(k)}(z_0)\|_{\mathcal{L}(X,Y)} r^k/k! < \infty$. It follows from Remark 1.6.2(e) that $\mathcal{R}(H(B_r(z_0))) \leq 2\rho_0$. Covering K by a finite set of such balls, the assertion follows.

Operator-Valued H^∞-Calculus and the Kalton–Weis Theorem

We now turn our attention to the operator-valued H^∞-functional calculus. Indeed, for a sectorial operator $A \in \mathcal{S}(X)$ consider an operator-valued function $F \in H^\infty(\Sigma_\phi; \mathcal{L}(X))$ satisfying

$$F(\lambda)(\mu - A)^{-1} = (\mu - A)^{-1}F(\lambda), \quad \mu \in \varrho(A), \lambda \in \Sigma_\phi$$

for some $\phi > \phi_A$. In this case we write $F \in H^\infty(\Sigma_\phi, \mathcal{A})$, where $\mathcal{A}$ denotes the subalgebra of all bounded operators acting on X, which commute with resolvents of A. Furthermore, we set

$$RH^\infty(\Sigma_\phi, \mathcal{A}) := \{F \in H^\infty(\Sigma_\phi, \mathcal{A}) : \{F(z) : z \in \Sigma_\phi\} \text{ is } \mathcal{R}\text{-bounded}\}.$$

As in the scalar-valued case, we denote by $RH_0^\infty(\Sigma_\phi; \mathcal{A})$ those elements of $RH^\infty(\Sigma_\phi; \mathcal{A})$ for which there exist constants $C, \varepsilon > 0$ such that

$$\|F(z)\|_{\mathcal{L}(X)} \leq C \left(\frac{|z|}{1 + |z|^2}\right)^\varepsilon, \quad z \in \Sigma_\phi.$$

For $F \in RH_0^\infty(\Sigma_\phi; \mathcal{L}(X))$, the integral

$$F(A) = \frac{1}{2\pi i} \int_{\partial \Sigma_{\phi'}} F(\lambda)(\lambda - A)^{-1} d\lambda \tag{1.6.5}$$

is well defined as a Bochner integral where $\phi_A < \phi' < \phi$.

We now show that formula (1.6.5) give rise to a bounded functional calculus on $RH^\infty(\Sigma_\phi, \mathcal{A})$ provided A admits a bounded H^∞-calculus. More precisely, the following theorem holds true.

Theorem 1.6.5 (Kalton–Weis) *Let $A \in \mathcal{H}^\infty(X)$. Assume that $F \in H^\infty(\Sigma_\phi, \mathcal{A})$ for some $\phi > \phi_A^\infty$ and such that $\mathcal{R}\{F(z) : z \in \Sigma_\phi\} \leq M$ for some $M > 0$. Then there exists a constant $C_A > 0$, depending only on A, such that*

$$F(A) \in \mathcal{L}(X) \text{ and } \|F(A)\|_{\mathcal{L}(X)} \leq C_A M.$$

Remark 1.6.6 It should be noted that in the situation of Theorem 1.6.5 there exists a bounded algebra homomorphism $\Phi_A : RH^\infty(\Sigma_\phi; \mathcal{A}) \to \mathcal{L}(X)$ with $\Phi_A(F) = F(A)$ as in (1.6.5) for $F \in RH_0^\infty(\Sigma_\phi; \mathcal{A})$ satisfying the following convergence property:
if (F_n) is a bounded sequence in $RH^\infty(\Sigma_\phi; \mathcal{A})$ such that $F_n(\lambda)x \to F(\lambda)x$ for some $F \in RH^\infty(\Sigma_\phi; \mathcal{A})$ and all $\lambda \in \Sigma_\phi, x \in X$, then

$$\Phi_A(F_n)x \to \Phi_A(F)x, \quad x \in X.$$

The proof of Theorem 1.6.5 is based on the following key lemma.

Lemma 1.6.7 *Let* $A \in \mathcal{H}^\infty(X)$ *and* $h \in \mathcal{H}_0^\infty(\Sigma_\phi)$ *for* $\phi > \phi_A^\infty$. *Then there exists a constant* $C > 0$ *such that*

$$\| \sum_{k \in \mathbb{Z}} \alpha_k h(2^k t A) \|_{\mathcal{L}(X)} \le C \sup_{k \in \mathbb{Z}} |\alpha_k|, \qquad \alpha_k \in \mathbb{C}, t > 0.$$

Proof The assumption $h \in \mathcal{H}_0^\infty(\Sigma_\phi)$ implies that there exist constants $c, \beta > 0$ such that

$$|h(z)| \le c \frac{|z|^\beta}{1 + |z|^{2\beta}}, \qquad z \in \Sigma_\phi.$$

Setting $f(z) := \sum_{k \in \mathbb{Z}} \alpha_k h(2^k t z)$ and observing that for $r = t|z|$

$$\sum_{k \in \mathbb{Z}} |h(2^k t z)| \le c \sum_{k \in \mathbb{Z}} \frac{(r2^k)^\beta}{1 + (r2^k)^{2\beta}} \le \frac{2c}{1 - 2^{-\beta}},$$

we see that

$$|f(z)| \le \sup_{k \in \mathbb{Z}} |\alpha_k| \sum_{k \in \mathbb{Z}} |h(2^k t z)| \le C \sup_{k \in \mathbb{Z}} |\alpha_k|$$

for some constant $C > 0$. Hence, the above series is absolutely converging and f defines a bounded and analytic function on Σ_ϕ. Since $A \in \mathcal{H}^\infty(X)$ and $\phi > \phi_A^\infty$ by assumption, we obtain

$$\| \sum_{k \in \mathbb{Z}} \alpha_k h(2^k t A) \|_{\mathcal{L}(X)} = \|f(A)\|_{\mathcal{L}(X)} \le C_A |f|_{H^\infty} \le C \sup_{k \in \mathbb{Z}} |\alpha_k|.$$

□

Proof of Theorem 1.6.5 We assume first that $F \in H_0^\infty(\Sigma_\phi; \mathcal{L}(X))$. Then

$$F(A) = \frac{1}{2\pi i} \int_\Gamma F(\lambda)(\lambda - A)^{-1} d\lambda = \frac{1}{2\pi i} \int_\Gamma F(\lambda)\lambda^{-1/2} A^{1/2} (\lambda - A)^{-1} d\lambda,$$

where Γ is defined by $\Gamma = \{re^{\pm i\theta} : r \ge 0\}$ for some $\theta \in (\phi_A^\infty, \phi)$. Without loss of generality we may assume in the following that $\lambda \in \Gamma^+ := \Gamma \cap \{\lambda \in \mathbb{C} : \mathrm{Im}\lambda \ge 0\}$. Due to our assumption that $F \in H_0^\infty(\Sigma_\phi; \mathcal{L}(X))$, the above integrals are absolutely convergent and for $\Gamma_N^+ := \{\lambda \in \Gamma^+ : 2^{-N} \le |\lambda| \le 2^N\}$ we may thus write

$$F(A) = \lim_{N \to \infty} \frac{1}{2\pi i} \int_{\Gamma_N^+} F(\lambda)\lambda^{-1/2} A^{1/2} (\lambda - A)^{-1} d\lambda = \lim_{N \to \infty} F_N(A),$$

where

$$F_N(A) = \frac{e^{i\theta/2}}{2\pi i} \int_{2^{-N}}^{2^N} F(re^{i\theta}) A^{1/2} (re^{i\theta} - A)^{-1} r^{-1/2} dr$$

$$= \frac{e^{i\theta/2}}{2\pi i} \int_{2^{-N}}^{2^N} F(re^{i\theta}) h(r^{-1}A) r^{-1} dr$$

$$= \frac{e^{i\theta/2}}{2\pi i} \sum_{k=-N}^{N-1} \int_1^2 F(2^k t e^{i\theta}) h(2^{-k} t^{-1} A) t^{-1} dt$$

$$= \frac{e^{i\theta/2}}{2\pi i} \int_1^2 H_N(t) t^{-1} dt,$$

and $h \in H_0^\infty$ is defined by $h(z) = z^{1/2}(e^{i\theta} - z)^{-1}$ and H_N is given by

$$H_N(t) := \sum_{k=-N}^{N-1} F(2^k t e^{i\theta}) h(2^{-k} t^{-1} A), \qquad t \in [1, 2].$$

For $x \in X$ and $x' \in X'$ we may estimate H_N by Lemma 1.6.7 as

$$|< H_N(t)x, x' >| = |\sum_{k=-N}^{N-1} < F(2^k t e^{i\theta}) h(2^{-k} t^{-1} A)x, x' >|$$

$$= \int_\Omega |\sum_{k=-N}^{N-1} \varepsilon_k^2 < F(2^k t e^{i\theta}) h^{1/2} (2^{-k} t^{-1} A)x, x' >|$$

$$= \int_\Omega |< \sum_{k=-N}^{N-1} \varepsilon_k F(2^k t e^{i\theta}) h^{1/2} (2^{-k} t^{-1} A)x, \sum_{k=-N}^{N-1} \varepsilon_k h^{1/2} (2^{-k} t^{-1} A^*)x' >|$$

$$\leq || \sum_{k=-N}^{N-1} \varepsilon_k F(2^k t e^{i\theta}) h^{1/2} (2^{-k} t^{-1} A)x ||_{L^2(\Omega; X)}$$

$$\times || \sum_{k=-N}^{N-1} \varepsilon_k h^{1/2} (2^{-k} t^{-1} A^*)x' ||_{L^2(\Omega; X')}$$

$$\leq \mathcal{R}(F(\Sigma_\phi)) || \sum_{k=-N}^{N-1} \varepsilon_k h^{1/2} (2^{-k} t^{-1} A)x ||_{L^2(\Omega; X)}$$

$$\times || \sum_{k=-N}^{N-1} \varepsilon_k h^{1/2} (2^{-k} t^{-1} A^*)x' ||_{L^2(\Omega; X')}$$

$$\leq C^2 \mathcal{R}(F(\Sigma_\phi)) ||x|| ||x'||.$$

Hence, $H_N(t)$ is uniformly bounded in $t \in [1, 2]$ and $N \in \mathbb{N}$, which shows that

$$\|F(A)\|_{\mathcal{L}(X)} \leq C^2 \mathcal{R}(F(\Sigma_\phi)).$$

Finally, in order to prove the assertion for general $F \in H^\infty(\Sigma_\phi; \mathcal{L}(X))$, we replace F by F_n, where F_n is given by $F_n(z) = F(z)\varrho_n$ and ϱ_n is the approximate identity introduced in Example 1.4.4.   □

$\mathcal{R}$-Bounded H^∞-Calculus and Property (α)

It is now a natural question to ask whether the set of operators produced by a bounded H^∞-calculus from a uniformly bounded set of functions is $\mathcal{R}$-bounded. We then say that $A \in \mathcal{H}^\infty(X)$ admits an $\mathcal{R}$-*bounded H^∞-calculus* provided the set

$$\{h(A) : h \in H^\infty(\Sigma_\theta), |h|_\infty^\theta \leq 1\}$$

is $\mathcal{R}$-bounded for some $\theta > 0$. The class of such operators is denoted by $\mathcal{RH}^\infty(X)$ and the $\mathcal{RH}^\infty$-angle $\phi_A^{\mathcal{R}\infty}$ of A is defined to be the infimum of such angles θ.

Note that we have the relations

$$\mathcal{RH}(X) \subset \mathcal{H}^\infty(X) \subset \mathcal{S}(X).$$

It is a very remarkable fact that, roughly speaking, the classes $\mathcal{H}^\infty(X)$ and $\mathcal{RH}^\infty(X)$ coincide for Banach spaces satisfying the so called property (α). A Banach space X is said to have *property* (α) if the contraction principle described in Remark 1.6.2(e) holds for random sequences $(r_i r_j)_{i,j \in \mathbb{N}}$, i.e. more precisely if there exists a constant $C > 0$ such that for each $n \in \mathbb{N}$, each subset $(x_{ij})_{i,j=1,...,n}$ of X and each subset $(\alpha_{ij})_{i,j=1,...,n}$ of $\mathbb{C}$ with $|\alpha_{ij}| = 1$

$$\int_0^1 \int_0^1 \| \sum_{i,j=1}^n r_i(s) r_j(t) \alpha_{ij} x_{ij} \|_X ds dt \leq C \int_0^1 \int_0^1 \| \sum_{i,j=1}^n r_i(s) r_j(t) \|_X ds dt.$$

We remark first that due to Khintchine's inequality, see Remark 1.6.2(d), $\mathbb{C}$ has property (α). Secondly, Fubini's theorem implies that $L^p(\Omega; X)$ has property (α) for $1 \leq p < \infty$ provided X enjoys this property and Ω is a σ-finite measure space. Thus, closed subspaces of $L^p(\Omega, \mathbb{C})$ have property (α) for the above values of p.

The precise statement about the relation between the classes $\mathcal{H}^\infty$ and $\mathcal{RH}^\infty$ reads as follows.

Proposition 1.6.8 *Let X be a Banach space with property* (α)*, $A \in \mathcal{H}^\infty(\Sigma_\phi)$, $\phi' > \phi$ and $R > 0$. Then the set*

$$\{f(A) : \|f\|_\infty^{\phi'} \leq R\}$$

is $\mathcal{R}$-bounded.

For a proof of this fact, we refer to [95].

Maximal L^p-Regularity

We finally consider the problem of maximal L^p-regularity for parabolic evolution equations. By this we mean the following. Consider the inhomogeneous Cauchy problem

$$(ACP_0) \qquad \begin{cases} u'(t) = Au(t) + f(t), & t \geq 0 \\ u(0) = u_0, \end{cases} \qquad (1.6.6)$$

where A is a sectorial operator on a Banach space X of spectral angle $\phi_A < \pi/2$. Let $1 < p < \infty$ and $0 < T \leq \infty$. We say that A admits *maximal L^p-regularity* on $[0, T)$ if for $u_0 = 0$ and all $f \in L^p([0, T); X)$ the solution of (1.6.6), given by

$$u(t) = \int_0^t e^{(t-s)A} f(s) ds,$$

is differentiable a.e., takes values in $D(A)$ a.e., and u' and Au belong to $L^p([0, T); X)$. It follows from the closed graph theorem that in this case there exists a constant C_p such that

$$\|u'\|_{L^p([0,T);X)} + \|Au\|_{L^p([0,T);X)} \leq C_p \|f\|_{L^p([0,T);X)}. \qquad (1.6.7)$$

Observe that by (1.6.6), $f = u' - Au$, and hence u' as well as Au cannot lie in a "better" function space as f. This explains the word "maximal".

Our aim is to show that such an estimate holds true in a certain class of Banach spaces if and only if A is a so called $\mathcal{R}$-sectorial operator . Here we call a sectorial operator A $\mathcal{R}$-*sectorial of angle* $\phi \in (0, \pi)$ provided $\sigma(A) \subset \overline{\Sigma_\phi}$ and for all $\phi' > \phi$ the set

$$\{\lambda R(\lambda, A) : \phi' \leq |\arg(\lambda)| \leq \pi\}$$

is $\mathcal{R}$-bounded. The infimum of all ϕ for which the above set is $\mathcal{R}$-bounded is called the $\mathcal{R}$-*angle* of A, denoted by $\phi_A^{\mathcal{R}}$.

The class of all $\mathcal{R}$-sectorial operators on X will be denoted by $\mathcal{RS}(X)$.

We are now in the position to state and prove the following result on the closedness of two resolvent commuting operators.

Theorem 1.6.9 (Kalton–Weis Sum Theorem) *Let A and B be two resolvent commuting sectorial operators on a Banach space X. Assume that $B \in \mathcal{H}^\infty(X)$ and that A is $\mathcal{R}$-sectorial with $\phi_B^\infty + \phi_A^{\mathcal{R}} < \pi$. Then $A + B$ is closed on $D(A) \cap D(B)$ and there exists a constant $C > 0$ such that*

$$\|Ax\| + \|Bx\| \leq C\|Ax + Bx\|, \quad x \in D(A) \cap D(B). \qquad (1.6.8)$$

Proof The idea of the proof is to show that $A(A + B)^{-1}$ and $B(A + B)^{-1}$ can be defined to be bounded operators on X by means of the operator-valued functional calculus developed in Theorem 1.6.5.

More precisely, let $\mathcal{B}$ be the subalgebra of $\mathcal{L}(X)$ consisting of all operators commuting with resolvents of B and choose $\phi_1 \in (\phi_B^\infty, \pi), \phi_2 \in (\phi_A^{\mathcal{R}}, \pi)$ such that $\phi_1 + \phi_2 < \pi$. Consider the functions F and G defined by

$$F(\lambda) := A(\lambda + A)^{-1}, \qquad G(\lambda) := (\lambda + A)\varrho_n(\lambda)^2 \varrho_n(A)^2$$

with (ϱ_n) as in Example 1.4.4. Then $F \in RH^\infty(\Sigma_{\phi_1}; \mathcal{B})$ and $G \in H_0^\infty(\Sigma_{\phi_1}; \mathcal{B})$ and $G(\lambda)F(\lambda) = \varrho_n(A)^2 A \varrho_n(\lambda)^2$. By the multiplicativity of the functional calculus,

$$(B + A)\varrho_n(B)^2 \varrho_n(A)^2 F(B) = G(B)F(B) = A\varrho_n(B)^2 \varrho(A)^2.$$

By Theorem 1.6.5, $F(B) \in \mathcal{L}(X)$ and there thus exists a constant $C > 0$ such that

$$\|\varrho_n(B)^2 \varrho_n(A)^2 Ax\| \leq C\|\varrho_n(B)^2 \varrho_n(A)^2 (B + A)x\|, \quad x \in D(A) \cap D(B).$$

Letting $n \to \infty$ gives

$$\|Ax\| \leq C\|(B + A)x\|, \quad x \in D(A) \cap D(B).$$

Since $\|Bx\| \leq \|Ax\| + \|(A + B)x\| \leq (C + 1)\|(A + B)x\|$ for $x \in D(A) \cap D(B)$, the estimate (1.6.8) follows.

In order to prove the closedness of $A + B$ on $D(A) \cap D(B)$, let $x_n \in D(A) \cap D(B)$ with $x_n \to x$ and $Ax_n + Bx_n \to y$ in X. By (1.6.8), (Ax_n) as well as (Bx_n) are Cauchy sequences so that $x \in D(A) \cap D(B)$ and $\lim_{n \to \infty}(Ax_n + Bx_n) = Ax + Bx = y$ by the closedness of A and B. The proof is complete. □

Let us come back to the maximal regularity estimate (1.6.7), i.e.

$$\|u'\|_{L^p([0,T);X)} + \|Au\|_{L^p([0,T);X)} \leq C_p \|f\|_{L^p([0,T);X)},$$

which we now rewrite as an inequality for operators on $L^p([0, T); X)$ for $1 < p < \infty$. Indeed, let $\tilde{B}$ be the derivative operator $\frac{d}{dt}$ on $\tilde{X} = L^p([0, T); X)$ and let $\tilde{A}$ be the extension of A to $\tilde{X}$ given by $(\tilde{A}f)(t) = -A(f(t))$. Then estimate (1.6.7) is equivalent to

$$\|\tilde{A}y\|_{\tilde{X}} + \|\tilde{B}y\|_{\tilde{X}} \leq C\|(\tilde{A} + \tilde{B})y\|_{\tilde{X}}, \quad y \in D(\tilde{A}) \cap D(\tilde{B}). \tag{1.6.9}$$

Of course, $\tilde{A}$ and $\tilde{B}$ are sectorial operators with commuting resolvents. We will show in the following that $\tilde{B}$ has a bounded H^∞-calculus on $\tilde{X} = L^p([0, T); X)$ for all angles $\phi > \pi/2$ provided X is a UMD-space. Recall that this class of spaces has been introduced in Definition 1.2.7 in Sect. 1.2.

Taking this for granted for the time being and assuming that A is $\mathcal{R}$-sectorial of angle $\phi_A^{\mathcal{R}} < \pi/2$, we obtain the main assertion of the following characterization theorem of maximal L^p-regularity.

Corollary 1.6.10 *Let A be the generator a bounded holomorphic semigroup on a UMD Banach space X. Then A has maximal L^p-regularity for one (all) $p \in (1, \infty)$ on $\mathbb{R}_+$ if and only if $-A$ is $\mathcal{R}$-sectorial of angle $\phi_A^{\mathcal{R}} < \pi/2$.*

For the proof of the fact that $\mathcal{R}$-sectoriality is necessary for maximal L^p-regularity, we refer to Proposition 3.17 of [36].

It remains to show that the derivative operator $\tilde{B}$ defined above has a bounded H^∞-calculus on $L^p(\mathbb{R}_+; X)$ for all angles $\phi > \pi/2$ provided X is a UMD-space. We will do this by applying an operator-valued Fourier multiplier theorem.

Operator-Valued Fourier Multipliers

Consider the Fourier transform $\mathcal{F}$ on $\mathcal{S}(\mathbb{R}^n; X)$, X, Y being arbitrary Banach spaces, as described in Sect. 1.2 and let $m : \mathbb{R}^n \to \mathcal{L}(X, Y)$ be a bounded and measurable function. It induces a map $T_m : \mathcal{S}(\mathbb{R}^n; X) \to L^\infty(\mathbb{R}^n; Y)$ by

$$T_m : \quad f \mapsto \mathcal{F}^{-1}(m(\cdot)[\widehat{f}(\cdot)]). \tag{1.6.10}$$

Given $1 < p < \infty$, we then call m an L^p-*Fourier multiplier* if there exists a constant C_p such that

$$\|T_m f\|_{L^p(\mathbb{R}^n; Y)} \leq C_p \|f\|_{L^p(\mathbb{R}^n; X)}, \quad f \in \mathcal{S}(\mathbb{R}^n; X). \tag{1.6.11}$$

In this case, T_m extends uniquely to an operator $T_m \in \mathcal{L}(L^p(\mathbb{R}^m; X), L^p(\mathbb{R}^n; Y))$ whose norm is the smallest constant C_p for which (1.6.11) holds.

The following theorem is the extension of the classical Mikhlin's theorem to the operator-valued situation.

Theorem 1.6.11 (Weis) *Assume that X and Y are UMD-Banach spaces and let $1 < p < \infty$. Suppose that $m \in C^1(\mathbb{R} \setminus \{0\}; \mathcal{L}(X, Y))$ satisfies the following conditions*

(i) $\mathcal{R}\{m(\xi) : \xi \in \mathbb{R} \setminus \{0\}\} =: M_0 < \infty$,
(ii) $\mathcal{R}\{\xi m'(\xi) : \xi \in \mathbb{R} \setminus \{0\}\} =: M_1 < \infty$.

Then the operator T_m defined by (1.6.10) is bounded from $L^p(\mathbb{R}; X)$ into $L^p(\mathbb{R}; Y)$ and there exists a constant $C_p > 0$ such that

$$\|T_m\|_{\mathcal{L}(L^p(\mathbb{R}; X), L^p(\mathbb{R}; Y))} \leq C_p(M_0 + M_1).$$

As discussed above we are in particular interested in the following example.

Example 1.6.12 Let X be a UMD-space, $1 < p < \infty$, and $\tilde{B} = \frac{d}{dt}$ on $L^p(\mathbb{R}; X)$. Then $\tilde{B}$ admits a bounded $\mathcal{H}^\infty(\Sigma_\phi)$-calculus for all $\phi > \pi/2$. This follows exactly in the same way as described in Example 1.4.9, now, however, with the classical Mikhlin's theorem being replaced by Theorem 1.6.11.

It is very interesting to note that we may strengthen Theorem 1.6.11 in the case where X and Y possess in addition property (α).

Theorem 1.6.13 *Assume that X and Y are UMD-Banach spaces with property (α), $1 < p < \infty$ and that $M \subset C^1(\mathbb{R}^n \setminus \{0\}; \mathcal{L}(X, Y))$ satisfies the conditions*

(i) $\mathcal{R}\{\{m(\xi) : \xi \in \mathbb{R} \setminus \{0\}, m \in M\}\} =: K_0 < \infty,$
(ii) $\mathcal{R}\{\{\xi m'(\xi) : \xi \in \mathbb{R} \setminus \{0\}, m \in M\}\} =: K_1 < \infty.$

Then the family of operators $\mathcal{T} := \{T_m : m \in M\} \subset \mathcal{L}(L^p(\mathbb{R}; X), L^p(\mathbb{R}; Y))$ defined by (1.6.10) is $\mathcal{R}$-bounded with $\mathcal{R}$-bound $\mathcal{R}\{\mathcal{T}\} \leq C(K_0 + K_1)$, where the constant $C > 0$ depends only on p, X, Y.

Turning our attention to the multi-dimensional case, we note that an operator-valued Fourier multiplier theorem of Lizorkin type can be deduced from Theorem 1.6.13 by induction.

Theorem 1.6.14 *Assume that X and Y are UMD-Banach spaces with property (α). Let $1 < p < \infty$ and $n \in \mathbb{N}$. Suppose that the family $M \subset C^n(\mathbb{R}^n \setminus \{0\}; \mathcal{L}(X, Y))$ satisfies*

$$\mathcal{R}\{\{\xi^\alpha D_\xi^\alpha m(\xi) : \xi \in \mathbb{R}^n \setminus \{0\}, \alpha \in 0, 1^n, m \in M\}\} =: K < \infty.$$

Then the family of operator $\mathcal{T} := \{T_m : m \in M\} \subset \mathcal{L}(L^p(\mathbb{R}^n; X), L^p(\mathbb{R}^n; Y))$ defined by (1.6.10) is $\mathcal{R}$-bounded with $\mathcal{R}$-bound $\mathcal{R}\{\mathcal{T}\} \leq CK$, where the constant $C > 0$ depends only on p, X, Y.

Further Results

Finally for $J = [0, T]$ with $T > 0$ consider the Cauchy problem

$$u'(t) - Au(t) = f, \qquad t \in J,$$
$$u(0) = u_0, \tag{1.6.12}$$

with initial data $u_0 \neq 0$. We would like to characterize those initial data u_0 for which the solution u of (1.6.12) has the maximal L^p-regularity property. To this end, we recall that the real interpolation space $(X, (D(A)))_{\theta, p}$ was introduced in Definition 1.3.16 and Remark 1.3.17 as $(X, D(A))_{\theta, p} = \{x \in X : \phi_{\theta, p}(x, \cdot) \in L^p(0, T)\}$ with norm

$$\|x\|_{\theta, p} := \|x\| + \|\phi_{\theta, p}(x, \cdot)\|_{L^p(0, T)}.$$

Thus, a solution of (1.6.12) with $f = 0$ satisfies $u \in L^p((0, T), D(A))$ if and only if $u_0 \in (X, D(A))_{1-1/p, p}$. In fact, $Au \in L^p((0, T); X)$ if and only if

$$\|\phi_{1-1/p, p}(u_0, \cdot)\|_{L^p(0, T)} = \left(\int_0^T t^{p(1-(1-1/p))-1} \|Ae^{tA}u_0\|_X^p dt \right)^{1/p}$$

$$= \left(\int_0^T \|Ae^{tA}u_0\|_X^p dt \right)^{1/p} < \infty.$$

Combining this observation with Corollary 1.6.10 we obtain the following theorem.

Theorem 1.6.15 *Let X be a UMD space, $1 < p < \infty$, and assume that $-A$ is $\mathcal{R}$-sectorial of angle $\phi_A^{\mathcal{R}} < \frac{\pi}{2}$. Then (1.6.12) has precisely one solution $u \in W^{1,p}(J; X) \cap L^p(J; D(A))$ if and only if*

$$f \in L^p(J; X) \text{ and } u_0 \in D_A(1 - 1/p, p)$$

The proof of characterization of the maximal L^p-regularity for solutions of the Cauchy problem (1.6.12) given in Corollary 1.6.10 and Theorem 1.6.15 was based on the Kalton–Weis sum Theorem 1.6.9. The latter theorem has many further consequences. We state here only the Mixed-Derivative theorem. Consider in the situation of Theorem 1.6.9 the function F given $F(\lambda) = \lambda^\alpha B^{1-\alpha}(\lambda + B)^{-1}$. The representation

$$F(\lambda) = \frac{1}{2\pi i} \int_\Gamma \frac{z^{-\alpha}}{1+z} \lambda z (\lambda z - B)^{-1} dz$$

for a suitable contour Γ shows that $F(\Sigma_\varphi)$ is $\mathcal{R}$-bounded provided B is $\mathcal{R}$-sectorial. We hence obtain the following result.

Proposition 1.6.16 (Mixed Derivative Theorem) *Let A and B be two resolvent commuting operators on a Banach space X. Assume that $A \in \mathcal{H}^\infty(X)$ and B is $\mathcal{R}$-sectorial with $\phi_A^\infty + \phi_B^{\mathcal{R}} < \pi$. Then $A^\alpha B^{1-\alpha}(A + B)^{-1}$ is a bounded operator on X for each $\alpha \in (0, 1)$. In particular, for each $\alpha \in (0, 1)$*

$$D(A) \cap D(B) = D(A + B) \hookrightarrow D(A^\alpha B^{1-\alpha}).$$

A typical situation is the following: Let $\Omega \subset \mathbb{R}^n$ be a bounded domain with smooth boundary, $X_0 = L^q(\Omega)$, $X_1 = H^{2,q}(\Omega)$ for $q \in (1, \infty)$ and for $T > 0$ and $p \in (1, \infty)$ set

$$\mathbb{E}_1(T) := L^p(0, T; X_1) \cap H^{1,p}(0, T; X_0).$$

Corollary 1.6.17 *If $\theta \in [0, 1]$, then*

$$\mathbb{E}_1(T) \hookrightarrow H^{\theta,p}(0, T; H^{2-2\theta,q}(\Omega)).$$

At this point one might ask the question whether all generators of analytic semigroups on $L^p(\Omega, \mu)$ where (Ω, μ) denotes a σ-finite measure space and $1 < p < \infty$ are $\mathcal{R}$-sectorial. Kalton and Lancien [90] gave a complete answer to this question.

Theorem 1.6.18 (Kalton–Lancien) *Assume that X has an unconditional basis and that all generators of analytic semigroups on X are $\mathcal{R}$-sectorial. Then X is isomorphic to a Hilbert space.*

We finish this section with a result saying that $\mathcal{R}$-sectorial operators are well behaved under perturbations. This fact will be used later on in various sections.

For $A \in \mathcal{S}(X)$ and $\theta > \phi_A$, we set

$$M(A)_\theta := \sup\{\|AR(\lambda, A)\| : \lambda \neq 0, \pi > |\arg(\lambda)| > \theta\};$$

if $A \in \mathcal{RS}(X)$ and $\theta > \phi_A^{\mathcal{R}}$, we set

$$R(A)_\theta := \mathcal{R}\{AR(\lambda, A) : \lambda \neq 0, \pi > |\arg(\lambda)| > \theta\}.$$

Proposition 1.6.19 *Let $A \in \mathcal{RS}(X)$ and $\theta > \phi_A^{\mathcal{R}}$. Assume that B is an operator with $D(A) \subset D(B)$ and*

$$\|Bx\| \leq a\|Ax\| + b\|x\|, \quad x \in D(A),$$

for some $a, b \geq 0$. If $a < (M(A)_\theta R(A)_\theta)^{-1}$, then there exists $v \geq 0$ such that $A + B + v$ is $\mathcal{R}$-sectorial. Moreover, $\phi_{A+B+v}^{\mathcal{R}} \leq \theta$ for this v.

The proof follows the lines of the perturbation Theorem 1.3.15 for sectorial operators, where now the arguments for the norm of $(A + B + \mu)^{-1}$ have to be replaced by $\mathcal{R}$-bounds. Details are left to the reader; see e.g. [36] or [95].

1.7 Quasilinear Evolution Equations

In this section we present several results on the local and global well-posedness for abstract quasilinear parabolic equations as well as on the dynamics of their solutions. These results will be employed in our investigations in various sections of Part B. They are mainly due to Prüss [120], Prüss and Simonett [121], Köhne et al. [94], and Prüss et al. [126]. A convenient reference for these results is the monograph by Prüss and Simonett [124], Chapter 5.

Consider the quasilinear problem

$$\dot{v} + A(v)v = F(v), \quad t > 0, \quad v(0) = v_0, \tag{1.7.1}$$

Assume that $(A, F) : V_\mu \to \mathcal{B}(X_1, X_0) \times X_0$ and $v_0 \in V_\mu$. Here X_1 and X_0 are Banach spaces such that $X_1 \hookrightarrow X_0$ with dense embedding and V_μ is an open subset of the real interpolation space

$$X_{\gamma,\mu} := (X_0, X_1)_{\mu-1/p,p}, \quad \mu \in (1/p, 1].$$

We are mainly interested in solutions v of (1.7.1) having *maximal L_p-regularity*, i.e.

$$v \in H_p^1(J; X_0) \cap L_p(J; X_1) =: \mathbb{E}_1(J), \quad \text{where } J = (0, T).$$

The trace space of this class of functions is given by $X_\gamma := X_{\gamma,1}$. However, to see and exploit the effect of parabolic regularization in the L_p-framework it is also useful to consider solutions in the class of weighted spaces

$$v \in H_{p,\mu}^1(J; X_0) \cap L_{p,\mu}(J; X_1) =: \mathbb{E}_{1,\mu}(J), \quad \text{which means } t^{1-\mu}v \in \mathbb{E}_1(J).$$

Here, for $\mu \in (1/p, 1]$ and a time interval $J \subset [0, \infty)$ we set

$$L_\mu^p(J; X) = \{u \in L_{loc}^1(J; X) : t^{1-\mu} u \in L^p(J; X)\},$$

$$H_\mu^{1,p}(J; X) = \{u \in L_{loc}^1(J; X) \cap H^{1,1}(J; X) : t^{1-\mu} u' \in L^p(J; X)\},$$

The trace space for this class of weighted spaces is given by $X_{\gamma,\mu}$. In our approach it is crucial to know that the operators $A(v)$ have the property of *maximal L_p-regularity*. We recall from Sect. 1.6 that an operator A_0 in X_0 with domain X_1 has maximal L_p-regularity, if the linear problem

$$\dot{v} + A_0 v = f, \quad t \in J, \ v(0) = 0, \tag{1.7.2}$$

admits a unique solution $v \in \mathbb{E}_1(J)$, for any given $f \in L_p(J; X_0) =: \mathbb{E}_0(J)$. In this case we write $A_0 \in \mathrm{MR}_p(X_0)$. It has been proved in [121] that in this case maximal regularity also holds in the weighted case. By this we mean that for each $f \in L_{p,\mu}(J; X_0)$ there exists a unique $u \in L_{p,\mu}(J; X_0)$ such that u', $Au \in L_{p,\mu}(J; X_0)$ and such that u solves (1.7.2). In this case we write $A_0 \in \mathrm{MR}_{p,\mu}(X_0)$.

Proposition 1.7.1 *Let $p \in (1, \infty)$ and $1/p < \mu \leq 1$. Then*

$$A_0 \in MR_p(X_0) \Longleftrightarrow A_0 \in MR_{p,\mu}(X_0)$$

Local Solutions

The local existence result for Eq. (1.7.1) reads as follows.

Theorem 1.7.2 *Let $p \in (1, \infty)$, $v_0 \in V_\mu$ be given and suppose that (A, F) satisfies*

$$(A, F) \in C^1(V_\mu; \mathcal{B}(X_1, X_0) \times X_0), \tag{1.7.3}$$

for some $\mu \in (1/p, 1]$. Assume in addition that $A(v_0)$ has maximal L_p-regularity.
Then there exist $a = a(v_0) > 0$ and $r = r(v_0) > 0$ with $\bar{B}_{X_{\gamma,\mu}}(v_0, r) \subset V_\mu$ such that problem (1.7.1) has a unique solution

$$v = v(\cdot, v_1) \in \mathbb{E}_{1,\mu}(0, a) \cap C([0, a]; V_\mu),$$

on $[0, a]$, for any initial value $v_1 \in \bar{B}_{X_{\gamma,\mu}}(v_0, r)$. In addition,

$$t \partial_t v \in \mathbb{E}_{1,\mu}(0, a),$$

in particular, for each $\delta \in (0, a)$ we have

$$v \in H_p^2((\delta, a); X_0) \cap H_p^1((\delta, a); X_1) \hookrightarrow C^1([\delta, a]; X_\gamma) \cap C^{1-1/p}([\delta, a]; X_1),$$

i.e. the solution regularizes instantly.

The next result provides information about the continuation of local solutions.

Corollary 1.7.3 *Let the assumptions of Theorem 1.7.2 be satisfied and assume that $A(v)$ has maximal L_p-regularity for all $v \in V_\mu$. Then the solution v of (1.7.1) has a maximal interval of existence $J(v_0) = [0, t_+(v_0))$, which is characterized by the following alternatives:*

(i) *Global existence: $t_+(v_0) = \infty$;*
(ii) $\liminf_{t \to t_+(v_0)} \text{dist}_{X_{\gamma,\mu}}(v(t), \partial V_\mu) = 0$;
(iii) $\lim_{t \to t_+(v_0)} v(t)$ *does not exist in $X_{\gamma,\mu}$.*

Stability of Equilibria and Global Solutions

Next we assume that there is an open set $V \subset X_\gamma$ such that

$$(A, F) \in C^1(V, \mathcal{B}(X_1, X_0) \times X_0). \qquad (1.7.4)$$

Let $\mathcal{E} \subset V \cap X_1$ denote the set of equilibrium solutions of (1.7.1), which means that

$$v \in \mathcal{E} \quad \text{if and only if} \quad v \in V \cap X_1 \text{ and } A(v)v = F(v).$$

Given an element $v_* \in \mathcal{E}$, we assume that v_* is contained in an m-dimensional *manifold of equilibria*. This means that there is an open subset $U \subset \mathbb{R}^m, 0 \in U$, and a C^1-function $\Psi : U \to X_1$, such that

- $\Psi(U) \subset \mathcal{E}$ and $\Psi(0) = v_*$,

- the rank of $\Psi'(0)$ equals m, and $\qquad (1.7.5)$

- $A(\Psi(\zeta))\Psi(\zeta) = F(\Psi(\zeta)), \quad \zeta \in U$.

We suppose that the operator $A(v_*)$ has the property of maximal L_p-regularity, and define the full linearization of (1.7.1) at v_* by

$$A_0 w = A(v_*)w + (A'(v_*)w)v_* - F'(v_*)w \quad \text{for } w \in X_1. \qquad (1.7.6)$$

After these preparations we can state the following result on convergence of solutions starting near v_* which is called the *generalized principle of linearized stability*.

Theorem 1.7.4 *Let $1 < p < \infty$. Suppose $v_* \in V \cap X_1$ is an equilibrium of (1.7.1), and suppose that the functions (A, F) satisfy (1.7.4). Suppose further that $A(v_*)$ has the property of maximal L_p-regularity and let A_0 be defined in (1.7.6). Suppose that v_* is* normally stable, *which means*

(i) *near v_* the set of equilibria $\mathcal{E}$ is a C^1-manifold in X_1 of dimension $m \in \mathbb{N}$,*
(ii) *the tangent space for $\mathcal{E}$ at v_* is isomorphic to $\mathsf{N}(A_0)$,*
(iii) *0 is a semi-simple eigenvalue of A_0, i.e. $\mathsf{N}(A_0) \oplus \mathsf{R}(A_0) = X_0$,*
(iv) $\sigma(A_0) \setminus \{0\} \subset \mathbb{C}_+ = \{z \in \mathbb{C} : \operatorname{Re} z > 0\}.$

Then v_* is stable in X_γ, and there exists $\delta > 0$ such that the unique solution v of (1.7.1) with initial value $v_0 \in X_\gamma$ satisfying $|v_0 - v_*|_\gamma \leq \delta$ exists on $\mathbb{R}_+$ and converges at an exponential rate in X_γ to some $v_\infty \in \mathcal{E}$ as $t \to \infty$.

The next result contains information on bounded solutions in the presence of compact embeddings and of a strict Lyapunov functional.

Theorem 1.7.5 *Let $p \in (1, \infty)$, $\mu \in (1/p, 1)$, $\bar{\mu} \in (\mu, 1]$, with $V_\mu \subset X_{\gamma,\mu}$ open. Assume that $(A, F) \in C^1(V_\mu; \mathcal{B}(X_1, X_0) \times X_0)$, and that the embedding $X_{\gamma,\bar{\mu}} \hookrightarrow X_{\gamma,\mu}$ is compact. Suppose furthermore that v is a maximal solution which is bounded in $X_{\gamma,\bar{\mu}}$ and satisfies*

$$\mathrm{dist}_{X_{\gamma,\mu}}(v(t), \partial V_\mu) \geq \eta > 0, \quad \text{for all } t \geq 0. \tag{1.7.7}$$

Suppose that $\Phi \in C(V_\mu \cap X_\gamma; \mathbb{R})$ is a strict Lyapunov functional for (1.7.1), which means that Φ is strictly decreasing along non-constant solutions.

Then $t_+(v_0) = \infty$, i.e. v is a global solution of (1.7.1). Its ω-limit set $\omega_+(v_0) \subset \mathcal{E}$ in X_γ is nonempty, compact and connected. If, in addition, there exists $v_* \in \omega_+(v_0)$ which is normally stable, then $\lim_{t \to \infty} v(t) = v_*$ in X_γ.

The Semilinear Case

We state here also a simplified version of the general result concerning semilinear evolution equations. It will then be applied to the primitive equations in Sect. 1.15. As above let X_0, X_1 be Banach spaces such that $X_1 \hookrightarrow X_0$ is densely embedded, and let $A: X_1 \to X_0$ be bounded. For $0 < T \leq \infty$ consider the semi-linear problem

$$u' + Au = F(u) + f, \quad 0 < t < T, \quad u(0) = u_0. \tag{1.7.8}$$

The subsequent existence and uniqueness result is based on the following assumptions. Here, for $\beta \in [0, 1]$ define X_β to be the complex interpolation space $[X_0, X_1]_\beta$.

(H1) A has maximal L^q-regularity for $q \in (1, \infty)$.
(H2) $F: X_\beta \to X_0$ satisfied the estimate

$$\|F(u_1) - F(u_2)\|_{X_0} \leq C(\|u_1\|_{X_\beta} + \|u_1\|_{X_\beta})(\|u_1 - u_2\|_{X_\beta})$$

for some $C > 0$ independent of u_1, u_2.
(H3) $\beta - (\mu - 1/q) \leq \frac{1}{2}(1 - (\mu - 1/q))$, that is $2\beta - 1 + 1/q \leq \mu$.
(S) X_0 is of class UMD, and the embedding

$$H^{1,q}(\mathbb{R}; X_0) \cap L^q(\mathbb{R}; X_1) \hookrightarrow H^{1-\beta,q}(\mathbb{R}; X_\beta)$$

is valid for each $\beta \in (0, 1)$ and $q \in (1, \infty)$.

Proposition 1.7.6 ([125]) *Assume that the assumptions* $(H1)$, $(H2)$, $(H3)$ *and* (S) *hold and let*

$$u_0 \in X_{\gamma,\mu} \quad \text{and} \quad f \in L^q(0, T; X_0).$$

Then there exists a time $T' = T'(u_0)$ *with* $0 < T' \leq T$ *such that problem* (1.7.8) *admits a unique solution*

$$u \in H^{1,q}_\mu(0, T'; X_0) \cap L^q_\mu(0, T'; X_1).$$

Furthermore, the solution u depends continuously on the data.

Remark 1.7.7 Let us remark that (S) holds true whenever X_0 is of class UMD and there is an operator $A_\# \in \mathcal{H}^\infty(X_0)$ with domain $D(A_\#) = X_1$ satisfying $\phi^\infty_{A_\#} < \pi/2$, see Remark 1.1 of [125].

When investigating the question of a global solution, we consider

$$t_+(u_0) := \sup\{T' > 0 : \text{Eq. (1.7.8) admits a solution on } (0, T')\}.$$

By the above Proposition 1.7.6, this set is non-empty, and we say that (1.7.8) has a *global solution* if for $f \in L^q(0, T; X_0)$ one has $t_+(u_0) = T$, where $0 < T \leq \infty$. Global existence results can be derived from suitable a priori bounds following [124, Theorem 5.7.1].

Proposition 1.7.8 ([124]) *Assume in addition to the assumptions of Proposition 1.7.6 that for* $\mu < \overline{\mu} \leq 1$ *the embedding*

$$X_{\gamma,\overline{\mu}} \hookrightarrow X_{\gamma,\mu}$$

is compact, and that for some $\tau \in (0, t_+(u_0))$ *the solution of* (1.7.8) *satisfies*

$$u \in C_b([\tau, t_+(u_0)); X_{\gamma,\overline{\mu}}),$$

then there is a global solution to (1.7.8)*, i.e.* $T' = T$.

Part B: Viscous Fluid Flows

In the second part of these lecture notes we investigate several problems from the theory of viscous fluid flows such as boundary layers, fluid structure interaction problems, free boundary value problems, liquid crystal flow and the primitive equations of geophysical flows. The methods and techniques established in Chapter I will play an important role in our investigation of these problems.

We start with a discussion of conservation laws for general heat conducting fluids and deduce, assuming certain constitutive laws, the fundamental equations of viscous fluid flows, the incompressible and compressible Navier–Stokes equations.

A natural step from here is the analysis of the Stokes equation in a half space $\mathbb{R}^n_+$ within the L^p-setting. The associated operator, the Stokes operator, is identified to be a sectorial operator on $L^p(\mathbb{R}^n_+)$. It generates in particular a bounded, holomorphic semigroup on $L^p_\sigma(\mathbb{R}^n_+)$ for all p satisfying $1 < p \le \infty$. For $1 < p < \infty$, the space $L^p_\sigma(\mathbb{R}^n_+)$ denotes the subspace consisting of all solenoidal vector fields in $L^p(\mathbb{R}^n_+)$. Whereas the case $1 < p < \infty$ is quite classical, it is not surprising that the proof for the case $p = \infty$ requires different methods. In this case the Helmholtz projection does not act as a bounded operator on $L^\infty(\mathbb{R}^n_+)$. We even show that the negative Stokes operator admits an $\mathcal{R}$-bounded H^∞-calculus on $L^p_\sigma(\mathbb{R}^n)$ for $1 < p < \infty$ and obtain as a consequence of the results in Sect. 1.6 the property of maximal L^p–L^q-regularity for the Stokes equation.

Section 1.10 deals with the H^∞-calculus as well as with the maximal L^p-regularity property for the classical Stokes as well as for the hydrostatic Stokes operator. The proof of these results are fairly involved and we thus present here only the key ideas of the proofs.

In the Sect. 1.11 we consider stability questions for Ekman boundary layers. These boundary layers are explicit stationary solutions of the Navier–Stokes equations in the rotational framework. Considering perturbations of the Ekman layer, we are interested in the question whether there exists a global, weak solution of the perturbed systems converging to zero as $t \to \infty$. This, of course, would imply the asymptotic stability of the Ekman layer. The methods developed in the Chapter I allow us then to prove that this is the case, provided the Reynolds number involved is small enough.

Sections 1.12 and 1.13 investigate moving and free boundary value problems. In fact, the fluid-rigid body interaction problem considered in Sect. 1.12 is a typical example of a moving domain problem, whereas the two-phase problem discussed in Sect. 1.13 represents a free boundary value problem. The fluid-rigid body interaction problem will be discussed for the situation of compressible fluids. Maximal regularity of the linearized equation in Lagragian coordinates allows us to prove a local well-posedness result for strong solutions. Strong solutions for the two-phase problem for generalized Newtonian fluids are again obtained by a fixed point argument in the associated space of maximal regularity.

The final two sections concern the primitive equations. These equations are a model for oceanic and atmospheric dynamics and are derived from the Navier–Stokes equations by assuming a hydrostatic balance for the pressure term. We show in Sect. 1.15 that these equations are globally strongly well-posedness for arbitrary large initial data lying in critical spaces, which in the given situation are given by the Besov spaces B^μ_{pq} for $p, q \in (1, \infty)$ with $1/p + 1/q \le \mu \le 1$. Finally, in Sect. 1.16 we show that the primitive equations may be obtained as the limit of anisotropically scaled Navier–Stokes equations. The proof relies on a maximal L^p-regularity estimate for the differences $(v_\varepsilon - v, \varepsilon(w_\varepsilon - w))$ by the aspect ration ε, where $(v_\varepsilon, w_\varepsilon)$ and (v, w) are solutions of the anisotropic Navier–Stokes equations and the primitive equations, respectively.

1.8 Balance Laws

In this section let $\Omega \subset \mathbb{R}^n$ be a domain with C^1-boundary.

First Principles

We begin with the balance laws of mass, momentum, and energy. They read as

$$\partial_t \rho + \operatorname{div}(\rho u) = 0 \qquad \text{in } \Omega,$$
$$\rho(\partial_t + u \cdot \nabla)u + \nabla \pi = \operatorname{div} S \qquad \text{in } \Omega, \qquad (1.8.1)$$
$$\rho(\partial_t + u \cdot \nabla)\epsilon + \operatorname{div} q = S : \nabla u - \pi \operatorname{div} u \qquad \text{in } \Omega,$$
$$u = 0, \quad q \cdot \nu = 0 \qquad \text{on } \partial\Omega.$$

Here ρ means density, u velocity, π pressure, ϵ internal energy, S extra stress and q heat flux. This immediately gives conservation of the total energy. In fact, we have

$$\rho(\partial_t + u \cdot \nabla)e + \operatorname{div}(q + \pi u - Su) = 0 \quad \text{in } \Omega,$$

where $e := |u|^2/2 + \epsilon$ means the total mass specific energy density (kinetic plus internal). The energy flux Φ_e is given by $\Phi_e := q + \pi u - Su$. Integrating over Ω yields

$$\partial_t \mathsf{E}(t) = 0, \quad \mathsf{E}(t) = \mathsf{E}_{kin}(t) + \mathsf{E}_{int}(t) = \int_\Omega \rho(t, x)e(t, x)dx,$$

provided

$$q \cdot \nu = u = 0 \quad \text{on } \partial\Omega. \qquad (1.8.2)$$

Hence, if (1.8.2) holds, total energy is preserved, independent of the particular choice of S and q.

Thermodynamics

Assume a given free energy ψ of the form $\psi = \psi(\rho, \theta, \tau)$, where θ denotes the (absolute) temperature and τ will be specified later. We then have the following thermodynamical relations:

$$\epsilon = \psi + \theta\eta \quad \text{internal energy,}$$
$$\eta = -\partial_\theta \psi \quad \text{entropy,} \qquad (1.8.3)$$
$$\kappa = \partial_\theta \epsilon = -\theta\partial_\theta^2 \psi \quad \text{heat capacity.}$$

Later on, for well-posedness of the heat problem, one requires $\kappa > 0$, i.e. ψ to be strictly concave with respect to $\theta \in (0, \infty)$.

In the classical case, where ψ depends only on ρ and θ, one has the *Clausius-Duhem equation*

$$\rho(\partial_t + u \cdot \nabla)\eta + \mathrm{div}(q/\theta) = S : \nabla u/\theta - q \cdot \nabla\theta/\theta^2 + (\rho^2\partial_\rho\psi - \pi)(\mathrm{div}\,u)/\theta \quad \text{in } \Omega.$$

Hence, in this case the entropy flux Φ_η is given by $\Phi_\eta := q/\theta$ and the entropy production by

$$\theta r := S : \nabla u - q \cdot \nabla\theta/\theta + (\rho^2\partial_\rho\psi - \pi)(\mathrm{div}\,u).$$

Employing the boundary conditions (1.8.2), we obtain for the total entropy N by integration over Ω

$$\partial_t \mathsf{N}(t) = \int_\Omega r(t, x)dx \geq 0, \quad \mathsf{N}(t) = \int_\Omega \rho(t, x)\eta(t, x)dx,$$

provided $r \geq 0$ in Ω. As div u has no sign we require

$$\pi = \rho^2\partial_\rho\psi, \tag{1.8.4}$$

which is the famous *Maxwell relation*. Further, as S and q are independent, this requirement leads to the classical conditions

$$S : \nabla u \geq 0 \quad \text{and} \quad q \cdot \nabla\theta \leq 0. \tag{1.8.5}$$

Summarizing, we see that whatever we choose for S and q, we always have conservation of energy and the total entropy is non-decreasing provided (1.8.5), (1.8.4) and (1.8.2) are satisfied. Thus, these conditions ensure the thermodynamic consistency of the model.

As an example for S and q consider the classical laws due to Newton and Fourier which are given by

$$S := S_N := 2\mu_s D + \mu_b \mathrm{div}\,u\,I, \quad 2D := (\nabla u + [\nabla u]^\mathsf{T}), \quad q := -\alpha_0\nabla\theta.$$

In this case, (1.8.5) is satisfied as soon as $\mu_s \geq 0$, $2\mu_s + n\mu_b \geq 0$ and $\alpha_0 \geq 0$ hold. Note that it does not matter at all whether μ_s, μ_b, α_0 are constants or whether they depend on ρ, θ, or on other variables.

Model Equations for the Isothermal, Incompressible Situation

In the following, we consider the isothermal situation, neglecting thus the influence of temperature variations in the flow. A fluid is said to be incompressible if

$$\operatorname{div} u = 0 \quad \text{for all } t > 0, x \in \Omega.$$

In this case, the density ϱ is constant along the trajectories associated with the velocity u. Given an incompressible fluid with homogeneous density ϱ at initial time, i.e. $x \mapsto \varrho(0, x) = \varrho_0$ is constant, the density ϱ remains constant, i.e. $\varrho(t, x) = \varrho_0$ for all $t > 0$ and all $x \in \Omega$.

For nonhomogeneous and incompressible fluids, the above balance laws yield the *nonhomogeneous, incompressible Navier–Stokes equations*. They read as

$$\begin{cases} \partial_t \varrho + \operatorname{div}(\varrho u) = 0 & \text{in } \Omega, \\ \partial_t(\varrho u) + \operatorname{div}(\varrho u \otimes u) - 2 \operatorname{div}(\mu(\varrho)D(u)) + \nabla \pi = 0 & \text{in } \Omega, \\ \operatorname{div} u = 0 & \text{in } \Omega. \end{cases} \quad (1.8.6)$$

Concerning boundary conditions, the homogeneous Dirichlet boundary condition $u = 0$ on $\partial \Omega$ is widely used. In case of nonhomogeneous fluids, we have to add side and boundary conditions for the density. In particular, in order to obtain physically meaningful solutions, we need to add the condition

$$\varrho(t, x) \geq 0 \quad \text{for all } t > 0 \text{ and all } x \in \Omega.$$

If the density ϱ of the fluid is homogeneous, then the mass balance equals the divergence free condition and we obtain the *homogeneous, incompressible Navier–Stokes equations*. Assuming $\varrho(0, x) = 1$ for all $x \in \Omega$ yields the equations

$$\begin{cases} \partial_t u + (u \cdot \nabla)u - \mu \Delta u + \nabla \pi = 0 & \text{in } \Omega, \\ \operatorname{div} u = 0 & \text{in } \Omega, \\ u = 0 & \text{on } \partial \Omega. \end{cases} \quad (1.8.7)$$

Of course, the above sets of equations have to be completed by imposing initial conditions for $u(0)$ and $\varrho(0)$, respectively.

Model Equations for the Isothermal, Compressible Situation

In the following, consider again the isothermal situation. A fluid is called *barotropic* if its pressure is a function of the density, i.e. $\pi = \pi(\varrho)$. The law $\pi = \pi(\varrho)$ serves as an constitutive equation for the barotropic fluid model. This is governed by the following system

$$
\begin{cases}
\partial_t \varrho + \text{div}\,(\varrho u) = 0 & \text{in } \Omega, \\
\varrho(\partial_t u + u \cdot \nabla u) - (\mu_b + \mu_s)\nabla(\text{div}\,u) - \mu_s \Delta u + \nabla \pi(\varrho) = 0 & \text{in } \Omega,
\end{cases}
\tag{1.8.8}
$$

Again, the above set of equations has to be completed by boundary and initial conditions as well as by compatibility conditions. The above system is called the *compressible Navier–Stokes system*. A typical boundary condition is of the form

$$
u = u_b \text{ and } (u_b | \nu) = 0 \text{ on } \partial \Omega,
$$

which guarantees no inflow or outflow.

1.9 The Stokes Equation in a Half Space

In this section, we consider the Stokes equation in a half space $H := \mathbb{R}^{n+1}_+$, i.e., we consider the set of equations

$$
\begin{aligned}
u_t - \Delta u + \nabla p &= f & \text{in } (0, \infty) \times H, \\
\text{div}\,u &= 0 & \text{in } (0, \infty) \times H, \\
u &= 0 & \text{on } (0, \infty) \times \partial H, \\
u(0) &= u_0,
\end{aligned}
\tag{1.9.1}
$$

where $u = (u_1, \ldots, u_n, u_{n+1})^T$ is interpreted as the velocity field and p as the pressure of the fluid, respectively.

Our aim is to show that the Stokes operator A associated in a natural way with the Stokes equation generates a bounded analytic semigroup on the solenoidal free subspace $L^p_\sigma(H)$ of $L^p(H)$ as well as on other function spaces related to bounded and continuous functions, respectively. Moreover, we prove that $-A$ admits an $\mathcal{R}$-bounded H^∞-calculus on these spaces. This implies in particular maximal L^p–L^q-estimates for the solution of the above Stokes equation by the results of the previous section.

Let us start with the L^p-theory of the solution of (1.9.1) for $1 \leq p \leq \infty$. If $1 < p < \infty$, one often considers the subspace $L^p_\sigma(H)$ of $L^p(H)$ consisting of all solenoidal functions f in $L^p(H)$. This space is defined by

$$
L^p_\sigma(H) := \overline{\{u \in C^\infty_c(H) : \text{div}\,u = 0 \text{ in } H\}}^{\|\cdot\|_{L^p(H)}}.
$$

Then $L^p(H)$ can be decomposed into

$$L^p(H) = L_\sigma^p(H) \oplus G_p(H),$$

where $G_p(H) := \{u \in L^p(H) : u = \nabla\pi \text{ for some } \pi \in H_{loc}^1(H)\}$ and there exists a unique projection $P : L^p(H) \to L_\sigma^p(H)$ having $G_p(H)$ as its null space. P is called the *Helmholtz projection* in $L^p(H)$. In fact, in the case of the whole space $\mathbb{R}^{n+1}$, $P_{\mathbb{R}^{n+1}}$ is given by $P_{\mathbb{R}^{n+1}} = (\delta_{ij} + R_i R_j)_{1 \le i,j, \le n+1}$, where R_i denotes the i-th Riesz transform introduced in Proposition 1.1.7. For the half space $H = \mathbb{R}_+^{n+1}$, P_H can be defined to be

$$P_H := RPE, \tag{1.9.2}$$

where the extension operator $E : L^p(H)^{n+1} \to L^p(\mathbb{R}^{n+1})^{n+1}$ is defined to be the even extension of f_k to $\mathbb{R}^{n+1}$ for $k = 1, \ldots, n$ and $(Ef)_{n+1}$ is set to be the odd extension of f_{n+1} to $\mathbb{R}^{n+1}$. Moreover, the restriction operator $R : L^p(\mathbb{R}^{n+1})^{n+1} \to L^p(H)^{n+1}$ is defined as $R := \frac{1}{2}E^*$, with E^* being the adjoint of E. It can be shown that

$$L_\sigma^p(H) = \{u \in L^p(H) : \text{div}\, u = 0, \gamma f_{n+1} = 0\}, \tag{1.9.3}$$

where γ is the trace operator and γf_{n+1} is interpreted for $f \in L^p(H)$ in the sence of traces.

We then associate with (1.9.1) the so-called Stokes operator A in $(L_\sigma^p(H))^{n+1}$ defined as

$$Au := P_H \Delta u \tag{1.9.4}$$

$$D(A) := (W^{2,p}(H) \cap W_0^{1,p}(H) \cap L_\sigma^p(H))^{n+1}. \tag{1.9.5}$$

Observe that the Helmholtz projection is neither bounded in $L^1(H)$ or $L^\infty(H)$. Thus the usual decomposition of $L^p(H)$ in $L_\sigma^p(H)$ and its orthogonal complement, which is true for $1 < p < \infty$, is no longer possible if $p = 1$ or $p = \infty$.

In order to treat also the case $p = \infty$, we will follow a *different strategy* and will solve the Stokes system by taking Fourier transforms and subsequent explicit calculations.

Our strategy to solve the Stokes system is the following: Taking Fourier transforms we obtain a representation of the solution of the corresponding resolvent equation as a sum of two terms, the first being the resolvent of the Dirichlet Laplacian on $L^p(\mathbb{R}_+^{n+1})$, the second one being a remainder term. We then derive pointwise upper bounds on the remainder term which allow to prove L^p-estimates for the solution of the corresponding resolvent problem for $1 < p \le \infty$, i.e., we obtain estimates of the form

$$\|u_\lambda\|_p \le \frac{M}{|\lambda|}\|f\|_p$$

for the resolvent equation defined below in (1.9.6) and λ belonging to a suitable sector of the complex plane.

The pointwise upper bound on the remainder term allows us also to show that for $1 < p < \infty$ the negative Stokes operator admits a bounded H^∞-calculus on $L_\sigma^p(H)^{n+1}$. In fact, even a stronger result is true: we prove that the Laplacian on $\mathbb{R}^n$ and the remainder term even admit an $\mathcal{R}$-bounded H^∞-calculus on $L^p(H)^{n+1}$ for $1 < p < \infty$. Thus the Stokes operator admits an $\mathcal{R}$-bounded H^∞-calculus on $L_\sigma^p(H)^{n+1}$. As a consequence, we obtain maximal L^p-L^q-regularity for the solution of the Stokes equation (1.9.1).

We recall that for $0 < \theta \leq \pi$ the sector Σ_θ in the complex plane is defined by $\Sigma_\theta := \{z \in \mathbb{C} \setminus \{0\}; \, |\arg z| < \theta\}$.

A Representation Formula and the Stokes Operator in the L^p-Setting for $1 < p < \infty$

Consider the resolvent equation for the Stokes problem in the half space $H = \mathbb{R}_+^{n+1} := \{(x, y) \in \mathbb{R}^{n+1}; \, x \in \mathbb{R}^n, \, y > 0\}$. Given $p \in (1, \infty]$, $\lambda \in \Sigma_\pi$ and $f \in L^p(\mathbb{R}_+^{n+1})^{n+1}$ with $\operatorname{div} f = 0$, we are aiming to find a velocity field $u = (u_1, \ldots, u_n, u_{n+1})^T$ and a pressure field p such that

$$\lambda u - \Delta u + \nabla p = f \quad \text{in } H, \tag{1.9.6}$$

$$\operatorname{div} u = 0 \quad \text{in } H,$$

$$u_{|\partial H} = 0.$$

For $x \in \mathbb{R}^n$ and $y > 0$ we write $u = (v, w)^T$ with $v = (v_1, \ldots, v_n)^T$ and $f = (f_v, f_w)^T$ with $f_v = ((f_v)_1, \ldots, (f_v)_n)^T$. Assume that $f_w(\xi, 0) = 0$ for all $\xi \in \mathbb{R}^n$. Applying the Fourier transform with respect to x we obtain

$$(\lambda + |\xi|^2)\hat{v}(\xi, y) - \partial_y^2 \hat{v}(\xi, y) = \hat{f}_v(\xi, y) - i\xi \cdot \hat{p}(\xi, y), \quad \xi \in \mathbb{R}^n, y > 0 \tag{1.9.7}$$

$$(\lambda + |\xi^2|)\hat{w}(\xi, y) - \partial_y^2 \hat{w}(\xi, y) = \hat{f}_w(\xi, y) - \partial_y \hat{p}(\xi, y), \quad \xi \in \mathbb{R}^n, y > 0 \tag{1.9.8}$$

$$i\xi \cdot \hat{v}(\xi, y) + \partial_y \hat{w}(\xi, y) = 0, \quad \xi \in \mathbb{R}^n, y > 0 \tag{1.9.9}$$

$$\hat{v}(\xi, 0) = 0, \quad \xi \in \mathbb{R}^n \tag{1.9.10}$$

$$\hat{w}(\xi, 0) = 0, \quad \xi \in \mathbb{R}^n. \tag{1.9.11}$$

Multiplying Eq. (1.9.7) by $i\xi$, applying ∂_y to (1.9.8) and adding them yields

$$|\xi|^2 \hat{p}(\xi, y) - \partial_y^2 \hat{p}(\xi, y) = -\partial_y \hat{f}_w(\xi, y) - i\xi \cdot \hat{f}_v(\xi, y) = -(\operatorname{div} f)\hat{\,}(\xi, y) = 0$$

for $\xi \in \mathbb{R}^n$ and $y > 0$, where we already took into account Eq. (1.9.9). Hence

$$\hat{p}(\xi, y) = e^{-|\xi|y} \hat{p}_0(\xi), \quad \xi \in \mathbb{R}^n, \quad y > 0$$

for some function $\hat{p}_0$. We thus obtain for $\hat{v}$ and $\hat{w}$ the following representations

$$\hat{v}(\xi, y) = \frac{1}{2\omega(\xi)} \int_0^\infty [e^{-\omega(\xi)|y-s|} - e^{-\omega(\xi)(y+s)}][\hat{f}_v(\xi, s) - i\xi e^{-|\xi|s} \hat{p}_0(\xi)]\, ds,$$

$$(1.9.12)$$

$$\hat{w}(\xi, y) = \frac{1}{2\omega(\xi)} \int_0^\infty [e^{-\omega(\xi)|y-s|} + e^{-\omega(\xi)(y+s)}][\hat{f}_w(\xi, s) + |\xi| e^{-|\xi|s} \hat{p}_0(\xi)]\, ds,$$

$$(1.9.13)$$

for $\xi \in \mathbb{R}^n$, $y > 0$ and where $\omega(\xi) := (|\lambda| + |\xi|^2)^{\frac{1}{2}}$ for $\xi \in \mathbb{R}^n$.
In order to determine $\hat{p}_0$, consider $\partial_y \hat{w}(\xi, y)$ at $y = 0$, i.e.

$$\partial_y \hat{w}(\xi, 0) = \int_0^\infty e^{-\omega(\xi)s}[\hat{f}_w(\xi, s) + |\xi| e^{-|\xi|s} \hat{p}_0(\xi)]\, ds = 0, \quad \xi \in \mathbb{R}^n.$$

This implies

$$\hat{p}_0(\xi) = -\frac{(\omega(\xi) + |\xi|)}{|\xi|} \int_0^\infty e^{-\omega(\xi)s} \hat{f}_w(\xi, s)\, ds, \quad \xi \neq 0. \qquad (1.9.14)$$

By assumption, $i\xi \cdot \hat{f}_v(\xi, y) + \partial_y \hat{f}_w(\xi, y) = 0$ for $\xi \in \mathbb{R}^n$ and $y > 0$. Integrating by parts yields

$$\hat{f}_w^L(\xi) := \int_0^\infty e^{-\omega(\xi)s} \hat{f}_w(\xi, s)\, ds = -\frac{i\xi}{\omega(\xi)} \int_0^\infty e^{-\omega(\xi)s} \hat{f}_v(\xi, s)\, ds$$

$$=: \frac{-i\xi}{\omega(\xi)} \hat{f}_v^L(\xi) \qquad (1.9.15)$$

for all $\xi \in \mathbb{R}^n$. Inserting (1.9.14) and (1.9.15) into (1.9.12) and (1.9.13) we obtain

$$\hat{v} = \hat{v}_1 + \hat{v}_2 \qquad (1.9.16)$$

$$\hat{w} = \hat{w}_1 + \hat{w}_2$$

with $\hat{v}_1, \hat{v}_2$ and $\hat{w}_1, \hat{w}_2$ given for $\xi \in \mathbb{R}^n$ and $y > 0$ by

$$\hat{v}_1(\xi, y) = \frac{1}{2\omega(\xi)} \int_0^\infty [e^{-\omega(\xi)|y-s|} - e^{-\omega(\xi)(y+s)}]\hat{f}_v(\xi, s)\, ds$$

$$\hat{v}_2(\xi, y) = \frac{1}{2\omega(\xi)} \int_0^\infty [e^{-\omega(\xi)|y-s|} - e^{-\omega(\xi)(y+s)}]\frac{(i\xi)}{|\xi|} e^{-|\xi|s}(\omega(\xi) + |\xi|)\, ds\, \hat{f}_w^L(\xi)$$

$$= \frac{|\xi|}{\omega(\xi)} \frac{1}{\omega(\xi) - |\xi|}[e^{-y|\xi|} - e^{-\omega(\xi)y}]\hat{f}_v^L(\xi)$$

$$\hat{w}_1(\xi, y) = \frac{1}{2\omega(\xi)} \int_0^\infty [e^{-\omega(\xi)|y-s|} - e^{-\omega(\xi)(y+s)}] \hat{f}_w(\xi, s) \, ds$$

$$\hat{w}_2(\xi, y) = \frac{-1}{\omega(\xi) - |\xi|} [e^{-y|\xi|} - e^{-\omega(\xi)y}] \hat{f}_w^L(\xi)$$

$$= \frac{i\xi}{\omega(\xi)} \cdot \frac{1}{\omega(\xi) - |\xi|} [e^{-y|\xi|} - e^{-\omega(\xi)y}] \hat{f}_v^L(\xi).$$

Observe that

$$v_1 = (\lambda - \Delta_D)^{-1} f_v, \quad w_1 = (\lambda - \Delta_D)^{-1} f_w,$$

where Δ_D denotes the Laplacian in $\mathbb{R}_+^{n+1}$ subject to homgeneous Dirichlet boundary conditions. It follows from Examples 1.3.13(c), (d) and Theorem 1.3.10 that for $1 < p \le \infty$ and $\lambda \in \Sigma_\theta$ with $\theta < \pi$ there exists a constant $M > 0$ such that

$$\|v_1\|_{L^p(\mathbb{R}_+^{n+1})^n} \le \frac{M}{|\lambda|} \|f_v\|_{L^p(\mathbb{R}_+^{n+1})^n}$$

$$\|w_1\|_{L^p(\mathbb{R}_+^{n+1})} \le \frac{M}{|\lambda|} \|f_w\|_{L^p(\mathbb{R}_+^{n+1})}.$$

Hence, in order to obtain L^p-estimates for v and w, we may restrict ourselves in the following to v_2 and w_2.

Let $\theta < \pi$ and define $\hat{r}_v : \mathbb{R}^n \times \mathbb{R}_+ \times \mathbb{R}_+ \times \Sigma_\theta \to \mathbb{C}$ by

$$\hat{r}_v(\xi, y, y', \lambda) := \frac{e^{-|\xi|y} - e^{-\omega(\xi)y}}{\omega(\xi) - |\xi|} \frac{|\xi|}{\omega(\xi)} e^{-\omega(\xi)y'}, \qquad (1.9.17)$$

where $\omega(\xi) = \sqrt{|\lambda| + |\xi|^2}$ and set

$$r_v(x, y, y', \lambda) := \frac{1}{(2\pi)^n} \int_{\mathbb{R}^n} e^{ix\cdot\xi} \hat{r}_v(\xi, y, y', \lambda) d\xi. \qquad (1.9.18)$$

Note that r_v is well defined since the above integral is absolutely convergent for $(y, y') \neq (0, 0)$. Observe first that by a scaling argument it suffices to consider the case $|\lambda| = 1$ and $|\arg \lambda| \le \theta < \pi$. In fact,

$$r_v(x, y, y', \lambda) = |\lambda|^{\frac{n-1}{2}} r_v(|\lambda|^{\frac{1}{2}}x, |\lambda|^{\frac{1}{2}}y, |\lambda|^{\frac{1}{2}}y', \frac{\lambda}{|\lambda|}).$$

Next, for $z \in \mathbb{C}$ consider the function ϕ given by $\phi(z) = (1 - e^{-z})z^{-1}$ for $z \in \mathbb{C}\backslash\{0\}$ and note that $|\phi(z)| \le \frac{C}{1+|z|}$ for $\operatorname{Re} z \ge 0$ and some suitable constant $C > 0$. Thus

$$\hat{r}_v(\xi, y, y', \lambda) = y e^{-|\xi|y} e^{-\omega(\xi)y'} \frac{|\xi|}{\omega(\xi)} \phi((\omega(\xi) - |\xi|)y).$$

Choose now a rotation Q in $\mathbb{R}^n$ such that $Qx = (|x|, 0, \ldots, 0)$ and write

$$Q\xi = (a, rb), \quad a \in \mathbb{R}, \quad r > 0, \quad b \in \mathbb{R}^{n-1}, \quad |b| = 1.$$

By this coordinate transformation and by a shift of the path of integration for a to the contour $a \to s + i\varepsilon(r + |s|)$, $s \in \mathbb{R}$, without changing the value of the integral thanks to Cauchy's theorem, for ε small enough, we obtain for a multiindex α

$$|(\partial_x)^\alpha r_v(x, y, y', \lambda)|$$

$$\leq M \int_0^\infty r^{n-2} \int_0^\infty e^{-c(r+s)(|x|+y+y')} y e^{-cy'} \frac{(r+s)^{1+|\alpha|}}{1+r+s+y} \, ds \, dr$$

$$= M y e^{-cy'} \int_0^\infty s^{n+|\alpha|} \frac{e^{-cs(|x|+y+y')}}{1+y+s} \, ds$$

for some constant $M > 0$ independent of $x \in \mathbb{R}^n$, $y, y' > 0$ and all $\lambda \in \mathbb{C}$ with $|\lambda| = 1$ and $|\arg \lambda| \leq \theta < \pi$. We thus have proved the following result.

Lemma 1.9.1 *Let $\theta \in (0, \pi)$ and α be a multiindex. Then there exist constants $M, c > 0$ such that*

$$|(\partial_x)^\alpha r_v(x, y, y', \lambda)| \leq M y e^{-cy'} \int_0^\infty \frac{s^{n+|\alpha|}}{1+y+s} e^{-cs(|x|+y+y')} \, ds,$$

where $x \in \mathbb{R}^n$, $y, y' > 0$ and $\lambda \in \mathbb{C}$ with $|\lambda| = 1$ and $|\arg \lambda| \leq \theta < \pi$.

Remark 1.9.2 *For $\theta < \pi$ we define $\hat{r}_w : \mathbb{R}^n \times \mathbb{R}_+ \times \mathbb{R}_+ \times \Sigma_\theta \to \mathbb{C}^n$ by*

$$\hat{r}_w(\xi, y, y', \lambda) := \frac{e^{-|\xi|y} - e^{-\omega(\xi)y}}{\omega(\xi) - |\xi|} \frac{i\xi}{\omega(\xi)} e^{-\omega(\xi)y'}. \qquad (1.9.19)$$

Copying the above proof we see that the assertion of Proposition 1.9.1 remains true if r_v is replaced by r_w, where

$$r_w(x, y, y', \lambda) = \frac{1}{(2\pi)^n} \int_{\mathbb{R}^n} e^{ix \cdot \xi} \hat{r}_w(\xi, y, y', \lambda) d\xi, \quad x \in \mathbb{R}^n, y, y' > 0. \qquad (1.9.20)$$

The kernel estimates given in Lemma 1.9.1 and Remark 1.9.2 allow us to derive L^p-estimates for v_2 and w_2 via the following simple lemma on L^p-continuity of integral operators acting in half spaces.

Lemma 1.9.3 *Suppose that $1 < p \leq \infty$ and let $\frac{1}{p} + \frac{1}{p'} = 1$. Let T be an integral operator in $L^p(\mathbb{R}_+^{n+1})$ of the form*

$$(Tf)(x, y) = \int_0^\infty \int_{\mathbb{R}^n} k(x - x', y, y') f(x', y') dx' dy', \quad x \in \mathbb{R}^n, y > 0,$$

where $k : \mathbb{R}^n \times \mathbb{R}_+ \times \mathbb{R}_+ \to \mathbb{C}$ is a measurable function.

(a) Let $1 < p < \infty$. If

$$\left(\int_0^\infty \left(\int_0^\infty \|k(\cdot, y, y')\|_1^{p'} dy' \right)^{\frac{p}{p'}} dy \right)^{\frac{1}{p}} =: M_1 < \infty,$$

then $T \in \mathcal{L}(L^p(\mathbb{R}_+^{n+1}))$ and $\|T\|_{\mathcal{L}(L^p(\mathbb{R}_+^{n+1}))} \le M_1$.

(b) Let $p = \infty$. If

$$\sup_{y>0} \int_0^\infty \|k(\cdot, y, y')\|_1 dy' =: M_2 < \infty,$$

then $T \in \mathcal{L}(L^\infty(\mathbb{R}_+^{n+1}))$ and $\|T\|_{\mathcal{L}(L^\infty(\mathbb{R}_+^{n+1}))} \le M_2$.

This follows immediately by applying Young's and Hölder's inequalities.

Combining the estimates obtained in Proposition 1.9.1 and Remark 1.9.2 with Lemma 1.9.3 we obtain the following estimates for v_2 and w_2.

Lemma 1.9.4 *Let $1 < p \le \infty$ and $\theta \in (0, \pi)$. Let v_2 and w_2 be defined as above. Then there exists a constant $M > 0$ such that*

$$\|v_2\|_{L^p(\mathbb{R}_+^{n+1})^n} \le \frac{M}{|\lambda|} \|f_v\|_{L^p(\mathbb{R}_+^{n+1})^n},$$

$$\|w_2\|_{L^p(\mathbb{R}_+^{n+1})} \le \frac{M}{|\lambda|} \|f_v\|_{L^p(\mathbb{R}_+^{n+1})^n}$$

for all $\lambda \in \Sigma_\theta$.

Proof Observe that for $\lambda \in \Sigma_\theta$,

$$v_2(x, y) = \int_0^\infty \int_{\mathbb{R}^n} r_v(x - x', y, y', \lambda) f_v(x', y') dx' dy', \quad x \in \mathbb{R}^n, y > 0,$$

where $r_v(x, y, y', \lambda) = |\lambda|^{\frac{n-1}{2}} r_v(|\lambda|^{\frac{1}{2}}x, |\lambda|^{\frac{1}{2}}y, |\lambda|^{\frac{1}{2}}y', \frac{\lambda}{|\lambda|}), x \in \mathbb{R}^n, y, y' > 0$ and r_v satisfies the estimate given in Lemma 1.9.1. Combining these bounds with Lemma 1.9.3, we obtain

$$\int_{\mathbb{R}^n} |(\partial_x)^\alpha r_v(x, y, y', \lambda)| dx \le C_n e^{-c|\lambda|^{\frac{1}{2}}y'} \frac{y}{1 + |\lambda|^{\frac{1}{2}}y} \cdot \frac{1}{(|\lambda|^{\frac{1}{2}}(y + y'))^{1+|\alpha|}}.$$

Hence, if $1 < p < \infty$, then

$$\int_0^\infty \|r_v(\cdot, y, y', \lambda)\|_1^{p'} dy' \le C_{n,p} |\lambda|^{-\frac{1}{2}} |\lambda|^{-\frac{p'}{2}} (1 + |\lambda|^{\frac{1}{2}}y)^{-p'}$$

and

$$\int_0^\infty \left(\int_0^\infty \|r_v(\cdot, y, y', \lambda)\|_1^p dy' \right)^{\frac{p}{p'}} dy \le C_{np} |\lambda|^{-p}.$$

If $p = \infty$, then $\int_0^\infty \int_{\mathbb{R}^n} |r_v(\cdot, y, y', \lambda)| dx dy' \leq C_n |\lambda|^{-1} (1 + |\lambda|^{\frac{1}{2}} y)^{-1}$ and hence

$$\sup_{y > 0} \int_0^\infty \|r_v(\cdot, y, y', \lambda)\|_1 dy' \leq \frac{C_n}{|\lambda|}.$$

The estimate for r_w follows in exactly the same way.   □

Summing up, we proved the following result:

Proposition 1.9.5 *Let $1 < p \leq \infty, 0 < \theta < \pi$ and $\lambda \in \Sigma_\theta$. Let $f \in L^p(\mathbb{R}_+^{n+1})^{n+1}$ such that $\mathrm{div}\, f = 0$ and $f_{n+1|_{\partial H}} = 0$. Let $u = (v, w)^T$ be defined as in (1.9.16). Then there exists a constant $M > 0$ such that*

$$\|u\|_{L^p(\mathbb{R}_+^{n+1})^{n+1}} \leq \frac{M}{|\lambda|} \|f\|_{L^p(\mathbb{R}_+^{n+1})^{n+1}}.$$

For $1 < p < \infty$ and $\lambda > 0$ consider the mapping

$$R(\lambda) : L_\sigma^p(H) \to L_\sigma^p(H), \quad f \mapsto u_\lambda,$$

where u_λ is defined as in (1.9.16). Let A be the Stokes operator in $L_\sigma^p(H)$ defined as in (1.9.4). Then $R(\lambda)(\lambda - A)f = f$ for all $f \in D(A)$ and $(\lambda - A)R(\lambda)f = f$ for all $f \in L_\sigma^p(H)$. Thus

$$R(\lambda) = (\lambda - A)^{-1}, \quad \lambda > 0.$$

Proposition 1.9.5 combined with Theorem 1.3.10 implies now the following fundamental result.

Theorem 1.9.6 *Let $1 < p < \infty$. Then the Stokes operator A defined as in (1.9.4) generates a bounded holomorphic semigroup on $L_\sigma^p(H)^{n+1}$.*

For different approaches to this basic result, we refer to [51, 62, 112, 135] and [151].

The Stokes Operator on Spaces of Bounded Functions

We now consider the Stokes equation in spaces of bounded functions, more precisely in $BUC_\sigma(H)$, $C_{0,\sigma}(H)$ and $L_\sigma^\infty(H)$. It is our aim to show that the Stokes operator, defined in the manner described below, is the generator of an holomorphic semigroup on these spaces (which is not strongly continuous in the case of $L_\sigma^\infty(H)$). To this end, define

$$BUC_\sigma(H) := \{f \in BUC(H);$$
$$\mathrm{div}\, f = 0, f(x_1, \ldots, x_n, 0) = 0 \quad \text{for all} \quad x_1, \ldots, x_n \in \mathbb{R}\}$$

and

$$C_{0,\sigma}(H) := \overline{\{f \in C_c^\infty(H); \operatorname{div} f = 0\}}^{\|\cdot\|_{L^\infty}}.$$

Let $X_\sigma(H)$ be one of the spaces $BUC_\sigma(H)$ or $C_{0,\sigma}(H)$. For $\theta \in (0, \pi)$ and $\lambda \in \Sigma_\theta$ consider the mapping

$$R(\lambda) : X_\sigma(H)^{n+1} \to L^\infty(H)^{n+1}, \quad f \mapsto u_\lambda,$$

where u_λ is the solution of the Stokes equation given in (1.9.16). Theorem 1.9.6 and a direct calculation show that $\{R(\lambda); \lambda > 0\}$ is a pseudo-resolvent in $X_\sigma(H)^{n+1}$. By this we mean a function $R : U \to \mathcal{L}(X_\sigma(H)^{n+1})$, U being a subset of $\mathbb{C}$, satisfying the resolvent equation given in (1.3.2).

Lemma 1.9.7 *Let* $f \in X_\sigma(H)^{n+1}$. *Then*

$$\lim_{\lambda \to \infty} \lambda R(\lambda) f = f.$$

Proof Notice first that

$$R(\lambda) f_v = (\lambda - \Delta_D)^{-1} f_v + v_2(\lambda),$$
$$R(\lambda) f_w = (\lambda - \Delta_D)^{-1} f_w + w_2(\lambda),$$

where Δ_D denotes the Dirichlet Laplacian and v_2, w_2 are defined as above. By Example 1.3.13, the Dirichlet Laplacian Δ_D generates a C_0-semigroup on $BUC(H)$ or $C_0(H)$, respectively. It hence follows that

$$\lim_{\lambda \to \infty} \lambda(\lambda - \Delta_D)^{-1} f = f \text{ for all } f \in X_\sigma(H).$$

It thus remains to prove that $\lim_{\lambda \to \infty} \lambda v_2(\lambda) = 0$ in $X_\sigma(H)^n$ and $\lim_{\lambda \to \infty} \lambda w_2(\lambda) = 0$ in $X_\sigma(H)$. In order to do so, note first that $\lim_{\lambda \to \infty} \lambda v_2(\lambda) = 0$ in $BUC_\sigma(H)^n$ if and only if

$$\lim_{\lambda \to \infty} \sup_{y>0} \int_0^\infty \int_{\mathbb{R}^n} \lambda^{(n+1)/2} r_v(\lambda^{1/2} x', \lambda^{1/2} y, \lambda^{1/2} y', 1) f_v(x - x', y') dx' dy' = 0.$$

$$(1.9.21)$$

Since the above inner integral equals $\widehat{r}_v(0, y, y', 1) = 0$, it follows that (1.9.21) is satisfied provided

$$\lim_{\lambda \to \infty} \sup_{y>0} \int_0^\infty \int_{\mathbb{R}^n} |\lambda^{(n+1)/2} r_v(\lambda^{1/2} x', \lambda^{1/2} y, \lambda^{1/2} y', 1)|$$

$$\times |f_v(x - x', y') - f_v(x, 0)| dx' dy' = 0.$$

Notice that the above double integral is dominated by

$$M \sup_{|x'| \le \frac{R}{\lambda^{1/2}}, |y'| \le \frac{S}{\lambda^{1/2}}} |f_v(x - x', y') - f_v(x, 0)| + 2|f_v|_\infty M(\frac{1}{S} + \frac{1}{R})$$

for some $M > 0$ and all $S, R > 0$ by the kernel estimates established above. This implies the assertion if $X_\sigma(H) = BUC_\sigma(H)$. The case where of $X_\sigma(H) = C_{0,\sigma}(H)$ is proved in a similar way.　　　　　　　　　　　　　　　□

The above lemma shows that $\ker R(\lambda) = 0$ for all $\lambda > 0$. Hence, by the following Lemma 1.9.8, (see e.g. [13], Prop. B.6, for a proof), there exists a closed, densely defined operator A_{X_σ} in $X_\sigma(H)^{n+1}$ such that

$$R(\lambda) = (\lambda - A_{X_\sigma})^{-1}, \lambda > 0.$$

We call this operator the *Stokes operator* A_{X_σ} on $X_\sigma(H)^{n+1}$.

Lemma 1.9.8 *Let X be a Banach space, $U \subset \mathbb{C}$ be open and $R : U \to \mathcal{L}(X)$ be a pseudo-resolvent. Then there is an operator A on X such that $R(\lambda) = R(\lambda, A)$ for all $\lambda \in U$ if and only if $\ker R(\lambda) = \{0\}$ for $\lambda \in U$.*

Proposition 1.9.5 implies now the following result, which was proved first in [40].

Proposition 1.9.9 *The Stokes operator A_{X_σ} generates a strongly continuous holomorphic semigroup on $X_\sigma(H)^{n+1}$.*

Finally, we consider the solution of the Stokes equation in $L_\sigma^\infty(H)$. This space is defined as follows: note that ∇ acts as a bounded operator from $\hat{W}^{1,1}(H)$ into $L^1(H)^{n+1}$, where $\hat{W}^{1,1}(H) = \{f \in L_{loc}^1(H); \nabla f \in L^1(H)\}$. Hence $Div := -\nabla^*$ is a bounded operator from $L^\infty(H)^{n+1}$ into $\hat{W}^{1,1}(H)^*$. We define

$$L_\sigma^\infty(H)^{n+1} := \ker Div.$$

Thus $f \in L_\sigma^\infty(H)$ if and only if $f \in L^\infty(H)$ and $\int_H \nabla \varphi f = 0$ for all $\varphi \in W^{1,1}(H)$. Consider the mapping $R(\lambda) : L_\sigma^\infty(H)^{n+1} \to L^\infty(H)^{n+1}$ defined as before. Proposition 1.9.5 and a direct calculation implies that $\{R(\lambda); \lambda > 0\}$ is a pseudo-resolvent. In contrast to the situation of $X_\sigma(H)$ we do *not* have that $\lim_{\lambda \to \infty} \lambda v_2(\lambda) = 0$ in $L_\sigma^\infty(H)$. However, the representation of the remainder term given above allows us to show that $\ker R(\lambda) = 0$ in $L_\sigma^\infty(H)$ for all $\lambda > 0$. Thus there exists a closed operator $A_{L_\sigma^\infty}$ in $L_\sigma^\infty(H)^{n+1}$ such that

$$R(\lambda) = (\lambda - A_{L_\sigma^\infty})^{-1}, \lambda > 0.$$

We call the operator $A_{L_\sigma^\infty}$ the *Stokes operator* in $L_\sigma^\infty(H)^{n+1}$. Note that $A_{L_\sigma^\infty}$ is not densely defined. However, the arguments given in the proof of Theorem 1.3.10 yield the following result, which we state with some abuse of language as follows. It was proved first in [40].

Theorem 1.9.10 *The Stokes operator $A_{L^\infty_\sigma}$ generates a holomorphic semigroup on $L^\infty_\sigma(H)^{n+1}$, which is not strongly continuous in 0.*

Remarks 1.9.11 It was shown recently by Abe and Giga [1] that the Stokes operator generates also a holomorphic semigroup on $L^\infty_\sigma(\Omega)$ provided $\Omega \subset \mathbb{R}^n$ is a bounded or exterior domain with smooth boundary $\partial\Omega$. Their approach is very different from the one given above and relies on certain blow up arguments.

H^∞-Calculus and $\mathcal{R}$-Bounded H^∞-Calculus for the Stokes Operator

Theorem 1.9.6 can be rephrased by saying that the negative Stokes operator $-A$ on $L^p_\sigma(H)^{n+1}$ is a sectorial operator for all $p \in (1, \infty)$. It is hence natural to ask whether $-A$ admits a bounded H^∞-calculus on $L^p(H)^{n+1}$ for every sector Σ_θ with $0 < \theta < \pi$.

In order to answer this question, recall that the resolvent $R(\lambda, A)$ can be represented as

$$R(\lambda, A)f = (\lambda - \Delta_D)^{-1}f + S(\lambda),$$

where $f = (f_v, f_w)^T$ and $S(\lambda)f = (v_2(\lambda), w_2(\lambda))^T$.

The following lemma on L^p-boundedness of integral operators will be useful in the following.

Lemma 1.9.12 *Let T be an integral operator of the form*

$$(Tf)(y) = \int_0^\infty k(y, y')f(y')dy', \quad y > 0, \tag{1.9.22}$$

where $k : \mathbb{R}_+ \times \mathbb{R}_+ \to \mathbb{C}$ is a measurable function such that the above integral is well defined. Suppose that for some $p \in (1, \infty)$ there exists a constant $M > 0$ such that

$$|(Tf)(y)| \le \frac{M}{y^{\frac{1}{p}}} \|f\|_{L^p(\mathbb{R}_+)}, \quad y > 0.$$

If $T \in \mathcal{L}(L^{q_0}(\mathbb{R}_+))$ for some $q_0 \in (p, \infty]$, then $T \in \mathcal{L}(L^q(\mathbb{R}_+))$ for all $q \in (p, q_0]$.

Proof By assumption, Tf is dominated pointwise by a function belonging to the weak L^p-space $L^p_w(\mathbb{R}_+)$. Thus $T : L^p(\mathbb{R}_+) \to L^p_w(\mathbb{R}_+)$ is a bounded operator. The assumption and the Marcienkiewicz interpolation theorem imply that $T \in \mathcal{L}(L^q(\mathbb{R}_+))$ for all $q \in (p, q_0]$. □

Corollary 1.9.13 *Let $k : \mathbb{R}_+ \times \mathbb{R}_+ \to \mathbb{C}$ be a measurable function. Suppose that there exists $M > 0$ such that*

$$|k(y, y')| \le \frac{M}{y + y'} \log\left(1 + \frac{y}{y'}\right), \quad y, y' > 0.$$

Let T be defined as in (1.9.22) and let $1 < p \le \infty$. Then $T \in \mathcal{L}(L^p(\mathbb{R}_+))$.

Proof Note first that

$$\int_0^\infty |k(y, y')| dy' \le M \int_0^\infty \frac{\log(1 + \frac{y}{y'})}{1 + \frac{y}{y'}} \frac{dy'}{y'} = M \int_0^\infty \frac{\log(1 + s)}{(1 + s)s} ds < \infty$$

which implies that $T \in \mathcal{L}(L^\infty(\mathbb{R}_+))$.

If $p > 1$ let $\frac{1}{p} + \frac{1}{p'} = 1$. For $f \in L^p(\mathbb{R}_+)$ we obtain by Hölder's inequality

$$|Tf(y)| \le M \left(\int_0^\infty \frac{\log^{p'}(1 + \frac{y}{y'}) dy'}{(y + y')^{p'}} \right)^{\frac{1}{p'}} \|f\|_{L^p(\mathbb{R}_+)} \le \frac{M}{y^{\frac{1}{p}}} \|f\|_{L^p(\mathbb{R}_+)}, \quad y > 0.$$

Thus the assertion follows from Lemma 1.9.12. □

Let now $h \in H_0^\infty(\Sigma_\theta)$, where $0 < \theta < \pi$ is fixed. Consider the function

$$k_{h,v}(x, y, y') = \frac{1}{2\pi i} \int_\Gamma h(\lambda) r_v(x, y, y', -\lambda) d\lambda, \quad x \in \mathbb{R}^n, y, y' > 0,$$

where r_v is defined as in (1.9.18) and $\Gamma := \{\rho e^{\pm i\varphi}, \rho \ge 0\}$ with $0 < \varphi < \theta$. The estimate for r_v given in Lemma 1.9.1 yields

$$|k_{h,v}(x, y, y')| \le C\|h\|_{H^\infty} \int_0^\infty |r(x, y, y', \rho e^{\pm i(\pi - \varphi)})| d\rho$$

$$\le C\|h\|_{H^\infty} \int_0^\infty y e^{-c\rho^{\frac{1}{2}} y'} \int_0^\infty \sigma^n \frac{e^{-c\sigma(|x| + y + y')}}{\rho^{\frac{1}{2}} + \rho y + \sigma} d\sigma d\rho$$

$$=: C\|h\|_{H^\infty} k_1(x, y, y'). \tag{1.9.23}$$

Now

$$\int_{\mathbb{R}^n} |k_1(x, y, y')| dx \le C \frac{y}{y + y'} \int_0^\infty \frac{e^{-csy'}}{1 + sy} ds, \quad y, y' > 0.$$

Splitting the latter integral at $s = \frac{1}{y'}$, we obtain

$$\int_{\mathbb{R}^n} |k_1(x, y, y')| dx \le \frac{C}{y + y'} \log(1 + \frac{y}{y'}), \quad y, y' > 0. \tag{1.9.24}$$

Define now the operator $T_{h,v}$ in $L^p(\mathbb{R}_+^{n+1})$ by

$$(T_{h,v} f)(x, y) := \int_0^\infty \int_{\mathbb{R}^n} k_{h,v}(x - x', y, y') f(x', y') dx' dy', \quad x \in \mathbb{R}^n, y > 0. \tag{1.9.25}$$

Then, by Young's inequality, (1.9.23), (1.9.24) and Corollary 1.9.13 we have

$$\int_0^\infty \int_{\mathbb{R}^n} |(T_{h,v}f)(x, y)|^p dx dy \leq \int_0^\infty \left(\int_0^\infty \|k_{h,v}(\cdot, y, y')\|_1 \|f(\cdot, y')\|_p dy' \right)^p dy$$

$$\leq C \|h\|_{H^\infty}^p \int_0^\infty \left(\int_0^\infty \log(1 + \frac{y}{y'}) \right.$$

$$\times \|f(\cdot, y')\|_p dy' \Big)^p dy$$

$$\leq C \|h\|_{H^\infty}^p \|f\|_{L^p(\mathbb{R}_+^{n+1})}^p. \tag{1.9.26}$$

Moreover, by Remark 1.9.2 the function r_w defined as in (1.9.20) satisfies also an estimate of the form given in Lemma 1.9.1. Thus, the function $k_{h,w}$ defined by

$$k_{h,w}(x, y, y') := \frac{1}{2\pi i} \int_\Gamma h(\lambda) r_w(x, y, y', -\lambda) d\lambda, \quad x \in \mathbb{R}^n, y, y' > 0$$

also satisfies

$$|k_{h,w}(x, y, y')| \leq C \|h\|_{H^\infty} k_1(x, y, y'), \quad x \in \mathbb{R}^n, y, y' > 0.$$

Define the operator $T_{h,w}$ as in (1.9.25) with $k_{h,v}$ replaced by $k_{h,w}$. We conclude that $T_{h,w}$ satisfies estimate (1.9.26).

Finally note that by Example 1.4.10 the operator $-\Delta_D$ admits a bounded $H^\infty(\Sigma_\theta)$-calculus on $L^p(\mathbb{R}_+^{n+1})$ for every $\theta \in (0, \pi)$. Summing up, we thus proved the following result.

Theorem 1.9.14 *Let* $1 < p < \infty$ *and let* A *be the Stokes operator in* $L_\sigma^p(\mathbb{R}_+^{n+1})^{n+1}$ *defined as in (1.9.4). Then* $-A$ *admits a bounded* $H^\infty(\Sigma_\theta)$-*calculus on* $L_\sigma^p(\mathbb{R}_+^{n+1})^{n+1}$ *for each* $\theta \in (0, \pi)$.

We finish this section by showing that $-A$ even admits a $\mathcal{R}$-bounded H^∞-calculus on $L_\sigma^p(\mathbb{R}_+^{n+1})^{n+1}$ for each $\theta \in (0, \pi)$. Note that this follows immediately by combing Theorem 1.9.14 with Proposition 1.6.8 since $L_\sigma^p(H)$ as a closed subspace of $L^p(H)$ has property (α).

In order to be selfcontained we, however, aim to give here a different proof which is based on the techniques developed above. To this end, let us consider first the Laplacian Δ on $L^p(\mathbb{R}^n)$.

Lemma 1.9.15 *Let* $1 < p < \infty$. *Then* $-\Delta$ *admits an* $\mathcal{R}$-bounded H^∞-*calculus on* $L^p(\mathbb{R}^n)$ *on the sector* Σ_θ *for* $0 < \theta < \pi$.

Proof Let $h \in H_0^\infty(\Sigma_\theta)$, where $0 < \theta < \pi$. Then the Fourier transform of $h(-\Delta)$ is given by $h(|\xi|^2)$ for $\xi \in \mathbb{R}^n$. The kernel $k_h(\cdot)$ corresponding to $h(|\cdot|^2)$ is given by

$$k_h(x) = \frac{1}{(2\pi)^n} \int_{\mathbb{R}^n} e^{ix\xi} h(|\xi|^2) d\xi, \quad x \in \mathbb{R}^n.$$

Choosing a rotation Q such that $Qx = (|x|, 0, \ldots, 0)$ and writing $Q\xi = (a, rb)$, with $a \in \mathbb{R}$, $r > 0$, $b \in \mathbb{R}^{n-1}$, $|b| = 1$ we obtain

$$k_h(x) = c_n \int_0^\infty r^{n-2} \int_{-\infty}^\infty h(\sqrt{r^2 + a^2}) e^{i|x|a} \, da \, dr, \quad x \in \mathbb{R}^n.$$

Next, deform the contour of integration via Cauchy's theorem to $a = s + i\epsilon(r + |s|)$ for $r > 0$ and $s \in \mathbb{R}$ and obtain

$$|D^\alpha k_h(x)| \le C_\alpha \|h\|_{H^\infty} \frac{1}{|x|^{n+|\alpha|}}, \quad x \in \mathbb{R}^n \setminus \{0\}$$

for each multiindex α. If $|x| \ge 2|y|$ we obtain

$$|k_h(x - y) - k_h(x)| = \left| \int_0^1 \frac{d}{dt} k_h(x - ty) \, dt \right|$$

$$\le |y| \int_0^1 \frac{dt}{|x - ty|^{n+1}} \|h\|_{H^\infty} \le C \frac{|y|}{|x|^{n+1}} \|h\|_{H^\infty}.$$

This implies that for $R > 0$ the uniform Hörmander condition is satisfied, i.e.:

$$\int_{|x| > 2|y|} \sup_{\|h\|_{H^\infty} \le R} |k_h(x - y) - k_h(x)| \, dx \le C|y|R \int_{|x| > 2|y|} \frac{dx}{|x|^{n+1}}$$

$$= C|y|R \int_{2|y|}^\infty \frac{dr}{r^2} = CR. \quad (1.9.27)$$

By Example 1.4.10a), $-\Delta \in \mathcal{H}^\infty(L^2(\mathbb{R}^n))$, and thus

$$\{h(A) : h \in H_0^\infty(\Sigma_\theta), \|h\|_{H^\infty(\Sigma_\theta)} \le R\} \subset \mathcal{L}(L^2(\mathbb{R}^n))$$

is uniformly bounded. By Remarks 1.6.2(c) and (d), there is a constant $C > 0$ such that

$$\left\| \left(\sum_{j=1}^N |H_j f_j|^2 \right)^{\frac{1}{2}} \right\|_{L^2(\mathbb{R}^n)} \le CR \left\| \left(\sum_{j=1}^N |f_j|^2 \right)^{\frac{1}{2}} \right\|_{L^2(\mathbb{R}^n)}$$

for $N \in \mathbb{N}$, $H_j := h_j(A), h_j \in H_0^\infty(\Sigma_\theta), \|h_j\|_{H^\infty(\Sigma_\theta)} \le R$ and $f_j \in L^2(\mathbb{R}^n)$. Set now $X := \mathbb{R}^N$ and define $K : L^p(\mathbb{R}^n, X) \to L^p(\mathbb{R}^n, X)$ by $(Kf)_i := H_i f_i$ for $i = 1, \ldots, N$. The uniform Hörmander condition (1.9.27) implies that

$$\int_{|x| > 2|y|} \|K(x - y) - K(x)\| \, dx \le CR.$$

Since K acts as a bounded operator on $L^2(\mathbb{R}^n; X)$, the Benedek-Calderon-Panzone Theorem 1.2.5 implies that K is L^p-bounded for $1 < p < \infty$. This means that there is a constant $C > 0$, depending only C given in (1.9.27), R and p such that

$$\|(\sum_{j=1}^{N} |H_j f_j|^2)^{\frac{1}{2}}\|_{L^p(\mathbb{R}^n)} \le C\|(\sum_{j=1}^{N} |f_j|^2)^{\frac{1}{2}}\|_{L^p(\mathbb{R}^n)}.$$

Remark 1.6.2(d) implies that the set

$$\{h(A) : h \in H_0^\infty(\Sigma_\theta), \|h\|_{H^\infty(\Sigma_\theta)} \le R\} \subset \mathcal{L}(L^p(\mathbb{R}^n))$$

is $\mathcal{R}$-bounded for all $R > 0$ and all $0 < \theta < \pi$.   □

Following the proof given in Example 1.4.10(b), it is now easy to deduce that $-\Delta_D$, where Δ_D denotes the Dirichlet Laplacian, admits an $\mathcal{R}$-bounded $H^\infty(\Sigma_\theta)$-calculus on $L^p(\mathbb{R}^{n+1}_+)$ for each $\theta \in (0, \pi)$.

Finally, combining (1.9.25) with (1.9.26), Corollary 1.9.13 and Lemma 1.6.3 we see that the sets

$$\{T_{h,v} : h \in H_0^\infty(\Sigma_\theta), \|h\|_{H^\infty(\Sigma_\theta)} \le R\} \subset \mathcal{L}(L^p(\mathbb{R}^{n+1}_+)),$$

$$\{T_{h,w} : h \in H_0^\infty(\Sigma_\theta), \|h\|_{H^\infty(\Sigma_\theta)} \le R\} \subset \mathcal{L}(L^p(\mathbb{R}^{n+1}_+))$$

are also $\mathcal{R}$-bounded.

Summarizing, we thus proved the following result.

Theorem 1.9.16 *Let* $1 < p < \infty$ *and let* A *be the Stokes operator in* $L_\sigma^p(\mathbb{R}^{n+1}_+)$. *Then* $-A$ *admits an* $\mathcal{R}$-*bounded* $H^\infty(\Sigma_\theta)$-*calculus on* $L_\sigma^p(\mathbb{R}^{n+1}_+)$ *for each* $\theta \in (0, \pi)$.

1.10   The Stokes and Hydrostatic Stokes Equations on Domains

Given an open set $\Omega \subset \mathbb{R}^n$, $n \ge 2$, and a time interval $J = (0, T)$ for $0 < T \le \infty$ the Stokes equation is given by the following set of equations

$$\begin{cases} \partial_t u - \Delta u + \nabla p = f & \text{in } J \times \Omega, \\ \operatorname{div} u = 0 & \text{in } J \times \Omega, \\ u = 0 & \text{on } J \times \partial\Omega, \\ u(0) = u_0 & \text{in } \Omega. \end{cases} \tag{1.10.1}$$

Here f as well as u_0 are given data and $u = 0$ on $\partial\Omega$ describes the classical Dirichlet boundary condition. An important tool in order to handle the incompressibility condition div $u = 0$ is the *Helmholtz projection* P, allowing to decompose the function space $L^p(\Omega)$ for $1 < p < \infty$ into the solenoidal space $L^p_\sigma(\Omega)$ and gradient fields for a large class of domains $\Omega \subset \mathbb{R}^n$.

The Helmholtz Decomposition

Given an open set $\Omega \subset \mathbb{R}^n$, the Helmholtz decomposition deals with the question whether $L^p(\Omega)$ can be decomposed into a direct sum of the space of solenoidal vector fields and the space of gradient fields. If such a decomposition holds true, the Stokes equation (1.10.1) can then be reformulated as an evolution equation in the L^p-setting. For $1 < p < \infty$ and $\Omega \subset \mathbb{R}^n$ being an arbitrary open set we set

$$G_p(\Omega) := \{u \in L^p(\Omega) : u = \nabla\pi \text{ for some } \pi \in W^{1,p}_{loc}(\Omega)\},$$

$$L^p_\sigma(\Omega) := \overline{\{u \in C^\infty_c(\Omega) : \text{div } u = 0 \text{ in } \Omega\}}^{\|\cdot\|_p}$$

Definition 1.10.1 Let $1 < p < \infty$ and $\Omega \subset \mathbb{R}^n$ be an open set. We say that the Helmholtz decomposition for $L^p(\Omega)$ exists whenever $L^p(\Omega)$ can be decomposed into

$$L^p(\Omega) = L^p_\sigma(\Omega) \oplus G_p(\Omega).$$

The unique projection operator $P_p : L^p(\Omega) \to L^p_\sigma(\Omega)$ having $G^p(\Omega)$ as its null space is called the *Helmholtz projection*.

If $\Omega \subset \mathbb{R}^n$ is an open set and $1 < p < \infty$, then the Helmholtz decomposition exists for $L^p(\Omega)$ if and only if it exists for $L^{p'}(\Omega)$ and we have $P'_p = P_{p'}$, i.e., P_2 is orthogonal.

The existence of the Helmholtz projection for $L^p(\Omega)$ is very strongly linked with the following weak Neumann problem: given $f \in L^p(\Omega)$, find $\pi \in \widehat{W}^{1,p}(\Omega)$ satisfying

$$\int_\Omega (\nabla\pi - f) \cdot \nabla\varphi = 0, \quad \varphi \in \widehat{W}^{1,p}(\Omega). \tag{1.10.2}$$

Note that if $\partial\Omega$ as well as f and π are smooth, then integration by parts yields

$$\Delta\pi = \text{div } f \text{ in } \Omega,$$

$$\partial_\nu \pi = f \cdot \nu \text{ on } \partial\Omega,$$

where ν denotes the outer normal on $\partial\Omega$, which explains that we call (1.10.2) the weak Neumann problem. The existence of the Helmholtz projection can be characterized in terms of the weak Neumann problem, see e.g. [57, 60, 133]

Lemma 1.10.2 *Let $\Omega \subset \mathbb{R}^n$ be an open set and $1 < p < \infty$. Then the Helmholtz decomposition exists for $L^p(\Omega)$ if and only if the weak Neumann problem (1.10.2) admits for all $f \in L^p(\Omega)^n$ a unique solution $\pi \in \widehat{W}^{1,p}(\Omega)$.*

Remarks 1.10.3 Let us remark that the Helmholtz decomposition exists for $L^2(\Omega)$ for any open set $\Omega \subset \mathbb{R}^n$ and that following Maslennikova and Bogovskii [110], there exist domains $\Omega \subset \mathbb{R}^n$ such that the Helmholtz decomposition does not hold for some $p \neq 2$.

In the following proposition we consider domains for which the Helmholtz decomposition is known to exist. Note that the bent half space H_h is defined by $H_h = \{x = (x', x_n) \in \mathbb{R}^n : x_n > h(x')\}$ for some bending function $h : \mathbb{R}^{n-1} \to \mathbb{R}$. We also say that Ω is a perturbed half space, if there exists an $R > 0$ such that $\Omega \setminus B_R = \mathbb{R}^n_+ \setminus B_R$, where B_R denotes the open ball centered at x with radius R.

Proposition 1.10.4 *Let $\Omega \subset \mathbb{R}^n$ be a domain and $1 < p < \infty$. Then the Helmholtz decomposition exists for $L^p(\Omega)$ provided*

(a) $\Omega = \mathbb{R}^n$ or $\Omega = \mathbb{R}^n_+$,
(b) Ω is a bent half space, i.e. for $\Omega = H_h$ provided $h \in C^1(\mathbb{R}^{n-1})$, $\|\nabla h\|_\infty \leq \delta$ and $\delta > 0$ is small enough,
(c) $\Omega \subset \mathbb{R}^n$ is bounded, exterior, or a perturbed half space of class C^1,
(d) $\Omega \subset \mathbb{R}^n$, $n \geq 2$, is a bounded and convex domain,
(e) $\Omega \subset \mathbb{R}^n$ is an aperture C^1-domain.
(f) $\Omega = \mathbb{R}^{n-1} \times (0, \delta)$, $n \geq 2$, $\delta > 0$, is a layer domain.

Moreover, there exists $\varepsilon > 0$ such that the Helmholtz decomposition for $L^p(\Omega)$ exists provided $\Omega \subset \mathbb{R}^3$ is bounded Lipschitz domain and $\frac{3}{2} - \varepsilon < p < 3 + \varepsilon$. The range of those p is sharp.

The Stokes Equation in Domains with Compact Boundaries
We consider the Stokes equation (1.10.1) on domains $\Omega \subset \mathbb{R}^n$ with smooth boundaries by means of a localization procedure. The procedure is explained for the resolvent problem

$$\begin{cases} \lambda u - \Delta u + \nabla p = f \text{ in } \Omega, \\ \operatorname{div} u = 0 \text{ in } \Omega, \\ u = 0 \text{ on } \partial\Omega, \end{cases} \tag{1.10.3}$$

where λ is assumed to lie in a suitable sector of the complex plane. For simplicity, we consider only domains with compact boundaries. By employing finite coverings, we then may include domains of the following type: given $n \geq 2$, a domain $\Omega \subset \mathbb{R}^n$ is called a *standard domain*, if Ω coincides with $\mathbb{R}^n$, $\mathbb{R}^n_+$, a bounded domain, an exterior domain, or a perturbed half-space.

The first step in the localization procedure consists of choosing a finite covering $(U_j)^m_{j=1}$ of $\Omega \subset \mathbb{R}^n$ with boundary of class C^2 and a partition of unity $(\varphi_j)^m_{j=1}$ subordinate to $(U_j)^m_{j=1}$. Multiplying (1.10.3) with φ_j leads to a localized perturbed version in U_j. Since the Stokes equations are invariant under rotations and translations, and by choosing U_j sufficiently small, one may assume that the localized version of Eq. (1.10.3) is either an equation on $\mathbb{R}^n$ or on a bent half-space

$$H_j := \{x \in \mathbb{R}^n : x_n > h_j(x')\}, \tag{1.10.4}$$

with a certain bending function $h_j : \mathbb{R}^{n-1} \to \mathbb{R}$. We further transforms the localized system on H_j by

$$v(x', x_n) := (u \circ \phi)(x', x_n) := u(x', x_n + h_j(x')), \quad (x', x_n) \in \mathbb{R}^n_+. \tag{1.10.5}$$

The resulting system for v then is an equation on $\mathbb{R}^n_+$. Summarizing, by this procedure the Stokes resolvent problem on a domain is reduced to finitely many equations on $\mathbb{R}^n_+$ or $\mathbb{R}^n$.

A fundamental problem arising in this approach is the fact that the condition div $u = 0$ is not preserved, neither under multiplication with a cut-off function nor by transformation (1.10.5). In order to overcome this difficulty, several strategies have been developed.

Strategy 1 We replace the transformation (1.10.5) by

$$v = Tu := u \circ \phi - \left(0, \ldots, 0, (\nabla' h_j, 0) \cdot (u \circ \phi)\right). \tag{1.10.6}$$

This transformation leaves the outer normal at the boundary invariant and implies div $v = 0$. The price one has to pay for this is a lift of the boundary smoothness from C^2 to C^3. This is due to the fact that $\nabla' h_j$ appears in the transformation and since the Stokes system is a second order system, we require existence of third order derivatives of the bending functions h_j. The transformation (1.10.6) is utilized e.g., in [59, 115, 135].

Strategy 2 Suppose the divergence problem div $w = g$, $w|_{x_n=0} = 0$ admits a solution given by a Bogovskii operator $B : g \mapsto w$. Then we may correct the lacking solenoidality by the term

$$Qv := v - B \operatorname{div} v. \tag{1.10.7}$$

The price to pay here is that we need to prove suitable mapping properties of Bogovskii type operators. This approach to the Stokes system is performed e.g., in [59] and [60].

Strategy 3 It is also possible to work with an inhomogeneous divergence condition right from the very beginning. In this case, the inhomogeneous Stokes systems in $\mathbb{R}^n$ and in $\mathbb{R}^n_+$ needs to be solved. Localizing the inhomogeneous equations produces then perturbation terms not only in the first n lines of (1.10.3), but also in the divergence condition.

We explain here strategy 1 in more detail. We consider first the Stokes resolvent problem on $\Omega = H$, where H is a bent half-space with h being a suitable bending function defined in (1.10.4). Since Ω is a C^3 domain, we may assume $h \in BUC^3(\mathbb{R}^{n-1})$. This implies that the transformation T defined in (1.10.6) is an isomorphism between the Sobolev spaces involved up to order two. In particular, it can be shown that

$$T \in \mathcal{L}_{is}(L^p_\sigma(H), L^p_\sigma(\mathbb{R}^n_+)) \cap \mathcal{L}_{is}(D(A_{H,p}), D(A_{\mathbb{R}^n_+,p})),$$

where $D(A_{\Omega,p})$ denotes the domain of the Stokes operator $A_{\Omega,p}$ on Ω. Hence, we may define

$$A_T := T A_{H,p} T^{-1},$$

which is an operator in $L^p_\sigma(\mathbb{R}^n_+)$ with domain $D(A_T) = D(A_{\mathbb{R}^n_+,p})$. The smallness of $\nabla' h$ implies that $B := A_T - A_{\mathbb{R}^n_+,p}$ is a relatively bounded perturbation of the Stokes operator $A_{\mathbb{R}^n_+,p}$. In the next step, we set

$$u := \sum_{j=1}^m \varphi_j u_j, \qquad p := \sum_{j=1}^m \varphi_j p_j,$$

where (u_j, p_j) is the restricted bent half-space solution to data $f_j = \phi_j f$ that corresponds to U_j. To be precise, modulo rotation and translation, one has

$$U_j \cap H_j = U_j \cap \Omega, \quad U_j \cap \partial H_j = U_j \cap \partial\Omega.$$

Then u solves the perturbed Stokes resolvent problem

$$\begin{cases} \lambda u - \Delta u + \nabla p = f + \sum_{j=1}^m (-u_j \Delta\varphi_j - \nabla u_j \nabla\varphi_j + p_j \nabla\varphi_j) & \text{in } \Omega, \\ \operatorname{div} u = \sum_{j=1}^m u_j \nabla\varphi_j & \text{in } \Omega, \\ u = 0 & \text{on } \partial\Omega. \end{cases}$$

$$(1.10.8)$$

It remains to prove that the remainder terms are of lower order so that they may be absorbed into the terms on the left hand side. Whereas the terms $u_j \Delta\varphi_j$ and $\nabla u_j \nabla\varphi_j$ are standard, a further difficulty is represented by the terms $p_j \nabla\varphi_j$. This relates to the fact that, a priori, only an estimate on ∇p of the form

$$\|\nabla p\|_p \leq C \|f\|_p$$

uniformly in the resolvent parameter λ is available, but no suitable estimate for the pressure itself. The following Lemma provides decay estimates for the pressure, see [115, Lemma 13].

Lemma 1.10.5 ([115]) *Let $\theta \in (0, \pi/2)$, $1 < p < \infty$ and $(u, p) \in D(A_{\Omega,p}) \times \widehat{W}^{1,p}(\Omega)$ the unique solution of the Stokes resolvent problem (1.10.3) on $L^p_\sigma(\Omega)$. For a bounded C^2-domain $G \subset \Omega$ set $p_G := p - \frac{1}{|G|} \int_G p\, dx \in L^p_0(G) = \{g \in L^p(G); \int_G g\, dx = 0\}$. Then for each $\alpha \in (0, 1/2p')$ and every bounded C^2-domain $G \subset \Omega$ there exists a constant $C > 0$, independent of λ and f, such that*

$$\|p_G\|_p \leq C |\lambda|^{-\alpha} \|f\|_p, \quad \lambda \in \Sigma_{\pi-\theta}, \ |\lambda| \geq 1. \qquad (1.10.9)$$

A sketch of the proof of the above lemma is as follows. Using $\nabla p_G = \nabla p$ and the fact that $\nabla p_G(x) = (I - P_\Omega)\Delta u(x)$ for $x \in \Omega$, we obtain

$$\int_\Omega p_G \eta \, dx = -\int_\Omega \nabla p_G \cdot \psi \, dx = \int_\Omega (-\Delta_D u)(I - P_\Omega)\psi \, dx$$

$$= \int_\Omega [(-\Delta_D)^{1-\alpha} u](-\Delta_D)^\alpha (I - P_\Omega)\psi \, dx.$$

By Proposition 1.4.12, $-\Delta_D$ admits a bounded H^∞-calculus on $L^{p'}(\Omega)$ and it thus follows from Theorem 1.5.6 that $D((-\Delta_D)^\alpha) = H^{2\alpha, p'}(\Omega)$. Note that for $\alpha \in (0, 1/2p')$ one has $H^{2\alpha, p'}(\Omega) = H_0^{2\alpha, p'}(\Omega)$, which implies that we may shift the part $(-\Delta_D)^\alpha$ of $-\Delta_D$ onto $(I - P_\Omega)\psi$ without getting boundary terms.

Summarizing, we obtain the following result.

Theorem 1.10.6 *Let $n \geq 2$, $1 < p, q < \infty$, $J = (0, T)$ for some $T > 0$ and assume that $\Omega \subset \mathbb{R}^n$ is a standard domain of class C^3. Then the Stokes operator defined by*

$$A_{\Omega, p} u := P\Delta u, \quad D(A_{\Omega, p}) := W^{2,p}(\Omega) \cap W_0^{1,p}(\Omega) \cap L_\sigma^p(\Omega) \qquad (1.10.10)$$

admits maximal L^q-regularity on $L_\sigma^p(\Omega)$. In particular, the solution u to the Cauchy problem

$$u'(t) - A_p u(t) = f(t), \quad t > 0, \quad u(0) = u_0,$$

satisfies the estimate

$$\|u'\|_{L^q(J; L^p(\Omega))} + \|A_p u\|_{L^q(J; L^p(\Omega))} \leq C\big(\|f\|_{L^q(J; L^p(\Omega))} + \|u_0\|_{X_\gamma}\big),$$

for some $C > 0$ independent of $f \in L^q(J; L_\sigma^p(\Omega))$ and $u_0 \in X_\gamma := \big(L_\sigma^p(\Omega), D(A_p)\big)_{1-1/q, q}$.

Moreover, A_p generates a bounded analytic C_0-semigroup on $L_\sigma^p(\Omega)$ and

(a) $\sigma(A_{\Omega, p}) = (-\infty, 0]$ if Ω is $\mathbb{R}^n$, $\mathbb{R}_+^n$ or an exterior domain,
(b) $\sigma(A_{\Omega, p}) = (-\infty, -\kappa]$ for some $\kappa = \kappa(\Omega) > 0$ provided Ω is bounded,
(c) $u_0 \in X_\gamma$ if and only if $u_0 \in B_{p,q}^{2-2/q}(\Omega) \cap L_\sigma^p(\Omega)$ and $u = 0$ on $\partial\Omega$.

Setting $\nabla\pi = (Id - P)\Delta(\lambda - A_p)^{-1}$, one obtains the following results for the Stokes equation (1.10.1) and its corresponding resolvent Eq. (1.10.3).

Corollary 1.10.7 *Given the assumptions of Theorem 1.10.6, the Stokes equation (1.10.1) admits a unique solution $(u, \pi) \in W^{1,q}(J; L^p(\Omega)) \cap L^q(J; W^{2,p}(\Omega) \cap W_0^{1,p}(\Omega) \cap L_\sigma^p(\Omega)) \times L^q(\widehat{W}^{1,p}(\Omega))$ and there exists a constant $C > 0$ such that*

$$\|u_t\|_{L^q(J; L^p(\Omega))} + \|u\|_{L^q(J; L^p(\Omega))} + \|\nabla^2 u\|_{L^q(J; L^p(\Omega))} + \|\nabla\pi\|_{L^q(J; L^p(\Omega))}$$

$$\leq C\big(\|f\|_{L^q(J; L^p(\Omega))} + \|u_0\|_{X_\gamma}\big).$$

First results on maximal L^p-regularity estimates for the instationary Stokes system (1.10.1) go back to the pioneering work of Solonnikov, see [135]. For a modern approach to his results, then also in the mixed $L^q - L^p$-context, based on the characterization of maximal L^p-regularity by the $\mathcal{R}$-boundedness property of the resolvent, see the work of Geissert et al. [59] and [60]. For the halfspace $\mathbb{R}^n_+$, results on the existence and analyticity of the Stokes semigroup go back to [40, 112, 151]. Giga and Sohr [64] proved for the first time global-in-time mixed $L^q - L^p$ maximal regularity estimates for smooth exterior domains by combining a result on the boundedness of the imaginary powers of the Stokes operator with the Dore–Venni theorem [44]. A different approach to maximal L^p-regularity of the Stokes equation (1.10.1) based on pseudo-differential methods was developed by Grubb and Solonnikov in [70].

An important tool in the investigation of nonlinear problems is the representation of the domain of the fractional powers of a sectorial operator in terms of suitable function spaces. Assuming that a sectorial operator A admits a bounded H^∞-calculus on a Banach space X, Theorem 1.5.6 tells us that in case

$$[X, D(A)]_\alpha = D(A^\alpha), \quad \alpha \in (0, 1). \tag{1.10.11}$$

The following result deals with the H^∞-calculus for the Stokes operator on $L^p_\sigma(\Omega)$.

Theorem 1.10.8 ([115]) *Let $n \geq 2$, $1 < p < \infty$, and assume that $\Omega \subset \mathbb{R}^n$ is a standard domain of class C^3. Then $-A_{\Omega,p}$ admits an $\mathcal{R}$-bounded H^∞-calculus on $L^p_\sigma(\Omega)$. In particular, relation (1.10.11) holds for $A = -A_{\Omega,p}$ and $X = L^p_\sigma(\Omega)$.*

The first proof of the boundedness of the imaginary powers of the Stokes operator on bounded domains with smooth boundaries goes back to Giga [63]. His proof is again based on Seeley's theorem. The case of an exterior domain, due to Giga and Sohr, was treated in [64]. A first proof of the fact that $-A_{\mathbb{R}^n_+,p}$ admits an $\mathcal{R}$-bounded H^∞-calculus on $L^p_\sigma(\mathbb{R}^n_+)$ was given by Desch, Hieber and Prüss in [40]. Moreover, the proof of the existence of an $\mathcal{R}$-bounded H^∞-calculus on standard domains in $L^p_\sigma(\Omega)$ is due to Noll and Saal [115]. For the case $n = 2$, see the work of Abels [3].

Remark 1.10.9 Let $\Omega \subset \mathbb{R}^n$ for $n \geq 3$ be a bounded Lipschitz domains. It was shown by Kunstmann and Weis [96] that the negative Stokes operator on $L^p_\sigma(\Omega)$ admits a bounded H^∞-calculus on $L^p_\sigma(\Omega)$ provided $|1/p - 1/2| \leq 1/2n$.

Finally, we consider the Stokes equation with inhomogeneous data of the form

$$\begin{cases} (\partial_t + \omega)u - \Delta u + \nabla\pi = f & \text{in } \mathbb{R}_+ \times \Omega, \\ \operatorname{div} u = g & \text{in } \mathbb{R}_+ \times \Omega, \\ u = h & \text{on } \mathbb{R}_+ \times \partial\Omega, \\ u(0) = u_0 & \text{in } \Omega, \end{cases} \tag{1.10.12}$$

where $\Omega \subset \mathbb{R}^n$ is a domain with compact boundary of class 3 and $\omega \in \mathbb{R}$, In the following theorem, the maximal $L^q - L^p$ regularity estimates for the solution (u, π) of (1.10.12) are characterized by conditions on the data (f, g, h, u_0). To this end, the set of conditions (D) is introduced:

Condition **(D)**:

(a) $f \in L^q(\mathbb{R}_+; L^p(\Omega))$, $u_0 \in B_{p,q}^{2-2/q}(\Omega)$,
(b) $g \in H^{1,q}(\mathbb{R}_+; \dot{H}^{-1,p}(\Omega)) \cap L^q(\mathbb{R}_+; H^{1,p}(\Omega))$, div $u_0 = g(0)$,
(c) $h \in F_{q,p}^{1-1/2p}(\mathbb{R}_+; L^p(\partial\Omega)) \cap L^q(\mathbb{R}_+; B_{p,p}^{2-1/p}(\partial\Omega))$ and $h(0) = u_0$ on $\partial\Omega$ if $q > 3/2$,
(d) $(g|h_\nu) \in H^{1,q}(\mathbb{R}_+; \dot{H}^{-1,p}(\Omega))$ and $h_\nu(0) = (\nu|u_0)$ on $\partial\Omega$.

Then the following theorem holds true.

Theorem 1.10.10 ([124]) *Let $\Omega \subset \mathbb{R}^n$ be a domain with compact boundary $\partial\Omega$ of class 3, let $1 < p, q < \infty$ and $q \neq 3, 3/2$. Then there exists $\omega_0 \in \mathbb{R}$ such that for each $\omega > \omega_0$ there exists a unique solution (u, π) of Eq. (1.10.12) within the class*

$$u \in H^{1,q}(\mathbb{R}_+; L^p(\Omega)) \cap L^q(\mathbb{R}_+, H^{2,p}(\Omega)) \text{ and } \pi \in L^q(\mathbb{R}_+; \dot{H}^{1,p}(\Omega))$$

if and only if the data (u_0, f, g, h) satisfy the above condition (D).

The proof of Theorem 1.10.10 is rather involved, see Section 7 of [124] for a very thorough study of the Stokes equation with inhomogeneous data. In addition, other types of boundary conditions as pure slip, outflow and free boundary conditions are studied there.

Primitive Equations
Consider the isothermal primitive equations of the form

$$\partial_t v + u \cdot \nabla v - \Delta v + \nabla_H \pi = f \quad \text{in } \Omega \times (0, T),$$

$$\partial_z \pi = 0 \quad \text{in } \Omega \times (0, T), \qquad (1.10.13)$$

$$\text{div } u = 0 \quad \text{in } \Omega \times (0, T),$$

$$v(0) = v_0.$$

Here $\Omega = G \times (-h, 0)$, where $G = (0, 1)^2$ and $h > 0$. The velocity u of the fluid is described by $u = (v, w)$ with $v = (v_1, v_2)$, and where v and w denote the horizontal and vertical components of u, respectively. Furthermore, π denotes the pressure of the fluid and f a given external force. The symbol $\nabla_H = (\partial_x, \partial_y)^T$ denotes the horizontal gradient, Δ the three dimensional Laplacian and ∇ and div the three dimensional gradient and divergence operators.

The above system is complemented by the set of boundary conditions

$$\partial_z v = 0, \quad w = 0 \quad \text{on } \Gamma_u \times (0, T),$$
$$v = 0, \quad w = 0 \quad \text{on } \Gamma_b \times (0, T), \qquad (1.10.14)$$
$$u, \pi \text{ are periodic on } \Gamma_l \times (0, T).$$

Here $\Gamma_u := G \times \{0\}$, $\Gamma_b := G \times \{-h\}$, $\Gamma_l := \partial G \times [-h, 0]$ denote the upper, bottom and lateral parts of the boundary $\partial \Omega$, respectively. Note that $w(x, y, z) = \int_z^0 \mathrm{div}_H\, v(x, y, \zeta)\, d\zeta$ for $(x, y) \in G$, $-h < z < 0$, and thus $\mathrm{div}_H\, \bar{v} = 0$ in G, where $\bar{v}$ stands for the average of v in the vertical direction, i.e.,

$$\bar{v}(x, y) := \frac{1}{h} \int_{-h}^0 v(x, y, z)\, dz, \quad (x, y) \in G.$$

Therefore, problem (1.10.13)–(1.10.14) is equivalent to finding a function $v : \Omega \to \mathbb{R}^2$ and a function $\pi : G \to \mathbb{R}$ satisfying the set of equations

$$\begin{aligned}
\partial_t v + v \cdot \nabla_H v + w \partial_z v - \Delta v + \nabla_H \pi &= f && \text{in } \Omega \times (0, T), \\
w &= \int_z^0 \mathrm{div}_H\, v\, d\zeta && \text{in } \Omega \times (0, T), \\
\mathrm{div}_H\, \bar{v} &= 0 && \text{in } G \times (0, T), \\
v(0) &= v_0,
\end{aligned}$$

(1.10.15)

as well as the boundary conditions

$$\begin{aligned}
\partial_z v &= 0 && \text{on } \Gamma_u \times (0, T), \\
v &= 0 && \text{on } \Gamma_b \times (0, T), \\
v \text{ and } \pi \text{ are periodic } &&& \text{on } \Gamma_l \times (0, T).
\end{aligned}$$

(1.10.16)

The Sobolev spaces equipped with periodic boundary conditions in the horizontal directions are defined by

$$W_{\mathrm{per}}^{m,p}(\Omega) := \{ f \in W^{m,p}(\Omega) \mid f \text{ is periodic of order } m - 1 \text{ on } \Gamma_l \},$$

$$W_{\mathrm{per}}^{m,p}(G) := \{ f \in W^{m,p}(G) \mid f \text{ is periodic of order } m - 1 \text{ on } \partial G \}.$$

We consider first the resolvent problem associated with the linearization of (1.10.14) within the L^p-setting. More precisely, let $\lambda \in \Sigma_{\pi-\varepsilon} = \{\lambda \in \mathbb{C} : |\arg \lambda| < \pi - \varepsilon\}$ for some $\varepsilon \in (0, \pi/2)$ and $f \in L^p(\Omega)^2$ for some $1 < p < \infty$ and consider the equation

$$\begin{aligned}
\lambda v - \Delta v + \nabla_H \pi &= f \text{ in } \Omega, \\
\mathrm{div}_H\, \bar{v} &= 0 \text{ in } G,
\end{aligned}$$

(1.10.17)

subject to the boundary conditions

$$\partial_z v = 0 \text{ on } \Gamma_u, \qquad v = 0 \text{ on } \Gamma_b, \qquad v \text{ and } \pi \text{ are periodic on } \Gamma_l. \quad (1.10.18)$$

The following resolvent estimate was deduced in [73].

Proposition 1.10.11 *Let* $\lambda \in \Sigma_{\pi-\varepsilon} \cup \{0\}$ *for some* $\varepsilon \in (0, \pi/2)$. *Moreover, let* $p \in (1, \infty)$ *and* $f \in L^p(\Omega)$. *Then Eqs.* (1.10.17) *and* (1.10.18) *admit a unique* solution $(v, \pi) \in W_{per}^{2,p}(\Omega) \times W_{per}^{1,p}(G) \cap L_0^p(G)$ and there exists a constant $C > 0$ such that

$$|\lambda| \, \|v\|_{L^p(\Omega)} + \|v\|_{W^{2,p}(\Omega)} + \|\pi\|_{W^{1,p}(G)} \leq C \|f\|_{L^p(\Omega)},$$

$$\lambda \in \Sigma_{\pi-\varepsilon} \cup \{0\}, \, f \in L^p(\Omega). \tag{1.10.19}$$

As in the case of the classical Helmholtz projection, the existence of the hydrostatic Helmholtz projection is closely related to the unique solvability of the Poisson problem in the weak sense. In the given situation, the equation $\Delta_H \pi = \mathrm{div}_H f$ in G, subject to periodic boundary conditions, plays an essential role.

Lemma 1.10.12 *Let* $p \in (1, \infty)$ *and* $f \in L^p(G)$. *Then there exists a unique* $\pi \in W_{per}^{1,p}(G) \cap L_0^p(G)$ *satisfying*

$$\langle \nabla_H \pi, \nabla_H \phi \rangle_{L^{p'}(G)} = \langle f, \nabla_H \phi \rangle_{L^{p'}(G)}, \quad \phi \in W_{per}^{1,p'}(G) \cap L_0^{p'}(G). \tag{1.10.20}$$

Furthermore, there exists a constant $C > 0$ *such that*

$$\|\pi\|_{W^{1,p}(G)} \leq C \|f\|_{L^p(G)}, \quad f \in L^p(G). \tag{1.10.21}$$

The above Lemma 1.10.12 allows to define the hydrostatic Helmholtz projection $P_p : L^p(\Omega) \to L^p(\Omega)$ as follows: given $v \in L^p(\Omega)$, let $\pi \in W_{per}^{1,p}(G) \cap L_0^p(G)$ be the unique solution of Eq. (1.10.20) with $f = \bar{v}$. One then sets

$$P_p v := v - \nabla_H \pi, \tag{1.10.22}$$

and calls P_p the *hydrostatic Helmholtz projection*. It follows from Lemma 1.10.12 that $P_p^2 = P_p$ and that thus P_p is indeed a projection. We define the closed subspace X_p of $L^p(\Omega)$ as $X_p := \mathrm{Rg} P_p$. This space plays the analogous role in the investigations of the primitive equations as the solenoidal space $L_\sigma^p(\Omega)$ plays in the theory of the Navier–Stokes equations.

The hydrostatic Helmholtz projection P_p defined as in (1.10.22) allows then to define the hydrostatic Stokes operator as follows. In fact, let $1 < p < \infty$ and X_p be defined as above. Then the *hydrostatic Stokes operator* A_p on X_p is defined as

$$\begin{cases} A_p v := P_p \Delta v, \\ D(A_p) := \{v \in W_{per}^{2,p}(\Omega)^2 : \mathrm{div}_H \, \bar{v} = 0 \text{ in } G, \, \partial_z v = 0 \text{ on } \Gamma_u, \, v = 0 \text{ on } \Gamma_b\}. \end{cases} \tag{1.10.23}$$

The resolvent estimates for Eqs. (1.10.17) and (1.10.18) given in Proposition 1.10.11 yield that A_p generates a bounded analytic semigroup on X_p. More precisely, one has the following result.

Proposition 1.10.13 *Let* $1 < p < \infty$. *Then the hydrostatic Stokes operator* A_p *generates a bounded analytic* C_0-*semigroup* T_p *on* X_p. *Moreover, there exist constants* $C, \beta > 0$ *such that*

$$\|T_p(t)f\|_{X_p} \le Ce^{-\beta t}\|f\|_{X_p}, \quad t > 0.$$

It was shown by Giga et al. [68] that $-A_p$ even admits an $\mathcal{R}$-bounded H^∞-calculus on X_p of angle 0, which implies in particular maximal $L^q - L^p$-estimates for the solution of the hydrostatic Stokes equation and allows further to characterize the domains $D(-A_p^\theta)$ of the fractional powers $-A_p^\theta$ for $0 < \theta < 1$ in terms of Sobolev spaces subject to the boundary conditions given.

Theorem 1.10.14 *Let* $p \in (1, \infty)$. *Then the operator* $-A_p$ *admits a bounded* $\mathcal{R}H^\infty$-*calculus on* X_p *with* $\phi_{A_p}^{\mathcal{R},\infty} = 0$.

Combining this result with a characterization of the complex interpolation spaces $[X_p, D(A_p)]_\theta$ proved in [83] allows then to characterize the domains $D(A_p^\theta)$ of the fractional powers $-A_p^\theta$ for $0 < \theta < 1$ as follows. For $p \in (1, \infty)$ and $s \in [0, \infty)$ the spaces $H_{per}^{s,p}(\Omega)$ are defined as $H_{per}^{s,p}(\Omega) := \overline{C_{per}^\infty(\overline{\Omega})}^{\|\cdot\|_{H^{s,p}(\Omega)}}$.

Corollary 1.10.15 ([68]) *Let* $1 < p < \infty$ *and* $\theta \in [0, 1]$ *with* $\theta \notin \{1/2p, 1/2 + 1/2p\}$. *Then*

$$D((A_p)^\theta) = \begin{cases} \{v \in H_{per}^{2\theta,p}(\Omega) \cap X_p : \partial_z v|_{\Gamma_N} = 0, \ v|_{\Gamma_D} = 0\}, & 1/2 + 1/2p < \theta \le 1, \\ \{v \in H_{per}^{2\theta,p}(\Omega) \cap X_p : v|_{\Gamma_D} = 0\}, & 1/2p < \theta < 1/2 + 1/2p, \\ \{v \in H_{per}^{2\theta,p}(\Omega)^2 \cap X_p, & \theta < 1/2p. \end{cases}$$

1.11   Nonlinear Stability of Ekman Boundary Layers

In this section we are considering the *nonlinear* Navier–Stokes equations in the rotational setting in the half-space $\mathbb{R}_+^3$ subject to homogeneous Dirichlet boundary conditions. More precisely, consider the set of equations

$$\begin{cases} \partial_t u - \nu \Delta u + \Omega e_3 \times u + (u \cdot \nabla)u + \nabla p = 0, & t > 0, \ x \in \mathbb{R}_+^3, \\ \operatorname{div} u = 0, & t > 0, \ x \in \mathbb{R}_+^3, \\ u(t, x_1, x_2, 0) = 0, & t > 0, \ x_1, x_2 \in \mathbb{R}, \\ u(0, x) = u_0, \ x \in \mathbb{R}_+^3, \end{cases} \quad (1.11.1)$$

where $u = (u^1, u^2, u^3)$ denotes the velocity field and p the pressure of an incompressible, viscous fluid. Here, e_3 denotes the unit vector in x_3-direction, $\nu > 0$ the viscosity of the fluid, and the constant $\Omega \in \mathbb{R}$ is called the Coriolis parameter, which is equal to twice of the frequency of the rotation around the x_3 axis.

It has been known for a long time that the above system has a stationary solution, which can be expressed explicitly as

$$u_E(x_3) = u_\infty(1 - e^{-x_3/\delta}\cos(x_3/\delta), e^{-x_3/\delta}\sin(x_3/\delta), 0)^T, \qquad (1.11.2)$$

$$p_E(x_2) = -\Omega u_\infty x_2, \qquad (1.11.3)$$

where δ is defined by $\delta := (\frac{2\nu}{\Omega})^{1/2}$ and $u_\infty \geq 0$ is a constant. This stationary solution of Eq. (1.11.1) is called in honour of the swedish oceanograph Ekman, the *Ekman spiral*; see also [47]. It describes mathematically rotating boundary layers in geophysical fluid dynamics between a geostrophic flow and a solid boundary at which the no slip boundary condition applies. Here, δ denotes the thickness of the layer, see also [66].

In the geostrophic flow region corresponding to large x_3, there is a uniform flow with velocity u_∞ in the x_1 direction. Associated with u_∞, there is a pressure gradient in the x_2-direction. The Ekman spiral in $\mathbb{R}^3_+$ matches this uniform velocity for large x_3 with the no slip boundary condition at $x_3 = 0$, i.e. we have $u_E(0) = 0$ and

$$u_E(x_3) \to (u_\infty, 0, 0) \quad \text{provided} \quad x_3 \to \infty.$$

In this section we are interested in stability questions for the Ekman spiral. The more general situation of stratified flows is described in detail in [93].

We consider hence perturbations of the Ekman spiral by functions u solving the above Eq. (1.11.1). To this end, set

$$w := u - u_E, \quad \text{and} \quad q := p - p_E.$$

Then, since (u_E, p_E) is a stationary solution of (1.11.1), the pair (w, q) satisfies the equations

$$\begin{cases} \partial_t w - \nu\Delta w + \Omega e_3 \times w + (u_E \cdot \nabla)w + w_3\partial_3 u_E + (w \cdot \nabla)w + \nabla q = 0, t > 0, x \in \mathbb{R}^3_+, \\ \operatorname{div} w = 0, t > 0, x \in \mathbb{R}^3_+, \\ w(x_1, x_2, 0) = 0, t > 0, x_1, x_2 \in \mathbb{R}, \\ w(0, x) = w_0, x \in \mathbb{R}^3_+, \end{cases}$$

$$(1.11.4)$$

where $w_0 = u_0 - u_E$. We are now interested in the following questions:

(a) does there exist a suitable notion of solutions (weak, mild, strong, etc.) such that for any initial data $w_0 \in L^2_\sigma(\mathbb{R}^3_+)$ the above set of Eq. (1.11.4) admits a solution for all $t \in (0, T]$, where $T > 0$ is arbitrary or for all $t \in [0, \infty)$?
(b) is the solution stable or asymptotically stable, i.e. does such a solution w satisfy

$$\|w(t)\|_2 \leq \|w_0\|_2 \text{ for all } t > 0 \text{ or } \lim_{t\to\infty} \|w(t)\|_2 = 0?$$

In the following, we give partial answers to the above questions. Indeed, we will show that there exists a global *weak solution* to the above Eq. (1.11.4) provided the Reynolds number Re given by $Re = u_\infty \delta \nu^{-1}$ is small enough. Secondly, assuming this condition, for every initial data $w_0 \in L^2_\sigma(\mathbb{R}^3_+)$, there exists at least one global weak solution w to (1.11.4) such that

$$\lim_{T \to \infty} \int_T^{T+1} \|w(t)\|_{H^1} dt = 0,$$

which shows in particular that the Ekman spiral is nonlinearly stable with respect to L^2-perturbations.

The Stokes-Coriolis-Ekman Operator

Rewriting the Eq. (1.11.4) as an evolution equation in $L^p_\sigma(\mathbb{R}^3_+)$ for $1 < p < \infty$ yields

$$\begin{cases} w' - A_{SCE}w + P(w \cdot \nabla)w = 0, & t > 0, \\ \qquad\qquad\qquad w(0) = w_0. \end{cases} \tag{1.11.5}$$

Here P denotes the Helmholtz projection from $L^p(\mathbb{R}^3_+)$ to $L^p_\sigma(\mathbb{R}^3_+)$ defined as in (1.9.2) and A_{SCE} denotes the Stokes-Coriolis-Ekman operator on $L^p_\sigma(\mathbb{R}^3_+)$ defined by

$$\begin{cases} A_{SCE}w := P(\nu\Delta w - \Omega e_3 \times w - [(u_E \cdot \nabla)w + w_3\partial_3 u_E]) =: (A_S + A_C + A_E)w \\ D(A_{SCE}) := W^{2,p}(\mathbb{R}^3_+) \cap W_0^{1,p}(\mathbb{R}^3_+) \cap L^p_\sigma(\mathbb{R}^3_+). \end{cases} \tag{1.11.6}$$

It now follows from Theorem 1.9.6 that the Stokes operator $A_S := P\Delta$ generates a bounded analytic semigroup e^{tA_S} on $L^p_\sigma(\mathbb{R}^3_+)$ for all $p \in (1, \infty)$. The perturbation Theorem 1.3.15 combined the standard interpolation theory implies that the Stokes-Coriolis-Ekman operator also generates also an holomorphic semigroup $e^{tA_{SCE}}$ on $L^p_\sigma(\mathbb{R}^3_+)$. Moreover, combing the Perturbation Theorem 1.6.19 for $\mathcal{R}$-sectorial operators with the fact that $-A_S$ is $\mathcal{R}$-sectorial, see Theorem 1.9.16, it follows from Corollary 1.6.10 that there exists a constant $\nu \geq 0$ such that $A_{SCE} + \nu$ has maximal L^p-regularity for all $p \in (1, \infty)$. We summarize our considerations in the following proposition.

Proposition 1.11.1 *Let* $1 < p < \infty$. *Then the operator* A_{SCE} *generates an holomorphic semigroup on* $L^p_\sigma(\mathbb{R}^3_+)$ *and there exists* $\nu \geq 0$ *such that* $-A_{SCE} + \nu$ *has maximal* L^p-*regularity on* $L^p_\sigma(\mathbb{R}^3_+)$.

The definition of a weak solution w for Eq. (1.11.4) given below requires that at least $w \in L^\infty((0, T); L^2_\sigma(\mathbb{R}^3_+))$ for all $T > 0$. We are hence interested in conditions implying that $e^{tA_{SCE}}$ is a bounded or a contraction semigroup on $L^2_\sigma(\mathbb{R}^3_+)$.

For this it is useful to note that, if $\alpha > 0$, then

$$\|e^{-(\cdot)/\alpha}v\|_{L^2(\mathbb{R}_+)} \le \frac{\alpha}{2}\|v'\|_{L^2(\mathbb{R}_+)} \tag{1.11.7}$$

for all $v \in H_0^1(\mathbb{R}_+)$. For smooth functions this follows easily from the fundamental theorem of calculus and Jensen's inequality; the general case is then implied by approximation.

Now, for $v_0 \in L_\sigma^2(\mathbb{R}_+^3)$ set $v(t) := e^{tA_{SCE}}v_0$. Then v satisfies

$$v'(t) - A_{SCE}v(t) = 0 \quad \text{for } t > 0 \text{ and } v(0) = v_0. \tag{1.11.8}$$

Multiplying with $v(t)$ and taking into account the skew symmetry of the second and third term of A_{SCE} yields

$$\frac{1}{2}\frac{d}{dt}\int_{\mathbb{R}_+^3}|v(t)|^2 dx + v\int_{\mathbb{R}_+^3}|\nabla v(t)|^2 dx + \int_{\mathbb{R}_+^3}v(t)\cdot(v_3(t)\cdot\partial_3 u_E)dx = 0, \quad t > 0.$$

Since

$$|\int_{\mathbb{R}_+^3}v(t)\cdot(v_3(t)\cdot\partial_3 u_E)dx| \le \sum_{j=1}^{2}\|e^{(\cdot)/2\delta}(\partial_3 u_E)_j v_3(t)\|_2\|e^{-(\cdot)/2\delta}v_j(t)\|_2,$$

and since

$$\partial_3 u_E(x_3) = \frac{u_\infty}{\delta}e^{-x_3/\delta}\begin{pmatrix}\cos(x_3/\delta) + \sin(x_3/\delta) \\ \cos(x_3/\delta) - \sin(x_3/\delta) \\ 0\end{pmatrix},$$

we see that

$$\|e^{(\cdot)/2\delta}(\partial_3 u_E)_j v_3(t)\|_2 \le \sqrt{2}\frac{u_\infty}{\delta}\|e^{-(\cdot)/2\delta}v_3(t)\|_2, \quad j = 1, 2. \tag{1.11.9}$$

Estimate (1.11.7) implies now

$$|\int_{\mathbb{R}_+^3}v(t)\cdot(v_3(t)\cdot\partial_3 u_E)dx| \le \sqrt{2}u_\infty\delta\|\nabla v(t)\|_2^2.$$

Thus, $\frac{d}{dt}\|v(t)\|_2^2 \le 0$ for all $t > 0$ provided $\sqrt{2}u_\infty\delta \le v$. Therefore $\|e^{tA_{SCE}}v_0\|_2 = \|v(t)\|_2 \le \|v_0\|_2$ for all $t > 0$ provided this condition is fulfilled. Setting

$$Re := u_\infty\delta v^{-1},$$

we thus proved the following result

Lemma 1.11.2 *The operator A_{SCE} generates a contraction semigroup $e^{tA_{SCE}}$ on $L^2_\sigma(\mathbb{R}^3_+)$ provided $\sqrt{2}Re \le 1$. Moreover, if*

$$\sqrt{2}Re < 1, \tag{1.11.10}$$

then there exists a constant $C > 0$ such that the solution v of (1.11.8) satisfies

$$\|v(t)\|_2^2 + C\int_0^t \|\nabla v(s)\|_2^2 ds \le \|v(0)\|_2^2, \qquad t \ge 0. \tag{1.11.11}$$

Remarks 1.11.3

(a) One easily sees that there exists also a constant $C > 0$ such that the solution v of (1.11.8) satisfies

$$\|v(t)\|_2^2 + C\int_s^t \|\nabla v(\tau)\|_2^2 ds \le \|v(s)\|_2^2, \qquad t \ge s. \tag{1.11.12}$$

for all $s \ge 0$.
(b) Combining Lemma 1.11.2 with Proposition 1.4.11, it follows that A_{SCE} admits a bounded $H^\infty(\Sigma_\theta)$-calculus on $L^2_\sigma(\mathbb{R}^3_+)$ for any angle $\theta > \pi/2$ provided condition (1.11.10) is satisfied. Furthermore, applying Theorem 1.5.6 yields

$$D(A_{SCE}^{1/2}) = H_0^1(\mathbb{R}^3_+) \cap L^2_\sigma(\mathbb{R}^3_+). \tag{1.11.13}$$

It thus follows that there exists a constant $C > 0$ such that for $u \in D(A_{SCE}^{1/2})$

$$\|A_{SCE}^{1/2}u\|_2 \le C(\|u\|_2 + \|\nabla u\|_2),$$

$$\|\nabla u\|_2 \le C(\|A_{SCE}^{1/2}u\|_2 + \|u\|_2). \tag{1.11.14}$$

(c) Assume (1.11.10). For $\alpha > 0$, let A_S^α and A_{SCE}^α be defined according to Theorem 1.5.3 and define q by $1/q = 1/2 - 2\alpha/3$. Then $\|u\|_q \le C\|A_S^\alpha u\|_2$ for $u \in D(A^\alpha)$ and one can show that

$$\|u\|_q \le C(\eta, \alpha)\|(A_{SCE} + \eta)^\alpha u\|_2, \quad u \in D(A_{SCE}^\alpha) \tag{1.11.15}$$

$$\|(A_{SCE}^* + \eta)^{-\alpha}u\|_2 \le C(\eta, \alpha)\|u\|_2, \quad u \in L^2_\sigma(\mathbb{R}^3_+) \tag{1.11.16}$$

(d) For $t \geq 0$ and $v_0 \in L_\sigma^2(\mathbb{R}_+^3)$ set $v^*(t) := e^{tA_{SCE}^*} v_0$. Then one can show as above that there exists a constant $C > 0$ such that

$$\|v^*(t)\|_2^2 + C \int_0^t \|\nabla v^*(s)\|_2^2 ds \leq \|v^*(0)\|_2^2, \qquad t \geq 0. \qquad (1.11.17)$$

The following result concerning the strong stability of $e^{tA_{SCE}}$ on $L_\sigma^2 \mathbb{R}_+^3$ will be one of the main ingredients for our stability estimates of weak solutions of system (1.11.5).

Proposition 1.11.4 *Assume that condition* (1.11.10) *is satisfied. Then*

$$\lim_{t \to \infty} \|e^{tA_{SCE}} v\|_2 = 0,$$

for every $v \in L_\sigma^2(\mathbb{R}_+^3)$, *i.e. the Stokes-Coriolis-Ekman semigroup on* $L_\sigma^2(\mathbb{R}_+^3)$ *is strongly stable.*

Proof Consider the operator $\partial_1 := \partial_1 Id_3$ on $X := L_\sigma^2(\mathbb{R}_+^3)^3$ with domain $D(\partial_1) = X \cap H^1(\mathbb{R}; L^2(\mathbb{R}_+^2))^3$. It can be shown that $\partial_1^* = -\partial_1$ and that ∂_1 has dense range in X.

First let $f \in Rg(\partial_1)$ and $g \in D(\partial_1)$ with $\partial_1 g = f$. By Lemma 1.11.2, $e^{tA_{SCE}}$ is a contraction semigroup on X and thus by Hölder's inequality

$$\|e^{tA_{SCE}} f\|_2 \leq \frac{1}{t} \int_0^t \|e^{sA_{SCE}} f\|_2 ds \leq \frac{1}{\sqrt{t}} \left(\int_0^t \|e^{sA_{SCE}} \partial_1 g\|_2^2 ds \right)^{1/2}.$$

Since the lower order terms of A_{SCE} do not depend on the x_1-variable, one can show that

$$\|e^{tA_{SCE}} \partial_1 g\|_2 = \|\partial_1 e^{tA_{SCE}} g\|_2, \qquad g \in D(\partial_1).$$

Hence, by the above energy inequality (1.11.12)

$$t\|e^{tA_{SCE}} f\|_2^2 \leq \int_0^t \|e^{sA_{SCE}} \partial_1 g\|_2^2 ds = \int_0^t \|\partial_1 e^{sA_{SCE}} g\|_2^2 ds$$

$$\leq \int_0^t \|\nabla e^{sA_{SCE}} g\|_2^2 ds \leq C\|g\|_2^2.$$

For general $f \in X$, note that since $Rg \, \partial_1$ is dense in X we may approximate f by $h \in Rg(\partial_1)$ and obtain thus the assertion. $\qquad \square$

Stability of Weak Solutions

We now state the definition of a weak solution to Eq. (1.11.4).

Definition 1.11.5 Let $w_0 \in L^2_\sigma(\mathbb{R}^3_+)$ and $T > 0$. A function w is called a *weak solution* of Eq. (1.11.4) if $w \in L^\infty((0, T); L^2_\sigma(\mathbb{R}^3_+)) \cap L^2((0, T); D(A^{1/2}_{SCE}))$ and

$$-\int_0^T \langle w, \phi \rangle h'(t)dt + \nu \int_0^T \langle \nabla w, \nabla \phi \rangle h(t)dt + \int_0^T \langle (u_E \cdot \nabla)w, \phi \rangle h(t)dt$$

$$+\int_0^T \langle w_3 \cdot \partial_3 u_E, \phi \rangle h(t)dt + \Omega \int_0^T \langle \mathbf{e}_3 \times w, \phi \rangle h(t)dt + \int_0^T \langle w \cdot \nabla w, \phi \rangle h(t)dt$$

$$= \langle w_0, \phi \rangle h(0),$$

holds for all $\phi \in D(A^{1/2}_{SCE})$ and all $h \in C^1([0, T], \mathbb{R})$ with $h(T) = 0$.

In the following we prove the existence of a global weak solution to the problem (1.11.4) provided the Reynolds number $Re = u_\infty \delta \nu^{-1}$ is small enough. More precisely, we assume from now on that condition (1.11.10) is satisfied. Note that our proof is inspired by a technique developed by Miyakawa and Sohr in [114].

We subdivide our considerations into three steps.

Step 1: Approximate Local Solutions
We first introduce smoothing operators J_k given by

$$J_k := k(k - A_{SCE})^{-1}, \quad k \in \mathbb{N},$$

on $X_p := L^p_\sigma(\mathbb{R}^3_+)$ for $1 < p < \infty$. They fulfill the following mapping properties.

Lemma 1.11.6

(a) $\|J_k\|_{\mathcal{L}(X_2)} \le 1$ for all $k \in \mathbb{N}$,
(b) $\|J_k\|_{\mathcal{L}(X_p)} \le C$ for some $C > 0$ and all $k \ge k_0$ and some $k_0 \ge 0$,
(c) $\|J_k u\|_{L^\infty} \le C(k)\|u\|_{X_2}$ for all $k \in \mathbb{N}$,
(d) $\|\nabla J_k u\|_{X_2} \le C(\|u\|_{X_2} + \|\nabla u\|_{X_2})$ for some $C > 0$ and all $u \in D(A^{1/2}_{SCE})$ and all $k \in \mathbb{N}$.

In fact, since A_{SCE} generates a contraction semigroup on X_2, see Lemma 1.11.2, the Hille-Yoshida Theorem 1.3.6 implies (a). Assertion (b) follows from Proposition 1.11.1. The Gagliardo-Nirenberg inequality implies $\|J_k u\|_\infty \le C\|u\|_2^{1/4}\|(A_{SCE} + 1)J_k u\|_2^{3/4}$ for $u \in X_2$ and (c) follows e.g. from the fact that $A_{SCE} \in \mathcal{H}^\infty(X_2)$; see Remark 1.11.3. Finally, (d) follows by combining the above assertions.

We now set

$$w_{0k} := J_k w_0 \text{ and } F_k w := -P(J_k w \cdot \nabla)w, \quad k \in \mathbb{N},$$

and construct approximate solutions w_k to Eq. (1.11.5) by solving the integral equations

$$w_k(t) = e^{tA_{SCE}}w_{0k} + \int_0^t e^{(t-s)A_{SCE}}F_k w_k(s)ds, \quad k \in \mathbb{N}. \tag{1.11.18}$$

To this end, consider for $T > 0$ the Banach space $X := C([0, T]; D(A_{SCE}^{1/2}))$ equipped with the norm

$$\|u\|_T := \sup_{0 \le t \le T} (\|u(t)\|_2 + \|A_{SCE}^{1/2}u(t)\|_2).$$

For $M > 0$ and $k \in \mathbb{N}$ consider the closed set

$$S(k, M, T) := \{u \in X, u(0) = w_{0k}, \|u\|_T \le M\},$$

as well as the nonlinear operator Γ_k defined on $S(k, M, T)$ given by

$$\Gamma_k u(t) := e^{tA_{SCE}}w_{0k} + \int_0^t e^{(t-s)A_{SCE}}F_k u(s)ds.$$

Note that by Lemma 1.11.6(c) and Remark 1.11.14

$$\|F_k u\|_2 \le C(k)(\|u\|_2^2 + \|A_{SCE}^{1/2}u\|_2^2), \quad k \in \mathbb{N}.$$

We thus may estimate $\Gamma_k u$ as follows

$$\|\Gamma_k u\|_T \le \|w_{0k}\|_2 + \|A_{SCE}^{1/2}w_{0k}\|_2 + \sup_{0 \le t \le T} \{C(k)\int_0^t \|u\|_2^2 + \|A_{SCE}^{1/2}u\|_2^2 ds\}$$

$$+ \sup_{0 \le t \le T} \{C(k)\int_0^t (t-s)^{-\frac{1}{2}} e^{\beta(t-s)}(\|u\|_2^2 + \|A_{SCE}^{1/2}u\|_2^2)ds\}$$

$$\le \|w_{0k}\|_2 + \|A_{SCE}^{1/2}w_{0k}\|_2 + C(k)M^2(T + e^{\beta T}T^{\frac{1}{2}})$$

for some $\beta \ge 0$. Similarly,

$$\|\Gamma_k u_1 - \Gamma_k u_2\|_T \le C(k)M(T + e^{\beta T}T^{\frac{1}{2}})\|u_1 - u_2\|_T.$$

Fix now M in such a way that $\|w_{0k}\|_2 + \|A_{SCE}^{1/2}w_{0k}\|_2 \le \frac{M}{2}$ and then $T^* < T$ such that $C(k)M^2(T + e^{\beta T}T^{\frac{1}{2}}) \le \frac{M}{2}$ and $C(k)M(T + e^{\beta T}T^{\frac{1}{2}}) < 1$. Then Γ_k is a strict contraction in $S(k, M, T^*)$ and by Banach fixed point theorem, there exists a unique w_k in $S(k, M, T^*)$ satisfying (1.11.18) for $t \in (0, T^*)$.

Step 2: A Priori Bounds for Fixed T
In the following we prove a priori bounds for $w_k(T)$ and $A_{SCE}^{1/2} w_k(T)$ for all $T > 0$.
To this end, note that $F_k w_k \in C([0, T^*]; X_2)$ and that w_k solves

$$w_k'(t) + A_{SCE} w_k = F_k w_k, \quad t \in (0, T^*). \tag{1.11.19}$$

Multiplying (1.11.19) with w_k and integrating by parts yields as above

$$\|w_k(T)\|_2^2 + C \int_0^T \|\nabla w_k(s)\|_2^2 ds \leq \|w_0\|_2^2. \tag{1.11.20}$$

In order to show that $\|A_{SCE}^{1/2} w_k(T)\|_2$ is bounded for all fixed $T > 0$, note that
by (1.11.14) it suffices to show this for $\|\nabla w_k(T)\|_2$. As above we see that

$$\|\nabla w_k\|_2^2 \leq C(k)\|w_0\|_2^2 + C \int_0^T \|w_k(s)\|_2^2 ds + C(k) \int_0^T (1 + \|w_k(s)\|_2^2)\|\nabla w_k(s)\|_2^2 ds.$$

Hence, by Gronwall's inequality there exists $C = C(\|w_0\|_2, T, k)$ such that
$\|\nabla w_k(T)\|_2 \leq C < \infty$. Combining this with (1.11.20), we see that

$$\sup_{0 \leq t \leq T} \{w_k(t)\|_2 + \|A_{SCE}^{1/2} w_k(t)\|_2\} \leq M < \infty.$$

Step 3: Weak Convergence
In this final step we show that the approximate global solutions (w_k) constructed
above, converge weakly to some weak solution w of system (1.11.4). Fix $T > 0$.
The above inequality (1.11.20) and (1.11.14) implies that

$$\int_0^T \|A_{SCE}^{1/2} w_k(s)\|_2^2 ds \leq C \int_0^T (\|w_k(s)\|_2^2 + \|\nabla w_k(s)\|_2^2) ds$$

$$\leq C(T + C)\|w(0)\|_2^2,$$

and hence that

$$w_k \in L^2(0, T; D(A_{SCE}^{1/2})) \cap L^\infty(0, T; L_\sigma^2(\mathbb{R}_+^3)) =: E =: E_1 \cap E_2, \quad k \in \mathbb{N}.$$

Thus (w_k) is a bounded sequence in E. Since E_1 is reflexive, there exists a
subsequence of (w_k) converging weakly in E_1. Further, (w_k) possesses a weak-star
convergent subsequence in E_2.

In order to investigate the strong convergence of (w_k) we write $w_k = w_k^1 + w_k^2$
with

$$w_k^{(1)}(t) := e^{t A_{SCE}} w_{0k}, \quad \text{and} \quad w_k^{(2)}(t) := \int_0^t e^{(t-s)A_{SCE}} F_k w_k(s) ds.$$

Performing the same calculations which led to (1.11.20), we obtain

$$\|w_k^{(1)}(T) - w_l^{(1)}(T)\|_2^2 + \int_0^T \|A_{SCE}^{1/2}(w_k^{(1)}(s) - w_l^{(1)}(s))\|_2^2 ds$$

$$\le C(T)\|(J_k - J_l)w_0\|_2^2 \to 0$$

as $k, l \to \infty$, since $J_k w_0 \to w_0$ in X_2. Hence $(w_k^{(1)})$ as well as $(w_k^{(2)})$ are bounded sequences in E.

Consider next the term $F_k w_k$ in $L_\sigma^r(\mathbb{R}_+^3)$ for $r = 5/4$. The Gagliardo–Nirenberg inequality yields

$$\|F_k w_k\|_r \le \|w_k\|_2^{2/5} \|\nabla J_k w_k\|_2^{3/5} \|\nabla w_k\|_2.$$

Since $\|\nabla J_k w_k\|_2 \le C(\|A_{SCE}^{1/2} w_k\|_2 + \|w_k\|_2)$, we see that by (1.11.20) that

$$\int_0^T \|F_k w_k\|_r^r dt \le C\|w_0\|_2.$$

Hence, $(F_k w_k)$ is a bounded sequence in $L^r(0, T; L_\sigma^r(\mathbb{R}_+^3))$. By construction, $w_k^{(2)}$ is the solution of the Cauchy problem

$$w_k'(t) + A_{SCE} w_k(t) = F_k w_k(t), \quad t > 0,$$
$$w(0) = 0.$$

By Proposition 1.11.1 there exists a constant $\nu_r > 0$ such that $-A_{SCE} + \nu_r$ admits maximal L^r-regularity on $L_\sigma^r(\mathbb{R}_+^3)$. A scaling argument shows not only $e^{-\nu_r t} w_k(t)$ but also

$$(w_k^2) \in L^r(0, T; D(A_{SCE,r})) \cap W^{1,r}(0, T; L_\sigma^r(\mathbb{R}_+^3))$$

is a bounded sequence in this space. By Lemma 1.11.6(b), $(J_k w_k^{(2)})_{k \ge k_0}$ is as well a bounded sequence in this space.l It thus follows from Theorem III.2.1 in [147] that $(w_k^{(2)})$ and $(J_k w_k^{(2)})_{k \ge k_0}$ are relatively compact in $L^2(K \times (0, T))$ for any fixed compact set $K \subset \mathbb{R}_+^3$. Hence, $(w_k^{(2)})$ and $(J_k w_k^{(2)})$ converge in $L^2(\mathbb{R}_+^3 \times (0, T))$. Therefore, $w_k(s) \to w(s)$ and $J_k w_k(s) \to w(s)$ for a.a. $s \in (0, T)$ for some function $w \in E$.

Finally, we need to verify that the function w constructed above is in fact a weak solution of our problem (1.11.4). We refrain, however, from giving details at this point.

We thus proved that for any initial value $w_0 \in L_\sigma^2(\mathbb{R}_+^3)$ there exists a weak solution w to (1.11.4) satisfying the energy inequality

$$\|w(t)\|_2^2 + C \int_s^t \|\nabla w(\tau)\|_2^2 d\tau \le \|w(s)\|_2^2 \tag{1.11.21}$$

for almost all $s \geq 0$, all $t > s$ and where $w(0) = w_0$ provided $\sqrt{2}Re < 1$. The advantage of our quite lengthy procedure is that we now have a representation formula for w, which allows us to deduce asymptotic properties of w.

In the following let $T_0 > 0$ and for $u \in D(A_{SCE}^{3/4})$ set $w := (A_{SCE} + 1)^{1/4}u$. Then, by the momentum inequality (1.5.1)

$$\|(A_{SCE} + 1)^{1/4}u\|_2 \leq C\|(A_{SCE} + 1)^{3/4}u\|_2^{1/3}\|u\|_2^{2/3}$$

for some $C > 0$. Hence, by (1.5.2) and (1.11.14)

$$\|w(t)\|_2 \leq C\|w(t)\|_{H^1}\|(A_{SCE} + 1)^{-1/4}w(t)\|_2^{2/3}.$$

Hölder's inequality combined with the energy inequality (1.11.21) yields

$$\int_{T_0}^{T_0+1} \|w(t)\|_2^2 dt \leq C\left(\int_{T_0}^{T_0+1} \|(A_{SCE} + 1)^{-1/4}w(t)\|_2^2 \, dt\right)^{2/3}$$

with C independent of T_0 and w.

Next, for $h > 0$ and s, t satisfying $0 \leq s \leq \tau \leq t < T_0$, consider

$$\Phi_h(\tau) := U(\tau)\int_s^t \varrho_h(\tau - \sigma)U(\sigma)w(\sigma)d\sigma,$$

where $U(\tau) = (A_{SCE} + 1)^{-1/4}e^{(t-\tau)A_{SCE}^*}$ and $\varrho_h(\tau) = h^{-1}\varrho(h^{-1}\tau)$ for a mollifier $\rho \in C^\infty(\mathbb{R})$. Since w is a weak solution of (1.11.4), integration by parts yields

$$< w(t), \Phi_h(t) > - < w(s), \Phi_h(s) >$$

$$= -\int_s^t < (w(\tau) \cdot \nabla)w(\tau), \Phi_h(\tau) > d\tau$$

$$+ \int_s^t < w(\tau), U(\tau)\int_s^t \frac{d}{d\tau}\varrho_h(\tau - \sigma)U(\sigma)w(\sigma)d\sigma > d\tau. \quad (1.11.22)$$

Now, Fubini's theorem as well as standard properties of mollifiers imply that

$$2 < w(s), \Phi_h(s) > \rightarrow \|e^{(t-s)A_{SCE}}(A_{SCE} + 1)^{-1/4}w(s)\|_2^2,$$

$$2 < w(t), \Phi_h(t) > \rightarrow \|(A_{SCE} + 1)^{-1/4}w(t)\|_2^2,$$

as $h \rightarrow 0$. Notice next that by symmetry of ϱ_h, the second term on the right hand side of (1.11.22) vanishes, whereas the first term on the right hand side of (1.11.22) can be estimated as

$$\left|\int_s^t < (w(\tau) \cdot \nabla)w(\tau), \Phi_h(\tau) > d\tau\right| \leq C\|w_0\|_2 \int_s^t \|\nabla w(\tau)\|_2^2 d\tau.$$

Indeed,

$$\left| \int_s^t < (w(\tau) \cdot \nabla)w(\tau), \Phi_h(\tau) > d\tau \right| \leq C \int_s^t \|w(\tau)\|_6 \|\nabla w(\tau)\|_2 \|\Phi_h\|_3 d\tau$$

$$\leq C \sup_{s \leq \tau \leq t} \{\|\Phi_h\|_3\} \int_s^t \|\nabla w(\tau)\|_2^2 d\tau$$

and the estimates given in Remark 1.11.3(c) imply that $\sup_{s \leq \tau \leq t} \|\Phi_h(\tau)\|_3 \leq C\|w_0\|_2$. Hence, letting $h \to 0$ in (1.11.22) yields

$$\|(A_{SCE}+1)^{-1/4}w(t)\|_2^2 \leq \|e^{(t-s)A_{SCE}}(A_{SCE}+1)^{-1/4}w(s)\|_2^2 + C \int_s^t \|\nabla w(\tau)\|_2^2 \, d\tau.$$

The energy inequality (1.11.21) implies that for $\varepsilon > 0$ there exists $m_0 \in \mathbb{N}$ such that $\|w(t_{m_2})\|_2^2 - \|w_{m_1}\|_2^2 \leq \varepsilon$. Thus, for given $\varepsilon > 0$ there exists $T_1 > 0$ such that

$$\int_{T_0}^T \|\nabla w(\tau)\|_2^2 d\tau < \varepsilon$$

for all $T > T_1$. For fixed $T_2 > 0$, Proposition 1.11.4 implies that

$$\|e^{(t-T_2)A_{SCE}}(A_{SCE}+1)^{-1/4}w(T_2)\|_2^2 \leq \varepsilon_0$$

provided $t > T_3$ for some $T_3 > 0$. Hence,

$$\int_{T_3}^{T_3+1} \|(A_{SCE}+1)^{-1/4}w(t)\|_2^2 \leq \int_{T_3}^{T_3+1} \|e^{(t-s)A_{SCE}}(A_{SCE}+1)^{-1/4}w(s)\|_2^2$$

$$+ C \int_{T_3}^{T_3+1} \int_{T_2}^t \|\nabla w(\tau)\|_2^2 \, d\tau dt < C\varepsilon_0.$$

Thus, there exists $T > 0$ such that

$$\int_T^{T+1} \|w(\tau)\|_{H^1} d\tau < \varepsilon.$$

Summing up, we proved the following result.

Theorem 1.11.7 *Assume that $\sqrt{2}Re < 1$. Then, given $w_0 \in L_\sigma^2(\mathbb{R}_+^3)$, there exists a least one weak solution w to Eq. (1.11.4) with $w(0) = w_0$ satisfying*

$$\lim_{T \to \infty} \int_T^{T+1} \|w(\tau)\|_{H^1} d\tau = 0.$$

For generalizations of this result to the situation of stratified flows we refer to the work of Koba [93]. We refer to the same reference for the existence of strong solutions for data being small in suitable norms.

1.12 Fluid-Rigid Body Interaction Problems for Compressible Fluids

The analysis of the movement of rigid or elastic bodies immersed in a fluid is a classical problem in fluid mechanics. In order to describe the problem, denote the bounded domain occupied by the body by $\mathcal{B}(t)$ and let $\mathcal{D}(t)$ be the exterior domain filled by the fluid, i.e. $\mathcal{D}(t) := \mathbb{R}^3 \backslash \overline{\mathcal{B}(t)}$. The interface between body and fluid is denoted by $\Gamma(t)$. In this section we are interested in the case where the body is a rigid body and the fluid is a compressible fluid within the barotropic regime. In this case, the motion of the fluid is given by the equations

$$\begin{aligned} \varrho_t + \operatorname{div}(\varrho v) &= 0 \quad \text{in } J_T \times \mathcal{D}(t) \\ \varrho(v_t + (v \cdot \nabla)v) - \operatorname{div} T(v, p) &= 0 \quad \text{in } J_T \times \mathcal{D}(t), \\ v &= v_{\mathcal{B}} \quad \text{in } J_T \times \Gamma(t), \\ v(0) &= v_0 \quad \text{in } \mathcal{D}(0), \\ \varrho(0) &= \varrho_0 \quad \text{in } \mathcal{D}(0), \end{aligned} \tag{1.12.1}$$

Here $J_T = (0, T)$ for some $T > 0$ and ϱ, v and p denote the density, velocity and pressure of the fluid, respectively. We assume that the fluid is of barotropic type, i.e. that the pressure $p = p(\varrho)$ satisfies the relation $p \in C^\infty(\mathbb{R}_+)$ and $p'(\varrho) > 0$ for all $\varrho > 0$. The stress tensor $T(v, p)$ is given by

$$T(v, p) = 2\mu D(v) + (\mu' - \mu) \operatorname{div} v I - pI.$$

Here $D(v) = \frac{1}{2}(\nabla v) + (\nabla v)^T$ denotes the deformation tensor, I is the 3×3 identity matrix, $\mu > 0$ and μ' are constants satisfying the relation $\mu + \mu' > 0$.

The fluid equations are coupled to the balance equations for the momentum and the angular momentum of the rigid body which read as

$$\begin{aligned} m\eta'(t) - \int_{\Gamma(t)} T(v, p)n(t, x)d\sigma &= F(t), \ t \in J_T, \\ (J\omega)'(t) - \int_{\Gamma(t)}(x - x_c) \times T(v, p)n(t, x)d\sigma &= M(t), \ t \in J_T, \\ \eta(0) &= \eta_0, \\ \omega(0) &= \omega_0, \end{aligned} \tag{1.12.2}$$

and which contain the drag force and the torque extended by the fluid onto the body. The constants m and J denote the body's mass and inertia tensor. Moreover, x_c is the position of its center of gravity and $\eta = \eta(t)$ and $\omega = \omega(t)$ denote its translational and angular velocity. Hence

$$v_B(t, x) := \eta(t) + \omega(t) \times (x - x_c(t)), \quad x \in \Gamma(t).$$

The functions F and M are external forces and torques.

Since the domain of the fluid $\mathcal{D}(t)$ depends on the motion of the rigid body, the problem is a so called *moving domain problem*. We start by rewriting Eqs. (1.12.1) and (1.12.2) into the following system of equations for the unknowns ϱ, v and η, ω

$$
\begin{cases}
\varrho_t + \operatorname{div}(\varrho v) = 0 & \text{in } J_T \times \mathcal{D}(t) \\
\varrho(v_t + (v \cdot \nabla)v) - \operatorname{div} T(v, p) = 0 & \text{in } J_T \times \mathcal{D}(t), \\
v(t, x) = \eta(t) + \omega(t) \times (x - x_c(t)) & \text{on } J_T \times \Gamma(t), \\
m\eta'(t) - \int_{\Gamma(t)} T(v, p)n(t, x)d\sigma = F(t) & t \in J_T, \\
(J\omega)'(t) - \int_{\Gamma(t)} (x - x_c) \times T(v, p)n(t, x)d\sigma = M(t) & t \in J_T, \\
v(0) = v_0, \quad \varrho(0) = \varrho_0 & \text{in } \mathcal{D}(0), \\
\eta(0) = \eta_0, \quad \omega(0) = \omega_0.
\end{cases}
$$

$$(1.12.3)$$

We note state our main result on the existence of unique, local strong solution to system (1.12.3). Let us denote by $\bar\varrho$ the mean value of ϱ_0 in $\mathcal{D}(0)$.

Theorem 1.12.1 *Let* $1 < p < \infty$ *and* $3 < q < \infty$. *Let* $T_0 > 0$ *and* $F, M \in L^p(J_{T_0}; \mathbb{R}^3)$ *and* $\mathcal{D}(0)$ *be an exterior domain of class* $C^{2,1}$. *Assume that*

$$\varrho_0 - \bar\varrho \in W^{1,q}(\mathcal{D}(0)), v_0 \in B_{q,p}^{2-2/p}(\mathcal{D}(0)) \text{ and } \eta_0, \omega_0 \in \mathbb{R}^3$$

are satisfying the compatibility condition $v_0 = \eta_0 + \omega_0 \times x$. *Then there exists* $T \in (0, T_0]$ *such that the system* (1.12.3) *admits a unique, strong solution* $(u, \varrho, \eta, \omega)$ *on* $J_T = (0, T)$ *within the class*

$$\varrho \in W^{1,p}(J_T; L^q(\mathcal{D}(\cdot))) \cap L^p(J_T; W^{1,q}(\mathcal{D}(\cdot))),$$

$$v \in W^{1,p}(J_T; L^q(\mathcal{D}(\cdot))^3) \cap L^p(J_T; W^{2,q}(\mathcal{D}(\cdot))^3),$$

$$(\eta, \omega) \in W^{1,p}(J_T; \mathbb{R}^6).$$

The above regularity properties for ϱ and v in Sobolev spaces on domains $\mathcal{D}(t)$ depending on t are understood in the following way: by $v \in W^{1,p}(J_T; L^q(\mathcal{D}(\cdot))^3) \cap L^p(J_T; W^{2,q}(\mathcal{D}(\cdot))^3)$ we mean that $X^*v \in W^{1,p}(J_T; L^q(\mathcal{D})^3) \cap L^p(J_T; W^{2,q}(\mathcal{D})^3)$, where $X^*v(t, y) := v(t, X(t, y))$ and X is being defined as the solution of Eq. (1.12.4) below. The regularity space for ϱ is defined in an analogous way.

In order to prove Theorem 1.12.1 it is natural to transform the original problem on $\mathcal{D}(t)$ to a problem on the fixed domain $\mathcal{D}(0)$. We will use a nonlinear, local change of coordinates for this, which only acts on a neighborhood of the rigid body.

This transformation goes back to Inoue and Wakimoto [89]. For compressible fluids it is also natural to switch, after this first change of coordinates, from Eulerian to Lagrangian coordinates. After these two transformations, our approach relies on maximal regularity estimates for the linearized transformed problem. Finally, we rewrite the nonlinear problem as a fixed point problem in the space of maximal regularity.

Let us start by introducing a diffeomorphism X between the time dependent and the fixed domain by considering the equation

$$\begin{cases} \partial_t X(t, y) = b(t, X(t, y)), & (0, T) \times \mathbb{R}^3, \\ X(0, y) = y, & y \in \mathbb{R}^3, \end{cases} \tag{1.12.4}$$

where b determines the modified velocity due to this change of coordinates. More precisely, we choose open balls $B_1, B_2 \subset \mathbb{R}^3$ such that $\overline{B} \subset B_1 \subset \overline{B_1} \subset B_2$ and define a cut-off function $\chi \in C^\infty(\mathbb{R}^3; [0, 1])$ by

$$\chi(y) := \begin{cases} 1 \text{ if } y \in \overline{B_1}, \\ 0 \text{ if } y \in \mathcal{D} \setminus B_2, \end{cases} \tag{1.12.5}$$

as well as a time-dependent vector field $b : [0, T] \times \mathbb{R}^3 \to \mathbb{R}^3$ by

$$b(t, x) := \chi(x - x_c(t))[m(t)(x - x_c(t)) + \eta(t)]. \tag{1.12.6}$$

Then $b \in W^{1,p}(0, T; C_c^\infty(\mathbb{R}^3))$ and that $b|_\Gamma = m(x - x_c) + \eta$. Given $\eta, \omega \in W^{1,p}(0, T)$, the Eq. (1.12.4) admits a unique solution $X \in C^1((0, T); C^\infty(\mathbb{R}^n))$ by the Picard–Lindelöf theorem. Denoting by J_X the Jacobian matrix of X, one can show that J_X is invertible provided $T > 0$ is small enough.

Performing this change of coordinates and switching also to Lagrangian coordinates in a second step, we obtain the system

$$\begin{cases} \theta_t + \bar{\varrho} \operatorname{div} u = f_0(u, \theta, \kappa, \Omega) & \text{in } J_T \times \mathcal{D}, \\ u_t - \operatorname{div} \mathcal{T}(u, \theta) = f_1(u, \theta, \kappa, \Omega) & \text{in } J_T \times \mathcal{D}, \\ u = 0 & \text{on } J_T \times \Gamma, \\ m\kappa' - \int_\Gamma \mathcal{T}(u, \theta) v d\sigma = g_0(u, \theta, \kappa, \Omega) & t \in J_T, \\ \mathcal{I}\Omega' - \int_\Gamma \xi \times \mathcal{T}(u, \theta) v d\sigma = g_1(u, \theta, \kappa, \Omega) & t \in J_T, \\ u(0) = w_0 & \text{in } \mathcal{D}, \\ \theta(0) = \varrho_0 - \bar{\varrho} & \text{in } \mathcal{D}, \\ \kappa(0) = \eta_0, \quad \Omega(0) = \omega_0, \end{cases} \tag{1.12.7}$$

where the right hand sides f_0, f_1, g_0 and g_1 collect all the nonlinear terms. Here $\mathcal{T}, v, \kappa, \Omega$ and $\mathcal{I}$ denote the transformed stress tensor, normal, translational, angular velocities and inertia tensor.

We now aim to prove a maximal regularity theorem for the linearized problem of (1.12.7), which reads as

$$
\begin{cases}
\varrho_t + \gamma_1 \operatorname{div} v = f_0 & \text{in } J_T \times \mathcal{D}, \\
v_t - \operatorname{div} T(v, \varrho) = f_1 & \text{in } J_T \times \mathcal{D}, \\
v = 0 & \text{on } J_T \times \Gamma, \\
m\kappa' - \int_\Gamma T(v, \varrho) v d\sigma = g_0 & t \in J_T, \\
\mathcal{I}\Omega' - \int_\Gamma \xi \times T(v, \varrho) v d\sigma = g_1 & t \in J_T, \\
v(0) = v_0 & \text{in } \mathcal{D}, \\
\varrho(0) = \varrho_0 - \bar\varrho & \text{in } \mathcal{D}, \\
\eta(0) = \eta_0, \\
\omega(0) = \omega_0.
\end{cases}
\tag{1.12.8}
$$

Here, $f = (f_0, f_1)$ and $g = (g_0, g_1)$ are given functions, $T(v, \varrho)$ is given by $T(v, \varrho) = 2\alpha D(v) + (\beta - \alpha) \operatorname{div} v I - \gamma_2 \varrho I$, where α, β, γ_1 and γ_2 are constants satisfying $\alpha > 0$, $\alpha + \beta > 0$ and $\gamma_1, \gamma_2 > 0$.

Step 1: Maximal Regularity for the Fluid Equation Consider the set of equations describing the compressible fluid

$$
\begin{cases}
\varrho_t + \gamma_1 \operatorname{div} v = f_0 & \text{in } J_T \times \mathcal{D}, \\
v_t - \operatorname{div} T(v, \varrho) = f_1 & \text{in } J_T \times \mathcal{D}, \\
v = 0 & \text{on } J_T \times \Gamma, \\
v(0) = v_0, & \text{in } \mathcal{D}, \\
\varrho(0) = \varrho_0 - \bar\varrho & \text{in } \mathcal{D}.
\end{cases}
\tag{1.12.9}
$$

We rewrite the above system (1.12.9) as an evolution equation on $W^{1,q}(\mathcal{D}) \times L^q(\mathcal{D})^3$, i.e. for $w = (\varrho, v)$, $f = (f_0, f_1)$ and $w_0 = (\varrho_0 - \bar\varrho, v_0)$ we consider the equation

$$
w_t + Aw = f, \quad w(0) = w_0
$$

where

$$
A := \begin{pmatrix} 0 & \gamma_1 \operatorname{div} \\ \gamma_2 \nabla & \mathcal{A}_D \end{pmatrix}
$$

with domain $D(A) := W^{1,q}(\mathcal{D}) \times D(\mathcal{A}_D)$ and where

$$
\mathcal{A}_D v := \alpha \Delta v + (\beta - \alpha) \nabla \operatorname{div} v \text{ for } v \in D(\mathcal{A}_D) \doteq (W^{2,q}(\mathcal{D}) \cap W_0^{1,q}(\mathcal{D}))^3.
$$

Let us decompose A as

$$A = \begin{pmatrix} 0 & \gamma_1 \operatorname{div} \\ \gamma_2 \nabla & \mathcal{A}_D \end{pmatrix} = \begin{pmatrix} 0 & \gamma_1 \operatorname{div} \\ 0 & \mathcal{A}_D \end{pmatrix} + \begin{pmatrix} 0 & 0 \\ \gamma_2 \nabla & 0 \end{pmatrix} =: A_0 + A_1.$$

Note that $\mathcal{A}_D$ is a parameter elliptic operator in sense of [36] subject to Dirichlet boundary conditions. Hence, by Theorem 8.2 of [36], $\mathcal{A}_D$ admits maximal regularity on $L^q(\mathcal{D})^3$. Moreover, since div is a bounded operator from $D(\mathcal{A}_D)$ to $W^{1,q}(\mathcal{D})$ it follows that A_0 admits maximal regularity on $W^{1,q}(\mathcal{D}) \times L^q(\mathcal{D})^3$. Furthermore, since ∇ maps $W^{1,q}(\mathcal{D})$ boundedly into $L^q(\mathcal{D})^3$, it follows that A is a bounded perturbation of A_0 and that thus A admits maximal regularity on $W^{1,q}(\mathcal{D}) \times L^q(\mathcal{D})^3$.

Step 2: Maximal Regularity for the Rigid Body Equations
  Let

$$\mathbb{I} = \begin{pmatrix} mI & 0 \\ 0 & I \end{pmatrix}$$

be the constant momentum matrix of our problem. For $0 < \varepsilon \leq 1 - 1/q$ we define the operator

$$\mathcal{J}_{\xi,\nu} : W_{\mathrm{loc}}^{\varepsilon+1/q,q}(\mathcal{D}, \mathbb{R}^{3\times3}) \to \mathbb{R}^6, \quad h \mapsto \begin{pmatrix} \int_\Gamma h\nu d\sigma \\ \int_\Gamma \xi \times h\nu d\sigma \end{pmatrix}.$$

The boundedness of the trace operator $\gamma : H_{\mathrm{loc}}^{\varepsilon/2+1/q,q}(\mathcal{D}) \to L^q(\partial\mathcal{D})$ implies that

$$|\mathcal{J}_{\xi,\nu}h| \leq C\|\gamma h\|_{L^q(\Gamma)} \leq C\|h\|_{W^{\varepsilon+1/q,q}(\mathcal{D})}, \quad h \in W^{\varepsilon+1/q,q}(\mathcal{D}). \tag{1.12.10}$$

Hence, the fourth and fifth line of the system (1.12.8) may be rewritten as

$$\mathbb{I}\begin{pmatrix} \kappa' \\ \Omega' \end{pmatrix} - \mathcal{J}_{\xi,\nu}T(v,\varrho) = \begin{pmatrix} g_0 \\ g_1 \end{pmatrix} \tag{1.12.11}$$

$$\kappa(0) = \eta_0, \quad \Omega(0) = \omega_0.$$

We then obtain

$$\|\mathcal{J}_{\xi,\nu}\varrho I]\|_{L_p} \leq C(T\|\varrho\|_{Y_{p,q}^T} + T^{1/p}\|\varrho_0 - \bar{\varrho}\|_{W^{1,q}(\mathcal{D})}).$$

By interpolation, $H^{\alpha,p}(J_T; H^{2-2\alpha,q}(\mathcal{D})) \hookrightarrow L^{2p}(J_T; W^{1+1/q+\varepsilon/2,q}(\mathcal{D}))$ provided $1/2p < \alpha < 1/2 - 1/2q$. Thus, by (1.12.10) and the mixed derivative Theorem 1.6.16

$$\|\mathcal{J}_{\xi,\nu}D(v)\|_{L_p} \leq C\|v\|_{L^p(J_T; W^{1+\varepsilon/2+1/q,q}(\mathcal{D}))}$$

$$\leq CT^{1/2p}\|v\|_{L^{2p}(J_T; W^{1+\varepsilon/2+1/q,q}(\mathcal{D}))}$$

$$\leq CT^{1/2p}\|v\|_{H^{\alpha,p}(J_T; H^{2-2\alpha,q}(\mathcal{D}))} \leq CT^{1/2p}\|v\|_{X_{p,q}^T}$$

provided $0 < \varepsilon < 2 - 2/q$. Similarly,

$$\|\mathcal{J}_{\xi,v}\operatorname{div} vI\|_p \le CT^\beta \|v\|_{X_{p,q}^T}$$

for some constant $\beta > 0$. We thus proved the following result.

Proposition 1.12.2 *Let $\mathcal{D}$ be an exterior domain with boundary class of $C^{2,1}$ and let $p, q \in (1, \infty)$ as well as $T > 0$. Let η_0, $\omega_0 \in \mathbb{R}^3$, $v_0 \in B_{p,q}^{2-2/p}(\mathcal{D})$ and $\varrho_0 - \bar{\varrho} \in W^{1,q}(\mathcal{D})$. Assume that $f_0 \in L^p(J_T; W^{1,q}(\mathcal{D}))$, $f_1 \in L^p(J_T; L^q(\mathcal{D})^3)$ and $g_0, g_1 \in L^p(J_T; \mathbb{R}^3)$. Then the system (1.12.8) admits a unique solution*

$$v \in X_{p,q}^T := W^{1,p}(J_T; L^q(\mathcal{D})^3) \cap L^p(J_T; W^{2,q}(\mathcal{D})^3),$$

$$\varrho \in Y_{p,q}^T := W^{1,p}(J_T; W^{1,q}(\mathcal{D})),$$

$$\kappa \in W^{1,p}(J_T; \mathbb{R}^6),$$

$$\Omega \in W^{1,p}(J_T; \mathbb{R}^6).$$

and there exists a constant $C > 0$ such that

$$\|v\|_{X_{p,q}^T} + \|\varrho\|_{Y_{p,q}^T} + \|\kappa\|_{W^{1,p}(J_T)} + \|\Omega\|_{W^{1,p}(J_T)}$$

$$\le C(\|f_0\|_{L^p(J_T; W^{1,q}(\mathcal{D}))} + \|f_1\|_{L_p(J_T; L_q(\mathcal{D}))} + \|g_0\|_{L_p(J_T)} + \|g_1\|_{L_p(J_T)}$$

$$+ \|v_0\|_{B_{p,q}^{2-2/p}(\mathcal{D})} + \|\varrho_0 - \bar{\varrho}\|_{W^{1,q}(\mathcal{D})} + |\eta_0| + |\omega_0|).$$

We finally solve (1.12.7) by an application of the contraction principle. To this end, we define the set $Z_{R,T}$ by

$$Z_{R,T} := \{(u, \theta, \kappa, \Omega) \in X_{p,q,0}^T \times Y_{p,q,0}^T \times W_0^{1,p}(J_T; \mathbb{R}^6) : \|u\|_{X_{p,q}^T}$$

$$+ \|\theta\|_{Y_{p,q}^T} + \|(\kappa, \Omega)\|_{L^p(J_T)} \le L\},$$

where $X_{p,q,0}^T = \{u \in X_{p,q}^T : u(0) = 0\}$, $Y_{p,q,0}^T = \{\theta \in Y_{p,q}^T : \theta(0) = 0\}$ and $W_0^{1,p}(J_T; \mathbb{R}^6) = \{(\kappa, \Omega) \in W^{1,p}(J_T; \mathbb{R}^6) : \eta(0) = \omega(0) = 0\}$. Given $(\tilde{u}, \tilde{\theta}, \tilde{\kappa}, \tilde{\Omega}) \in Z_{R,T}$, there exists a unique solution $(u, \theta, \kappa, \Omega)$ to the linear problem (1.12.8) with initial data $u_0 = 0$, $\varrho_0 - \bar{\varrho} = 0$, $\eta_0 = \omega_0 = 0$ and right hand sides $\hat{f}_i(\tilde{u}, \tilde{\theta}, \tilde{\kappa}, \tilde{\Omega})$, $\hat{g}_i(\tilde{u}, \tilde{\theta}, \tilde{\kappa}, \tilde{\Omega})$ for $i = 0, 1$. Set $\hat{u} := u - u^*$, $\hat{\theta} := \theta - \theta^*$, $\hat{\kappa} := \kappa - \kappa^*$ and $\hat{\Omega} := \Omega - \Omega^*$, where $(u^*, \theta^*, \kappa^*, \Omega^*)$ is the unique solution to system (1.12.8) with $f_0 = f_1 = g_0 = g_1 = 0$ and $\gamma_1 = \bar{\varrho}$. For $(\tilde{u}, \tilde{\theta}, \tilde{\kappa}, \tilde{\Omega}) \in Z_{R,T}$ we define the map Φ by

$$\Phi(\tilde{u}, \tilde{\theta}, \tilde{\kappa}, \tilde{\Omega}) := (\hat{u}, \hat{\theta}, \hat{\kappa}, \hat{\Omega}),$$

and show that Φ is a contraction from $Z_{R,T}$ into itself. For this, we need to estimate the nonlinear terms $\hat{f}_i(\tilde{u}, \tilde{\theta}, \tilde{\kappa}, \tilde{\Omega})$ and $\hat{g}_i(\tilde{u}, \tilde{\theta}, \tilde{\kappa}, \tilde{\Omega})$. We will not do this here in detail, but refer to [74]. The unique solution of Eq. (1.12.7) is then transferred back to the unique solution of the original system (1.12.3) within the regularity class stated in Theorem 1.12.1. The proof of Theorem 1.12.1 is complete.

1.13 Two-Phase Free Boundary Value Problems for a Class of Non-Newtonian Fluids

In this section we consider a two-phase free boundary value problem for generalized Newtonian fluids. The problem reads as follows: let $n \geq 2$ and $\Gamma_0 \subset \mathbb{R}^n$ be a hypersurface that separates a region $\Omega_1(0)$ filled with a viscous, incompressible fluid from $\Omega_2(0)$, the complement of $\overline{\Omega_1(0)}$ in $\mathbb{R}^n$. The region $\Omega_2(0)$ is also occupied with a second incompressible, viscous fluid and it is assumed that the two fluids are immiscible. Denoting by $\Gamma(t)$ the position of Γ_0 at time t, $\Gamma(t)$ is then the interface separating the two fluids occupying the regions $\Omega_1(t)$ and $\Omega_2(t)$.

We recall from Sect. 1.8 that an incompressible fluid is subject to the set of equations

$$\rho(\partial_t u + u \cdot \nabla u) = \operatorname{div} T,$$

$$\operatorname{div} u = 0,$$

where ρ denotes the density of the fluid and where the stress tensor T can be decomposed as $T = \tau - qI$. Here q denotes the pressure and τ the tangential part of the stress tensor of the fluid. For a Newtonian fluid, τ is given by $\tau = 2\mu D(u)$, where $D(u) = [\nabla u + (\nabla u)^T]/2$ denotes the deformation tensor and μ the viscosity coefficient of the fluid.

A Two-Phase Problem for Generalized Newtonian Fluids
In this section, we consider a class of non-Newtonian fluids, where τ as above is replaced by

$$\tau = 2\mu(|D(u)|^2)D(u)$$

for some function μ satisfying

$$\mu \in C^3([0, \infty)) \quad \text{and} \quad \mu(0) > 0. \tag{1.13.1}$$

In the special case of power law fluids, one has

$$\mu(|D(u)|^2) = \nu + \beta|D(u)|^{d-2}$$

for some $d \geq 1$ and constants $\nu, \beta \geq 0$. If $d < 2$, the fluid is then called a shear thinning fluid, if $d > 2$ it is called a shear thickening fluid. Fluids of this type are special cases of so called *Stokesian fluids*, which were investigated analytically for fixed domains by Amann in [7] and [9].

The motion of the two immiscible, incompressible and viscous fluids is then governed by the following set of equations

$$\begin{cases} \rho(\partial_t v + v \cdot \nabla v) = \operatorname{div} T - \rho \gamma_a e_N, & \text{in } \Omega(t), \\ \operatorname{div} = 0 & \text{in } \Omega(t), \\ -[\![Tn_\Gamma]\!] = \sigma H_\Gamma n_\Gamma & \text{on } \Gamma(t), \\ [\![v]\!] = 0 & \text{on } \Gamma(t), \\ V_\Gamma = v \cdot n_\Gamma & \text{on } \Gamma(t), \\ v|_{t=0} = v_0 & \text{in } \Omega_0, \\ \Gamma|_{t=0} = \Gamma_0, & \end{cases} \qquad (1.13.2)$$

where $t > 0$ and where we assumed that the unknown $\Gamma(t)$ is determined by an unknown scalar function $h = h(t, x')$ for $x' \in \mathbb{R}^{n-1}$, that is

$$\Gamma(t) = \{ (x', x_n) : x' \in \mathbb{R}^{n-1}, x_n = h(t, x') \}$$

and $\Omega(t) = \Omega_1(t) \cup \Omega_2(t)$ with

$$\Omega_n(t) = \{ (x', x_n) : x' \in \mathbb{R}^{n-1}, (-1)^n(x_n - h(t, x')) > 0 \}, \quad n = 1, 2.$$

Here, $e_n = (0, \ldots, 0, 1)^T$ and the normal field on $\Gamma(t)$, pointing from $\Omega_1(t)$ into $\Omega_2(t)$, is denoted by $n_\Gamma(t, \cdot)$. Moreover, $V_\Gamma(t, \cdot)$ and $H_\Gamma(t, \cdot)$ denote the normal velocity and mean curvature of $\Gamma(t)$, respectively. Furthermore, γ_a denotes the gravitational acceleration and σ the coefficient of the surface tension.

We suppose that the stress tensor T is given by the generalized Newtonian type described above, that is, for given scalar functions $\mu_1, \mu_2 : [0, \infty) \to \mathbb{R}$,

$$T = \chi_{\Omega_1(t)} T_1(v, q) + \chi_{\Omega_2(t)} T_2(v, q),$$

$$T_j(v, \pi) = -qI + 2\mu_j(|D(v)|^2) D(v), \quad j = 1, 2,$$

and where $|D(u)|^2 = \sum_{i,j=1}^{N} (D_{ij}(u))^2$. The function χ_D denotes the indicator function of a set $D \subset \mathbb{R}^n$, and the density ρ is defined by $\rho = \chi_{\Omega_1(t)} \rho_1 + \chi_{\Omega_2(t)} \rho_2$ for the densities $\rho_j > 0$ of the j-th fluid.

The system is complemented by the initial fluid velocity v_0 and the given initial height function h_0, as well as $\Omega_0 = \Omega_1(0) \cup \Omega_2(0)$ and Γ_0 given by

$$\Omega_j(0) = \{ (x', x_n) : x' \in \mathbb{R}^{n-1}, (-1)^j(x_n - h_0(x')) > 0 \}, \quad j = 1, 2,$$

$$\Gamma_0 = \{ (x', x_n) : x' \in \mathbb{R}^{n-1}, x_n = h_0(x') \}.$$

In addition, we denote the unit normal field, pointing from $\Omega_1(0)$ into $\Omega_2(0)$, on Γ_0 by n_0. The quantity $[\![f]\!] = [\![f]\!](t, x)$ is the jump of the quantity f, which is defined on $\Omega(t)$ across the interface $\Gamma(t)$ as

$$[\![f]\!](t, x) = \lim_{\varepsilon \to 0+} \{f(t, x + \varepsilon n_\Gamma) - f(t, x - \varepsilon n_\Gamma)\} \quad \text{for } x \in \Gamma(t).$$

The problem then is to find functions v, q, and h solving the Eq. (1.13.2).

Well-posedness results for the above system (1.13.2) in the case of *Newtonian fluids* and in the special case of one-phase flows with or without surface tension were first obtained by Solonnikov [136–138], Shibata and Shimizu [130, 132]. The case of an ocean of infinite extend and which is bounded below by a solid surface and bounded above by a free surface was treated first by Beale [15] and Tani and Tanaka [146]

Besides the articles cited already above, the two-phase problem for Newtonian fluids was studied by Denisova in [34] and [35], and by Tanaka in [145] using *Lagrangian coordinates*. Indeed, Denisova proved local wellposedness in the Newtonian case in $W_2^{r,r/2}$ for $r \in (5/2, 3)$ for the case where one of the domains is bounded and Tanaka obtained wellposedness (including thermo-capillary convection) in $W_2^{r,r/2}$ for $r \in (7/2, 4)$.

Prüss and Simonett were using in [124] a different approach by transforming problem (1.13.2) to a problem on a fixed domain via the *Hanzawa transform* and applying then an optimal regularity approach for the linearized equations. Like this they proved wellposedness of the above problem in the case of Newtonian fluids.

Problems of the above kind for *non-Newtonian fluids* were treated by Abels in [4] in the context of measure-valued varifold solutions. His result covers in particular the situation where $\mu_j(s) = \nu_j s^{(d-2)/2}$ for $j = 1, 2$ and $d \in (1, \infty)$. Note, however, that his approach does not give the uniqueness of a solution. We refer also to [5].

In the main result of this section we show that system (1.13.2) admits a unique, strong solution on $(0, T)$ for arbitrary $T > 0$ provided the viscosity functions μ_n fulfill (1.13.1) and the initial data are sufficiently small in their natural norms. More precisely, we have the following result.

Theorem 1.13.1 *Let $p \in (n + 2, \infty)$ and $J = (0, T)$ for some $T > 0$. Suppose that $\rho_1 > 0$, $\rho_2 > 0$, $\gamma_a \geq 0$, $\sigma > 0$ and that*

$$\mu_j \in C^3([0, \infty)) \quad \text{and} \quad \mu_j(0) > 0, \quad j = 1, 2.$$

Then there exists $\varepsilon_0 > 0$ such that for $(v_0, h_0) \in W_p^{2-2/p}(\Omega_0)^n \times W_p^{3-2/p}(\mathbb{R}^{n-1})$ satisfying the compatibility conditions

$$[\![\mu(|D(v_0)|^2)D(v_0)n_0 - \{n_0 \cdot \mu(|D(v_0)|^2)D(v_0)n_0\}n_0]\!] = 0 \quad \text{on } \Gamma_0,$$

$$\text{div}_0 = 0 \quad \text{in } \Omega_0, \quad [\![v_0]\!] = 0 \quad \text{on } \Gamma_0,$$

with $\mu(|D(v_0)|^2) = \chi_{\Omega_1(0)}\mu_1(|D(v_0)|^2) + \chi_{\Omega_2(0)}\mu_2(|D(v_0)|^2)$ *as well as the smallness condition*

$$\|v_0\|_{W_p^{2-2/p}(\Omega_0)} + \|h_0\|_{W_p^{3-2/p}(\mathbb{R}^{n-1})} < \varepsilon_0,$$

the system (1.13.2) *admits a unique solution* (v, q, h) *within the class*

$$v \in (H_p^1(J, L_p(\Omega(t)) \cap L_p(J, H_p^2(\Omega(t)))))^n,$$

$$q \in L_p(J, \dot{H}_p^1(\Omega(t))),$$

$$h \in W_p^{2-1/(2p)}(J, L_p(\mathbb{R}^{n-1})) \cap H_p^1(J, W_p^{2-1/p}(\mathbb{R}^{n-1}))$$

$$\cap W_p^{1/2-1/(2p)}(J, H_p^2(\mathbb{R}^{n-1})) \cap L_p(J, W_p^{3-1/p}(\mathbb{R}^{n-1})).$$

Remarks 1.13.2

(a) Some remarks on our notation are in order at this point. Setting

$$\dot{\mathbb{R}}^n = \mathbb{R}^n \setminus \mathbb{R}_0^n, \quad \mathbb{R}_0^n = \{(x', x_n) : x' \in \mathbb{R}^{n-1}, \; x_n = 0\},$$

we mean by $v \in H_p^1(J, L_p(\Omega(t))) \cap L_p(J, H_p^2(\Omega(t)))^N$ that

$$\Theta^* v = v \circ \Theta \in (H_p^1(J, L_p(\dot{\mathbb{R}}^n)) \cap L_p(J, H_p^2(\dot{\mathbb{R}}^n)))^n,$$

where Θ and Θ^* are defined below by (1.13.4) and (1.13.5), respectively. The regularity statement for q is understood in the same way.

(b) The assumption $p > n + 2$ implies that

$$h \in BUC(J, BUC^2(\mathbb{R}^{n-1})), \quad \partial_t h \in BUC(J, BUC^1(\mathbb{R}^{n-1})),$$

which means that the condition on the free interface can be understood in the classical sense.

(c) Typical examples of viscosity functions μ satisfying our conditions are given by

$$\mu(s) = \nu\left(1 + s^{\frac{d-2}{2}}\right) \quad \text{with } d = 2, 4, 6, \text{ or } d \geq 8,$$

$$\mu(s) = \nu(1 + s)^{\frac{d-2}{2}} \quad \text{with } 1 \leq d < \infty$$

for $\nu > 0$. For more information and details we refer e.g. to the work of [43] and [108]. Obviously, if $d = 2$, the above viscosity functions correspond to the Newtonian situation.

Let us remark at this point that our proof of Theorem 1.13.1 is inspired by the work by Prüss and Simonett in [122] and [123]. Our strategy may be described as follows: we first transform the system (1.13.2) to a problem on a fixed domain. Then maximal regularity properties of the associated linearized problem due to Prüss and Simonett [123] enable us prove Theorem 1.13.1 by the contraction principle.

Transformation to a Fixed Domain

Let us start the proof of Theorem 1.13.1 by calculating the divergence of the stress tensor, i.e. by calculating explicitly

$$\operatorname{div}\{\mu_N(|D(u)|^2)D(u)\} \quad \text{for} \quad N = 1, 2.$$

Let us remark first that given a vector u of length m for $m \geq 2$, we denote by u_i its i-th component and by u' its tangential component, i.e. $u = (u_1, \ldots, u_m)^T$ and $u' = (u_1, \ldots, u_{m-1})^T$. We then obtain

$$(\operatorname{div}\{\mu_N(|D(u)|^2)D(u)\})_i = \frac{1}{2} \sum_{j,k,l=1}^{n} \{2\dot{\mu}_N(|D(u)|^2)D_{ij}(u)D_{kl}(u)$$

$$+ \mu_N(|D(u)|^2)\delta_{ik}\delta_{jl}\}(\partial_j\partial_k u_l + \partial_j\partial_l u_k).$$

For vectors u, v we set $A_N(u)v := (A_{n,1}(u)v, \ldots, A_{n,N}(u)v)^T$, where

$$A_{N,i}(u)v := - \sum_{j,k,l=1}^{n} A_{N,i}^{j,k,l}(D(u))(\partial_j\partial_k v_l + \partial_j\partial_l v_k) \quad \text{and}$$

$$A_{N,i}^{j,k,l}(D(u)) := \frac{1}{2}\left(2\dot{\mu}_N(|D(u)|^2)D_{ij}(u)D_{kl}(u) + \mu_N(|D(u)|^2)\delta_{ik}\delta_{jl}\right)$$

for $N = 1, 2$ and $i = 1, \ldots, n$. We then have

$$A_N(u)u = - \operatorname{div}\{\mu_N(|D(u)|^2)D(u)\} \quad \text{and} \quad A_N(0)u = -\mu_N(0)(\Delta u + \nabla \operatorname{div} u).$$

In addition, we set

$$A(u)v := \chi_{\Omega_1(t)}A_1(u)v + \chi_{\Omega_2(t)}A_2(u)v \quad \text{and} \quad \tilde{q} := q + \rho\gamma_a x_n.$$

Then system (1.13.2) may be rewritten as

$$\begin{cases} \rho(\partial_t v + v \cdot \nabla v) - \mu(0)\Delta v + \nabla\tilde{q} = -(A(v) - A(0))v & \text{in } \Omega(t), \\ \operatorname{div} = 0 & \text{in } \Omega(t), \\ -[\![\tilde{T}n_\Gamma]\!] = \sigma H_\Gamma n_\Gamma + [\![\rho]\!]\gamma_a x_N n_\Gamma & \text{on } \Gamma(t), \\ [\![v]\!] = 0 & \text{on } \Gamma(t), \\ V_\Gamma = v \cdot n_\Gamma & \text{on } \Gamma(t), \\ v|_{t=0} = v_0 & \text{in } \Omega_0, \\ \Gamma|_{t=0} = \Gamma_0, \end{cases}$$

$$(1.13.3)$$

where $\tilde{T} = \chi_{\Omega_1(t)}T_1(v, \tilde{q}) + \chi_{\Omega_2(t)}T_2(v, \tilde{q})$ and $\mu(0) = \chi_{\Omega_1(t)}\mu_1(0) + \chi_{\Omega_2(t)}\mu_2(0)$.

Our first aim is to transform the problem (1.13.3) to a problem on the fixed domain $\dot{\mathbb{R}}^n$. To this end, we define a transformation Θ on $J \times \dot{\mathbb{R}}^n$ for $J = (0, T)$ with $T > 0$ as

$$\Theta : J \times \dot{\mathbb{R}}^n \ni (\tau, \xi', \xi_n) \mapsto (t, x', x_n) \in \bigcup_{s \in J} \{s\} \times \Omega(s),$$

$$\text{with } t = \tau, \ x' = \xi', \ x_n = \xi_n + h(\tau, \xi') \tag{1.13.4}$$

for some scalar-valued function h. Note that $\det \mathcal{J}\Theta = 1$, where $\mathcal{J}\Theta$ denotes the Jacobian matrix of Θ. We now define

$$u(\tau, \xi) := \Theta^* v(t, x) := v(\Theta(\tau, \xi)), \quad \pi(\tau, \xi) := \Theta^* \widetilde{q}(t, x), \tag{1.13.5}$$

as well as

$$\Theta_* f(\tau, \xi) := f(\Theta^{-1}(t, x)) \quad \text{for } f : \dot{\mathbb{R}}^n \to \mathbb{R}^n, \tag{1.13.6}$$

where Θ^{-1} is given by $\Theta^{-1}(t, x) = (t, x', x_n - h(t, x'))$. Hence, the system (1.13.3) is reduced to the following problem on $\dot{\mathbb{R}}^n$

$$\begin{cases} \rho \partial_\tau u - \mu(0) \Delta u + \nabla \pi = F(u, \pi, h) & \text{in } \dot{\mathbb{R}}^n, \\ \operatorname{div} u = F_d(u, h) & \text{in } \dot{\mathbb{R}}^n, \\ -[\![\mu(0)(D_N u_j + D_j u_n)]\!] = G_j(u, [\![\pi]\!], h) & \text{on } \mathbb{R}_0^n, \\ [\![\pi]\!] - 2[\![\mu(0) D_N u_N]\!] - ([\![\rho]\!]\gamma_a + \sigma \Delta')h = G_N(u, h) & \text{on } \mathbb{R}_0^n, \\ [\![u]\!] = 0 & \text{on } \mathbb{R}_0^n, \\ \partial_\tau h - u_n = G_h(u', h) & \text{on } \mathbb{R}_0^n, \\ u|_{t=0} = u_0 & \text{on } \dot{\mathbb{R}}^n, \\ h|_{t=0} = h_0 & \text{on } \mathbb{R}^{n-1}, \end{cases}$$

$$\tag{1.13.7}$$

where $j = 1, \ldots, n - 1$ and $F = (F_1, \ldots, F_n)^T$. The terms on the right hand side of (1.13.7) are given by

$$F_i(u, \pi, h) := \rho\{(\partial_\tau h)D_n u_i - (u \cdot \nabla)u_i + (u' \cdot \nabla' h)D_n u_i\}$$

$$-\mu(0) \sum_{j=1}^n \mathcal{F}_{jj}(h)u_i + (D_i h)D_N \pi + \mathcal{A}_i(u, h)$$

$$G_j(u, [\![\pi]\!], h) := \sigma \mathcal{H}(h)D_j h - \{([\![\rho]\!]\gamma_a + \sigma \Delta')h\}D_j h + [\![\pi]\!]D_j h + \mathcal{B}_j(u, h)$$

$$G_n(u, h) := -\sigma \mathcal{H}(h) + \mathcal{B}_n(u, h),$$

$$F_d(u, h) := (D_n u') \cdot \nabla' h = D_n(u' \cdot \nabla' h),$$

$$G_h(u, h) := -u' \cdot \nabla' h.$$

Here $\mathcal{A}_i(u, h)$, $\mathcal{B}_j(u, h)$ and $\mathcal{B}_n(u, h)$ are given by

$$\mathcal{A}_i(u, h) := \sum_{j,k,\ell=1}^{n} \left(A_i^{j,k,\ell}(E(u, h)) - A_i^{j,k,\ell}(0) \right)(D_j D_k u_\ell + D_j D_\ell u_k)$$

$$- \sum_{j,k,\ell=1}^{N} \left(A_i^{j,k,\ell}(E(u, h)) - A_i^{j,k,\ell}(0) \right)$$

$$\times (\mathcal{F}_{jk}(h)u_\ell + \mathcal{F}_{j\ell}(h)u_k), \quad i = 1, \ldots, n,$$

$$\mathcal{B}_j(u, h) := - [\![\mu(|E(u, h)|^2)D_n u_n]\!]D_j h$$

$$+ [\![\{\mu(|E(u, h)|^2) - \mu(0)\}(D_n u_j + D_j u_n)]\!]$$

$$- \sum_{k=1}^{n-1} [\![\mu(|E(u, h)|^2)(D_j u_k + D_k u_j)]\!]D_k h$$

$$+ \sum_{k=1}^{n-1} [\![\mu(|E(u, h)|^2)(D_N u_j D_k h$$

$$+ D_n u_k D_j h)]\!]D_k h, \quad j = 1, \ldots, n-1,$$

$$\mathcal{B}_n(u, h) := -2[\![\{\mu(|E(u, h)|^2) - \mu(0)\}D_n u_n]\!] + [\![\mu(|E(u, h)|^2)D_n u_n]\!]|\nabla' h|^2$$

$$- \sum_{k=1}^{n-1} [\![\mu(|E(u, h)|^2)(D_n u_k + D_k u_n)]\!]D_k h,$$

where $A_i^{j,k,l}(E(u, h)) := \chi_{\mathbb{R}^n_-} A_{i,1}^{j,k,l}(E(u, h)) + \chi_{\mathbb{R}^n_+} A_{i,2}^{j,k,l}(E(u, h))$. In particular,

$$\mu(|E(u, h)|^2) = \chi_{\mathbb{R}^n_-}\mu_1(|E(u, h)|^2) + \chi_{\mathbb{R}^n_+}\mu_2(|E(u, h)|^2),$$

$$\rho = \chi_{\mathbb{R}^n_-}\rho_1 + \chi_{\mathbb{R}^n_+}\rho_2,$$

$$\mu(0) = \chi_{\mathbb{R}^n_-}\mu_1(0) + \chi_{\mathbb{R}^n_+}\mu_2(0).$$

Finally, in order to simplify our notation, we set

$$G(u, [\![\pi]\!], h) := (G_1(u, [\![\pi]\!], h), \ldots, G_{n-1}(u, [\![\pi]\!], h), G_n(u, h))^T$$

$$\mathcal{A}(u, h) := (\mathcal{A}_1(u, h), \ldots, \mathcal{A}_n(u, h))^T,$$

$$\mathcal{B}(u, h) := (\mathcal{B}_1(u, h), \ldots, \mathcal{B}_n(u, h))^T.$$

The above set of Eq. (1.13.7) leads to the following associated linear problem

$$
\begin{cases}
\rho \partial_t u - \nu \Delta u + \nabla \pi = f & \text{in } \dot{\mathbb{R}}^n, \\
\operatorname{div} u = f_d & \text{in } \mathbb{R}^n, \\
-[\![\nu(D_n u_j + D_j u_n)]\!] = g_j & \text{on } \mathbb{R}_0^n, \\
[\![\pi]\!] - 2[\![\mu D_n u_n]\!] - ([\![\rho]\!]\gamma_a + \sigma \Delta')h = g_n & \text{on } \mathbb{R}_0^n, \\
[\![u]\!] = 0 & \text{on } \mathbb{R}_0^n, \\
\partial_t h - u_n = g_h & \text{on } \mathbb{R}_0^n, \\
u|_{t=0} = u_0 & \text{in } \dot{\mathbb{R}}^n, \\
h|_{t=0} = h_0 & \text{on } \mathbb{R}^{n-1},
\end{cases}
\tag{1.13.8}
$$

where $j = 1, \ldots, n-1$ and set $g = (g_1, \ldots, g_n)^T$. Here,

$$
\rho = \rho_1 \chi_{\mathbb{R}_-^n} + \rho_2 \chi_{\mathbb{R}_+^n}, \qquad \nu = \nu_1 \chi_{\mathbb{R}_-^n} + \nu_2 \chi_{\mathbb{R}_+^n}
$$

with positive constants ρ_N, ν_N for $N = 1, 2$.

Characterization of Maximal Regularity by Data

The optimal regularity property of the solution of the above problem (1.13.8) will be of central importance in the following. The following result due to Prüss and Simonett [122] characterizes the set of right-hand sides of Eq. (1.13.8) which yield a unique solution of (1.13.8) in the space of maximal regularity.

Proposition 1.13.3 (Prüss–Simonett [122, 123]) *Let* $1 < p < \infty$, $p \neq 3/2, 3$, $a > 0$ *and* $J = (0, a)$. *Suppose that*

$$
\rho_N > 0, \ \nu_N > 0, \ \gamma_a \geq 0 \text{ and } \sigma > 0, \quad N = 1, 2.
$$

Then, Eq. (1.13.8) admits a unique solution (u, π, h) *with regularity*

$$
u \in (H_p^1(J, L_p(\dot{\mathbb{R}}^n)) \cap L_p(J, H_p^2(\dot{\mathbb{R}}^n)))^n,
$$

$$
\pi \in L_p(J, \dot{H}_p^1(\dot{\mathbb{R}}^n)),
$$

$$
[\![\pi]\!] \in W_p^{1/2 - 1/(2p)}(J, L_p(\mathbb{R}^{n-1})) \cap L_p(J, W_p^{1-1/p}(\mathbb{R}^{n-1})),
$$

$$
h \in W_p^{2-1/(2p)}(J, L_p(\mathbb{R}^{n-1})) \cap H_p^1(J, W_p^{2-1/p}(\mathbb{R}^{n-1})) \cap L_p(J, W_p^{3-1/p}(\mathbb{R}^{n-1}))
$$

if and only if the data $(f, f_d, g, g_h, u_0, h_0)$ *satisfy the following regularity and compatibility conditions:*

$$
f \in L_p(J, L_p(\dot{\mathbb{R}}^n))^n,
$$

$$
f_d \in H_p^1(J, \dot{H}_p^{-1}(\mathbb{R}^n)) \cap L_p(J, H_p^1(\dot{\mathbb{R}}^n)),
$$

$$g \in (W_p^{1/2-1/(2p)}(J, L_p(\mathbb{R}^{n-1})) \cap L_p(J, W_p^{1-1/p}(\mathbb{R}^{n-1})))^n,$$

$$g_h \in W_p^{1-1/(2p)}(J, L_p(\mathbb{R}^{n-1})) \cap L_p(J, W_p^{2-1/p}(\mathbb{R}^{n-1})),$$

$$u_0 \in W_p^{2-2/p}(\dot{\mathbb{R}}^n)^n,$$

$$h_0 \in W_p^{3-2/p}(\mathbb{R}^{n-1}),$$

$$f_d|_{t=0} = \operatorname{div} u_0 \quad \text{in } \dot{\mathbb{R}}^n \quad \text{and} \quad [\![u_0]\!] = 0 \quad \text{on } \mathbb{R}^{N-1} \text{ if } p > 3/2,$$

$$g_j|_{t=0} = -[\![\nu(D_n u_{0j} + D_j u_{0n})]\!] \quad \text{on } \mathbb{R}^{n-1} \text{ if } p > 3$$

for all $j = 1, \ldots, n - 1$. *Moreover, the solution map* $[(f, f_d, g, g_h, u_0, h_0) \mapsto (u, \pi, [\![\pi]\!], h)]$ *is continuous between the corresponding spaces.*

The Nonlinear Problem
For $a > 0$ let $J = (0, a)$ and set

$$\mathbb{E}_1(a) := \{u \in (H_p^1(J, L_p(\dot{\mathbb{R}}^n)) \cap L_p(J, H_p^2(\dot{\mathbb{R}}^n)))^N \mid [\![u]\!] = 0\},$$

$$\mathbb{E}_2(a) := L_p(J, \dot{H}_p^1(\dot{\mathbb{R}}^n)),$$

$$\mathbb{E}_3(a) := W_p^{1/2-1/(2p)}(J, L_p(\mathbb{R}^{n-1})) \cap L_p(J, W_p^{1-1/p}(\mathbb{R}^{n-1})),$$

$$\mathbb{E}_4(a) := W_p^{2-1/(2p)}(J, L_p(\mathbb{R}^{n-1})) \cap H_p^1(J, W_p^{2-1/p}(\mathbb{R}^{n-1}))$$

$$\cap W_p^{1/2-1/(2p)}(J, H_p^2(\mathbb{R}^{n-1})) \cap L_p(J, W_p^{3-1/p}(\mathbb{R}^{n-1}))$$

as well as

$$\mathbb{F}_1(a) := L_p(J, L_p(\dot{\mathbb{R}}^n))^N,$$

$$\mathbb{F}_2(a) := H_p^1(J, \dot{H}_p^{-1}(\mathbb{R}^n)) \cap L_p(J, H_p^1(\dot{\mathbb{R}}^n)),$$

$$\mathbb{F}_3(a) := (W_p^{1/2-1/(2p)}(J, L_p(\mathbb{R}^{n-1})) \cap L_p(J, W_p^{1-1/p}(\mathbb{R}^{n-1})))^N,$$

$$\mathbb{F}_4(a) := W_p^{1-1/(2p)}(J, L_p(\mathbb{R}^{n-1})) \cap L_p(J, W_p^{2-1/p}(\mathbb{R}^{n-1})).$$

The solution space $\mathbb{E}(a)$ and the data space $\mathbb{F}(a)$ are defined for $a > 0$ by

$$\mathbb{E}(a) := \{(u, \pi, q, h) \in \mathbb{E}_1(a) \times \mathbb{E}_2(a) \times \mathbb{E}_3(a) \times \mathbb{E}_4(a) \mid [\![\pi]\!] = q\},$$

$$\mathbb{F}(a) := \mathbb{F}_1(a) \times \mathbb{F}_2(a) \times \mathbb{F}_3(a) \times \mathbb{F}_4(a).$$

and are endowed with their natural norms. Finally, for $(u, \pi, q, h) \in \mathbb{E}(a)$ consider the nonlinear mapping $\mathcal{N}$ defined by

$$\mathcal{N}(u, \pi, q, h) := (F(u, \pi, h), F_d(u, h), G(u, q, h), G_h(u, h)). \tag{1.13.9}$$

The mapping $\mathcal{N}$ possesses the following properties.

Lemma 1.13.4 *Let* $n + 2 < p < \infty$, $a > 0$ *and* $r > 0$. *Suppose that* $\mu_N(s) \in$ $C^3([0, \infty))$ *for* $N = 1, 2$ *and in addition that* $\rho_1 > 0$, $\rho_2 > 0$, $\gamma_a \geq 0$ *and* $\sigma > 0$ *are constants. Then*

$$\mathcal{N} \in C^1(B_{\mathbb{E}(a)}(r), \mathbb{F}(a)), \quad \mathcal{N}(0) = 0 \text{ and } D\mathcal{N}(0) = 0.$$

Finally, let us return to the nonlinear problem (1.13.7). Let us define the space of initial data $\mathbb{I}$ by

$$\mathbb{I} := W_p^{2-2/p}(\dot{\mathbb{R}}^n)^n \times W_p^{3-2/p}(\mathbb{R}^{n-1}).$$

The following result shows that problem (1.13.7) on the fixed domain admits a unique strong solution provided the data u_0 and h_0 are sufficiently small in their corresponding norms.

Proposition 1.13.5 *Let* $n+2 < p < \infty$ *and* $a > 0$. *Suppose that* $\mu_N \in C^3([0, \infty))$ *and that*

$$\rho_N > 0, \mu_N(0) > 0, \ \gamma_a \geq 0 \text{ and } \sigma > 0 \text{ for } N = 1, 2.$$

Then there exist positive constants ε_0 *and* δ_0 *(depending on* a *and* p*), such that system (1.13.7) admits a unique solution* $(u, \pi, [\![\pi]\!], h)$ *in* $B_{\mathbb{E}(a)}(\delta_0)$ *provided the initial data* $(u_0, h_0) \in \mathbb{I}$ *satisfy the smallness condition*

$$\|(u_0, h_0)\|_{\mathbb{I}} < \varepsilon_0,$$

as well as the compatibility conditions

$$[\![\mu(|E(u_0, h_0)|^2)E(u_0, h_0)n_0 - \{n_0 \cdot \mu(|E(u_0, h_0)|^2)E(u_0, h_0)n_0\}n_0]\!] = 0 \text{ on } \mathbb{R}_0^n,$$
$$(1.13.10)$$

$$\operatorname{div} u_0 = F_d(u_0, h_0) \quad \text{in } \dot{\mathbb{R}}^n, \quad [\![u_0]\!] = 0 \quad \text{on } \mathbb{R}_0^n.$$

For $z = (u, \pi, q, h) \in \mathbb{E}(a)$, the nonlinear problem (1.13.7) can be restated as

$$Lz = \mathcal{N}(z), \quad (u, h)|_{t=0} = (u_0, h_0), \tag{1.13.11}$$

where L denotes the linear operator on the left side of (1.13.8) with $\nu = \mu(0)$ and $\mathcal{N}$ is defined in (1.13.9). We divide the proof into several steps.

Step 1 We introduce an auxiliary function $z^* \in \mathbb{E}(a)$ that satisfies

$$Lz^* = (0, f_d^*, g^*, g_h^*), \quad (u^*, h^*)|_{t=0} = (u_0, h_0),$$

where

$$f_d^* \in \mathbb{F}_2(a) \quad \text{and} \quad f_d^*|_{t=0} = F_d(u_0, h_0),$$

$$g^* \in \mathbb{F}_3(a) \quad \text{and} \quad g^*|_{t=0} = G(u_0, [\![\pi_0]\!], h_0), \qquad (1.13.12)$$

$$g_h^* \in \mathbb{F}_4(a) \quad \text{and} \quad g_h^*|_{t=0} = G_h(u_0', h_0),$$

and where $\|z^*\|_{\mathbb{E}(a)} \le C_0 \|(u_0, h_0)\|_{\mathbb{I}}$ with a positive constant C_0 independent of (u_0, h_0). The last inequality then yields

$$\|Lz^*\|_{\mathbb{F}(a)} \le C_1 C_0 \|(u_0, h_0)\|_{\mathbb{I}} \qquad (1.13.13)$$

for some positive constant C_1.

Step 2 In order to proceed with our proof, we define closed subspaces $_0\mathbb{E}(a)$, $_0\mathbb{F}(a)$ of $\mathbb{E}(a)$, $\mathbb{F}(a)$ as

$$_0\mathbb{E}(a) := \{z = (u, \pi, q, h) \in \mathbb{E}(a) : (u, q, h)|_{t=0} = (0, 0, 0)\},$$

$$_0\mathbb{F}(a) := \{(f, f_d, g, g_h) \in \mathbb{F}(a) : (f_d, g, g_h)|_{t=0} = (0, 0, 0)\},$$

respectively. We then replace z by $z + z^*$ in (1.13.11) in order to obtain

$$Lz = \mathcal{N}(z + z^*) - Lz^* =: K_0(z), \quad z \in {_0\mathbb{E}(a)}.$$

This leads to the fixed point equation

$$z = L_0^{-1} K_0(z), \quad z \in {_0\mathbb{E}(a)}, \qquad (1.13.14)$$

where L_0 denotes the restriction of L to $_0\mathbb{E}(a)$. By (1.13.12), we have $K_0(z) \in {_0\mathbb{F}(a)}$ for any $z \in {_0\mathbb{E}(a)}$. In addition, it follows from Lemma 1.13.4 and (1.13.13) that $K_0 \in C^1(B_{_0\mathbb{E}(a)}(r), {_0\mathbb{F}(a)})$ for $r > 0$. Consequently, $L_0^{-1} K_0 : {_0\mathbb{E}(a)} \to {_0\mathbb{E}(a)}$ is well-defined.

Step 3 In this last step, we prove that $\Phi := L_0^{-1} K_0$ is a contraction mapping in $B_{_0\mathbb{E}(a)}(r_0)$ for some positive number r_0. By Lemma 1.13.4, we may choose $r_0 > 0$ small enough such that

$$\sup_{z \in B_{\mathbb{E}(a)}(2r_0)} \|D\mathcal{N}(z)\|_{\mathcal{L}(\mathbb{E}(a), \mathbb{F}(a))} \le \frac{1}{4\|L_0^{-1}\|_{\mathcal{L}(_0\mathbb{F}(a), _0\mathbb{E}(a))}}.$$

In addition, we choose ε_0 sufficiently small and satisfying

$$0 < \varepsilon_0 < \min\left\{ (\frac{r_0}{C_0}, \frac{r_0}{2C_1 C_0 \|L_0^{-1}\|_{\mathcal{L}(_0\mathbb{F}(a), _0\mathbb{E}(a))}} \right\}.$$

For $z \in B_{_0\mathbb{E}(a)}(r_0)$, the mean value theorem and (1.13.13) imply that

$$\|\Phi(z)\|_{_0\mathbb{E}(a)} \le \|L_0^{-1}\|_{\mathcal{L}(_0\mathbb{F}(a),\,_0\mathbb{E}(a))}\|K_0(z)\|_{_0\mathbb{F}(a)}$$

$$\le \|L_0^{-1}\|_{\mathcal{L}(_0\mathbb{F}(a),\,_0\mathbb{E}(a))}\left(\|\mathcal{N}(z+z^*) - \mathcal{N}(0)\|_{\mathbb{F}(a)} + \|Lz^*\|_{\mathbb{F}(a)}\right)$$

$$\le \|L_0^{-1}\|_{\mathcal{L}(_0\mathbb{F}(a),\,_0\mathbb{E}(a))}\left(\sup_{\bar{z}\in B_{\mathbb{E}(a)}(2r_0)}\|D\mathcal{N}(\bar{z})\|_{\mathcal{L}(\mathbb{E}(a),\mathbb{F}(a))}\|z+z^*\|_{\mathbb{E}(a)}\right.$$

$$\left. + C_1 C_0 \varepsilon_0\right) \le \frac{r_0}{2} + \frac{r_0}{2} = r_0,$$

which yields that Φ is a mapping from $B_{_0\mathbb{E}(a)}(r_0)$ into itself. Given $z_1, z_2 \in B_{_0\mathbb{E}(a)}(r_0)$ and noting that

$$\Phi(z_1) - \Phi(z_2) = L_0^{-1}\left(K_0(z_1) - K_0(z_2)\right) = L_0^{-1}\left(\mathcal{N}(z_1+z^*) - \mathcal{N}(z_2+z^*)\right),$$

we obtain by the mean value theorem

$$\|\Phi(z_1) - \Phi(z_2)\|_{_0\mathbb{E}(a)} \le \|L_0^{-1}\|_{\mathcal{L}(_0\mathbb{F}(a),\,_0\mathbb{E}(a))}\|\mathcal{N}(z_1+z^*) - \mathcal{N}(z_2+z^*)\|_{_0\mathbb{F}(a)}$$

$$\le \|L_0^{-1}\|_{\mathcal{L}(_0\mathbb{F}(a),\,_0\mathbb{E}(a))}\left(\sup_{\bar{z}\in B_{\mathbb{E}(a)}(2r_0)}\|D\mathcal{N}(\bar{z})\|_{\mathcal{L}(\mathbb{E}(a),\mathbb{F}(a))}\right)\|z_1 - z_2\|_{_0\mathbb{E}(a)}$$

$$\le \frac{1}{4}\|z_1 - z_2\|_{_0\mathbb{E}(a)}.$$

This implies that Φ is a contraction on $B_{_0\mathbb{E}(a)}(r_0)$ and we thus obtain the existence of a unique solution $z \in B_{_0\mathbb{E}(a)}(r_0)$ of (1.13.14). The proof of Proposition 1.13.5 is complete.   □

Proof of Theorem 1.13.1 Observe that the compatibility conditions stated in of Theorem 1.13.1 are satisfied if and only if (1.13.10) is satisfied. The mapping Θ_{h_0} given by

$$\Theta_{h_0}(\xi', \xi_n) := (\xi', \xi_n + h_0(\xi')) \quad \text{for } (\xi', \xi_n) \in \dot{\mathbb{R}}^n$$

defines for $h_0 \in W_p^{3-2/p}(\mathbb{R}^{n-1})$ a C^2-diffeomorphism from $\dot{\mathbb{R}}^n$ onto Ω_0 with inverse $\Theta_{h_0}^{-1}(x', x_n) = (x', x_n - h_0(x'))$. Thus there exists a constant $C(h_0)$ such that

$$C(h_0)^{-1}\|v_0\|_{W_p^{2-2/p}(\Omega_0)} \le \|u_0\|_{W_p^{2-2/p}(\dot{\mathbb{R}}^n)} \le C(h_0)\|v_0\|_{W_p^{2-2/p}(\Omega_0)}.$$

Hence, the smallness condition in Theorem 1.13.1 implies the smallness condition in Proposition 1.13.5 and the latter then yields a unique solution $(u, \pi, [\![\pi]\!], h) \in B_{\mathbb{E}(a)}(\delta_0)$ of (1.13.7). Finally, setting

$$(v, q) = (\Theta_* u, \Theta_* \pi) = (u \circ \Theta^{-1}, \pi \circ \Theta^{-1}),$$

where Θ_* is defined as in (1.13.6), we obtain a unique solution (v, q, h) of the original problem (1.13.2) with the regularities stated in Theorem 1.13.1. The proof is complete.   □

1.14 Nematic Liquid Crystals

The Ericksen-Leslie model describing the hydrodynamic flow of nematic liquid crystals was derived by Ericksen [49] from the classical balance laws for mass, linear and angular momentum and by considering it as an anisotropic fluid. The system was completed by Leslie [97] with the addition of constitutive relations. For a derivation of this model based on thermodynamical principles we refer to [77].

The non-isothermal Ericksen-Leslie model for incompressible fluids subject to general Leslie stress S_L and isotropic free energy $\psi(\theta, \tau)$ developed and described in [77] reads as

$$
\begin{cases}
\rho \mathcal{D}_t u + \nabla \pi = \operatorname{div} S & \text{in } \Omega, \\
\operatorname{div} u = 0 & \text{in } \Omega, \\
\rho \mathcal{D}_t \epsilon + \operatorname{div} q = S : \nabla u + \operatorname{div}(\lambda \nabla d \mathcal{D}_t d) & \text{in } \Omega, \\
\gamma \mathcal{D}_t d - \mu_V V d - \operatorname{div}[\lambda \nabla]d = \lambda |\nabla d|^2 d + \mu_D P_d D d & \text{in } \Omega, \\
u = 0, \quad q \cdot \nu = 0, \quad \partial_\nu d = 0 & \text{on } \partial\Omega, \\
\rho(0) = \rho_0, \quad u(0) = u_0, \quad \theta(0) = \theta_0, \quad d(0) = d_0 & \text{in } \Omega.
\end{cases}
$$

$$(1.14.1)$$

The unknown variables u, π, θ, d denote velocity, pressure, (absolute) temperature, and director, respectively, $\mathcal{D}_t = \partial_t + u \cdot \nabla$ the Lagrangian derivative, and $P_d := I - d \otimes d$. We also impose the condition

$$|d| = 1 \quad \text{in } (0, T) \times \Omega. \tag{1.14.2}$$

Here *incompressible* means that the density ρ is constant and *isotropic* means that the free energy ψ is a function only of θ and $\tau = |\nabla d|_2^2/2$. These equations have to be supplemented by the thermodynamical laws for the internal energy ϵ, entropy η, heat capacity κ and Ericksen tension λ, according to

$$\epsilon = \psi + \theta\eta, \quad \eta = -\partial_\theta\psi, \quad \kappa = \partial_\theta\epsilon, \quad \lambda = \rho\partial_\tau\psi, \tag{1.14.3}$$

and by the constitutive laws

$$
\begin{cases}
S = S_N + S_E + S_L^{stretch} + S_L^{diss}, \quad D = (\nabla u + [\nabla u]^\mathsf{T})/2, \quad q = -\alpha\nabla\theta, \\
S_N = 2\mu_s D, \quad S_E = -\lambda\nabla d[\nabla d]^\mathsf{T} \\
S_L^{stretch} = \frac{\mu_D + \mu_V}{2\gamma} \mathsf{n} \otimes d + \frac{\mu_D - \mu_V}{2\gamma} d \otimes \mathsf{n}, \quad \mathsf{n} = \mu_V V d + \mu_D P_d D d - \gamma\mathcal{D}_t d, \\
S_L^{diss} = \frac{\mu_P}{\gamma}(\mathsf{n} \otimes d + d \otimes \mathsf{n}) + \frac{\gamma\mu_L + \mu_P^2}{2\gamma}(P_d D d \otimes d + d \otimes P_d D d) + \mu_0(Dd|d)d \otimes d.
\end{cases}
$$

$$(1.14.4)$$

All coefficients μ_j, α_j and γ are functions of θ, τ, in accordance with the principle of equi-presence. For thermodynamic consistency and well-posedness we require

$$\mu_s > 0, \quad \alpha > 0, \quad \mu_0, \mu_L \geq 0, \quad \kappa, \gamma > 0, \quad \lambda, \lambda + 2\tau\partial_\tau\lambda > 0. \qquad (1.14.5)$$

It is convenient to write the equation for the internal energy as an equation for the temperature θ. It reads as

$$\rho\kappa\mathcal{D}_t\theta + \operatorname{div} q = (S - (1 - \theta\partial_\theta\lambda/\lambda)S_E) : \nabla u + \operatorname{div}(\lambda\nabla)d \cdot \mathcal{D}_t d + (\theta\partial_\theta\lambda)\nabla d : \nabla\mathcal{D}_t d.$$

The Simplified System

Due to the complexity of the system (1.14.1), (1.14.3)–(1.14.5) we analyze first a simplified version, which reads as

$$\begin{cases} \partial_t u + (u \cdot \nabla)u - \nu\Delta u + \nabla\pi = -\lambda\operatorname{div}([\nabla d]^\mathsf{T}\nabla d) & \text{in } (0, T) \times \Omega, \\ \partial_t d + (u \cdot \nabla)d = \gamma(\Delta d + |\nabla d|^2 d) & \text{in } (0, T) \times \Omega, \\ \operatorname{div} u = 0 & \text{in } (0, T) \times \Omega, \\ (u, \partial_\nu d) = (0, 0) & \text{on } (0, T) \times \partial\Omega, \\ (u, d)_{|t=0} = (u_0, d_0) & \text{in } \Omega. \end{cases}$$
$$(1.14.6)$$

As above, the function $u : (0, \infty) \times \Omega \to \mathbb{R}^n$ describes the velocity field, $\pi : (0, \infty) \times \Omega \to \mathbb{R}$ the pressure, and $d : (0, \infty) \times \Omega \to \mathbb{R}^n$ represents the macroscopic molecular orientation of the liquid crystal. We impose again the condition

$$|d| = 1 \quad \text{in } (0, T) \times \Omega. \qquad (1.14.7)$$

We will show in the following that this condition is indeed preserved by the above system. In the simplified model the parameters $\nu > 0, \lambda > 0$ and $\gamma > 0$ are constants and for simplicity we set $\nu = \lambda = \gamma = 1$.

Our main idea is to consider (1.14.6) not as a semilinear equation as done in other approaches but as a *quasilinear* evolution equation of the form

$$z'(t) + A(z(t))z(t) = F(z(t)), \quad t \in J, \quad z(0) = z_0. \qquad (1.14.8)$$

We thus incorporate the term $\operatorname{div}([\nabla d]^\mathsf{T}\nabla d)$ into the quasilinear operator A given by

$$A(d) = \begin{bmatrix} \mathcal{A} & \mathbb{P}\mathcal{B}(d) \\ 0 & \mathcal{D} \end{bmatrix},$$

where $\mathcal{A}$ denotes the Stokes operator, $\mathcal{D}$ the Neumann Laplacian, and $\mathcal{B}$ is given by

$$[\mathcal{B}(d)h]_i := \partial_i d_l \Delta h_l + \partial_k d_l \partial_k \partial_i h_l, \qquad (1.14.9)$$

for which we employ the sum convention. Note that $\mathcal{B}(d)d = \text{div}([\nabla d]^{\mathsf{T}}\nabla d)$. We now reformulate (1.14.6) equivalently as a quasilinear parabolic evolution equation for the unknown $z = (u, d)$. To this end, for $1 < q < \infty$ define the Banach space X_0 by

$$X_0 := L_{q,\sigma}(\Omega) \times L_q(\Omega)^n,$$

where $\Omega \subset \mathbb{R}^n$ is a bounded domain with boundary $\partial\Omega \in C^2$. The Neumann-Laplacian $\mathcal{D}_q$ in $L_q(\Omega)$ is defined by $\mathcal{D}_q = -\Delta$ with domain $D(\mathcal{D}_q) := \{d \in H_q^2(\Omega)^n : \partial_\nu d = 0 \text{ on } \partial\Omega\}$. It follows from Proposition 1.4.12 that $\mathcal{D}_q$ has the property of L_p-maximal regularity. Denoting by $\mathbb{P} : L_q(\Omega)^n \to L_{q,\sigma}(\Omega)$ the Helmholtz projection defined in Sect. 1.10, we consider the Stokes Operator $\mathcal{A}_q = -\mathbb{P}\Delta$ in $L_{q,\sigma}(\Omega)$ with domain $D(\mathcal{A}_q) = \{u \in H_q^2(\Omega)^n : \text{div } u = 0 \text{ in } \Omega, u = 0 \text{ on } \partial\Omega\}$. By Theorem 1.10.6, $\mathcal{A}_q$ has the property of L_p-maximal regularity. Next, we define the space X_1 by $X_1 := D(\mathcal{A}_q) \times D(\mathcal{D}_q)$, equipped with its canonical norms. Then $X_1 \overset{d}{\hookrightarrow} X_0$ densely. The quasilinear part $A(z)$ of (1.14.8) is given by the tri-diagonal matrix

$$A(z) = \begin{bmatrix} \mathcal{A}_q & \mathbb{P}\mathcal{B}_q(d) \\ 0 & \mathcal{D}_q \end{bmatrix},$$

where the operator $\mathcal{B}_q$ is given as in (1.14.9). By the tri-diagonal structure of $A(z)$ and by the regularity of $\mathcal{B}_q$ we see that $A(z)$ also has the property of L_p-maximal regularity, for each $z \in C^1(\overline{\Omega})^{2n}$. Note that the conditions (A) and (F) of Theorem 1.7.2 hold, as soon as we have the embedding $X_{\gamma,\mu} \hookrightarrow C^1(\overline{\Omega})^{2n}$. The space X_γ is given by

$$X_\gamma = (X_0, X_1)_{1-1/p,p} = D_{\mathcal{A}_q}(1 - 1/p, p) \times D_{\mathcal{D}_q}(1 - 1/p, p);$$

see [8, 38]. The trace space of the class $z \in H_{p,\mu}^1(J; X_0) \cap L_{p,\mu}(J; X_1)$ reads as

$$X_{\gamma,\mu} = (X_0, X_1)_{\mu-1/p,p} = D_{\mathcal{A}_q}(\mu - 1/p, p) \times D_{\mathcal{D}_q}(\mu - 1/p, p),$$

provided $p \in (1, \infty)$ and $\mu \in (1/p, 1]$.

In order to obtain the embeddings $X_\gamma \hookrightarrow C^1(\overline{\Omega})^{2n}$ and more generally $X_{\gamma,\mu} \hookrightarrow C^1(\overline{\Omega})^{2n}$ we impose on $p, q \in (1, \infty)$ the conditions

$$\frac{2}{p} + \frac{n}{q} < 1, \quad \frac{1}{2} + \frac{1}{p} + \frac{n}{2q} < \mu \leq 1. \tag{1.14.10}$$

Standard Sobolev embedding theorems can then be applied.

Furthermore, we recall from [148, Theorem 4.3.3] and [10, Theorem 3.4], respectively, the following characterizations of the interpolation spaces involved,

$$d \in D_{\mathcal{D}_q}(\mu - 1/p, p) \quad \Leftrightarrow \quad d \in B_{qp}^{2\mu-2/p}(\Omega)^n, \ \partial_\nu d = 0 \text{ on } \partial\Omega,$$

and

$$u \in D_{\mathcal{A}_q}(\mu - 1/p, p) \quad \Leftrightarrow \quad u \in B_{qp}^{2\mu - 2/p}(\Omega)^n \cap L_{q,\sigma}(\Omega), \ u = 0 \text{ on } \partial\Omega.$$

Observe that both of these characterizations make sense, since the condition (1.14.10) guarantees the existence of the trace.

Applying Theorem 1.7.2 we obtain the following result on local well-posedness of (1.14.6).

Theorem 1.14.6 *Let p, q, μ be subject to (1.14.10), and assume $z_0 = (u_0, d_0) \in X_{\gamma,\mu}$, which means that $u_0, d_0 \in B_{qp}^{2\mu - 2/p}(\Omega)^n$ satisfy the compatibility conditions*

$$\text{div } u_0 = 0 \text{ in } \Omega, \quad u_0 = \partial_\nu d_0 = 0 \text{ on } \partial\Omega.$$

Then for some $a = a(z_0) > 0$, there is a unique solution

$$z \in H_{p,\mu}^1(J, X_0) \cap L_{p,\mu}(J; X_1), \quad J = [0, a],$$

of (1.14.6) on J. Moreover,

$$z \in C([0, a]; X_{\gamma,\mu}) \cap C((0, a]; X_\gamma),$$

i.e. the solution regularizes instantly in time. It depends continuously on z_0 and exists on a maximal time interval $J(z_0) = [0, t^+(z_0))$. Therefore problem (1.14.6), i.e. (1.14.8), generates a local semi-flow in its natural state space $X_{\gamma,\mu}$.

The following lemma tells that the condition (1.14.7) is preserved by (1.14.6).

Lemma 1.14.7 *Suppose that μ, p, q are satisfying (1.14.10) and let $z_0 = (u_0, d_0) \in X_{\gamma,\mu}$ with $|d_0| \equiv 1$, $a > 0$. Let $z \in H_{p,\mu}^1(J; X_0) \cap L_{p,\mu}(J; X_1)$ be a solution of (1.14.6) on the interval $J = [0, a]$. Then $|d(t)| \equiv 1$ holds for all $t \in [0, a]$.*

We consider next the linearization of (1.14.6) at $z_* \in \mathcal{E}_0$, which is given by the linear evolution equation

$$\dot{z} + A_* z = f, \quad z(0) = z_0,$$

in X_0, where

$$A_* = \text{diag}(\mathcal{A}_q, \mathcal{D}_q), \quad D(A_*) = X_1.$$

As Ω is bounded, the spectrum $\sigma(\mathcal{A}_q)$ consists only of positive eigenvalues and $0 \notin \sigma(\mathcal{A}_q)$. On the other hand, $\mathcal{D}_q$ has 0 as an eigenvalue, which is semi-simple and the remaining part of $\sigma(\mathcal{D}_q)$ consist only of positive eigenvalues. Thus $\sigma(A_*) \backslash \{0\} \subset [\delta, \infty)$ for some $\delta > 0$ and the kernel of A_* is given by

$$N(A_*) = \{0\} \times \mathbb{R}^n,$$

which equals the tangent space. Verifying that the equilibrium is normally stable, we obtain by Theorem 1.7.4 the following result on global existence for data close to an equilibrium point.

Theorem 1.14.8 *Let p, q satisfy the first inequality in* (1.14.10). *Then each equilibrium $z_* \in \{0\} \times \mathbb{R}^n$ is stable in X_γ, i.e. there exists $\epsilon > 0$ such that a solution $z(t)$ of* (1.14.6) *with initial value $z_0 \in X_\gamma$, $|z_0 - z_*|_{X_\gamma} \leq \epsilon$, exists globally and converges exponentially to some $z_\infty \in \{0\} \times \mathbb{R}^n$ in X_γ, as $t \to \infty$.*

The Ericksen-Leslie System with General Leslie Stress

The Ericksen-Leslie equations with general Lesie stress lead to a mixed order system. For this reason, in order to formulate our main well-posedness result in this situation, we introduce a functional analytic setting as follows. Denote the principal variable by $v = (u, \theta, d)$. Then v belongs to the base space X_0 defined by

$$X_0 := L_{q,\sigma}(\Omega) \times L_q(\Omega; \mathbb{R}) \times H_q^1(\Omega; \mathbb{R}^n),$$

where $1 < p, q < \infty$ and σ indicates solenoidal vector fields. Following Theorem 1.10.6, Proposition 1.4.12 and Remark 1.4.13, the regularity space will be

$$X_1 := \{u \in H_q^2(\Omega; \mathbb{R}^n) \cap L_{q,\sigma}(\Omega) : u = 0 \text{ on } \partial\Omega\} \times Y_1,$$

with

$$Y_1 := \{(\theta, d) \in H_q^2(\Omega) \times H_q^3(\Omega; \mathbb{R}^n) : \partial_\nu \theta = \partial_\nu d = 0 \text{ on } \partial\Omega\}.$$

We consider solutions within the class

$$v \in H_{p,\mu}^1(J; X_0) \cap L_{p,\mu}(J; X_1),$$

where $J = (0, a)$ with $0 < a \leq \infty$ is an interval and $\mu \in (1/p, 1]$ indicates a time weight, as before. The time trace space of this class is given by

$$X_{\gamma,\mu} = \{u \in B_{qp}^{2(\mu-1/p)}(\Omega)^n \cap L_{q,\sigma}(\Omega) : u = 0 \text{ on } \partial\Omega\} \times Y_{\gamma,\mu},$$

where

$$Y_{\gamma,\mu} = \{(\theta, d) \in B_{qp}^{2(\mu-1/p)}(\Omega) \times B_{qp}^{1+2(\mu-1/p)}(\Omega; \mathbb{R}^n) : \partial_\nu \theta = \partial_\nu d = 0 \text{ on } \partial\Omega\},$$

whenever the boundary traces exist. It satisfies

$$X_{\gamma,\mu} \hookrightarrow B_{qp}^{2(\mu-1/p)}(\Omega)^{n+1} \times B_{qp}^{1+2(\mu-1/p)}(\Omega)^n \hookrightarrow C(\overline{\Omega})^{n+1} \times C^1(\overline{\Omega})^n,$$

provided

$$\frac{1}{p} + \frac{n}{2q} < \mu \le 1. \tag{1.14.11}$$

For brevity we set $X_\gamma := X_{\gamma,1}$, as before. The state manifold of the problem is defined by

$$\mathcal{SM} = \{v \in X_\gamma : \theta(x) > 0, \ |d(x)|_2 = 1 \text{ in } \Omega\}.$$

The fundamental well-posedness result regarding the general Ericksen-Leslie system reads then as follows.

Theorem 1.14.9 *Let $J = (0, a)$, $1 < p, q < \infty$, $1 \ge \mu > 1/2 + 1/p + n/2q$, and assume that $\psi \in C^4((0, \infty) \times [0, \infty))$ as well as $\alpha, \mu_j, \gamma \in C^2((0, \infty) \times [0, \infty))$, $j = S, V, D, P, L, 0$, and the positivity conditions (1.14.5). Then the following assertions are valid:*

(i) *(Local Well-Posedness)*
  Let $v_0 \in X_{\gamma,\mu}$. Then for some $a = a(v_0) > 0$, there is a unique solution

$$v \in H^1_{p,\mu}(J; X_0) \cap L_{p,\mu}(J; X_1)$$

  of (1.14.1)–(1.14.4) on J. Moreover,

$$v \in C([0, a]; X_{\gamma,\mu}) \cap C((0, a]; X_\gamma),$$

  i.e., the solution regularizes instantly in time. It depends continuously on v_0 and exists on a maximal time interval $J(v_0) = [0, t^+(v_0))$. Moreover,

$$t\partial_t v \in H^1_{p,\mu}(J; X_0) \cap L_{p,\mu}(J; X_1), \quad a < t^+(v_0),$$

  and $\mathsf{E}(t) \equiv \mathsf{E}_0$ and $-\mathsf{N}$ is a strict Lyapunov functional. Furthermore, the problem (1.14.1), (1.14.3), (1.14.4) generates a local semi-flow in its natural state manifold $\mathcal{SM}$.
(ii) *(Stability of Equilibria)*
  Any equilibrium $v_* \in \mathcal{E}$ of (1.14.1)–(1.14.4) is stable in X_γ. Moreover, for each $v_* \in \mathcal{E}$ there is $\varepsilon > 0$ such that if $v_0 \in \mathcal{SM}$ with $|v_0 - v_*|_{X_{\gamma,\mu}} \le \varepsilon$, then the solution v of (1.14.1)–(1.14.4) with initial value v_0 exists globally in time and converges at an exponential rate in X_γ to some $v_\infty \in \mathcal{E}$.
(iii) *(Long-Time Behaviour)*

  (a) *Suppose that*

$$\sup_{t \in (0, t^+(v_0))} \left[\|v(t)\|_{X_{\gamma,\mu}} + \|1/\theta(t)\|_{L_\infty} \right] < \infty.$$

  Then $t^+(v_0) = \infty$ and v is a global solution.

(b) *If v is a global solution, bounded in $X_{\gamma,\mu}$ and with $1/\theta$ bounded, then v converges exponentially in $\mathcal{SM}$ to an equilibrium $v_\infty \in \mathcal{E}$ of (1.14.1)– (1.14.4), as $t \to \infty$.*

It is remarkable that the above theorem holds true *without any structural assumptions* on the Leslie coefficients, except for Condition 1.14.5. In particular, the above well-posedness results hold true *without* assuming Parodi's relation and no conditions for μ_V, μ_D, μ_P are needed.

Remarks 1.14.10

(a) A related class of models also dealing with the non-isothermal situation was presented by Feireisl et al. [53] as well as by Feireisl, Frémond, Rocca and Schimperna in [54]. Their model includes stretching as well as rotational terms and is consistent with the fundamental laws of Thermodynamics. The equation for the director d, however, is given in the penalized form. They show that the presence of the term $|\nabla d|_2^2$ in the internal energy as well as the stretching term $d \cdot \nabla u$ give rise, in order to respect the laws of Thermodynamics, to two new non dissipative contributions in the stress tensor S and in the flux q.

(b) Wu, Xu and Liu reconsidered in [155] the isothermal penalized Ericksen-Leslie model. Their main result says that under certain assumptions on the data and the Leslie coefficients, the penalized Ericksen-Leslie system admits a unique, global solution provided the viscosity is large enough. Wang et al. [153] proved local well-posedness of the isothermal general Ericksen-Leslie system as well as global well-posedness for small initial data under various conditions on the Leslie coefficients, which ensure that the energy of the system is dissipated.

(c) Let us mention here also the well-posedness results concerning the non-isothermal situation obtained in [78] and the very recent work of De Anna and Liu [33], in which the non-isothermal compressible Ericksen-Leslie system is investigated, too. In contrast to the approach in [33], the approach described here does *not* impose any structural assumptions on the Leslie-coefficients.

Key ideas of the proof of Theorem 1.14.9 We recall that the parameter functions are having the regularity properties

$$\mu_j, \alpha, \gamma \in C^2((0, \infty) \times [0, \infty)), \quad \psi \in C^4((0, \infty) \times [0, \infty)). \tag{1.14.12}$$

and that furthermore the positivity conditions (1.14.5) are assumed to be fulfilled.

Step 1: Linearization We linearize Eq. (1.14.1) at an initial value $v_0 = [u_0, \theta_0, d_0]^\mathsf{T}$ and drop all terms of lower order. This yields the principal linearization

$$\begin{cases} \mathcal{L}_\pi(\partial_t, \nabla)v_\pi = f & \text{in } J \times \Omega, \\ u = \partial_\nu\theta = \partial_\nu d = 0 & \text{on } J \times \partial\Omega, \\ u = \theta = d = 0 & \text{on } \{0\} \times \Omega. \end{cases} \tag{1.14.13}$$

Here $J = (0, a)$, $v_\pi = [u, \pi, \theta, d]^\mathsf{T}$ is the unknown, and $f = [f_u, f_\pi, f_\theta, f_d]^\mathsf{T}$ are the given data. The differential operator $\mathcal{L}_\pi(\partial_t, \nabla)$ is defined via its symbol $\mathcal{L}_\pi(z, i\xi)$, which is given by

$$\mathcal{L}_\pi(z, i\xi) = \begin{bmatrix} M_u(z, \xi)\, i\xi & 0 & izR_1(\xi)^\mathsf{T} \\ i\xi^\mathsf{T} & 0 & 0 & 0 \\ 0 & 0 & m_\theta(z, \xi) & -iz\theta_0 ba(\xi) \\ -iR_0(\xi) & 0 & -iba(\xi) & M_d(z, \xi) \end{bmatrix}, \tag{1.14.14}$$

with $b = \partial_\theta \lambda$, and $\lambda_1 = \partial_\tau \lambda$. We also introduce the parabolic part of this symbol by dropping pressure gradient and divergence, i.e.

$$\mathcal{L}(z, i\xi) = \begin{bmatrix} M_u(z, \xi) & 0 & izR_1(\xi)^\mathsf{T} \\ 0 & m_\theta(z, \xi) & iz\theta_0 ba(\xi) \\ -iR_0(\xi) & iba(\xi) & M_d(z, \xi) \end{bmatrix}. \tag{1.14.15}$$

The entries of these matrices are given by

$$m_\theta = \rho \kappa z + \alpha |\xi|^2, \quad a(\xi) = \xi \cdot \nabla d_0,$$

$$M_d = \gamma z + \lambda |\xi|^2 + \lambda_1 a(\xi) \otimes a(\xi) = m_d(z, \xi) + \lambda_1 a(\xi) \otimes a(\xi),$$

$$R_0 = \frac{\mu_D + \mu_V}{2} P_0 \xi \otimes d_0 + \frac{\mu_D - \mu_V}{2} (\xi | d_0) P_0,$$

$$R_1 = (\frac{\mu_D + \mu_V}{2} + \mu_P) P_0 \xi \otimes d_0 + (\frac{\mu_D - \mu_V}{2} + \mu_p)(\xi | d_0) P_0,$$

$$M_u = \rho z + \mu_s |\xi|^2 + \mu_0 (\xi | d_0)^2 d_0 \otimes d_0 + a_1 (\xi | d_0) P_0 \xi \otimes d_0,$$

$$+ a_2 (\xi | d_0)^2 P_0 + a_3 |P_0 \xi|^2 d_0 \otimes d_0 + a_4 (\xi | d_0) d_0 \otimes P_0 \xi.$$

Here $P_0 = P_{d_0} = I - d_0 \otimes d_0$, and a_j are certain coefficients.

Step 2: Maximal L^p-Regularity Let $1 < p, q < \infty$ and assume that (1.14.12) holds. Then (1.14.13) admits a unique solution $v_\pi = [u, \pi, \theta, d]^\mathsf{T}$ satisfying

$$(u, \theta) \in {}_0H_p^1(J; L_q(\Omega))^{n+1} \cap L_p(J; H_q^2(\Omega))^{n+1},$$

$$\pi \in L_p(J; \dot{H}_q^1(\Omega)),$$

$$d \in {}_0H_p^1(J; H_q^1(\Omega))^n \cap L_p(J; H_q^3(\Omega))^n,$$

if and only if

$$(f_u, f_\theta) \in L_p(J; L_q(\Omega))^{n+1}, \quad f_d \in L_p(J; H_q^1(\Omega))^n,$$

$$f_\pi \in {}_0H_p^1(J; {}_0H_q^{-1}(\Omega)) \cap L_p(J; H_q^1(\Omega)).$$

In order to prove this, we set $J := \text{diag}(I, 1/\theta_0, zI)$ and show first that the symbol $\bar{J}\mathcal{L}$ is accretive for $\text{Re}\, z > 0$, i.e. the associated system is strongly elliptic. Observe that for proving this we do not need any conditions for the coefficients $\mu_D, \mu_V, \mu_P, \partial_\theta\lambda$. Next, we perform a Schur reduction to reduce the above symbol to a symbol depending only for u. This implies that the resulting generalized Stokes symbol for (u, π) is strongly elliptic. Then we may apply a result due to Bothe and Prüss [18] to prove maximal L^p regularity in the case of $\mathbb{R}^n$. In a second step we verify the Lopatinskii-Shapiroo condition to obtain the corresponding result on the half space. A localization procedure finishes the proof of the above assertion.

Step 3: Local Existence We rewrite Problem (1.14.1) as an abstract quasi-linear evolution equation of the form

$$\dot{v} + A(v)v = F(v), \quad t > 0, \quad v(0) = v_0, \tag{1.14.16}$$

replacing $\mathcal{D}_t d$ appearing in the equations for u and θ by the equation for d. Here $v = (u, \theta, d)$ and the *Helmholtz projection* P is applied to the equation for u. The base space will be

$$X_0 := L_{q,\sigma}(\Omega) \times L_q(\Omega) \times H_q^1(\Omega; \mathbb{R}^n).$$

Then, by Theorem 1.7.2 for some $a = a(z_0) > 0$ there is a unique solution

$$v \in H_{p,\mu}^1(J, X_0) \cap L_{p,\mu}(J; X_1), \quad J = [0, a],$$

of (1.14.16), i.e., (1.14.1) on J. Moreover,

$$t[\frac{d}{dt}]v \in H_{p,\mu}^1(J; X_0) \cap L_{p,\mu}(J; X_1),$$

and it can be shown that $|d(t, x)|_2 \equiv 1$, $\mathsf{E}(t) \equiv \mathsf{E}_0$, and $-\mathsf{N}$ is a strict Lyapunov functional; see [76, 77] for details. Furthermore, the problem (1.14.16) generates a local semi-flow in its natural state manifold $\mathcal{SM}$.

Step 4: Dynamics The linearization of (1.14.1) at an equilibrium $v_* = (0, \theta_*, d_*)$ is given by the operator $A_* = A(v_*)$ in X_0. This operator has maximal L_p-regularity, it is the negative generator of a *compact analytic C_0-semigroup*, and it has *compact resolvent*. So its spectrum consists only of countably many eigenvalues of finite multiplicity, which have all positive real parts, hence are stable, except for 0. The eigenvalue 0 is semi-simple. Its eigenspace is given by

$$\mathsf{N}(A_*) = \{(0, \vartheta, \mathsf{d}) : \vartheta \in \mathbb{R}, \mathsf{d} \in \mathbb{R}^n\},$$

hence it coincides with the set of constant equilibria $\bar{\mathcal{E}}$, when ignoring the constraint $|d|_2 = 1$ and conservation of energy. Therefore each such equilibrium is normally

stable and Theorem 1.7.4 implies assertion (ii). Finally, assertion (iii) follows from
Theorem 1.7.8.   □

1.15 Global Strong Well-Posedness of the Primitive Equations

The primitive equations for the ocean and atmosphere are considered to be a
fundamental model for geophysical flows which is derived from Navier–Stokes
equations assuming a hydrostatic balance for the pressure term in the vertical
direction. The mathematical analysis of the primitive equations was pioneered by
Lions, Teman and Wang in a series of articles [103, 104]; for a survey of known
results and further references, we refer to an article by Li and Titi [99].

In contrast to Navier–Stokes equations, the $3D$ primitive equations admit a
unique, global, strong solution for arbitrary large data in H^1. This breakthrough
result was proved by Cao and Titi [25] in 2007 using energy methods. A different
approach to the primitive equations, based on methods of evolution equations, has
been presented in [73]. There a Fujita-Kato type iteration scheme was developed in
addition to H^2-a priori bounds for the solution.

In this section we present an alternative approach to the primitive equations based
on techniques from maximal L^q-regularity. This approach has several advantages
compared to the two other approaches. In fact, the regularizing effect of the solution
plays an important role when extending local solutions to global ones by means
of certain a priori bounds. We mention here the H^2-a priori bounds used in [73,
83]. In the following, we show that a priori bounds in the maximal regularity space
$L^2(0, T; H^2) \cap H^1(0, T; L^2)$ are already sufficient to prove the global existence of
a solution within the L^q–L^p-class.

Secondly, our approach allows to prove the existence and uniqueness of a global,
strong solution for initial values lying in critical spaces, which in the given situation
are the Besov spaces B_{pq}^μ for $p, q \in (1, \infty)$ and with $1/p + 1/q \le \mu \le 1$. Here, we
use in an essential way the concept of time weights for maximal L^p-regularity; see
Sect. 1.7 and [124] as well as [125] for details and proofs. Choosing in particular
$p = q = 2$ and $\mu = 1$ and noting that $B_{22}^1 = H^1$, we rediscover in particular the
celebrated result by Cao and Titi [25].

The precise formulation of the primitive equations reads as follows: Consider a
cylindrical domain $\Omega = G \times (-h, 0) \subset \mathbb{R}^3$ with $G = (0, 1) \times (0, 1)$, $h > 0$ and
denote by $v: \Omega \to \mathbb{R}^2$ the vertical velocity of the fluid and by $\pi_s: G \to \mathbb{R}$ its
surface pressure. The primitive equations are then given by

$$\begin{cases} \partial_t v + v \cdot \nabla_H v + w(v) \cdot \partial_z v - \Delta v + \nabla_H \pi_s = f, & \text{in } \Omega \times (0, T), \\ \qquad\qquad\qquad\qquad\qquad \text{div}_H \bar{v} = 0, & \text{in } \Omega \times (0, T), \\ \qquad\qquad\qquad\qquad\qquad\quad v(0) = v_0, & \text{in } \Omega, \end{cases} \qquad (1.15.1)$$

where $x, y \in G$ denote the horizontal coordinates and $z \in (-h, 0)$ the vertical one. We use the notations

$$\Delta = \partial_x^2 + \partial_y^2 + \partial_z^2, \quad \nabla_H = (\partial_x, \partial_y)^T,$$

$$\mathrm{div}_H v = \partial_x v_1 + \partial_y v_2 \quad \text{and} \quad \overline{v} := \frac{1}{h} \int_{-h}^{0} v(\cdot, \cdot, \xi) d\xi.$$

Here the horizontal velocity $w = w(v)$ is given by

$$w(v)(x, y, z) = -\int_{-h}^{z} \mathrm{div}_H v(x, y, \xi) d\xi,$$

where $w(x, y, -h) = w(x, y, 0) = 0$.

The Eq. (1.15.1) are supplemented by mixed boundary conditions on

$$\Gamma_u = G \times \{0\}, \quad \Gamma_b = G \times \{-h\} \quad \text{and} \quad \Gamma_l = \partial G \times (-h, 0),$$

i.e., the upper, bottom and lateral parts of the boundary $\partial \Omega$, respectively, defined by

$$v, \pi_s \text{ are periodic on } \Gamma_l \times (0, \infty),$$

$$v = 0 \text{ on } \Gamma_D \times (0, \infty) \quad \text{and} \quad \partial_z v = 0 \text{ on } \Gamma_N \times (0, \infty). \tag{1.15.2}$$

The Dirichlet, Neumann and mixed boundary conditions are comprised by the notation

$$\Gamma_D \in \{\emptyset, \Gamma_u, \Gamma_b, \Gamma_u \cup \Gamma_b\} \quad \text{and} \quad \Gamma_N = (\Gamma_u \cup \Gamma_b) \setminus \Gamma_D.$$

Given $p \in (1, \infty)$, the space $L_\sigma^p(\Omega)$ of *hydrostatically solenoidal vector fields* is given as in Sect. 1.10 as the subspace of $L^p(\Omega)^2$ defined by

$$L_\sigma^p(\Omega) := \overline{\{v \in C_{per}^\infty(\overline{\Omega})^2 : \mathrm{div}_H \overline{v} = 0\}}^{\|\cdot\|_{L^P(\Omega)^2}}. \tag{1.15.3}$$

Here horizontal periodicity is modeled by the function spaces $C_{per}^\infty(\overline{\Omega})$ and $C_{per}^\infty(\overline{G})$ consists of all smooth functions, which are periodic with respect to x, y coordinates but not necessarily in the z coordinate. Furthermore, following Sect. 1.10 there exists a continuous projection P_p, called the *hydrostatic Helmholtz projection*, from $L^p(\Omega)^2$ onto $L_\sigma^p(\Omega)$. In particular, P_p annihilates the pressure term $\nabla_H \pi_s$.

The Hydrostatic Stokes Operator

For $p \in (1, \infty)$ and $s \in [0, \infty)$ define the spaces

$$H^{s,p}_{per}(\Omega) := \overline{C^{\infty}_{per}(\overline{\Omega})}^{\|\cdot\|_{H^{s,p}(\Omega)}} \quad \text{and} \quad H^{s,p}_{per}(G) := \overline{C^{\infty}_{per}(\overline{G})}^{\|\cdot\|_{H^{s,p}(G)}},$$

where $H^{0,p}_{per} := L^p$. Here $H^{s,p}(\Omega)$ denotes the Bessel potential spaces, which are defined as restrictions of Bessel potential spaces on the whole space to Ω, compare e.g. [148, Definition 3.2.2.]. For $p, q \in (1, \infty)$ and $s \in [0, \infty)$ we also define the periodic Besov spaces

$$B^{s}_{pq,per}(\Omega) := \overline{C^{\infty}_{per}(\overline{\Omega})}^{\|\cdot\|_{B^{s}_{pq}(\Omega)}} \quad \text{and} \quad B^{s}_{pq,per}(G) := \overline{C^{\infty}_{per}(\overline{G})}^{\|\cdot\|_{B^{s}_{pq}(G)}},$$

where B^{s}_{pq} denotes Besov spaces, which are defined as restrictions of Besov spaces defined on the whole space $B^{s}_{p,q}(\mathbb{R}^3)$, compare e.g. [148, Definitions 3.2.2].

Following Sect. 1.10, the *hydrostatic Stokes operator* A_p in $L^p_{\overline{\sigma}}(\Omega)$ is defined as

$$A_p v := P_p \Delta v, \quad D(A_p) := \{v \in H^{2,p}_{per}(\Omega)^2 : \partial_z v\big|_{\Gamma_N} = 0, \ v\big|_{\Gamma_D} = 0\} \cap L^p_{\overline{\sigma}}(\Omega).$$

By Theorem 1.10.14, $A_p \in H^{\infty}(L^p_{\overline{\sigma}}(\Omega))$ with $\Phi^{\infty}_A = 0$. In particular, A_p admits the property of maximal L^q-regularity. Following Proposition 1.7.1 this is equivalent to maximal L^q-regularity of A_p in time-weighted spaces. Recall that these spaces were defined in Sect. 1.7 for $\mu \in (1/q, 1]$ by

$$L^q_{\mu}(J; D(A_p)) = \{v \in L^1_{loc}(J; D(A_p)): t^{1-\mu}v \in L^q(J; D(A_p))\},$$

$$H^{1,q}_{\mu}(J; L^p_{\overline{\sigma}}(\Omega)) = \{v \in L^q_{\mu}(J; L^p_{\overline{\sigma}}(\Omega)) \cap H^{1,1}(J; L^p_{\overline{\sigma}}(\Omega)): t^{1-\mu}v_t \in L^q(J; L^p_{\overline{\sigma}}(\Omega))\}.$$

The natural trace spaces of these spaces are determined by real interpolation $(\cdot, \cdot)_{\theta,q}$ for $\theta \in (0, 1)$ and $p, q \in (1, \infty)$. Thanks to Theorem 1.10.14 these spaces can be identified explicitly as described in the following lemma.

Lemma 1.15.1 *Let* $\theta \in (0, 1)$, $p, q \in (1, \infty)$ *and* $X_{\theta,q} := (L^p_{\overline{\sigma}}(\Omega), D(A_p))_{\theta,q}$. *Then*

$$X_{\theta,q} = \begin{cases} \{v \in B^{2\theta}_{pq,per}(\Omega) \cap L^p_{\overline{\sigma}}(\Omega): \partial_z v\big|_{\Gamma_N} = 0, \ v\big|_{\Gamma_D} = 0\}, & \frac{1}{2} + \frac{1}{2p} < \theta < 1, \\ \{v \in B^{2\theta}_{pq,per}(\Omega) \cap L^p_{\overline{\sigma}}(\Omega): v\big|_{\Gamma_D} = 0\}, & \frac{1}{2p} < \theta < \frac{1}{2} + \frac{1}{2p}, \\ B^{2\theta}_{pq,per}(\Omega) \cap L^p_{\overline{\sigma}}(\Omega), & 0 < \theta < \frac{1}{2p}. \end{cases}$$

Strong Global Well-Posedness

The main result of this section concerns the global, strong well-posedness of the primitive equations for arbitrarily large data in critical Besov spaces.

Theorem 1.15.2 (Global Well-Posedness) *Let $p, q \in (1, \infty)$ such that $1/p + 1/q \leq 1$. For $0 < T < \infty$ let $\mu \in [1/p + 1/q, 1]$ and assume*

$$v_0 \in X_{\mu - 1/q, q} \quad \text{and} \quad P_p f \in H_\mu^{1,q}(0, T; L_\sigma^p(\Omega)) \cap H^{1,2}(\delta, T; L_\sigma^2(\Omega))$$

for some sufficiently small $\delta > 0$. Then there exists a unique, strong solution v to the primitive Eq. (1.15.1) satisfying

$$v \in H_\mu^{1,q}(0, T; L_\sigma^p(\Omega)) \cap L_\mu^q(0, T; D(A_p)).$$

Considering in particular the case $p = q = 2$, we rediscover the global existence result due to Cao and Titi [25].

In order to prove Theorem 1.15.2 we introduce a setting as described in Sect. 1.7. In fact, for $1 < p < \infty$ set

$$X_0 := L_\sigma^p(\Omega) \quad \text{and} \quad X_1 := D(A_p),$$

where $L_\sigma^p(\Omega)$ is defined as in (1.15.3). We then define the bilinear map F_p by

$$F_p(v, v') := P_p(v \cdot \nabla_H v' + w(v)\partial_z v'),$$

and set $F_p(v) := F_p(v, v)$. Since $u' = (v', w(v'))$ is divergence-free, we also obtain the representation

$$F_p(v, v') = P_p \text{div}(u' \otimes v).$$

The mapping F_p satisfies then the following properties.

Lemma 1.15.3 *Let $p \in (1, \infty)$ and $s \geq 0$. Then there exists a constant $C > 0$ (depending only on Ω and p, s) such that*

$$\|F_p(v, v')\|_{H^{s,p}(\Omega)^2} \leq C \|v\|_{H^{s+1+1/p,p}} \|v'\|_{H^{s+1+1/p,p}},$$

for all $v, v' \in H^{s+1+1/p,p}(\Omega)^2$, i.e., the mapping

$$F_p(\cdot, \cdot): H^{s+1+1/p,p}(\Omega)^2 \times H^{s+1+1/p,p}(\Omega)^2 \to H^{s,p}(\Omega)^2$$

is a continuous bilinear map.

Denoting by $X_\beta = [L_\sigma^p(\Omega), D(A_p)]_\beta$ the complex interpolation space of order $\beta \in [0, 1]$, we verify that $[L_\sigma^p(\Omega), D(A_p)]_{(1+1/p)/2} \subset H^{1+1/p,p}(\Omega)^2$, which for $\beta = \frac{1}{2}(1 + 1/p)$ yields the existence a constant $C > 0$, independent of v, v', such that

$$\|F_p(v) - F_p(v')\|_{L_\sigma^p(\Omega)} \leq C \left(\|v\|_{X_\beta} + \|v'\|_{X_\beta} \right) \|v - v'\|_{X_\beta}. \tag{1.15.4}$$

The local well-posedness of the primitive equations follows now from Proposition 1.7.6 by using Theorem 1.10.14 and the remark after Proposition 1.7.6 for condition (S) and $(H1)$ and estimate (1.15.4) for the conditions $(H2)$ and $(H3)$. More precisely, the following result holds true.

Proposition 1.15.4 (Local Well-Posedness) *Let $p, q \in (1, \infty)$ with $1/p + 1/q \leq 1$, $\mu \in [1/p + 1/q, 1]$ and $T > 0$. Assume that*

$$v_0 \in X_{\mu - 1/q, q} \quad and \quad P_p f \in L_\mu^q(0, T; L_{\bar\sigma}^p(\Omega)).$$

Then there exists $T' = T'(v_0)$ with $T' \in (0, T]$ and a unique, strong solution

$$v \in H_\mu^{1,q}(0, T'; L_{\bar\sigma}^p(\Omega)) \cap L_\mu^q(0, T'; D(A_p))$$

to (1.15.1) on $(0, T')$.

The proof of global well-posedness of the primitive equations is based on the following a priori bound in $H^1(0, T; L^2) \cap L^2(0, T; H^2)$.

Theorem 1.15.5 (A Priori Bounds) *There exists a continuous function B satisfying the following property: any solution of (1.15.1) fulfilling for $0 < T < \infty$ the conditions*

$$v \in H^1(0, T; L_{\bar\sigma}^2(\Omega))) \cap L^2(0, T; D(A_2)),$$

$$v_0 \in \{H^1 \cap L_{\bar\sigma}^2(\Omega) : v|_{\Gamma_D} = 0\}, \quad P_2 f \in L^2(0, T; L_{\bar\sigma}^2(\Omega))$$

satisfies

$$\|v\|_{H^1(0,T;L_{\bar\sigma}^2(\Omega))) \cap L^2(0,T;D(A_2))} \leq B\left(\|v_0\|_{H^1(\Omega)}, \|P_2 f\|_{L^2(0,T;L^2(\Omega))}, T\right).$$

The proof of Theorem 1.15.2 is based on Proposition 1.15.4, the above a priori bound and the following result within the L^2-setting.

Proposition 1.15.6 *Let $0 < T < \infty$ and $v_0 \in \{H^1 \cap L_{\bar\sigma}^2(\Omega) : v|_{\Gamma_D} = 0\}$.*

(a) If $P_2 f \in L^2(0, T; L_{\bar\sigma}^2(\Omega))$, then there exists a unique, strong solution v to the primitive Eq. (1.15.1) within the class

$$v \in H^1(0, T; L_{\bar\sigma}^2(\Omega))) \cap L^2(0, T; D(A_2)).$$

(b) If in addition $t \mapsto t \cdot P_2 f_t(t) \in L^2(0, T; L_{\bar\sigma}^2(\Omega))$, then

$$t \cdot v_t \in H^1(0, T; L_{\bar\sigma}^2(\Omega))) \cap L^2(0, T; D(A_2)).$$

Before giving the proof of Proposition 1.15.6 we note the following elementary fact about extending regularity of solutions from $(0, T')$ for any $0 < T' < T$ to $(0, T)$.

Lemma 1.15.7 *Let* $v \in \mathbb{E}_{1,\mu}(0, T')$ *for any* $0 < T' < T$, *and* $\sup\limits_{0<T'<T} \|v\|_{\mathbb{E}_{1,\mu}(0,T')}$
$< C$ *for some constant* $C > 0$. *Then* $v \in \mathbb{E}_{1,\mu}(0, T)$.

Proof of Proposition 1.15.6 In order to prove assertion (a) consider

$$t_+(v_0) := \sup\{T' > 0 \colon \text{Eq.} (1.15.1) \text{ has a solution in } \mathbb{E}_{1,1}(0, T')\}.$$

Proposition 1.15.4 yields $t_+(v_0) > 0$ and that the solutions in $\mathbb{E}_{1,1}(0, T')$ are unique. Assume now that $t_+(v_0) < T$. By Theorem 1.15.5, $\|v\|_{\mathbb{E}_{1,1}(0,T')} \leq B\big(\|v_0\|_{H^1(\Omega)}, \|P_2 f\|_{L^2(0,T;L^2(\Omega))}, t_+(v_0)\big)$ for any $0 < T' < t_+(v_0)$. Hence, by Lemma 1.15.7 we have $v \in \mathbb{E}_{1,1}(0, t_+(v_0))$. Since the trace in $\mathbb{E}_{1,1}(0, t_+(v_0))$ is well-defined, $v(t_+(v_0))$ can be taken as new initial value, thus extending the solution beyond $t_+(v_0)$ and contradicting the assumption. Hence, $t_+(v_0) = T$ and combing again Theorem 1.15.5 with Lemma 1.15.7 yields $v \in \mathbb{E}_{1,1}(0, T)$. This proves part (a).

Assertion (b) is proved by Angenent's method. For details see [65]. □

Proof of Theorem 1.15.2 Using Angenent's method we verify that the local solution v obtained in Proposition 1.15.4 enjoys additional time regularity. In particular, we deduce that $v \in H^{1,q}(\delta, T; D(A_p)) \hookrightarrow C^0(\delta, T; D(A_p))$ for some $0 < \delta \leq T'$ and $0 < T' < T$. Now, using $v(T')$ as new initial value, and taking advantage of the embedding $D(A_q) \subset (L^2_\sigma(\Omega), D(A_2))_{1/2,q}$ for $q \in [6/5, \infty)$ and the additional assumption $P_2 f \in W^{1,2}(\delta, T; L^2_\sigma(\Omega))$, we see that v is also an L^2 solution, at least for $\delta > 0$. This holds for $q \in [6/5, \infty)$, and for $q \in (1, 6/5)$ by a bootstrapping argument; see [83].

By Proposition 1.15.6 there exists now a global L^2 solution $v \in C_b(\delta, D(A_2))$. Lemma 1.15.1 and classical embedding results yield

$$D(A_2) \hookrightarrow X_{\overline{\mu},q} \quad \text{for } 0 \leq \overline{\mu} - \mu < 2 - \tfrac{2}{p},$$

and the compactness of the embedding $X_{\overline{\mu},q} \hookrightarrow X_{\mu,q}$ for $1/p < \mu < \overline{\mu} < 1$. Since

$$\|v\|_{C_b(\delta,T;X_{\overline{\mu},q})} \leq C \|v\|_{C_b(\delta,T;D(A_2))},$$

Proposition 1.7.8 applies and the solution v exists hence globally, that is for any $T > 0$. □

1.16 Justification of the Hydrostatic Approximation for the Primitive Equations by Scaled Navier–Stokes Equations

In this section we show that the primitive equations can be obtained as the limit of anisotropically scaled Navier–Stokes equations. The scaling parameter $\varepsilon > 0$

represents the ratio of the depth to the horizontal width. Such an approximation is motivated by the fact that for large-scale oceanic dynamics, this aspect ratio ε is rather small and implies anisotropic viscosity coefficients. For an aspect ratio ε, i.e., in the case where the spacial domain can be represented as $\Omega_\varepsilon = G \times (-\varepsilon, +\varepsilon)$ for some $G \subset \mathbb{R}^2$, and a horizontal and vertical eddy viscosity 1 and ε^2, respectively, the system can be rescaled into the form

$$\begin{cases} \partial_t v_\varepsilon + u_\varepsilon \cdot \nabla v_\varepsilon - \Delta v_\varepsilon + \nabla_H p_\varepsilon = 0, \\ \varepsilon(\partial_t w_\varepsilon + u_\varepsilon \cdot \nabla w_\varepsilon - \Delta w_\varepsilon) + \frac{1}{\varepsilon}\partial_z p_\varepsilon = 0, \\ \operatorname{div} u_\varepsilon = 0, \end{cases} \qquad (1.16.1)$$

in the time-space domain $(0, T) \times \Omega_1$, which is *independent* of the aspect ratio. We refer to [98] for more details on this rescaling procedure. Here the horizontal and vertical velocities v_ε and w_ε describe the three-dimensional velocity u_ε, while p_ε denotes the pressure of the fluid.

In the following, we show convergence results of the above system in the strong sense within the L^p–L^q-setting hereby extending a previous result due to Li and Titi [98] to a more general setting. Our method is very different from the one introduced by [98]. Whereas they rely on second order energy estimates, our approach is based on maximal L^p–L^q-regularity estimates for the heat or Stokes equation.

Consider the cylindrical domain $\Omega := (0, 1)^2 \times (-1, 1)$. Let $u = (v, w)$ be the solution of the primitive equations

$$\begin{cases} \partial_t v + u \cdot \nabla v - \Delta v + \nabla_H p = 0 & \text{in } (0, T) \times \Omega, \\ \partial_z p = 0 & \text{in } (0, T) \times \Omega, \\ \operatorname{div} u = 0 & \text{in } (0, T) \times \Omega, \\ p \text{ periodic in } x, y \\ v, w \text{ periodic in } x, y, z, \text{ even and odd } \text{in } z, \\ u(0) = u_0 & \text{in } \Omega, \end{cases} \qquad \text{(PE)}$$

and $u_\varepsilon = (v_\varepsilon, w_\varepsilon)$ be the solution of the anisotropic Navier–Stokes equations

$$\begin{cases} \partial_t v_\varepsilon + u_\varepsilon \cdot \nabla v_\varepsilon - \Delta v_\varepsilon + \nabla_H p_\varepsilon = 0 & \text{in } (0, T) \times \Omega, \\ \partial_t w_\varepsilon + u_\varepsilon \cdot \nabla w_\varepsilon - \Delta w_\varepsilon + \frac{1}{\varepsilon^2}\partial_z p_\varepsilon = 0 & \text{in } (0, T) \times \Omega, \\ \operatorname{div} u_\varepsilon = 0 & \text{in } (0, T) \times \Omega, \\ p_\varepsilon \text{ periodic in } x, y, z, \text{ even} & \text{in } z, \\ v_\varepsilon, w_\varepsilon \text{ periodic in } x, y, z, \text{ even and odd} & \text{in } z, \\ u_\varepsilon(0) = u_0 & \text{in } \Omega. \end{cases} \qquad \text{(NS}_\varepsilon)$$

Here v and v_ε denote the (two-dimensional) horizontal velocities, w and w_ε the vertical velocities, and p and p_ε denote the pressure term for the primitive equations as well as the Navier–Stokes equations, respectively. These are functions of three space variables $x, y \in (0, 1)$, $z \in (-1, 1)$. Since w is odd, the divergence free condition for the primitive equation translates into $\operatorname{div}_H \overline{v} = 0$, where

$\overline{v}(x, y) = \frac{1}{2} \int_{-1}^{1} v(x, y, z) \, dz$, and

$$w(\cdot, \cdot, z) = -\int_{-1}^{z} \operatorname{div}_H v(\cdot, \cdot, \zeta) \, d\zeta.$$

The Setting

For $p, q \in (1, \infty)$ and $s \in [0, \infty)$ let the Bessel potential spaces $H^{s,p}_{per}(\Omega) = \overline{C^{\infty}_{per}(\overline{\Omega})}^{\|\cdot\|_{H^{s,p}}}$ and the Besov spaces $B^{s}_{p,q,per}(\Omega) = \overline{C^{\infty}_{per}(\overline{\Omega})}^{\|\cdot\|_{B^{s}_{p,q}}}$ be defined as in Sect. 1.15. The space $H^{s,p}(\Omega)$ denotes the Bessel potential space of order s, with norm $\|\cdot\|_{H^{s,p}}$ defined via the restriction of the corresponding space defined on the whole space to Ω, see e.g. [148, Definition 3.2.2.]. Similarly, $B^{s}_{p,q}(\Omega)$ denotes a Besov space on Ω, which is defined again by restrictions of functions on the whole space to Ω, see again [148, Definition 3.2.2.]. Note that $L^{p}(\Omega) = H^{0,p}_{per}(\Omega)$ and $B^{s}_{p,2,per}(\Omega) = H^{s}_{p,per}(\Omega)$. The anisotropic structure of the primitive equations motivates the definition of the Bessel potential spaces $H^{s,p}_{xy} := H^{s,p}((0,1)^2)$ and $H^{s,p}_{z} := H^{s,p}(-1,1)$ for the horizontal and vertical variables, respectively. Similarly as above we write $L^{p}_{xy} := H^{0,p}_{xy}$ and $L^{p}_{z} := H^{0,p}_{z}$ and set $H^{s,p}_{xy} H^{r,q}_{z} := H^{s,p}((0,1)^2; H^{r,q}_{z})$.

The divergence free conditions in the above sets of equations can be encoded into the space of solenoidal functions

$$L^{p}_{\sigma}(\Omega) = \overline{\{u \in C^{\infty}_{per}(\overline{\Omega})^3 : \operatorname{div} u = 0\}}^{\|\cdot\|_{L^p}} \quad \text{and}$$

$$L^{p}_{\overline{\sigma}}(\Omega) = \overline{\{v \in C^{\infty}_{per}(\overline{\Omega})^2 : \operatorname{div}_H \overline{v} = 0\}}^{\|\cdot\|_{L^p}}.$$

For given $p, q \in (1, \infty)$ we set

$$X_0 := L^{q}(\Omega), \quad X_1 := H^{2,q}_{per}(\Omega),$$

$$X^{v}_0 := \{v \in L^{q}_{\overline{\sigma}}(\Omega) : v \text{ even in } z\},$$

$$X^{v}_1 := \{v \in H^{2,q}_{per}(\Omega)^2 \cap L^{q}_{\overline{\sigma}}(\Omega) : v \text{ even in } z\},$$

$$X^{u}_0 := \{(v_1, v_2, w) \in L^{q}_{\sigma}(\Omega) : v_1, v_2 \text{ even } w \text{ odd in } z\},$$

$$X^{u}_1 := \{(v_1, v_2, w) \in H^{2,q}_{per}(\Omega)^3 \cap L^{q}_{\sigma}(\Omega) : v_1, v_2 \text{ even } w \text{ odd in } z\},$$

and consider the trace space X_γ defined by $X_\gamma = (X^{u}_0, X^{u}_1)_{1-1/p, p}$. Given $p, q \in (1, \infty)$ and following [68], the trace space X_γ can be characterized as

$$X_\gamma = \begin{cases} \{(v_1, v_2, w) \in B^{2-2/p}_{p,q,per}(\Omega)^3 \cap L^{q}_{\sigma}(\Omega) : v = (v_1, v_2) \text{ even}, w \text{ odd in } z, \\ \qquad\qquad\qquad (\partial_z v, w) = 0 \text{ at } z = -1, 0, 1\}, \quad 1 > \frac{2}{p} + \frac{1}{q}, \\ \{(v_1, v_2, w) \in B^{2-2/p}_{p,q,per}(\Omega)^3 \cap L^{q}_{\sigma}(\Omega) : v = (v_1, v_2) \text{ even}, w \text{ odd in } z, \\ \qquad\qquad\qquad w = 0 \text{ at } z = -1, 0, 1\}, \qquad 1 < \frac{2}{p} + \frac{1}{q}. \end{cases}$$

For $p, q \in (1, \infty)$ and $T \in (0, \infty]$ we define the maximal regularity spaces

$$\mathbb{E}_0(T) := L^p(0, T; X_0), \quad \mathbb{E}_1(T) := L^p(0, T; X_1) \cap H^{1,p}(0, T; X_0),$$

and analogously $\mathbb{E}_0^v(T)$, $\mathbb{E}_1^v(T)$ and $\mathbb{E}_0^u(T)$, $\mathbb{E}_0^u(T)$ with respect to X_0^v, X_1^v and X_0^u, X_1^u, respectively. In order to simplify our notation we sometimes omit the subscripts u and v and write only $\mathbb{E}_0(T)$ and $\mathbb{E}_1(T)$. Finally, we say that $u = (v, w)$ is a *strong solution to the primitive equations* (in the L^p–L^q-setting, if $v \in \mathbb{E}_1^v$ and (PE) holds almost everywhere. We say that u_ε is a *strong solution to the Navier–Stokes equations*, if $u \in \mathbb{E}_1^u$ and (NS$_\varepsilon$) holds almost everywhere.

The idea of our approach consists of controling the maximal regularity norm of the differences $(v_\varepsilon - v, \varepsilon(w_\varepsilon - w))$ by the aspect ratio ε. To this end, we introduce the difference equations of (NS$_\varepsilon$) and (PE): setting $V_\varepsilon := v_\varepsilon - v$, $W_\varepsilon := w_\varepsilon - w$, $U_\varepsilon := (V_\varepsilon, W_\varepsilon)$ and $P_\varepsilon = p_\varepsilon - p$, we obtain

$$\begin{cases} \partial_t V_\varepsilon - \Delta V_\varepsilon = F_H(V_\varepsilon, W_\varepsilon) - \nabla_H P_\varepsilon & \text{in } (0, T) \times \Omega, \\ \partial_t \varepsilon W_\varepsilon - \Delta \varepsilon W_\varepsilon = \varepsilon F_z(V_\varepsilon, W_\varepsilon) - \frac{1}{\varepsilon}\partial_z P_\varepsilon & \text{in } (0, T) \times \Omega, \\ \operatorname{div} U_\varepsilon = 0 & \text{in } (0, T) \times \Omega, \\ P_\varepsilon \text{ periodic in } x, y, z, \text{ even} & \text{in } z, \\ V_\varepsilon, W_\varepsilon \text{ periodic in } x, y, z, \text{ even and odd} & \text{in } z, \\ U_\varepsilon(0) = 0 & \text{in } \Omega, \end{cases} \tag{1.16.2}$$

where the forcing terms F_H and F_z are given by

$$F_H(V_\varepsilon, W_\varepsilon) := -U_\varepsilon \cdot \nabla v - u \cdot \nabla V_\varepsilon - U_\varepsilon \cdot \nabla V_\varepsilon,$$

$$F_z(V_\varepsilon, W_\varepsilon) := -U_\varepsilon \cdot \nabla w - u \cdot \nabla W_\varepsilon - U_\varepsilon \cdot \nabla W_\varepsilon - \partial_t w - u \cdot \nabla w + \Delta w.$$

Our strategy is as follows: Applying the maximal regularity estimate given below in Proposition 1.16.3 to (1.16.2), we estimate $\|U_\varepsilon\|_{\mathbb{E}_1}$ in terms of the right hand sides. Like this we obtain a quadratic inequality for the norm of the differences (see Corollary 1.16.6) and for proving the main result of this section we need to ensure that the constant term as well as the coefficient in front of the linear term are sufficiently small. This can be achieved provided the aspect ratio ε is small enough and provided the vertical and horizontal solution of the primitive equations exist globally in the maximal regularity class.

Main Result and Proof

Introducing assumption (A)

(A): Let $q \in \left(\frac{4}{3}, \infty\right)$ and $p \geq \max\left\{\frac{q}{q-1}; \frac{2q}{3q-4}\right\}$, i.e., $1 \geq \begin{cases} \frac{1}{p} + \frac{1}{q}, \text{if } q \geq 2, \\ \frac{2}{3p} + \frac{4}{3q}, \text{if } q \leq 2, \end{cases}$

we are in the position to state the main result of this section.

Theorem 1.16.1 *Assume that Assumption (A) is satisfied. Let $u_0 \in X_\gamma$, $T > 0$ and (v, w) and $(v_\varepsilon, w_\varepsilon)$ be solutions of (PE) and (NS$_\varepsilon$), respectively. Then there exists a* *constant $C > 0$, independent of ε, such that*

$$\|(V_\varepsilon, \varepsilon W_\varepsilon)\|_{\mathbb{E}_1(T)} \le C\varepsilon$$

for sufficiently small ε. In particular,

$$(v_\varepsilon, \varepsilon w_\varepsilon) \to (v, 0) \text{ in } L^p(0, T; H^{2,q}(\Omega)) \cap H^{1,p}(0, T; L^q(\Omega))$$

as $\varepsilon \to 0$ with convergence rate $\mathcal{O}(\varepsilon)$.

Remarks 1.16.2

(a) If the solution $u = (v, w)$ of the primitive equations exists globally in time, the convergence rate is uniform for all $T \in (0, \infty]$, see Remark 4.10.b) in [55]. For example, if $p = q = 2$ and the initial data are mean value free, one can show that the solution to the primitive equations exists globally in $\mathbb{E}_1(T)$ with $T = \infty$.

(b) Let us note that the above result covers in particular the case $p = q = 2$, which was investigated before in [98].

(c) The scaled Navier–Stokes equations are locally well-posed in the maximal regularity space of the torus and the parity conditions are preserved. The above Theorem 1.16.1 yields that for each $T > 0$ there exists an $\varepsilon > 0$ such that the solution exists on $(0, T)$.

(d) The primitive equations are well-posed for all times $T > 0$ provided $u_0 \in X_\gamma$, see Sect. 1.15.

We start the proof of Theorem 1.16.1 by considering the linearization of (1.16.2). More specifically, given $F \in \mathbb{E}_0(T)$ and $U_0 \in X_\gamma$, we consider the linear problem

$$\begin{cases} \partial_t U - \Delta U = F - \nabla_\varepsilon P & \text{in } (0, T) \times \Omega, \\ \nabla_\varepsilon \cdot U = 0 & \text{in } (0, T) \times \Omega, \\ U, P \text{ periodic} & \text{in } x, y, z, \\ U(0) = U_0 & \text{in } \Omega, \end{cases} \qquad (1.16.3)$$

where $\nabla_\varepsilon := (\partial_x, \partial_y, \varepsilon^{-1}\partial_z)^T$. The following proposition gives a maximal regularity estimate for U, where the constants are *independent* of the aspect ratio and the pressure gradient.

Proposition 1.16.3 *Let $p, q \in (1, \infty)$, $T > 0$, $F \in \mathbb{E}_0(T)$, $U_0 \in X_\gamma$ and $\varepsilon > 0$. Then there is a unique solution U, P to the Eq. (1.16.3) with $U \in \mathbb{E}_1(T)$ and $\nabla_\varepsilon P \in \mathbb{E}_0(T)$, where P is unique up to a constant. Moreover, there exist constants $C > 0$ and $C_T > 0$, independent of ε, such that*

$$\|U\|_{\mathbb{E}_1(T)} \le C\|F\|_{\mathbb{E}_0(T)} + C_T\|U_0\|_{X_\gamma}.$$

In order to prove Proposition 1.16.3 we define the ε-dependent Helmholtz projection $\mathbb{P}_\varepsilon$ by

$$\mathbb{P}_\varepsilon := \mathrm{Id} - \nabla_\varepsilon \Delta_\varepsilon^{-1} \mathrm{div}_\varepsilon, \quad \text{where } \Delta_\varepsilon = \nabla_\varepsilon \cdot \nabla_\varepsilon.$$

The following lemma shows that $\mathbb{P}_\varepsilon$ is a bounded projection with uniform norm bound independent of ε.

Lemma 1.16.4 *Let $\varepsilon > 0$, $q \in (1, \infty)$ and assume that $F = (f_H, f_z) \in L^q(\Omega)$ and $P \in H^{1,q}_{per}(\Omega)$ are satisfying the equation $-(\Delta_H + \varepsilon^{-2}\partial_z^2)P = \mathrm{div}(f_H, \varepsilon^{-1}f_z)$ for $\varepsilon > 0$. Then there exists a constant $C > 0$, independent of ε, such that*

$$\|(\nabla_H P, \varepsilon^{-1}\partial_z P)\|_q \le C\|F\|_q.$$

Proof For $k_\varepsilon = (k_1, k_2, \varepsilon^{-1}k_3)^T$ and $m(k) = -\frac{k \otimes k}{|k|^2} \in \mathbb{R}^{3 \times 3}$ set $m_\varepsilon(k) = m(k_\varepsilon)$. Then $(\nabla_H, \varepsilon^{-1}\partial_z)P = \mathcal{F}^{-1}m_\varepsilon \mathcal{F} F$ and $k\nabla m_\varepsilon(k) = k_\varepsilon \nabla m(k_\varepsilon)$. Hence,

$$\sup_{\gamma \in \{0,1\}^3} \sup_{k \neq 0} |k^\gamma D^\gamma m_\varepsilon(k)| = \sup_{\gamma \in \{0,1\}^3} \sup_{k_\varepsilon \neq 0} |k_\varepsilon^\gamma D^\gamma m(k_\varepsilon)| = 1,$$

and Markincinkiewicz's theorem, see Corollary 1.1.9, implies that m_ε is an L^p-Fourier multiplier satisfying $\|\mathcal{F}^{-1}m_\varepsilon\mathcal{F}\|_{\mathcal{L}(L^q(\Omega))} \le C$ for some $C = C(q) > 0$ independent of ε. □

Applying Lemma 1.16.4 to (1.16.3) and taking into account the equalities $\mathbb{P}_\varepsilon \Delta = \Delta \mathbb{P}_\varepsilon$ and $\mathbb{P}_\varepsilon \nabla_\varepsilon P = 0$, Eq. (1.16.3) reduces to the heat equation with right hand side $\mathbb{P}_\varepsilon F$. Maximal L^p-regularity estimates of the three-dimensional Laplacian in the periodic setting yield

$$\|U\|_{\mathbb{E}_1(T)} \le C\|\mathbb{P}_\varepsilon F\|_{\mathbb{E}_0(T)} + C_T\|U_0\|_{X_\gamma} \le C\|F\|_{\mathbb{E}_0(T)} + C_T\|U_0\|_{X_\gamma}. \quad □$$

Finally, we state that the solution $u = (v, w)$ of the primitive equations belongs to the maximal regularity class $\mathbb{E}_1^u(T)$. For a detailed proof, see [55], Prop. 4.8.

Proposition 1.16.5 *Let p, q fulfill Assumption (A) and let v be the strong solution of the primitive equations associated to v_0 satisfying $(v_0, w_0) \in X_\gamma$. Then*

$$u = (v, w) \in \mathbb{E}_1^u(T) \text{ for all } T > 0.$$

Corollary 1.16.6 *Let $\mathcal{T} > 0$ and $p, q \in (1, \infty)$ such that $1/p + 1/q \le 1$. Let $(V_\varepsilon, W_\varepsilon) \in \mathbb{E}_1^u(\mathcal{T})$ denote the solution of Eq. (1.16.2) for some $u = (v, w) \in \mathbb{E}_1(\mathcal{T})$ and initial data $U_0 \in X_\gamma$ and set $X_\varepsilon(T) := \|(V_\varepsilon, \varepsilon W_\varepsilon)\|_{\mathbb{E}_1(T)}$. Then, for any*

$\eta \in [0, 1 - 1/p - 1/q]$ *there exists a constant* $C > 0$, *independent of* ε, *such that*

$$X_\varepsilon(T) \le CT^\eta \left[X_\varepsilon(T) \|u\|_{\mathbb{E}_1(T)} + X_\varepsilon^2(T) \right]$$

$$+ \varepsilon C \left[\|u\|_{\mathbb{E}_1(T)} + T^\eta \|u\|_{\mathbb{E}_1(T)}^2 \right] + C \|U_0\|_{X_\gamma},$$

for all $T \in [0, \mathcal{T}]$.

Proof of Theorem 1.16.1 Fix $\mathcal{T} > 0$ and let u be the solution of Eq. (PE). Then Proposition 1.16.5 implies $u \in \mathbb{E}_1(\mathcal{T})$. We show that

$$X_\varepsilon(T) = \|(V_\varepsilon, \varepsilon W_\varepsilon)\|_{\mathbb{E}_1(T)} \le \varepsilon C(\|(v, w)\|_{\mathbb{E}_1(\cdot)}, \mathcal{T}, p, q)$$

for all $T \in [0, \mathcal{T}]$ and $\varepsilon > 0$ small enough. To this end, by the uniform continuity of the mapping $T \mapsto \|u\|_{\mathbb{E}_1(T)}$ on $[0, \mathcal{T}]$, there is a $T^* \in [0, \mathcal{T}]$ such that $\|u\|_{\mathbb{E}_1(T+T^*)}^p - \|u\|_{\mathbb{E}_1(T)}^p \le (2C\mathcal{T}^\eta)^{-p}$ for all $T \in [0, \mathcal{T} - T^*]$ and where C denotes the constant given in Corollary 1.16.6. The latter with $U_0 = 0$ implies

$$CX_\varepsilon^2(T) - \tfrac{1}{2}X_\varepsilon(T) + \varepsilon \ge 0, \quad T \in [0, T^*]. \tag{1.16.4}$$

Since $X_\varepsilon(0) = 0$ we may solve the above quadratic inequality for $\varepsilon < (16C)^{-1}$ and obtain $X_\varepsilon \le 2\varepsilon$ on $[0, T^*]$.

Observe next that inequality (1.16.4) holds on a time interval independent of ε. More specifically, if one replaces T^* by $T_\varepsilon < T^*$, where T_ε is the maximal existence time of (NS$_\varepsilon$) with initial data u_0, then similarly as above we obtain $X_\varepsilon \le 2\varepsilon$ on $[0, T_\varepsilon]$, which yields a contradiction to the maximality of the existence time.

Assume now that there exists some $m \in \mathbb{N}$ such that $mT^* < \mathcal{T}$ and $X_\varepsilon \le \varepsilon 2K_m$ in $[0, mT^*]$, where $K_1 = 1$ and $K_m = 2^{1/p} [(2Cc_{mT^*} + 1) K_{m-1} + 1]$ and c_T denotes the embedding constant of $\mathbb{E}_1(T) \hookrightarrow L^\infty(0, T; X_\gamma)$. Let $(\tilde{V}_\varepsilon, \varepsilon \tilde{W}_\varepsilon)(T) = (V_\varepsilon, \varepsilon W_\varepsilon)(T + mT^*)$ be the unique solution of problem (1.16.2) with respect to $\tilde{u}(T) = u(T + mT^*)$ and initial data $U_0 = (V_\varepsilon, \varepsilon W_\varepsilon)(mT^*)$. Setting

$$\tilde{X}_\varepsilon^p(T) := \|(\tilde{V}_\varepsilon, \varepsilon \tilde{W}_\varepsilon)\|_{\mathbb{E}_1(T)}^p = X_\varepsilon^p(T + mT^*) - X_\varepsilon^p(mT^*),$$

Corollary 1.16.6 and the argument about the ε-independency of the time interval given above imply

$$C\tilde{X}_\varepsilon^2(T) - \tfrac{1}{2}\tilde{X}_\varepsilon(T) + \varepsilon + C\|U_0\|_{X_\gamma} \ge 0, \quad T \in [0, \min\{T^*; \mathcal{T} - mT^*\}].$$

By assumption $\|U_0\|_{X_\gamma} \le c_{mT^*} X_\varepsilon(mT^*) \le c_{mT^*} \varepsilon 2K_m$. Since $\tilde{X}_\varepsilon(0) = 0$ and $\tilde{X}_\varepsilon$ is continuous in $[0, \min\{T^*; \mathcal{T} - mT^*\}]$, we may solve the quadratic inequality for $\varepsilon < (16C(1 + 2Cc_{mT^*}K_m))^{-1}$ and obtain $\tilde{X}_\varepsilon \le \varepsilon 2(1 + 2Cc_{mT^*}K_m)$ in

[0, min{T^*; $\mathcal{T} - mT^*$}]. Hence, by the assumption on m

$$X_\varepsilon^p(T) \le (\varepsilon 2(1 + 2Cc_{mT^*}K_m))^p + X_\varepsilon^p(mT^*)$$

$$\le 2\left[2\varepsilon(1 + 2Cc_{mT^*}K_m) + \varepsilon 2K_m\right]^p = (\varepsilon 2K_{m+1})^p,$$

for all $T \in [mT^*, \min\{(m+1)T^*; \mathcal{T}\}]$. The assumption on m implies $X_\varepsilon \le \varepsilon 2K_{m+1}$ in $[0, \min\{(m+1)T^*; \mathcal{T}\}]$. By induction we get $X_\varepsilon \le \varepsilon 2K_M$ in $[0, \mathcal{T}]$ with $M = [\frac{\mathcal{T}}{T^*}]$. The proof of Theorem 1.16.1 is complete. $\square$

1.17   Notes

Sections 1.1 and 1.2   The content of Sect. 1.1 is very standard and can be found in many books, for example [86] or [141]. Excellent sources for more information on Fourier transforms, Fourier multipliers and the Hilbert transform are [45, 69, 140] and [14]. The corresponding vector-valued setting is also well studied and documented. In fact, basic properties on Bochner integrals and Banach space valued L^p-spaces can be found e.g. in Section 1.1 of [13]. Properties of Banach space valued distributions and Fourier transforms are described in detail e.g. in [8, 12, 95] or [87]. It is remarkable that the fundamental theorem on scalar valued singular integrals based on Calderon-Zygmund theory extends without difficulties to the Banach space valued setting described in Theorem 1.2.3.

A very thorough study of UMD-spaces can be found in the monograph by Hytönen et al. [87]. Their significant role in the theory of scalar- or vector-valed L^p-Fourier multipliers was recognized by Bourgain in [22] and is described in detail in Chapter I of [95], see also [88]. We only note here that UMD spaces are necessarily reflexive spaces and that the converse is not true. Due to the work of Burkholder [24] it is known that a Banach space X is a UMD-space if and only if X is ζ-convex in the sense that there exists a symmetric, biconvex function $\zeta : X \times X \to \mathbb{R}$ satisfying $\zeta(0, 0) > 0$ and $\zeta(x, y) \le \|x + y\|$ for $x, y \in X$ with $\|x\| = \|y\| = 1$. The abbreviation UMD stands for the property of unconditional martingale differences and it was shown by Bourgain [20] that spaces possessing this property coincide with UMD-spaces as defined in Sect. 1.2. There are many assertions in vector-valued harmonic analysis and probability theory which are equivalent to the UMD property, see e.g. [23] and [87].

Section 1.3   The theory of semigroups is well described in various books. The approach described in Sect. 1.3 based on Laplace transforms is described in detail in [13]. For other approaches see e.g. [48] or [117]. The books [8] and [106] concentrate on analytic semigroups and their properties. A short introduction to the theory of semigroups can be found e.g. in [50].

Section 1.4 The concept of sectorial operators is again a classical one and was already used many authors such as Hille and Phillips, Triebel [148] and Tanabe [144] and Kato [92].

Operators admitting a bounded H^∞-functional calculus have been introduced by McIntosh [113], where also the convergence lemma, Proposition 1.4.5, is proved. Excellent sources for further properties of the H^∞-calculus are [71, 95] and [88].

The examples presented in Example 1.4.9 and 1.4.10 are standard and are based on Mikhlin's theorem. More complicated examples of operators admitting a bounded H^∞-calculus are discussed in Proposition 1.4.11. Its proof is based on the following vector-valued version of the transference principle from harmonic analysis. Note that for an operator-valued kernel $k \in L^1(\mathbb{R}, \mathcal{L}(X))$ the convolution operator $S_k : L^p(\mathbb{R}; X) \to L^p(\mathbb{R}; X)$ is defined for $1 \leq q \leq \infty$ by

$$S_k f(t) = \int_{-\infty}^{\infty} k(s)[f(t-s)]ds, \quad f \in L^p(\mathbb{R}; X).$$

Let U be a C_0-group of bounded operators on a Banach space X with

$$\sup_{t \in \mathbb{R}} \|U(t)\|_{\mathcal{L}(X)} \leq C_U < \infty,$$

and assume that $k \in L^1(\mathbb{R}; \mathcal{L}(X))$ is satisfying $k(t)U(s) = U(s)k(t)$ for all $s, t \in \mathbb{R}$. Then, given $x \in X$, the mapping $T_k : x \mapsto \int_{\infty}^{\infty} k(t)[U(t)]x dt$ defines a bounded operator on X such that

$$\|T_k\|_{\mathcal{L}(X)} \leq C_U^2 \|S_k\|_{\mathcal{L}(L^p(X))}.$$

For details see [77] and [88].

Another important class of operators admitting a bounded H^∞-calculus is closely related to boundary value problems of order $2m$ subject to general boundary conditions. They are of the form

$$\lambda u + \mathcal{A}(x, D)u = f \quad \text{in } \Omega,$$
$$\mathcal{B}_j(x, D)u = g_j \quad \text{on } \partial\Omega, \quad j = 1, \ldots, m.$$

Here $\Omega \subset \mathbb{R}^{n+1}$ is a domain with compact, smooth boundary, f and g_j are given functions, $\mathcal{A}$ is a differential operator of order $2m$ of the form $\mathcal{A}(x, D) = \sum_{|\alpha| \leq 2m} a_\alpha(x) D^\alpha$ and $\mathcal{B}_j(x, D) = \sum_{|\beta| \leq m_j} b_{j,\beta}(x) D^\beta$ with $m_j \leq 2m$ for $j = 1, \ldots m$. We denote by A_B the realization of $\mathcal{A}(x, D)$ in $L^p(\Omega)$ with domain

$$D(A_B) = \{u \in H_p^{2m}(\Omega) : \mathcal{B}_j(x, D)u = 0 \text{ on } \partial\Omega, j = 1, \ldots, m\}.$$

Assume that the top-order coefficients of $\mathcal{A}(x, D)$ are Hölder continuous and that the above boundary value problem satifies certain ellipticity, smoothness as well as the Lopatinskii-Shapiro condition. It was then shown in [37] that, roughly speaking, there exists $\mu > 0$ such that $A_B + \mu$ admits a bounded H^∞-calculus on $L^p(\Omega)$. For the precise assumptions and the precise assertion, we refer to Theorem 2.3 of [37].

At this point we refer further to a result by Duong and Li [46] on bounded H^∞-calculus for elliptic operators on $\mathbb{R}^n$ satisfying only very weak smoothness assumption on the coefficients, more precisely only a smallness condition on the BMO-norm of the top-order coefficients.

Section 1.5 For general expositions concerning fractional powers of operators we refer e.g. to the books of Triebel [148] and Amann [8]. The approach described in Sect. 1.5 uses an extended H^∞-calculus which is described in more detail e.g. in Section I.2 of [36] or in Appendix B of [95]. We follow here closely the presentation in [95] in which one finds also additional information on representation formulas for A^α and on interpolation and extrapolation scales.

The characterization of the domain $D(A^\alpha)$ of a sectorial operators A in terms of complex interpolation spaces is important in many applications. It should be noted that Theorem 1.5.6 holds true under the weaker assumption that A admits bounded imaginary powers. Observe that a sectorial operator is said to admit *bounded imaginary powers* if $A^{is} \in \mathcal{L}(X)$ for all $s \in \mathbb{R}$ and there exists a constant $C > 0$ such that $\|A^{is}\| \leq C$ for all s with $|s| \leq 1$. It seems that Theorem 1.5.6 goes back to [148]. The reiteration theorem in interpolation theory yields in particular

$$[X_\alpha, X_\beta]_\theta = X_{\alpha(1-\theta)+\theta\beta}, \qquad 0 \leq \alpha < \beta \leq 1, \theta \in (0,1),$$

for the fractional power spaces associated to operators A having bounded imaginary powers. The latter class of operators plays a central role in the Dore–Venni approach to maximal L^p-regularity of evolution equations.

Section 1.6 The operator-valued H^∞-calculus in its general form as presented in Sect. 1.6 was developed in a fundamental article by Kalton and Weis [91] in 2001. Their approach uses in an essential way the notation of $\mathcal{R}$-boundedness of a set of bounded operators. It is shown that the functional calculus for scalar-valued bounded holomorphic functions described in Sect. 1.4 can be extended to operator-valued functions F provided the range $\{F(\lambda) : \lambda \in \Sigma_\theta\}$ of the function F is not only bounded but $\mathcal{R}$-bounded. The notion of $\mathcal{R}$-boundedness is implicitly already contained in the article [21]. In [17], this property was called "Riesz-property"; the name "randomized boundedness" appears in [28] and it this property is also often also known as "Rademacher boundedness". For detailed proofs of the assertions given in Remark 6.2, Lemmas 6.3 and 6.4, we refer to [36, 95] or [88].

The Key Lemma 1.6.7 is also being used in a crucial way in the proof of the boundedness of H^∞-calculus for operators associated to general boundary value problems subject to general boundary conditions describred above.

Proposition 1.6.8 asserts that the classes $\mathcal{H}^\infty(X)$ and $\mathcal{R}\mathcal{H}^\infty(X)$ coincide for Banach spaces satisfying the so-called property (α). Banach spaces having this property were introduced by Pisier in 1978, see [119]. The property (α) also plays a crucial role in the theory of vector-valued Fourier multipliers. This was already observed by Zimmermann in [156] in 1989. In particular, the vector-valued Fourier multiplier theorems stated in Theorem 1.6.13 and 1.6.14 require the additional assumption that X possesses property (α). For more information on this and on Fourier multipliers we refer to [11, 36, 124, 154] and [88].

The definition of an $\mathcal{R}$-bounded H^∞-calculus is a natural extension of the concept of the bounded H^∞-calculus. It appeared in an article by Desch, Hieber and Prüss [40] on the Stokes problem in the half space $\mathbb{R}_+^n$ and in the abstract setting by Kalton and Weis in [91]. The fact that the Laplacian or more generally, parameter elliptic operators of angle $\varphi_A < \pi$, admit an $\mathcal{R}$-bounded H^∞-calculus on $L^p(\mathbb{R}^n)$ was shown in [40] by combining kernel estimates with Theorem 1.2.3. Of course, taking into account the property (α) of $L^p(\mathbb{R}^n)$, this follows immediately also from Proposition 1.6.8.

Our approach to the problem of maximal L^p-regularity for evolution equations of the form $u'(t) + Au(t) = f(t), t > 0, u(0) = 0$ is based on the beautiful Kalton–Weis sum theorem which was proved in [91]. The problem whether the sum of two resolvent commuting operators is closed was already investigated in 1975 by Da Prato and Grisvard in [32]. They proved a maximal L^p-regularity result for real interpolation spaces between X and $D(A)$ via the sum method. By the same method, Dore and Venni [44] succeeded in 1987 in proving that $A + B$ is closed provided X is a UMD space and A as well as B admit bounded imaginary powers with power angles θ_A and θ_B such that $\theta_A + \theta_B < \pi$. The Kalton–Weis theorem requires thus less for A, namely $\mathcal{R}$-Boundedness instead of bounded imaginary powers for A, but more on B, namely bounded H^∞-calculus instead of bounded imaginary powers. In order to deduce a maximal L^p-regularity estimate of the form

$$\|u'\|_{L^p((0,T);X)} + \|Au\|_{L^p((0,T);X)} \le C_p \|f\|_{L^p((0,T);X)},$$

from the sum theorem, we need to show that the derivative operator $B = d/dt$ has a bounded H^∞-calculus on $L^p(\mathbb{R}_+; X)$. Assuming that X has the UMD property, it follows easily from the vector-valued Fourier multiplier Theorem 1.6.11 that B has a bounded H^∞-calculus on $L^p(\mathbb{R}_+; X)$ for all angles $\phi > \pi/2$.

It is interesting to note that the UMD property of X is not only sufficient for this assertion but also necessary. Thus, the UMD property may also be characterized in terms of spectral theory.

The notion of $\mathcal{R}$-sectorial operators goes back to Clément and Prüss [29] and Weis [154].

The operator-valued version of Mikhlin's theorem was proved first by Weis in [154]. For detailed treatments also of the n-dimensional situation, we refer to Chapter I.3 and I.4 of [36], Chapters 1.3 and 1.4 of [95] and [124].

Let us finally mention a maximal $L^1(0, T; \dot{B}^s_{p,1}(\mathbb{R}^n))$-regularity result for the solution of the heat equation on $\mathbb{R}^n$ for certain homogeneous Besov spaces which is due to Danchin and Mucha, see [30, 31]. More precisely, let $\varphi : \mathbb{R}^n \to [0, 1]$ be a function with support in $\{\xi \in \mathbb{R}^n : 1/2 \le |\xi| \le 2\}$ such that $\sum_{j \in \mathbb{Z}} \varphi(2^{-j}\xi) = 1$ for all $\xi \ne 0$. Then the Littlewood-Paley decomposition $((\dot{\Delta})_j)_{j \in \mathbb{Z}}$ over $\mathbb{R}^n$ is given by

$$\dot{\Delta}_j u := \mathcal{F}^{-1}(\varphi(2^{-j}\cdot)\mathcal{F}u),$$

where $\mathcal{F}$ denotes the Fourier transform. Moreover, we denote by $\mathcal{S}_h'(\mathbb{R}^n)$ the set of all tempered distributions u over $\mathbb{R}^n$ such that for all functions $\theta \in C_c^\infty(\mathbb{R}^n)$ one has $\lim_{\lambda \to \infty} \theta(\lambda D)u = 0$ in $L^\infty(\mathbb{R}^n)$. For $u \in \mathcal{S}_h(\mathbb{R}^n)$, $s \in \mathbb{R}$ and $1 \le p, r \le \infty$ set

$$\|u\|_{\dot{B}_{p,r}^s(\mathbb{R}^n)} := \|2^{sj}\|\dot{\Delta}_j u\|_{L^p(\mathbb{R}^n)}\|_{l^r(\mathbb{Z})}$$

Then the *homogeneous Besov space* $\dot{B}_{p,r}^s(\mathbb{R}^n)$ is defined as

$$\dot{B}_{p,r}^s(\mathbb{R}^n) := \{u \in \mathcal{S}_h'(\mathbb{R}^n) : \|u\|_{\dot{B}_{p,r}^s(\mathbb{R}^n)} < \infty\}.$$

Consider the Gaussian semigroup $T(t) = e^{t\Delta}$ introduced in Example 1.3.13b). Then it is not difficult to show that there exist two constants $c, C > 0$ such that for all $j \in \mathbb{Z}$

$$\|e^{t\Delta}\dot{\Delta}_j h\|_{L^p(\mathbb{R}^n)} \le Ce^{-c2^{2j}}\|\dot{\Delta}_j h\|_{L^p(\mathbb{R}^n)}.$$

Hence, by the variation of constant formula, the solution u of the inhomogeneous heat equation

$$u_t - \Delta u = f, \quad t \in J := (0, T), x \in \mathbb{R}^n, \quad u(0) = u_0 \qquad (1.17.1)$$

satisfies

$$\|\dot{\Delta}_j u\|_{L^\infty(J;L^p(\mathbb{R}^n))} + 2^{2j}\|\dot{\Delta}_j u\|_{L^1(J;L^p(\mathbb{R}^n))} \le C\big(\|\dot{\Delta}_j u_0\|_{L^p(\mathbb{R}^n)} + \|\dot{\Delta}_j f\|_{L^1(J;L^p(\mathbb{R}^n))}\big)$$

Multiplying with 2^{js} and summing over j yields the following maximal regularity estimate

$$\|u\|_{L^\infty(J;\dot{B}_{p,1}^s(\mathbb{R}^n))} + \|u'\|_{L^1(J;\dot{B}_{p,1}^s(\mathbb{R}^n))} + \|\Delta u\|_{L^1(J;\dot{B}_{p,1}^s(\mathbb{R}^n))}$$

$$\le C\big(\|f\|_{L^1(J;\dot{B}_{p,1}^s(\mathbb{R}^n))} + \|u_0\|_{\dot{B}_{p,1}^s(\mathbb{R}^n)}\big). \qquad (1.17.2)$$

We thus proved the following result: Let $s \in \mathbb{R}$, $p \in [1, \infty]$ and assume that $f \in L^1(J; \dot{B}_{p,1}^s(\mathbb{R}^n))$ and $u_0 \in \dot{B}_{p,1}^s(\mathbb{R}^n)$. Then the heat Eq. (1.17.1) admits a unique solution in $C([0, T); \dot{B}_{p,1}^s(\mathbb{R}^n))$ satisfying

$$u_t, \Delta u \in L^1(J; \dot{B}_{p,1}^s(\mathbb{R}^n))$$

and there is a constant $C > 0$ such that (1.17.2) is satisfied. We finally note that an estimate of the form (1.17.2) *cannot* hold true if the space $\dot{B}_{p,1}^s(\mathbb{R}^n)$ is replaced by any reflexive space and in particular by an L^p space for $1 < p < \infty$. Let us emphasize that estimates of the form (1.17.2) can be deduced from a more abstract point of view also from the classical Da Prato-Grisvard theorem [32], which is valid also for the case $p = 1$.

Section 1.7 There is a huge amount of articles dealing with abstract quasilinear evolution equations. Early work by the Japanese school founded by Kato and Tanaba and the Russian school founded by Sobolevskii was continued by Amann, von Wahl, Da Prato, Lundardi and many others. We refer here e.g. to the monographs [8, 106] and [144] and to a series of articles by Amann started with [6]. The results described in this section go back to Clément and Li [27] and [120]. For a very thorough study of quasilinear parabolic problems (including the theory of time weights and the principle of generalized linearized stability due to Prüss, Simonett and Zacher [126]) we refer to the excellent monograph by Prüss and Simonett [124].

Section 1.8 The content of this section is rather standard and can be found in many textbooks, for example [116]. For a recent comprehensive investigation, see [109].

Section 1.9 The Stokes equation in the halfspace $\mathbb{R}^n_+$ often serves as a model problem for this equation in bounded, exterior or more complicated types of domains. Indeed, the localization procedure described in Sect. 1.10 is based on a good understanding of this or the corresponding resolvent equation. Following [40], we describe an approach which differs from the ones given e.g. by Solonnikov [135], McCracken [112] and Ukai [151]. The advantage of our approach is twofold: first, the representation formula for the solution of the Stokes resolvent problem developed in Sect. 1.9 allows to show by fairly easy means that $-A$ admits an $\mathcal{R}$-bounded H^∞-calculus on $L^p_\sigma(\mathbb{R}^n_+)$, where A denotes the Stokes operator in $L^p_\sigma(\mathbb{R}^n_+)$ defined as in (1.9.4).

Secondly, our representation allows to deduce L^∞-estimates for the solution of the Stokes resolvent problem which imply that the Stokes operator in $L^\infty_\sigma(\mathbb{R}^n_+)$ generates a holomorphic semigroup on this space (which, of course, is not strongly continuous in 0). Note that the usual localization procedure does not work in this setting, which means that generator results for the Stokes semigroup on $L^\infty_\sigma(\Omega)$, Ω a bounded or exterior domain with smooth boundary, are much more difficult to obtain. Recently this problem was solved by Abe and Giga [1]. Their proof is based on a contradiction argument. For an approach based on resolvent estimates which is inspired by the Masuda-Stewart technique for elliptic operator, see [2].

Section 1.10 There is a vast amount of literature concerning the linear Stokes equations. An excellent sourse of information are the monographs by Galdi [57], Sohr [134], Robinson et al. [128] and Tsai [150]. Pioneering results in the context of L^p-spaces are due to Solonnikov [135], Giga [62, 63], Giga and Sohr [64], Farwig and Sohr [51] and Shibata and Shimizu [130] and Abels [3]. For a recent survey on this topic we refer to the work of Hieber and Saal [79].

Section 1.11 An excellent source for mathematical models and their analysis arising in geophysics is the book by Chemin et al. [26]. They proved the following very remarkable fact concerning the Navier–Stokes equations in the rotational setting in

$\mathbb{R}^3$: given $u(0) \in L^2(\mathbb{R}^2)^3 + H^{1/2}(\mathbb{R}^3)^3$, there exists $\Omega_0 > 0$ such that for every $\Omega \in \mathbb{R}$ with $|\Omega| \geq \Omega_0$, the equation

$$\begin{cases} \partial_t u - \nu \Delta u + \Omega e_3 \times u + (u \cdot \nabla)u + \nabla p = 0, & t > 0, \ x \in \mathbb{R}^3, \\ \qquad\qquad\qquad\qquad\qquad \text{div} u = 0, & t > 0, \ x \in \mathbb{R}^3, \\ \qquad\qquad\qquad\qquad\quad u(0, x) = u_0, \ x \in \mathbb{R}^3, \end{cases} \tag{1.17.3}$$

admits a unique, global solution. Their proof is based on dispersive estimates for certain terms in the solution operator of the linear part due to the Coriolis force.

Global well-posedness results for the above equation for arbitrary Ω but initial data being small with respect to $H^{1/2}$ or other function spaces go back to Hieber and Shibata [81] and Giga et al. [67].

For more information on geophysical fluids we refer also to the book by Pedlosky [118].

It is a remarkable fact that the Navier–Stokes equations in the rotational seeting, i.e. with an additional term representing the Coriolis force, admit an explicit stationary solution. This stationary solution is called the *Ekman spiral* and the study of its stability properties is an important task, see e.g. [26, 42].

In this context, the Stokes-Coriolis-Ekman semigroup arises naturally. It seems to be unknown whether the Stokes-Coriolis-Ekman semigroup defined as in (1.11.2) remains a bounded semigroup on $L^2_\sigma(\mathbb{R}^n_+)$ for large Reynolds numbers. An answer to this question would be helpful for clarifying questions related to the stability/instability of the Ekman spiral for large Reynolds numbers. Our presentation here follows [72] and [93]. The latter booklet studies more generally stationary solutions of the rotating Navier–Stokes-Boussinesq equations with stratification effects, i.e.

$$\begin{cases} \partial_t u - \nu \Delta u + \Omega e_3 \times u + (u \cdot \nabla)u + \nabla p = G e_3, & t > 0, \ x \in \mathbb{R}^3_+, \\ \qquad\quad \partial_t \theta - \kappa \Delta \theta + (u \cdot \nabla)\theta = -N^2 u_3, & t > 0, \ x \in \mathbb{R}^3_+, \\ \qquad\qquad\qquad\qquad\qquad\qquad \text{div} u = 0, & t > 0, \ x \in \mathbb{R}^3_+, \\ \qquad\quad u(t, x_1, x_2, 0) = (a_1, b_1, 0), & t > 0, \ x_1, x_2 \in \mathbb{R}, \\ \qquad\qquad \theta(t, x_1, x_2, 0) = c_1, & t > 0, \ x_1, x_2 \in \mathbb{R}, \end{cases}$$

$$\tag{1.17.4}$$

subject to initial data. Here G and N represent the gravitational force and the Brunt-Väisälä frequency.

For further results concerning Ekman layers in rotating fluids we refer to the work of Masmoudi [111] and Rousset [129].

Hieber and Stannat considered in [82] also the situation of the stochastic Navier–Stokes-Coriolis equation in $\mathbb{T}^2 \times (0, b)$. They proved, as a stochastic analogue of the deterministic stability result described in Theorem 1.11.7, stochastic stability of the Ekman spiral by considering stationary martingale solutions.

Section 1.12 The analysis of the motion of a body immersed in a liquid is a classical problem in fluid mechanics. The methods used depend on the assumptions whether the body is considered to be rigid or elastic or whether the fluid is incompressible, compressible, viscolelastic or has a complex stress tensor. For a survey of known results and methods concerning the stationary and instationary problem for Newtonian fluids and for prescribed, steady, self-propelled and free movements, see e.g. the articles by Galdi [56] and Galdi and Neustupa [58].

There are several possibilities to transform this *moving domain problem* to a probem on a fixed domain. One possible transformation was introduced by Galdi [56]: it is linear and the whole space is rotated and shifted back to its original position at every time $t > 0$. This transformation generates in the fluid equations a drift term with unbounded coefficients, i.e. the new fluid operator is of the form

$$Lu := P(\Delta u + (\omega \times x \cdot \nabla)u - \omega \times u)$$

in the purely rotational case. Here P denotes the Helmholtz projection and ω the rotational velocity. One of the fundamental difficulties of this approach is that the transformed problem is *no* longer parabolic.

A second approach is characterized by a non-linear, "local" change of coordinates which only acts in a suitable bounded neighborhood of the obstacle. Tucsnak, Cumsille and Takahashi used this transform going back to Inoue and Wakimoto [89] and showed the existence of a unique, local strong L^2-solution to the fluid rigid body problem in two and three space dimensions, see [142, 143]. The L^p-theory of this approach was developed in [61] for Newtonian and also for generalized Newtonian fluids. For weak solutions, see e.g. [41] and [52]. Section 1.12 follows the approach based on this second change of variables, however now in the situation of a compressibe fluid, see [74]. For a different approach, see [19].

The methods for investigating the case of an elastic body are again different. We refer here e.g. to the fundamental article by Raymond and Vanninathan [127].

Section 1.13 There is an enormous amount of literature concerning free boundary value problems for viscous fluids. The monograph [124] is an excellent source of information about the state of the art concerning strong solutions and sharp interfaces. Pioneering results for Newtonian fluids in the one phase setting go back to Solonnikov [136, 137], Beale [15], Tani and Tanaka [146] and Shibata and Shimizu [131]; for the two-phase situation we refer to Denisova [34] and Prüss and Simonett [122]. For the spin-coating problem, see [39].

Problems of this kind for non-Newtonian fluids were treated by Abels [4] in the context of measure-valued varifold solutions. The approach presented here for generalized non-Newtonian fluids follows [80].

Section 1.14 The Ericksen-Leslie models describing the hydrodynamic flow of nematic liquid crystal flow was pioneered by Ericksen [49]. Starting from the classical balance laws for mass, linear and angular momentum he introduced the a set of equation describing the evolution of liquid crystals. The system was completed by Leslie [97] with the addition of constitutive equations. For more information see the book by Virga [152] and the survey article [102] by Lin and Wang. For a derivation of this model from thermodynamical principles we refer to the survey paper [77]. The latter and [152] contain many further results and references. The mathematical analysis of the simplified system started with the work of Lin and Liu, see [100] and [101]. The approach to the simplified system presented here is taken form [84].

The non-isothermal Ericksen-Leslie model for incompressible fluids subject to general Leslie stress S_L and general Ericksen tensor can be found e.g. in [77]. It is shown to be thermodynamically consistent. The well-posedness results concerning the situation of general Leslie stress go back to [76] and [78], see also [77]. Observe that these results do not make any structural assumptions on the Leslie coefficients and that Parodi's condition is not assumed in contrast to previous work, see e.g. [105] and [33].

For results dealing with anisotropic elasticity, i.e. with the *general Ericksen tensor S_E* but with vanishing Leslie tensor S_L, in particular without stretching, see the recent work of Hong et al. [85] and Ma, Li and Gong [107]. Let us mention here also that in the case of $\Omega = \mathbb{R}^n$, the local existence result given in Theorem 1.14.9 extends to the situation of general Ericksen stress tensor, see [78].

Sections 1.15 and 1.16 The mathematical analysis of the primitive equations was initiated by Lions, Teman and Wang in a series of articles [103, 104]. For a survey of known results and further references, we refer to an survey article by Li and Titi [99]. A breakthrough result in this contex was proved by Cao and Titi [25] in 2007. They proved that the primitive equations admit a unique, global, strong solution for arbitrary large data in H^1. A different approach to the primitive equations based on methods of evolution equations has been initiated in [73]. The results presented in Sects. 1.15 and 1.16 go back to the work of Giga et al. [65] and Furukawa et al. [55].

References

1. K. Abe, Y. Giga, Analyticity of the Stokes semigroup in spaces of bounded functions. Acta Math. **211**, 1–46 (2013)
2. K. Abe, Y. Giga, M. Hieber, Stokes resolvent estimates in spaces of bounded functions. Ann. Sci. Ec. Norm. Super. **48**, 521–543 (2015)
3. H. Abels, Bounded imaginary powers and H_∞-calculus for the Stokes operator in two-dimensional exterior domains. Math. Z. **251**, 589–605 (2005)
4. H. Abels, On generalized solutions of two-phase flows for viscous incompressible fluids. Interfaces Free Bound. **9**, 31–65 (2007)
5. H. Abels, L. Dienig, Y. Terasawa, Existence of weak solutions for a diffusive interface models of non-Newtonian two-phase flows, Nonlinear Anal. Ser. B, **15**, 149–157 (2014)
6. H. Amann, Dynamic theory of quasilinear parabolic equations I: abstract evolution equations. Nonlinear Anal. **12**, 895–919 (1988)
7. H. Amann, Stability of the rest state of a viscous incompressible fluid. Arch. Ratational Mech. Anal. **126**, 231–242 (1994)
8. H. Amann, *Linear and Quasilinear Parabolic Problems*, vol. I (Birkhäuser, Basel, 1995)

9. H. Amann, Stability and bifurcation in viscous incompressible fluids. Zapiski Nauchn. Seminar. POMI **233**, 9–29 (1996)
10. H. Amann, On the strong solvability of the Navier–Stokes equations. J. Math. Fluid Mech. **2**, 16–98 (2000)
11. H. Amann, *Anisotropic Function Spaces and Maximal Regularity for Parabolic Problems.* Necas Center for Mathematical Modeling Lecture Notes (Prague, 2009)
12. H. Amann, *Linear and Quasilinear Parabolic Problems*, vol. II (Birkhäuser, Basel, 2019)
13. W. Arendt, Ch. Batty, M. Hieber, F. Neubrander, *Vector-Valued Laplace Transforms and Cauchy Problems*, 2nd edn. (Birkhauser, Basel, 2011)
14. H. Bahouri, J, Chemin, R. Danchin, *Fourier Analysis and Nonlinear Partial Differential Equations*, vol. 343 (Springer, Grundlehren, 2011)
15. J. T. Beale, Large-time regularity of viscous surface waves. Arch. Rational Mech. Anal. **84**, 307–352 (1983–1984)
16. A. Benedek, A.P. Calderón, R. Panzone, Convolution operators on Banach space valued functions. Proc. Nat. Acad. Sci. USA **48**, 356–365 (1962)
17. E. Berkson, T. Gillespie, Spectral decompositions and harmonic analysis on UMD spaces. Studia Math. **112**, 13–49 (1994)
18. D. Bothe, J. Prüss, L^p-theory for a class of non-Newtonian fluids. SIAM J. Math. Anal. **39**, 379–421 (2007)
19. M. Boulakia, S. Guerrero, A regularity result for a solid-fluid system associated to the compressible Navier–Stokes equations. Ann. Inst. H. Poincaré Anal. Non Linéaire **26**, 777–813 (2009)
20. J. Bourgain, Some remarks on Banach spaces in which martingale difference sequences are unconditional. Ark. Mat. **21**, 163–168 (1983)
21. J. Bourgain, Vector-valued singular integrals and the $H^1 - BMO$ duality, in *Probabilty Theory and Harmonic Analysis* (Marcel-Dekker, New York, 1986), pp. 1–19
22. J. Bourgain, Vector-valued Hausdorff-Young inequalities and applications, in *Geometric Aspects of Functional Analysis* (Springer, Berlin, 1988), pp. 239–249
23. D. Burkholder, Martingale transforms and the geometry of Banach spaces, in *Probability in Banach Spaces, III* (Springer, Berlin, 1981), pp. 35–50
24. D. Burkholder, A geometrical condition that implies the existence of certain singular integrals of Banach-space-valued functions, in ed. by W. Beckner, A.P. Calderón, R. Fefferman, P.W. Jones. Confernece Harm Anal., 1981, pp. 270–286, Wadsworth, 1983.
25. Ch. Cao, E. Titi, Global well–posedness of the three-dimensional viscous primitive equations of large scale ocean and atmosphere dynamics. Ann. Math. **166**, 245–267 (2007)
26. J. Chemin, B. Desjardins, I. Gallagher, E. Grenier, *Mathemtical Geophysics*. Oxford Lecture Series in Mathematics and Its Applications, vol. 32 (Oxford University Press, Oxford, 2006)
27. P. Clément, S. Li, Abstract parabolic quasilinear equations and application to a groundwater flow problem. Adv. Math. Sci. Appl. **3**, 17–32 (1993–1994)
28. P. Clément, B. de Pagter, F. Sukochev, H. Witvliet, Schauder decomposition and multiplier theorems. Studia Math. **138**, 135–163 (2000)
29. P. Clément, J. Prüss, An operator-valued transference principle and maximal regularity on vector-valued L^p-sapces, in *Evolution Equations*, ed. by G. Lumer, L. Weis. Lectures Notes Pure Applied Mathematics, vol. 215 (Marcel Dekker, New York, 2001), pp. 67–87
30. R. Danchin, P. Mucha, A critical functional framework for the inhomogeneous Navier–Stokes equations in the half space. J. Funct. Anal. **256**, 881–927 (2009)
31. R. Danchin, P. Mucha. Critical functional framework and maximal regularity in action on systems of incompressible flows. Memoirs Soc. Math. France **143**, 151 (2015)
32. G. Da Prato, P. Grisvard, Sommes d'opérateurs linéaires et équations différentielles opérationelles. J. Math. Pures Appl. **54**, 305–387 (1975)
33. F. De Anna, C. Liu, Non-isothermal general Ericksen-Leslie system: derivation, analysis and thermodynamic consistency. Arch. Rational Mech. Anal. **231**, 637–717 (2019)
34. I. Denisova, A priori estimates for the solution of the linear nonstationary problem connected with the motion of a drop in a liquid medium, Proc. Stekhlov Inst. Math. **3**, 1–24 (1991)

35. I. Denisova, Problem of the motion of two viscous incompressible fluids separated by a closed free interface, Acta Appl. Math. **37**, 31–40 (1994)

36. R. Denk, M. Hieber, J. Prüss, $\mathcal{R}$-Boundedness, Fourier multipliers and problems of elliptic and parabolic type. Mem. Amer. Math. Soc. **166**, 144 (2003)

37. R. Denk, G. Dore, M. Hieber, J. Prüss, A. Venni, New thoughts on old ideas of R.T. Seeley. Math. Annalen, **166**, 545–583 (2004)

38. R. Denk, M. Hieber, J. Prüss, Optimal L^p-L^q-estimates for parabolic boundary value problem with inhomogeneous data. Math. Z. **257**, 193–224 (2007)

39. R. Denk, M. Geissert, M. Hieber, J. Saal, O. Sawada, The spin-coating process: analysis of the free boundary value problem. Comm. Partial Differ. Equ. **36**, 1145–1192 (2011)

40. W. Desch, M. Hieber, J. Prüss, L^p-theory of the Stokes equation in a half-space. J. Evol. Equ. **1**, 115–142 (2001)

41. B. Desjardins, M. Esteban, On weak solutions for fluid rigid structure interaction: compressible and incompressible models. Comm. Partial Differ. Equ. **25**, 1399–1413 (2000)

42. B. Desjardins, E. Dormy, E. Grenier, Stability of mixed Ekman-Hartmann boundary layers. Nonlinearity **12**, 181–199 (1999)

43. L. Diening, M. Ruzicka, Strong solutions for generalized Newtonian fluids. J. Math. Fluid Mech. **7**, 413–450 (2005)

44. G. Dore, A. Venni, On the closedness of the sum of two closed operators. Math. Z. **196**, 189–201 (1987)

45. J. Duistermaat, J. Kolk, *Distributions*. Cornerstones Series (Birkhäuser, New York, 2011)

46. X. Duong, L. Yan, Bounded holomorphic functional calculus for non-divergence form differential operators. Diff. Int. Equ. **15**, 709–730 (2002)

47. V.W. Ekman, On the influence of the earth's rotation on ocean currents. Arkiv Matem. Astr. Fysik, (Stockholm) **11**, 1–52 (1905)

48. K.-J. Engel, R. Nagel, *One-Parameter Semigroups for Linear Evolution Equations* (Springer, Berlin, 2000)

49. J. L. Ericksen, Hydrostatic theory of liquid crystals. Arch. Ration. Mech. Anal. **9**, 371–378 (1962)

50. L.C. Evans, *Partial Differential Equations* (American Mathematical Society, Providence, 1998)

51. R. Farwig, H. Sohr, Generalized resolvent estimates for the Stokes operator in bounded and unbounded domains. J. Math. Soc. Japan **46**, 607–643 (1994)

52. E. Feireisl, On the motion of rigid bodies in a viscous compressible fluid. Arch. Ration. Mech. Anal. **167**, 281–308 (2003)

53. E. Feireisl, E. Rocca, G. Schimperna, On a non-isothermal model for nematic liquid crystals. Nonlinearity **24**, 243–257 (2011)

54. E. Feireisl, M. Frémond, E. Rocca, G. Schimperna, A new approach to non-isothermal models for nematic liquid crystals. Arch. Ration. Mech. Anal. **205**, 651–672 (2012)

55. K. Furukawa, Y. Giga, M. Hieber, A. Hussein, T. Kashiwabara, M. Wrona, Rigorous justification of the hydrostatic approximation for the primitive equations by scaled Navier–Stokes equations. Submitted.

56. G.P. Galdi, On the motion of a rigid body in a viscous liquid: a mathematical analysis with applications, in *Handbook of Mathematical Fluid Dynamics*, ed. by S. Friedlander, D. Serre, vol. I (Elsevier, North-Holland, 2002), pp. 653–791

57. G.P. Galdi, *An Introduction to the Mathematical Theory of the Navier–Stokes Equations. Steady State Problems*, 2nd edn. (Springer, New York, 2011)

58. G.P. Galdi, J. Neustupa, Steady state Navier–Stokes flow around a moving body, in *Handbook of Mathematical Analysis in Mechanics of Viscous Fluids*, ed. by Y. Giga, A. Novotny, vol. 1 (Springer, Beriln, 2018), pp. 341–418

59. M. Geissert, M. Hess, M. Hieber, C. Schwarz, K. Stavrakidis, Maximal L^p–L^q-estimates for the Stokes equation: a short proof of Solonnikov's theorem. J. Math. Fluid Mech., **12**, 47–60 (2010)
60. M. Geissert, H. Heck, M. Hieber, O. Sawada, Weak Neumann implies Stokes. J. Reine Angew. Math. **669**, 75–100 (2012)
61. M. Geissert, K. Götze, M. Hieber, L^p-theory fro strong solutions to fluid-rigid body interaction in Newtonian and generalized Newtonian fluids. Trans. Amer. Math. Soc. **365**, 1393–1439 (2013)
62. Y. Giga, Analyticity of the semigroup generated by the Stokes operator in L^p-spaces. Math. Z. **178**, 297–329 (1981)
63. Y. Giga, Domains of fractional powers of the Stokes operator in L_r spaces. Arch. Ration. Mech. Anal. **89**, 251–265 (1985)
64. Y. Giga, H. Sohr, Abstract L^p-estimates for the Cauchy problem with applications to the Navier–Stokes equations in exterior domains. J. Funct. Anal. **102**, 72–94 (1991)
65. Y. Giga, M. Gries, M. Hieber, A. Hussein, T. Kashiwabara, Analyticity of solutions to the primitive equations. Math. Nachrichten, to appear.
66. Y. Giga, K. Inui, A. Mahalov, S. Matsui, J. Saal, Rotating Navier–Stokes equations in $\mathbb{R}^3_+$ with initial data nondecreasing at infinity: the Ekman boundary layer problem. Arch. Ration. Mech. Anal. **186**, 177–224 (2007)
67. Y. Giga, K. Inui, A. Mahalov, S. Matsui, J. Saal, Uniform global solvability of the rotating Navier–Stokes equations for nondecaying initial data. Indiana Univ. Math. J. **57**, 2775–2791 (2008)
68. Y. Giga, M. Gries, M. Hieber, A. Hussein, T. Kashiwabara, Bounded H^∞-calculus for the Hydrostatic Stokes operator on L^p-spaces and applications. Proc. Amer. Math. Soc. **145**, 3865–3876 (2017)
69. L. Grafakos, *Classical Fourier Analysis* (Springer, Berlin, 2008)
70. G. Grubb, V.A. Solonnikov, Boundary value problems for the nonstationary Navier–Stokes equations treated by pseudo-differential methods. Math. Scand. **69**, 217–290 (1991)
71. M. Haase, *The Functional Calculus for Sectorial Operators* (Birkhäuser, Basel, 2006)
72. M. Hess, M. Hieber, A. Mahalov, J. Saal, Nonlinear stability of the Ekman boundary layers. Bull. London Math. Soc. **42**, 691–706 (2010)
73. M. Hieber, T. Kashiwabara, Global strong well–posedness of the three dimensional primitive equations in L^p–spaces. Arch. Ration. Mech. Anal.**221**, 1077–1115 (2016)
74. M. Hieber, M. Murata, The L^p-approach to the fluid rigid body interaction problem for compressible fluids. Evol. Equ. Contr. Theory **4**, 69–87 (2015)
75. M. Hieber, J. Prüss, Functional calculi for linear operators in vector-valued L^p-spaces via the transference principle. Adv. Diff. Equ. **3**, 847–872 (1998)
76. M. Hieber, J. Prüss, Dynamics of the Ericksen-Leslie equations with general Leslie stress I: the incompressible isotropic case. Math. Ann. **369**, 977–996 (2017)
77. M. Hieber, J. Prüss, Modeling and analysis of the Ericksen-Leslie equations for nematic liquid crystal flow, in ed. by Y. Giga, A. Novotny. *Handbook of Mathematical Analysis in Mechanics of Viscous Fluids*, vol. 2 (Springer, Berlin, 2018), pp. 1057–1134
78. M. Hieber, J. Prüss, Dynamics of the Ericksen-Leslie equations with general Leslie stress II: the compressible isotropic case. Arch. Ration. Mech. Anal. *233*, 1441–1468 (2019)
79. M. Hieber, J. Saal, The Stokes equation in the L^p-setting: well-posedness and regularity properties. in *Handbook of Mathematical Analysis in Mechanics of Viscous Fluids*. ed. by Y. Giga, A. Novotny, vol. 1 (Springer, Berlin, 2018), pp. 117–206
80. M. Hieber, H. Saito, Strong solutions for two-phase free boundary problems for a class of non-Newtonian fluids. J. Evolut. Equ. **17**, 335–358 (2017)
81. M. Hieber, Y. Shibata, The Fujita-Kato approach to the Navier–Stokes equation in the rotational framework. Math. Z. **265**, 481–491 (2010)
82. M. Hieber, W. Stannat, Stochastic stability of the Ekman spiral. Ann. Sc. Norm. Super. Pisa **XII**, 189–208 (2013)

83. M. Hieber, A. Hussein, T, Kashiwabara, Global strong L^p well-posedness of the 3D primitive equations with heat and salinity diffusion. J. Diff. Equ. **261**, 6950–6981 (2016)
84. M. Hieber, M. Nesensohn, J. Prüss, K. Schade, Dynamics of nematic lquid crystals: the quasilinear approach. Ann. Inst. H. Poincaré Anal. Non Linéaire **33**, 397–408 (2016)
85. M. Hong, J. Li, Z. Xin, Blow-up criteria of strong solutions to the Ericksen-Leslie system in $\mathbb{R}^3$. Comm. Partial Differ. Equ. **39**, 1284–1328 (2014)
86. L. Hörmander, *The Analysis of Linear Partial Differential Operators*, vol. I, II (Springer, Berlin, 1983)
87. T. Hytönen, J. van Neerven, M. Veraar, L. Weis, *Analysis in Banach Spaces*, vol. I (Springer, Berlin, 2016)
88. T. Hytönen, J. van Neerven, M. Veraar, L. Weis, *Analysis in Banach Spaces*, vol. II (Springer, Berlin, 2017)
89. A. Inoue, M. Wakimoto, On existence of solutions of the Navier–Stokes equation in a time dependent domain. J. Fac. Sci. Univ. Tokyo Sect. IA **24**, 303–319 (1977)
90. N. Kalton, G. Lancien, A solution to the problem of L_p-maximal regularity. Math. Z. **235**, 559–568 (2000)
91. N. Kalton, L. Weis, The H^∞-calculus and sums of closed operators. Math. Ann. **321**, 319–345 (2001)
92. T. Kato, *Perturbation Theory of Linear Operators* (Springer, Berlin, 1966)
93. H. Koba, Nonlinear stability of Ekman boundary layers in rotating stratified fluids. Memoirs Amer. Math. Soc. **228**, 1 (2014)
94. M. Köhne, J. Prüss, M. Wilke, On quasilinear parabolic evolution equations in weighted L_p-spaces. J. Evol. Equ. **10**, 443–463 (2010)
95. P. Kunstmann, L. Weis, Maximal L^p-regularity for parabolic equations, Fourier multiplier theorems and H^∞-functional calculus, in *Functional Analytic Methods for Evolution Equations*. ed. by M. Ianelli, R. Nagel, S. Piazzera. Lecture Notes in Mathematics, vol. 1855 (Springer, Berlin, 2004), pp. 65–311
96. P. Kunstmann, L. Weis, New criteria for the H^∞-calculus and the Stokes operator on bounded Lipschitz domains. J. Evol. Equ. **17**, 387–409 (2017)
97. F. M. Leslie, Some constitutive equations for liquid crystals. Arch. Ration. Mech. Anal. **28**, 265–283 (1968)
98. J. Li, E. Titi, The primitive equations as the small aspect ratio limit of the Navier–Stokes equations: rigorous justification of the hydrostatic approximation (2017). arXiv:1706.08885
99. J. Li, E. Titi, Recent advances concerning certain classes of geophysical flows, in *Handbook of Mathematical Analysis in Mechanics of Viscous Fluids*. ed. by Y. Giga, A. Novotny, vol. 1 (Springer, Berlin, 2018), pp. 933–972
100. F. Lin, Nonlinear theory of defects in nematic liquid crystals: phase transition and flow phenomena. Comm. Pure Appl. Math. **42**, 789–814 (1989)
101. F. Lin, Ch. Liu, Nonparabolic dissipative systems modeling the flow of liquid crystals. Comm. Pure Appl. Math. **48**, 501–537 (1995)
102. F. Lin, C. Wang, Recent developments of analysis for hydrodynamic flow of nematic liquid crystals. Philos. Trans. R. Soc. Lon. Ser. A, Math. Phys. Eng. Sci. **372**, 20130361 (2014)
103. J.L. Lions, R. Temam, Sh.H. Wang, New formulations of the primitive equations of atmosphere and applications. Nonlinearity **5**, 237–288 (1992)
104. J.L. Lions, R. Temam, Sh.H. Wang, On the equations of the large-scale ocean. Nonlinearity **5**, 1007–1053 (1992)
105. C. Liu, H. Wu, X. Xu, On the general Ericksen-Leslie system: Parodi's relation, well-posedness and stability. Arch. Ration. Mech. Anal. **208**, 59–107 (2013)
106. A. Lunardi, *Analytic Semigroups and Optimal Regularity in Parabolic Problems* (Birkhäuser, Basel, 1995)
107. W. Ma, H. Gong, J. Li, Global strong solutions to incompressible Ericksen-Leslie system in $\mathbb{R}^3$. Nonlinear Anal. **109**, 230–235 (2014)

108. J. Málek, J. Necas, M. Ruzicka, On weak solutions to a class of non-Newtonian incompressible fluids in bounded three-dimensional domains: the case $p \geq 2$. Adv. Differ. Equ. **6**, 257–302 (2001)

109. J. Malek, V. Prusa, Derivation of equations for incmpressible and compressible fluids, in ed. by Y. Giga, A. Novotny. *Handbook of Mathematical Analysis in Mechanics of Viscous Fluids*, vol. 1 (Springer, Berlin, 2018), pp. 3–72

110. V. Maslennikova, M. Bogovskii, Elliptic boundary values in unbounded domains with noncompact and nonsmooth boundaries. Rend. Sem. Mat. Fis. Milano **56**, 125–138 (1986)

111. N. Masmoudi, Ekman layers of rotating fluids: the case of general initial data. Comm. Pure Appl. Math. **53**, 432–483 (2000)

112. M. McCracken, The resolvent problem for the Stokes equation on half spaces in L_p. SIAM J. Math. Anal. **12**, 201–228 (1981)

113. A. McIntosh, Operators which have an H_∞-calculus, in Miniconference on operator theory and partial differential equations. ed. by B. Jefferies, A. McIntosh, W. Ricker. Proceeding Center Mathematica Analysis A.N.U., vol. 14 (1986), pp.210–231

114. T. Miyakawa, H. Sohr, Weak solutions of Navier–Stokes equations. Math. Z. **199**, 455–478 (1988)

115. A. Noll, J. Saal, H^∞-calculus for the Stokes operator on L_q-spaces. Math. Z. **244**, 651–688 (2003)

116. M. Padula, *Asymptotic Stability of Steady Compressible Fluids*. Lecture Notes Mathematics, vol. 2024 (Springer, Berlin, 2010)

117. A. Pazy, *Semigroups of Linear Operators and Applications to Partial Differential Equations* (Springer, New York, 1983)

118. J. Pedlovsky, *Geophysical Fluid Dynamics* (Springer, New York, 1987)

119. G. Pisier, Some results on Banach spaces without local unconditional structure. Compos. Math. **37**, 3–19 (1978)

120. J. Prüss, Maximal regularity for evolution equations in L_p-spaces. In: *Conf. Semin. Mat. Univ. Bari*, **(2002)**(285), (2003), 1–39.

121. J. Prüss, G. Simonett, Maximal regularity for evolution equations in weighted L_p-spaces. Arch. Math. **82**, 415–431 (2004)

122. J. Prüss, G. Simonett, On the two-phase Navier–Stokes equations with surface tension. Inter. Free Bound. **12**, 311–345 (2010)

123. J. Prüss, G. Simonett, Analytic solutions for the two-phase Navier–Stokes equations with surface tension and gravity, in *Parabolic Problems*, Progress. Nonlinear Differential Equations Applications, vol. 80, (Birkhäuser, Basel, 2011), pp. 507–540

124. J. Prüss, G. Simonett, *Moving Interfaces and Quasilinear Parabolic Evolution Equations*. Monographs in Mathematics, vol. 105 (Birkhäuser, Basel, 2016)

125. J. Prüss, M. Wilke. Addendum to the paper "On quasilinear parabolic evolution equations in weighted L_p-spaces II". J. Evol. Equ. **17**, 1381–1388 (2017)

126. J. Prüss, G. Simonett, R. Zacher, On convergence of solutions to equilibria for quasilinear parabolic problems. J. Diff. Equ. **246**, 3902–3931 (2009)

127. J.-P. Raymond, M. Vanninathan, A fluid-structure model coupling the Navier–Stokes equations and the Lamé system. J. Math. Pures Appl. **102**, 546–596 (2014)

128. J. Robinson, J. Rodrigo, W. Sadowski, *The Three-Dimensional Navier–Stokes Equations*. Cambridge Studies in Advanced Mathematicals, vol. 157 (Cambridge University Press, Cambridge, 2016)

129. F. Rousset, Stability of large Ekman boundary layers in rotating fluids. Arch. Ration. Mech. Anal. **172**, 213–245 (2004)

130. Y. Shibata, S. Shimizu, On a free boundary value problem for the Navier–Stokes equations. Differ. Integral Equ. **20**, 241–276 (2007)

131. Y. Shibata, S. Shimizu, Report on a local in time solvability of free surface problems for the Navier–Stokes equations with surface tension. Appl. Anal. **90**, 201–214 (2011)

132. Y. Shibata, S. Shimizu, Maximal L_p-L_q regularity for the two-phase Stokes equations; model problems. J. Differ. Equ. **251**, 373–419 (2011)

133. C. Simader, H. Sohr, *The Dirichlet Problem for the Laplacian in Bounded and Unbounded Domains*. Pitman Research Notes in Mathematicals, vol. 360 (CRC Press, Boca Raton, 1997)
134. H. Sohr, *The Navier–Stokes Equations. An Elementary Functional Analytic Approach* (Birkhäuser, Basel, 2001)
135. V.A. Solonnikov, Estimates for solutions of nonstationary Navier–Stokes equations. J. Soviet Math. **8**, 213–317 (1977)
136. V.A. Solonnikov, Solvability of a problem of evolution of an isolated amount of a viscous incompressible capillary fluid. Zap. Nauchn. Sem. LOMI **140**, 179–186 (1984); English transl. in J. Soviet Math. **37** (1987)
137. V.A. Solonnikov, On the quasistationary approximation in the problem of motion of a capillary drop, in *Topics in Nonlinear Analysis. The Herbert Amann Anniversary Volume.* ed. by J. Escher, G. Simonett (Birkhäuser, Basel, 1999), pp. 641–671
138. V.A. Solonnikov, On the stability of nonsymmetric equilibrium figures of a rotating viscous imcompressible liquid. Inter. Free Bound. **6**, 461–492 (2004)
139. E.M. Stein, *Topics in Harmonic Analysis Related to Littlewood-Paley Theory* (Princeton University Press, Princeton, 1970)
140. E.M. Stein, *Harmonic Analysis: Real-Variables Methods, Orthogonality and Oscillatory Integrals* (Princeton University Press, Princeton, 1993)
141. E.M. Stein, R. Shakarchi, *Fourier Analysis: An Introduction* (Princeton University Press, Princeton, 2003)
142. T. Takahashi, Analysis of strong solutions for the equations modeling the motion of a rigid-fluid system in a bounded domain. Adv. Differ. Equ. **8**, 1499–1532 (2003)
143. T. Takahashi, M. Tucsnak, Global strong solutions for the two-dimensional motion of an infinite cylinder in a viscous fluid. J. Math. Fluid Mech. **6**, 63–77 (2004)
144. H. Tanabe, *Equations of Evolution* (Pitman, London, 1979)
145. N. Tanaka, Two-phase free boundary problem for viscous incompressible thermocapillary convection. Jpn. J. Math. **21**, 1–42 (1995)
146. A. Tani, N. Tanaka, Large-time existence of surface waves in incompressible viscous fluids with or without surface tension. Arch. Ration. Mech. Anal. **130**, 303–314 (1995)
147. H. Temam, *Navier Stokes Equations*. Monographs in Mathematics, vol. 78 (Birkhäuser, Basel, 1992)
148. H. Triebel, *Interpolation Theory, Function Spaces, Differential Operators* (North-Holland, Amsterdam, 1978)
149. H. Triebel, *Theory of Function Spaces*, (Reprint of 1983 edition) (Springer, Berlin, 2010)
150. T.-P. Tsai, Lectures on the Navier–Stokes equations, in *Graduate Studies in Mathematics* (American Mathematical Society,, Providence, 2018)
151. S. Ukai, A solution formula for the Stokes equation in $\mathbb{R}_3^+$, Comm. Pure Appl. Math. **11**, 611–621 (1987)
152. E. G. Virga, *Variational Theories for Liquid Crystals* (Chapman-Hall, London, 1994)
153. W. Wang, P. Zhang, Z. Zhang, Well-posedness of the Ericksen-Leslie system. Arch. Ration. Mech. Anal. **210**, 837–855 (2013)
154. L. Weis, Operator valued Fourier multiplier theorems and maximal L^p-regularity. Math. Ann. **319**, 735–758 (2001)
155. H. Wu, X. Xu, Ch. Liu, On the general Ericksen-Leslie system: Parodi's relation, well-posedness and stability. Arch. Ration. Mech. Anal. **208**, 59–107 (2013)
156. F. Zimmermann, On vector-valued Fourier multiplier theorems. Studia Math. **89**, 201–222 (1989)

Chapter 2
Partial Regularity for the 3D Navier–Stokes Equations

James C. Robinson

2.1 Introduction

These notes give a relatively quick introduction to some of the main results for the three-dimensional Navier–Stokes equations, concentrating in particular on 'partial regularity' results that limit the size of the set of (potential) singularities, both in time (Sect. 2.2.5) and in space-time (Sect. 2.4).

We will consider the Navier–Stokes equations

$$\partial_t u - \Delta u + (u \cdot \nabla)u + \nabla p = 0, \qquad \nabla \cdot u = 0, \tag{2.1}$$

on the three-dimensional torus $\mathbb{T}^3$, i.e. $[0, 2\pi)^3$ with 'periodic boundary conditions'. We choose this domain for simplicity, since it is both bounded and boundaryless. Often we will consider the initial-value problem, i.e. we solve (2.1) for $t > 0$ given $u(x, 0) = u_0(x)$.

The existence of weak solutions (solutions with finite kinetic energy) on the whole space has been known since the work of Leray [22], but whether or not smooth solutions exist for all time is still unresolved, and is one of the Clay Foundation's Million-Dollar Millennium Problems [12]. The first 'partial regularity' results were due to Scheffer who showed that the 1-dimensional Hausdorff measure of the set of spatial singularities is finite at the first singular time [33], that the Hausdorff dimension of the set of space-time singularities is no larger than 5/3 [35], and that the 1/2-dimensional Hausdorff measure of the set of singular times is zero ([32]; cf. our Proposition 2.12).

J. C. Robinson (✉)
Department of Mathematics, University of Warwick, Coventry, UK
e-mail: j.c.robinson@warwick.ac.uk

© Springer Nature Switzerland AG 2020   147
G. P. Galdi, Y. Shibata (eds.), *Mathematical Analysis of the Navier-Stokes Equations*, Lecture Notes in Mathematics 2254,
https://doi.org/10.1007/978-3-030-36226-3_2

The ingredients that go into known partial regularity results are usually (i) a conditional regularity result, i.e. a guarantee of local regularity of solutions under a local smallness condition and (ii) a global bound on solutions that involves the same quantity as the conditional regularity result. Then (ii) can be used to show that the condition in (i) cannot be violated at too many points, which serves to limit the size of the singular set.

A space-time point (x, t) is termed 'regular' if $u \in L^\infty(U)$ for some space-time neighbourhood U of (x, t). That this is a reasonable definition of a regular point is a consequence of a local conditional regularity result due to Serrin [37]: local boundedness is sufficient for spatial smoothness. We given a sketch of this result in Sect. 2.3 (in fact we show that if $u \in L^\alpha(U)$ for any $\alpha > 5$ then u is spatially smooth within U; and we also indicate how to extend this result to the case $u \in L^5(U)$).

The most well known partial regularity result for the Navier–Stokes equations is due to Caffarelli et al. [3] [hereafter CKN] and guarantees that the 1-dimensional Hausdorff measure of the set of space-time singularities is zero; in particular the Hausdorff dimension of this set is no larger than one. The hard part of their proof is a local conditional regularity result. In fact they proved two such results; we only prove the first in these notes (and then in a simplified form, neglecting the pressure), but this is perhaps the harder of the two: from what we prove it is relatively straightforward to show that the box-counting (Minkowski) dimension of the set of space-time singularities is bounded by $5/3$.

We then present a result to due Robinson and Sadowski [30] that makes crucial use of this bound on the dimension of the singular set: we show that given any weak solution u, the 'particle trajectories', i.e. solutions of $\dot{X} = u(X, t)$, exist and are unique for almost every choice of initial condition, so that it makes sense to consider even such an irregular solution from a Lagrangian point of view. The point of including this result is two-fold: it provides an application of the partial regularity theory, and demonstrates that, despite the lack of a complete theory of existence and uniqueness, one can prove interesting results about solutions of the 3D Navier–Stokes equations that are valid without invoking any unproved assumptions.

2.1.1 Notation and Preliminaries

We will use

$$A \lesssim B$$

to mean that $A \leq cB$ for some absolute constant c; and

$$A \lesssim_r B$$

to mean that $A \leq c(r)B$. Similarly we use $A \approx B$ to mean that $c_1 A \leq B \leq c_2 B$ for some constants c_1 and c_2.

We write $\nabla = (\partial_1, \partial_2, \partial_3)$, where $\partial_j = \frac{\partial}{\partial x_j}$, so that

$$\nabla \cdot u = \operatorname{div} u, \qquad \nabla \times u = \operatorname{curl} u, \qquad \text{and} \qquad [(u \cdot \nabla)u]_i = \sum_{j=1}^{3} (u_j \partial_j) u_i.$$

For functions on the torus we can easily define the norms in the Sobolev spaces $H^s(\mathbb{T}^3)$ using Fourier series: if

$$u = \sum_{k \in \mathbb{Z}^3} \hat{u}_k e^{ik \cdot x}$$

then

$$\|u\|_{H^s}^2 := \sum_k (1 + |k|^2)^s |\hat{u}_k|^2$$

and

$$\|u\|_{\dot{H}^s}^2 := \sum_k |k|^{2s} |\hat{u}_k|^2.$$

Note that if u has zero average then $\hat{u}_0 = 0$, and so $\|u\|_{H^s} \approx \|u\|_{\dot{H}^s}$. Also note that from these expressions it is easy to obtain the generalised Poincaré inequality $\|u\|_{H^{s_1}} \lesssim \|u\|_{H^{s_2}}$ for all $s_2 \geq s_1$.

It is useful to recall that H^s is an algebra if $s > 3/2$, i.e.

$$\|fg\|_{H^s} \lesssim \|f\|_{H^s} \|g\|_{H^s} \qquad f, g \in H^s, \ s > 3/2. \tag{2.2}$$

Let

$$C_\sigma^\infty(\mathbb{T}^3) = \left\{ \varphi \in [C^\infty(\mathbb{T}^3)]^3 : \nabla \cdot \varphi = 0, \int_{\mathbb{T}^3} \varphi = 0 \right\},$$

i.e. divergence-free smooth periodic functions with zero average; the subscript σ is used to indicate that the functions are divergence free. We set

$$H := \text{completion of } C_\sigma^\infty(\mathbb{T}^3) \text{ in the norm of } L^2(\mathbb{T}^3),$$

denoting the L^2 norm by $\| \cdot \|$, and

$$V := \text{completion of } C_\sigma^\infty(\mathbb{T}^3) \text{ in the norm of } H^1(\mathbb{T}^3) = H \cap H^1(\mathbb{T}^3).$$

When we omit the domain from a spatial integral it should be understood to be $\mathbb{T}^3$.

We will often use 'Lebesgue interpolation' (which follows from Hölder's inequality), in particular the inequality

$$\|u\|_{L^4} \leq \|u\|_{L^2}^{1/4}\|u\|_{L^6}^{3/4}.$$

Coupled with the Sobolev embedding $H^1(\mathbb{T}^3) \subset L^6(\mathbb{T}^3)$ this yields (when $\int u = 0$) the Ladyzhenskaya inequality

$$\|u\|_{L^4} \lesssim \|u\|_{L^2}^{1/4}\|\nabla u\|_{L^2}^{3/4}. \tag{2.3}$$

2.2 Basic Existence and Uniqueness Results

We will not discuss in detail here the proof of the existence of solutions, referring instead to the relevant sections of Robinson et al. [31], hereafter RRS. However, we show how the regularity of weak and strong solutions (and their existence) is to be expected, by using a priori estimates obtained by manipulating the equations as if all the terms were smooth.

2.2.1 Weak Solutions

Both the definition of weak solutions and the proof of their existence are based on the following a priori energy estimate for smooth solutions. If we take the inner product of (2.1) with u and integrate then two terms vanish: for the nonlinear term we have[1]

$$\int u_i(\partial_i u_j)u_j = -\int u_i u_j(\partial_i u_j)$$

and so this term vanishes, and for the pressure term

$$\int (\partial_i p)u_i = -\int p(\partial_i u_i) = 0,$$

since $\partial_i u_i = \text{div}\, u = 0$.

---

[1] This is a particular case of the very useful anti-symmetry property

$$\langle (u \cdot \nabla)v, w \rangle = -\langle (u \cdot \nabla)w, v \rangle, \tag{2.4}$$

which we will use from time to time in what follows.

Therefore

$$\frac{1}{2}\frac{d}{dt}\|u\|^2 + \|\nabla u\|^2 = 0$$

from which it follows that

$$\frac{1}{2}\|u(t)\|^2 + \int_0^t \|\nabla u(s)\|^2 \, ds = \frac{1}{2}\|u_0\|^2;$$

this estimate is at the basis of the proof that for any $u_0 \in H$ there exists at least one weak solution $u \in L^\infty(0, T; L^2) \cap L^2(0, T; H^1)$ for every $T > 0$.
[One can perform the same estimates for the Euler equations (when $v = 0$) and show that the kinetic energy is constant. The required level of regularity for solutions to ensure that this holds is the subject of the *Onsager Conjecture*, resolved by work of Constantin et al. [6] and Isett [18]; see also Buckmaster et al. [2].]

We now define the notion of a weak solution more precisely. Let

$$\mathcal{D}_\sigma = \{\varphi \in C_c^\infty(\mathbb{T}^3 \times [0, \infty))^3 : \operatorname{div}\varphi(t) = 0 \text{ for all } t \in [0, \infty)\}.$$

Definition 2.1 A weak solution of (2.1) is a function u such that

$$u \in L^\infty(0, T; H) \cap L^2(0, T; V) \quad \text{for every } T > 0$$

and

$$\int_0^s -\langle u, \partial_t\varphi\rangle + \int_0^s \langle \nabla u, \nabla\varphi\rangle + \int_0^s \langle (u \cdot \nabla)u, \varphi\rangle = \langle u_0, \varphi(0)\rangle - \langle u(s), \varphi(s)\rangle$$

$$(2.5)$$

for all $\varphi \in \mathcal{D}_\sigma$ and almost every $s > 0$.

[Note that this definition is distinct from requiring that u satisfies the equations 'in the sense of distributions'. However, one can prove that given a weak solution u there exists a corresponding pressure such that the original equations hold in this sense, see Chapter 5 of RRS (Section 5.2).]

Exercise Derive (2.5) as a consequence of (2.1) when u is smooth by taking the inner product with $\varphi \in \mathcal{D}_\sigma$, integrating in both space and time, and integrating by parts.

The existence of weak solutions on $\mathbb{T}^3$ can be proved rigorously via a Galerkin procedure,[2] see Chapter 4 of RRS, for example, which also yields a solution that

---

[2]Leray's 1934 paper treats the equations on the whole space and does not use the Galerkin approach (see Ożański and Pooley [25], for a modern treatment of the methods in his paper). A 'Galerkin-like' argument for the equations on the whole space can be found in the book by Chemin et al. [4]. The first proof of existence of solutions on bounded domains was due to Hopf [17].

satisfies the strong energy inequality [(2.6), below]. These are known as Leray–Hopf weak solutions.

Theorem 2.2 *Given* $u_0 \in H$ *there exists at least one weak solution that satisfies the strong energy inequality*

$$\frac{1}{2}\|u(t)\|^2 + \int_s^t \|\nabla u(\tau)\|^2 \, d\tau \leq \frac{1}{2}\|u(s)\|^2 \tag{2.6}$$

for every $t \geq s$, *for a set of* s *of full measure that includes* $s = 0$.

The 'energy inequality' is (2.6) allowing only $s = 0$.

It is not known whether Leray–Hopf weak solutions are unique. Very recently, Buckmaster and Vicol [1] have shown that if one does not require the energy inequality then distributional solutions are not unique, and there are reasons to suspect that this non-uniqueness extends to Leray–Hopf weak solutions [19].

2.2.2 Strong Solutions

We can show that if $u_0 \in V$ then smoother solutions exist with

$$u \in L^\infty(0, T; V) \cap L^2(0, T; H^2), \tag{2.7}$$

at least for some $T > 0$, and if the norm of u_0 in V is sufficiently small then the solutions exist for all $t \geq 0$. A weak solution with the regularity in (2.7) is called a strong solution on $[0, T]$.

Taking the L^2 inner product of (2.1) with Δu we obtain

$$\frac{1}{2}\frac{d}{dt}\|\nabla u\|^2 + \|\Delta u\|^2 \leq \int [(u \cdot \nabla)u] \cdot \Delta u$$

$$\leq \|u\|_\infty \|\nabla u\| \|\Delta u\| \tag{2.8}$$

$$\lesssim \|\nabla u\|^{3/2} \|\Delta u\|^{3/2}$$

$$\leq \frac{1}{2}\|\Delta u\|^2 + C\|\nabla u\|^6.$$

Exercise Use Fourier series to show that

$$\|u\|_{L^\infty} \lesssim \|\nabla u\|^{1/2} \|\Delta u\|^{1/2} \tag{2.9}$$

whenever $u \in H^2(\mathbb{T}^3)$.

Therefore

$$\frac{d}{dt}\|\nabla u\|^2 + \|\Delta u\|^2 \lesssim \|\nabla u\|^6. \tag{2.10}$$

In particular

$$\frac{d}{dt}\|\nabla u\|^2 \leq c\|\nabla u\|^6,$$

from which it follows that

$$\|\nabla u(t)\|^2 \leq \frac{\|\nabla u_0\|^2}{\sqrt{1 - 2ct\|\nabla u_0\|^4}}. \tag{2.11}$$

These calculations suggest that if $u_0 \in H^1$ there exists $T = T(\|\nabla u_0\|)$ [for example, $T = \|\nabla u_0\|^{-4}/4c$] such that

$$u \in L^\infty(0, T; H^1).$$

Once we have $u \in L^\infty(0, T; H^1)$ we can return to (2.10) and integrate from 0 to T to show that $u \in L^2(0, T; H^2)$:

$$\|\nabla u(T)\|^2 + \int_0^T \|\Delta u(s)\|^2 \, ds \lesssim \|\nabla u(0)\|^2 + \int_0^T \|\nabla u(s)\|^6 \, ds.$$

Again, to make this rigorous one can argue using Galerkin approximations (Chapter 6 in RRS).

Since $u \in L^\infty(0, T; V) \cap L^2(0, T; H^2)$ it follows that $\partial_t u \in L^2(0, T; H)$. Indeed, choosing $\varphi \in H$ we have, using (2.9),

$$|(\partial_t u, \varphi)| \leq |(\Delta u, \varphi)| + |(u \cdot \nabla u, \varphi)|$$

$$\leq \|\Delta u\|\|\varphi\| + \|u\|_\infty \|\nabla u\|\|\varphi\|$$

$$\lesssim [\|\Delta u\| + \|\nabla u\|^{3/2}\|\Delta u\|^{1/2}]\|\varphi\|, \tag{2.12}$$

and the time integrability of $\|\partial_t u\|$ follows from the regularity of u. This implies, in particular, that $u \in C^0([0, T]; V)$ (see Theorem 4 in Section 5.9.2 in Evans [9] or Theorem 7.2 in Robinson [27] for details).

Furthermore, strong solutions are unique (actually, something stronger is true, as we will soon see in the next section). If u and v are strong solutions and we consider their difference $w = v - u$ then

$$\partial_t w - \Delta w + \underbrace{(v \cdot \nabla)v - (u \cdot \nabla)u}_{(v \cdot \nabla)w + (w \cdot \nabla)u} + \nabla(p - q) = 0,$$

and if we take the inner product with w then

$$\frac{1}{2}\frac{d}{dt}\|w\|^2 + \|\nabla w\|^2 = -\int (w \cdot \nabla)u \cdot w \tag{2.13}$$

$$\leq \|w\|_{L^4}^2 \|\nabla u\|$$

$$\lesssim \|w\|^{1/2} \|\nabla w\|^{3/2} \|\nabla u\|,$$

using the Ladyzhenskaya inequality $\|w\|_{L^4} \lesssim \|w\|_{L^2}^{1/4} \|\nabla w\|_{L^2}^{3/4}$ (see (2.3)). Now we use Young's inequality to split the right-hand side, and absorb the $\|\nabla w\|^2$ using the left-hand side, to obtain

$$\frac{1}{2}\frac{d}{dt}\|w\|^2 \lesssim \|\nabla u\|^4 \|w\|^2; \tag{2.14}$$

since $\nabla u \in L^\infty(0, T; V)$ we can use Gronwall's inequality to deduce that $w = 0$ for all $t \in [0, T]$, i.e. that $u = v$.

The inequality (2.10) can also be used to prove the global-in-time existence of strong solutions when the initial condition is sufficiently small, if we use the Poincaré inequality $\|\nabla u\| \lesssim \|\Delta u\|$. Indeed, we then have

$$\frac{d}{dt}\|\nabla u\|^2 \leq c\|\nabla u\|^6 - \lambda\|\nabla u\|^2; \tag{2.15}$$

if we take $\|\nabla u_0\|^4 < \varrho_0^4 := \lambda/c$ then $\|\nabla u(t)\|$ is non-increasing and so the solution remains bounded in H^1 for all positive times.

Putting all these facts together we have the following result on the existence and uniqueness of strong solutions.

Theorem 2.3 *If $u_0 \in V$ then there exists a time $T = T(\|\nabla u_0\|)$ such that there exists a unique strong solution $u \in L^\infty(0, T; V) \cap L^2(0, T; H^2)$. Furthermore there exists $\varrho_0 > 0$ such that whenever $\|\nabla u_0\| < \varrho_0$ this strong solution exists for all $t \geq 0$.*

It turns out that strong solutions are also smooth in space (while they exist): so if u is a strong solution on $(0, T)$ it is smooth on $(\varepsilon, T]$ for any $\varepsilon > 0$. We can prove this inductively, following Constantin and Foias [5].

Lemma 2.4 *If $u_0 \in H^k$ and u is a strong solution on $(0, T)$ then*

$$u \in L^\infty(0, T; H^k) \cap L^2(0, T; H^{k+1}).$$

Proof We will show that if $s \geq 1$ then

$$u \in L^\infty(0, T; H^s) \cap L^2(0, T; H^{s+1}) \text{ and } u_0 \in H^{s+1}$$

implies that

$$u \in L^\infty(0, T; H^{s+1}) \cap L^2(0, T; H^{s+2}).$$

If we take the inner product of the equation with u in H^{s+1} we obtain

$$\frac{1}{2}\frac{d}{dt}\|u\|_{H^{s+1}}^2 + \|\nabla u\|_{H^{s+1}}^2 \leq |\langle (u \cdot \nabla)u, u\rangle_{H^{s+1}}|$$

$$\lesssim \|u\|_{H^{s+1}}^2 \|\nabla u\|_{H^{s+1}}$$

$$\leq \frac{1}{2}\|\nabla u\|_{H^{s+1}}^2 + c\|u\|_{H^{s+1}}^4,$$

where we have used the fact that H^{s+1} is an algebra (see (2.2)) and Young's inequality to split the right-hand side. Therefore

$$\frac{d}{dt}\|u\|_{H^{s+1}}^2 + \|\nabla u\|_{H^{s+1}}^2 \lesssim \|u\|_{H^{s+1}}^4 = \|u\|_{H^{s+1}}^2 \|u\|_{H^{s+1}}^2.$$

Since by assumption $u \in L^2(0, T; H^{s+1})$ and $u_0 \in H^{s+1}$ it follows—dropping the $\|\nabla u\|_{H^{s+1}}^2$ term and integrating in time—that in fact $u \in L^\infty(0, T; H^{s+1})$. Now we can integrate again, but this time retaining the ∇u term, to see that $u \in L^2(0, T; H^{s+2})$.

We can start the induction with $s = 1$ since for any strong solution we have $u \in L^\infty(0, T; H^1) \cap L^2(0, T; H^2)$.   □

With this result it is easy to prove the smoothness of strong solutions.

Proposition 2.5 *If $u_0 \in V$ and $u \in L^\infty(0, T; H^1) \cap L^2(0, T; H^2)$ is a strong solution then $u \in L^\infty(\varepsilon, T; H^k)$ for every $k \in \mathbb{N}$ and $\varepsilon > 0$.*

Proof Fix $\varepsilon > 0$. Since $u \in L^\infty(0, T; V) \cap L^2(0, T; H^2)$, there exists a time $t_1 \in (0, \varepsilon)$ such that $u(t_1) \in H^2$. Since strong solutions are unique, the strong solution $v(t)$ with initial condition $u(t_1)$ coincides with $u(t_1 + t)$, so it follows from Lemma 2.4 that $u \in L^\infty(t_1, T; H^2) \cap L^2(t_1, T; H^3)$. Now we can find $t_2 \in (t_1, \varepsilon)$ such that $u(t_2) \in H^3$; it follows from Lemma 2.4 (again using the uniqueness of strong solutions) that $u \in L^\infty(t_2, T; H^3) \cap L^2(t_2, T; H^4)$. Continuing in this way we obtain the result as stated.   □

Corollary 2.6 *If $u_0 \in V$ and $u \in L^\infty(0, T; H^1) \cap L^2(0, T; H^2)$ is a strong solution then $u \in C^0((0, T]; H^k)$ for every $k \geq 0$. In particular the solution is smooth in space: $u(t) \in C^\infty(\mathbb{T}^3)$ for every $t \in (0, T]$.*

Proof It follows from an estimate similar to that in (2.12) that we have $\partial_t u \in L^\infty(0, T; H^k)$ for every $k \geq 0$, and so $u \in C^0((0, T]; H^k)$ for every $k \geq 0$ [see Theorem 4 in Section 5.9.2 in Evans [9] or Theorem 7.2 in Robinson [27]]. Thus for every $t \in (0, T]$ we have $u(t) \in H^k$ for every $k \geq 0$, and hence $u(t) \in C^\infty$.   □

2.2.3 Weak-Strong Uniqueness

One very useful property of strong solutions is that they are unique in the class of weak solutions that satisfy the energy inequality ('weak-strong uniqueness'). The following argument gives an indication of why this is true, but it is not rigorous since it requires more smoothness of the two solutions than we actually have.

Suppose that u is a strong solution and v is a weak solution, and consider, as above, the difference $w = u - v$. If we neglect any smoothness considerations and follow our previous calculations that led to (2.14) then we obtain

$$\frac{1}{2}\frac{d}{dt}\|w\|^2 \lesssim \|\nabla u\|^4 \|w\|^2;$$

noticing that the smoothness of v does not play a role here, we can again use the fact that $\nabla u \in L^\infty(0, T; V)$ to deduce that $w = 0$ for all $t \in [0, T]$, i.e. that $u = v$.

However, this argument is not valid, since the time derivative of a *weak* solution does not have sufficient regularity to ensure that $\langle \partial_t w, w \rangle$ makes sense. Indeed, if v is a weak solution then the regularity of $\partial_t v$ is essentially determined by that of $(v \cdot \nabla)v$, and

$$|\langle (v \cdot \nabla)v, \varphi \rangle| = |\langle (v \cdot \nabla)\varphi, v \rangle| \le \|u\|_{L^4}^2 \|\nabla \varphi\| \le \|v\|_{L^2}^{1/2} \|\nabla v\|^{3/2} \|\nabla \varphi\|,$$

which shows that $\partial_t v \in L^{4/3}(0, T; H^{-1})$ [we used the anti-symmetry property of the nonlinear term (2.4)]. This is not enough to allow us to pair $\partial_t w$ (whose regularity is limited by that of $\partial_t v$) with w, which is only in $L^2(0, T; H^1)$.

A rigorous proof of the weak-strong uniqueness property essentially proceeds as follows: (i) a strong solution u is sufficiently regular that the NSE hold as an equality in $L^2(0, T; H)$, so one can take the inner product with a weak solution $v \in L^2(0, T; H)$; (ii) a strong solution u has sufficient regularity to be used as a test function in the definition of a weak solution v; (iii) add the two equations from (i) and (ii) and use the fact that v satisfies the energy inequality to deduce that $w = u - v$ satisfies

$$\frac{1}{2}\|w(t)\|^2 + \int_0^t \|\nabla w\|^2 \le \left| \int_0^t \langle (w \cdot \nabla)w, u \rangle \right|, \tag{2.16}$$

(this is the rigorous version of (2.13) in this context) and from here one can argue to show that $w \equiv 0$. Indeed, it follows that

$$\frac{1}{2}\|w(t)\|^2 + \int_0^t \|\nabla w\|^2 \lesssim \int_0^t \|w\|^{1/2} \|\nabla w\|^{3/2} \|\nabla u\|$$

$$\le c \int_0^t \|\nabla u\|^4 \|w\|^2 + \frac{1}{2}\int_0^t \|\nabla w\|^2$$

and so we obtain

$$\|w(t)\|^2 + \int_0^t \|\nabla w(s)\|^2 \, ds \lesssim \int_0^t \|\nabla u(s)\|^4 \|w(s)\|^2 \, ds.$$

Since u is a strong solution we have $u \in L^\infty(0, T; H^1)$, and since $u(0) = v(0)$ we also have $w(0) = 0$, from which it follows that $w(t) \equiv 0$ for all $t \in (0, T)$. See Section 4 of Galdi [15] or Chapter 6 in RRS for details.

Now let us consider what implications this has for weak solutions.

One is that all Leray–Hopf weak solutions are eventually regular. We know that such solutions satisfy the energy inequality

$$\frac{1}{2}\|u(t)\|^2 + \int_0^t \|\nabla u(s)\|^2 \, ds \le \frac{1}{2}\|u_0\|^2,$$

and we have seen that if $\|\nabla u_0\| < \varrho_0$ then the solution is strong for all $t \ge 0$ (see (2.15)). From the energy inequality it follows that if $T \ge \frac{1}{2\varrho_0^2}\|u_0\|^2$ then there exists a $t_0 \in [0, T]$ such that $\|\nabla u(t_0)\| < \varrho_0$; weak-strong uniqueness now guarantees that u coincides with the strong solution through $u(t_0)$ for all $t \ge t_0$.

Since any weak solution satisfies $u \in L^2(0, T; H^1)$, we have $u(t_1) \in H^1$ for almost every t_1; it follows from weak-strong uniqueness that for $t \ge t_1$ the weak solution u coincides with the strong solution with initial data $u(t_1)$; it therefore follows from (2.11) that

$$\|\nabla u(t_2)\|^2 \le \frac{\|\nabla u(t_1)\|^2}{\sqrt{1 - 2c(t_2 - t_1)\|\nabla u(t_1)\|^4}} \qquad t_1 \le t_2 \le T. \tag{2.17}$$

while expression within the square root in the denominator remains positive.

2.2.4 Regular and Singular Times

Let us call a time $t \in (0, \infty)$ a *regular time* if $\|\nabla u\| \in L^\infty(U)$ for some neighbourhood U of $\{t\}$.

Let $\mathcal{R}$ denote the set of all regular times. This is clearly open, so we can write $\mathcal{R}$ as a disjoint union of open intervals:

$$\mathcal{R} = \bigcup_{i=1}^\infty (a_i, b_i); \tag{2.18}$$

since weak (Leray–Hopf) solutions are eventually strong (and so bounded) we can take $b_1 = \infty$.

We now show that $u(t)$ is regular if $t \in \mathcal{R}$.

Lemma 2.7 *If $t \in \mathcal{R}$ then $u(t) \in C^\infty$.*

Proof If $t \in \mathcal{R}$ then $u \in L^\infty((t - \delta, t + \delta), V)$ for some $\delta > 0$. In particular there exists $t_0 \in (t - \delta, t - \delta/2)$ such that $u(t_0) \in V$; the regularity result of Corollary 2.6 now implies that $u(t) \in C^\infty$.   □

We will call a time a *singular time* if it is not regular. Thus, if T is a singular time then there exists a sequence of times $t_j \to T$ such that $\|\nabla u(t_j)\| \geq j$, and so, by rearranging[3] (2.17) we obtain, for any $t < T$,

$$\|\nabla u(t)\|^4 \geq \frac{\|\nabla u(t_j)\|^4}{1 + 2c(t_j - t)\|\nabla u(t_j)\|^4}; \tag{2.19}$$

hence, on letting $j \to \infty$, it follows that

$$\|\nabla u(t)\|^2 \geq \frac{1}{\sqrt{2c(T - t)}}. \tag{2.20}$$

In particular, any singular time T must actually be a 'blowup time' with $\|\nabla u(t)\| \to \infty$ as $t \to T^-$. [While the H^1 norm must blow up, it is not known whether any weak solution v that coincides with u on $(0, T)$ must satisfy $\limsup_{t \to T^+} \|\nabla v(t)\| = \infty$.]

Alternatively we can interpret (2.20) as a 'local regularity' result. If we start from (2.20) and integrate from $T - r$ to T then we obtain

$$\int_{T-r}^T \|\nabla u(t)\|^2 \, dt \geq \sqrt{\frac{1}{2c}} \int_{T-r}^T \frac{1}{\sqrt{T - t}} \, dt = \sqrt{\frac{2}{c}} r^{1/2}.$$

If we set $\varepsilon_* = \sqrt{2/c}$ then we have the following conditional 'ε-regularity' result.

Lemma 2.8 *There exists an absolute constant ε_* such that if*

$$r^{-1/2} \int_{T-r}^T \|\nabla u(s)\|^2 \, ds < \varepsilon_*$$

then u is regular at time T.

---

[3] A more careful argument proceeds by contradiction: if (2.19) does not hold then $\|\nabla u(t)\|^4 < \text{RHS}$ of (2.19). In this case it follows from (2.17) that a strong solution v with $v(t) = u(t)$ exists on the time interval $[t, t_j]$ and $\|\nabla v(t_j)\| < \|\nabla u(t_j)\|$. But by weak-strong uniqueness we must have $u = v$ on $[t, t_j]$, which yields a contradiction.

2.2.5   'Partial Regularity': The Set of Singular Times

All the partial regularity results that we will discuss rely on two components: an ε-regularity result, like Lemma 2.8, and a global bound on the same quantity, such as $\nabla u \in L^2(\Omega \times (0, T))$. These two ingredients can be combined to limit the size of 'singular' points of u: there cannot be too many of these, otherwise the integrability would be violated.

2.2.5.1   Dimensions

To describe the 'size' of a set we will use its box-counting dimension (also called the Minkowski dimension). There are a number of equivalent definitions of this dimension; we focus on two here (which also work for subsets of infinite-dimensional spaces) and will use both in what follows. For more details see Falconer [11] or Robinson [28].

Definition in Terms of Coverings
Let X be compact set. We let $N(X, \varepsilon)$ be the minimum number of balls of radius ε whose centres lie in X needed to cover X, and define the box-counting dimension of X to be

$$\dim_B(X) := \limsup_{\varepsilon \to 0} \frac{\log N(X, \varepsilon)}{-\log \varepsilon}.$$

Essentially this extracts the exponent d from $N(X, \varepsilon) \sim \varepsilon^{-d}$.

If $d > \dim_B(X)$ then $N(X, \varepsilon) \le \varepsilon^{-d}$ for ε sufficiently small; while for each $d < \dim_B(X)$ there is a sequence $\varepsilon_j \to 0$ such that $N(X, \varepsilon_j) > \varepsilon_j^{-d}$.

Definition in Terms of ε-Separated Points
Let $M(X, \varepsilon)$ be the maximal number of ε-separated points in X, and use the same definition as above with $N(X, \varepsilon)$ replaced by $M(X, \varepsilon)$. This gives the same quantity.

[To show that these definitions are equivalent: (1) $M(X, \varepsilon) \le N(X, \varepsilon)$ since none of the ε-separated balls cover the centres of the others (2) We have $N(X, \varepsilon) \le M(X, \varepsilon/3)$—if some $x \in X$ does not belong to an ε-ball about an $\varepsilon/3$-separated point then $B(x, \varepsilon/3)$ and $B(y, \varepsilon/3)$ are disjoint for every $\varepsilon/3$-separated y, giving a new $\varepsilon/3$-separated point x.]

As above, if $d < \dim_B(X)$ then there is a sequence $\varepsilon_j \to 0$ for which it is possible to find a set of at least ε_j^{-d} points that are ε_j separated.

It can be useful to mix these definitions, as the following simple result shows.

Lemma 2.9 *The set $A_\alpha := \{n^{-\alpha} : n \in \mathbb{N}\} \cup \{0\}$ has box-counting dimension $1/(1 + \alpha)$.*

Proof Points in A_α are a distance ε apart for

$$n^{-\alpha} - (n+1)^{-\alpha} \sim \alpha n^{-(1+\alpha)} \geq \varepsilon,$$

i.e. for $n \lesssim \varepsilon^{-1/(1+\alpha)}$. It follows that $M(X, \varepsilon/2) \gtrsim \varepsilon^{-1/(1+\alpha)}$.

However, n intervals of length ε will certainly cover these points, and the remaining points are contained in the interval $[0, \varepsilon^{\alpha/(1+\alpha)}]$, which requires no more than $\varepsilon^{-1/(1+\alpha)}$ intervals of length ε to cover: $N(A_\alpha, \varepsilon) \lesssim \varepsilon^{-1/(1+\alpha)}$.

These upper and lower bounds imply that $\dim_B(A_\alpha) = 1/(1+\alpha)$ as claimed. $\qquad\square$

Exercise Suppose that (e_n) is an orthonormal subset of a Hilbert space H, and define $H_\alpha := \{n^{-\alpha} e_n : n \in \mathbb{N}\} \cup \{0\}$. Show that $\dim_B(H_\alpha) = 1/\alpha$.

The following consequence of the covering definition will be useful later (and can itself serve as the basis for yet another equivalent definition of the box-counting dimension in Euclidean spaces).

Lemma 2.10 *If X is a compact subset of $\mathbb{R}^n$ and $d > \dim_B(X)$ then there exists $C > 0$ such that*

$$\mu(O(X, \varepsilon)) \leq C \varepsilon^{n-d},$$

where

$$O(X, \varepsilon) = \{x + y : x \in X, |y| < \varepsilon\} \qquad (2.21)$$

is the ε-neighbourhood of X.

Proof By definition, for any $d > \dim_B(X)$, we have $N(X, \varepsilon) < C\varepsilon^{-d}$ for some $C > 0$. Since X can be covered by $C\varepsilon^{-d}$ balls of radius ε, it follows that the ε-neighbourhood of X can be covered by $C\varepsilon^{-d}$ balls of radius 2ε, of total measure no more than $2^n C \varepsilon^{n-d}$. $\qquad\square$

Exercise Show that if $X \subset \mathbb{R}^n$ then

$$\dim_B(X) = n - \sup\{s : \mu(O(X, \varepsilon)) \leq C\varepsilon^s \text{ for some } C > 0\}.$$

Corollary 2.11 *If $X \subset \mathbb{R}^n$ and $\dim_B(X) < n$ then $\mu(X) = 0$.*

Proof Choose d with $\dim_B(X) < d < n$. Since $X \subset O(X, \varepsilon)$ for every $\varepsilon > 0$, it follows that

$$\mu(X) \leq \mu(O(X, \varepsilon)) \leq C\varepsilon^{n-d}$$

and so $\mu(X) = 0$. $\qquad\square$

We can now use this definition to deduce a 'partial regularity' result concerning the singular times of a Navier–Stokes solution.

2.2.5.2   Dimension of the Set of Singular Times

We now bound the box-counting dimension of the set of singular times. The result in this form is due to Robinson and Sadowski [29], but the proof we give is based on Kukavica [21].

Proposition 2.12 *If*

$$\mathcal{T} = \{t \geq 0 : \|\nabla u(t)\| \text{ is unbounded in a neighbourhood of } t\}$$

then $\dim_B(\mathcal{T}) \leq 1/2$.

Proof It follows from Lemma 2.8 that if t is a singular time then

$$\int_{t-r}^{t} \|\nabla u(s)\|^2 \, ds \geq \varepsilon_0 r^{1/2}$$

for every $r > 0$. Suppose now, for a contradiction, that the set of all singular times $\mathcal{T}$ satisfies $\dim_B(\mathcal{T}) > d > 1/2$. Then there is a sequence $\varepsilon_j \to 0$ for which one can find a collection $\mathcal{T}_j$ of at least ε_j^{-d} points that are ε_j separated. Then the intervals $(t - \varepsilon_j/2, t)$ are disjoint for distinct points t in $\mathcal{T}_j$, and so

$$\int_0^T \|\nabla u(s)\|^2 \, ds \gtrsim \varepsilon_j^{-d} (\varepsilon_0 \varepsilon_j^{1/2}) = \varepsilon_0 \varepsilon_j^{-(d-1/2)};$$

letting $j \to \infty$ contradicts the integrability of $\|\nabla u\|^2$.   □

Using Corollary 2.11 it follows that the set of singular times has zero measure, i.e. almost every time is regular.

Exercise While the fact that any weak solution satisfies $u \in L^2(0, T; H^1)$ implies that $u(t) \in H^1$ for almost every t, this is not the same as saying that almost every time is regular; to see this construct a function $f \in L^2(0, 1)$ for which the set of singular times,

$$\text{sing}(f) := \{t \in [0, 1] : f \text{ is unbounded in a neighbourhood of } t\},$$

has full measure.

The proof of Proposition 2.12 relies only on two ingredients, namely that the quantity $X(t) := \|\nabla u(t)\|^2$ satisfies

$$\dot{X} \lesssim X^3 \qquad \text{and} \qquad \int_0^T X(s) \, ds < \infty. \tag{2.22}$$

Using only these two facts we can do no better.

Lemma 2.13 *For each $0 < d < 1/2$ there exists a function $X_d : [0, 1] \to \mathbb{R}$ that satisfies (2.22) and for which*

$$\dim_B(\text{sing}(X_d)) = d.$$

Proof Let A_α be the set from Lemma 2.9; if $\alpha = (1 - d)/d$ then $\dim_B(A_\alpha) = d$, and $d < 1/2$ corresponds to choosing some $\alpha > 1$. Now let

$$X_d = \left(n^{-\alpha} - t\right)^{-1/2} \qquad \text{for} \qquad (n+1)^{-\alpha} < t < n^{-\alpha}$$

with $X_d(n^{-\alpha})$ chosen arbitrarily. Then

$$\dot{X}_d(t) = \frac{1}{2}\left(n^{-\alpha} - t\right)^{-3/2} = \frac{1}{2}X_d(t)^3$$

for almost every $t \in [0, 1]$, and

$$\int_0^1 X_d(t)\,dt = \sum_{n=1}^{\infty} \int_{(n+1)^{-\alpha}}^{n^{-\alpha}} (n^{-\alpha} - t)^{-1/2}\,dt$$

$$= \sum_{n=1}^{\infty} \frac{1}{2}\left(\frac{1}{n^\alpha} - \frac{1}{(n+1)^\alpha}\right)^{1/2}$$

$$\leq \sum_{n=1}^{\infty} \left(\frac{\alpha}{n^{1+\alpha}}\right)^{1/2} < \infty$$

since $(1 + \alpha)/2 > 1$. □

The above result shows that it is not possible to improve the bound on the box-counting dimension using only the ingredients from (2.22) that went into the proof of Proposition 2.12. But with a new definition one can prove something a little stronger. The following argument is due to Kukavica [21]. Given $r > 0$, let

$$\mathcal{F}^{s,r}(X) := N(X, r)r^s$$

$$= \inf\{nr^s : \text{covers of } X \text{ by } n \text{ balls of radius } r, \ n \in \mathbb{N}\}$$

and set $\mathcal{F}^s(X) := \limsup_{r \to 0} \mathcal{F}^{s,r}(X)$. This quantity has the following properties:

(i) $\dim_B(X) = \inf\{s \geq 0 : \mathcal{F}^s(X) = 0\}$;
(ii) $\mathcal{H}^s(X) \leq \mathcal{F}^s(X)$, where $\mathcal{H}^s$ is the s-dimensional Hausdorff measure.

Since the inequality in (ii) can be strict, the following result improves on the bound in Lemma 2.12 and on the result of Scheffer [34] that $\mathcal{H}^{1/2}(\mathcal{T}) = 0$.

Proposition 2.14 $\mathcal{F}^{1/2}(\mathcal{T}) = 0$.

Proof Fix $\varepsilon > 0$ and choose $t_i \in \mathcal{T}$ such that $(t_i - r/3, t_i + r/3)$ form a maximal family of disjoint intervals with centres in $\mathcal{T}$. The expanded intervals $(t_i - r, t_i + r)$ then form a cover $\mathcal{O}$ all of $\mathcal{T}$.

Since $\dim_B(\mathcal{T}) \leq 1/2$ we can choose r small enough that the measure of $\mathcal{O}$ is small enough that

$$\int_{\mathcal{O}} \|\nabla u\|^2 < \varepsilon.$$

Since $t_i \in \mathcal{T}$ we have

$$\int_{t_i - r/3}^{t_i} \|\nabla u(s)\|^2 \, ds \geq \varepsilon_*(r/3)^{1/2};$$

since the intervals $(t_i - r/3, t_i)$ are disjoint we have

$$\varepsilon_* N(\mathcal{T}, r/3)(r/3)^{1/2} \leq \sum_i \int_{t_i - r/3}^{t_i} \|\nabla u\|^2 \, ds$$

$$\leq \int_{\mathcal{O}} \|\nabla u(s)\|^2 \, ds < \varepsilon$$

and the result follows. $\square$

2.3 Serrin's Local Regularity Result for $u \in L^{5+}(Q^*)$

In this section we prove a local conditional regularity result, due to Serrin [37]. We take a solution that solves the Navier–Stokes equations only on some region U of space-time, and show that if $u \in L^\alpha(U)$ for some $\alpha > 5$ then in fact u is smooth in the spatial variables within U. We then give a sketch of the proof of the same result when $u \in L^5(U)$.

[Two comments are in order here. First, Serrin's result is more general, allowing for different integrability in space and time: if $u \in L^r(a, b; L^s(\Omega))$ with

$$\frac{2}{r} + \frac{3}{s} < 1 \tag{2.23}$$

then u is smooth in the spatial variables on $\Omega \times (a, b)$. The same result allowing for equality in the 'Serrin condition' (2.23) (which therefore means that $u \in L^5(U)$ is in fact sufficient for spatial smoothness) was shown later by Fabes et al. [10] when $3 < s < \infty$; the endpoint case $r = 2, s = \infty$ was covered by Struwe [38] and the case $r = \infty, s = 3$ by Escauriaza et al. [8].]

If (u, p) solve the Navier–Stokes equations on $\mathbb{R}^3$, then so do the rescaled pair

$$u_\lambda(x, t) := \lambda u(\lambda x, \lambda^2 t) \qquad p_\lambda(x, t) := \lambda^2 p(\lambda x, \lambda^2 t) \tag{2.24}$$

for any $\lambda > 0$. The means that in space-time the 'natural' set to consider is not a ball in $\mathbb{R}^4$, but rather the 'cylinder'

$$Q_r^*(x, t) = B_r(x) \times (t - r^2/2, t + r^2/2) \tag{2.25}$$

(we choose the factor $1/2$ in the time direction for later convenience).

By 'solving the Navier–Stokes equations on U' we mean that u is a weakly divergence-free distributional solution of the equations on U, i.e. that[4]

$$\int_U \langle u, \partial_t \varphi \rangle + \langle u, \Delta \varphi \rangle + \langle u \otimes u : \nabla \varphi \rangle \, dx \, dt = 0, \tag{2.26}$$

for every $\varphi \in \mathscr{D}_\sigma(U)$, where

$$\mathscr{D}_\sigma(U) := \{\varphi \in [C_c^\infty(U)]^3 : \nabla \cdot \varphi(x, t) = 0\}$$

and $(u \otimes v)_{ij} = u_i v_j$.

In terms of 'local solutions' within parabolic cylinders, it is easy to see that if (u, p) solve the Navier–Stokes equations in $Q_1^*(0)$, then the rescaled (u_λ, p_λ) solve the equations on $Q_{1/\lambda}^*(0)$.

In this section we give a (sketch) proof of the following local conditional regularity theorem.

Theorem 3.1 *Suppose that u solves the Navier–Stokes equations on $Q_r^*(x, t)$ and $u \in L^\alpha(Q_r^*(x, t))$ for some $\alpha > 5$. Then in fact u is smooth in the spatial variables in $Q_{r/2}^*(x, t)$.*

The following is an almost immediate corollary. Note that in particular this shows that the above theorem is true with the conclusion holding in $Q_r^*(x, t)$.

Corollary 3.2 *Let U be an open set in space-time, and u a solution of the Navier–Stokes equations on U such that $u \in L^\alpha(U)$ with $\alpha > 5$. Then u is smooth in the spatial variables within U.*

Proof Take $(x, t) \in U$, choose $r > 0$ such that $Q_r^*(x, t) \in U$, and apply Theorem 3.1 to show that u is smooth in the spatial variables within $Q_{r/2}^*(x, t)$. □

We will show that if $u \in L^\alpha(Q_1^*)$ then u is spatially smooth in $Q_{1/2}^*$; Theorem 3.1 then follows using the rescaling in (2.24).

---

[4]Here we use a colon for the matrix product, i.e. $A : B = \sum_{i,j=1}^3 A_{ij} B_{ij}$.

The key idea is to work with the vorticity form of the Navier–Stokes equations: if u is smooth then we can define $\omega = \nabla \times u$ (the vorticity), and then ω satisfies the vorticity equation[5]

$$\omega_t - \Delta\omega + (u \cdot \nabla)\omega - (\omega \cdot \nabla)u = 0.$$

If we rewrite this as

$$\omega_t - \Delta\omega = (\omega \cdot \nabla)u - (u \cdot \nabla)\omega = \mathrm{div}(\omega \otimes u - u \otimes \omega) \qquad (2.27)$$

then we have recast the equation as a heat equation for ω, although admittedly the right-hand side depends on ω and on u (which itself also depends on ω—we will discuss recovering u from ω later).

[With a little care, it is possible to show something similar based on the formulation of what it means for u to be a 'local weak solution' in (2.26): for any test function $\phi \in [C_c^\infty(U)]^3$ we take $\varphi = \mathrm{curl}\,\phi$, which is divergence free and so an element of $\mathscr{D}_\sigma(U)$. We can therefore use this φ as a test function in (2.26) to give

$$\int_U \langle u, \partial_t(\nabla \times \phi)\rangle + \langle u, \Delta(\nabla \times \phi)\rangle + \langle u \otimes u, \nabla(\nabla \times \phi)\rangle \, dx \, dt = 0,$$

which after manipulations similar to those leading from the Navier–Stokes equations to (2.27) yields the weak form of the vorticity equation

$$\int_U \langle \omega, \partial_t \phi\rangle + \langle \omega, \Delta\phi\rangle + \langle (u \otimes \omega) - (\omega \otimes u), \nabla\phi\rangle \, dx \, dt = 0,$$

see RRS for details.]

The formulation in (2.27) is convenient, since it does not contain the pressure and properties of solutions of the heat equation are well understood. In particular if we set

$$K(x, t) = \begin{cases} ct^{-3/2}e^{-|x|^2/4t} & t > 0 \\ 0 & \text{otherwise} \end{cases}$$

---

[5]The vorticity equation follows on taking the curl of the Navier–Stokes equations and using the two vector identities

$$\frac{1}{2}\nabla|u|^2 = (u \cdot \nabla)u + u \times \omega \quad \text{and} \quad \nabla \times (a \times b) = a(\nabla \cdot b) - b(\nabla \cdot a) + (b \cdot \nabla)a - (a \cdot \nabla)b$$

along with the fact that both u and ω are divergence free.

(K is the 'heat kernel') and the support of f is contained in $Q_r^*(a, s)$ then $K \star f$, defined by setting

$$K \star f(x, t) := \int_{s-r^2/2}^{s+r^2/2} \int_{B(a,r)} K(x - \xi, t - \tau) f(\xi, \tau) \, d\xi \, d\tau,$$

is the solution for $(x, t) \in Q_r^*(a, s)$ of

$$\omega_t - \Delta\omega = f(x, t) \qquad \text{with} \quad \omega(x, s - r^2/2) = 0.$$

We can use this to write down a representation formula for ω within Q_r^*, $r < 1$, in terms of $g = \omega \otimes u - u \otimes \omega$:

$$\begin{aligned}
\omega &= K \star (\text{div } g) + H(x, t) \\
&= \nabla K \star g + H(x, t), \qquad\qquad (2.28)
\end{aligned}$$

where $H(x, t)$ is a solution of the heat equation (and hence smooth in both space and time). The term $H(x, t)$ therefore plays only a minimal role in what follows. [Note that there are also some subtleties here that we have ignored. For example, are u and ω (and so g) actually smooth enough to use the representation in (2.28)? Serrin's 'rigorous' derivation is very short on details, but the idea is relatively standard: find a way to mollify Eq. (2.26) and take limits. One method is to start with a test function $\phi \in \mathcal{D}_\sigma$ and then use $\varphi = \psi_\varepsilon * \phi$ in (2.26), where ψ_ε is a standard mollifying function.]

In order to exploit this representation formula, we will use Young's inequality for convolutions, which we recall here. Note that $r = \infty$ is included: we obtain $f * g \in L^\infty$ if $p^{-1} + q^{-1} \le 1$, where $*$ denotes convolution over the whole space.

Lemma 3.3 (Young's Inequality) *Let $1 \le p, q, r \le \infty$ satisfy*

$$\frac{1}{r} + 1 = \frac{1}{p} + \frac{1}{q}.$$

Then for all $f \in L^p$, $g \in L^q$, we have $f * g \in L^r$ with

$$\|f * g\|_{L^r} \le \|f\|_{L^p} \|g\|_{L^q}. \qquad\qquad (2.29)$$

2.3.1 Step 1: Show that $\omega \in L^\infty(Q_s^*)$, $s < 1$

If we apply Young's inequality using ∇K as the kernel, then we obtain the following, which is fundamental to the first part of Serrin's argument. In the statement we use Q_ρ^* to denote any space-time cylinder.

Lemma 3.4 *If* $g \in L^q(Q^*_\rho)$ *for some* $q \le 5$ *then*

$$\|\nabla K \star g\|_{L^r(Q^*_\rho)} \le c\|g\|_{L^q(Q^*_\rho)}$$

for any r such that

$$\frac{1}{r} > \frac{1}{q} - \frac{1}{5}; \tag{2.30}$$

if $g \in L^q(Q^*_\rho)$ *for some* $q > 5$ *then* $\nabla K \star g \in L^\infty(Q^*_\rho)$.

Proof We have $\nabla K \in L^p((0, \rho^2) \times B_{2r}(0))$ for any $1 \le p < 5/4$. To see this, first observe that

$$|\nabla K(x, t)| \le c|x|t^{-5/2}e^{-|x|^2/4t}.$$

Now one can calculate

$$\|\nabla K\|^p_{L^p((0,\rho^2) \times B_{2\rho}(0))} \le \int_0^{\rho^2} \int_{B_{2\rho}(0)} c|x|^p t^{-5p/2} e^{-p|x|^2/4t} \, dx \, dt$$

$$= c \int_0^{\rho^2} t^{-5p/2} t^{p/2+3/2} \left(\int_{B_{\rho\sqrt{2p/t}}(0)} |y|^p e^{-|y|^2} \, dy \right) dt$$

$$\le c' \int_0^{\rho^2} t^{-2p+(3/2)} \, dt < \infty,$$

provided that $1 \le p < 5/4$. The result now follows from Lemma 3.3. □

Corollary 3.5 *If* $\omega \in L^p(Q^*_r)$ *and* $u \in L^\alpha(Q^*_r)$, *then if*

$$\frac{1}{\alpha} + \frac{1}{\rho} \ge \frac{1}{5} \tag{2.31}$$

we have $\omega \in L^\sigma(Q^*_r)$ *provided that*

$$\frac{1}{\sigma} > \frac{1}{\alpha} + \frac{1}{\rho} - \frac{1}{5}. \tag{2.32}$$

If

$$\frac{1}{\alpha} + \frac{1}{\rho} < \frac{1}{5} \tag{2.33}$$

then $\omega \in L^\infty(Q^*_r)$.

Proof Since $u \in L^\alpha(Q_r^*)$ and $\omega \in L^\rho(Q_r^*)$ we have

$$g = u \otimes \omega + \omega \otimes u \in L^{\rho'}(Q_r^*)$$

with

$$\frac{1}{\rho'} = \frac{1}{\alpha} + \frac{1}{\rho}.$$

Now we can use Lemma 3.4 to show that if $\rho' \leq 5$, i.e. under condition (2.31), then $\omega = (\nabla K \star g) + H \in L^\sigma(Q_r^*)$ for any σ satisfying (2.32). Similarly, if $\rho' > 5$, i.e. under condition (2.33), it follows that $\omega \in L^\infty(Q_r^*)$. □

If u is a weak solution we know that $u \in L^2(0, T; H^1)$; it follows that $\nabla u \in L^2(0, T; L^2)$, so in particular $\omega \in L^2(Q_1^*)$. We now iterate the above argument to improve the regularity of ω and show that $\omega \in L^\infty(Q_r^*)$ for every $r < 1$.

Proposition 3.6 *If* $\omega \in L^2(Q_r^*)$ *and* $u \in L^\alpha(Q_r^*)$ *for some* $\alpha > 5$ *then in fact* $\omega \in L^\infty(Q_r^*)$.

Proof We apply Corollary 3.5 repeatedly: it is enough to show that $\omega \in L^\rho(Q_r^*)$ with ρ satisfying (2.33), since the next step of the iteration then shows that $\omega \in L^\infty(Q_r^*)$.

If $\omega \in L^{\rho_n}(Q_r^*)$ then Corollary 3.5 implies that while $\alpha^{-1} + \rho_n^{-1} \geq 5^{-1}$ we have $\omega \in L^{\rho_{n+1}}(Q_r^*)$ if

$$\frac{1}{\alpha} + \frac{1}{\rho_{n+1}} > \left[\frac{1}{\alpha} + \frac{1}{\rho_n}\right] - \left[\frac{1}{5} - \frac{1}{\alpha}\right].$$

Since $\alpha > 5$ this shows that $\alpha^{-1} + \rho_n^{-1}$ decreases by a constant factor (anything less than $1/5 - 1/\alpha$) with each iteration. It follows that after a finite number of iterations condition (2.33) will be satisfied, and then $u \in L^\infty(Q_r^*)$ as claimed. □

[As an example, if we assume that $u \in L^\infty(Q_r^*)$ and start with $\omega \in L^2(Q_r^*)$ then on the first iteration we obtain $\omega \in L^p$ for any $p < 10/3$; on the second $\omega \in L^p$ for any $p < 10$; and then on the third $\omega \in L^\infty$.]

2.3.2 Show that $u \in C_x^\infty(Q_s^*)$ for any $s < 1$

Proving smoothness of u is based on an iterative process consisting of two stages. It turns out that we have to keep track of the smoothness in time and in space separately. To do this we adopt the notation

$$L_t^p L_x^q(Q_r^*) \qquad \text{and} \qquad L_t^p C_x^\alpha(Q_r^*)$$

to consist of functions f in

$$L^p((-r^2/2, r^2/2); L^q(B_r)) \quad \text{and} \quad L^p((-r^2/2, r^2/2); C^\alpha(B_r)),$$

respectively.

The first stage in this iteration is the reconstruction of u from ω via the Biot–Savart Law: if we take the curl of the relation $\omega = \nabla \times u$ then we obtain

$$-\Delta u = \nabla \times \omega, \tag{2.34}$$

where we have used the vector identity

$$\nabla \times (\nabla \times u) = \nabla(\nabla \cdot u) - \Delta u$$

and the fact that u is divergence free.

On the whole space it follows, by inverting the Laplacian in (2.34), that

$$u = -\frac{1}{4\pi} \int_{\mathbb{R}^3} \frac{(x - y) \times \omega(y)}{|x - y|^3} \, dy;$$

if we only know ω on a subset U of $\mathbb{R}^3$ then instead we have

$$u(x) = -\frac{1}{4\pi} \int_U \frac{(x - y) \times \omega(y)}{|x - y|^3} \, dy + H(x), \tag{2.35}$$

where H is harmonic ($\Delta H = 0$). For proofs of these facts see Chapter 12 in RRS.

As a consequence of this representation formula, it is possible to show (in line with what one would guess from (2.34)) that u is 'one derivative better' than ω, i.e.

$$\omega \in L^\infty(t_1, t_2; C^{k,\alpha}(B_R)) \quad \Rightarrow \quad u \in L^\infty(t_1, t_2; C^{k+1,\alpha'}(B_{R'})) \tag{2.36}$$

for $R' < R$ and for any $0 < \alpha' < \alpha$ (see Theorem 12.6 in RRS).

In order to improve the regularity of ω, since we know that ω satisfies

$$\omega_t - \Delta\omega = \text{div}(\omega u - u\omega)$$

we can appeal to 'standard' results for the heat equation: if

$$\omega_t - \Delta\omega = \text{div} g \quad \text{in } Q_R^* \tag{2.37}$$

then for any $R' < R$ and any $0 < \alpha < 1$ we have

$$g \in L_t^\infty L_x^\infty(Q_R^*) \quad \Rightarrow \quad \omega \in L_t^\infty C_x^{0,\alpha}(Q_R^*) \tag{2.38}$$

and (by considering derivatives of the Eq. (2.37))

$$\partial^k g \in L_t^\infty L_x^\infty(Q_R^*) \qquad \Rightarrow \qquad \omega \in L_t^\infty C_x^{k,\alpha}(Q_{R'}^*), \tag{2.39}$$

where by $\partial^k g \in L_t^\infty L_x^\infty$ we mean that all derivatives of order up to k are bounded (note that this is a little weaker than $\omega \in L_t^\infty C_x^k$); we also have

$$g \in L_t^\infty C_x^{k,\alpha}(Q_R^*) \qquad \Rightarrow \qquad \partial^{k+1}\omega \in L_t^\infty L_x^\infty(Q_{R'}^*). \tag{2.40}$$

[Full proofs can be found in Appendix D of RRS.]

We now use these to reach a degree of regularity at which we can establish a more systematic induction (throughout what follows we take α, α' with $0 < \alpha' < \alpha < 1$) and $R > R' > R'' > R''' > R^{(4)} > \cdots > R/2$:

$$u, \omega \in L_t^\infty L_x^\infty(Q_R^*) \qquad \Rightarrow \qquad \omega u - u\omega \in L_t^\infty L_x^\infty(Q_R^*)$$

$$\omega u - u\omega \in L_t^\infty L_x^\infty(Q_R^*) \qquad \Rightarrow \qquad \omega \in L_t^\infty C_x^{0,\alpha}(Q_R^*) \qquad \text{by (2.38)}$$

$$\omega \in L_t^\infty C_x^{0,\alpha}(Q_R^*) \qquad \Rightarrow \qquad u \in L_t^\infty C_x^{1,\alpha'}(Q_{R'}^*) \qquad \text{by (2.36)}$$

$$u, \omega \in L_t^\infty C_x^{0,\alpha}(Q_R^*) \qquad \Rightarrow \qquad \omega u - u\omega \in L_t^\infty C_x^{0,\alpha}(Q_R^*)$$

$$\omega u - u\omega \in L_t^\infty C_x^{0,\alpha}(Q_R^*) \qquad \Rightarrow \qquad \partial\omega \in L_t^\infty L_x^\infty(Q_{R'}^*) \qquad \text{by (2.39)}.$$

We now have $u \in L_t^\infty C_x^{1,\alpha'}(Q_{R'}^*)$ and $\partial\omega \in L_t^\infty L_x^\infty(Q_{R'}^*)$ and from here the induction process to improve from $u \in L_t^\infty C_x^{k,\alpha}$, $\partial^k\omega \in L_t^\infty L_x^\infty$ on $Q_{R(2k+1)}^*$ to $u \in L_t^\infty C_x^{k+1,\alpha'}$, $\partial^{k+1}\omega \in L_t^\infty L_x^\infty$ on $Q_{R(2k+3)}^*$ is more systematic:

$$\left.\begin{array}{l} u \in L_t^\infty C_x^k(Q_{R'}^*) \\ \partial^k\omega \in L_t^\infty L_x^\infty(Q_{R'}^*) \end{array}\right\} \quad \Rightarrow \quad \partial^k(\omega u - u\omega) \in L_t^\infty L_x^\infty(Q_{R'}^*)$$

$$\partial^k(\omega u - u\omega) \in L_t^\infty L_x^\infty(Q_{R'}^*) \quad \Rightarrow \quad \omega \in L_t^\infty C_x^{k,\alpha}(Q_{R''}^*) \qquad \text{by (2.39)}$$

$$u, \omega \in L_t^\infty C_x^{k,\alpha'}(Q_{R'}^*) \quad \Rightarrow \quad \omega u - u\omega \in L_t^\infty C_x^{k,\alpha'}(Q_{R''}^*)$$

$$\omega u - u\omega \in L_t^\infty C_x^{k,\alpha'}(Q_{R''}^*) \quad \Rightarrow \quad \partial^{k+1}\omega \in L_t^\infty L_x^\infty(Q_{R'''}^*) \qquad \text{by (2.40)}$$

$$\omega \in L_t^\infty C_x^{k,\alpha}(Q_{R''}^*) \quad \Rightarrow \quad u \in L_t^\infty C_x^{k+1,\alpha'}(Q_{R'''}^*) \qquad \text{by (2.36)}.$$

It is clear that continuing in this way leads to $u \in L_t^\infty C_x^k$ for every $k \in \mathbb{N}$, and so u is spatially smooth as claimed.

Note, in particular, that the bounds on the spatial derivatives of u are uniform with respect to time within $Q_{R/2}^*$. This will be useful in the final proof in these notes (of the almost-everywhere uniqueness of particle trajectories), where this will ensure that u is Lipschitz continuous in space with a Lipschitz constant that is uniform in time.

2.3.3   The Case $u \in L^5(Q^*)$

The following argument[6] gives an indication of how one can prove regularity for the boundary case $u \in L^5$ (and more generally when $u \in L^r(a, b; L^s(\Omega))$ with equality in the condition (2.23)).

The first observation is that it is sufficient to prove the result under a smallness condition on $\|u\|_{L^5(Q^*_r)}$, since if $u \in L^5(U)$ then it follows that for any $\varepsilon > 0$, for each point (x, t) there is an $r > 0$ such that $\|u\|_{L^5(Q^*_r)} < \varepsilon$.

Starting from the equality

$$\omega = \nabla K \star g + H(x, t), \tag{2.41}$$

choose any $r < \infty$. Then, recalling that $g = \omega u - u \omega$, we have

$$\|g\|_{L^m} \lesssim \|u\|_{L^5} \|\omega\|_{L^r}, \qquad \text{where} \qquad \frac{1}{m} = \frac{1}{5} + \frac{1}{r},$$

and so using Lemma 3.4 we obtain[7]

$$\|\omega\|_{L^r} \leq C \|u\|_{L^5} \|\omega\|_{L^r} + \|H\|_{L^r}. \tag{2.42}$$

If $\|u\|_{L^5} < \varepsilon := 1/2C$ then this inequality yields

$$\|\omega\|_{L^r} \leq 2\|H\|_{L^r}. \tag{2.43}$$

The Biot–Savart Law in (2.35) allows us to recover u from ω (at each fixed time t); since this is essentially a convolution with a kernel of the order of $|x|^{-2}$ at infinity, the kernel is in the weak Lebesgue space $L^{3/2,\infty}$. Away from the endpoint values $(1 < p, q, r < \infty)$ Young's convolution inequality (2.29) still holds when the L^p norm on the right-hand side replaced by the norm in $L^{p,\infty}$,

$$\|f * g\|_{L^r} \lesssim \|f\|_{L^{p,\infty}} \|g\|_{L^q},$$

---

[6]This is the basis of the proof given in the paper by Takahashi [39], although rather than following exactly the argument here he works with the equation for $\phi\omega$, where ϕ is a cutoff function.

[7]Here our argument has what is potentially a fatal flaw: in the inequality (2.42) we have assumed that $\omega \in L^r$, which is in fact what we want to prove (the inequality is trivially true if $\omega \notin L^r$ since then both sides are infinite). However, this can be circumvented by considering estimates for the equation

$$\partial_t W^\varepsilon - \Delta W^\varepsilon = \text{div}(W^\varepsilon u_\varepsilon - u_\varepsilon W^\varepsilon),$$

where u_ε is a mollified version of the original function u, and then taking limits as $\varepsilon \to 0$.

and so it follows that if $\omega \in L_x^q$ for some $q > 3$ then $u \in L_x^\infty$ with

$$\|u\|_{L^\infty} \lesssim \|\omega\|_{L^q}.$$

In particular if we use (2.43) to bound ω in $L^6(U)$ (for example) then for each t we have $\|u\|_{L^\infty} \lesssim \|\omega\|_{L^6}$ and hence

$$u \in L_t^\infty L_x^6 \subset L^6(U).$$

This finishes our sketch proof, since we have already shown that if $u \in L^6(U)$ then u is spatially smooth in U.

2.4   Space-Time Partial Regularity

We will now develop—following Caffarelli et al. [3]—a theory that to some extent parallels that for regular times, but now for space-time points. We will call a space-time point (x, t) is *regular* if it has a space-time neighbourhood U such that $u \in L^\infty(U)$. It then follows from the results of Serrin sketched in the previous section that u is smooth (in space, at least) in U.

If (x, t) is not regular then it is *singular*, and we set

$$S := \{(x, t) : \mathbb{T}^3 \times (0, \infty) : (x, t) \text{ is a singular point}\}.$$

We again respect the 'parabolic scaling' of the Eq. (2.24); we now take our basic domain to be the 'non-anticipating' parabolic cylinder

$$Q_r(a, s) := \{(x, t) : |x - a| < r, s - r^2 < t < s\}.$$

Note that (a, s) is the 'centre-right' point of $Q_r(a, s)$ and is *not* contained in $Q_r(a, s)$. We have $Q_r^*(a, s) = Q_r(a, s + r^2/2)$, where Q_r^* is the centred cylinder we defined earlier in (2.25).

We will show that there exists an absolute constant $\varepsilon_0 > 0$ such that if for some $r > 0$ we have

$$\frac{1}{r^2} \int_{Q_r(a,s)} |u|^3 + |p|^{3/2} < \varepsilon_0$$

then u is bounded on $Q_{r/2}(a, s)$.

2.4.1   The Navier–Stokes Inequality

The Navier–Stokes equations themselves require two equalities: one expressing conservation of momentum and the other the incompressibility condition. We now replace the Navier–Stokes 'equality' by the following inequality:

$$u \cdot (\partial_t u - \Delta u + (u \cdot \nabla)u + \nabla p) \leq 0.$$

This is essentially the pointwise form of the energy equality, changed to an inequality. If we require this equation to hold weakly, then we obtain (iii) of the following definition. We retain the incompressibility condition in (i) and the relation between the pressure and u is given in (ii).

Definition 4.1 A pair (u, p) is a weak solution of the Navier–Stokes inequality (NSI) on $U \times (t_1, t_2)$ if

(i) $u \in L^\infty(t_1, t_2; L^2(U))$, $\nabla u \in L^2(U \times (t_1, t_2))$ with $\nabla \cdot u = 0$ almost everywhere, and $p \in L^{5/3}(U \times (t_1, t_2))$;
(ii) $-\Delta p = \partial_i \partial_j (u_i u_j)$ for a.e. $t \in (t_1, t_2)$; and
(iii) the local energy inequality

$$\int_U |u(t)|^2 \varphi + 2 \int_{t_1}^t \int_U |\nabla u|^2 \varphi \leq \int_{t_1}^t \int_U |u|^2 (\partial_t \varphi + \Delta \varphi) + \int_{t_1}^t \int_U (|u|^2 + 2p) u \cdot \nabla \varphi$$

is valid for every $t \in (t_1, t_2)$ for every choice of non-negative test function $\varphi \in C_c^\infty(U \times (t_1, t_2))$.

Exercise Derive the local energy equality under the assumption that u is smooth.

CKN prove something very like the following theorem, although their condition on the pressure is a little different since they only assumed the regularity $p \in L^{5/4}(U \times (t_1, t_2))$, which was the best known to hold in bounded domains when they wrote their paper. Note that the theorem does not require (u, p) to be a weak solution of the Navier–Stokes equations, only of the Navier–Stokes inequality.

Theorem 4.2 *There exists an absolute constant $\varepsilon_0 > 0$ such that if (u, p) is a weak solution of the Navier–Stokes inequality on $Q_1(0, 0)$ and*

$$\int_{Q_1(0,0)} |u|^3 + |p|^{3/2} \leq \varepsilon_0 \tag{2.44}$$

then $u \in L^\infty(Q_{1/2}(0, 0))$.

2.4.2 Partial Regularity

It is easy to show that if (u, p) solves the NSI on $Q_r(0, 0)$ then (u_r, p_r), where

$$u_r(x, t) = ru(rx, r^2t) \quad \text{and} \quad p_r(x, t) = r^2 p(rx, r^2t),$$

satisfies the NSI on Q_1 and

$$\int_{Q_1} |u_r|^3 + |p_r|^{3/2} = \frac{1}{r^2} \int_{Q_r} |u|^3 + |p|^{3/2}.$$

This observation yields a more useful version of Theorem 4.2.

Theorem 4.3 *If (u, p) is a weak solution of the Navier–Stokes inequality on Q_r and*

$$\frac{1}{r^2} \int_{Q_r} |u|^3 + |p|^{3/2} < \varepsilon_0$$

then $u \in L^\infty(Q_{r/2})$.

From here we can easily obtain an upper bound on the dimension of the singular set; but it is useful first to recast Theorem 4.3 in what is actually a slightly weaker form. Note that since $(x, t) \notin Q_r(x, t)$, Theorem 4.3 does not actually give a condition to guarantee that (x, t) is a regular point, but this is easily remedied. Recall that previously we defined the 'centred' cylinder $Q_r^*(x, t)$ about (x, t) to be

$$Q_r^*(x, t) = B_r(x) \times (t - r^2/2, t + r^2/2);$$

we can deduce the following from Theorem 4.3.

Corollary 4.4 *If (u, p) is a weak solution of the Navier–Stokes inequality on $Q_r^*(x, t)$ and*

$$\frac{1}{r^2} \int_{Q_r^*(x,t)} |u|^3 + |p|^{3/2} < \varepsilon_0/4 \tag{2.45}$$

then $u \in L^\infty(Q_{r/4}^*(x, t))$; in particular, (x, t) is a regular point.

Proof Since

$$Q_r^*(x, t) = Q_r(x, t + r^2/2) \supset Q_{r/2}(x, t + r^2/32)$$

it follows from (2.45) that

$$\frac{1}{(r/2)^2} \int_{Q_{r/2}(x,t+r^2/32)} |u|^3 + |p|^{3/2} < \varepsilon_0,$$

and we can use Theorem 4.3 to guarantee that $u \in L^\infty(Q_{r/4}(x, t + r^2/32))$. Since

$$Q_{r/4}(x, t + r^2/32) = Q^*_{r/4}(x, t)$$

the theorem follows as stated.   □

In order to use this result to bound the dimension of the singular set we first obtain some space-time integrability of u and p from the regularity we know for weak solutions, namely $u \in L^\infty(0, T; L^2) \cap L^2(0, T; L^6)$ (since $H^1 \subset L^6$). We use Lebesgue interpolation to show that

$$\|u\|_{L^{10/3}(U)} \leq \|u\|^{2/5}_{L^2(U)} \|u\|^{3/5}_{L^6(U)}$$

and hence

$$\|u\|^{10/3}_{L^{10/3}(U)} \leq \|u\|^{4/3}_{L^2(U)} \|u\|^2_{L^6(U)},$$

from which it follows that

$$\int_0^T \int_U |u|^{10/3} < \infty,$$

i.e. that $u \in L^{10/3}(U \times (0, T))$. Note that $p \in L^{5/3}(U \times (0, T))$ by assumption.

Now, using Hölder's inequality we have

$$\int_{Q_r} |u|^3 \leq \left\{ \int_{Q_r} |u|^{10/3} \right\}^{9/10} \left\{ \int_{Q_r} 1 \right\}^{1/10}$$

which yields

$$\int_{Q_r} |u|^{10/3} \gtrsim r^{-5/9} \left\{ \int_{Q_r} |u|^3 \right\}^{10/9}.$$

At an irregular point we cannot have (2.45), and so either

$$\int_{Q_r} |u|^{10/3} \gtrsim r^{5/3} \quad \text{or} \quad \int_{Q_r} |p|^{5/3} \gtrsim r^{5/3}. \tag{2.46}$$

[The second lower bound on the pressure term follows in a similar way, again using Hölder's inequality and (2.45).]

We now combine these for a dimension estimate; one can find the basis of the following argument in CKN, although they treat only the Hausdorff dimension.

Proposition 4.5 ([30]) *If (u, p) is a weak solution of the Navier–Stokes inequality on $U \times (a, b)$ and S denotes the set of space-time singularities then*

$$\dim_B(S \cap K) \le 5/3$$

for any compact subset K of $U \times (a, b)$.

Proof Suppose that $d = \dim_B(S \cap K) > 5/3$ and take δ with $5/3 < \delta < d$. Then there exists a sequence $\epsilon_j \to 0$, so that for each j there is a maximal collection $\{z_n^{(j)}\}$ of at least $\epsilon_j^{-\delta}$ points in $S \cap K$ that are $2\epsilon_j$ separated. For each j it follows that the ϵ_j-balls centred at the z_n are disjoint, and from Corollary 4.4 and (2.46)—assuming that the lower bound on the u integral holds for infinitely many points—we have

$$\int_K |u|^{10/3} \ge \sum_j \int_{Q_{\epsilon_j}^*} |u|^{10/3} \gtrsim \epsilon_j^{5/3} \epsilon_j^{-\delta},$$

which on letting $j \to \infty$ contradicts the fact that $u \in L^{10/3}(U \times (a, b))$. A similar contradiction occurs if the lower bound on the p integral holds for infinitely many points, since $p \in L^{5/3}(U \times (a, b))$. $\square$

Lemma 2.11 now implies that almost-every space-time point is regular.

Exercise Suppose that (u, p) is a weak solution of the Navier–Stokes inequality on $\mathbb{R}^3$. Show that the set of space-time singularities is bounded. (Use a contradiction argument again.)

2.4.3 Idea of the Proof of Theorem 4.2

In order to prove Theorem 4.2 we will use an inductive argument to show that for every $(a, s) \in Q_{1/2}(0, 0)$ and every $r_n := 2^{-n}$ we have

$$\frac{1}{r_n} \sup_{s-r_n^2 < t \le s} \int_{B_{r_n}(a)} |u(t)|^2 + \frac{1}{r_n} \int_{Q_{r_n}(a,s)} |\nabla u|^2 \lesssim \varepsilon_0^{2/3} r_n^2; \tag{2.47}$$

since in particular this shows that

$$\frac{1}{r_n^3} \int_{B_{r_n}(a)} |u(s)|^2 \lesssim \varepsilon_0^{2/3}$$

it follows from the Lebesgue Differentiation Theorem that $u \in L^\infty(Q_{1/2}(0, 0))$.

The bound in (2.47) immediately yields on a bound on $\int_{Q_{r_n}} |u|^3$ using the interpolation inequality

$$\frac{1}{r^2} \int_{Q_r(a,s)} |u|^3 \lesssim \left[\frac{1}{r} \sup_{s-r^2 < t < s} \int_{B_r(a)} |u(t)|^2 + \frac{1}{r} \int_{Q_r(a,s)} |\nabla u|^2 \right]^{3/2} \qquad (2.48)$$

which we will prove shortly (Sect. 2.4.4). We can then use the local energy inequality

$$\int |u(t)|^2 \varphi + 2 \int_{t_1}^{t} \int |\nabla u|^2 \varphi \leq \int_{t_1}^{t} \int |u|^2 (\partial_t \varphi + \Delta \varphi) + \int_{t_1}^{t} \int (|u|^2 + 2p) u \cdot \nabla \varphi,$$
$$(2.49)$$

whose right-hand side is essentially cubic in u (as $p \sim u^2$), to obtain (2.47) at the next 'level' (n replaced by $n + 1$).

We will not give the details of how one deals with the pressure, since this involves many of the more technical parts of the argument and, while crucial, can obscure the overall structure of the proof.

2.4.4 Interpolation Inequality

The interpolation inequality (2.48) is just a combination of the Lebesgue interpolation inequality $\|u\|_{L^3} \leq \|u\|_{L^2}^{1/2} \|u\|_{L^6}^{1/2}$, and the Sobolev embedding $H^1(B_r) \subset L^6(B_r)$. We prove the result on Q_1; inequality (2.48) follows by rescaling.

Lemma 4.6 *If $u \in L^\infty(-1, 0; L^2(B_1(0)))$ and $\nabla u \in L^2(Q_1(0, 0))$ then we have $u \in L^3(Q_1(0, 0))$ and*

$$\int_{Q_1(0,0)} |u|^3 \lesssim \left[\sup_{-1 < t < 0} \int_{B_1(0)} |u(t)|^2 + \int_{Q_1(0,0)} |\nabla u|^2 \right]^{3/2}.$$

Proof First we use Lebesgue interpolation and the embedding $H^1 \subset L^6$ in the spatial variables:

$$\|u\|_{L^3}^2 \leq \|u\|_{L^2} \|u\|_{L^6}$$

$$\lesssim \|u\|_{L^2} \|u\|_{H^1}$$

$$\lesssim \|u\|_{L^2}^2 + \|u\|_{L^2} \|\nabla u\|_{L^2}.$$

It follows that

$$\int_{B_1} |u(t)|^3 \lesssim \left[\int_{B_1} |u(t)|^2\right]^{3/2} + \left[\int_{B_1} |u(t)|^2\right]^{3/4} \left[\int_{B_1} |\nabla u(t)|^2\right]^{3/4}.$$

Now we integrate in time:

$$\int_{Q_1} |u|^3 \lesssim \left[\sup_t \int_{B_1} |u(t)|^2\right]^{3/2} + \left[\int_{Q_1} |\nabla u|^2\right]^{3/4} \left[\int_{-1}^{0} \left(\int_{B_1} |u(t)|^2\right)^3 \, dt\right]^{1/4}$$

$$\lesssim \left[\sup_t \int_{B_1} |u(t)|^2\right]^{3/2} + \left[\int_{Q_1} |\nabla u|^2\right]^{3/4} \left[\sup_t \int_{B_1} |u(t)|^2\right]^{3/4}$$

$$\lesssim \left[\sup_t \int_{B_1} |u(t)|^2 + \int_{Q_1} |\nabla u|^2\right]^{3/2}. \qquad \square$$

2.4.5 Inductive Argument

We will move back and forth between two sets of inductive hypotheses[8]: hypothesis (A_n)

$$\frac{1}{r_k^2} \int_{Q_{r_k}(a,s)} |u|^3 \leq \varepsilon_0^{2/3} r_k^3 \qquad k = 1, \ldots, n \tag{2.50}$$

and hypothesis (B_n)

$$\frac{1}{r_n} \sup_{s-r_n^2 < t \leq s} \int_{B_{r_n}(a)} |u(t)|^2 + \frac{1}{r_n} \int_{Q_{r_n}(a,s)} |\nabla u|^2 \lesssim \varepsilon_0^{2/3} r_n^2. \tag{2.51}$$

Now we fix $(a, s) \in Q_{1/2}(0, 0)$, and omit the (a, s) and a arguments on $Q(a, s)$ and $B(a)$.

---

[8] Wojciech Ożański recently remarked to me that the pressure term usually included in hypothesis (A_n) (as in CKN or RRS, for example) is not in fact necessary: the pressure estimates required in the course of the proof rely only on the estimates for u in our (2.50) and (2.51) and on the initial smallness assumption on p in (2.44). A very elegant version of the full inductive argument, including the pressure, is presented in his monograph [24]).

(A_1) follows given the original assumption (2.44), since

$$4 \int_{Q_{1/2}(a,s)} |u|^3 \leq 4 \int_{Q_1(0,0)} |u|^3 \leq 4\varepsilon_0 \leq 2^{-3}\varepsilon_0^{2/3}$$

provided that ε_0 is chosen sufficiently small.

The main part of the proof involves using the local energy inequality to show that (B_n) follows from (A_n). To do this we use a particular choice of test function in the local energy inequality, namely a function φ_n such that $|\partial_t \varphi_n + \Delta \varphi_n| \lesssim r_n^2$, with

$$\varphi_n \approx r_n^{-1} \quad \text{on} \quad Q_{r_n}(a,s),$$

$$|\nabla \varphi_n| \lesssim \begin{cases} r_n^{-2} & \text{on } Q_{r_n}(a,s) \\ r_n^2 r_k^{-4} & \text{on } Q_{r_k}(a,s) \setminus Q_{r_{k+1}}(a,s), \ k = 1, \ldots, n-1, \end{cases}$$

and $\operatorname{supp} \varphi_n \cap Q_1 \subset Q_{1/3}$.

[We choose

$$\varphi_n = r_n^2 \chi_0(x) \chi_n(t) \psi_n(x,t),$$

where the χs are cutoff functions:

$$\chi_0(x) := \begin{cases} 1 & x \in B_{1/4} \\ 0 & x \notin B_{1/3}, \end{cases} \qquad \chi_n(t) := \begin{cases} 1 & -1/16 \leq t \leq 0 \\ 0 & t < -1/9, \ t > r_n^2/2, \end{cases}$$

and ψ_n is a solution of the backwards heat equation satisfying

$$\partial_t \psi + \Delta \psi = 0, \qquad \psi(x, r_n^2) = \delta(x),$$

i.e. for $t < r_n^2$ we have

$$\psi_n(x,t) = \frac{1}{(r_n^2 - t)^{3/2}} \exp\left(\frac{-|x|^2}{4(r_n^2 - t)}\right).$$

The properties above are checked by direct calculation.]

If we use the test function φ_n in the local energy inequality we obtain, for any choice of $t \in (s - r_n^2, s]$,

$$\frac{1}{r_n} \int_{B_{r_n}(a)} |u(t)|^2 + \frac{1}{r_n} \int_{Q_{r_n}(a,s)} |\nabla u|^2 \lesssim r_n^2 \int_{Q_{1/2}(a,s)} |u|^2 + \int_{Q_{1/2}(a,s)} |u|^3 |\nabla \varphi_n|$$

$$+ 2 \int_{Q_{1/2}(a,s)} pu \cdot \nabla \varphi_n.$$

Since we are neglecting the pressure we only have two terms on the right-hand side to deal with.

The first is easy:

$$r_n^2 \int_{Q_{1/2}} |u|^2 \leq r_n^2 \left(\int_{Q_{1/2}} |u|^3 \right)^{2/3} \left(\int_{Q_{1/2}} 1 \right)^{1/3} \lesssim r_n^2 \varepsilon_0^{2/3}.$$

The second requires a little more work. We estimate this term in 'rings':

$$\int_{Q_{1/2}(a,s)} |u|^3 |\nabla \varphi_n| = \sum_{k=1}^{n-1} \int_{Q_{r_k} \setminus Q_{r_{k+1}}} |u|^3 |\nabla \varphi_n| + \int_{Q_{r_n}} |u|^3 |\nabla \varphi_n|$$

$$\lesssim r_n^2 \sum_{k=1}^{n-1} r_k^{-4} \int_{Q_{r_k}} |u|^3 + r_n^{-2} \int_{Q_{r_n}} |u|^3$$

$$= r_n^2 \sum_{k=1}^{n} \left(r_k^{-4} \int_{Q_{r_k}} |u|^3 \right)$$

$$\lesssim r_n^2 \sum_{k=1}^{n} \left(r_k^{-4} r_k^5 \varepsilon_0^{2/3} \right) = r_n^2 \left(\sum_{k=1}^{n} r_k \right) \varepsilon_0^{2/3}$$

$$\approx r_n^2 \varepsilon_0^{2/3}.$$

Combining these gives, for each choice of $t \in (s - r_n^2, s]$,

$$\frac{1}{r_n} \int_{B_{r_n}(a)} |u(t)|^2 + \frac{1}{r_n} \int_{Q_{r_n}(a,s)} |\nabla u|^2 \lesssim r_n^2 \varepsilon_0^{2/3},$$

from which we easily obtain

$$\frac{1}{r_n} \sup_{s-r_n^2 < t \leq s} \int_{B_{r_n}} |u(t)|^2 + \frac{1}{r_n} \int_{Q_{r_n}} |\nabla u|^2 \lesssim r_n^2 \varepsilon_0^{2/3},$$

i.e. (B_n).

[The inclusion of the pressure term in the local energy inequality is dealt with in a similar way to the term that is cubic in u. The trick to observe that if u is divergence free and φ is compactly supported in U then

$$\int_U \alpha u \cdot \nabla \varphi = - \int_U \alpha (\nabla \cdot u) \varphi = 0.$$

In this way the final term can be decomposed into 'rings' as

$$\sum_{k=1}^{n-1} \int_{Q_{r_k} \setminus Q_{r_{k+1}}} (p - (p)_{r_k})u \cdot \nabla\varphi_n + \int_{Q_{r_n}} (p - (p)_{r_n})u \cdot \nabla\varphi_n.$$

One now has to bound the terms involving the pressure.

This can be done by using the relationship between the pressure and the velocity—property (ii) in the definition of a weak solution of the Navier–Stokes inequality.

The key estimate is

$$\frac{1}{r^{3/2}} \int_{Q_r} |p - (p)_r|^{3/2} \lesssim \frac{1}{r^{3/2}} \int_{Q_{2r}} |u|^3 + r^3 \int_{Q_{1/2}} |u|^3 + |p|^{3/2}$$

$$+ r^5 \left\{ \sup_{s-r^2 < t < s} \int_{2r < |y| < 1/2} \frac{|u(t)|^2}{|y|^4} \, dy \right\}^{3/2}.$$

The trick is to take a cutoff function ϕ with support in Q_r, and to write

$$\phi p = (-\Delta)^{-1}(-\Delta)(\phi p) = (-\Delta)^{-1}(\phi \Delta p + \text{other terms});$$

then the first term in the estimate comes from using the Calderón–Zygmund Theorem; the other terms are estimated in different ways. Details can be found in Chapter 15 of RRS; the proof there is slightly simpler than that in CKN, since it assumes slightly more regularity for p.]

Now, (A_{n+1}) follows easily given (A_n) and (B_n) using the interpolation inequality (2.48):

$$\frac{1}{r_{n+1}^2} \int_{Q_{r_{n+1}}} |u|^3 \lesssim \frac{1}{r_n^2} \int_{Q_{r_n}} |u|^3$$

$$\lesssim \left[\frac{1}{r_n} \sup_{s-r_n^2 < t < s} \int_{B_{r_n}} |u(t)|^2 + \frac{1}{r_n} \int_{Q_{r_n}} |\nabla u|^2 \right]^{3/2}$$

$$\lesssim \left[\varepsilon_0^{2/3} r_n^2 \right]^{3/2} = \varepsilon_0 r_n^3 \le \varepsilon_0^{2/3} r_{n+1}^3,$$

provided that ε_0 is small enough is counteract the constants implicit in the $\lesssim$ inequality.

We have shown that under the condition

$$\int_{Q_1(0,0)} |u|^3 + |p|^{3/2} \le \varepsilon_0$$

it follows that for every $(a, s) \in Q_{1/2}(0, 0)$ we have

$$\frac{1}{r_k^3} \int_{B_{r_k}(a)} |u(s)|^2 \lesssim \varepsilon_0^{2/3}$$

for every $k \in \mathbb{N}$. It now follows from the Lebesgue Differentiation Theorem that $u \in L^\infty(Q_{1/2}(0, 0))$, which is what we set out to prove.

CKN also have a second regularity theorem, which yields an improved bound on the Hausdorff dimension of the singular set. This theorem is given in terms of a condition on the gradient of u.

Theorem 4.7 *There exists an absolute constant $\varepsilon_1 > 0$ such that if (u, p) is a weak solution of the Navier–Stokes inequality on $Q_r(x, t)$ and*

$$\limsup_{r \to 0} \frac{1}{r} \int_{Q_r(a,s)} |\nabla u|^2 \leq \varepsilon_1$$

then $u \in L^\infty(Q_\rho(a, s))$ for some ρ with $0 < \rho < R$.

For details of this result see Chapter 16 of RRS; the proof there, which uses ideas from the paper of Lin [23] and Kukavica [20], is significantly simpler than that in CKN.

A consequence of this result is that $\dim_H(S) \leq 1$ (the proof uses a version of the Vitali Covering Lemma). Scheffer [36] has shown that using only the ingredients in the CKN proof this cannot be improved: for any $\gamma < 1$ he constructed a weak solution of the Navier–Stokes inequality that has a singular set with $\dim_H(S) \geq \gamma$. This is the space-time analogue our Lemma 2.13. (Scheffer's papers are very dense and quite hard to unravel: for a more accessible version of his construction see Ożański [24].)

2.5 Lagrangian Trajectories

Finally, we will investigate the existence of Lagrangian trajectories, or 'particle paths', associated with weak solutions of the Navier–Stokes equations. The problem, essentially, is that given a particular weak solution u, we want to solve the ODE

$$\dot{X} = u(X, t), \qquad X(0) = a. \tag{2.52}$$

First we show that solutions are volume preserving when u is smooth, and that they are unique for every $a \in \mathbb{T}^3$ when u is a strong solution. We then show when u is a weak solution that satisfies the local energy inequality the solutions are still unique for almost every $a \in \mathbb{T}^3$. The results in this section are due to Robinson and Sadowski [30]; see also Chapter 17 of RRS, which gives more of the details.

Local existence and uniqueness of solutions of (2.52) can be guaranteed using standard arguments when, for example, u is locally Lipschitz with a Lipschitz constant that is uniform in time, i.e. when

$$|u(X, t) - u(Y, t)| \le L \qquad \text{for all } X, Y \in B(a, r), \ t \in [0, \tau)$$

for some $r > 0$, $\tau > 0$. Existence can be shown by finding a fixed point of the mapping

$$X(\cdot) \mapsto a + \int_0^t u(X(s), s) \, ds,$$

while uniqueness follows from the differential inequality

$$\frac{1}{2} \frac{d}{dt} |W|^2 \le L |W|^2, \qquad W(0) = 0$$

satisfied by the difference $W(t) = X_1(t) - X_2(t)$ of any two solutions of (2.52). For details see Hartman [16], for example.

2.5.1 Volume-Preserving Solutions When u is Smooth

When u is sufficiently smooth, e.g. $u \in C^1([0, T]; C^2(\mathbb{T}^2))$, we can use the fact that u is divergence free to show that the corresponding flow preserves volumes.

To see this, given $V(0) \in \mathbb{T}^3$ we set

$$V(t) = \{X(t; a) : a \in V(0)\};$$

then by a change of variables we have

$$\mu(V(t)) = \int_{V(t)} 1 \, dx = \int_{V(0)} |\det(\nabla X(t; y))| \, dy. \tag{2.53}$$

Since

$$\frac{d}{dt} \nabla X(t; a) = \nabla u(X(t; a), t) \nabla X(t; a)$$

it follows (for details see Section 17.1 in RRS) that

$$\det \nabla X(t) = \exp\left(\int_0^t \operatorname{tr}(\nabla u(s)) \, ds\right) \det \nabla X(0)$$

$$= \exp\left(\int_0^t \nabla \cdot u(s) \, ds\right) = 1,$$

since $\nabla X(0) = \mathrm{Id}$ and $\nabla \cdot u(s) = 0$. From (2.53) we obtain $\mu(V(t)) = \mu(V(0))$.

2.5.2 Unique Solutions When u is a Strong Solution

First note that if u is a strong solution, i.e. if $u \in L^\infty(0, T; H^1) \cap L^2(0, T; H^2)$, then by taking the inner product[9] with tu in $\dot{H}^2$ we obtain

$$\frac{1}{2}\frac{\mathrm{d}}{\mathrm{d}t}\left(t\|u\|_{H^2}^2\right) + t\|\nabla u\|_{H^2}^2 = \frac{1}{2}\|u\|_{H^2}^2 - t\langle(u \cdot \nabla)u, u\rangle_{H^2}$$

$$\leq \frac{1}{2}\|u\|_{H^2}^2 + t\|(u \cdot \nabla)u\|_{H^2}\|u\|_{H^2}$$

$$\lesssim \|u\|_{H^2}^2 + t\|u\|_{H^2}\|\nabla u\|_{H^2}\|u\|_{H^2} \qquad (2.54)$$

(since $H^2(\mathbb{T}^3)$ is an algebra) which yields

$$\frac{\mathrm{d}}{\mathrm{d}t}\left(t\|u\|_{H^2}^2\right) + t\|\nabla u\|_{H^2}^2 \lesssim \|u\|_{H^2}^2 + \|u\|_{H^2}^2\left(t\|u\|_{H^2}^2\right).$$

Integrating in time shows that $\sqrt{t}u \in L^2(0, T; H^3)$; it follows in particular that

$$\int_0^T \|u\|_{H^{11/4}} \lesssim \int_0^T \|u\|_{H^1}^{1/8} t^{-7/16} t^{7/16} \|u\|_{H^3}^{7/8}\,\mathrm{d}t$$

$$\lesssim \left(\int_0^T t^{-7/9}\|u\|_{H^1}^{2/9}\,\mathrm{d}t\right)^{9/16}\left(\int_0^T t\|u\|_{H^3}^2\,\mathrm{d}t\right)^{7/16},$$

i.e. that $u \in L^1(0, T; H^{11/4})$. Since $H^{11/4} \subset C^{0,1}$ on a three-dimensional domain, this is enough to ensure that

$$|u(x, t) - u(y, t)| \leq C(t)|x - y|$$

for some $C(\cdot) \in L^1(0, T)$, which in turn is enough to ensure that solutions of (2.52) are unique on $[0, T]$. (In fact it is sufficient to take $u_0 \in H^{1/2}$; on their interval of existence the resulting solutions have enough regularity to ensure that solutions of (2.52) are unique, see Dashti and Robinson [7].)

2.5.3 Existence of Trajectories When u is a Weak Solution

However, a weak solution u is not smooth enough to solve the ODE (2.52) directly, so we need to reinterpret this as the integral equation

$$X(t) = a + \int_0^t u(X(s), s)\,\mathrm{d}s. \qquad (2.55)$$

---

[9]We define $\langle f, g\rangle_{H^2} = \sum_k (1 + |k|^2)^2 \hat{f}_k \hat{g}_k$.

The key property of weak solutions that allows us to make sense of this integral equation is the fact that $u \in L^1(0, T; L^\infty)$. We will prove this via an auxiliary estimate on the H^2 norm, a proof due to Foias et al. [13].

Lemma 5.1 *Let u be a weak solution of the NSE. Then for every $T > 0$*

$$\int_0^T \|\Delta u\|^{2/3} \, ds < \infty, \tag{2.56}$$

and consequently $u \in L^1(0, T; L^\infty)$.

Proof Returning to (2.10) we have

$$\frac{d}{dt} \|\nabla u\|^2 + \|\Delta u\|^2 \lesssim \|\nabla u\|^6;$$

we divide both sides by $\|\nabla u\|^4 \neq 0$ to obtain

$$\frac{1}{\|\nabla u\|^4} \frac{d}{dt} \|\nabla u\|^2 + \frac{\|\Delta u\|^2}{\|\nabla u\|^4} \lesssim \|\nabla u\|^2.$$

Now integrate over an interval of regularity (a_i, b_i) [see (2.18)] to obtain

$$\frac{1}{\|\nabla u(a_i)\|^2} + \int_{a_i}^{b_i} \frac{\|\Delta u(s)\|^2}{\|\nabla u(s)\|^4} \, ds \lesssim \int_{a_i}^{b_i} \|\nabla u(s)\|^2 \, ds,$$

since $\lim_{t \to b_i^-} \|\nabla u(t)\| = \infty$. Since the set of regular times has full measure, summing over i in the previous inequality it follows that

$$\int_0^T \frac{\|\Delta u(s)\|^2}{\|\nabla u(s)\|^4} \, ds \lesssim \int_0^T \|\nabla u(s)\|^2 \, ds < \infty.$$

Now, observe that

$$\int_0^T \|\Delta u(s)\|^{2/3} \, ds = \int_0^T \frac{\|\Delta u(s)\|^{2/3}}{\|\nabla u(s)\|^{4/3}} \|\nabla u(s)\|^{4/3} \, ds$$

$$\leq \left(\int_0^T \frac{\|\Delta u(s)\|^2}{\|\nabla u(s)\|^4} \, ds \right)^{1/3} \left(\int_0^T \|\nabla u(s)\|^2 \, ds \right)^{2/3},$$

yielding (2.56).

The L^∞ bound follows using

$$\|u\|_{L^\infty} \lesssim \|\nabla u\|^{1/2} \|\Delta u\|^{1/2}; \tag{2.57}$$

integrating from 0 to T and using Hölder's inequality in time

$$\int_0^T \|u\|_{L^\infty} \leq \left(\int_0^T \|\nabla u\|^2 \right)^{1/4} \left(\int_0^T \|\Delta u\|^{2/3} \right)^{3/4}. \tag{2.58}$$

This inequality will be useful in itself shortly.   □

We now show that there is at least one solution of (2.55) when u is a weak solution.

Proposition 5.2 *If u is any weak solution then there is at least one solution of*

$$X(t) = a + \int_0^t u(X(s), s) \, ds.$$

Proof We approximate u by the smooth function u_n obtained by truncating the Fourier expansion of u after n terms. Then

$$\|\nabla u_n\| \leq \|\nabla u\|, \qquad \|\Delta u_n\| \leq \|\Delta u\|,$$

and $u_n \to u$ in $L^2(0, T; H^1)$ and in $L^{2/3}(0, T; H^2)$ (using the Dominated Convergence Theorem and (2.56)). It follows from (2.58) that $u_n \to u$ in $L^1(0, T; L^\infty)$ also, and we now take a subsequence (which we relabel) so that $u_n(s) \to u(s)$ in L^∞ for almost every s. Note that, uniformly in n,

$$\|u_n(t)\|_{L^\infty} \lesssim \|\nabla u_n(t)\|^{1/2} \|\Delta u_n(t)\|^{1/2} \leq \|\nabla u(t)\|^{1/2} \|\Delta u(t)\|^{1/2} =: \theta(t),$$

where $\theta \in L^1(0, T)$.
   Now solve

$$\dot{X}_n = u_n(X_n, t), \qquad X_n(0) = a.$$

The solution is bounded, since u_n is uniformly bounded in $L^1(0, T; L^\infty)$, and the functions X_n are equicontinuous since

$$|X_n(t) - X_n(s)| \leq \int_s^t \|u_n(t)\|_{L^\infty} \lesssim \int_s^t \theta(r) \, dr.$$

We can use the Arzelà–Ascoli Theorem to find a subsequence of the (X_n) that converges uniformly on $[0, T]$ to some $X(\cdot)$. To show that $X(\cdot)$ satisfies the required integral equation, we want to pass to the limit in

$$X_n(t) = a + \int_0^t u_n(X_n(s), s) \, ds.$$

Since $|u_n(X_n(s), s)| \leq \|u_n(s)\|_{L^\infty} \lesssim \theta(s)$, using the Dominated Convergence Theorem it suffices to show that $u_n(X_n(s), s) \to u(X(s), s)$ for almost every s. We can write

$$|u_n(X_n(s), s) - u(X(s), s)|$$
$$\leq |u_n(X_n(s), s) - u(X_n(s), s)| + |u(X_n(s), s) - u(X(s), s)|$$
$$\leq \|u_n(s) - u(s)\|_{L^\infty} + |u(X_n(s), s) - u(X(s), s)|.$$

We know that $u_n(s) \to u(s)$ in L^∞ for almost every $s \in [0, T]$, and since u is continuous for almost every s we have the required convergence and $X(\cdot)$ satisfies

$$X(t) = a + \int_0^t u(X(s), s) \, ds$$

as required.   □

2.5.4 Existence of a Flow When u is a Weak Solution

Remarkably, it is possible to 'put together' a selection of the solutions whose existence is guaranteed by Proposition 5.2 in such a way that the resulting flow mapping preserves volumes. This result is due to Foias et al. [14]; full details of the proof can be found in RRS.

Theorem 5.3 ([14]) *For any Leray–Hopf weak solution u there exists at least one solution mapping $\Phi \colon \mathbb{T}^3 \times [0, T] \to \mathbb{T}^3$ such that for each $a \in \mathbb{T}^3$, $X_a(\cdot) := \Phi(a, \cdot)$ satisfies*

$$X_a(t) = a + \int_0^t u(X_a(s), s) \, ds \tag{2.59}$$

and the mapping $\Phi(\cdot, t)$ is volume preserving in the sense that for any $t > 0$ and any Borel set $B \subset \mathbb{T}^3$ we have

$$\mu[\Phi(\cdot, t)^{-1}(B)] = \mu(B).$$

Proof Proposition 5.2 guarantees the existence of at least one solution of the integral equation (2.59) for each $a \in \mathbb{T}^3$. It is possible to choose one of these solutions for each a in such a way that the resulting solution mapping Φ is measurable (this requires a 'measurable selection theorem' due to Robertson [26] whose proof can be found in Appendix E of RRS). The volume-preserving property is proved by considering the absolutely continuous function $J(t) := \int_{\mathbb{T}^3} g(\Phi(a, t)) \, da$ for $g \in C^\infty(\mathbb{T}^3)$ and showing that $\dot{J} = 0$ almost everywhere; this is possible since u is smooth except on a set of zero measure.   □

2.5.5 Properties of the Flow Mapping

We now show that if u satisfies the local energy inequality—so that we know that the set of space-times singularities Σ has $\dim_B(\Sigma) \leq 5/3$—the trajectories that form the flow map avoid the singular set for almost every $a \in \mathbb{T}^3$. The following result is in fact a little more general.

Theorem 5.4 ([30]) *Suppose that $\Omega \subset \mathbb{R}^d$ and that $\Phi \colon \Omega \times [0, T] \to \Omega$ is a volume-preserving solution mapping corresponding to a vector field u with $u \in L^1(0, T; L^\infty(\Omega))$ for every $T > 0$. If Σ is a compact subset of Ω with $\dim_B(\Sigma) < d - 1$ then for almost every initial condition $a \in \Omega$, $\Phi(a, t) \notin \Sigma$ for all $t \geq 0$.*

Proof Choose $T > 0$, fix $N \in \mathbb{N}$, write $t_j = jT/N$, and consider the problem of whether the trajectory $X_a(t) := \Phi(a, t)$ starting at a avoids the set Σ on the time interval $[t_k, t_{k+1}]$ for some $k \in \{0, \ldots, N - 1\}$. Since

$$X_a(t) - X_a(s) = \int_s^t u(X_a(\tau), \tau) \, d\tau,$$

it follows that for all $t \in [t_k, t_{k+1}]$,

$$|X_a(t) - X_a(t_k)| \leq \delta_k := \int_{t_k}^{t_{k+1}} \|u(s)\|_\infty \, ds.$$

So if $X_a(t) \in \Sigma$ for some $t \in [t_k, t_{k+1}]$, we must have

$$X_a(t_k) \in O(\Sigma, \delta_k)$$

[recall that $O(\Sigma, \delta_k)$ denotes the (open) δ_k-neighbourhood of Σ, see (2.21)].

Since $\dim_B(\Sigma) < d - 1$, Lemma 2.10 guarantees that

$$\mu(O(\Sigma, \delta)) \leq c_d \delta^\alpha \quad \text{for some } \alpha > 1.$$

Since Φ is measure preserving,

$$\mu\{a : X_a(t_k) \in O(\Sigma, \delta_k)\} \leq c_d \delta_k^\alpha.$$

Thus for any choice of N, the measure of the set Ω_Σ of initial conditions for which $X_a(t) \in \Sigma$ for some $t \in [0, T]$ is bounded by

$$c_d \sum_{k=0}^{N-1} \delta_k^\alpha = c_d \sum_{k=0}^{N-1} \left(\int_{t_k}^{t_{k+1}} \|u(s)\|_\infty \, ds \right)^\alpha. \tag{2.60}$$

Since $u \in L^1(0, T; L^\infty)$ the integral of $\|u\|_\infty$ exists and is absolutely continuous. In particular, given any $\epsilon > 0$ there exists an $N \in \mathbb{N}$ such that for each $k \in$

$\{0, \ldots, N - 1\}$,

$$\int_{t_k}^{t_{k+1}} \|u(s)\|_\infty \, ds < \epsilon.$$

Using this in (2.60) it follows that

$$\mu(\Omega_\Sigma) \le c_d \epsilon^{\alpha-1} \int_0^T \|u(s)\|_\infty \, ds.$$

Since $\alpha > 1$ and both ϵ and T are arbitrary this completes the proof. $\square$

An important property of the box-counting dimension is that it is non-increasing under Lipschitz mappings: we will use this to deduce that the dimension of the set

$$\{x \in \mathbb{T}^3 : (x, t) \in S \text{ for some } t > 0\}$$

(a subset of $\mathbb{T}^3$) is no larger than the dimension of S (a subset of $\mathbb{T}^3 \times (0, \infty)$).

Lemma 5.5 *Suppose that* $X \subset \mathbb{R}^N$ *and* $f : \mathbb{R}^N \to \mathbb{R}^m$ *is Lipschitz. Then* $\dim_B(f(X)) \le \dim_B(X)$.

Proof Suppose that $|f(x) - f(x')| \le L|x - x'|$. If X can be covered with $N(X, \varepsilon)$ balls of radius ε then the image of these balls under f cover $f(X)$. Since the image under f of a ball of radius ε is contained in a ball of radius $L\varepsilon$, it follows that $f(X)$ can be covered by $N(X, \varepsilon)$ balls of radius $L\varepsilon$ in $\mathbb{R}^m$. The result is now a consequence of the definition of the box-counting dimension. $\square$

Theorem 5.6 *Let* u *be a weak solution that arises from an initial condition* $u_0 \in H \cap H_0^1(\Omega)$ *and satisfies the local energy inequality. If* Φ *is a corresponding solution mapping then*

$$\Phi(a, t) \notin S \text{ for all } t \ge 0 \qquad \text{for almost every } a \in \Omega.$$

Proof Fix $T > 0$ and let $\Omega \times (0, T) = \bigcup_{n=1}^\infty K_n$ with K_n compact.

If P denotes the projection onto the spatial component then

$$\dim_B(P[S \cap K_n]) \le \dim_B(S \cap K_n) \le 5/3;$$

we have used the fact that $\dim_B$ is non-increasing under the Lipschitz mapping P (Lemma 5.5) and the bound on the box-counting dimension of S from Theorem 4.5.

Theorem 5.4 now guarantees that almost every trajectory avoids the set $P(S \cap K_n)$ for all $t \ge 0$, and so in particular avoids $S \cap K_n$ for every n, so almost every trajectory avoids S. $\square$

Finally, we can deduce that the Lagrangian trajectories are unique for almost every initial condition. To do this we need to use the fact that u is smooth away from the space-time singular set, which follows from the local regularity result due to Serrin from Sect. 2.3: if $u \in L^{\infty}(U \times (a, b))$ then in fact u is smooth in the spatial variables on $U \times (a, b)$.

By a suitable weak solution we mean a Leray–Hopf weak solution that is also a weak solution of the Navier–Stokes inequality.

Corollary 5.7 *If u is a suitable weak solution corresponding to an initial condition $u_0 \in V$ then almost every initial condition $a \in \mathbb{T}^3$ gives rise to a unique particle trajectory.*

Proof Taking $u_0 \in H^1(\mathbb{T}^3)$ implies that trajectories are unique on some $[0, T)$, so the issue is uniqueness for $t \geq T$, in which range of times the partial regularity results of Sect. 2.4 serve to limit the dimension of the set of space-time singularities.

Let $X_a(\cdot)$ be a trajectory that avoids the singular set for all $t \geq 0$, and suppose (for a contradiction) that there are two trajectories that pass through the space-time point $(X_a(t), t)$ for some particular $t > 0$.

Since $(X_a(t), t) \notin S$ it is a regular point; by definition this means that $u \in L^{\infty}(U)$ for some space-time neighbourhood $U = W \times (t - \varepsilon, t + \varepsilon)$ of $(X_a(t), t)$, and then the result of Serrin in Theorem 3.1 guarantees that u is (spatially) smooth in U, and in particular Lipschitz continuous on W with a constant that is uniform over $(t - \varepsilon, t + \varepsilon)$. It follows (see the discussion at the beginning of this section) that the solution of $\dot{X} = u(X, t)$ is unique at $(X_a(t), t)$, so there cannot be two trajectories passing through this point.   □

References

1. T. Buckmaster, V. Vicol, Nonuniquenness of weak solutions to the Navier–Stokes equation. arXiv:1709.10033 (2017)
2. T. Buckmaster, C. De Lellis, L. Székelyhidi Jr., V. Vicol, Onsager's conjecture for admissible weak solutions. arXiv:1701.08678 (2017)
3. L. Caffarelli, R. Kohn, L. Nirenberg, Partial regularity of suitable weak solutions of the Navier–Stokes equations. Comm. Pure. Appl. Math. **35**, 771–931 (1982)
4. J.-Y. Chemin, B. Desjardins, I. Gallagher, E. Grenier, *Mathematical Geophysics* (Oxford University Press, Oxford, 2006)
5. P. Constantin, C. Foias, *Navier–Stokes Equations* (University of Chicago Press, Chicago, 1988)
6. P. Constantin, E. Weinan, E.S. Titi, Onsager's conjecture on the energy conservation for solutions of Euler's equation. Commun. Math. Phys. **165**, 207–209 (1994)
7. M. Dashti, J.C. Robinson, A simple proof of uniqueness of the particle trajectories for solutions of the Navier–Stokes equations. Nonlinearity **22**, 735–746 (2009)
8. L. Escauriaza, G. Seregin, V. Šverák, $L_{3,\infty}$-solutions of Navier–Stokes equations and backward uniqueness. Russ. Math. Surv. **58**, 211–250 (2003)
9. L.C. Evans, *Partial Differential Equations* (American Mathematical Society, Providence, 2010)
10. E.B. Fabes, B.F. Jones, N.M. Rivière, The initial value problem for the Navier–Stokes equations with data in L^p. Arch. Ration. Mech. Anal. **45**, 222–240 (1972)
11. K.J. Falconer, *Fractal Geometry* (Wiley, Chichester, 1990)
12. C.L. Fefferman, Existence and smoothness of the Navier–Stokes equation, in *The Millennium Prize Problems* (Clay Mathematics Institute, Cambridge, 2000), pp. 57–67

13. C. Foias, C. Guillopé, R. Temam, New a priori estimates for Navier–Stokes equations in dimension 3. Comm. Partial Diff. Equ. **6**, 329–359 (1981)
14. C. Foias, C. Guillopé, R. Temam, Lagrangian representation of a flow. J. Diff. Equ. **57**, 440–449 (1985)
15. G.P. Galdi, An introduction to the Navier–Stokes initial-boundary value problem, in ed. by G.P. Galdi, J.G. Heywood, R. Rannacher. *Fundamental Directions in Mathematical Fluid Dynamics* (Birkhäuser, Basel, 2000), pp. 1–70
16. P. Hartman, *Ordinary Differential Equations* (Wiley, Baltimore, 1973)
17. E. Hopf, Über die Aufgangswertaufgave für die hydrodynamischen Grundliechungen. Math. Nachr. **4**, 213–231 (1951)
18. P. Isett, A proof of Onsager's conjecture. Ann. Math. **188**, 871–963 (2018)
19. H. Jia, V. Šverák, Are the incompressible 3d Navier–Stokes equations locally ill-posed in the natural energy space? J. Func. Anal. **268**, 3734–3766 (2015)
20. I. Kukavica, Partial regularity results for solutions of the Navier–Stokes system, in ed. by J.C. Robinson, J.L. Rodrigo, *Partial Differential Equations and Fluid Mechanics* (Cambridge University Press, Cambridge, 2009), pp. 121–145
21. I. Kukavica, The fractal dimension of the singular set for solutions of the Navier–Stokes system. Nonlinearity **22**, 2889–2900 (2009)
22. J. Leray, Essai sur le mouvement d'un liquide visqueux emplissant l'espace. Acta Math. **63**, 193–248 (1934)
23. F. Lin, A new proof of the Caffarelli–Kohn–Nirenberg theorem. Comm. Pure Appl. Math. **51**, 241–257 (1998)
24. W. Ożański, *The Partial Regularity Theory of Caffarelli, Kohn, and Nirenberg and Its Sharpness*. Lecture Notes in Mathematical Fluid Mechanics (Birkhäuser/Springer, 2019)
25. W. Ożański, B. Pooley, Leray's fundamental work on the Navier–Stokes equations: a modern review of "Sur le mouvement d'un liquid visqueux emplissant l'espace". in ed. by C.L. Fefferman, J.L. Rodrigo, J.C. Robinson. *Partial Differential Equations in Fluid Mechanics*. LMS Lecture Notes (Cambridge University Press, Cambridge, 2018)
26. A.P. Robertson, On measurable selections. Proc. Roy. Soc. Edinburgh Sect. A **73**, 1–7 (1974)
27. J.C. Robinson, *Infinite-Dimensional Dynamical Systems* (Cambridge University Press, Cambridge, 2001)
28. J.C. Robinson, *Dimensions, Embeddings, and Attractors* (Cambridge University Press, Cambridge, 2011)
29. J.C. Robinson, W. Sadowski, Decay of weak solutions and the singular set of the three-dimensional Navier–Stokes equations. Nonlinearity **20**, 1185–1191 (2007)
30. J.C. Robinson, W. Sadowski, Almost everywhere uniqueness of Lagrangian trajectories for suitable weak solutions of the three-dimensional Navier–Stokes equations. Nonlinearity **22**, 2093–2099 (2009)
31. J.C. Robinson, J.L. Rodrigo, W. Sadowski, *The Three-Dimensional Navier–Stokes Equations* (Cambridge University Press, Cambridge, 2016)
32. V. Scheffer, Turbulence and Hausdorff dimension, in *Turbulence and Navier–Stokes Equation, Orsay 1975*. Springer Lecture Notes in Mathematics, vol. 565 (Springer, Berlin, 1976), pp. 174–183
33. V. Scheffer, Partial regularity of solutions to the Navier–Stokes equations. Pacific J. Math. **66**, 535–552 (1976)
34. V. Scheffer, Hausdorff measure and the Navier–Stokes equations. Comm. Math. Phys. **55**, 97–112 (1977)
35. V. Scheffer, The Navier–Stokes equations on a bounded domain. Comm. Math. Phys. **73**, 1–42 (1980)
36. V. Scheffer, Nearly one-dimensional singularities of solutions to the Navier–Stokes inequality. Comm. Math. Phys. **110**, 525–551 (1987)

37. J. Serrin, On the interior regularity of weak solutions of the Navier–Stokes equations. Arch. Rat. Mech. Anal. **9**, 187–195 (1962)
38. M. Struwe, On partial regularity result for the Navier–Stokes equations. Comm. Pure Appl. Math. **41**, 437–458 (1988)
39. S. Takahashi, On interior regularity criteria for weak solutions of the Navier–Stokes equations. Manuscripta Math. **69**, 237–254 (1990)

Chapter 3
$\mathcal{R}$ Boundedness, Maximal Regularity and Free Boundary Problems for the Navier Stokes Equations

Yoshihiro Shibata

3.1 Introduction

3.1.1 Free Boundary Problem for the Navier–Stokes Equations

In this chapter, we study free boundary problem for the Navier–Stokes equations, and $\mathcal{R}$ bounded solution operators and L_p–L_q maximal regularity theorem for the Stokes equations with free boundary conditions. Typical problem for the free boundary problem of the Navier–Stokes equations are (P1) the motion of an isolated liquid mass and (P2) the motion of a viscous incompressible fluid contained in an ocean of infinite extent.

The mathematical problem for the free boundary problem of the Navier–Stokes equations is to find a time dependent domain Ω_t, t being time variable, in the N-dimensional Euclidean space $\mathbb{R}^N$, the velocity vector field, $\mathbf{v}(x,t) = {}^\top(v_1(x,t), \ldots, v_N(x,t))$, where ${}^\top M$ denotes the transposed M, and the pressure

Y. Shibata (✉)
Department of Mathematics, Waseda University, Tokyo, Japan

© Springer Nature Switzerland AG 2020
G. P. Galdi, Y. Shibata (eds.), *Mathematical Analysis of the Navier-Stokes Equations*, Lecture Notes in Mathematics 2254,
https://doi.org/10.1007/978-3-030-36226-3_3

field $\mathfrak{p} = \mathfrak{p}(x, t)$ satisfying the Navier–Stokes equations in Ω_t:

$$
\begin{cases}
\partial_t \mathbf{v} + (\mathbf{v} \cdot \nabla)\mathbf{v} - \mathrm{Div}\,(\mu \mathbf{D}(\mathbf{v}) - \mathfrak{p}\mathbf{I}) = 0 & \text{in } \bigcup_{0<t<T} \Omega_t \times \{t\}, \\[2mm]
\mathrm{div}\,\mathbf{v} = 0 & \text{in } \bigcup_{0<t<T} \Omega_t \times \{t\}, \\[2mm]
(\mu \mathbf{D}(\mathbf{v}) - \mathfrak{p}\mathbf{I})\mathbf{n}_t = \sigma H(\Gamma_t)\mathbf{n}_t - p_0 \mathbf{n}_t & \text{on } \bigcup_{0<t<T} \Gamma_t \times \{t\}, \\[2mm]
V_n = \mathbf{v} \cdot \mathbf{n}_t & \text{on } \bigcup_{0<t<T} \Gamma_t \times \{t\}, \\[2mm]
\mathbf{v}|_{t=0} = \mathbf{v}_0 \quad \text{in } \Omega_0, \quad \Omega_t|_{t=0} = \Omega_0,
\end{cases}
\tag{3.1}
$$

where Γ_t is the boundary of Ω_t and $\mathbf{n}_t$ is the unit outer normal to Γ_t. As for the remaining notation in (3.1), $\partial_t = \partial/\partial t$, $\mathbf{v}_0 = {}^\top(v_{01}, \ldots, v_{0N})$ is a given initial velocity field, $\mathbf{D}(\mathbf{v}) = \nabla \mathbf{v} + {}^\top \nabla \mathbf{v}$ the doubled deformation tensor with (i, j)th component $D_{ij}(\mathbf{v}) = \partial_i v_j + \partial_j v_i$, $\partial_i = \partial/\partial x_i$, $\mathbf{I}$ the $N \times N$ identity matrix, $H(\Gamma_t)$ the $N - 1$ fold mean curvature of Γ_t given by $H(\Gamma_t)\mathbf{n}_t = \Delta_{\Gamma_t}x$ ($x \in \Gamma_t$), where Δ_{Γ_t} is the Laplace–Beltrami operator on Γ_t, V_n the velocity of the evolution of free surface Γ_t in the direction of $\mathbf{n}_t$, p_0 the outside pressure, and μ and σ are positive constants representing the viscous coefficient and the coefficient of the surface tension, respectively. Moreover, for any matrix field $\mathbf{K}$ with (i, j)th component K_{ij}, the quantity $\mathrm{Div}\,\mathbf{K}$ is an N-vector with ith component $\sum_{j=1}^{N} \partial_j K_{ij}$, and for any vector of functions $\mathbf{w} = {}^\top(w_1, \ldots, w_N)$, we set $\mathrm{div}\,\mathbf{w} = \sum_{j=1}^{N} \partial_j w_j$ and $(\mathbf{w} \cdot \nabla)\mathbf{w}$ is an N-vector with ith component $\sum_{j=1}^{N} w_j \partial_j w_i$. The domain $\Omega_0 = \Omega$ is given, Γ denotes the boundary of Ω, and $\mathbf{n}$ the unit outer normal to Γ. For simplicity, we assume that the mass density equals one in this chapter.

Concerning the outside pressure p_0, an equilibrium state is that $\mathbf{v} = 0$, and so from the first equation in (3.1) it follows that $\nabla \mathfrak{p} = 0$, that is the pressure $\mathfrak{p}$ is a constant. Moreover, $\mathbf{n}_t = \mathbf{n}$ and $\sigma H(\Gamma_t) = \sigma H(\Gamma)$ for any $t \geq 0$. Thus,

$$
p_0 = \sigma H(\Gamma) + \text{constant.} \tag{3.2}
$$

Since Ω_t is unknown, we transform Ω_t to some fixed domain Ω, and the system of linearized equations of the nonlinear equations on Ω has the following forms:

$$
\begin{cases}
\partial_t \mathbf{u} - \mathrm{Div}\,(\mu \mathbf{D}(\mathbf{u}) - \mathfrak{p}\mathbf{I}) = \mathbf{f} & \text{in } \Omega^T, \\[2mm]
\mathrm{div}\,\mathbf{u} = g = \mathrm{div}\,\mathbf{g} & \text{in } \Omega^T, \\[2mm]
\partial_t \eta + \,<\mathbf{a} \mid \nabla'_\Gamma \eta> -\mathbf{n} \cdot \mathbf{u} = d & \text{on } \Gamma^T, \\[2mm]
(\mu \mathbf{D}(\mathbf{u}) - \mathfrak{p}\mathbf{I})\mathbf{n} - \sigma((\Delta_\Gamma + b)\eta)\mathbf{n} = \mathbf{h} & \text{on } \Gamma^T, \\[2mm]
(\mathbf{u}, \eta)|_{t=0} = (\mathbf{u}_0, \eta_0) & \text{in } \Omega \times \Gamma.
\end{cases}
\tag{3.3}
$$

Here, η is an unknown scalar function obtained by linearization of kinematic equation $V_n = \mathbf{v} \cdot \mathbf{n}_t$, Γ is the boundary of Ω, $\nabla'_\Gamma \eta$ denotes the tangential derivative of η on Γ, $\mathbf{a}$ and b are given functions defined on Γ, and $\mathbf{f}$, g, $\mathbf{g}$, d, $\mathbf{h}$, $\mathbf{u}_0$ and η_0 are prescribed functions. Moreover, we set $\Omega^T = \Omega \times (0, T)$ and $\Gamma^T = \Gamma \times (0, T)$.

The main topics for Eq. (3.3) is to prove the maximal L_p–L_q regularity and the decay properties of solutions. To prove these properties, we consider the corresponding resolvent problem:

$$
\begin{cases}
\lambda \mathbf{u} - \mathrm{Div}\, (\mu \mathbf{D}(\mathbf{u}) - \mathfrak{p}\mathbf{I}) = \mathbf{f} & \text{in } \Omega, \\
\mathrm{div}\, \mathbf{u} = g = \mathrm{div}\, \mathbf{g} & \text{in } \Omega, \\
\lambda\eta + < \mathbf{a} \mid \nabla'_\Gamma \eta > - \mathbf{n} \cdot \mathbf{u} = d & \text{on } \Gamma, \\
(\mu \mathbf{D}(\mathbf{u}) - \mathfrak{p}\mathbf{I})\mathbf{n} - \sigma((\Delta_\Gamma + b)\eta)\mathbf{n} = \mathbf{h} & \text{on } \Gamma.
\end{cases}
\tag{3.4}
$$

The main issue of Eq. (3.4) is to prove the existence of $\mathcal{R}$ bounded solution operators, which, combined with the Weis operator valued Fourier multiplier theorem [74], yields the maximal L_p–L_q regularity for Eq. (3.1). Moreover, using some spectral analysis of solutions to Eq. (3.4), we derive decay properties of solutions to Eq. (3.3).

As was stated in the abstract, the $\mathcal{R}$ bounded solution operator for problem (3.4) derives the L_p–L_q maximal regularity as well as the generation of C_0 analytic semigroup. Moreover, by the transference theorem, the $\mathcal{R}$ bounded solution operator derives the maximal L_p–L_q regularity for the periodic solutions. This is not a topics of this chapter but will be explained elsewhere. Thus, the $\mathcal{R}$ bounded solution operator is a key issue in the study of the system of parabolic equations including the equations appearing in the fluid mechanics, MHD, the mathematical theory of liquid crystal, the theory of multi-component mixture flow, and so on.

This chapter is organized as follows. In the rest of Sect. 3.1 we derive the boundary condition in Eq. (3.1) in view of the conservation of mass and momentum. Moreover, we derive the conservation of angular momentum. To end Sect. 3.1, we give a short history of mathematical studies of Eq. (3.1), and further notation used throughout this chapter. In Sect. 3.2, we give the definition of uniformly C^k domains ($k \geq 2$), the weak Dirichlet problem, and the Laplace–Beltrami operators. The weak Dirichlet problem is used to define the reduced Stokes equations obtained by eliminating the pressure term and the Laplace–Beltrami operator plays an essential role to describe the surface tension. In Sect. 3.3, we explain some transformation of Ω_t to a fixed domain and derive the system of nonlinear equations on this fixed domain. In Sect. 3.4, we state the maximal L_p–L_q regularity theorems for the linearized equations (3.3) and also the existence theorem of $\mathcal{R}$-bounded solution operator for the resolvent problem (3.4). Moreover, using the $\mathcal{R}$ bounded solution operator and the Weis operator valued Fourier multiplier theorem, we prove the maximal L_p–L_q regularity theorem. In Sect. 3.5, we prove the existence of $\mathcal{R}$-bounded solution operators by dividing the studies into the following subsections: Sect. 3.5.1 is devoted to a model problem in $\mathbb{R}^N$, Sect. 3.5.2 perturbed problem in $\mathbb{R}^N$, Sect. 3.5.3 a model problem in $\mathbb{R}^N_+$, and Sect. 3.5.4 a problem in a bent half space. And then, in the rest of subsections, putting together the results obtained in previous subsections and using the partition of unity, we prove the existence of $\mathcal{R}$-bounded solution operators and also we prove the uniqueness of solutions. Since the pressure term is non-local, it does not fit the usual localization argument using in the parameter elliptic problem, and so according to an idea due to Grubb-Solonnikov [21], we consider the reduced Stokes equation which is defined in Sect. 3.4.3. In Sect. 3.6, we prove the local well-posedness of Eq. (3.1). Here, we use the Hanzawa transform to transform Eq. (3.1) to nonlinear equations on a fixed domain. Thus, one of main difficulties is to treat the nonlinear term of the form $< \mathbf{u} \mid \nabla'_\Gamma \rho >$ on the boundary which arises from the kinematic equation $V_n = \mathbf{v} \cdot \mathbf{n}_t$. If we assume that the initial data of both $\mathbf{u}$ and ρ are small, then this term is harmless. But, if we want to avoid the smallness condition on the initial velocity field $\mathbf{u}_0 = \mathbf{u}|_{t=0}$, we approximate initial data $\mathbf{u}_0$ by a family of functions $\{\mathbf{u}_\kappa\}$ such that $\|\mathbf{u}_\kappa - \mathbf{u}_0\|_{L_q} \sim \kappa^\alpha$ and $\|\mathbf{u}_\kappa\|_{H_q^2} \sim \kappa^{-\beta}$ as $\kappa \to 0$ with some positive constants α and β. In Sect. 3.7, we prove the global well-posedness of Eq. (3.1) under the assumption that the initial domain $\Omega_t|_{t=0} = \Omega$ is closed to a ball and initial data are small. We use the Hanzawa transform whose center is the barycenter of Ω_t which is crucial, and then, the shape of domain becomes a ball when time goes infinity. In the final section, Sect. 8, we consider Eq. (3.1) in the case where $\sigma = 0$, that is the surface tension is not taken into account, and we prove the global well-posedness of Eq. (3.1) in the case where $\Omega = \Omega_t|_{t=0}$ is an exterior domain in $\mathbb{R}^N$ ($N \geq 3$). Since we consider the without surface tension case, we can not represent the free surface by using some representing function like Hanzawa transform, because we can not obtain enough regularity of such functions. Thus, we use the local Lagrange transform to transform Γ_t to $\Gamma_t|_{t=0} = \Gamma$, which is identity on the outside of some large ball.

3.1.2 Derivation of Boundary Conditions

We drive the boundary conditions which guarantee the conservation of mass and the conservation of momentum. For a while, the mass density ρ is assumed to be a function of (x, t). Let $\phi_t : \Omega \to \mathbb{R}^N$ be a smooth injective map with suitable regularity for each time $t \geq 0$ such that $\phi_t(y)|_{t=0} = y$, and let

$$\Omega_t = \{x = \phi_t(y) \mid y \in \Omega\}.$$

Let ρ and $\mathbf{v}$ satisfy the Navier–Stokes equations:

$$\partial_t \rho + \operatorname{div}(\rho \mathbf{v}) = 0 \quad \text{in} \bigcup_{0 < t < T} \Omega_t \times \{t\}, \tag{3.5}$$

$$\rho(\partial_t \mathbf{v} + \mathbf{v} \cdot \nabla \mathbf{v}) - \operatorname{Div} \mathbf{S}(\mathbf{v}, \mathfrak{p}) = 0 \quad \text{in} \bigcup_{0 < t < T} \Omega_t \times \{t\}. \tag{3.6}$$

We say that Eq. (3.5) is the equation of mass and Eq. (3.6) the equation of momentum. Let

$$\mathbf{S}(\mathbf{v}, \mathfrak{p}) = \mu \mathbf{D}(\mathbf{v}) + (\nu - \mu)\text{div } \mathbf{v}\mathbf{I} - \mathfrak{p}\mathbf{I}, \tag{3.7}$$

which denotes the stress tensor, where μ and ν are positive constants describing the first and second viscosity coefficients, respectively. If we assume that ρ is a positive constant, then by (3.5) div $\mathbf{v} = 0$, which, combined with (3.7), leads to $\mathbf{S}(\mathbf{v}, \mathfrak{p}) = \mu \mathbf{D}(\mathbf{v}) - \mathfrak{p}\mathbf{I}$. This is the situation of Eq. (3.1).

Let

$$\mathbf{S}(\mathbf{v}, \mathfrak{p}) = \begin{pmatrix} \mathbf{S}_1(\mathbf{v}, \mathfrak{p}) \\ \vdots \\ \mathbf{S}_N(\mathbf{v}, \mathfrak{p}) \end{pmatrix}.$$

Where $\mathbf{S}_i(\mathbf{v}, \mathfrak{p})$ are N row vectors of functions whose j-th component is $\mu D_{ij}(\mathbf{v}) + (\nu - \mu)\text{div } \mathbf{v}\delta_{ij} - \mathfrak{p}\delta_{ij}$, δ_{ij} being the Kronecker delta symbols, that is, $\delta_{ii} = 1$ and $\delta_{ij} = 0$ $(i \neq j)$. Then, Eq. (3.6) is written componentwise as

$$\rho(\partial_t v_i + \sum_{j=1}^{N} v_j \partial_j v_i) - \text{div } \mathbf{S}_i(\mathbf{v}, \mathfrak{p})$$

$$= \rho(\partial_t v_i + \sum_{j=1}^{N} v_j \partial_j v_i) - \mu \Delta v_i - \nu \partial_i \text{div } \mathbf{v} + \partial_i \mathfrak{p} = 0 \quad (i = 1, \ldots, N),$$

where $\Delta v_i = \sum_{j=1}^{N} \dfrac{\partial^2 v_i}{\partial x_j^2}$.

To drive boundary conditions that preserve the conservation of mass and the conservation of momentum, we use the following theorem.

Theorem 3.1.1 (Reynolds Transport Theorem) *Let $y = \phi_t^{-1}(x)$ be the inverse map of $x = \phi_t(y)$. Let $J = J(y, t)$ be the Jacobian of the transformation: $x = \phi_t(y)$ and $\mathbf{w}(x, t) = (\partial_t \phi_t)(\phi_t^{-1}(x))$. Then,*

$$\frac{\partial}{\partial t} J(y, t) = (\text{div}_x \mathbf{w}(x, t)) J(y, t).$$

Remark 3.1.2 Reynolds transport theorem will be proved in (3.87) of Sect. 3.2.8 below.

Let $f = f(x, t)$ be a function defined on $\overline{\Omega}_t$, and then

$$\frac{d}{dt} \int_{\Omega_t} f(x, t) \, dx = \int_{\Omega_t} \partial_t f(x, t) + \operatorname{div}(f(x, t)\mathbf{w}(x, t)) \, dx. \tag{3.8}$$

In fact, by Reynolds transport theorem,

$$\frac{d}{dt} \int_{\Omega_t} f(x, t) \, dx = \frac{d}{dt} \int_{\Omega} f(\phi_t(y), t) J(y, t) \, dy$$

$$= \int_{\Omega} \{(\partial_t f + (\nabla f) \cdot \partial_t \phi_t) J + f \partial_t J\} \, dy = \int_{\Omega_t} (\partial_t f + \operatorname{div}(f\mathbf{w})) \, dx.$$

We first consider the conservation of mass:

$$\frac{d}{dt} \int_{\Omega_t} \rho \, dx = 0. \tag{3.9}$$

By (3.8) and the equation of mass, Eq. (3.5),

$$\frac{d}{dt} \int_{\Omega_t} \rho \, dx = \int_{\Omega_t} \partial_t \rho + \operatorname{div}(\rho\mathbf{w}) \, dx = \int_{\Omega_t} \operatorname{div}(\rho(\mathbf{w} - \mathbf{v})) \, dx$$

$$= \int_{\Gamma_t} \rho(\mathbf{w} - \mathbf{v}) \cdot \mathbf{n}_t \, d\omega,$$

where $d\omega$ is the area element on Γ_t. Thus, in order that the mass conservation (3.9) holds, it is sufficient to assume that

$$\rho(\mathbf{w} - \mathbf{v}) \cdot \mathbf{n}_t = 0. \tag{3.10}$$

Since ρ is the mass density, we may assume that $\rho > 0$, and so, we have

$$\mathbf{w} \cdot \mathbf{n}_t = \mathbf{v} \cdot \mathbf{n}_t \quad \text{on } \Gamma_t. \tag{3.11}$$

Let $V_{\Gamma_t} = \mathbf{w} \cdot \mathbf{n}_t$, which represents the velocity of the evolution of the free surface Γ_t in the $\mathbf{n}_t$ direction, and then (3.11) is written as

$$V_{\Gamma_t} = \mathbf{v} \cdot \mathbf{n}_t \quad \text{on } \Gamma_t \tag{3.12}$$

for $t \in (0, T)$. This is called a **kinematic condition**, or non-slip condition. Let us assume that Γ_t is represented by $F(x, t) = 0$ for $x \in \Gamma_t$ locally. Since $\Gamma_t = \{x = \phi_t(y) \mid y \in \Gamma\}$ for $t \in (0, T)$, we have $F(\phi_t(y), t) = 0$ for $y \in \Gamma$. Differentiation of this formula with respect to t yields that

$$\partial_t F + (\partial_t \phi_t) \cdot \nabla_x F = \partial_t F + \mathbf{w} \cdot \nabla F = 0.$$

Since $\mathbf{n}_t = \nabla_x F / |\nabla_x F|$, it follows from (3.11) that $\mathbf{w} \cdot \nabla_x F = \mathbf{v} \cdot \nabla_x F$ on Γ_t, and so

$$\partial_t F + \mathbf{v} \cdot \nabla F = 0 \quad \text{on} \quad \bigcup_{0<t<T} \Gamma_t \times \{t\}. \tag{3.13}$$

This expresses the fact that the free surface Γ_t consists for all $t > 0$ of the same fluid particles, which do not leave it and are not incident on it from inside Ω_t.

Secondly, we consider the conservation of momentum:

$$\frac{d}{dt} \int_{\Omega_t} \rho v_i \, dx = 0 \quad (i = 1, \dots, N). \tag{3.14}$$

As $\partial_t \rho + \operatorname{div}(\rho \mathbf{v}) = 0$ (cf. (3.5)), Eq. (3.6) is rewritten as

$$\partial_t (\rho v_i) + \operatorname{div}(\rho v_i \mathbf{v}) - \operatorname{div} \mathbf{S}_i(\mathbf{v}, \mathfrak{p}) = 0.$$

And then, by (3.8)

$$\frac{d}{dt} \int_{\Omega_t} \rho v_i \, dx = \int_{\Omega_t} \partial_t (\rho v_i) + \operatorname{div}(\rho v_i \mathbf{w}) \, dx$$

$$= \int_{\Omega_t} \operatorname{div}(\rho v_i (\mathbf{w} - \mathbf{v})) + \operatorname{div} \mathbf{S}_i(\mathbf{v}, \mathfrak{p}) \, dx$$

$$= \int_{\Gamma_t} \rho v_i (\mathbf{w} - \mathbf{v}) \cdot \mathbf{n}_t \, d\omega + \int_{\Gamma_t} \mathbf{S}_i(\mathbf{v}, \mathfrak{p}) \cdot \mathbf{n}_t \, d\omega.$$

Thus, by (3.10),

$$\frac{d}{dt} \int_{\Omega_t} \rho v_i \, dx = \int_{\Gamma_t} \mathbf{S}_i(\mathbf{v}, \mathfrak{p}) \cdot \mathbf{n}_t \, d\omega.$$

To obtain (3.14), it suffices to assume that

$$\int_{\Gamma_t} \mathbf{S}_i(\mathbf{v}, \mathfrak{p}) \cdot \mathbf{n}_t \, d\omega = 0. \quad (i = 1, \dots, N).$$

Let $H(\Gamma_t)$ and Δ_{Γ_t} be the $N - 1$ fold mean curvature of Γ_t and the Laplace–Beltrami operator on Γ_t. We know that

$$H(\Gamma_t)\mathbf{n}_t = \Delta_{\Gamma_t} x \quad (x \in \Gamma_t). \tag{3.15}$$

In this lecture note, our boundary conditions are

$$\mathbf{S}(\mathbf{v}, \mathfrak{p})\mathbf{n}_t = \sigma H(\Gamma_t)\mathbf{n}_t - p_0 \mathbf{n}_t, \quad V_{\Gamma_t} = \mathbf{v} \cdot \mathbf{n}_t \quad \text{on} \ \Gamma_t \tag{3.16}$$

for $t \in (0, T)$. Here, $\mathbf{S}(\mathbf{v}, \mathfrak{p})\mathbf{n}_t = {}^\top(\mathbf{S}_1(\mathbf{v}, \mathfrak{p}) \cdot \mathbf{n}_t, \ldots, \mathbf{S}_N(\mathbf{v}, \mathfrak{p}) \cdot \mathbf{n}_t)$, and by (3.16) $\mathbf{S}_i(\mathbf{v}, \mathfrak{p}) \cdot \mathbf{n}_t = \sigma \Delta_{\Gamma_t} x_i$ for $i = 1, \ldots, N$. Outside pressure $p_0 \mathbf{n}_t$ is an external force, and so to obtain the conservation of momentum and angular momentum, we assume that $p_0 = 0$ below. Note that if p_0 is a constant, then $\mathfrak{p} - p_0$ is a new pressure, and so we can reduce the situation to the case where $p_0 = 0$. When $p_0 = 0$,

$$\int_{\Gamma_t} \mathbf{S}_i(\mathbf{v}, \mathfrak{p}) \cdot \mathbf{n}_t \, d\omega = \int_{\Gamma_t} \Delta_{\Gamma_t} x_i \, d\omega = 0.$$

Thus, the conservation of momentum (3.14) holds.

We will show two more identities. By (3.5), (3.6), and (3.16), we have

$$\frac{d}{dt} \int_{\Omega_t} \rho(x_i v_j - x_j v_i) \, dx = 0, \tag{3.17}$$

$$\frac{d}{dt} \int_{\Omega_t} \rho x_i \, dx = \int_{\Omega_t} \rho v_i \, dx. \tag{3.18}$$

The formula (3.17) is called the conservation of angular momentum. In fact, by (3.8)

$$\frac{d}{dt} \int_{\Omega_t} \rho(x_i v_j - x_j v_i) \, dx$$

$$= \int_{\Omega_t} \{\partial_t(\rho(x_i v_j - x_j v_i)) + \mathrm{div}\,(\rho(x_i v_j - x_j v_i)\mathbf{w})\} \, dx$$

$$= \int_{\Omega_t} \{(\partial_t \rho)(x_i v_j - x_j v_i) + x_i \rho \partial_t v_j - x_j \rho \partial_t v_i + \mathrm{div}\,(\rho(x_i v_j - x_j v_i)\mathbf{w})\} \, dx$$

$$= \int_{\Omega_t} \{-(\mathrm{div}\,(\rho \mathbf{v}))(x_i v_j - x_j v_i) - \rho(x_i \mathbf{v} \cdot \nabla v_j - x_j \mathbf{v} \cdot \nabla v_i)$$

$$\quad + x_i \mathrm{div}\, \mathbf{S}_j(\mathbf{v}, \mathfrak{p}) - x_j \mathrm{div}\, \mathbf{S}_i(\mathbf{v}, \mathfrak{p}) + \mathrm{div}\,(\rho(x_i v_j - x_j v_i)\mathbf{w})\} \, dx$$

$$= \int_{\Omega_t} [\mathrm{div}\, \{\rho(x_i v_j - x_j v_i)(\mathbf{w} - \mathbf{v})\} + \mathrm{div}\,(x_i \mathbf{S}_j(\mathbf{v}, \mathfrak{p})) - \mathrm{div}\,(x_j \mathbf{S}_i(\mathbf{v}, \mathfrak{p}))$$

$$\quad + \sum_{k=1}^{N} (\delta_{ik} S_{jk}(\mathbf{v}, \mathfrak{p}) - \delta_{jk} S_{ik}(\mathbf{v}, \mathfrak{p})] \, dx,$$

where $S_{jk}(\mathbf{v}, \mathfrak{p}) = D_{jk}(\mathbf{v}) + ((\nu - \mu)\mathrm{div}\, \mathbf{v} - \mathfrak{p})\delta_{jk}$. By (3.11), we have

$$\int_{\Omega_t} \mathrm{div}\, \{\rho(x_i v_j - x_j v_i)(\mathbf{w} - \mathbf{v})\} \, dx = \int_{\Gamma_t} \rho(x_i v_j - x_j v_i)(\mathbf{w} - \mathbf{v}) \cdot \mathbf{n}_t \, d\omega = 0.$$

By (3.15) and (3.16), we have

$$\int_{\Omega_t} \mathrm{div}\,(x_i \mathbf{S}_j(\mathbf{v}, \mathfrak{p})) - \mathrm{div}\,(x_j \mathbf{S}_i(\mathbf{v}, \mathfrak{p}))\,dx$$

$$= \int_{\Gamma_t} \{x_i \mathbf{S}_j(\mathbf{v}, \mathfrak{p}) \cdot \mathbf{n}_t - x_j \mathbf{S}_i(\mathbf{v}, \mathfrak{p}) \cdot \mathbf{n}_t\}\,d\omega$$

$$= \sigma \int_{\Gamma_t} \{x_i \Delta_{\Gamma_t} x_j - x_j \Delta_{\Gamma_t} x_i\}\,d\omega$$

$$= -\sigma \int_{\Gamma_t} \{\nabla_{\Gamma_t} x_i \cdot \nabla_{\Gamma_t} x_j - \nabla_{\Gamma_t} x_j \cdot \nabla_{\Gamma_t} x_i\}\,d\omega = 0.$$

Since $S_{ij}(\mathbf{v}, \mathfrak{p}) = S_{ji}(\mathbf{v}, \mathfrak{p})$, we have

$$\int_{\Omega_t} \sum_{k=1}^{N} (\delta_{ik} S_{jk}(\mathbf{v}, \mathfrak{p}) - \delta_{jk} S_{ik}(\mathbf{v}, \mathfrak{p}))\,dx = 0.$$

Thus, we have (3.17).

Analogously, by (3.8), (3.5), and (3.11), we have

$$\frac{d}{dt} \int_{\Omega_t} \rho x_i\,dx = \int_{\Omega_t} \{\partial_t(\rho x_i) + \mathrm{div}\,(\rho x_i \mathbf{w})\}\,dx$$

$$= \int_{\Omega_t} \{(\partial_t \rho) x_i + \mathrm{div}\,(\rho x_i \mathbf{w})\}\,dx = \int_{\Omega_t} \{-(\mathrm{div}\,(\rho \mathbf{v})) x_i + \mathrm{div}\,(\rho x_i \mathbf{w})\}\,dx$$

$$= \int_{\Omega_t} \{\mathrm{div}\,(x_i \rho(\mathbf{w} - \mathbf{v})) + \rho v_i\}\,dx = \int_{\Gamma_t} \rho x_i(\mathbf{w} - \mathbf{v}) \cdot \mathbf{n}_t\,d\omega + \int_{\Omega_t} \rho v_i\,dx$$

$$= \int_{\Omega_t} \rho v_i\,dx.$$

Thus, we have (3.18).

Finally, assuming that $\rho = 1$ and adding the initial conditions:

$$\mathbf{v}|_{t=0} = \mathbf{v}_0 \quad \text{in } \Omega, \quad \Omega_t|_{t=0} = \Omega, \tag{3.19}$$

we have the set of equations in Eq. (3.1).

3.1.3 Short History

The problems (P1) and (P2) have been studied by many mathematicians. In case of (P1), the initial domain, Ω, is bounded. In the case that $\sigma > 0$, the local in time unique existence theorem was proved by Solonnikov [59, 62–66], and Padula and Solonnikov [34] in the L_2 Sobolev-Slobodetskii space, by Schweizer [41] in the semigroup setting, by Moglilevskiĭ and Solonnikov [28, 29, 65] in the Hölder spaces. In the case that $\sigma = 0$, the local in time unique existence theorem was proved by Solonnikov [61] and Mucha and Zajączkowski [30] in the L_p Sobolev-Slobodetskii space, and by Shibata and Shimizu [52, 53] in the L_p in time and L_q in space setting.

The global in time unique existence theorem was proved, in the case that $\sigma = 0$, by Solonnikov [61] in the L_p Sobolev-Slobodetskii space, and by Shibata and Shimizu [52, 53] in the L_p in time and L_q in space under the assumption that the initial velocities are small and orthogonal to the rigid space, $\{\mathbf{u} \mid \mathbf{D}(\mathbf{u}) = 0\}$, and, in the case that $\sigma > 0$, was proved by Solonnikov [60] in the L_2 Sobolev-Slobodetskii space, by Padula and Solonnikov [33] in the Hölder spaces, and by Shibata [47] in the L_p in time and L_q in space setting under the assumptions that the initial domain, Ω, is sufficiently close to a ball and the initial velocities are small and orthogonal to the rigid space.

In case of (P2), the initial domain, Ω, is a perturbed layer given by $\Omega = \{x \in \mathbb{R}^N \mid -b < x_N < \eta(x'), \; x' = (x_1, \ldots, x_N) \in \mathbb{R}^{N-1}\}$. The local in time unique existence theorem was proved by Beale [5, 6], Allain [3] and Tani [71] in the L_2 Sobolev-Slobodetskii space when $\sigma > 0$ and by Abels [1] in the L_p Sobolev-Slobodetskii space when $\sigma = 0$. The global in time unique existence theorem for small initial velocities was proved in the L_2 Sobolev-Slobodetskii space by Beale [5, 6] and Tani and Tanaka [72] in the case that $\sigma > 0$ and by Saito [38] in the L_p in time and L_q in space setting in the case that $\sigma = 0$. The decay rate was studied by Beale and Nishida [7] (cf. also [24] for the detailed proof), Lynn and Sylvester [27], Hataya [23] and Hataya-Kawashima [25] in the L_2 framework with polynomial decay, and by Saito [38] with exponential decay.

Recently, the local well-posedness in general unbounded domains was proved by Shibata [44] and [55] under the assumption that the initial domain is uniformly C^3 and the weak Dirichlet problem is uniquely solvable in the initial domain. To transform Eq. (3.1) to the problem in the reference domain, in [44] the Lagrange transform was used and in [55] the Hanzawa transform was used. Moreover, in the case that $\Omega_t = \{x \in \mathbb{R}^N \mid x_N < \eta(x', t) \; x' = (x_1, \ldots, x_{N-1}) \in \mathbb{R}^{N-1}\}$, which is corresponding to the ocean problem without bottom physically, the global in time unique existence theorem for Eq. (3.1) was proved by Saito and Shibata [39, 40], and in the case that Ω is an exterior domain in $\mathbb{R}^N$ ($N \geq 3$), the local and global in time unique existence theorems were proved by Shibata [46, 48, 49]. In such unbounded domains case, we can only show that the L_q space norm of solutions of the Stokes equations with free boundary conditions decay polynomially, and so we have to choose different exponents p and q to guarantee the L_p integrability on time interval $(0, \infty)$. This is one of the reasons why the maximal L_p–L_q regularity theory with freedom of choice of p and q is necessary in the study of the free boundary problems. In Sect. 3.8 below, it is explained how to prove the global in time unique existence theorem by combining the decay properties of the lower order terms and the maximal L_p–L_q regularity of the highest order terms. Finally, the author is honored to refer the readers to the lecture note written by Solonnikov, who is a pioneer of the study of free boundary problems of the Navier- Stokes equations, and Denisova [67], where it is given an overview of studies achieved by Solonnikov and his followers about the free boundary problems of the Navier–Stokes equations in a bounded domain and related topics.

We remark that the two phase fluid flows separated by sharp interface problem has been studied by Abels [2], Denisova and Solonnikov [12–15], Giga and Takahashi [20, 69], Nouri and Poupaud [32], Pruess, Simonett et al. [26, 35–37, 58], Shibata and Shimizu [51], Shimizu [56, 57], Tanaka [70] and references therein.

3.1.4 Further Notation

This section is ended by explaining further notation used in this lecture note. We denote the sets of all complex numbers, real numbers, integers, and natural numbers by $\mathbb{C}$, $\mathbb{R}$, $\mathbb{Z}$, and $\mathbb{N}$, respectively. Let $\mathbb{N}_0 = \mathbb{N} \cup \{0\}$. For any multi-index $\alpha = (\alpha_1, \ldots, \alpha_N) \in \mathbb{N}_0^N$ we set $\partial_x^\alpha h = \partial_1^{\alpha_1} \cdots \partial_N^{\alpha_N} h$ with $\partial_i = \partial/\partial x_i$. For any scalor function f, we write

$$\nabla f = (\partial_1 f, \ldots, \partial_N f), \quad \bar\nabla f = (f, \partial_1 f, \ldots, \partial_N f),$$
$$\nabla^n f = (\partial_x^\alpha f \mid |\alpha| = n), \quad \bar\nabla^n f = (\partial_x^\alpha f \mid |\alpha| \le n) \quad (n \ge 2),$$

where $\partial_x^0 f = f$. For any m-vector of functions, $\mathbf{f} = {}^\top(f_1, \ldots, f_m)$, we write

$$\nabla \mathbf{f} = (\nabla f_1, \ldots, \nabla f_m), \quad \bar\nabla \mathbf{f} = (\bar\nabla f_1, \ldots, \bar\nabla f_m),$$
$$\nabla^n \mathbf{f} = (\nabla^n f_1, \ldots, \nabla^n f_m), \quad \bar\nabla^n \mathbf{f} = (\bar\nabla^n f_1, \ldots, \bar\nabla^n f_m).$$

For any N-vector of functions, $\mathbf{u} = {}^\top(u_1, \ldots, u_N)$, sometime $\nabla \mathbf{u}$ is regarded as an $N \times N$ matrix of functions whose (i, j)th component is $\partial_i u_j$, that is

$$\nabla \mathbf{u} = \begin{pmatrix} \partial_1 u_1 & \partial_2 u_1 & \cdots & \partial_N u_1 \\ \partial_1 u_2 & \partial_2 u_2 & \cdots & \partial_N u_2 \\ \vdots & \vdots & \ddots & \vdots \\ \partial_1 u_N & \partial_2 u_N & \cdots & \partial_N u_N \end{pmatrix}.$$

For any m-vector $V = (v_1, \ldots, v_m)$ and n-vector $W = (w_1, \ldots, w_n)$, $V \otimes W$ denotes an (m, n) matrix whose (i, j)th component is $V_i W_j$. For any (mn, N) matrix $A = (A_{ij,k} \mid i = 1, \ldots, m, j = 1, \ldots, n, k = 1, \ldots, N)$, $AV \otimes W$ denotes an N column vector whose ith component is the quantity: $\sum_{j=1}^m \sum_{j=1}^n A_{jk,i} v_j w_k$.

For any N vector $\mathbf{a}$, $\mathbf{a}_i$ denotes the ith component of $\mathbf{a}$ and for any $N \times N$ matrix $\mathbf{A}$, $\mathbf{A}_{ij}$ denotes the (i, j)th component of $\mathbf{A}$, and moreover, the $N \times N$ matrix whose

(i, j)th component is K_{ij} is written as (K_{ij}). Let δ_{ij} be the Kronecker delta symbol, that is $\delta_{ii} = 1$ and $\delta_{ij} = 0$ for $i \neq j$. In particular, $\mathbf{I} = (\delta_{ij})$ is the $N \times N$ identity matrix. Let $\mathbf{a} \cdot \mathbf{b} = < \mathbf{a}, \mathbf{b} > = \sum_{j=1}^{N} \mathbf{a}_j \mathbf{b}_j$ for any N-vectors $\mathbf{a}$ and $\mathbf{b}$. For any N-vector $\mathbf{a}$, let $\Pi_0 \mathbf{a} = \mathbf{a}_\tau := \mathbf{a} - < \mathbf{a}, \mathbf{n} > \mathbf{n}$. For any two $N \times N$ matrices $\mathbf{A}$ and $\mathbf{B}$, the quantity $\mathbf{A} : \mathbf{B}$ is defined by $\mathbf{A} : \mathbf{B} = \mathrm{tr}\mathbf{AB} = \sum_{i,j=1}^{N} \mathbf{A}_{ij} \mathbf{B}_{ji}$. Given $1 < q < \infty$, let $q' = q/(q-1)$. For $L > 0$, let $B_L = \{x \in \mathbb{R}^N \mid |x| < L\}$ and $S_L = \{x \in \mathbb{R}^N \mid |x| = L\}$. Moreover, for $L < M$, we set $D_{L,M} = \{x \in \mathbb{R}^N \mid L < |x| < M\}$.

For any domain G in $\mathbb{R}^N$, let $C_0^\infty(G)$ be the set of all C^∞ functions whose supports are compact and contained in G. Let $(\mathbf{u}, \mathbf{v})_G = \int_G \mathbf{u} \cdot \mathbf{v} \, dx$ and $(\mathbf{u}, \mathbf{v})_{\partial G} = \int_{\partial G} \mathbf{u} \cdot \mathbf{v} \, ds$, where ds denotes the surface element on ∂G and ∂G is the boundary of G. For $T > 0$, $G \times (0, T) = \{(x, t) \mid x \in G, t \in (0, T)\}$ is written simply by G^T.

For $1 \leq q \leq \infty$, let $L_q(G)$, $H_q^m(G)$, and $B_{q,p}^s(G)$ be the standard Lebesgue, Sobolev, and Besov spaces on G, and let $\| \cdot \|_{L_q(G)}$, $\| \cdot \|_{H_q^m(G)}$, and $\| \cdot \|_{B_{q,p}^s(G)}$ denote their respective norms. We write $L_q(G)$ as $H_q^0(G)$, and $B_{q,q}^s(G)$ simply as $W_q^s(G)$. For any Banach space X with norm $\| \cdot \|_X$, let $X^d = \{(f_1, \ldots, f_d) \mid f_i \in X \ (i = 1, \ldots, d)\}$, and write the norm of X^d simply as $\| \cdot \|_X$, which is defined by $\|f\|_X = \sum_{j=1}^{d} \|f_j\|_X$ for $f = (f_1, \ldots, f_d) \in X^d$.

For $1 \leq p \leq \infty$, $L_p((a, b), X)$ and $H_p^m((a, b), X)$ denote the standard Lebesgue and Sobolev spaces of X-valued functions defined on an interval (a, b), and their respective norms are denoted by $\| \cdot \|_{L_p((a,b),X)}$ and $\| \cdot \|_{H_p^m((a,b),X)}$. For $\theta \in (0, 1)$, $H_p^\theta(\mathbb{R}, X)$ denotes the standard X-valued Bessel potential space defined by

$$H_p^\theta(\mathbb{R}, X) = \{f \in L_p(\mathbb{R}, X) \mid \|f\|_{H_p^\theta(\mathbb{R},X)} < \infty\},$$

$$\|f\|_{H_p^\theta(\mathbb{R},X)} = \left(\int_\mathbb{R} \|\mathcal{F}^{-1}[(1 + \tau^2)^{\theta/2} \mathcal{F}[f](\tau)](t)\|_X^p \, dt \right)^{1/p},$$

where $\mathcal{F}$ and $\mathcal{F}^{-1}$ denote the Fourier transform and the inverse Fourier transform, respectively.

For $1 \leq q < \infty$, we define homogeneous spaces $\hat{H}_q^1(G)$ and $\hat{H}_{q,0}^1(G)$ by setting

$$\hat{H}_q^1(G) = \{u \in L_{q,\mathrm{loc}}(\Omega) \mid \nabla u \in L_q(G)^N\},$$

$$\hat{H}_{q,0}^1(G) = \{u \in L_{q,\mathrm{loc}}(\Omega) \mid \nabla u \in L_q(G)^N, \quad u|_{\partial G} = 0\},$$

For $1 < q < \infty$, the solenoidal space $J_q(G)$ on G of this chapter is defined by setting

$$J_q(G) = \{\mathbf{u} \in L_q(G)^N \mid (\mathbf{u}, \nabla \varphi)_G = 0 \quad \text{for any } \varphi \in \hat{H}_{q',0}^1(G)\}.$$

For two Banach spaces X and Y, $X + Y = \{x + y \mid x \in X, y \in Y\}$, $\mathcal{L}(X, Y)$ denotes the set of all bounded linear operators from X into Y and $\mathcal{L}(X, X)$ is written simply as $\mathcal{L}(X)$. For a domain U in $\mathbb{C}$, Hol $(U, \mathcal{L}(X, Y))$ denotes the set of all $\mathcal{L}(X, Y)$-valued holomorphic functions defined on U. Let $\mathcal{R}_{\mathcal{L}(X,Y)}(\{\mathcal{T}(\lambda) \mid \lambda \in U\})$ be the $\mathcal{R}$ norm of the operator family $\mathcal{T}(\lambda) \in$ Hol $(U, \mathcal{L}(X, Y))$. Let

$$\Sigma_{\epsilon_0} = \{\lambda \in \mathbb{C} \setminus \{0\} \mid |\arg \lambda| \leq \pi - \epsilon_0\}, \quad \Sigma_{\epsilon_0,\lambda_0} = \{\lambda \in \Sigma_{\epsilon_0} \mid |\lambda| \geq \lambda_0\},$$

$$\mathbb{C}_{+,\lambda_0} = \{\lambda \in \mathbb{C} \mid \operatorname{Re} \lambda \geq \max(0, \lambda_0)\}.$$

Let $\mathbb{R}_+^N = \{x = (x_1, \ldots, x_N) \mid x_N > 0\}$ and $\mathbb{R}_0^N = \{x = (x_1, \ldots, x_N) \mid x_N = 0\}$. The letter C denotes a generic constant and $C_{a,b,c,\ldots}$ denotes that the constant $C_{a,b,c,\ldots}$ depends on $a, b, c, \cdots$. The value of C and $C_{a,b,c,\ldots}$ may change from line to line.

3.2 Preliminaries

In this section, we study a uniform C^k domain, a uniform C^k domain whose inside has a finite covering, the weak Dirichlet problem, Besov spaces on the boundary, and the Laplace- Beltrami operator on the boundary.

3.2.1 Definitions of Domains

We first introduce the definition of a uniform C^k ($k = 2$ or 3) domain.

Definition 3.2.1 Let $k \in \mathbb{N}$. We say that Ω is a uniform C^k domain, if there exist positive constants a_1, a_2, and A such that the following assertion holds: For any $x_0 = (x_{01}, \ldots, x_{0N}) \in \Gamma$ there exist a coordinate number j and a C^k function $h(x')$ defined on $B'_{a_1}(x'_0)$ such that $\|h\|_{H^k_\infty(B'_{a_1}(x'_0))} \leq A$ and

$$\Omega \cap B_{a_2}(x_0) = \{x \in \mathbb{R}^N \mid x_j > h(x') \ (x' \in B_{a_1}(x'_0))\} \cap B_{a_2}(x_0),$$

$$\Gamma \cap B_{a_2}(x_0) = \{x \in \mathbb{R}^N \mid x_j = h(x') \ (x' \in B'_{a_1}(x'_0))\} \cap B_{a_2}(x_0).$$

Here, we have set

$$y' = (y_1, \ldots, y_{j-1}, y_{j+1}, \ldots, y_N) \ (y \in \{x, x_0\}),$$

$$B'_{a_1}(x'_0) = \{x' \in \mathbb{R}^{N-1} \mid |x' - x'_0| < a_1\},$$

$$B_{a_2}(x_0) = \{x \in \mathbb{R}^N \mid |x - x_0| < a_2\}.$$

The uniform C^k domains are characterized as follows.

Proposition 3.2.2 *Let $k = 2$ or $k = 3$. Let Ω be a uniform C^k domain in $\mathbb{R}^N$. Then, for any $M_1 \in (0, 1)$, there exist constants $M_2 > 0$ and $0 < r_0 < 1$, at most countably many N-vector of functions $\Phi_j \in C^k(\mathbb{R}^N)^N$ and points $x_j^0 \in \Omega$ and $x_j^1 \in \Gamma$ such that the following assertions hold:*

(i) *The maps: $\mathbb{R}^N \ni x \mapsto \Phi_j(x) \in \mathbb{R}^N$ are bijections satisfying the following conditions: $\nabla \Phi_j = \mathcal{A}_j + B_j$, $\nabla(\Phi_j)^{-1} = \mathcal{A}_{j,-} + B_{j,-}$, where $\mathcal{A}_j$ and $\mathcal{A}_{j,-}$ are $N \times N$ constant orthogonal matrices, and B_j and $B_{j,-}$ are $N \times N$ matrices of $C^{k-1}(\mathbb{R}^N)$ functions defined on $\mathbb{R}^N$ satisfying the conditions: $\|(B_j, B_{j,-})\|_{L_\infty(\mathbb{R}^N)} \le M_1$, and $\|\nabla(B_j, B_{j,-})\|_{L_\infty(\mathbb{R}^N)} \le C_A$, where C_A is a constant depending on constants A, a_1 and a_2 appearing in Definition 3.2.1. Moreover, if $k = 3$, then $\|\nabla^2(B_j, B_{j,-})\|_{L_\infty(\mathbb{R}^N)} \le M_2$.*

(ii) *$\Omega = \left(\bigcup_{j=1}^\infty B_{r_0}(x_j^0)\right) \cup \left(\bigcup_{j=1}^\infty (\Phi_j(\mathbb{R}_+^N) \cap B_{r_0}(x_j^1))\right)$, $B_{r_0}(x_j^0) \subset \Omega$, $\Phi_j(\mathbb{R}_0^N) \cap B_{r_0}(x_j^1) = \Gamma \cap B_{r_0}(x_j^1)$.*

(iii) *There exist C^∞ functions ζ_j^i and $\tilde{\zeta}_j^i$ ($i = 0, 1$, $j \in \mathbb{N}$) such that*

$$0 \le \zeta_j^i, \ \tilde{\zeta}_j^i \le 1, \ \ \operatorname{supp} \zeta_j^i \subset \operatorname{supp} \tilde{\zeta}_j^i \subset B_{r_0}(x_j^i), \ \ \tilde{\zeta}_j^i = 1 \text{ on } \operatorname{supp} \zeta_j^i,$$

$$\sum_{i=0}^1 \sum_{j=1}^\infty \zeta_j^i = 1 \text{ on } \overline{\Omega}, \sum_{j=1}^\infty \zeta_j^1 = 1 \text{ on } \Gamma, \|\nabla \zeta_j^i\|_{H_\infty^{k-1}(\mathbb{R}^N)}, \|\nabla \tilde{\zeta}_j^i\|_{H_\infty^{k-1}(\mathbb{R}^N)} \le M_2.$$

(iv) *Below, for the notational simplicity, we set $B_j^i = B_{r_0}(x_j^i)$. For each j, let ℓ_k^{ij} ($k = 1, \ldots, m_j^i$) be numbers for which $B_j^i \cap B_{\ell_k^{ij}}^i \ne \emptyset$ and $B_j^i \cap B_m^i = \emptyset$ for $m \notin \{\ell_k^{ij} \mid k = 1, \ldots, m_j^i\}$. Then, there exists an $L \ge 2$ independent of M_1 such that $m_j^i \le L$.*

Proof Proposition 3.2.2 was essentially proved by Enomoto-Shibata [18, Appendix], except for $\|\nabla(B_j^i, B_{j,-}^i)\|_{L_\infty(\mathbb{R}^N)} \le C_A$. There, it was proved that

$$\|\nabla(B_j^i, B_{j,-}^i)\|_{L_\infty(\mathbb{R}^N)} \le CM_2,$$

that is the estimate of $\nabla(B_j^i, B_{j,-}^i)$ depends on M_2. Since the proof of Proposition 3.2.2 is almost the same as in [18, Appendix], we shall give an idea how to improve this point below. Let $x_0 = (x_0', x_{0N}) \in \Gamma$ and we assume that

$$\Omega \cap B_\beta(x_0) = \{x \in \mathbb{R}^N \mid x_N > h(x') \ (x' \in B_\alpha'(x_0'))\} \cap B_\beta(x_0),$$

$$\Gamma \cap B_\beta(x_0) = \{x \in \mathbb{R}^N \mid x_N = h(x') \ (x' \in B_\alpha'(x_0'))\} \cap B_\beta(x_0).$$

We only consider the case where $k = 3$. In fact, by the same argument, we can improve the estimate in the case where $k = 2$. We assume that $h \in C^3(B_{a_1}'(x_0'))$,

$\|h\|_{H^3_\infty(B'_{a_1}(x'_0))} \leq A$, and $x_{0N} = h(x'_0)$. Below, C denotes a generic constant depending on A, a_1 and a_2 but independent of ϵ. Let $\rho(y)$ be a function in $C^\infty_0(\mathbb{R}^N)$ such that $\rho(y) = 1$ for $|y'| \leq 1/2$ and $|y_N| \leq 1/2$ and $\rho(y) = 0$ for $|y'| \geq 1$ or $|y_N| \geq 1$. Let $\rho_\epsilon(y) = \rho(y/\epsilon)$. We consider a C^∞ diffeomorphism:

$$x_j = \Phi^\epsilon_j(y) = x_{0j} + \sum_{k=1}^N t_{j,k} y_k + \sum_{k,\ell=1}^N s_{j,k\ell} y_k y_\ell \rho_\epsilon(y).$$

Here, $t_{j,k}$ and $s_{j,k\ell}$ are some constants satisfying the conditions (3.22), (3.20), and (3.21), below. Let

$$G_\epsilon(y) = \Phi^\epsilon_N(y) - h(\Phi^\epsilon_1(y), \ldots, \Phi^\epsilon_{N-1}(y)).$$

Notice that $G_\epsilon(0) = x_{0N} - h(x'_0) = 0$. We choose $t_{j,k}$ and $s_{\ell,mn}$ in such a way that

$$\frac{\partial G_\epsilon}{\partial y_N}(0) = t_{N,N} - \sum_{k=1}^{N-1} \frac{\partial h}{\partial x_k}(x'_0)t_{k,N} \neq 0,$$

$$\frac{\partial G_\epsilon}{\partial y_j}(0) = t_{N,j} - \sum_{k=1}^{N-1} \frac{\partial h}{\partial x_k}(x'_0)t_{k,j} = 0,$$

(3.20)

$$\frac{\partial^2 G_\epsilon}{\partial y_\ell \partial y_m}(0) = s_{N,\ell m} + s_{N,m\ell} - \sum_{k=1}^{N-1} \frac{\partial h}{\partial x_k}(x'_0)(s_{k,\ell m} + s_{k,m\ell})$$

$$- \sum_{j,k=1}^{N-1} \frac{\partial^2 G_\epsilon}{\partial x_j \partial x_k}(x'_0)t_{j,\ell}t_{k,m} = 0.$$

(3.21)

Moreover, setting

$$T = \begin{pmatrix} t_{1,1} & t_{2,1} & \cdots & t_{N,1} \\ t_{1,2} & t_{2,2} & \cdots & t_{N,2} \\ \vdots & \vdots & \ddots & \vdots \\ t_{1,N} & t_{2,N} & \cdots & t_{N,N} \end{pmatrix},$$

we assume that T is an orthogonal matrix, that is

$$\sum_{\ell=1}^N t_{\ell,m}t_{\ell,n} = \delta_{mn} = \begin{cases} 1 & \text{for } m = n, \\ 0 & \text{for } m \neq n. \end{cases}$$

(3.22)

We write $\dfrac{\partial h}{\partial x_j}(x_0')$ simply by h_j and set

$$H_j = \sqrt{1 + \sum_{\ell=j}^{N-1} h_\ell^2} = \sqrt{1 + h_j^2 + h_{j+1}^2 + \cdots + h_{N-1}^2}.$$

Let

$$t_{N,N-j} = \frac{h_{N-j}}{H_{N-j} H_{N+1-j}}, \quad t_{N-k,N-j} = -\frac{h_{N-k} h_{N-j}}{H_{N-j} H_{N+1-j}}$$

for $k = 1, \ldots, j - 1$, and

$$t_{N-j,N-j} = \frac{H_{N+1-j}}{H_{N-1}}, \quad t_{k,N-j} = 0,$$

for $k = 1, \ldots, N - j - 1$ and $j = 1, \ldots, N - 1$, and

$$t_{i,N} = -\frac{h_i}{H_1}, \quad t_{N,N} = \frac{1}{H_1}$$

for $i = 1, \ldots, N - 1$. Then, we see that such $t_{j,k}$ satisfy (3.20) and (3.22). In particular,

$$\frac{\partial G_\epsilon}{\partial y_N}(0) = \frac{1}{H_1}. \tag{3.23}$$

Moreover, assuming the symmetry: $s_{\ell,jk} = s_{\ell,kj}$, we have

$$s_{N,jk} = \frac{1}{2H_2} \sum_{m,n=1}^{N-1} \frac{\partial^2 h}{\partial x_m \partial x_n}(x_0') t_{m,j} t_{n,k},$$

$$s_{i,jk} = -\frac{h_i}{2H_1^2} \sum_{m,n=1}^{N-1} \frac{\partial^2 h}{\partial x_m \partial x_n}(x_0') t_{m,j} t_{n,k}.$$

By successive approximation, we see that there exists a constant $\epsilon_0 > 0$ such that for any $\epsilon \in (0, \epsilon_0)$ there exists a function $\psi_\epsilon \in C^3(B_\epsilon'(0))$ satisfying the following conditions:

$$\psi_\epsilon(0) = \partial_i \psi_\epsilon(0) = \partial_i \partial_j \psi_\epsilon(0) = 0,$$

$$\|\psi_\epsilon\|_{L_\infty(B'_\epsilon(0))} \le C\epsilon^2, \quad \|\partial_i \psi_\epsilon\|_{L_\infty(B'_\epsilon(0))} \le C\epsilon,$$

$$\|\partial_i \partial_j \psi_\epsilon\|_{L_\infty(B'_\epsilon(0))} \le C, \quad \|\partial_i \partial_j \partial_k \psi_\epsilon\|_{L_\infty(B'_\epsilon(0))} \le C\epsilon^{-1},$$

$$G_\epsilon(y', \psi_\epsilon(y')) = 0 \quad \text{for } y' \in B'_\epsilon(0), \tag{3.24}$$

where i, j and k run from 1 through $N - 1$. Notice that

$$
\begin{aligned}
x_N - h_\epsilon(x') &= G_\epsilon(y) \\
&= G_\epsilon(y', \psi_\epsilon(y')) \\
&\quad + \int_0^1 (\partial_N G_\epsilon)(y', \psi_\epsilon(y') + \theta(y_N - \psi_\epsilon(y'))) \, d\theta(y_N - \psi_\epsilon(y')) \\
&= ((\partial_N G_\epsilon)(0) + \tilde{G}_\epsilon(y))(y_N - \psi_\epsilon(y')),
\end{aligned}
\tag{3.25}
$$

where we have used $G_\epsilon(y', \psi_\epsilon(y')) = 0$ and

$$
\begin{aligned}
\tilde{G}_\epsilon(y) = \int_0^1 \int_0^1 \Big\{ &\sum_{\ell=1}^{N-1} (\partial_\ell \partial_N G_\epsilon)(\tau y', \tau(\psi_\epsilon(y') + \theta(y_N - \psi_\epsilon(y')))) y_\ell \\
&+ (\partial_N G_\epsilon)(\tau y', \tau(\psi_\epsilon(y') + \theta(y_N - \psi_\epsilon(y'))))(\psi_\epsilon(y') \\
&+ \theta(y_N - \psi_\epsilon(y'))) \Big\} \, d\theta d\tau.
\end{aligned}
$$

Since $(\partial_N G_\epsilon)(0) = 1/H_1$, choosing $\epsilon_0 > 0$ so small that $|\tilde{G}_\epsilon(y)| \le 1/(2H_1)$ for $|y| \le \epsilon_0$, we see that $x_N - h(x') \ge 0$ and $y_N - \psi_\epsilon(y') \ge 0$ are equivalent.

Let ω be a function in $C_0^\infty(\mathbb{R}^{N-1})$ such that $\omega(y') = 1$ for $|y'| \le 1/2$ and $\omega(y') = 0$ for $|y'| \ge 1$ and set $\omega_\epsilon(y') = \psi_\epsilon(y')\omega(y'/\epsilon)$. Then, by (3.24) we have

$$\|\omega_\epsilon\|_{L_\infty(\mathbb{R}^{N-1})} \le C\epsilon^2, \quad \|\partial_i \omega_\epsilon\|_{L_\infty(\mathbb{R}^{N-1})} \le C\epsilon,$$

$$\|\partial_i \partial_j \omega_\epsilon\|_{L_\infty(\mathbb{R}^{N-1})} \le C, \quad \|\partial_i \partial_j \partial_k \omega_\epsilon\|_{L_\infty(\mathbb{R}^{N-1})} \le C\epsilon^{-1}. \tag{3.26}$$

where i, j, and k run from 1 through $N - 1$. Setting $\Psi^\epsilon(z) = \Phi^\epsilon(z', z_N + \omega_\epsilon(z'))$, that is $y_N = z_N + \omega_\epsilon(z')$, and $y_j = z_j$ for $j = 1, \ldots, N - 1$, we see that there exists an $\epsilon_0 > 0$ such that for any $\epsilon \in (0, \epsilon_0)$, the map: $z \to x = \Psi^\epsilon(z)$ is a diffeomorphism of C^3 class from $\mathbb{R}^N$ onto $\mathbb{R}^N$. Since

$$\frac{\partial x_m}{\partial z_k} = t_{m,k} + b_{m,k}, \quad \frac{\partial x_m}{\partial z_N} = t_{m,N} + b_{n,N}$$

where we have set

$$b_{m,k} = \sum_{i,j=1} \frac{\partial}{\partial z_k}(s_{m,ij} y_i y_j \rho_\epsilon(y))$$

$$+ \{t_{m,N} + \sum_{i,j=1}^N \frac{\partial}{\partial z_k}(s_{m,ij} y_i y_j \rho_\epsilon(y))\} \frac{\partial \omega_\epsilon}{\partial z_k}(z'),$$

$$b_{n,N} = \sum_{i,j=1} \frac{\partial}{\partial z_N}(s_{m,ij} y_i y_j \rho_\epsilon(y)),$$

let $\mathcal{A}$ and $\mathcal{B}$ be the $N \times N$ matrices whose (m, n)th components are $t_{m,n}$ and $b_{m,n}$, respectively. Then, by (3.26), $\mathcal{A}$ is an orthogonal matrix and B satisfies the estimates:

$$\|B\|_{L_\infty(\mathbb{R}^N)} \leq C\epsilon, \quad \|\nabla B\|_{L_\infty(\mathbb{R}^N)} \leq C, \quad \|\nabla^2 B\|_{L_\infty(\mathbb{R}^N)} \leq C\epsilon^{-1}.$$

Moreover, by (3.25) we have

$$x_N - h(x') = (1/H_1 + \tilde{G}_\epsilon(z', z_N + \omega_\epsilon(z')))(z_N + (\omega(z'/\epsilon) - 1)\psi_\epsilon(z')),$$

which shows that when $|z'| \leq \epsilon/2$, $x_N \geq h(x')$ and $z_N \geq 0$ are equivalent. We can construct the sequences of $C_0^\infty(\mathbb{R}^N)$ functions, $\{\zeta_j^i\}$, $\{\tilde{\zeta}_j^i\}$, by standard manner (cf. Enomoto-Shibata [18, Appendix]). This completes the proof of Proposition 3.2.2.

$\square$

To show the a priori estimates for the Stokes equations with free boundary condition in a general domain, we need some restriction of the domains. In this chapter note, we adopt the following conditions.

Definition 3.2.3 Let $k = 2$ or 3 and let Ω be a domain in $\mathbb{R}^N$. We say that Ω is a uniformly C^k domain whose inside has a finite covering if Ω is a uniformly C^k domain in the sense of Definition 3.2.1 and the following assertion hold:
(v) Let ζ_j^i be the partition of unity given in Proposition 3.2.2 (iii) and set $\psi^0 = \sum_{j=1}^\infty \zeta_j^0$. Let $\mathcal{O} = \text{supp} \nabla \psi^0 \cup (\bigcup_{j=1}^\infty \text{supp} \nabla \zeta_j^1)$. Then, there exists a finite number of subdomains $\mathcal{O}_j$ $(j = 1, \ldots, \iota)$ such that $\mathcal{O} \subset \bigcup_{j=1}^\iota \mathcal{O}_j$ and each $\mathcal{O}_j$

satisfies one of the following conditions:

(a) There exists an $R > 0$ such that $\mathcal{O}_j \subset \Omega_R$, where $\Omega_R = \{x \in \Omega \mid |x| < R\}$.

(b) There exist a translation τ, a rotation $\mathcal{A}$, a domain $D \subset \mathbb{R}^{N-1}$, a coordinate function $a(x')$ defined for $x' \in D$, and a positive constant b such that $0 \leq a(x') < b$ for $x \in D$,

$$\mathcal{A} \circ \tau(\mathcal{O}_j) \subset \{x = (x', x_N) \mid x' \in D, \ a(x') \leq x_N \leq b\} \subset \mathcal{A} \circ \tau(\Omega),$$

$$\{x = (x', x_N) \in \mathbb{R}^N \mid x' \in D, \ x_N = a(x')\} \subset \mathcal{A} \circ \tau(\Gamma).$$

Here, for any subset E of $\mathbb{R}^N$, $\mathcal{A}(E) = \{Ax \mid x \in E\}$ with some orthogonal matrix A and $\tau(E) = \{x + y \mid x \in E\}$ with some $y \in \mathbb{R}^N$.

Example 3.2.4 Let Ω be a domain whose boundary Γ is a C^k hypersurface. If Ω satisfies one of the following conditions, then Ω is a uniform C^k domain whose inside has a finite covering.

(1) Ω is bounded, or Ω is an exterior domain, that is, $\Omega = \mathbb{R}^N \setminus \overline{\mathcal{O}}$ with some bounded domain $\mathcal{O}$.

(2) $\Omega = \mathbb{R}^N_+$ (half space), or Ω is a perturbed half space, that is, there exists an $R > 0$ such that $\Omega \cap B^R = \mathbb{R}^N_+ \cap B^R$, where $B^R = \{x \in \mathbb{R}^N \mid |x| > R\}$.

(3) Ω is a layer L or a perturbed layer, that is, there exists an $R > 0$ such that $\Omega \cap B^R = L \cap B^R$. Here $L = \{x = (x', x_N) \in \mathbb{R}^N \mid x' = (x_1, \ldots, x_{N-1}) \in \mathbb{R}^{N-1}, \ a < x_N < b\}$ for some constants a and b for which $a < b$.

(4) Ω is a tube, that is, there exists a bounded domain D in $\mathbb{R}^{N-1}$ such that $\Omega = D \times \mathbb{R}$.

(5) There exist an $R > 0$ and several orthogonal transforms, $\mathcal{R}_i$ ($i = 1, \ldots, M$), such that $\Gamma \cap B^R = \left(\bigcup_{i=1}^{M} \mathcal{R}_i \mathbb{R}^N_0 \right) \cap B^R$.

(6) There exist an $R > 0$, half tubes, T_i ($i = 1, \ldots, M$), and orthogonal transforms, $\mathcal{R}_i$ ($i = 1, \ldots, M$), such that $\Omega \cap B^R = \left(\bigcup_{i=1}^{M} \mathcal{R}_i T_i \right) \cap B^R$, where what T_i is a half tube means that $T_i = D_i \times [0, \infty)$ with some bounded domain D_i of $\mathbb{R}^{N-1}$.

In the following, we write $B_{r_0}(x_j^i)$, $\Phi_j(\mathbb{R}^N_+)$, and $\Phi_j(\mathbb{R}^N_0)$ simply by B_j^i, Ω_j and Γ_j, respectively. In view of Proposition 3.2.2 (ii), we have $\Omega_j \cap B_j^1 = \Omega \cap B_j^1$ and $\Gamma_j \cap B_j^1 = \Gamma \cap B_j^1$. By the finite intersection property stated in Proposition 3.2.2 (iv), for any $r \in [1, \infty)$ there exists a constant $C_{r,L}$ such that

$$\left[\sum_{j=1}^{\infty} \| f \|^r_{L_r(\Omega \cap B_j^i)} \right]^{\frac{1}{r}} \leq C_{r,L} \| f \|_{L_r(\Omega)} \quad \text{for any } f \in L_r(\Omega). \tag{3.27}$$

Let $n \in \mathbb{N}_0$, $f \in H_q^n(\Omega)$, and let η_j^i be functions in $C_0^\infty(B_j^i)$ with $\|\eta_j^i\|_{H_\infty^n(\mathbb{R}^N)} \le c_0$ for some constant c_0 independent of $j \in \mathbb{N}$. Since $\Omega \cap B_j^1 = \Omega_j \cap B_j^1$, by (3.27)

$$\sum_{j=1}^\infty \|\eta_j^0 f\|_{H_q^n(\mathbb{R}^N)}^q + \sum_{j=1}^\infty \|\eta_j^1 f\|_{H_q^n(\Omega_j)}^q \le C_q \|f\|_{H_q^n(\Omega)}^q. \tag{3.28}$$

3.2.2 Besov Spaces on Γ

We now define Besov spaces on Γ. Before turning to it, we recall some basic facts. Let f be a function defined on Γ such that supp $f \subset \Gamma \cap B_j^1 \cap B_k^1$. Let $f_j = f \circ \Phi_j^{-1}$, and then by Proposition 3.2.2 (iv), we see that for any $s \in [-1, 3]$ and $j, k \in \mathbb{N}$,

$$\|f_j\|_{W_q^s(\mathbb{R}_0^N)} \le C_{s,q} \|f_k\|_{W_q^s(\mathbb{R}_0^N)}. \tag{3.29}$$

In fact, in the case that $s = 0, 1, 2, 3$, noting $W_q^s = H_q^s$, we see that the inequality (3.29) follows from the direct calculations. When $s = -1$, it follows from duality argument. Finally, in the case that $s \notin \mathbb{Z}$, the inequality (3.29) follows from real interpolation. Let $\Gamma_j = \Phi_j(\mathbb{R}_0^N)$ be spaces given in Proposition 3.2.2 and the space $W_q^s(\Gamma_j)$ and its norm $\|\cdot\|_{W_q^s(\Gamma_j)}$ are defined by

$$W_q^s(\Gamma_j) = \{f \mid f \circ \Phi_j \in W_q^s(\mathbb{R}_0^N)\}, \quad \|f\|_{W_q^s(\Gamma_j)} = \|f \circ \Phi_j^{-1}\|_{W_q^s(\mathbb{R}_0^N)}.$$

In view of (3.29), if supp $f \subset \Gamma \cap B_j^1 \cap B_k^1$, then

$$\|f\|_{W_q^s(\Gamma_j)} \le C_{s,q} \|f\|_{W_q^s(\Gamma_k)}. \tag{3.30}$$

For $s \in [-1, 3]$, we now define $W_q^s(\Gamma)$ by

$$W_q^s(\Gamma) = \{f = \sum_{j=1}^\infty f_j \mid \text{supp } f_j \subset \Gamma \cap B_j^1, \quad f_j \in W_q^s(\Gamma_j),$$

$$\|f\|_{W_q^s(\Gamma)} = \left\{ \sum_{j=1}^\infty \|f_j\|_{W_q^s(\Gamma_j)}^q \right\}^{1/q} < \infty\}. \tag{3.31}$$

Since each $W_q^s(\Gamma_j)$ is a Banach space, so is $W_q^s(\Gamma)$. Given $f \in W_q^s(\Gamma)$, we have

$$\tilde{\zeta}_j f \in W_q^s(\Gamma_j), \quad \sum_{j=1}^{\infty} \|\tilde{\zeta}_j f\|_{W_q^s(\Gamma_j)}^q \leq (c_0 C_{s,q} 3^N)^q \|f\|_{W_q^s(\Gamma)}^q. \tag{3.32}$$

In fact, let $f = \sum_{j=1}^{\infty} f_j \in W_q^s(\Gamma)$, where f_j satisfy (3.31). For each j, let ℓ_k^{1j} ($k = 1, \ldots, m_j^1$) be the numbers given in Proposition 3.2.2 (iv) for which $B_j^1 \cap B_{\ell_k^{1j}}^1 \neq \emptyset$ and $B_j^1 \cap B_m^1 = \emptyset$ for $m \notin \{\ell_k^{1j} \mid k = 1, \ldots, m_j^1\}$, and then

$$\tilde{\zeta}_j f = \sum_{k=1,\ldots,m_j^1} \tilde{\zeta}_j f_{\ell_k^{1j}}.$$

Since $\mathrm{supp}\, \tilde{\zeta}_j f_{\ell_k^{1j}} \subset \Gamma \cap B_{\ell_k^{1j}}^1 \cap B_j^1$, by (3.30)

$$\|\tilde{\zeta}_j f\|_{W_q^s(\Gamma_j)} \leq \sum_{k=1,2,\ldots,m_j^1} \|\tilde{\zeta}_j f_{\ell_k^{1j}}\|_{W_q^s(\Gamma_j)} \leq C_{s,q} \sum_{k=1,2,\ldots,m_j^1} \|\tilde{\zeta}_j f_{\ell_k^{1j}}\|_{W_q^s(\Gamma_{\ell_k^{1j}})}$$

$$\leq c_0 C_{s,q} \sum_{k=1,2,\ldots,m_j^1} \|f_{\ell_k^{1j}}\|_{W_q^s(\Gamma_{\ell_k^{1j}})} < \infty,$$

where c_0 is the number appearing in Proposition 3.2.2. Thus, we have $\tilde{\zeta}_j f \in W_q^s(\Gamma_j)$. Since $m_j^1 \leq L$ with some constant L independent of M_1 as follows from Proposition 3.2.2, we have

$$\sum_{j=1}^{\infty} \|\tilde{\zeta}_j f\|_{W_q^s(\Gamma_j)}^q \leq \sum_{j=1}^{\infty} (c_0 C_{s,q})^q (m_j^1)^{q-1} \sum_{k=1,\ldots,m_j^1} \|f_{\ell_k^{1j}}\|_{W_q^s(\Gamma_{\ell_k^{1j}})}^q$$

$$\leq (c_0 C_{s,q} L)^q \sum_{n=1}^{\infty} \|f_n\|_{W_q^s(\Gamma_n)}^q,$$

which yields (3.32).

3.2.3 The Weak Dirichlet Problem

Let $\hat{H}_{q,0}^1(\Omega)$ be the homogeneous Sobolev space defined by letting

$$\hat{H}_{q,0}^1(\Omega) = \{\varphi \in L_{q,\mathrm{loc}}(\Omega) \mid \nabla\varphi \in L_q(\Omega)^N, \quad \varphi|_\Gamma = 0\}. \tag{3.33}$$

Let $1 < q < \infty$. The variational equation:

$$(\nabla u, \nabla \varphi)_\Omega = (\mathbf{f}, \nabla \varphi)_\Omega \quad \text{for all } \varphi \in \hat{H}^1_{q',0}(\Omega) \tag{3.34}$$

is called the weak Dirichlet problem, where $q' = q/(q-1)$.

Definition 3.2.5 We say that the weak Dirichlet problem (3.34) is uniquely solvable in $\hat{H}^1_{q,0}(\Omega)$ if for any $\mathbf{f} \in L_q(\Omega)^N$, problem (3.34) admits a unique solution $u \in \hat{H}^1_{q,0}(\Omega)$ possessing the estimate: $\|\nabla u\|_{L_q(\Omega)} \leq C\|\mathbf{f}\|_{L_q(\Omega)}$.

We define an operator $\mathcal{K}$ acting on $\mathbf{f} \in L_q(\Omega)^N$ by $u = \mathcal{K}(\mathbf{f})$.

Remark 3.2.6

(1) When $q = 2$, the weak Dirichlet problem is uniquely solvable for any domain Ω, which is easily proved by using the Hilbert space structure of the space $\hat{H}^1_{2,0}(\Omega)$. But, for any $q \neq 2$, speaking generally, we do not know whether the weak Dirichlet problem is uniquely solvable.

(2) Given $\mathbf{f} \in L_q(\Omega)^N$ and $g \in W^{1-1/q}_q(\Gamma)$, we consider the weak Dirichlet problem:

$$(\nabla u, \nabla \varphi)_\Omega = (\mathbf{f}, \nabla \varphi)_\Omega \quad \text{for every } \varphi \in \hat{H}^1_{q',0}(\Omega), \tag{3.35}$$

subject to $u = g$ on Γ. Let G be an extension of g to Ω such that $G = g$ on Γ and $\|G\|_{H^1_q(\Omega)} \leq C\|g\|_{W^{1-1/q}_q(\Gamma)}$ for some constant $C > 0$. Let $v \in \hat{H}^1_{q,0}(\Omega)$ be a solution of the weak Dirichlet problem:

$$(\nabla v, \nabla \varphi)_\Omega = (\mathbf{f} - \nabla G, \nabla \varphi)_\Omega \quad \text{for every } \varphi \in \hat{H}^1_{q',0}(\Omega).$$

Then, $u = G + v \in H^1_q(\Omega) + \hat{H}^1_{q,0}(\Omega)$ is a unique solution of Eq. (3.35) possessing the estimate:

$$\|\nabla u\|_{L_q(\Omega)} \leq C(\|g\|_{W^{1-1/q}_q(\Gamma)} + \|\mathbf{f}\|_{L_q(\Omega)})$$

for some constant $C > 0$.

Example 3.2.7 When Ω is a bounded domain, an exterior domain, half space, a perturbed half space, layer, a perturbed layer, and a tube, then the weak Dirichlet problem is uniquely solvable for $q \in (1, \infty)$.

Theorem 3.2.8 *Let $1 < q < \infty$. Let Ω be a uniform C^2 domain. Given $\mathbf{f} \in L_q(\Omega)^N$, let $u \in \hat{H}^1_{q,0}(\Omega)$ be a unique solution of the weak Dirichlet problem (3.34) possessing the estimate: $\|\nabla u\|_{L_q(\Omega)} \leq C\|\mathbf{f}\|_{L_q(\Omega)}$. If we assume that $\mathrm{div}\,\mathbf{f} \in L_q(\Omega)$ in addition, then $\nabla^2 u \in L_q(\Omega)$ and*

$$\|\nabla^2 u\|_{L_q(\Omega)} \leq C_{M_2,q}(\|\mathrm{div}\,\mathbf{f}\|_{L_q(\Omega)} + \|\mathbf{f}\|_{L_q(\Omega)}).$$

Proof This theorem will be proved in Sect. 3.2.6, after some preparations for weak Laplace problem in $\mathbb{R}^N$ in Sect.3.2.4 and weak Dirichlet problem in the half space in Sect. 3.2.5 below.   □

Remark 3.2.9 From Theorem 3.2.8, we know the existence of solutions of the strong Dirichlet problem:

$$\Delta u = \operatorname{div} \mathbf{f} \quad \text{in } \Omega, \quad u|_\Gamma = 0. \tag{3.36}$$

But, the uniqueness of solutions of Eq. (3.36) does not hold generally. For example, let $B_1 = \{x \in \mathbb{R}^N \mid |x| > 1\}$ for $N \geq 2$, and let $f(x)$ be a function defined by

$$f(x) = \begin{cases} \log |x| & \text{when } N = 2, \\ |x|^{-(N-2)} - 1 & \text{when } N \geq 3. \end{cases}$$

Then, $f(x)$ satisfies the homogeneous Dirichlet problem: $\Delta f = 0$ in B_1 and $f|_{\partial B_1} = 0$, where $\partial B_1 = \{x \in \mathbb{R}^N \mid |x| = 1\}$.

3.2.4 The Weak Laplace Problem in $\mathbb{R}^N$

In this subsection, we consider the following weak Laplace problem in $\mathbb{R}^N$:

$$(\nabla u, \nabla \varphi)_{\mathbb{R}^N} = (\mathbf{f}, \nabla \varphi)_{\mathbb{R}^N} \quad \text{for any } \varphi \in \hat{H}^1_{q'}(\mathbb{R}^N). \tag{3.37}$$

We shall prove the following theorem.

Theorem 3.2.10 *Let* $1 < q < \infty$. *Then, for any* $\mathbf{f} \in L_q(\mathbb{R}^N)^N$, *the weak Laplace problem (3.37) admits a unique solution* $u \in \hat{H}^1_q(\mathbb{R}^N)$ *possessing the estimate:* $\|\nabla u\|_{L_q(\mathbb{R}^N)} \leq C\|\mathbf{f}\|_{L_q(\mathbb{R}^N)}$.

Moreover, if we assume that $\operatorname{div} \mathbf{f} \in L_q(\mathbb{R}^N)$ *in addition, then* $\nabla^2 u \in L_q(\mathbb{R}^N)^{N^2}$ *and*

$$\|\nabla^2 u\|_{L_q(\mathbb{R}^N)} \leq C\|\operatorname{div} \mathbf{f}\|_{L_q(\mathbb{R}^N)}.$$

Proof To prove the theorem, we consider the strong Laplace equation:

$$\Delta u = \operatorname{div} \mathbf{f} \quad \text{in } \mathbb{R}^N. \tag{3.38}$$

Let

$$H^1_{q,\operatorname{div}}(D) = \{\mathbf{f} \in L_q(D)^N \mid \operatorname{div} \mathbf{f} \in L_q(D)\},$$

where D is any domain in $\mathbb{R}^N$. Since $C_0^\infty(\mathbb{R}^N)^N$ is dense both in $L_q(\mathbb{R}^N)^N$ and $H^1_{q,\text{div}}(\mathbb{R}^N)$, we may assume that $\mathbf{f} \in C_0^\infty(\mathbb{R}^N)^N$. Let $\mathcal{F}[f] = \hat{f}$ and $\mathcal{F}^{-1}$ denote respective the Fourier transform of f and the Fourier inverse transform. We then set

$$u = -\mathcal{F}^{-1}\left[\frac{\mathcal{F}[\text{div}\,\mathbf{f}](\xi)}{|\xi^2|}\right] = -\mathcal{F}^{-1}\left[\frac{\sum_{j=1}^N i\xi_j \mathcal{F}[f_j](\xi)}{|\xi|^2}\right]$$

for $\mathbf{f} = {}^\top(f_1, \ldots, f_N)$. By the Fourier multiplier theorem, we have

$$
\begin{aligned}
\|\nabla u\|_{L_q(\mathbb{R}^N)} &\leq C\|\mathbf{f}\|_{L_q(\mathbb{R}^N)}, \\
\|\nabla^2 u\|_{L_q(\mathbb{R}^N)} &\leq C\|\text{div}\,\mathbf{f}\|_{L_q(\mathbb{R}^N)}.
\end{aligned}
\tag{3.39}
$$

Of course, u satisfies Eq. (3.38).

We now prove that u satisfies the weak Laplace equation (3.38). For this purpose, we use the following lemma.

Lemma 3.2.11 *Let $1 < q < \infty$ and let*

$$d_q(x) = \begin{cases} (1 + |x|^2)^{1/2} & \text{for } N \neq q, \\ (1 + |x|^2)^{1/2} \log(2 + |x|^2)^{1/2} & \text{for } N = q. \end{cases}$$

Then, for any $\varphi \in \hat{H}_q^1(\mathbb{R}^N)$, there exists a constant c for which

$$\left\|\frac{\varphi - c}{d_q}\right\|_{L_q(\mathbb{R}^N)} \leq C\|\nabla\varphi\|_{L_q(\mathbb{R}^N)}$$

with some constant independent of φ and c.

Proof For a proof, see Galdi [19, Chapter II]. $\square$

To use Lemma 3.2.11, we use a cut-off function, ψ_R, of Sobolev's type defined as follows: Let ψ be a function in $C^\infty(\mathbb{R})$ such that $\psi(t) = 1$ for $|t| \leq 1/2$ and $\psi(t) = 0$ for $|t| \geq 1$, and set

$$\psi_R(x) = \psi\left(\frac{\ln\ln|x|}{\ln\ln R}\right).$$

Notice that

$$|\nabla\psi_R(x)| \leq \frac{c}{\ln\ln R}\frac{1}{|x|\ln|x|}, \quad \text{supp}\,\nabla\psi_R \subset D_R, \tag{3.40}$$

where we have set $D_R = \{x \in \mathbb{R}^N \mid e^{\sqrt{\ln R}} \le |x| \le R\}$. Noting that $\mathbf{f} \in C_0^\infty(\mathbb{R}^N)^N$, by (3.38) for large $R > 0$ and $\varphi \in \hat{H}_{q'}^1(\mathbb{R}^N)$ we have

$$
\begin{aligned}
(\mathbf{f}, \nabla\varphi)_{\mathbb{R}^N} &= (\mathbf{f}, \nabla(\varphi - c))_{\mathbb{R}^N} = -(\operatorname{div}\mathbf{f}, \varphi - c)_{\mathbb{R}^N} \\
&= -(\psi_R \operatorname{div}\mathbf{f}, \varphi - c)_{\mathbb{R}^N} = -(\psi_R \Delta u, \varphi - c)_{\mathbb{R}^N} \\
&= ((\nabla\psi_R)\cdot(\nabla u), \varphi - c)_{\mathbb{R}^N} + (\psi_R \nabla u, \nabla\varphi)_{\mathbb{R}^N},
\end{aligned}
\tag{3.41}
$$

where c is a constant for which

$$
\left\| \frac{\varphi - c}{d_{q'}} \right\|_{L_{q'}(\mathbb{R}^N)} \le C \|\nabla\varphi\|_{L_{q'}(\mathbb{R}^N)}
\tag{3.42}
$$

with some constant $C > 0$. By (3.40) and (3.42), we have

$$
\begin{aligned}
|((\nabla\psi_R)\cdot(\nabla u), \varphi - c)_{\mathbb{R}^N}| &\le \|d_{q'}(\nabla\psi_R)\cdot(\nabla u)\|_{L_q(\mathbb{R}^N)} \left\| \frac{\varphi - c}{d_{q'}} \right\|_{L_{q'}(\mathbb{R}^N)} \\
&\le \frac{C}{\ln\ln R} \|\nabla u\|_{L_q(D_R)} \|\nabla\varphi\|_{L_{q'}(\mathbb{R}^N)} \to 0
\end{aligned}
\tag{3.43}
$$

as $R \to \infty$. By (3.41) and (3.43) we see that u satisfies the weak Dirichlet problem (3.37). The uniqueness follows from the existence theorem just proved for the dual problem. Moreover, if $\operatorname{div}\mathbf{f} \in L_q(\mathbb{R}^N)$ in addition, then $\nabla^2 u \in L_q(\mathbb{R}^N)^{N^2}$, and so by (3.39) we complete the proof of Theorem 3.2.10. $\qquad\square$

3.2.5 The Weak Dirichlet Problem in the Half Space Case

In this subsection, we consider the following weak Dirichlet problem in $\mathbb{R}_+^N$:

$$
(\nabla u, \nabla\varphi)_{\mathbb{R}_+^N} = (\mathbf{f}, \nabla\varphi)_{\mathbb{R}_+^N} \quad \text{for any } \varphi \in \hat{H}_{q',0}^1(\mathbb{R}_+^N).
\tag{3.44}
$$

We shall prove the following theorem.

Theorem 3.2.12 *Let $1 < q < \infty$. Then, for any $\mathbf{f} \in L_q(\mathbb{R}_+^N)^N$, the weak Dirichlet problem (3.44) admits a unique solution $u \in \hat{H}_{q,0}^1(\mathbb{R}_+^N)$ possessing the estimate:* $\|\nabla u\|_{L_q(\mathbb{R}_+^N)} \le C\|\mathbf{f}\|_{L_q(\mathbb{R}_+^N)}$.

Moreover, if we assume that $\operatorname{div}\mathbf{f} \in L_q(\mathbb{R}_+^N)$ in addition, then $\nabla^2 u \in L_q(\mathbb{R}_+^N)^{N^2}$ and

$$
\|\nabla^2 u\|_{L_q(\mathbb{R}_+^N)} \le C\|\operatorname{div}\mathbf{f}\|_{L_q(\mathbb{R}_+^N)}.
$$

Proof We may assume that $\mathbf{f} = {}^\top(f_1, \ldots, f_N) \in C_0^\infty(\mathbb{R}_+^N)^N$ in the following, because $C_0^\infty(\mathbb{R}_+^N)$ is dense both in $L_q(\mathbb{R}_+^N)^N$ and $H_{q,\mathrm{div}}^1(\mathbb{R}_+^N)$. We first consider the strong Dirichlet problem:

$$\Delta u = \mathrm{div}\,\mathbf{f} \quad \text{in } \mathbb{R}_+^N, \qquad u|_{x_N=0} = 0. \tag{3.45}$$

For any function, $f(x)$, defined in $\mathbb{R}_+^N$, let f^e and f^o be the even extension and the odd extension of f defined by letting

$$f^e(x) = \begin{cases} f(x', x_N) & x_N > 0, \\ f(x', -x_N) & x_N < 0, \end{cases} \qquad f^o(x) = \begin{cases} f(x', x_N) & x_N > 0, \\ -f(x', -x_N) & x_N < 0, \end{cases} \tag{3.46}$$

where $x' = (x_1, \ldots, x_{N-1}) \in \mathbb{R}^{N-1}$ and $x = (x', x_N) \in \mathbb{R}^N$.

Noting that $(\mathrm{div}\,\mathbf{f})^o = \sum_{j=1}^{N-1} \partial_j(f_j)^o + \partial_N(f_N)^e$, we define u by letting

$$u = -\mathcal{F}^{-1}\left[\frac{\mathcal{F}[(\mathrm{div}\,\mathbf{f})^o](\xi)}{|\xi|^2}\right]$$

$$= -\mathcal{F}^{-1}\left[\frac{\sum_{j=1}^{N-1} i\xi_j \mathcal{F}[(f_j)^o](\xi) + i\xi_N \mathcal{F}[(f_N)^e](\xi)}{|\xi|^2}\right].$$

We then have

$$\|\nabla u\|_{L_q(\mathbb{R}^N)} \leq C\|\mathbf{f}\|_{L_q(\mathbb{R}_+^N)}, \quad \|\nabla^2 u\|_{L_q(\mathbb{R}^N)} \leq C\|\mathrm{div}\,\mathbf{f}\|_{L_q(\mathbb{R}_+^N)}, \tag{3.47}$$

and moreover u satisfies Eq. (3.45).

We next prove that u satisfies the weak Dirichlet problem Eq. (3.44). For this purpose, instead of Lemma 3.2.11, we use the Hardy type inequality:

$$\left(\int_0^\infty \left(\int_0^x f(y)\,dy\right)^p x^{-r-1}\,dy\right)^{1/p} \leq (p/r)\left(\int_0^\infty (yf(y))^p y^{-r-1}\,dy\right)^{1/p}, \tag{3.48}$$

where $f \geq 0$, $p \geq 1$ and $r > 0$ (cf. Stein [68, A.4 p.272]). Of course, using zero extension of f suitably, we can replace the interval $(0, \infty)$ by (a, b) for any $0 \leq a < b < \infty$ in (3.48). Let $D_{R,2R} = \{x \in \mathbb{R}_+^N \mid R \leq |x| \leq 2R\}$. Using (3.48), we see that for any $\varphi \in \hat{H}_{q',0}^1(\mathbb{R}_+^N)$

$$\lim_{R\to\infty} R^{-1}\|\varphi\|_{L_{q'}(D_{R,2R})} = 0. \tag{3.49}$$

In fact, using $\varphi|_{x_N=0} = 0$, we write $\varphi(x', x_N) = \int_0^{x_N} (\partial_s \varphi)(x', s)\, ds$. Thus, by (3.48) we have

$$\int_a^b |\varphi(x', x_N)|^{q'}\, dx_N \le \left(\frac{bq'}{q'-1}\right)^{q'} \int_a^b |(\partial_N \varphi)(x', x_N)|^{q'}\, dx_N$$

for any $0 < a < b$. Let

$$E_R^1 = \{x \in \mathbb{R}^N \mid |x'| \le 2R, \ R/2 \le x_N < 2R\},$$

$$E_R^2 = \{x \in \mathbb{R}^N \mid 0 \le x_N \le 2R, \ R/2 \le |x'| \le 2R\},$$

and then $D_{R,2R} \subset E_R^1 \cup E_R^2$. Thus, by (3.48),

$$\left(\int_{D_{R,2R}} |\varphi(x)|^{q'}\, dx\right)^{1/q'} \le \left(\frac{Rq'}{q'-1}\right)\left\{ \int_{|x'|\le R} \int_{R/2}^{2R} |\partial_N \varphi(x)|^{q'}\, dx_N dx' \right.$$

$$\left. + \int_{R/2 \le |x'| \le 2R} \int_0^{2R} |\partial_N \varphi(x)|^{q'}\, dx_N dx' \right\}^{1/q'},$$

which leads to (3.49).

Let ω be a function in $C_0^\infty(\mathbb{R}^N)$ such that $\omega(x) = 1$ for $|x| \le 1$ and $\varphi(x) = 0$ for $|x| \ge 2$, and we set $\omega_R(x) = \omega(x/R)$. For any $\varphi \in \hat{H}_{q',0}^1(\mathbb{R}_+^N)$ and for large $R > 0$, we have

$$(\mathrm{div}\,\mathbf{f}, \varphi)_{\mathbb{R}_+^N} = (\omega_R \mathrm{div}\,\mathbf{f}, \varphi)_{\mathbb{R}_+^N} = (\omega_R \Delta u, \varphi)_{\mathbb{R}_+^N}$$

$$= -((\nabla \omega_R) \cdot (\nabla u), \varphi)_{\mathbb{R}_+^N} - (\omega_R \nabla u, \nabla \varphi)_{\mathbb{R}_+^N}. \tag{3.50}$$

By (3.49)

$$|((\nabla \omega_R) \cdot (\nabla u), \varphi)_{\mathbb{R}_+^N}| \le R^{-1} \|\nabla u\|_{L_q(D_{R,2R})} \|\varphi\|_{L_{q'}(D_{R,2R})} \to 0$$

as $R \to \infty$. On the other hand, $(\mathrm{div}\,\mathbf{f}, \varphi)_{\mathbb{R}_+^N} = -(\mathbf{f}, \nabla \varphi)_{\mathbb{R}_+^N}$, where we have used $\mathbf{f} \in C_0^\infty(\mathbb{R}_+^N)^N$. Thus, by (3.50) we have

$$(\nabla u, \nabla \varphi)_{\mathbb{R}_+^N} = (\mathbf{f}, \nabla \varphi)_{\mathbb{R}_+^N}$$

for any $\varphi \in \hat{H}_{q',0}^1(\mathbb{R}_+^N)$. This shows that u is a solution of the weak Dirichlet problem. The uniqueness follows from the existence of solutions for the dual problem, which completes the proof of Theorem 3.2.12. □

3.2.6 Regularity of the Weak Dirichlet Problem

In this subsection, we shall prove Theorem 3.2.8 in Sect. 3.2.3. Let ζ_j^i ($i = 0, 1$, $j \in \mathbb{N}$) be cut-off functions given in Proposition 3.2.2. We first consider the regularity of $\zeta_j^0 u$. For this purpose, we use the following lemma.

Lemma 3.2.13 Let Ω be a uniformly C^2 domain in $\mathbb{R}^N$. Then, there exists a constant $c_1 > 0$ independent of $j \in \mathbb{N}$ such that

$$\|\varphi\|_{H_q^1(\Omega_j \cap B_j^1)} \leq c_1 \|\nabla\varphi\|_{L_q(\Omega_j \cap B_j^1)} \quad \text{for any } \varphi \in \hat{H}_{q,0}^1(\Omega_j),$$

$$\|\psi\|_{H_q^1(\Omega \cap B_j^1)} \leq c_1 \|\nabla\psi\|_{L_q(\Omega \cap B_j^1)} \quad \text{for any } \psi \in \hat{H}_{q,0}^1(\Omega),$$

$$\|\varphi - c_j^0(\varphi)\|_{H_q^1(B_j^0)} \leq c_1 \|\nabla\varphi\|_{L_q(B_j^0)} \quad \text{for any } \varphi \in \hat{H}_q^1(\mathbb{R}^N),$$

$$\|\psi - c_j^0(\psi)\|_{H_q^1(B_j^0)} \leq c_1 \|\nabla\psi\|_{L_q(B_j^0)} \quad \text{for any } \psi \in \hat{H}_q^1(\Omega).$$

Here, $c_j^0(\varphi)$ and $c_j^0(\psi)$ are suitable constants depending on φ and ψ, respectively.

Proof For a proof, see Shibata [42, in the proofs of Lemma 3.4 and Lemma 3.5].
□

Continuation of Proof of Theorem 3.2.8 Let $c_j^0 = c_j^0(\varphi)$ be a constant in Lemma 3.2.13 such that

$$\|u - c_j^0\|_{L_q(B_j^0)} \leq c_1 \|\nabla u\|_{L_q(B_j^0)}. \tag{3.51}$$

For any $\varphi \in \hat{H}_{q'}^1(\mathbb{R}^N)$, we have

$$(\nabla(\zeta_j^0(u - c_j^0)), \nabla\varphi)_{\mathbb{R}^N}$$

$$= ((\nabla\zeta_j^0)(u - c_j^0), \nabla\varphi)_{\mathbb{R}^N} + (\nabla u, \nabla(\zeta_j^0\varphi))_{\mathbb{R}^N} - ((\nabla u) \cdot (\nabla\zeta_j^0), \varphi)_{\mathbb{R}^N}$$

$$= -((\Delta\zeta_j^0)(u - c_j^0) + 2(\nabla\zeta_j^0) \cdot (\nabla u) + \zeta_j^0\text{div}\,\mathbf{f}, \varphi)_{\mathbb{R}^N},$$

where we have used $(\nabla u, \nabla(\zeta_j^0\varphi))_{\mathbb{R}^N} = (\mathbf{f}, \nabla(\zeta_j^0\varphi))_{\mathbb{R}^N} = -(\zeta_j^0\text{div}\,\mathbf{f}, \varphi)_{\mathbb{R}^N}$. Let $f = (\Delta\zeta_j^0)(u - c_j^0) + 2(\nabla\zeta_j^0) \cdot (\nabla u) + \zeta_j^0\text{div}\,\mathbf{f}$. Since $C_0^\infty(\mathbb{R}^N) \subset \hat{H}_q^1(\mathbb{R}^N)$, for any $\varphi \in C_0^\infty(\mathbb{R}^N)$ we have

$$(\Delta(\zeta_j^0(u - c_j^0)), \varphi)_{\mathbb{R}^N} = (f, \varphi)_{\mathbb{R}^N},$$

which yields that

$$\Delta(\zeta_j^0(u - c_j^0)) = f \quad \text{in } \mathbb{R}^N \tag{3.52}$$

in the sense of distribution. By Lemma 3.2.13, $f \in L_q(\mathbb{R}^N)$ and

$$\|f\|_{L_q(\mathbb{R}^N)} \le C(\|\operatorname{div} \mathbf{f}\|_{L_q(B_j^0)} + \|\nabla u\|_{L_q(B_j^0)}). \tag{3.53}$$

From (3.52) it follows that

$$\partial_k \partial_\ell \Delta(\zeta_j^0(u - c_j^0)) = \partial_k \partial_\ell f$$

for any $k, \ell \in \mathbb{N}$. Since both sides are compactly supported distributions, we can apply the Fourier transform and the inverse Fourier transform. We then have

$$\partial_k \partial_\ell (\zeta_j^0(u - c_j^0)) = \mathcal{F}^{-1}\left[\frac{\xi_k \xi_\ell}{|\xi|^2} \mathcal{F}[f](\xi)\right].$$

By the Fourier multiplier theorem, we have

$$\|\partial_k \partial_\ell (\zeta_j^0(u - c_j^0))\|_{L_q(\mathbb{R}^N)} \le C \|f\|_{L_q(\mathbb{R}^N)}.$$

Since $\partial_k \partial_\ell (\zeta_j^0(u - c_j^0)) = \zeta_j^0 \partial_k \partial_\ell u + (\partial_k \zeta_j^0)\partial_\ell u + (\partial_\ell \zeta_j^0)\partial_k u + (\partial_k \partial_\ell \zeta_j^0)(u - c_j^0)$, by (3.51) and (3.53) we have $\zeta_j^0 \nabla^2 u \in L_q(\mathbb{R}^N)^N$ and

$$\|\zeta_j^0 \nabla^2 u\|_{L_q(\Omega)} \le C_{M2}(\|\operatorname{div} \mathbf{f}\|_{L_q(B_j^0)} + \|\nabla u\|_{L_q(B_j^0)}). \tag{3.54}$$

We next consider $\zeta_j^1 u$. For any $\varphi \in \hat{H}_{q',0}^1(\Omega_j)$, we have

$$(\nabla(\zeta_j^1 u), \nabla \varphi)_{\Omega_J} = (g, \varphi)_{\Omega_j}, \tag{3.55}$$

where we have set $g = -(\zeta_j^1 \operatorname{div} \mathbf{f} + 2(\nabla u) \cdot (\nabla \zeta_j^1) + (\Delta \zeta_j^1)u)$. By Lemma 3.2.13, $g \in L_q(\Omega_j)$ and

$$\|g\|_{L_q(\Omega_j)} \le C(\|\operatorname{div} \mathbf{f}\|_{L_q(\Omega \cap B_j^1)} + \|\nabla u\|_{L_q(\Omega \cap B_j^1)}). \tag{3.56}$$

We use the symbols given in Proposition 3.2.2. Let $a_{k\ell}$ and $b_{k\ell}$ be the (k, ℓ)th component of $N \times N$ matrices $\mathcal{A}_j$ and B_j given in Proposition 3.2.2. By the change of variables: $y = \Phi_j(x)$, the variational Eq. (3.55) is transformed to

$$\sum_{k,\ell=1}^N ((\delta_{k\ell} + A_{k\ell})\partial_k v, \partial_\ell \varphi)_{\mathbb{R}_+^N} = (h, \varphi)_{\mathbb{R}_+^N}. \tag{3.57}$$

Here, we have set

$$v = \zeta_j^1 u \circ \Phi_j, \quad h = g \circ \Phi_j, \quad J = \det(\mathcal{A}_j + B_j) = 1 + J^0,$$

$$A_{k\ell} = \sum_{m=1}^{N} \{a_{\ell m} b_{km} + a_{\ell m} J_0 (a_{k\ell} + b_{k\ell}) + b_{\ell m} J (a_{km} + b_{km})\}.$$

By Proposition 3.2.2 and (3.56), we have

$$\|A_{k\ell}\|_{L_\infty(\mathbb{R}^N)} \leq CM_1, \quad \|\nabla A_{k\ell}\|_{L_\infty(\mathbb{R}^N)} \leq C_A,$$

$$\|h\|_{L_q(\mathbb{R}_+^N)} \leq C(\|\operatorname{div} \mathbf{f}\|_{L_q(B_j^1 \cap \Omega)} + \|\nabla u\|_{L_q(B_j^1 \cap \Omega)}), \tag{3.58}$$

where C_A is a constant depending a_1, a_2 and A appearing in Definition 3.2.1. Since $\operatorname{supp} g \subset \Phi^{-1}(B_j^1) \cap \Omega$, by Lemma 3.2.13

$$|(h, \varphi)_{\mathbb{R}_+^N}| \leq \|h\|_{L_q(B_j^1 \cap \mathbb{R}_+^N)} \|\varphi\|_{L_{q'}(B_j^1 \cap \mathbb{R}_+^N)} \leq C\|g\|_{L_q(B_j^1 \cap \Omega)} \|\nabla \varphi\|_{L_{q'}(\mathbb{R}_+^N)}$$

for any $\varphi \in \hat{H}_{q',0}^1(\mathbb{R}_+^N)$, where C is a constant independent of $j \in \mathbb{N}$. Thus, by the Hahn-Banach theorem, there exists an $\mathbf{h} \in L_q(\mathbb{R}_+^N)^N$ such that $\|\mathbf{h}\|_{L_q(\mathbb{R}_+^N)} \leq C\|h\|_{L_q(B_j^1 \cap \mathbb{R}_+^N)}$ and $(\mathbf{h}, \nabla\varphi)_{\mathbb{R}_+^N} = (h, \varphi)_{\mathbb{R}_+^N}$ for any $\varphi \in \hat{H}_{q',0}^1(\mathbb{R}_+^N)$. In particular, $\operatorname{div} \mathbf{h} = -h \in L_q(\mathbb{R}_+^N)$. Thus, the variational problem (3.57) reads

$$\sum_{k,\ell=1}^{N} ((\delta_{k\ell} + A_{k\ell})\partial_k v, \partial_\ell \varphi)_{\mathbb{R}_+^N} = (\mathbf{h}, \nabla\varphi)_{\mathbb{R}_+^N} \quad \text{for any } \varphi \in \hat{H}_{q',0}^1(\mathbb{R}_+^N).$$

We now prove that if $M_1 \in (0, 1)$ is small enough, then for any $\mathbf{g} \in L_q(\mathbb{R}_+^N)^N$, there exists a unique solution $w \in \hat{H}_{q,0}^1(\mathbb{R}_+^N)$ of the variational problem:

$$\sum_{k,\ell=1}^{N} ((\delta_{k\ell} + A_{k\ell})\partial_k w, \partial_\ell \varphi)_{\mathbb{R}_+^N} = (\mathbf{g}, \nabla\varphi)_{\mathbb{R}_+^N} \quad \text{for any } \varphi \in \hat{H}_{q',0}^1(\mathbb{R}_+^N), \tag{3.59}$$

possessing the estimate:

$$\|\nabla w\|_{L_q(\mathbb{R}_+^N)} \leq C\|\mathbf{g}\|_{L_q(\mathbb{R}_+^N)}. \tag{3.60}$$

Moreover, if $\operatorname{div} \mathbf{g} \in L_q(\mathbb{R}_+^N)$, then $\nabla w \in H_q^1(\mathbb{R}_+^N)^N$ and

$$\|\nabla^2 w\|_{L_q(\mathbb{R}_+^N)} \leq C\|\operatorname{div} \mathbf{g}\|_{L_q(\mathbb{R}_+^N)} + C_A\|\mathbf{g}\|_{L_q(\mathbb{R}_+^N)}. \tag{3.61}$$

In fact, we prove the existence of w by the successive approximation. Let $w_1 \in \hat{H}^1_{q,0}(\mathbb{R}^N_+)$ be a solution of the weak Dirichlet problem:

$$(\nabla w_1, \nabla \varphi)_{\mathbb{R}^N_+} = (\mathbf{g}, \nabla \varphi)_{\mathbb{R}^N_+} \quad \text{for any } \varphi \in \hat{H}^1_{q',0}(\mathbb{R}^N_+). \tag{3.62}$$

By Theorem 3.2.12, w_1 uniquely exists and satisfies the estimate:

$$\|\nabla w_1\|_{L_q(\mathbb{R}^N_+)} \leq C \|\mathbf{g}\|_{L_q(\mathbb{R}^N_+)}. \tag{3.63}$$

Additionally, we assume that div $\mathbf{g} \in L_q(\mathbb{R}^N_+)$, and then $\nabla^2 w_1 \in H^1_q(\mathbb{R}^N_+)^N$ and

$$\|\nabla^2 w_1\|_{L_q(\mathbb{R}^N_+)} \leq C \|\text{div } \mathbf{g}\|_{L_q(\mathbb{R}^N_+)}. \tag{3.64}$$

Given $w_j \in \hat{H}^1_{q,0}(\mathbb{R}^N_+)$, let $w_{j+1} \in \hat{H}^1_{q,0}(\mathbb{R}^N_+)$ be a solution of the weak Dirichlet problem:

$$(\nabla w_{j+1}, \nabla \varphi)_{\mathbb{R}^N_+} = (\mathbf{g}, \nabla \varphi)_{\mathbb{R}^N_+} - \sum_{k,\ell=1}^{N} (A_{k\ell} \partial_k w_j, \partial_\ell \varphi)_{\mathbb{R}^N_+} \tag{3.65}$$

for any $\varphi \in \hat{H}^1_{q',0}(\mathbb{R}^N_+)$. By Theorem 3.2.12 and (3.58), w_{j+1} exists and satisfies the estimate:

$$\|\nabla w_{j+1}\|_{L_q(\mathbb{R}^N_+)} \leq C(\|\mathbf{g}\|_{L_q(\Omega_+)} + M_1 \|\nabla w_j\|_{L_q(\mathbb{R}^N_+)}). \tag{3.66}$$

Applying Theorem 3.2.12 and (3.58) to the difference $w_{j+1} - w_j$, we have

$$\|\nabla(w_{j+1} - w_j)\|_{L_q(\mathbb{R}^N_+)} \leq CM_1 \|\nabla(w_j - w_{j-1})\|_{L_q(\mathbb{R}^N_+)}. \tag{3.67}$$

Choosing $CM_1 \leq 1/2$ in (3.67), we see that $\{w_j\}_{j=1}^{\infty}$ is a Cauchy sequence in $\hat{H}^1_{q,0}(\mathbb{R}^N_+)$, and so the limit $w \in H^1_{q,0}(\mathbb{R}^N_+)$ exists and satisfies the weak Dirichlet problem (3.59). Moreover, taking the limit in (3.66), we have

$$\|\nabla w\|_{L_q(\mathbb{R}^N_+)} \leq C \|\mathbf{g}\|_{L_q(\Omega_+)} + CM_1 \|\nabla w\|_{L_q(\mathbb{R}^N_+)}.$$

Since $CM_1 \leq 1/2$, we have $\|\nabla w\|_{L_q(\mathbb{R}^N_+)} \leq 2C \|\mathbf{g}\|_{L_q(\Omega_+)}$. Thus, we have proved that the weak Dirichlet problem (3.59) admits at least one solution $w \in \hat{H}^1_{q,0}(\Omega_+)$ possessing the estimate (3.60). The uniqueness follows from the existence of solutions to the dual problem. Thus, we have proved the unique existence of solutions of Eq. (3.59).

We now prove that $\nabla w \in H_q^1(\mathbb{R}_+^N)^N$ provided that $\operatorname{div} \mathbf{g} \in L_q(\mathbb{R}_+^N)$. By Theorem 3.2.12, $\nabla w_1 \in H_q^1(\mathbb{R}_+^N)^N$ and w_1 satisfies the estimate:

$$\|\nabla^2 w_1\|_{L_q(\mathbb{R}_+^N)} \leq C\|\operatorname{div} \mathbf{g}\|_{H_q^1(\Omega_+)}.$$

Additionally, we assume that $\nabla^2 w_j \in L_q(\mathbb{R}_+^N)^N$, and then applying Theorem 3.2.12 to (3.65) and using (3.58) give that $\nabla^2 w_{j+1} \in L_q(\mathbb{R}_+^N)^{N^2}$ and

$$\|\nabla^2 w_{j+1}\|_{L_q(\mathbb{R}_+^N)} \leq C\|\operatorname{div} \mathbf{g}\|_{L_q(\mathbb{R}_+^N)} + CM_1\|\nabla^2 w_j\|_{L_q(\mathbb{R}_+^N)} + C_A\|\nabla w_j\|_{L_q(\mathbb{R}_+^N)}.$$

$$(3.68)$$

And also, applying Theorem 3.2.12 to the difference $w_{j+1} - w_j$ and using (3.58), we have

$$\|\nabla^2(w_{j+1} - w_j)\|_{L_q(\mathbb{R}_+^N)}$$
$$\leq CM_1\|\nabla^2(w_j - w_{j-1})\|_{L_q(\mathbb{R}_+^N)} + C_A\|\nabla(w_j - w_{j-1})\|_{L_q(\mathbb{R}_+^N)},$$

which, combined with (3.67), leads to

$$\|\nabla^2(w_{j+1} - w_j)\|_{L_q(\mathbb{R}_+^N)} + \|\nabla(w_j - w_{j-1})\|_{L_q(\mathbb{R}_+^N)}$$
$$\leq CM_1\|\nabla^2(w_j - w_{j-1})\|_{L_q(\mathbb{R}_+^N)} + C(C_A + 1)M_1\|\nabla(w_{j-1} - w_{j-2})\|_{L_q(\mathbb{R}_+^N)}.$$

Choosing $M_1 > 0$ so small that $CM_1 \leq 1/2$ and $(C_A + 1)M_1 \leq 1/2$, then we have

$$\|\nabla^2(w_{j+1} - w_j)\|_{L_q(\mathbb{R}_+^N)} + \|\nabla(w_j - w_{j-1})\|_{L_q(\mathbb{R}_+^N)} \leq (1/2)^{j-1} L$$

with $L = \|\nabla^2(w_3 - w_2)\|_{L_q(\mathbb{R}_+^N)} + \|\nabla(w_2 - w_1)\|_{L_q(\mathbb{R}_+^N)}$. From this it follows that $\{\nabla^2 w_j\}_{j=1}^\infty$ is a Cauchy sequence in $L_q(\Omega)$, which yields that $\nabla^2 w \in L_q(\mathbb{R}_+^N)^{N^2}$. Moreover, taking the limit in (3.68) and using (3.60) gives that

$$\|\nabla^2 w\|_{L_q(\mathbb{R}_+^N)} \leq C\|\operatorname{div} \mathbf{g}\|_{L_q(\mathbb{R}_+^N)} + (1/2)\|\nabla^2 w\|_{L_q(\mathbb{R}_+^N)} + C_A\|\mathbf{g}\|_{L_q(\mathbb{R}_+^N)}.$$

which leads to (3.61).

Applying what we have proved and using the estimate:

$$\|\operatorname{div} \mathbf{h}\|_{L_q(\mathbb{R}_+^N)} + \|\mathbf{h}\|_{L_q(\mathbb{R}_+^N)} \leq C\|h\|_{L_q(\mathbb{R}_+^N)} \leq C(\|\operatorname{div} \mathbf{f}\|_{L_q(B_j^1)} + \|\nabla u\|_{L_q(B_j^1)}),$$

which follows from (3.58), we have $\nabla v = \nabla(\zeta_j^1 u \circ \Phi_j) \in H_q^1(\mathbb{R}_+^N)^N$ and

$$\|\nabla(\zeta_j^1 u \circ \Phi_j)\|_{H_q^1(\mathbb{R}_+^N)} \leq C(\|\operatorname{div} \mathbf{f}\|_{L_q(B_j^1 \cap \Omega)} + \|\nabla u\|_{L_q(B_j^1 \cap \Omega)}).$$

Since $\|u\|_{L_q(B_j^1 \cap \Omega)} \leq c_1 \|\nabla u\|_{L_q(B_j^1 \cap \Omega)}$ as follows from Lemma 3.2.13, we have

$$\|\zeta_j^1 \nabla^2 u\|_{L_q(\Omega)} \leq C(\|\operatorname{div} \mathbf{f}\|_{L_q(B_j^1 \cap \Omega)} + \|\nabla u\|_{L_q(B_j^1 \cap \Omega)}). \tag{3.69}$$

Combining (3.54), (3.69) and (3.28) gives

$$\|\nabla^2 u\|_{L_q(\Omega)} \leq C(\|\operatorname{div} \mathbf{f}\|_{L_q(\Omega)} + \|\nabla u\|_{L_q(\Omega)}) \leq C(\|\operatorname{div} \mathbf{f}\|_{L_q(\Omega)} + \|\mathbf{f}\|_{L_q(\Omega)}),$$

which completes the proof of Theorem 3.2.8.

3.2.7 Laplace–Beltrami Operator

In this subsection, we introduce the Laplace–Beltrami operators and some important formulas from differential geometry.

Let Γ be a hypersurface of class C^3 in $\mathbb{R}^N$. Let Γ be parametrized as $p = \phi(\theta) = {}^\top(\phi_1(\theta), \ldots, \phi_N(\theta))$ locally at $p \in \Gamma$, where $\theta = (\theta_1, \ldots, \theta_{N-1})$ runs through a domain $\Theta \subset \mathbb{R}^{N-1}$. Let

$$\tau_i = \tau_i(p) = \frac{\partial}{\partial \theta_i} \phi(\theta) = \partial_i \phi \quad (i = 1, \ldots, N-1), \tag{3.70}$$

which forms a basis of the tangent space $T_p \Gamma$ of Γ at p. Let $\mathbf{n} = \mathbf{n}(p)$ denote the outer unit normal of Γ at p. Notice that

$$< \tau_i, \mathbf{n} > = 0. \tag{3.71}$$

Here and in the following, $< \cdot, \cdot >$ denotes a standard inner product in $\mathbb{R}^N$. To introduce the formula of $\mathbf{n}$, we notice that

$$\det \begin{pmatrix} \frac{\partial \phi_1}{\partial \theta_1} & \cdots & \frac{\partial \phi_1}{\partial \theta_{N-1}} & \frac{\partial \phi_1}{\partial \theta_k} \\ \vdots & \ddots & \vdots & \vdots \\ \frac{\partial \phi_N}{\partial \theta_1} & \cdots & \frac{\partial \phi_N}{\partial \theta_{N-1}} & \frac{\partial \phi_N}{\partial \theta_k} \end{pmatrix} = 0$$

for any $k = 1, \ldots, N-1$. Thus, to satisfy (3.71), $\mathbf{n}$ is defined by

$$\mathbf{n} = \frac{{}^\top(h_1, \ldots, h_N)}{H} \tag{3.72}$$

with $H = \sqrt{\sum_{j=1}^{N} h_i^2}$ and

$$h_i = \frac{\partial(\phi_1, \ldots, \hat{\phi}_i, \ldots, \phi_N)}{\partial(\theta_1, \ldots, \theta_{N-1})} = (-1)^{N+i} \det \begin{pmatrix} \frac{\partial \phi_1}{\partial \theta_1} & \cdots & \frac{\partial \phi_1}{\partial \theta_{N-1}} \\ \vdots & \ddots & \vdots \\ \frac{\partial \phi_{i-1}}{\partial \theta_1} & \cdots & \frac{\partial \phi_{i-1}}{\partial \theta_{N-1}} \\ \frac{\partial \phi_{i+1}}{\partial \theta_1} & \cdots & \frac{\partial \phi_{i+1}}{\partial \theta_{N-1}} \\ \vdots & \ddots & \vdots \\ \frac{\partial \phi_N}{\partial \theta_1} & \cdots & \frac{\partial \phi_N}{\partial \theta_{N-1}} \end{pmatrix}.$$

For example, when $N = 3$,

$$\mathbf{n} = \frac{\frac{\partial \phi}{\partial \theta_1} \times \frac{\partial \phi}{\partial \theta_2}}{\left| \frac{\partial \phi}{\partial \theta_1} \times \frac{\partial \phi}{\partial \theta_2} \right|} = H^{-1\top}(h_1, h_2, h_3) \quad h_1 = \frac{\partial \phi_2}{\partial \theta_1} \frac{\partial \phi_3}{\partial \theta_2} - \frac{\partial \phi_3}{\partial \theta_1} \frac{\partial \phi_2}{\partial \theta_2},$$

$$h_2 = \frac{\partial \phi_3}{\partial \theta_1} \frac{\partial \phi_1}{\partial \theta_2} - \frac{\partial \phi_1}{\partial \theta_1} \frac{\partial \phi_3}{\partial \theta_2}, \quad h_3 = \frac{\partial \phi_1}{\partial \theta_1} \frac{\partial \phi_2}{\partial \theta_2} - \frac{\partial \phi_2}{\partial \theta_1} \frac{\partial \phi_1}{\partial \theta_2}, \quad H = \sqrt{h_1^2 + h_2^2 + h_3^2}.$$

Let

$$g_{ij} = g_{ij}(p) = <\tau_i, \tau_j> \quad (i, j = 1, \ldots, N-1),$$

and let G be an $(N-1) \times (N-1)$ matrix whose (i, j)th components are g_{ij}. The matrix G is called the **first fundamental form** of Γ. In the following, we employ Einstein's summation convention, which means that equal lower and upper indices are to be summed. Since for any $\xi \in \mathbb{R}^{N-1}$ with $\xi \neq 0$, $< G\xi, \xi > = g_{ij}\xi^i \xi^j = < \xi^i \tau_i, \xi^j \tau_j > = |\xi^i \tau_i|^2 > 0$, G is a positive symmetric matrix, and therefore G^{-1} exists. Let g^{ij} be the (i, j)th component of G^{-1} and let

$$\tau^i = g^{ij} \tau_j.$$

Using $g^{ik} g_{kj} = g_{ik} g^{kj} = \delta_j^i$, where δ_j^i are the Kronecker delta symbols defined by $\delta_i^i = 1$ and $\delta_j^i = 0$ for $i \neq j$, we have

$$<\tau_i, \tau^j> = <\tau^i, \tau_j> = \delta_j^i. \tag{3.73}$$

In fact, $<\tau_i, \tau^j> = <\tau_i, g^{jk} \tau_k> = g^{jk} <\tau_i, \tau_k> = g^{jk} g_{ki} = \delta_i^j$. Thus, $\{\tau^j\}_{j=1}^{N-1}$ is a dual basis of $\{\tau_i\}_{i=1}^{N-1}$. In particular, we have

$$\tau^i = g^{ij} \tau_j, \quad \tau_i = g_{ij} \tau^j. \tag{3.74}$$

For any $\mathbf{a} \in T_p \Gamma$, we write

$$\mathbf{a} = a^i \tau_i = a_i \tau^i.$$

By (3.73), we have $< \mathbf{a}, \tau^i > = < a^j \tau_j, \tau^i > = a^j < \tau_j, \tau^i > = a^i$ and $< \mathbf{a}, \tau_i > = a_j < \tau^j, \tau_i > = a_i$, and so

$$\mathbf{a} = < \mathbf{a}, \tau^i > \tau_i = < \mathbf{a}, \tau_i > \tau^i. \tag{3.75}$$

In particular, by (3.73) and (3.74), we have

$$a_i = g_{ij} a^j, \quad a^i = g^{ij} a_j$$

with $a_i = < \mathbf{a}, \tau_i >$ and $a^i = < \mathbf{a}, \tau^i >$. Notice that $\{\tau_1, \ldots, \tau_{N-1}, \mathbf{n}\}$ forms a basis of $\mathbb{R}^N$. Namely, for any N-vector $\mathbf{b} \in \mathbb{R}^N$ we have

$$\mathbf{b} = b^i \tau_i + < \mathbf{b}, \mathbf{n} > \mathbf{n} = b_i \tau^i + < \mathbf{b}, \mathbf{n} > \mathbf{n}$$

with $b^i = < \mathbf{b}, \tau^i >$ and $b_i = < \mathbf{b}, \tau_i >$.

We next consider $\tau_{ij} = \partial_i \partial_j \phi = \partial_j \tau_i$. Notice that $\tau_{ij} = \tau_{ji}$. Let

$$\Lambda_{ij}^k = < \tau_{ij}, \tau^k >, \quad \ell_{ij} = < \tau_{ij}, \mathbf{n} >, \tag{3.76}$$

and then, we have

$$\tau_{ij} = \Lambda_{ij}^k \tau_k + \ell_{ij} \mathbf{n}. \tag{3.77}$$

Let L be an $N - 1 \times N - 1$ matrix whose (i, j)th component is ℓ_{ij}, which is called the **second fundamental form** of Γ. Let

$$\mathcal{H}(\Gamma) = \frac{1}{N-1} \text{tr} \, (G^{-1} L) = \frac{1}{N-1} g^{ij} \ell_{ij}. \tag{3.78}$$

The $\mathcal{H}(\Gamma)$ is called the **mean curvature** of $\mathcal{H}(\Gamma)$.

Let $g = \det G$. One of the most important formulas in this section is

$$\partial_i (\sqrt{g} g^{ij} \tau_j) = \sqrt{g} g^{ij} \ell_{ij} \mathbf{n}. \tag{3.79}$$

This formula will be proved below.

The Λ_{ij}^k is called **Christoffel symbols**. We know the formula:

$$\Lambda_{ij}^r = \frac{1}{2} g^{rk} (\partial_i g_{jk} + \partial_j g_{ki} - \partial_k g_{ij}). \tag{3.80}$$

In fact,

$$\partial_i g_{jk} = \partial_i < \tau_j, \tau_k > = < \tau_{ij}, \tau_k > + < \tau_j, \tau_{ki} >,$$

$$\partial_j g_{ki} = \partial_j < \tau_k, \tau_i > = < \tau_{jk}, \tau_i > + < \tau_k, \tau_{ij} >,$$

$$\partial_k g_{ij} = \partial_k < \tau_i, \tau_j > = < \tau_{ki}, \tau_j > + < \tau_i, \tau_{jk} >,$$

and so we have

$$\partial_i g_{jk} + \partial_j g_{ki} - \partial_k g_{ij} = 2 < \tau_{ij}, \tau_k >,$$

which implies (3.80).

We now prove (3.79) by studying several steps. We first prove

$$\partial_k g_{ij} = g_{jr} \Lambda_{ki}^r + g_{ir} \Lambda_{kj}^r. \tag{3.81}$$

In fact, by (3.77) and $< \tau_i, \mathbf{n} > = 0$,

$$\partial_k g_{ij} = < \tau_{ki}, \tau_j > + < \tau_i, \tau_{kj} > = < \Lambda_{ki}^r \tau_r, \tau_j > + < \tau_i, \Lambda_{kj}^r \tau_r >$$

$$= g_{jr} \Lambda_{ki}^r + g_{ir} \Lambda_{kj}^r.$$

We next prove

$$\partial_k g^{ij} = -g^{ir} \Lambda_{rk}^j - g^{jr} \Lambda_{rk}^i. \tag{3.82}$$

In fact, by $g^{ij} g_{j\ell} = \delta_\ell^i$, we have

$$0 = \partial_k(g^{ij} g_{jk}) = (\partial_k g^{ij}) g_{j\ell} + g^{ij} \partial_k g_{j\ell}.$$

Using (3.81), we have

$$(\partial_k g^{ij}) g_{j\ell} = -g^{ij} \partial_k g_{j\ell} = -g^{ij}(g_{\ell r} \Lambda_{kj}^r + g_{jr} \Lambda_{k\ell}^r)$$

$$= -g^{ij} g_{\ell r} \Lambda_{kj}^r - \delta_r^i \Lambda_{k\ell}^r = -g^{ij} g_{\ell r} \Lambda_{kj}^r - \Lambda_{k\ell}^i,$$

and therefore

$$\partial_k g^{im} = (\partial_k g^{ij}) \delta_j^m = (\partial_k g^{ij}) g_{j\ell} g^{\ell m} = -g^{\ell m}(g^{ij} g_{\ell r} \Lambda_{kj}^r + \Lambda_{k\ell}^i)$$

$$= -\delta_r^m g^{ij} \Lambda_{kj}^r - g^{\ell m} \Lambda_{k\ell}^i = -g^{ij} \Lambda_{kj}^m - g^{m\ell} \Lambda_{k\ell}^i = -g^{i\ell} \Lambda_{k\ell}^m - g^{m\ell} \Lambda_{k\ell}^i.$$

Setting $m = j$ and $\ell = r$ in the above formula, we have (3.82), because $\Lambda_{k\ell}^m = \Lambda_{\ell k}^m$ and $\Lambda_{k\ell}^i = \Lambda_{\ell k}^i$.

We next prove

$$\partial_i g = 2\Lambda_{ij}^j g. \tag{3.83}$$

Recall that $g = \det G$ and the (i, j)th component of G is $< \tau_i, \tau_j >$. From the definition of differentiation, we have

$$\frac{\det G(x + \mathbf{e}_i \Delta x_i) - \det G(x)}{\Delta x_i}$$

$$= \frac{\det(G(x) + \partial_i G(x)\Delta x_i) - \det G(x)}{\Delta x_i} + O(\Delta x_i)$$

$$= \frac{\det G(x)(\det(I + \partial_i G(x)G^{-1}(x)\Delta x_i) - 1)}{\Delta x_i} + O(\Delta x_i)$$

$$= \det G(x) \operatorname{tr}(\partial_i G(x)G^{-1}(x)) + O(\Delta x_i).$$

Thus, we have

$$\partial_i g = \det G \operatorname{tr}(\partial_i G G^{-1}).$$

Using (3.77) and $< \tau_i, \mathbf{n} >= 0$, we have

$$\operatorname{tr}(\partial_i G G^{-1}) = \partial_i(< \tau_j, \tau_k >)g^{kj} = (< \tau_{ij}, \tau_k > + < \tau_j, \tau_{ik} >)g^{kj}$$

$$= ((\Lambda_{ij}^r \tau_r, \tau_k > + < \Lambda_{ik}^r \tau_r, \tau_j >)g^{kj} = (g_{rk}\Lambda_{ij}^r + g_{rj}\Lambda_{ik}^r)g^{kj}$$

$$= 2\Lambda_{ij}^j$$

Putting these two formulas gives (3.83).
We next prove

$$\partial_i(\sqrt{g}g^{ij}) = -\sqrt{g}g^{ik}\Lambda_{ik}^j. \tag{3.84}$$

In fact, by (3.82) and (3.83),

$$\frac{1}{\sqrt{g}}\partial_i(\sqrt{g}g^{ij}) = \frac{1}{2g}(\partial_i g)g^{ij} + \partial_i g^{ij} = \frac{1}{2g}2\Lambda_{i\ell}^\ell g g^{ij} - g^{ir}\Lambda_{ri}^j - g^{jr}\Lambda_{ri}^i$$

$$= g^{jr}\Lambda_{r\ell}^\ell - g^{ir}\Lambda_{ri}^j - g^{jr}\Lambda_{\ell r}^\ell = -g^{ik}\Lambda_{ik}^j.$$

Thus, we have (3.84).

We now prove (3.79). By (3.77) and (3.84),

$$
\begin{aligned}
\partial_i(\sqrt{g}\,g^{ij}\tau_j) &= \partial_i(\sqrt{g}\,g^{ij})\tau_j + \sqrt{g}\,g^{ij}\partial_i\tau_{ij} \\
&= -\sqrt{g}\,g^{ik}\Lambda_{ik}^j\tau_j + \sqrt{g}\,g^{ij}(\Lambda_{ij}^k\tau_k + \ell_{ij}\mathbf{n}) \\
&= -\sqrt{g}\,g^{ij}\Lambda_{ij}^k\tau_k + \sqrt{g}\,g^{ij}\Lambda_{ij}^k\tau_k + \sqrt{g}\,g^{ij}\ell_{ij}\mathbf{n} \\
&= \sqrt{g}\,g^{ij}\ell_{ij}\mathbf{n}.
\end{aligned}
$$

Thus, we have (3.79).

We now introduce the **Laplace–Beltrami operator** Δ_Γ on Γ, which is defined by

$$
\Delta_\Gamma f = \frac{1}{\sqrt{g}}\partial_i(\sqrt{g}\,g^{ij}\partial_j f).
$$

By (3.84), we have

$$
\Delta_\Gamma f = g^{ij}\partial_i\partial_j f - g^{ik}\Lambda_{ik}^j\partial_j f. \tag{3.85}
$$

By (3.78) and (3.79), we have

$$
\Delta_\Gamma \phi = (N-1)\mathcal{H}(\Gamma)\mathbf{n}.
$$

Usually, we put $H(\Gamma) = (N-1)\mathcal{H}(\Gamma)$, and so we have

$$
\Delta_\Gamma x = H(\Gamma)\mathbf{n} \quad \text{for } x \in \Gamma. \tag{3.86}
$$

One fundamental result for the Laplace–Beltrami operator is the following.

Lemma 3.2.14 *Let $1 < q < \infty$. Assume that Ω is a uniform C^3 domain and let Γ be the boundary of Ω. Then, there exists a large number $m > 0$ such that for any $f \in W_q^{-1/q}(\Gamma)$, there exists a unique $v \in W_q^{2-1/q}(\Gamma)$ such that v satisfies the equation*

$$
(m - \Delta_\Gamma)v = f \quad \text{on } \Gamma
$$

and the estimate

$$
\|v\|_{W_q^{2-1/q}(\Gamma)} \leq C\|f\|_{W_q^{-1/q}(\Gamma)}
$$

for some constant $C > 0$.

Remark 3.2.15

(1) Lemma 3.2.14 was proved by Amann–Hieber–Simonett [4, Theorem 10.3] in the case where Γ has finite covering, and by Shibata [44, Theorem 2.2] in the case where Γ is the boundary of a uniform C^2 domain.
(2) Let $(m - \Delta_\Gamma)^{-1}$ be defined by $(m - \Delta_\Gamma)^{-1} f = v$.

3.2.8 Parametrized Surface

Let $\phi_t : \Omega \to \mathbb{R}^N$ be an injection map with suitable regularity for each time $t \geq 0$. Let Γ be a C^3 hypersurface $\subset \Omega$ and set

$$\Gamma_t = \{x = \phi_t(y) \mid y \in \Gamma\}.$$

First we prove Theorem 3.1.1. Let $J(t)$ be the Jacobian of the map $x = \phi(y)$. Let $\mathbf{w}(x, t) = (\partial_t \phi_t)(\phi^{-1}(x))$. We shall prove that

$$\frac{\partial}{\partial t} J(t) = (\operatorname{div}_x \mathbf{w}(x, t)) J(t). \tag{3.87}$$

This formula is called a Reynolds transport theorem. In fact, in view of the definition of differentiation, we consider

$$\frac{J(t + \Delta t) - J(t)}{\Delta t}.$$

Writing $\phi_t = {}^\top(x_1(y, t), \ldots, x_N(y, t))$, we have $J(t) = \det X(t)$ with

$$X(t) = \begin{pmatrix} \frac{\partial x_1}{\partial y_1} & \cdots & \frac{\partial x_1}{\partial y_N} \\ \vdots & \ddots & \vdots \\ \frac{\partial x_N}{\partial y_1} & \cdots & \frac{\partial x_N}{\partial y_N} \end{pmatrix}.$$

By mean value theorem, we write $X(t + \Delta t) = X(t) + X'(t)\Delta t + O((\Delta t)^2)$ with

$$X'(t) = \begin{pmatrix} \frac{\partial^2 x_1}{\partial y_1 \partial t} & \cdots & \frac{\partial^2 x_1}{\partial y_N \partial t} \\ \vdots & \ddots & \vdots \\ \frac{\partial^2 x_N}{\partial y_1 \partial t} & \cdots & \frac{\partial^2 x_N}{\partial y_N \partial t} \end{pmatrix}.$$

Using this symbol, we write

$$
\begin{aligned}
\frac{J(t+\Delta t)-J(t)}{\Delta t} &= \frac{\det X(t+\Delta t)-\det X(t)}{\Delta t} \\
&= \frac{\det(X(t)+\Delta t\,X'(t)+O((\Delta t)^2))-\det X'(t)}{\Delta t} \\
&= \frac{\det X(t)(\det(\mathbf{I}+\Delta t\,X'(t)X^{-1}(t)+O((\Delta t)^2))-1)}{\Delta t} \\
&= (\det X(t))\mathrm{tr}(X'(t)X^{-1}(t)) + O(\Delta t).
\end{aligned}
$$

Since the (i, j)th component of $X'(t)X(t)^{-1}$ is

$$
\sum_{k=1}^{N} \frac{\partial^2 x_i}{\partial y_k \partial t}\frac{\partial y_k}{\partial x_j} = \frac{\partial w_i}{\partial x_j},
$$

where $\mathbf{w}(x) = (\partial_t \phi)(\phi^{-1}(x)) = {}^\top(w_1, \ldots, w_N)$, we have $\mathrm{tr}\,(X'(t)X^{-1})(t) = \mathrm{div}_x \mathbf{w}$, which shows (3.87).

We also prove that

$$
\frac{\partial}{\partial t}|\Gamma_t| = -\int_{\Gamma_t} H(\Gamma_t) <\mathbf{n}_t, \dot\phi > d\sigma, \tag{3.88}
$$

where $|\Gamma_t|$ is the area of Γ_t, $\dot\phi = \partial_t \phi$, and $d\sigma$ is the surface element of Γ_t. The formula $<\mathbf{n}_t, \dot\phi >$ denotes the velocity of the evolution of Γ_t with respect to $\mathbf{n}_t$.

To prove (3.88), we have to parametrize Γ_t. Let Γ be parametrized as $y = y(\theta)$ for $\theta \in \Theta \subset \mathbb{R}^{N-1}$, and then the first fundamental form of Γ_t is given by $G_t = (g_{ij}(t))$ with $g_{ij}(t) =< \tau_i(t), \tau_j(t) >$, where

$$
\tau_i(t) = \frac{\partial \phi_t(y(\theta))}{\partial \theta_i}.
$$

Let $g(t) = \det G_t$ and $G_t^{-1} = (g^{ij}(t))$. Since the surface element of Γ_t is given by $d\sigma = \sqrt{g}\,d\theta$, we have

$$
|\Gamma(t)| = \int_{\Gamma_t} d\sigma = \int_{\Theta} \sqrt{g(t)}\, d\theta.
$$

Thus,

$$
\frac{d}{dt}|\Gamma(t)| = \int_{\Theta} \frac{d}{dt}\sqrt{g(t)}\, d\theta = \int_{\Theta} \frac{\dot g}{2\sqrt{g}}\, d\theta
$$

To find $\dot{g}$, we calculate $(g(t + \Delta t) - g(t))/\Delta t$ as follows:

$$\frac{g(t + \Delta t) - g(t)}{\Delta t}$$

$$= \frac{\det G(t + \Delta t) - \det G(t)}{\Delta t}$$

$$= \frac{\det(G(t) + \partial_t G(t)\Delta t + O((\Delta t)^2)) - \det G(t)}{\Delta t}$$

$$= \frac{\det(G(t)(\det(\mathbf{I} + \partial_t G(t)G(t)^{-1}\Delta t + O((\Delta t)^2)) - 1)}{\Delta t}$$

$$= \det G(t))\mathrm{tr}(\partial_t G(t)G(t)^{-1}) + O(\Delta t).$$

Thus, we have

$$\dot{g} = g\,\mathrm{tr}\,(\partial_t G(t)G(t)^{-1}) = g(< \dot{\tau}_i, \tau_j > + < \tau_i, \dot{\tau}_j >)g^{ji}$$

$$= g(< \dot{\tau}_i, \tau_j > g^{ji} + < \dot{\tau}_i, \tau_j > g^{ij}) = 2gg^{ij} < \dot{\tau}_i, \tau_j >,$$

where we have used $g^{ij} = g^{ji}$. Putting these formulas together gives

$$\frac{d}{dt}|\Gamma(t)| = \int_\Theta \sqrt{g}g^{ij} < \dot{\tau}_i, \tau_j > d\theta$$

$$= \int_\Theta \partial_i(\sqrt{g}g^{ij} < \dot{\phi}, \tau_j >)\,d\theta - \int_\Theta < \partial_i(\sqrt{g}g^{ij}\tau_j), \dot{\phi} > d\theta$$

$$= -\int_\Theta \frac{1}{\sqrt{g}} < \partial_i(\sqrt{g}g^{ij}\partial_j\phi), \dot{\phi} > \sqrt{g}\,d\theta = -\int_{\Gamma_t} < \Delta_{\Gamma_t}\phi, \dot{\phi} > d\sigma$$

$$= -\int_{\Gamma_t} H(\Gamma_t) < \mathbf{n}_t, \dot{\phi} > d\sigma,$$

where we have used (3.86). This shows (3.88).

3.2.9 Example of Mean Curvature

Let Γ be a closed hypersurface in $\mathbb{R}^3$ defined by $|x| = r(\omega)$ for $\omega \in S_1 = \{\omega \in \mathbb{R}^3 \mid |\omega| = 1\}$. We introduce the polar coordinates:

$$\omega_1 = \cos\varphi\sin\theta, \quad \omega_2 = \sin\varphi\sin\theta, \quad \omega_3 = \cos\theta$$

for $\varphi \in [0, 2\pi)$ and $\theta \in [0, \pi)$. And then, Γ is represented by

$$x_1 = r(\varphi, \theta) \cos\varphi \sin\theta, \quad x_2 = r(\varphi, \theta) \sin\varphi \sin\theta, \quad x_3 = r(\varphi, \theta) \cos\theta.$$

Let $H(\Gamma)$ be the doubled mean curvature of Γ. In the sequel, we will show the following well-known formula (cf. [59]):

$$
H(\Gamma) = \frac{1}{r \sin\theta} \left\{ \frac{\partial}{\partial \varphi} \left(\frac{r_\varphi}{\sin\theta \sqrt{r^2 + |\nabla r|^2}} \right) + \frac{\partial}{\partial \theta} \left(\frac{\sin\theta \, r_\theta}{\sqrt{r^2 + |\nabla r|^2}} \right) \right\} \\
- \frac{2}{\sqrt{r^2 + |\nabla r|^2}},
\tag{3.89}
$$

where $\nabla r = (r_\theta, r/\sin\theta)$, and so $|\nabla r|^2 = r_\theta^2 + (r_\varphi/\sin\theta)^2$.

From the definition, we have

$$H(\Gamma) = \sum_{ij=1}^{2} g^{ij} \ell_{ij},$$

where g^{ij} is the (i, j)th component of the inverse matrix of the first fundamental form and ℓ_{ij} is the (i, j)th component of the second fundamental form. We have to find g^{ij} and ℓ_{ij}. We first find g_{ij}. Since

$$
\frac{\partial x}{\partial \varphi} = r_\varphi \begin{pmatrix} \cos\varphi \sin\theta \\ \sin\varphi \sin\theta \\ \cos\theta \end{pmatrix} + r \begin{pmatrix} -\sin\varphi \sin\theta \\ \cos\varphi \sin\theta \\ 0 \end{pmatrix},
$$
$$
\frac{\partial x}{\partial \theta} = r_\theta \begin{pmatrix} \cos\varphi \sin\theta \\ \sin\varphi \sin\theta \\ \cos\theta \end{pmatrix} + r \begin{pmatrix} \cos\varphi \cos\theta \\ \sin\varphi \cos\theta \\ -\sin\theta \end{pmatrix},
\tag{3.90}
$$

the first fundamental form G is given by $G = \begin{pmatrix} g_{11} & g_{12} \\ g_{12} & g_{22} \end{pmatrix}$ with

$$g_{11} = \frac{\partial x}{\partial \varphi} \cdot \frac{\partial x}{\partial \varphi} = r_\varphi^2 + r^2 \sin^2\theta,$$

$$g_{12} = \frac{\partial x}{\partial \varphi} \cdot \frac{\partial x}{\partial \theta} = r_\varphi r_\theta,$$

$$g_{22} = \frac{\partial x}{\partial \theta} \cdot \frac{\partial x}{\partial \theta} = r_\theta^2 + r^2.$$

And then,

$$g = \det G = r^2 \sin^2 \theta (r^2 + r_\theta^2 + (r_\varphi / \sin \theta)^2) = r^2 \sin^2 \theta (r^2 + |\nabla r|^2),$$

where $|\nabla r|^2 = r_\theta^2 + (r_\varphi / \sin \theta)^2$. Thus,

$$G^{-1} = \frac{1}{g} \begin{pmatrix} r_\theta^2 + r^2 & -r_\varphi r_\theta \\ -r_\varphi r_\theta & r_\varphi^2 + r^2 \sin^2 \theta \end{pmatrix}.$$

and so

$$g^{11} = \frac{r_\theta^2 + r^2}{g}, \quad g^{12} = g^{21} = -\frac{r_\varphi r_\theta}{g}, \quad g^{22} = \frac{r_\varphi^2 + r^2 \sin^2 \theta}{g}. \tag{3.91}$$

We next calculate the second fundamental form of Γ. For this purpose, we first calculate the unit outer normal $\mathbf{n}_\Gamma$ to Γ, which is given by

$$\mathbf{n}_\Gamma = \frac{1}{\sqrt{g}} \begin{pmatrix} r_\varphi r \sin \varphi - r_\theta r \cos \varphi \sin \theta \cos \theta + r^2 \cos \varphi \sin^2 \theta \\ -r_\varphi r \cos \varphi - r_\theta r \sin \varphi \sin \theta \cos \theta + r^2 \sin \varphi \sin^2 \theta \\ r_\theta r \sin^2 \theta + r^2 \sin \theta \cos \theta \end{pmatrix}. \tag{3.92}$$

We next find the second fundamental form. For this, first we calculate the second derivatives of x as follows:

$$\frac{\partial^2 x}{\partial \varphi^2} = r_{\varphi\varphi} \begin{pmatrix} \cos \varphi \sin \theta \\ \sin \varphi \sin \theta \\ \cos \theta \end{pmatrix} + 2r_\varphi \begin{pmatrix} -\sin \varphi \sin \theta \\ \cos \varphi \sin \theta \\ 0 \end{pmatrix} + r \begin{pmatrix} -\cos \varphi \sin \theta \\ -\sin \varphi \sin \theta \\ 0 \end{pmatrix};$$

$$\frac{\partial^2 x}{\partial \varphi \partial \theta} = r_{\varphi\theta} \begin{pmatrix} \cos \varphi \sin \theta \\ \sin \varphi \sin \theta \\ \cos \theta \end{pmatrix} + r_\varphi \begin{pmatrix} \cos \varphi \cos \theta \\ \sin \varphi \cos \theta \\ -\sin \theta \end{pmatrix} + r_\theta \begin{pmatrix} -\sin \varphi \sin \theta \\ \cos \varphi \sin \theta \\ 0 \end{pmatrix} \tag{3.93}$$

$$+ r \begin{pmatrix} -\sin \varphi \cos \theta \\ \cos \varphi \cos \theta \\ 0 \end{pmatrix};$$

$$\frac{\partial^2 x}{\partial \theta^2} = r_{\theta\theta} \begin{pmatrix} \cos \varphi \sin \theta \\ \sin \varphi \sin \theta \\ \cos \theta \end{pmatrix} + 2r_\theta \begin{pmatrix} \cos \varphi \cos \theta \\ \sin \varphi \cos \theta \\ -\sin \theta \end{pmatrix} + r \begin{pmatrix} -\cos \varphi \sin \theta \\ -\sin \varphi \sin \theta \\ -\cos \theta \end{pmatrix}. \tag{3.94}$$

Thus, we have

$$\ell_{11} = < \frac{\partial^2 x}{\partial \varphi^2}, \mathbf{n}_r >$$

$$= \frac{1}{\sqrt{g}} (r^2 r_{\varphi\varphi} \sin \theta - 2rr_\varphi^2 \sin \theta + r^2 r_\theta \sin^2 \theta \cos \theta - r^3 \sin^3 \theta),$$

$$\ell_{12} = < \frac{\partial^2 x}{\partial \varphi \partial \theta}, \mathbf{n}_r >$$

$$\hspace{6cm} (3.95)$$

$$= \frac{1}{\sqrt{g}} (r^2 r_{\varphi\theta} \sin \theta - 2rr_\varphi r_\theta \sin \theta - r^2 r_\varphi \cos \theta),$$

$$\ell_{22} = < \frac{\partial^2 x}{\partial \theta^2}, \mathbf{n}_r >$$

$$= \frac{1}{\sqrt{g}} (r^2 r_{\theta\theta} \sin \theta - 2r_\theta^2 r \sin \theta - r^3 \sin \theta).$$

We now calculate $H(\Gamma)$. Noting that $g^{12} = g^{21}$ and $\ell_{12} = \ell_{21}$, by (3.91) and (3.95), we have

$$H(\Gamma) = g^{11} \ell_{11} + 2g^{12} \ell_{12} + g^{22} \ell_{22}$$

$$= \frac{1}{g^{3/2}} \{ ((r_\theta^2 + r^2) r_{\varphi\varphi} - 2r_\varphi r_\theta r_{\varphi\theta} + r_\varphi^2 r_{\theta\theta} - 3rr_\varphi^2) r^2 \sin \theta$$

$$\hspace{6cm} (3.96)$$

$$+ (rr_{\theta\theta} - 3r_\theta^2 - 2r^2) r^3 \sin^3 \theta + (r_\theta + r^2) r_\theta r^2 \sin^2 \theta \cos \theta$$

$$+ 2r_\varphi^2 r_\theta r^2 \cos \theta \}.$$

On the other hand, we have

$$\frac{1}{r \sin \theta} \frac{\partial}{\partial \varphi} \frac{r_\varphi}{\sin \theta \sqrt{r^2 + |\nabla r|^2}} = \frac{r^2 \sin \theta ((r^2 + r_\theta^2) r_{\varphi\varphi} - rr_\varphi^2 - r_\varphi r_\theta r_{\theta\varphi})}{r^3 \sin^3 \theta (r^2 + |\nabla r|^2)^{3/2}};$$

$$\frac{1}{r \sin \theta} \frac{\partial}{\partial \theta} \frac{\sin \theta r_\theta}{\sqrt{r^2 + |\nabla r|^2}} = \frac{1}{r^3 \sin^3 \theta (r^2 + |\nabla r|^2)^{3/2}}$$

$$\times \{ r^2 (r^2 r_{\theta\theta} - r^3 r_\theta^2) \sin^3 \theta$$

$$+ r^2 (r_\varphi^2 r_{\theta\theta} - r_\varphi r_\theta r_{\theta\varphi}) \sin \theta$$

$$+ r^2 (r^2 + r_\theta^2) r_\theta \sin^2 \theta \cos \theta + 2r^2 r_\theta r_\varphi^2 \cos \theta \}.$$

Thus, we have

$$\frac{1}{r\sin\theta}\left\{\frac{\partial}{\partial\varphi}\left(\frac{r_\varphi}{\sin\theta\sqrt{r^2+|\nabla r|^2}}\right)+\frac{\partial}{\partial\theta}\left(\frac{\sin\theta\, r_\theta}{\sqrt{r^2+|\nabla r|^2}}\right)\right\}-\frac{2}{\sqrt{r^2+|\nabla r|^2}}$$

$$=\frac{B}{r^3\sin^3\theta(r^2+|\nabla r|^2)^{3/2}}$$

with

$$\begin{aligned}
B &= r^2\sin\theta((r^2+r_\theta^2)r_{\varphi\varphi}-rr_\varphi^2-r_\varphi r_\theta r_{\theta\varphi})+r^2(r^2 r_{\theta\theta}-rr_\theta^2)\sin^3\theta\\
&\quad +r^2(r_\varphi^2 r_{\theta\theta}-r_\varphi r_\theta r_{\theta\varphi})\sin\theta+r^2(r^2+r_\theta^2)r_\theta\sin^2\theta\cos\theta+2r^2 r_\theta r_\varphi^2\cos\theta\\
&\quad -2r^3\sin^3\theta(r^2+r_\theta^2+r_\varphi^2\sin^{-2}\theta)\\
&= r^2((r^2+r_\theta^2)r_{\varphi\varphi}+r_\varphi^2 r_{\theta\theta}-3rr_\varphi^2-2r_\varphi r_\theta r_{\varphi\theta})\sin\theta\\
&\quad +r^3\sin^3\theta(rr_{\theta\theta}-3r_\theta^2-2r^2)+r^2(r^2+r_\theta^2)r_\theta\sin^2\theta\cos\theta\\
&\quad +r^2 r_\theta r_\varphi^2\cos\theta,
\end{aligned}$$

which, combined with (3.96), leads to (3.89).

3.3 Free Boundary Problem with Surface Tension

In this section, we consider the case where σ is a positive constant in (3.1), and we shall transform Eq. (3.1) to some problem formulated on a fixed domain by using the Hanzawa transform.

3.3.1 Hanzawa Transform

As Ω_t is unknown, we have to transform Ω_t to a fixed domain Ω. In the following, we assume that the reference domain Ω is a uniform C^3 domain. Let Γ be the boundary of Ω and $\mathbf{n}$ the unit outer normal to Γ. We may assume that $\mathbf{n}$ is an N vector of C^3 functions defined on $\mathbb{R}^N$ satisfying the condition: $\|\mathbf{n}\|_{H^3_\infty(\mathbb{R}^N)} < \infty$. We assume that Γ_t is given by

$$\Gamma_t = \{x = y + \rho(y,t)\mathbf{n} + \xi(t) \mid y \in \Gamma\} \quad (t \in (0,T)) \tag{3.97}$$

with an unknown function $\rho(x, t)$, where $\xi(t)$ is a function depending solely on t. Let $H_\rho(y, t)$ be a suitable extension of $\rho(y, t)$ to $\mathbb{R}^N$ such that for each $t \in (0, T)$ $\rho(y, t) = H_\rho(y, t)$ for $y \in \Gamma$ and

$$\|H_\rho(\cdot, t)\|_{H_q^k(\mathbb{R}^N)} \leq C\|\rho(\cdot, t)\|_{W_q^{k-1/q}(\Gamma)},$$
$$\|\partial_t H_\rho(\cdot, t)\|_{H_q^\ell(\mathbb{R}^N)} \leq C\|\partial_t \rho(\cdot, t)\|_{W_q^{\ell-1/q}(\Gamma)} \tag{3.98}$$

with some constant $C > 0$ for $k = 1, 2, 3$ and $\ell = 1, 2$. We then define Ω_t by

$$\Omega_t = \{x = y + \omega(y)H_\rho(y, t)\mathbf{n}(y) + \xi(t) \mid y \in \Omega\} \quad (t \in (0, T)),$$

where $\omega(y)$ is a C^∞ function which equals 1 near Γ and zero far from Γ. For the notational simplicity, we set $\Psi_\rho(y, t) = \omega(y)H_\rho(y, t)\mathbf{n}(y)$. The transformation:

$$x = y + \Psi_\rho(y, t) + \xi(t) \tag{3.99}$$

is called the Hanzawa transform, which was originally introduced by Hanzawa [22] to treat classical solutions of the Stefan problem. Usually, $\xi(t)$ is also unknown functions and to prove the local well-posedness, we set $\xi(t) = 0$.

In the following, we study how to change the equations and boundary conditions under such transformation. Assume that

$$\sup_{t \in (0, T)} \|\Psi_\rho(\cdot, t)\|_{H_\infty^1(\mathbb{R}^N)} \leq \delta, \tag{3.100}$$

where δ is a small positive number determined in such a way that several conditions stated below will be satisfied. We first choose $0 < \delta < 1$, and then the Hanzawa transform (3.99) is injective for each $t \in (0, T)$. In fact, let $x_i = y_i + \Psi_\rho(y_i, t) + \xi(t)$ $(i = 1, 2)$. We then have

$$|x_1 - x_2| = |y_1 + \Psi_\rho(y_1, t) - (y_2 + \Psi_\rho(y_2, t)|$$
$$\geq |y_1 - y_2| - \|\nabla\Psi_\rho(\cdot, t)\|_{L_\infty(\mathbb{R}^N)}|y_1 - y_2| \geq (1 - \delta)|y_1 - y_2|.$$

Thus, the condition $0 < \delta < 1$ implies if $y_1 \neq y_2$ then $x_1 \neq x_2$, that is, the Hanzawa transform is injective.

We now set

$$\Omega_t = \{x = y + \Psi_\rho(y, t) + \xi(t) \mid y \in \Omega\} \quad (t \in (0, T)).$$

Notice that the Hanzawa transform maps Ω onto Ω_t injectively.

3.3.2   Transformation of Equations and the Divergence Free Condition

In this subsection, for the latter use we consider more general transformation: $x = y + \Psi(y, t)$, where $\Psi(y, t)$ satisfies the condition:

$$\sup_{t \in (0,T)} \|\Psi(\cdot, t)\|_{H^1_\infty(\mathbb{R}^N)} \leq \delta$$

with some small $\delta > 0$. Moreover, we assume that $\partial_t \Psi$ exists.

Let $\mathbf{v}$ and $\mathfrak{p}$ be solutions of Eq. (3.1), and let

$$\mathbf{u}(y, t) = \mathbf{v}(y + \Psi(y, t), t), \quad \mathfrak{q}(y, t) = \mathfrak{p}(y + \Psi(y, t), t).$$

We now show that the first equation in (3.1) is transformed to

$$\partial_t \mathbf{u} - \mathrm{Div}\,(\mu \mathbf{D}(\mathbf{u}) - \mathfrak{q}\mathbf{I}) = \mathbf{f}(\mathbf{u}, \Psi) \quad \text{in } \Omega^T, \tag{3.101}$$

and the divergence free condition: $\mathrm{div}\,\mathbf{v} = 0$ in (3.1) is transformed to

$$\mathrm{div}\,\mathbf{u} = g(\mathbf{u}, \Psi) = \mathrm{div}\,\mathbf{g}(\mathbf{u}, \Psi) \quad \text{in } \Omega^T. \tag{3.102}$$

Here, $\mathbf{f}(\mathbf{u}, \Psi)$, $g(\mathbf{u}, \Psi)$ and $\mathbf{g}(\mathbf{u}, \Psi)$ are suitable non-linear functions with respect to $\mathbf{u}$ and $\nabla\Psi$ given in (3.113) and (3.108), below.

Let $\partial x / \partial y$ be the Jacobi matrix of the transformation (3.99), that is,

$$\frac{\partial x}{\partial y} = \mathbf{I} + \nabla\Psi(y, t) = (\delta_{ij} + \frac{\partial \Psi_j}{\partial y_i}), \quad \nabla\Psi = \begin{pmatrix} \partial_1\Psi_1 & \partial_2\Psi_1 & \dots & \partial_N\Psi_1 \\ \partial_1\Psi_2 & \partial_2\Psi_2 & \dots & \partial_N\Psi_2 \\ \vdots & \vdots & \ddots & \vdots \\ \partial_1\Psi_N & \partial_2\Psi_N & \dots & \partial_N\Psi_N \end{pmatrix}$$

where $\Psi(y, t) = {}^\top(\Psi_1(y, t), \dots, \Psi_N(y, t))$, and $\partial_i \Psi_j = \dfrac{\partial \Psi_j}{\partial y_i}$. If $0 < \delta < 1$, then

$$\left(\frac{\partial x}{\partial y}\right)^{-1} = \mathbf{I} + \sum_{k=1}^{\infty}(-\nabla\Psi(y, t))^k$$

exists, and therefore there exists an $N \times N$ matrix $\mathbf{V}_0(\mathbf{k})$ of C^∞ functions defined on $|\mathbf{k}| < \delta$ such that $\mathbf{V}_0(0) = 0$ and

$$\left(\frac{\partial x}{\partial y}\right)^{-1} = \mathbf{I} + \mathbf{V}_0(\nabla\Psi(y, t)). \tag{3.103}$$

Here and in the following, $\mathbf{k} = (k_{ij})$ and k_{ij} are the variables corresponding to $\partial_i \Psi_j$.

Let $V_{0ij}(\mathbf{k})$ be the (i, j)th component of $\mathbf{V}_0(\mathbf{k})$. We then have

$$\nabla_x = (\mathbf{I} + \mathbf{V}_0(\mathbf{k}))\nabla_y, \quad \frac{\partial}{\partial x_i} = \sum_{j=1}^{N}(\delta_{ij} + V_{0ij}(\mathbf{k}))\frac{\partial}{\partial y_j}, \tag{3.104}$$

where $\nabla_z = {}^\top(\partial/\partial z, \ldots, \partial/\partial z_N)$ for $z = x$ and y. By (3.104), we can write $\mathbf{D}(\mathbf{v})$ as $\mathbf{D}(\mathbf{v}) = \mathbf{D}(\mathbf{u}) + \mathcal{D}_{\mathbf{D}}(\mathbf{k})\nabla\mathbf{u}$ with

$$\mathbf{D}(\mathbf{u})_{ij} = \frac{\partial u_i}{\partial y_j} + \frac{\partial u_j}{\partial y_i},$$

$$(\mathcal{D}_{\mathbf{D}}(\mathbf{k})\nabla\mathbf{u})_{ij} = \sum_{k=1}^{N}\left(V_{0jk}(\mathbf{k})\frac{\partial u_i}{\partial y_k} + V_{0ik}(\mathbf{k})\frac{\partial u_j}{\partial y_k}\right). \tag{3.105}$$

We next consider div $\mathbf{v}$. By (3.104), we have

$$\operatorname{div}_x\mathbf{v} = \sum_{j=1}^{N}\frac{\partial v_j}{\partial x_j} = \sum_{j,k=1}^{N}(\delta_{jk} + V_{0jk}(\mathbf{k}))\frac{\partial u_j}{\partial y_k} = \operatorname{div}_y\mathbf{u} + \mathbf{V}_0(\mathbf{k}) : \nabla\mathbf{u}. \tag{3.106}$$

Let J be the Jacobian of the transformation (3.99). Choosing $\delta > 0$ small enough, we may assume that $J = J(\mathbf{k}) = 1 + J_0(\mathbf{k})$, where $J_0(\mathbf{k})$ is a C^∞ function defined for $|\mathbf{k}| < \sigma$ such that $J_0(0) = 0$.

To obtain another representation formula of $\operatorname{div}_x\mathbf{v}$, we use the inner product $(\cdot, \cdot)_{\Omega_t}$. For any test function $\varphi \in C_0^\infty(\Omega_t)$, we set $\psi(y) = \varphi(x)$. We then have

$$(\operatorname{div}_x\mathbf{v}, \varphi)_{\Omega_t} = -(\mathbf{v}, \nabla\varphi)_{\Omega_t} = -(J\mathbf{u}, (\mathbf{I} + \mathbf{V}_0)\nabla_y\psi)_\Omega$$

$$= (\operatorname{div}((\mathbf{I} + {}^\top\mathbf{V}_0)J\mathbf{u}), \psi)_\Omega = (J^{-1}\operatorname{div}((\mathbf{I} + {}^\top\mathbf{V}_0)J\mathbf{u}), \varphi)_{\Omega_t},$$

which, combined with (3.106), leads to

$$\operatorname{div}_x\mathbf{v} = \operatorname{div}_y\mathbf{u} + \mathbf{V}_0(\mathbf{k}) : \nabla\mathbf{u} = J^{-1}(\operatorname{div}_y\mathbf{u} + \operatorname{div}_y(J^\top\mathbf{V}_0(\mathbf{k})\mathbf{u})). \tag{3.107}$$

Recalling that $J = J(\mathbf{k}) = 1 + J_0(\mathbf{k})$, we define $g(\mathbf{u}, \Psi)$ and $\mathbf{g}(\mathbf{u}, \Psi)$ by letting

$$g(\mathbf{u}, \Psi) = -(J_0(\mathbf{k})\operatorname{div}\mathbf{u} + (1 + J_0(\mathbf{k}))\mathbf{V}_0(\mathbf{k}) : \nabla\mathbf{u}),$$

$$\mathbf{g}(\mathbf{u}, \Psi) = -(1 + J_0(\mathbf{k}))^\top\mathbf{V}_0(\mathbf{k})\mathbf{u}, \tag{3.108}$$

and then by (3.107) we see that the divergence free condition: $\operatorname{div}\mathbf{v} = 0$ is transformed to Eq. (3.102). In particular, it follows from (3.107) that

$$J_0(k)\operatorname{div}\mathbf{u} + J(k)\mathbf{V}_0(\mathbf{k}) : \nabla\mathbf{u} = \operatorname{div}(J(k)^\top\mathbf{V}_0(\mathbf{k})\mathbf{u}). \tag{3.109}$$

To derive Eq. (3.101), we first observe that

$$\sum_{j=1}^{N} \frac{\partial}{\partial x_j}(\mu \mathbf{D}(\mathbf{v})_{ij} - \mathfrak{p}\delta_{ij})$$

$$= \sum_{j,k=1}^{N} \mu(\delta_{jk} + V_{0jk})\frac{\partial}{\partial y_k}(\mathbf{D}(\mathbf{u})_{ij} + (\mathcal{D}_{\mathbf{D}}(\mathbf{k})\nabla\mathbf{u})_{ij}) - \sum_{j=1}^{N}(\delta_{ij} + V_{0ij})\frac{\partial \mathfrak{q}}{\partial y_j},$$

$$(3.110)$$

where we have used (3.105). Since

$$\frac{\partial}{\partial t}[v_i(y + \Psi(y,t),t)] = \frac{\partial v_i}{\partial t}(x,t) + \sum_{j=1}^{N}\frac{\partial \Psi_j}{\partial t}\frac{\partial v_i}{\partial x_j}(x,t),$$

we have

$$\frac{\partial v_i}{\partial t} = \frac{\partial u_i}{\partial t} - \sum_{j,k=1}^{N}\frac{\partial \Psi_j}{\partial t}(\delta_{jk} + V_{0jk})\frac{\partial u_i}{\partial y_k},$$

and therefore,

$$\frac{\partial v_i}{\partial t} + \sum_{j=1}^{N}v_j\frac{\partial v_i}{\partial x_j} = \frac{\partial u_i}{\partial t} + \sum_{j,k=1}^{N}(u_j - \frac{\partial \Psi_j}{\partial t})(\delta_{jk} + V_{0jk}(\mathbf{k}))\frac{\partial u_i}{\partial y_k}. \quad (3.111)$$

Putting (3.110) and (3.111) together gives

$$0 = \left(\frac{\partial u_i}{\partial t} + \sum_{j,k=1}^{N}(u_j - \frac{\partial \Psi_j}{\partial t})(\delta_{jk} + V_{0jk}(\mathbf{k}))\frac{\partial u_i}{\partial y_k}\right)$$

$$- \mu\sum_{j,k=1}^{N}(\delta_{jk} + V_{0jk}(\mathbf{k}))\frac{\partial}{\partial y_k}(\mathbf{D}(\mathbf{u})_{ij} + (\mathcal{D}_{\mathbf{D}}(\mathbf{k})\nabla\mathbf{u})_{ij})$$

$$- \sum_{j=1}^{N}(\delta_{ij} + V_{0ij}(\mathbf{k}))\frac{\partial \mathfrak{q}}{\partial y_j}.$$

Since $(\mathbf{I} + \nabla\Psi)(\mathbf{I} + \mathbf{V}_0) = (\partial x/\partial y)(\partial y/\partial x) = \mathbf{I}$,

$$\sum_{i=1}^{N}(\delta_{mi} + \partial_m\Psi_i)(\delta_{ij} + V_{0ij}(\mathbf{k})) = \delta_{mj}, \quad (3.112)$$

and so we have

$$
0 = \sum_{i=1}^{N} (\delta_{mi} + \partial_m \Psi_i) \left(\frac{\partial u_i}{\partial t} + \sum_{j,k=1}^{N} (u_j - \frac{\partial \Psi_i}{\partial t})(\delta_{jk} + V_{0jk}(\mathbf{k})) \frac{\partial u_i}{\partial y_k} \right)
$$

$$
- \mu \sum_{i,j,k=1}^{N} (\delta_{mi} + \partial_m \Psi_i)(\delta_{jk} + V_{0jk}(\mathbf{k})) \frac{\partial}{\partial y_k} (\mathbf{D}(\mathbf{u})_{ij} + (\mathcal{D}_{\mathbf{D}}(\mathbf{k})\nabla\mathbf{u})_{ij})
$$

$$
- \frac{\partial q}{\partial y_m}.
$$

Thus, changing i to ℓ and m to i in the formula above, we define an N-vector of functions $\mathbf{f}(\mathbf{u}, \Psi)$ by letting

$$
\mathbf{f}(\mathbf{u}, \Psi)|_i = - \sum_{j,k=1}^{N} (u_j - \frac{\partial \Psi_j}{\partial t})(\delta_{jk} + V_{0jk}(\mathbf{k})) \frac{\partial u_i}{\partial y_k}
$$

$$
- \sum_{\ell=1}^{N} \partial_i \Psi_\ell \left(\frac{\partial u_\ell}{\partial t} + \sum_{j,k=1}^{N} (u_j - \frac{\partial \Psi_j}{\partial t})(\delta_{jk} + V_{0jk}(\mathbf{k})) \frac{\partial u_\ell}{\partial y_k} \right)
$$

$$
+ \mu \left(\sum_{j=1}^{N} \frac{\partial}{\partial y_j} (\mathcal{D}_{\mathbf{D}}(\mathbf{k})\nabla\mathbf{u})_{ij} + \sum_{j,k=1}^{N} V_{0jk}(\mathbf{k}) \frac{\partial}{\partial y_k} (\mathbf{D}(\mathbf{u})_{ij} + (\mathcal{D}_{\mathbf{D}}(\mathbf{k})\nabla\mathbf{u})_{ij}) \right.
$$

$$
\left. + \sum_{j,k,\ell=1}^{N} \partial_i \Psi_\ell (\delta_{jk} + V_{0jk}(\mathbf{k})) \frac{\partial}{\partial y_k} (\mathbf{D}(\mathbf{u})_{\ell j} + (\mathcal{D}_{\mathbf{D}}(\mathbf{k})\nabla\mathbf{u})_{\ell j}) \right), \qquad (3.113)
$$

where $\mathbf{f}(\mathbf{u}, \Psi)|_i$ denotes the ith comploment of $\mathbf{f}(\mathbf{u}, \Psi)$. We then see that Eq. (3.6) is transformed to Eq. (3.101).

3.3.3   *Transformation of the Boundary Conditions*

In this subsection, we consider the Hanzawa transform given in Sect.3.3.1. To represent Γ locally, we use the local coordinates near $x_\ell^1 \in \Gamma$ such that

$$
\Omega \cap B_\ell^1 = \{y = \Phi_\ell(p) \mid p \in \mathbb{R}_+^N\} \cap B_\ell^1,
$$
$$
\Gamma \cap B_\ell^1 = \{y = \Phi_\ell(p', 0) \mid (p', 0) \in \mathbb{R}_0^N\} \cap B_\ell^1. \qquad (3.114)
$$

Let $\{\zeta_\ell^1\}_{\ell \in \mathbb{N}}$ be a partition of unity given in Proposition 3.2.2. In the following we use the formula:

$$f = \sum_{\ell=1}^{\infty} \zeta_\ell^1 f \quad \text{in } \Gamma$$

for any function, f, defined on Γ.

We write $\rho = \rho(y(p_1, \ldots, p_{N-1}, 0), t)$ in the following. By the chain rule, we have

$$\frac{\partial \rho}{\partial p_i} = \frac{\partial}{\partial p_i} \Psi_\rho(\Phi_\ell(p_1, \ldots, p_{N-1}, 0), t) = \sum_{m=1}^{N} \frac{\partial \Psi_\rho}{\partial y_m} \frac{\partial \Phi_{\ell,m}}{\partial p_i}\bigg|_{p_N=0}, \tag{3.115}$$

where we have set $\Phi_\ell = {}^{\top}(\Phi_{\ell,1}, \ldots, \Phi_{\ell,N})$, and so, $\partial \rho/\partial p_i$ is defined in B_ℓ^1 by letting

$$\frac{\partial \rho}{\partial p_i} = \sum_{m=1}^{N} \frac{\partial \Psi_\rho}{\partial y_m} \circ \Phi_\ell \frac{\partial \Phi_{\ell,m}}{\partial p_i}. \tag{3.116}$$

We first represent $\mathbf{n}_t$. Recall that Γ_t is given by $x = y + \rho(y, t)\mathbf{n} + \xi(t)$ for $y \in \Gamma$ (cf. (3.97)). Let

$$\mathbf{n}_t = a\Big(\mathbf{n} + \sum_{i=1}^{N-1} b_i \tau_i\Big) \quad \text{with } \tau_i = \frac{\partial}{\partial p_i} y = \frac{\partial}{\partial p_i} \Phi_\ell(p', 0).$$

These vectors τ_i $(i = 1, \ldots, N-1)$ form a basis of the tangent space of Γ at $y = y(p_1, \ldots, p_{N-1})$. Since $|\mathbf{n}_t|^2 = 1$, we have

$$1 = a^2\Big(1 + \sum_{i,j=1}^{N-1} g_{ij} b_i b_j\Big) \quad \text{with } g_{ij} = \tau_i \cdot \tau_j \tag{3.117}$$

because $\tau_i \cdot \mathbf{n} = 0$. The vectors $\dfrac{\partial x}{\partial p_i}$ $(i = 1, \ldots, N-1)$ form a basis of the tangent space of Γ_t, and so $\dfrac{\partial x}{\partial p_i} \cdot \mathbf{n}_t = 0$. Thus, we have

$$0 = a\Big(\mathbf{n} + \sum_{j=1}^{N-1} b_j \tau_j\Big) \cdot \Big(\frac{\partial y}{\partial p_i} + \frac{\partial \rho}{\partial p_i} \mathbf{n} + \rho \frac{\partial \mathbf{n}}{\partial p_i}\Big). \tag{3.118}$$

Since $\mathbf{n} \cdot \dfrac{\partial y}{\partial p_i} = \mathbf{n} \cdot \tau_i = 0$, $\dfrac{\partial \mathbf{n}}{\partial p_i} \cdot \mathbf{n} = 0$ (because of $|\mathbf{n}|^2 = 1$), and $\dfrac{\partial y}{\partial p_i} \cdot \dfrac{\partial y}{\partial p_j} = \tau_i \cdot \tau_j = g_{ij}$, by (3.118) we have

$$\frac{\partial \rho}{\partial p_i} + \sum_{i=1}^{N-1} (g_{ij} + \rho \frac{\partial \mathbf{n}}{\partial p_i} \cdot \tau_j) b_j = 0. \tag{3.119}$$

Let H be an $(N-1) \times (N-1)$ matrix whose (i, j)th component is $\dfrac{\partial \mathbf{n}}{\partial p_i} \cdot \tau_j$. Since $\mathbf{n}$ is defined in $\mathbb{R}^N$ as an N-vector of C^2 functions with $\|\mathbf{n}\|_{H^2_\infty(\mathbb{R}^N)} < \infty$, we can write

$$\frac{\partial \mathbf{n}}{\partial p_i} \cdot \tau_j = \sum_{m=1}^{N} \frac{\partial \mathbf{n}}{\partial y_m} \circ \Phi_\ell \frac{\partial \Phi_{\ell,m}}{\partial p_i} \cdot \frac{\partial \Phi_\ell}{\partial p_j}, \tag{3.120}$$

and then H is defined in $\mathbb{R}^N$ and $\|H\|_{H^2_\infty(\mathbb{R}^N)} \leq C$ with some constant C independent of $\ell \in \mathbb{N}$. Under the assumption (3.100), we may assume that the inverse of $\mathbf{I} + \rho H G^{-1}$ exists. Let $\nabla'_\Gamma \rho = (\dfrac{\partial \rho}{\partial p_1}, \ldots, \dfrac{\partial \rho}{\partial p_{N-1}})$ with $\dfrac{\partial \rho}{\partial p_i} = \dfrac{\partial}{\partial p_i} \rho(\Phi_\ell(p', 0))$, and then by (3.119) we have

$$b = -(G + \rho H)^{-1} \nabla'_\Gamma \rho = -G^{-1} (\mathbf{I} + \rho H G^{-1})^{-1} \nabla'_\Gamma \rho. \tag{3.121}$$

Putting (3.121) and (3.117) together gives

$$a = (1 + <Gb, b>)^{-1/2}$$
$$= (1 + <(\mathbf{I} + \rho H G^{-1})^{-1} \nabla'_\Gamma \rho, G^{-1} (\mathbf{I} + \rho H G^{-1})^{-1} \nabla'_\Gamma \rho >)^{-1/2}.$$

Thus, we have

$$\mathbf{n}_t = \mathbf{n} - \sum_{i,j=1}^{N-1} g^{ij} \tau_i \frac{\partial \rho}{\partial p_j} + \mathbf{V}_\Gamma(\rho, \nabla'_\Gamma \rho) \tag{3.122}$$

where we have set

$$\mathbf{V}_\Gamma(\rho, \nabla'_\Gamma \rho) = - <G^{-1}((\mathbf{I} + \rho H G^{-1})^{-1} - \mathbf{I}) \nabla'_\Gamma \rho, \tau >$$
$$+ \{(1 + <(\mathbf{I} + \rho H G^{-1})^{-1} \nabla'_\Gamma \rho,$$
$$G^{-1} (\mathbf{I} + \rho H G^{-1})^{-1} \nabla'_\Gamma \rho >)^{-1/2} - 1\}$$
$$\times (\mathbf{n} - <G^{-1} (\mathbf{I} + \rho H G^{-1})^{-1} \nabla'_\Gamma \rho, \tau >).$$

From (3.116), $\nabla'_\Gamma \rho$ is extended to $\mathbb{R}^N$ by letting $\nabla'_\Gamma \rho = (\nabla \Phi_\ell) \nabla \Psi_\rho \circ \Phi_\ell := \Xi_{\rho,\ell}$, and so $\mathbf{V}_\Gamma(\rho, \nabla'_\Gamma \rho)$ can be extended to B^1_ℓ by

$$
\begin{aligned}
\mathbf{V}_\Gamma(\rho, \nabla'_\Gamma \rho) = & - < G^{-1}((\mathbf{I} + \Psi_\rho H G^{-1})^{-1} - \mathbf{I}) \Xi_{\rho,\ell}, \tau > \\
& + \{(1 + < (\mathbf{I} + \Psi_\rho H G^{-1})^{-1} \Xi_{\rho,\ell}, \\
& \quad G^{-1}(\mathbf{I} + \rho H G^{-1})^{-1} \Xi_{\rho,\ell} >)^{-1/2} - 1\} \\
& \times (\mathbf{n} - < G^{-1}(\mathbf{I} + \rho H G^{-1})^{-1} \Xi_{\rho,\ell}, \tau >),
\end{aligned}
\tag{3.123}
$$

where H is an $(N-1) \times (N-1)$ matrix whose (i,j)th component, H_{ij}, is given by

$$
H_{ij} = \sum_{m=1}^N \frac{\partial \mathbf{n}}{\partial y_m} \circ \Phi_\ell \frac{\partial \Phi_{\ell,m}}{\partial p_i} \cdot \frac{\partial \Phi_\ell}{\partial p_j}.
$$

Thus, we may write

$$
\mathbf{V}_\Gamma(\rho, \nabla'_\Gamma \rho) = \mathbf{V}_{\Gamma,\ell}(\bar{\mathbf{k}}) \bar{\nabla} \Psi_\rho \otimes \bar{\nabla} \Psi_\rho
$$

on B^1_ℓ with some function $\mathbf{V}_{\Gamma,\ell}(\bar{\mathbf{k}}) = \mathbf{V}_{\Gamma,\ell}(y, \bar{\mathbf{k}})$ defined on $B^1_\ell \times \{\bar{\mathbf{k}} \mid |\bar{\mathbf{k}}| \le \delta\}$ with $\mathbf{V}_{\Gamma,\ell}(0) = 0$ possessing the estimate

$$
\|(\mathbf{V}_{\Gamma,\ell}(\bar{\mathbf{k}}), \partial_{\bar{\mathbf{k}}} \mathbf{V}_{\Gamma,\ell}(\cdot, \bar{\mathbf{k}}))\|_{H^1_\infty(B^1_\ell)} \le C
$$

with some constant C independent of ℓ. Here and in the following $\bar{\mathbf{k}}$ are the corresponding variables to $\bar{\nabla} \Psi_\rho = (\Psi_\rho, \nabla \Psi_\rho)$. In view of (3.122), we have

$$
\mathbf{n}_t = \mathbf{n} - \sum_{i,j=1}^{N-1} g^{ij} \tau_i \frac{\partial \rho}{\partial y_j} + \mathbf{V}_{\Gamma,\ell}(\bar{\mathbf{k}}) \bar{\nabla} \Psi_\rho \otimes \bar{\nabla} \Psi_\rho \quad \text{on } B^1_\ell \cap \Gamma.
\tag{3.124}
$$

Let

$$
\mathbf{V}_\Gamma(\bar{\mathbf{k}}) = \sum_{\ell=1}^\infty \zeta^1_\ell \mathbf{V}_{\Gamma,\ell}(\bar{\mathbf{k}}),
\tag{3.125}
$$

and then we have

$$
\|(\mathbf{V}_\Gamma(\bar{\mathbf{k}}), \partial_{\bar{\mathbf{k}}} \mathbf{V}_\Gamma(\cdot, \bar{\mathbf{k}}))\|_{H^1_\infty(\Omega)} \le C
\tag{3.126}
$$

for $|\bar{\mathbf{k}}| \le \delta$ with some constant $C > 0$.

In view of (3.116) and (3.122), the unit outer normal $\mathbf{n}_t$ is also represented by

$$\mathbf{n}_t = \mathbf{n} - \sum_{i,j=1}^{N-1} \sum_{m=1}^{N} g^{ij} \tau_i \frac{\partial \Psi_\rho}{\partial y_m} \circ \Phi_\ell \frac{\partial \Phi_{\ell,m}}{\partial p_i} + \mathbf{V}_{\Gamma,\ell}(\bar{\mathbf{k}}) \bar{\nabla} \Psi_\rho \otimes \bar{\nabla} \Psi_\rho$$

on B_ℓ^1, and so we may write

$$\mathbf{n}_t = \mathbf{n} + \tilde{\mathbf{V}}_{\Gamma,\ell}(\bar{\nabla}\Psi_\rho)\bar{\nabla}\Psi_\rho$$

for some functions $\tilde{\mathbf{V}}_{\Gamma,\ell}(\bar{\mathbf{k}}) = \tilde{\mathbf{V}}_{\Gamma,\ell}(y, \bar{\mathbf{k}})$ defined on $B_\ell^1 \times \{\bar{\mathbf{k}} \mid |\bar{\mathbf{k}}| \leq \delta\}$ possessing the estimate:

$$\|(\tilde{\mathbf{V}}_{\Gamma,\ell}, \partial_{\bar{\mathbf{k}}}\tilde{\mathbf{V}}_{\Gamma,\ell})(\cdot, \bar{\mathbf{k}}))\|_{H_\infty^1(B_\ell^1)} \leq C \quad \text{for } |\bar{\mathbf{k}}| \leq \delta \tag{3.127}$$

with some constant C independent of $\ell \in \mathbb{N}$. Thus, setting

$$\tilde{\mathbf{V}}_\Gamma(\bar{\mathbf{k}}) = \sum_{\ell=1}^{\infty} \zeta_\ell^1 \tilde{\mathbf{V}}_{\Gamma,\ell}(\bar{\mathbf{k}}), \tag{3.128}$$

we have

$$\mathbf{n}_t = \mathbf{n} + \tilde{\mathbf{V}}_\Gamma(\bar{\nabla}\Psi_\rho)\bar{\nabla}\Psi_\rho \tag{3.129}$$

and

$$\|(\tilde{\mathbf{V}}_\Gamma(\cdot, \bar{\mathbf{k}}), \partial_{\bar{\mathbf{k}}}\mathbf{V}_\Gamma(\cdot, \bar{\mathbf{k}}))\|_{H_\infty^1(\Omega)} \leq C \quad \text{for } |\bar{\mathbf{k}}| \leq \delta. \tag{3.130}$$

We now consider the kinematic equation: $V_{\Gamma_t} = \mathbf{v} \cdot \mathbf{n}_t$ in (3.1). Since $x = y + \rho(y, t)\mathbf{n} + \xi(t)$, by (3.124) we have

$$\begin{aligned}
V_{\Gamma_t} &= \frac{\partial x}{\partial t} \cdot \mathbf{n}_t \\
&= \sum_{\ell=1}^{\infty} \zeta_\ell^1 < (\partial_t \rho)\mathbf{n} + \xi'(t), \mathbf{n} - \sum_{i,j=1}^{N-1} g^{ij} \tau_i \frac{\partial \rho}{\partial p_j} + \mathbf{V}_{\Gamma,\ell}(\bar{\mathbf{k}}) \bar{\nabla} \Psi_\rho \otimes \bar{\nabla} \Psi_\rho > \\
&= \partial_t \rho + \xi'(t) \cdot \mathbf{n} + \partial_t \rho < \mathbf{n}, \mathbf{V}_\Gamma(\bar{\mathbf{k}}) \bar{\nabla} \Psi_\rho \otimes \bar{\nabla} \Psi_\rho > \\
&\quad - < \xi'(t) \mid \bar{\nabla}'_\Gamma \rho > + < \xi'(t), \mathbf{V}_\Gamma(\bar{\mathbf{k}}) \bar{\nabla} \Psi_\rho \otimes \bar{\nabla} \Psi_\rho > .
\end{aligned}$$

Here, for any N-vector of functions, $\mathbf{d}$, we have set

$$< \mathbf{d} \mid \nabla'_\Gamma \rho > = \sum_{i,j=1}^{N-1} g^{ij} < \tau_i, \mathbf{d} > \frac{\partial \rho}{\partial p_j}. \tag{3.131}$$

On the other hand,

$$
\mathbf{v} \cdot \mathbf{n}_t = \sum_{\ell=1}^{\infty} \zeta_\ell^1 \mathbf{u} \cdot (\mathbf{n} - \sum_{i,j=1}^{N-1} g^{ij} \tau_i \frac{\partial \rho}{\partial p_j} + \mathbf{V}_{\Gamma,\ell}(\bar{\mathbf{k}}) \bar{\nabla} \Psi_\rho \otimes \bar{\nabla} \Psi_\rho)
$$

$$
= \mathbf{n} \cdot \mathbf{u} - <\mathbf{u} \mid \nabla'_\Gamma \rho> + \mathbf{u} \cdot \mathbf{V}_\Gamma(\bar{\mathbf{k}}) \bar{\nabla} \Psi_\rho \otimes \bar{\nabla} \Psi_\rho.
$$

From the consideration above, the kinematic equation is transformed to the following equation:

$$
\partial_t \rho + \xi'(t) \cdot \mathbf{n} - \mathbf{n} \cdot \mathbf{u} + <\mathbf{u} \mid \nabla'_\Gamma \rho> = d(\mathbf{u}, \rho) \quad \text{on } \Gamma \times (0, T) \tag{3.132}
$$

with

$$
d(\mathbf{u}, \rho) = \mathbf{u} \cdot \mathbf{V}_\Gamma(\bar{\mathbf{k}}) \bar{\nabla} \Psi_\rho \otimes \bar{\nabla} \Psi_\rho - \partial_t \rho < \mathbf{n}, \mathbf{V}_\Gamma(\bar{\mathbf{k}}) \bar{\nabla} \Psi_\rho \otimes \bar{\nabla} \Psi_\rho >
$$
$$
+ <\xi'(t) \mid \nabla'_\Gamma \rho> - <\xi'(t), \mathbf{V}_\Gamma(\bar{\mathbf{k}}) \bar{\nabla} \Psi_\rho \otimes \bar{\nabla} \Psi_\rho > . \tag{3.133}
$$

We next consider the boundary condition:

$$
(\mu \mathbf{D}(\mathbf{v}) - \mathfrak{p}\mathbf{I})\mathbf{n}_t = \sigma H(\Gamma_t)\mathbf{n}_t - p_0 \mathbf{n}_t \tag{3.134}
$$

in Eq. (3.1). It is convenient to divide the formula in (3.134) into the tangential part and normal part on Γ_t as follows:

$$
\mathbf{\Pi}_t \mu \mathbf{D}(\mathbf{v})\mathbf{n}_t = 0, \tag{3.135}
$$

$$
< \mu \mathbf{D}(\mathbf{v})\mathbf{n}_t, \mathbf{n}_t > -\mathfrak{p} = \sigma < H(\Gamma_t)\mathbf{n}_t, \mathbf{n}_t > -p_0. \tag{3.136}
$$

Here, $\mathbf{\Pi}_t$ is defined by $\mathbf{\Pi}_t \mathbf{d} = \mathbf{d} - <\mathbf{d}, \mathbf{n}_t > \mathbf{n}_t$ for any N-vector of functions, $\mathbf{d}$.
By (3.129) we can write

$$
\mathbf{\Pi}_t \mathbf{d} = \mathbf{\Pi}_0 \mathbf{d} + < \tilde{\mathbf{V}}_\Gamma(\bar{\nabla}\Psi_\rho)\bar{\nabla}\Psi_\rho, \mathbf{d} > \mathbf{n} + <\mathbf{n}, \mathbf{d} > \tilde{\mathbf{V}}_\Gamma(\bar{\nabla}\Psi_\rho)\bar{\nabla}\Psi_\rho
$$
$$
+ < \tilde{\mathbf{V}}_\Gamma(\bar{\nabla}\Psi_\rho)\bar{\nabla}\Psi_\rho, \mathbf{d} > \tilde{\mathbf{V}}_\Gamma(\bar{\nabla}\Psi_\rho)\bar{\nabla}\Psi_\rho, \tag{3.137}
$$

for any $\mathbf{d} \in \mathbb{R}^N$. Thus, recalling (3.105), we see that the boundary condition (3.136) is transformed to the following formula:

$$
(\mu \mathbf{D}(\mathbf{u})\mathbf{n})_\tau = \mathbf{h}'(\mathbf{u}, \Psi_\rho) \quad \text{on } \Gamma \times (0, T), \tag{3.138}
$$

where we have set

$$
\mathbf{h}'(\mathbf{u}, \Psi_\rho) = -\mu\{\mathbf{\Pi}_0 \mathcal{D}_D(\mathbf{k})\nabla\mathbf{u} + < \tilde{\mathbf{V}}_\Gamma(\bar{\mathbf{k}})\bar{\mathbf{k}}, \mathbf{D}(\mathbf{u}) + \mathcal{D}_D(\mathbf{k})\nabla\mathbf{u} > \mathbf{n}
$$
$$
+ < \mathbf{n}, \mathbf{D}(\mathbf{u}) + \mathcal{D}_D(\mathbf{k})\nabla\mathbf{u} > \tilde{\mathbf{V}}_\Gamma(\bar{\mathbf{k}})\bar{\mathbf{k}}
$$
$$
+ < \tilde{\mathbf{V}}_\Gamma(\bar{\mathbf{k}})\bar{\mathbf{k}}, \mathbf{D}(\mathbf{u}) + \mathcal{D}_D(\mathbf{k})\nabla\mathbf{u} > \tilde{\mathbf{V}}_{\Gamma,\ell}(\bar{\mathbf{k}})\bar{\mathbf{k}}\} \tag{3.139}
$$

with $\mathbf{k} = \nabla \Psi_\rho$ and $\bar{\mathbf{k}} = (\Psi_\rho, \nabla \Psi_\rho)$. Here and in the following, for any N-vector $\mathbf{a}$, we set

$$\mathbf{a}_\tau = \mathbf{a} - <\mathbf{a}, \mathbf{n}> \mathbf{n}.$$

To consider the transformation of the boundary condition (3.136), we first consider the Laplace–Beltrami operator on Γ_t. From (3.85), we have

$$\Delta_{\Gamma_t} f = g_t^{ij} \partial_i \partial_j f - g_t^{ij} \Lambda_{tik}^j \partial_j f, \tag{3.140}$$

where Λ_{tik}^j are Christoffel symbols of Γ_t defined by (3.76). Recall $x = y + \rho\mathbf{n} + \xi(t)$. The first fundamental form of $\Gamma_t = (g_{tij})$ is given by

$$g_{tij} = <\frac{\partial x}{\partial x_i}, \frac{\partial x}{\partial x_j}> = g_{ij} + \tilde{g}_{ij}\rho + \frac{\partial \rho}{\partial p_i} \frac{\partial \rho}{\partial p_j} \tag{3.141}$$

with

$$\tilde{g}_{ij} = <\tau_i, \frac{\partial \mathbf{n}}{\partial p_j}> + <\tau_j, \frac{\partial \mathbf{n}}{\partial p_i}> + <\frac{\partial \mathbf{n}}{\partial p_i}, \frac{\partial \mathbf{n}}{\partial p_j}> .$$

In view of (3.114), we can write τ_i and $\partial_j\mathbf{n}$ as

$$\tau_i = \frac{\partial y}{\partial p_i} = \sum_{j=1}^{N-1} \frac{\partial \Phi_\ell(p)}{\partial p_j}, \quad \frac{\partial \mathbf{n}}{\partial p_j} = \sum_{k=1}^{N} \frac{\partial \mathbf{n}}{\partial y_k} \frac{\partial \Phi_{\ell k}(p)}{\partial p_j}. \tag{3.142}$$

Let H be an $(N-1) \times (N-1)$ matrix whose (i,j)th components are $\tilde{g}_{ij}$, and then

$$G_t = G + \rho H + \nabla_p \rho \otimes \nabla_p \rho = G(\mathbf{I} + \rho G^{-1} H + G^{-1} \nabla_p \rho \otimes \nabla_p \rho).$$

Here, $\nabla_p \rho = (\frac{\partial \rho}{\partial p_1}, \ldots, \frac{\partial \rho}{\partial p_{N-1}})$. Choosing $\delta > 0$ small enough in (3.100), we know that the inverse of $\mathbf{I} + \rho G^{-1} H + \nabla_p \rho \otimes \nabla_p \rho$ exists, and so

$$G_t^{-1} = (\mathbf{I} + \rho G^{-1} H + G^{-1}\nabla_p \rho \otimes \nabla_p \rho))^{-1} G^{-1} = G^{-1} - \rho G^{-1} H G^{-1} + O_2,$$

that is,

$$g_t^{ij} = g^{ij} + \rho h^{ij} + O_2,$$

where h^{ij} denotes the (i,j)th component of $-G^{-1}HG^{-1}$. Here and in the following, O_2 denotes some nonlinear function with respect to ρ and $\nabla_p \rho$ of the form:

$$O_2 = a_0 \Psi_\rho^2 + \sum_{j=1}^{N} a_j \Psi_\rho \frac{\partial \Psi_\rho}{\partial y_j} + \sum_{j,k=1}^{N} b_{jk} \frac{\partial \Psi_\rho}{\partial y_j} \frac{\partial \Psi_\rho}{\partial y_k} \tag{3.143}$$

with suitable functions a_j, and b_{jk} possessing the estimates

$$|(a_j, b_{jk})| \leq C, \quad |\nabla(a_j, b_{jk})| \leq C(|\nabla \Psi_\rho| + |\nabla^2 \Psi_\rho|), \tag{3.144}$$

provided that (3.100) holds. Moreover, the Christoffel symbols are given by $\Lambda_{tij}^k = g_t^{k\ell} < \tau_{tij}, \tau_{t\ell} >$, where $\tau_{ti} = \frac{\partial x}{\partial i}$ and $\tau_{tij} = \frac{\partial^2 x}{\partial p_i \partial p_j}$. Since

$$\tau_{tij} = \tau_{ij} + \partial_i \partial_j(\rho \mathbf{n}) = \tau_{ij} + \rho \partial_i \partial_j \mathbf{n} + (\partial_i \rho) \partial_j \mathbf{n} + (\partial_j \rho) \partial_i \mathbf{n} + (\partial_i \partial_j \rho) \mathbf{n},$$

we have

$$< \tau_{tij}, \tau_{t\ell} > = < \tau_{ij}, \tau_\ell > + \rho(< \partial_i \partial_j \mathbf{n}, \tau_\ell > + < \tau_{ij}, \partial_\ell \mathbf{n} >)$$
$$+ \partial_j \rho < \partial_i \mathbf{n}, \tau_\ell > + \partial_i \rho < \partial_j \mathbf{n}, \tau_\ell > .$$

Thus,

$$\Lambda_{tij}^k = (g^{k\ell} + h^{k\ell}\rho + O_2)(< \tau_{ij}, \tau_\ell > + \rho(< \partial_i \partial_j \mathbf{n}, \tau_\ell > + < \tau_{ij}, \partial_\ell \mathbf{n} >)$$
$$+ \partial_j \rho < \partial_i \mathbf{n}, \tau_\ell > + \partial_i \rho < \partial_j \mathbf{n}, \tau_\ell >)$$
$$= \Lambda_{ij}^k + \rho g^{k\ell}(< \partial_i \partial_j \mathbf{n}, \tau_\ell > + < \tau_{ij}, \partial_\ell \mathbf{n} >)$$
$$+ \partial_j \rho \, g^{k\ell} < \partial_i \mathbf{n}, \tau_\ell > + \partial_i \rho \, g^{k\ell} < \partial_j \mathbf{n}, \tau_\ell > + O_2,$$

and so we may write

$$\Lambda_{tij}^k = \Lambda_{ij}^k + \rho A_{ij}^k + \partial_i \rho B_j^k + \partial_j \rho B_i^k + O_2 \tag{3.145}$$

with

$$A_{ij}^k = g^{k\ell}(< \partial_i \partial_j \mathbf{n}, \tau_\ell > + < \tau_{ij}, \partial_\ell \mathbf{n} >),$$
$$B_j^k = g^{k\ell} < \partial_i \mathbf{n}, \tau_\ell >, \quad B_i^k = g^{k\ell} < \partial_j \mathbf{n}, \tau_\ell > .$$

Combining these formulas with (3.140) gives

$$\Delta_{\Gamma_t} f = \Delta_\Gamma f + \sum_{i,j=1}^{N-1} (h^{ij}\rho + O_2)\partial_i \partial_j f + \sum_{k=1}^{N-1} (h^k + O_2)\partial_k f \tag{3.146}$$

with

$$h^k = - \sum_{i,j=1}^{N-1} (g^{ij} A_{ij}^k \rho + g^{ij} B_j^k \partial_i \rho + g^{ij} B_i^k \partial_j \rho + \Lambda_{ij}^k h^{ij} \rho).$$

Thus, we have

$$H(\Gamma_t)\mathbf{n}_t = \Delta_{\Gamma_t}(y + \rho\mathbf{n})$$

$$= \Delta_\Gamma y + \rho \Delta_\Gamma \mathbf{n} + \mathbf{n} \Delta_\Gamma \rho + g^{ij}(\partial_i \rho \partial_j \mathbf{n} + \partial_j \rho \partial_i \mathbf{n})$$

$$+ \rho h^{ij} \partial_i \partial_j y + (\rho h^{ij} \partial_i \partial_j \rho)\mathbf{n} + h^k \partial_k y + O_2 \partial_i \partial_j \rho + O_2,$$

which, combined with (3.122) and $< \mathbf{n}, \partial_j \mathbf{n} >= 0$, leads to

$$< H(\Gamma_t)\mathbf{n}_t, \mathbf{n}_t > =< \Delta_\Gamma y, \mathbf{n} > +\rho < \Delta_\Gamma \mathbf{n}, \mathbf{n} > +\Delta_\Gamma \rho + \rho h^{ij} < \partial_i \partial_j y, \mathbf{n} >$$

$$+ (h^{ij}\rho + O_2)\partial_i \partial_j \rho - g^{ij} < \Delta_\Gamma y, \tau_i > \partial_j \rho + O_2.$$

Noting that $\Delta_\Gamma y = H(\Gamma)\mathbf{n}$, we have

$$< H(\Gamma_t)\mathbf{n}_t, \mathbf{n}_t > = \Delta_\Gamma \rho + H(\Gamma) + \rho < \Delta_\Gamma \mathbf{n}, \mathbf{n} >$$

$$+ \rho h^{ij} < \partial_i \partial_j y, \mathbf{n} > +(h^{ij}\rho + O_2)\partial_i \partial_j \rho + O_2.$$

Recalling (3.143) and (3.144), we see that there exists a function $\mathbf{V}'_\Gamma(\bar{\mathbf{k}})$ defined on $\bar{\Omega} \times \{\bar{\mathbf{k}} \mid |\bar{\mathbf{k}}| \leq C\}$ satisfying the estimate:

$$\sup_{|\bar{\mathbf{k}}| \leq \delta} \|(\mathbf{V}'_\Gamma(\cdot, \bar{\mathbf{k}}), \partial_{\bar{\mathbf{k}}} \mathbf{V}'_\Gamma(\cdot, \bar{\mathbf{k}}))\|_{H^1_\infty(\Omega)} \leq C$$

with some constant C, where $\partial_{\bar{\mathbf{k}}}$ denotes the partial derivatives with respect to variables $\bar{\mathbf{k}}$, for which

$$\sum_{\ell=1}^{\infty} \zeta_\ell^1 ((h^{ij}\rho + O_2)\partial_i \partial_j \rho + O_2) = \mathbf{V}'_\Gamma(\bar{\mathbf{k}})\bar{\mathbf{k}} \otimes \bar{\bar{\mathbf{k}}}, \tag{3.147}$$

where $\bar{\mathbf{k}} = (\Psi_\rho, \nabla\Psi_\rho)$ and $\bar{\bar{\mathbf{k}}} = (\Psi_\rho, \nabla\Psi_\rho, \nabla^2\Psi_\rho)$. Therefore, we have

$$< H(\Gamma_t)\mathbf{n}_t, \mathbf{n}_t > = \Delta_\gamma \rho + H(\Gamma) + \mathbb{B}\rho + \mathbf{V}'_\Gamma(\bar{\nabla}\Psi_\rho)\bar{\nabla}\Psi_\rho \otimes \bar{\nabla}^2\Psi_\rho, \tag{3.148}$$

where $\bar{\nabla}^2\Psi_\rho = (\Psi_\rho, \nabla\Psi_\rho, \nabla^2\Psi_\rho)$ and we have set

$$\mathbb{B}\rho = \{< \Delta_\Gamma \mathbf{n}, \mathbf{n} > + \sum_{\ell=1}^{\infty} \zeta_\ell^1 (\sum_{i,j=1}^{N-1} h^{ij} < \partial_i \partial_j y, \mathbf{n} >)\}\rho. \tag{3.149}$$

To turn to Eq. (3.136), in view of (3.105), (3.129), and (3.128), we write

$$
\begin{aligned}
< \mathbf{n}, \mu \mathbf{D}(\mathbf{v})\mathbf{n}_t > = &< \mathbf{n}, \mu \mathbf{D}(\mathbf{u})\mathbf{n} > + < \mathbf{n}, \mu \mathbf{D}(\mathbf{u})\mathbf{V}_\Gamma(\bar{\mathbf{k}})\bar{\mathbf{k}} > \\
&+ < \mathbf{n}, \mu (\mathcal{D}_{\mathbf{D}}(\mathbf{k})\nabla \mathbf{u})(\mathbf{n} + \mathbf{V}_\Gamma(\bar{\mathbf{k}})\bar{\mathbf{k}}) >,
\end{aligned} \tag{3.150}
$$

where $\mathbf{k} = \nabla \Psi_\rho$ and $\bar{\mathbf{k}} = (\Psi_\rho, \nabla \Psi_\rho)$. Thus, from (3.148), (3.150) and (3.2) it follows that the boundary condition (3.136) is transformed to

$$
< \mathbf{n}, \mu \mathbf{D}(\mathbf{u})\mathbf{n} > -(\mathfrak{q} + c_0) - \sigma(\Delta_\Gamma \rho + \mathbb{B}\rho) = h_N(\mathbf{u}, \Psi_\rho), \tag{3.151}
$$

where c_0 is a constant and we have set

$$
\begin{aligned}
h_N(\mathbf{u}, \Psi_\rho) = &- < \mathbf{n}, \mu \mathbf{D}(\mathbf{u})\mathbf{V}_\Gamma(\bar{\mathbf{k}})\bar{\mathbf{k}} > \\
&- < \mathbf{n}, \mu (\mathcal{D}_{\mathbf{D}}(\mathbf{k})\nabla \mathbf{u})(\mathbf{n} + \mathbf{V}_\Gamma(\bar{\mathbf{k}})\bar{\mathbf{k}}) > + \sigma \mathbf{V}'_\Gamma(\bar{\mathbf{k}})(\bar{\mathbf{k}}, \bar{\bar{\mathbf{k}}}),
\end{aligned} \tag{3.152}
$$

where $\mathbf{k} = \nabla \Psi_\rho$, $\bar{\mathbf{k}} = \bar{\nabla} \Psi_\rho$, and $\bar{\bar{\mathbf{k}}} = \bar{\nabla}^2 \Psi_\rho$. Setting $\mathfrak{q} + c_0 = \mathfrak{q}'$, we have $\nabla \mathfrak{q} = \nabla \mathfrak{q}'$, and so we may assume that $c_0 = 0$.

3.3.4 Linearization Principle

In this subsection, we give a linearization principle[1] of Eq. (3.1) for the local well-posedness and the global well-posedness. Putting (3.101), (3.102), (3.132), (3.138), and (3.151) together, we see that Eq. (3.1) is transformed to the equations:

$$
\begin{cases}
\partial_t \mathbf{u} - \mathrm{Div}\,(\mu \mathbf{D}(\mathbf{u}) - \mathfrak{q}\mathbf{I}) = \mathbf{f}(\mathbf{u}, \Psi_\rho) & \text{in } \Omega^T, \\[4pt]
\mathrm{div}\,\mathbf{u} = g(\mathbf{u}, \Psi_\rho) = \mathrm{div}\,\mathbf{g}(\mathbf{u}, \Psi_\rho) & \text{in } \Omega^T, \\[4pt]
\partial_t \rho + \xi'(t) \cdot \mathbf{n} - \mathbf{u} \cdot \mathbf{n} + < \mathbf{u} \mid \nabla'_\Gamma \rho > = d(\mathbf{u}, \rho) & \text{on } \Gamma^T, \\[4pt]
(\mu \mathbf{D}(\mathbf{u})\mathbf{n})_\tau = \mathbf{h}'(\mathbf{u}, \Psi_\rho) & \text{on } \Gamma^T, \\[4pt]
< \mu \mathbf{D}(\mathbf{u})\mathbf{n}, \mathbf{n} > -(\mathfrak{q} + \sigma H(\Gamma)) - \sigma(\Delta_\Gamma + \mathbb{B})\rho = h_N(\mathbf{u}, \Psi_\rho) & \text{on } \Gamma^T, \\[4pt]
\mathbf{u}|_{t=0} = \mathbf{u}_0 \text{ in } \Omega, \quad \rho|_{t=0} = \rho_0 \text{ on } \Gamma.
\end{cases} \tag{3.153}
$$

---

[1]The linearization principle means how to divide a nonlinear equation into a linear part and a non-linear part.

Local Well-Posedness

To prove the local well-posedness, the unique existence of local in time solutions, we take $\xi(t) = 0$. For the reference body Ω, we choose the boundary Γ_0 of the initial domain in such a way that

$$\Gamma_0 = \{x = y + h_0(y)\mathbf{n} \mid y \in \Gamma\}$$

and $\|h_0\|_{B_{q,p}^{3-1/p-1/q}(\Gamma)}$ is small enough. But, we do not want to have any restriction on the size of the initial velocity, $\mathbf{u}_0$. Thus, we approximate $\mathbf{u}_0$ by $\mathbf{u}_\kappa$ defined by letting

$$\mathbf{u}_\kappa = \frac{1}{\kappa} \int_0^\kappa T(s)\tilde{\mathbf{u}}_0 \, ds,$$

where $\tilde{\mathbf{u}}_0$ is a suitable extension of $\mathbf{u}_0$ to $\mathbb{R}^N$ satisfying the condition:

$$\|\tilde{\mathbf{u}}_0\|_{B_{q,p}^{2(1-1/p)}(\mathbb{R}^N)} \leq C\|\mathbf{u}_0\|_{B_{q,p}^{2(1-1/p)}(\Omega)}, \tag{3.154}$$

and $T(s)$ is some analytic semigroup defined on $\mathbb{R}^N$ satisfying the estimate:

$$\|T(\cdot)\tilde{\mathbf{u}}_0\|_{L_\infty((0,\infty), B_{q,p}^{2(1-1/p)}(\mathbb{R}^N))} \leq C\|\mathbf{u}_0\|_{B_{q,p}^{2(1-1/p)}(\Omega)},$$

$$\|T(\cdot)\tilde{\mathbf{u}}_0\|_{L_p((0,\infty), H_q^2(\mathbb{R}^N))} + \|T(\cdot)\tilde{\mathbf{u}}_0\|_{H_p^1((0,\infty), L_q(\mathbb{R}^N))} \leq C\|\mathbf{u}_0\|_{B_{q,p}^{2(1-1/p)}(\Omega)}. \tag{3.155}$$

By (3.155)

$$\|\mathbf{u}_\kappa\|_{B_{q,p}^{2(1-1/p)}(\Omega)} \leq C\|\mathbf{u}_0\|_{B_{q,p}^{2(1-1/p)}(\Omega)},$$

$$\|\mathbf{u}_\kappa\|_{H_q^2(\Omega)} \leq C\|\mathbf{u}_0\|_{B_{q,p}^{2(1-1/p)}(\Omega)}\kappa^{-1/p}. \tag{3.156}$$

The original idea to use $\mathbf{u}_\kappa$ goes back to Padula and Solonnikov [34]. From Eq. (3.153), the linearization principle of Eq. (3.1) is the following equations:

$$\begin{cases} \partial_t \mathbf{u} - \mathrm{Div}\,(\mu\mathbf{D}(\mathbf{u}) - \mathfrak{q}\mathbf{I}) = \mathbf{f}(\mathbf{u}, \Psi_\rho) & \text{in } \Omega^T \\ \mathrm{div}\,\mathbf{u} = g(\mathbf{u}, \Psi_\rho) = \mathrm{div}\,\mathbf{g}(\mathbf{u}, \Psi_\rho) & \text{in } \Omega^T \\ \partial_t \rho + <\mathbf{u}_\kappa \mid \nabla_\Gamma' \rho> -\mathbf{u}\cdot\mathbf{n} = d(\mathbf{u}, \Psi_\rho) + <\mathbf{u}_\kappa - \mathbf{u} \mid \nabla_\Gamma' \rho> & \text{in } \Gamma^T \\ (\mu\mathbf{D}(\mathbf{u})\mathbf{n})_\tau = \mathbf{h}'(\mathbf{u}, \Psi_\rho) & \text{in } \Gamma^T \\ <\mu\mathbf{D}(\mathbf{u})\mathbf{n}, \mathbf{n}> -\mathfrak{q} - \sigma(\Delta_\Gamma + \mathbb{B})\rho = h_N(\mathbf{u}, \Psi_\rho) & \text{in } \Gamma^T \\ \mathbf{u}|_{t=0} = \mathbf{u}_0 \text{ in } \Omega, \quad \rho|_{t=0} = \rho_0 \text{ on } \Gamma. \end{cases} \tag{3.157}$$

Global Well-Posedness

In this lecture note, we only treat the global well-posedness in the case where the boundary is compact, that is Ω is a bounded domain or an exterior domain. The case where the boundary is unbounded was treated, for example, by Saito and Shibata [39, 40] in the L_p–L_q framework.

When Ω is a bounded domain, roughly speaking, the global well-posedess holds in the maximal L_p–L_q regularity class provided that initial data are small and orthogonal to the rigid motion and the reference domain Ω is very closed to a ball, and moreover the solutions decay exponentially, which will be proved in Sect. 3.7 below.

When Ω is an exterior domain, we consider (3.1) with $\sigma = 0$, that is without surface tension. In this case, instead of the Hanzawa transform, we use the partial Lagrange transform. Because, if we use the Hanzawa transform, we can not obtain necessary regularity of ρ representing Γ_t. Roughly speaking, we have the global wellposedness provided that initial data are small, and moreover the solutions decay with polynomial order. For detailed, see Sect. 3.8 below.

3.4 Maximal L_p–L_q Regularity

To prove the local and global well-posedness for Eq. (3.1), in this lecture note the maximal regularity theorem for the linear part (the left hand side of Eq. (3.157)) plays an essential role. In this section, we study the maximal L_p–L_q regularity theorem and the generation of C_0 analytic semigroup associated with the Stokes equations with free boundary conditions.

3.4.1 Statement of Maximal Regularity Theorems

In this subsection, we shall state the maximal L_p–L_q regularity for the following three problems:

$$\begin{cases} \partial_t \mathbf{u} - \text{Div}\,(\mu \mathbf{D}(\mathbf{u}) - \mathfrak{p}\mathbf{I}) = \mathbf{f}, & \text{div}\,\mathbf{u} = g = \text{div}\,\mathbf{g} & \text{in } \Omega^T, \\ (\mu \mathbf{D}(\mathbf{u}) - \mathfrak{p}\mathbf{I})\mathbf{n} = \mathbf{h} & & \text{on } \Gamma^T, \\ \mathbf{u} = \mathbf{u}_0 & & \text{in } \Omega; \end{cases} \quad (3.158)$$

$$\begin{cases} \partial_t \mathbf{v} - \text{Div}\,(\mu \mathbf{D}(\mathbf{v}) - \mathfrak{q}\mathbf{I}) = 0, \quad \text{div}\,\mathbf{v} = 0 & \text{in } \Omega^T, \\[4pt] \qquad\qquad \partial_t h - \mathbf{n} \cdot \mathbf{v} + \mathcal{F}_1 \mathbf{v} = d & \text{on } \Gamma^T, \\[4pt] (\mu \mathbf{D}(\mathbf{v}) - \mathfrak{q}\mathbf{I})\mathbf{n} - (\mathcal{F}_2 h + \sigma \Delta_\Gamma h)\mathbf{n} = 0 & \text{on } \Gamma^T, \\[4pt] \qquad\qquad\qquad (\mathbf{v}, h)|_{t=0} = (0, h_0) & \text{in } \Omega \times \Gamma; \end{cases} \tag{3.159}$$

$$\begin{cases} \partial_t \mathbf{w} - \text{Div}\,(\mu \mathbf{D}(\mathbf{w}) - \mathfrak{r}\mathbf{I}) = 0, \quad \text{div}\,\mathbf{w} = 0 & \text{in } \Omega^T, \\[4pt] \partial_t \rho + A_\kappa \cdot \nabla'_\Gamma \rho - \mathbf{w} \cdot \mathbf{n} + \mathcal{F}_1 \mathbf{w} = d & \text{on } \Gamma^T, \\[4pt] (\mu \mathbf{D}(\mathbf{w}) - \mathfrak{r}\mathbf{I})\mathbf{n} - (\mathcal{F}_2 \rho + \sigma \Delta_\Gamma \rho)\mathbf{n} = 0 & \text{on } \Gamma^T, \\[4pt] \qquad\qquad\qquad (\mathbf{w}, \rho)|_{t=0} = (0, 0) & \text{in } \Omega \times \Gamma. \end{cases} \tag{3.160}$$

Here, $\mathcal{F}_1$ and $\mathcal{F}_2$ are linear operators such that

$$\|\mathcal{F}_1 \mathbf{v}\|_{W_q^{2-1/q}(\Gamma)} \le M_0 \|\mathbf{v}\|_{H_q^1(\Omega)}, \quad \|\mathcal{F}_2 h\|_{H_q^1(\Omega)} \le C_0 \|h\|_{H_q^2(\Omega)} \tag{3.161}$$

with some constant M_0. If we consider the total problem:

$$\begin{cases} \partial_t U - \text{Div}\,(\mu \mathbf{D}(U) - P\mathbf{I}) = \mathbf{f}, \quad \text{div}\,U = g = \text{div}\,\mathbf{g} & \text{in } \Omega^T, \\[4pt] \partial_t H + A_\kappa \cdot \nabla'_\Gamma H - U \cdot \mathbf{n} + \mathcal{F}_1 U = d & \text{on } \Gamma^T, \\[4pt] (\mu \mathbf{D}(U) - P\mathbf{I})\mathbf{n} - (\mathcal{F}_2 H + \sigma \Delta_\Gamma H)\mathbf{n} = \mathbf{h} & \text{on } \Gamma^T, \\[4pt] \qquad\qquad\qquad (U, H)|_{t=0} = (\mathbf{u}_0, h_0) & \text{in } \Omega \times \Gamma, \end{cases} \tag{3.162}$$

then, U, P and H are given by $U = \mathbf{u} + \mathbf{v} + \mathbf{w}$ and $P = \mathfrak{p} + \mathfrak{q} + \mathfrak{r}$ and $H = h + \rho$, where $(\mathbf{u}, \mathfrak{p})$ is a solution of Eq. (3.158), $(\mathbf{v}, \mathfrak{q}, h)$ a solution of Eq. (3.159) with $d = 0$, and $(\mathbf{w}, \mathfrak{r}, \rho)$ a solution of Eq. (3.160) by replacing d by $d + \mathbf{n} \cdot \mathbf{u} - \mathcal{F}_1 \mathbf{u} - A_\kappa \cdot \nabla'_\Gamma h$.

We now introduce a solenoidal space $J_q(\Omega)$ defined by

$$J_q(\Omega) = \{\mathbf{f} \in L_q(\Omega)^N \mid (\mathbf{f}, \nabla \varphi)_\Omega = 0 \quad \text{for any } \varphi \in \hat{H}_{q',0}^1(\Omega)\}. \tag{3.163}$$

Before stating the maximal L_p–L_q regularity theorems for the three equations given above, we introduce the assumptions on μ, σ, and A_κ.

Assumption on μ, σ and A_κ

There exist positive constants m_0, m_1, m_2, m_3, a and b for which,

$$m_0 \leq \mu(x), \sigma(x) \leq m_1, \quad |\nabla(\mu(x), \sigma(x))| \leq m_1 \quad \text{for any } x \in \overline{\Omega},$$

$$|A_\kappa(x)| \leq m_2, \quad |A_\kappa(x) - A_\kappa(y)| \leq m_2|x - y|^a \quad \text{for any } x, y \in \Gamma,$$

$$\|A_\kappa\|_{W_r^{2-1/r}(\Gamma)} \leq m_3\kappa^{-b} \quad (\kappa \in (0, 1)), \tag{3.164}$$

where r is an exponent in (N, ∞).

Moreover, in view of (3.164), for $B_j^i = B_{r_0}(x_j^i)$ given in Proposition 3.2.2 we assume that

$$|\mu(x) - \mu(x_j^0)| \leq M_1, \quad \text{for } x_j^0 \in B_j^i, \tag{3.165}$$

$$|\mu(x) - \mu(x_j^1)| \leq M_1, \quad |\sigma(x) - \sigma(x_j^1)| \leq M_1, \quad |A_\kappa(x) - A_\kappa(x_j^1)| \leq M_1 \tag{3.166}$$

for any $x \in B_j^1$. Since $H_q^1(\Omega)$ is usually not dense in $\hat{H}_q^1(\Omega)$, it does not hold that $\operatorname{div} \mathbf{u} = \operatorname{div} \mathbf{g}$ implies $(\mathbf{u}, \nabla\varphi)_\Omega = (\mathbf{g}, \nabla\varphi)_\Omega$ for all $\varphi \in \hat{H}_q^1(\Omega)$. Of course, the opposite direction holds. Thus, finally we introduce the following definition.

Definition 3.4.1 For $\mathbf{u}, \mathbf{g} \in L_q(\Omega)^N$, we say that $\operatorname{div} \mathbf{u} = \operatorname{div} \mathbf{g}$ in Ω if $\mathbf{u} - \mathbf{g} \in J_q(\Omega)$.

To solve the divergence equation $\operatorname{div} \mathbf{u} = g$ in Ω, it is necessary to assume that g is given by $g = \operatorname{div} \mathbf{g}$ for some $\mathbf{g}$, and so we define the space $DI_q(G)$ by

$$DI_q(G) = \{(g, \mathbf{g}) \mid g \in H_q^1(G), \ \mathbf{g} \in L_q(G), \ g = \operatorname{div} \mathbf{g} \text{ in } G\},$$

where G is any domain in $\mathbb{R}^N$.

Before stating the main results in this section, we give a definition of the uniqueness. For Eq. (3.158), the uniqueness is defined as follows:

- If $\mathbf{v}$ and $\mathfrak{p}$ with

$$\mathbf{v} \in L_p((0, T), H_q^2(\Omega)^N) \cap H_p^1((0, T), L_q(\Omega)^N),$$

$$\mathfrak{p} \in L_p((0, T), H_q^1(\Omega) + \hat{H}_{q,0}^1(\Omega))$$

satisfy the homogeneous equations:

$$\begin{cases} \partial_t \mathbf{v} - \operatorname{Div}(\mu\mathbf{D}(\mathbf{v}) - \mathfrak{p}\mathbf{I}) = 0, & \operatorname{div} \mathbf{v} = 0 \quad \text{in } \Omega^T, \\ (\mu\mathbf{D}(\mathbf{v}) - \mathfrak{p}\mathbf{I})\mathbf{n} = 0 & \text{on } \Gamma^T, \\ \mathbf{v}|_{t=0} = 0 & \text{on } \Omega, \end{cases} \tag{3.167}$$

then, $\mathbf{v} = 0$ and $\mathfrak{p} = 0$.

For Eq. (3.159) and (3.160), the uniqueness is defined as follows:

- If $\mathbf{v}$, $\mathfrak{p}$ and ρ with

$$\mathbf{v} \in L_p((0, T), H_q^2(\Omega)^N) \cap H_p^1((0, T), L_q(\Omega)^N),$$

$$\mathfrak{p} \in L_p((0, T), H_q^1(\Omega) + \hat{H}_{q,0}^1(\Omega)),$$

$$\rho \in L_p((0, T), W_q^{3-1/q}(\Gamma)) \cap H_p^1((0, T), W_q^{2-1/q}(\Gamma))$$

satisfy the homogeneous equations:

$$\begin{cases} \partial_t \mathbf{v} - \mathrm{Div}\,(\mu \mathbf{D}(\mathbf{v}) - \mathfrak{p}\mathbf{I}) = 0, \quad \mathrm{div}\,\mathbf{v} = 0 & \text{in } \Omega^T, \\ \partial_t \rho + A_\kappa \cdot \nabla_\Gamma' \rho - \mathbf{v} \cdot \mathbf{n} + \mathcal{F}_1 \mathbf{v} = 0 & \text{on } \Gamma^T, \\ (\mu \mathbf{D}(\mathbf{v}) - \mathfrak{p}\mathbf{I} - ((\mathcal{F}_2 + \sigma \Delta_\Gamma)\rho)\mathbf{I})\mathbf{n} = 0 & \text{on } \Gamma^T, \\ (\mathbf{v}, \rho)|_{t=0} = (0, 0) & \text{on } \Omega \times \Gamma, \end{cases} \tag{3.168}$$

then, $\mathbf{v} = 0$, $\mathfrak{p} = 0$, and $\rho = 0$.

Notice that when $A_\kappa = 0$, the uniqueness is stated in the same manner as in (3.168). We now state the maximal L_p–L_q regularity theorem.

Theorem 3.4.2 *Let* $1 < p, q < \infty$ *with* $2/p + 1/q \neq 1$. *Assume that* Ω *is a uniform* C^2 *domain and that the weak Dirichlet problem is uniquely solvable for index* q. *Then, there exists a* $\gamma_0 > 0$ *such that the following assertion holds: Let* $\mathbf{u}_0 \in B_{q,p}^{2(1-1/p)}(\Omega)^N$ *be initial data for Eq. (3.158) and let* $\mathbf{f}$, g, $\mathbf{g}$, *and* $\mathbf{h}$ *be functions in the right side of Eq. (3.158) such that* $\mathbf{f} \in L_p((0, T), L_q(\Omega)^N)$, *and*

$$e^{-\gamma t}\mathbf{g} \in H_p^1(\mathbb{R}, L_q(\Omega)^N), \quad e^{-\gamma t}g \in L_p(\mathbb{R}, H_q^1(\Omega)) \cap H_p^{1/2}(\mathbb{R}, L_q(\Omega)),$$

$$e^{-\gamma t}\mathbf{h} \in L_p(\mathbb{R}, H_q^1(\Omega)^N) \cap H_p^{1/2}(\mathbb{R}, L_q(\Omega)^N) \tag{3.169}$$

for any $\gamma \geq \gamma_0$. *Assume that the compatibility condition:*

$$\mathbf{u}_0 - \mathbf{g}|_{t=0} \in J_q(\Omega) \quad \text{and} \quad \mathrm{div}\,\mathbf{u}_0 = g|_{t=0} \quad \text{in } \Omega \tag{3.170}$$

holds. In addition, we assume that the compatibility condition:

$$(\mu \mathbf{D}(\mathbf{u}_0)\mathbf{n})_\tau = (\mathbf{h}|_{t=0})_\tau \quad \text{on } \Gamma \tag{3.171}$$

holds for $2/p + 1/q < 1$. *Then, problem (3.158) admits solutions* $\mathbf{u}$ *and* $\mathfrak{p}$ *with*

$$\mathbf{u} \in L_p((0, T), H_q^2(\Omega)^N) \cap H_p^1((0, T), L_q(\Omega)^N),$$

$$\mathfrak{p} \in L_p((0, T), H_q^1(\Omega) + \hat{H}_{q,0}^1(\Omega))$$

possessing the estimate:

$$\|\mathbf{u}\|_{L_p((0,T),H_q^2(\Omega))} + \|\partial_t \mathbf{u}\|_{L_p((0,T),L_q(\Omega))}$$

$$\leq Ce^{\gamma T}(\|\mathbf{u}_0\|_{B_{q,p}^{2(1-1/p)}(\Omega)} + \|\mathbf{f}\|_{L_p((0,T),L_q(\Omega))} + \|e^{-\gamma t}\mathbf{g}\|_{H_p^1(\mathbb{R},L_q(\Omega))}$$

$$+ \|e^{-\gamma t}(g,\mathbf{h})\|_{L_p(\mathbb{R},H_q^1(\Omega))} + \|e^{-\gamma t}(g,\mathbf{h})\|_{H_p^{1/2}(\mathbb{R},L_q(\Omega))})$$

for some constant $C > 0$ and any $\gamma \geq \gamma_0$, where C is independent of γ.

Moreover, if we assume that the weak Dirichlet problem is uniquely solvable for $q' = q/(q-1)$ in addition, then the uniqueness for Eq. (3.158) holds.

Theorem 3.4.3 *Let $1 < p < \infty$, $1 < q \leq r$ and $2/p + 1/q \neq 1$. Assume that Ω is a uniform C^3 domain and that the weak Dirichlet problem is uniquely solvable for q. Let $h_0 \in B_{q,p}^{3-1/p-1/q}(\Gamma)$ be initial data for Eq. (3.159) and let d be a function in the right side of Eq. (3.159) such that*

$$d \in L_p((0,T), W_q^{2-1/q}(\Gamma)). \tag{3.172}$$

Then, problem (3.159) admits solutions $\mathbf{v}$, $\mathfrak{q}$ and h with

$$\mathbf{v} \in L_p((0,T), H_q^2(\Omega)^N) \cap H_p^1((0,T), L_q(\Omega)^N),$$

$$\mathfrak{q} \in L_p((0,T), H_q^1(\Omega) + \hat{H}_{q,0}^1(\Omega)),$$

$$h \in L_p((0,T), W_q^{3-1/q}(\Gamma)) \cap H_p^1((0,T), W_q^{2-1/q}(\Gamma))$$

possessing the estimate:

$$\|\mathbf{v}\|_{L_p((0,T),H_q^2(\Omega))} + \|\partial_t \mathbf{v}\|_{L_p((0,T),L_q(\Omega))}$$

$$+ \|h\|_{L_p((0,T),W_q^{3-1/q}(\Gamma))} + \|\partial_t h\|_{L_p((0,T),W_q^{2-1/q}(\Gamma))}$$

$$\leq Ce^{\gamma T}(\|h_0\|_{B_{q,p}^{3-1/p-1/q}(\Gamma)} + \|d\|_{L_p((0,T),W_q^{2-1/q}(\Gamma))})$$

for some constants $C > 0$ and any $\gamma \geq \gamma_0$, where C is independent of γ.

Moreover, if we assme that the weak Dirichlet problem is uniquely solvable for $q' = q/(q-1)$ in addition, then the uniqueness for Eq. (3.159) holds.

Theorem 3.4.4 *Let $1 < p, q < \infty$. Assume that Ω is a uniform C^3 domain and that the weak Dirichlet problem is uniquely solvable for q. Let d be a function in the right side of Eq. (3.160) such that*

$$d \in L_p((0,T), W_q^{2-1/q}(\Gamma)). \tag{3.173}$$

Then, problem (3.160) *admits unique solutions* **w**, $\mathfrak{t}$ *and* ρ *with*

$$\mathbf{w} \in L_p((0, T), H_q^2(\Omega)^N) \cap H_q^1((0, T), L_q(\Omega)^N),$$

$$\mathfrak{t} \in L_p((0, T), H_q^1(\Omega) + \hat{H}_{q,0}^1(\Omega)),$$

$$\rho \in L_p((0, T), W_q^{3-1/q}(\Gamma)) \cap H_p^1((0, T), W_q^{2-1/q}(\Gamma))$$

possessing the estimate:

$$\|\mathbf{w}\|_{L_p((0,T), H_q^2(\Omega))} + \|\partial_t \mathbf{w}\|_{L_p((0,T), L_q(\Omega))}$$

$$+ \|\rho\|_{L_p((0,T), W_q^{3-1/q}(\Gamma))} + \|\partial_t \rho\|_{L_p((0,T), W_q^{2-1/q}(\Gamma))}$$

$$\leq C e^{\gamma \kappa^{-b} T} \|d\|_{L_p((0,T), W_q^{2-1/q}(\Gamma))}$$

for some constants $C > 0$ *and any* $\gamma \geq \gamma_0$ *with some* γ_0, *where* C *is independent of* γ *and* $\kappa \in (0, 1)$, *where* $\kappa \in (0, 1)$ *and* b *is the constant appearing in* (3.164).

Moreover, if we assume that Ω *is a uniform* C^3 *domain whose inside has a finite covering and that the weak Dirichlet problem is uniquely solvable for* $q' = q/(q-1)$ *in addition, then the uniqueness for Eq.* (3.160) *holds.*

Remark 3.4.5 The uniqueness follows from the existence of solutions to the dual problem in the case of Eqs. (3.158) and (3.159). But, the uniqueness for the Eq. (3.160) follows from the a priori estimates, because we can not find a suitable dual problem for Eq. (3.160). Thus, in addition, we need the assumption that the inside of Ω has a finite covering.

Applying Theorems 3.4.2–3.4.4, we have the following corollary.

Corollary 3.4.6 *Let* $1 < p < \infty$, $1 < q \leq r$ *and* $2/p + 1/q \neq 1$. *Assume that* Ω *is a uniform* C^3 *domain and that the weak Dirichlet problem is uniquely solvable for* q. *Then, there exists a* γ_0 *for which the following assertion holds: Let* $\mathbf{u}_0 \in B_{q,p}^{2(1-1/p)}(\Omega)^N$ *and* $h_0 \in B_{q,p}^{3-1/p-1/q}(\Gamma)$ *be initial data for Eq.* (3.162) *and let* **f**, g, **g**, d, *and* **h** *be functions appearing in the right side of Eq.* (3.162) *and satisfying the conditions:*

$$\mathbf{f} \in L_p((0, T), L_q(\Omega)^N), \quad d \in L_p((0, T), W_q^{2-1/q}(\Gamma)),$$

$$e^{-\gamma t} g \in L_p(\mathbb{R}, H_q^1(\Omega)) \cap H_p^{1/2}(\mathbb{R}, L_q(\Omega)), \quad e^{-\gamma t} \mathbf{g} \in H_p^1(\mathbb{R}, L_q(\Omega)^N),$$

$$e^{-\gamma t} \mathbf{h} \in L_p(\mathbb{R}, H_q^1(\Omega)^N) \cap H_p^{1/2}(\mathbb{R}, L_q(\Omega)^N).$$

for any $\gamma \geq \gamma_0$. Assume that the compatibility conditions (3.170) and (3.171) are satisfied. Then, problem (3.162) admits solutions U P, and H with

$$U \in L_p((0, T), H_q^2(\Omega)^N) \cap H_p^1((0, T), L_q(\Omega)^N),$$

$$P \in L_p((0, T), H_q^1(\Omega) + H_{q,0}^1(\Omega)),$$

$$H \in L_p((0, T), W_q^{3-1/q}(\Gamma)) \cap H_p^1((0, T), W_q^{2-1/q}(\Gamma))$$

possessing the estimate:

$$\|U\|_{L_p((0,T),H_q^2(\Omega))} + \|\partial_t U\|_{L_p((0,T),L_q(\Omega))} + \|H\|_{L_p((0,T),W_q^{3-1/q}(\Gamma))}$$

$$+ \|\partial_t H\|_{L_p((0,T),W_q^{2-1/q}(\Gamma))}$$

$$\leq Ce^{2\gamma\kappa^{-b}T}\{\|\mathbf{u}_0\|_{B_{q,p}^{2(1-1/p)}(\Omega)} + \kappa^{-b}\|h_0\|_{B_{q,p}^{3-1/p-1/q}(\Gamma)}$$

$$+ \|\mathbf{f}\|_{L_p((0,T),L_q(\Omega))} + \|e^{-\gamma t}\partial_t g\|_{L_p(\mathbb{R},L_q(\Omega))} + \|e^{-\gamma t}(g, \mathbf{h})\|_{L_p(\mathbb{R},H_q^1(\Omega))}$$

$$+ \|e^{-\gamma t}(g, \mathbf{h})\|_{H_p^{1/2}(\mathbb{R},L_q(\Omega))} + \|d\|_{L_p((0,T),W_q^{2-1/q}(\Gamma))}\}$$

for any $\gamma \geq \gamma_0$ with some constant C independent of γ and κ, where $\kappa \in (0, 1)$, and b is the constant appearing in (3.164).

Proof As was mentioned after Eq. (3.162), U, P and H are given by $U = \mathbf{u}+\mathbf{v}+\mathbf{w}$, $P = \mathfrak{p}+\mathfrak{q}+\mathfrak{r}$ and $H = h+\rho$. Since $\mathbf{u}$ and $\mathfrak{p}$ are solutions of Eq. (3.158), by Theorem 3.4.2 we have

$$\|\mathbf{u}\|_{L_p((0,T),H_q^2(\Omega))} + \|\partial_t\mathbf{u}\|_{L_p((0,T),L_q(\Omega))}$$

$$\leq Ce^{\gamma T}\{\|\mathbf{u}_0\|_{B_{q,p}^{2(1-1/p)}(\Omega)} + \|\mathbf{f}\|_{L_p((0,T),L_q(\Omega))} + \|e^{-\gamma t}\partial_t g\|_{L_p(\mathbb{R},L_q(\Omega))}$$

$$+ \|e^{-\gamma t}(g, \mathbf{h})\|_{L_p(\mathbb{R},H_q^1(\Omega))} + \|e^{-\gamma t}(g, \mathbf{h})\|_{H_p^{1/2}(\mathbb{R},L_q(\Omega))}\}.$$

$$(3.174)$$

Since $\mathbf{v}$, $\mathfrak{q}$ and h are solutions of Eq. (3.159) with $d = 0$, applying Theorem 3.4.3 with $d = 0$ yields that

$$\|\mathbf{v}\|_{L_p((0,T),H_q^2(\Omega))} + \|\partial_t\mathbf{v}\|_{L_p((0,T),L_q(\Omega))} + \|h\|_{L_p((0,T),W_q^{3-1/q}(\Gamma))}$$

$$+ \|\partial_t h\|_{L_p((0,T),W_q^{2-1/q}(\Gamma))} \leq Ce^{\gamma T}\|h_0\|_{B_{q,p}^{3-1/p-1/q}(\Gamma)}.$$

$$(3.175)$$

Finally, recalling that $\mathbf{w}$, $\mathfrak{r}$ and ρ are solutions of Eq. (3.160) with d replaced by $d + \mathbf{n} \cdot \mathbf{u} - \mathcal{F}_1\mathbf{u} - A_\kappa \cdot \nabla_\Gamma' h$, applying Theorem 3.4.4 and using the estimate

$$\|A_\kappa \cdot \nabla_\Gamma' h\|_{W_q^{2-1/q}(\Gamma)} \leq C\|A_\kappa\|_{H_r^2(\Omega)}\|h\|_{W_q^{3-1/q}(\Gamma)},$$

which follows from the Sobolev imbedding theorem and the assumption: $N < r < \infty$, we have

$$\|\mathbf{w}\|_{L_p((0,T),H_q^2(\Omega))} + \|\partial_t \mathbf{w}\|_{L_p((0,T),L_q(\Omega))}$$

$$+ \|\rho\|_{L_p((0,T),W_q^{3-1/q}(\Gamma))} + \|\partial_t \rho\|_{L_p((0,T),W_q^{2-1/q}(\Gamma))}$$

$$\leq C e^{\gamma \kappa^{-b} T} \{ \|h_0\|_{B_{q,p}^{3-1/p-1/q}(\Gamma)}$$

$$+ \|d + \mathbf{n} \cdot \mathbf{u} - \mathcal{F}_1 \mathbf{u} - A_\kappa \cdot \nabla'_\Gamma h\|_{L_p((0,T),W_q^{2-1/q}(\Gamma))} \}$$

$$\leq C e^{\gamma \kappa^{-b} T} \{ \kappa^{-b} \|h\|_{L_p((0,T),W_q^{3-1/q}(\Gamma))} + \|d\|_{L_p((0,T),W_q^{2-1/q}(\Gamma))}$$

$$+ \|\mathbf{u}\|_{L_p((0,T),H_q^2(\Omega))} \},$$

which, combined with (3.174) and (3.175), leads to the required estimate. This completes the proof of Corollary 3.4.6.   □

3.4.2 $\mathcal{R}$ Bounded Solution Operators

To prove Theorems 3.4.2–3.4.4, we use $\mathcal{R}$ bounded solution operators associated with the following generalized resolvent problems:

$$\begin{cases} \lambda \mathbf{u} - \mathrm{Div}\,(\mu \mathbf{D}(\mathbf{u}) - \mathfrak{p}\mathbf{I}) = \mathbf{f}, & \mathrm{div}\,\mathbf{u} = g = \mathrm{div}\,\mathbf{g} & \text{in } \Omega, \\ & (\mu \mathbf{D}(\mathbf{u}) - \mathfrak{p}\mathbf{I})\mathbf{n} = \mathbf{h} & \text{on } \Gamma; \end{cases} \tag{3.176}$$

$$\begin{cases} \lambda \mathbf{v} - \mathrm{Div}\,(\mu \mathbf{D}(\mathbf{v}) - \mathfrak{q}\mathbf{I}) = \mathbf{f}, & \mathrm{div}\,\mathbf{v} = g = \mathrm{div}\,\mathbf{g} & \text{in } \Omega, \\ \lambda \rho - \mathbf{v} \cdot \mathbf{n} + \mathcal{F}_1 \mathbf{v} = d & \text{on } \Gamma, \\ (\mu \mathbf{D}(\mathbf{v}) - \mathfrak{q}\mathbf{I})\mathbf{n} - (\mathcal{F}_2 \rho + \sigma \Delta_\Gamma \rho)\mathbf{n} = \mathbf{h} & \text{on } \Gamma; \end{cases} \tag{3.177}$$

$$\begin{cases} \lambda \mathbf{v} - \mathrm{Div}\,(\mu \mathbf{D}(\mathbf{v}) - \mathfrak{q}\mathbf{I}) = \mathbf{f}, & \mathrm{div}\,\mathbf{v} = g = \mathrm{div}\,\mathbf{g} & \text{in } \Omega, \\ \lambda \rho + A_\kappa \cdot \nabla'_\Gamma \rho - \mathbf{v} \cdot \mathbf{n} + \mathcal{F}_1 \mathbf{v} = d & \text{on } \Gamma, \\ (\mu \mathbf{D}(\mathbf{v}) - \mathfrak{q}\mathbf{I})\mathbf{n} - (\mathcal{F}_2 \rho + \sigma \Delta_\Gamma \rho)\mathbf{n} = \mathbf{h} & \text{on } \Gamma. \end{cases} \tag{3.178}$$

In the following, we consider Eq. (3.177) and (3.178) at the same time. For this, we set $A_0 = 0$, and then Eq. (3.177) is represented by Eq. (3.178) with $\kappa = 0$.

We make a definition.

Definition 3.4.7 Let X and Y be two Banach spaces. A family of operators $\mathcal{T} \subset \mathcal{L}(X, Y)$ is called $\mathcal{R}$ bounded on $\mathcal{L}(X, Y)$, if there exist constants $C > 0$ and $p \in [1, \infty)$ such that for each $n \in \mathbb{N}$, $\{T_j\}_{j=1}^n \subset \mathcal{T}$, and $\{f_j\}_{j=1}^n \subset X$, we have

$$\| \sum_{k=1}^n r_k T_k f_k \|_{L_p((0,1),Y)} \leq C \| \sum_{k=1}^n r_k f_k \|_{L_p((0,1),X)}.$$

Here, the Rademacher functions r_k, $k \in \mathbb{N}$, are given by $r_k : [0, 1] \to \{-1, 1\}$ $t \mapsto \mathrm{sign}\,(\sin 2^k \pi t)$. The smallest such C is called $\mathcal{R}$-bound of $\mathcal{T}$ on $\mathcal{L}(X, Y)$, which is denoted by $\mathcal{R}_{\mathcal{L}(X,Y)} \mathcal{T}$.

We introduce the definition of the uniqueness of solutions. For Eq. (3.176), the uniqueness is defined as follows:

- Let $\lambda \in U \subset \mathbb{C}$. If $\mathbf{u}$ and $\mathfrak{p}$ with

$$\mathbf{u} \in H_q^2(\Omega)^N, \quad \mathfrak{p} \in H_q^1(\Omega) + \hat{H}_{q,0}^1(\Omega)$$

satisfy the homogeneous equations:

$$\begin{cases} \lambda \mathbf{u} - \mathrm{Div}\,(\mu \mathbf{D}(\mathbf{u}) - \mathfrak{p}\mathbf{I}) = 0, & \mathrm{div}\,\mathbf{u} = 0 \quad \text{in } \Omega, \\ (\mu \mathbf{D}(\mathbf{u}) - \mathfrak{p}\mathbf{I})\mathbf{n} = 0 & \text{on } \Gamma, \end{cases} \tag{3.179}$$

then $\mathbf{u} = 0$ and $\mathfrak{p} = 0$.

For Eq. (3.177) and (3.178), the uniqueness is defined as follows:

- Let $\lambda \in U \subset \mathbb{C}$. If $\mathbf{v}$, $\mathfrak{q}$ and ρ with

$$\mathbf{v} \in H_q^2(\Omega)^N, \quad \mathfrak{q} \in H_q^1(\Omega) + \hat{H}_{q,0}^1(\Omega), \quad \rho \in W_q^{3-1/q}(\Gamma)$$

satisfy the homogeneous equations:

$$\begin{cases} \lambda \mathbf{v} - \mathrm{Div}\,(\mu \mathbf{D}(\mathbf{v}) - \mathfrak{q}\mathbf{I}) = 0, & \mathrm{div}\,\mathbf{v} = 0 \quad \text{in } \Omega, \\ \lambda \rho + A_\kappa \cdot \nabla_\Gamma' \rho - \mathbf{v} \cdot \mathbf{n} + \mathcal{F}_1 \mathbf{v} = 0 & \text{on } \Gamma, \\ (\mu \mathbf{D}(\mathbf{v}) - \mathfrak{q}\mathbf{I} - ((\mathcal{F}_2 + \sigma \Delta_\Gamma)\rho)\mathbf{I})\mathbf{n} = 0 & \text{on } \Gamma, \end{cases} \tag{3.180}$$

then $\mathbf{u} = 0$, $\mathfrak{q} = 0$ and $\rho = 0$.

We have the following theorems.

Theorem 3.4.8 *Let* $1 < q < \infty$ *and* $0 < \epsilon_0 < \pi/2$. *Assume that* Ω *is a uniform* C^3
domain and that the weak Dirichlet problem is uniquely solvable for q. *Let* $A_0 = 0$
and let A_κ ($\kappa \in (0, 1)$) *be an* $N - 1$ *vector of real valued functions satisfying*
(3.164). Assume that $1 < q \leq r$ *when* $\kappa \in (0, 1)$.

(1) *(Existence) Let*

$$X_q(\Omega) = \{\mathbf{F} = (\mathbf{f}, d, \mathbf{h}, g, \mathbf{g}) \mid \mathbf{f} \in L_q(\Omega)^N, \quad (g, \mathbf{g}) \in DI_q(\Omega),$$

$$\mathbf{h} \in H_q^1(\Omega)^N, \quad d \in W_q^{2-1/q}(\Gamma)\},$$

$$\mathcal{X}_q(\Omega) = \{F = (F_1, F_2, \ldots, F_7) \mid F_1, F_3, F_7 \in L_q(\Omega)^N, \quad F_4 \in H_q^1(\Omega)^N,$$

$$F_5 \in L_q(\Omega), \quad F_6 \in H_q^1(\Omega), \quad F_2 \in W_q^{2-1/q}(\Gamma)\},$$

$$\Lambda_{\kappa,\lambda_0} = \begin{cases} \Sigma_{\epsilon_0,\lambda_0} & \text{for } \kappa = 0, \\ \mathbb{C}_{+,\lambda_0} & \text{for } \kappa \in (0, 1), \end{cases} \qquad \gamma_\kappa = \begin{cases} 1 & \text{for } \kappa = 0, \\ \kappa^{-b} & \text{for } \kappa \in (0, 1). \end{cases}$$

Then, there exist a constant $\lambda_0 > 0$ *and operator families* $\mathcal{A}_1(\lambda)$, $\mathcal{P}_1(\lambda)$ *and*
$\mathcal{H}_1(\lambda)$ *with*

$$\mathcal{A}_1(\lambda) \in Hol\,(\Lambda_{\kappa,\lambda_0\gamma_\kappa}, \mathcal{L}(\mathcal{X}_q(\Omega), H_q^2(\Omega)^N)),$$

$$\mathcal{P}_1(\lambda) \in Hol\,(\Lambda_{\kappa,\lambda_0\gamma_\kappa}, \mathcal{L}(\mathcal{X}_q(\Omega), H_q^1(\Omega) + \hat{H}_{q,0}^1(\Omega))),$$

$$\mathcal{H}_1(\lambda) \in Hol\,(\Lambda_{\kappa,\lambda_0\gamma_\kappa}, \mathcal{L}(\mathcal{X}_q(\Omega), H_q^3(\Omega)))$$

such that for every $\lambda = \gamma + i\tau \in \Lambda_{\kappa,\lambda_0\gamma_\kappa}$ *and* $(\mathbf{f}, d, \mathbf{h}, g, \mathbf{g}) \in X_q(\Omega)$, $\mathbf{v} = \mathcal{A}_1 F_\lambda$ *and* $\mathfrak{q} = \mathcal{P}_1(\lambda)F_\lambda$, *and* $\rho = \mathcal{H}_1(\lambda)F_\lambda$ *are solutions of Eq. (3.178), where*

$$F_\lambda = (\mathbf{f}, d, \lambda^{1/2}\mathbf{h}, \mathbf{h}, \lambda^{1/2}g, g, \lambda\mathbf{g}).$$

Moreover, we have

$$\mathcal{R}_{\mathcal{L}(\mathcal{X}_q(\Omega), H_q^{2-j}(\Omega)^N)}(\{(\tau\partial_\tau)^\ell(\lambda^{j/2}\mathcal{A}_1(\lambda)) \mid \lambda \in \Lambda_{\kappa,\lambda_0\gamma_\kappa}\}) \leq r_b;$$

$$\mathcal{R}_{\mathcal{L}(\mathcal{X}_q(\Omega), L_q(\Omega)^N)}(\{(\tau\partial_\tau)^\ell(\nabla\mathcal{P}_1(\lambda)) \mid \lambda \in \Lambda_{\kappa,\lambda_0\gamma_\kappa}\}) \leq r_b;$$

$$\mathcal{R}_{\mathcal{L}(\mathcal{X}_q(\Omega), H_q^{3-k}(\Omega))}(\{(\tau\partial_\tau)^\ell(\lambda^k\mathcal{H}_1(\lambda)) \mid \lambda \in \Lambda_{\kappa,\lambda_0\gamma_\kappa}\}) \leq r_b$$

for $\ell = 0, 1$, $j = 0, 1, 2$, *and* $k = 0, 1$ *with some constant* $r_b > 0$.

(2) *(Uniqueness)*

(i) *When* $\kappa = 0$, *if we assume that the weak Dirichlet problem is uniquely*
solvable for $q' = q/(q-1)$ *in addition, then the uniqueness for Eq. (3.177)*
holds.

(ii) *When $\kappa \in (0, 1)$, if we assume that Ω is a uniformly C^3 domain whose inside has a finite covering and that the weak Dirichlet problem is uniquely solvable for $q' = q/(q-1)$ in addition, then the uniqueness for Eq. (3.178) holds.*

Remark 3.4.9

(1) F_1, F_2, F_3, F_4, F_5, F_6, and F_7 are variables corresponding to $\mathbf{f}$, d, $\lambda^{1/2}\mathbf{h}$, $\mathbf{h}$, $\lambda^{1/2}g$, g, and λg, respectively.

(2) We define the norms $\| \cdot \|_{X_q(\Omega)}$ and $\| \cdot \|_{\mathcal{X}_q(\Omega)}$ by

$$\|(\mathbf{f}, d, \mathbf{h}, g, \mathbf{g})\|_{X_q(\Omega)} = \|(\mathbf{f}, \mathbf{g})\|_{L_q(\Omega)} + \|(g, \mathbf{h})\|_{H_q^1(\Omega)} + \|d\|_{W_q^{2-1/q}(\Gamma)};$$

$$\|(F_1, \ldots, F_7)\|_{\mathcal{X}_q(\Omega)} = \|(F_1, F_3, F_7)\|_{L_q(\Omega)} + \|(F_4, F_6)\|_{H_q^1(\Omega)}$$

$$+ \|F_2\|_{W_q^{2-1/q}(\Gamma)}.$$

Theorem 3.4.10 *Let $1 < q < \infty$ and $0 < \epsilon_0 < \pi/2$.*

(1) *(Existence) Assume that Ω is a uniform C^2 domain and that the weak Dirichlet problem is uniquely solvable for q. Let*

$$\tilde{X}_q(\Omega) = \{\tilde{\mathbf{F}} = (\mathbf{f}, \mathbf{h}, g, \mathbf{g}) \mid \mathbf{f} \subset L_q(\Omega)^N, \ (g, \mathbf{g}) \in DI_q(\Omega), \ \mathbf{h} \in H_q^1(\Omega)^N\},$$

$$\tilde{\mathcal{X}}_q(\Omega) = \{\tilde{F} = (F_1, F_3, \ldots, F_7) \mid F_1, F_3, F_7 \in L_q(\Omega)^N, \ F_5 \in L_q(\Omega),$$

$$F_4 \in H_q^1(\Omega)^N, \ F_6 \in H_q^1(\Omega)\}.$$

Then, there exist a constant $\lambda_0 > 0$ and operator families $\mathcal{A}_2(\lambda)$ and $\mathcal{P}_2(\lambda)$ with

$$\mathcal{A}_2(\lambda) \in Hol\left(\Sigma_{\epsilon_0, \lambda_0}, \mathcal{L}(\tilde{\mathcal{X}}_q(\Omega), H_q^2(\Omega)^N)\right),$$

$$\mathcal{P}_2(\lambda) \in Hol\left(\Sigma_{\epsilon_0, \lambda_0}, \mathcal{L}(\tilde{\mathcal{X}}_q(\Omega), H_q^1(\Omega) + \hat{H}_{q,0}^1(\Omega))\right)$$

such that for every $\lambda \in \Sigma_{\epsilon_0, \lambda_0}$ and $(\mathbf{f}, \mathbf{h}, g, \mathbf{g}) \in \tilde{X}_q(\Omega)$, $\mathbf{u} = \mathcal{A}_2(\lambda)\tilde{\mathbf{F}}_\lambda$ and $\mathfrak{p} = \mathcal{P}_2(\lambda)\tilde{\mathbf{F}}_\lambda$ are solutions of Eq. (3.176), where we have set

$$\tilde{\mathbf{F}}_\lambda = (\mathbf{f}, \lambda^{1/2}\mathbf{h}, \mathbf{h}, g, \lambda^{1/2}g, \lambda g).$$

Moreover, we have

$$\mathcal{R}_{\mathcal{L}(\tilde{\mathcal{X}}_q(\Omega), H_q^{2-j}(\Omega)^N)}(\{(\tau \partial_\tau)^\ell (\lambda^{j/2} \mathcal{A}_2(\lambda)) \mid \lambda \in \Sigma_{\epsilon_0, \lambda_0}\}) \leq r_b;$$

$$\mathcal{R}_{\mathcal{L}(\tilde{\mathcal{X}}_q(\Omega), L_q(\Omega)^N)}(\{(\tau \partial_\tau)^\ell (\nabla \mathcal{P}_2(\lambda)) \mid \lambda \in \Sigma_{\epsilon_0, \lambda_0}\}) \leq r_b$$

for $\ell = 0, 1$, $j = 0, 1, 2$ with some constant $r_b > 0$.

(Uniqueness) *If we assume that the weak Dirichlet problem is uniquely solvable for $q' = q/(q-1)$ in addition, then the uniqueness for Eq. (3.176) holds.*

3.4.3 Stokes Operator and Reduced Stokes Operator

Since the pressure term has no time evolution, sometimes it is convenient to eliminate the pressure terms, for example to formulate the problem in the semigroup setting. For $\mathbf{u} \in H_q^2(\Omega)^N$ and $h \in H_q^3(\Omega)$, we introduce functionals $K_0(\mathbf{u})$ and $K(\mathbf{u}, h)$. Let $K_0(\mathbf{u}) \in H_q^1(\Omega) + \hat{H}_{q,0}^1(\Omega)$ be a unique solution of the weak Dirichlet problems:

$$(\nabla K_0(\mathbf{u}), \nabla\varphi)_\Omega = (\text{Div}\,(\mu\mathbf{D}(\mathbf{u})) - \nabla\text{div}\,\mathbf{u}, \nabla\varphi)_\Omega \tag{3.181}$$

for any $\varphi \in \hat{H}_{q',0}^1(\Omega)$, subject to

$$K_0(\mathbf{u}) = \mu < \mathbf{D}(\mathbf{u})\mathbf{n}, \mathbf{n} > -\text{div}\,\mathbf{u} \quad \text{on } \Gamma.$$

And, let $K(\mathbf{u}, h) \in H_q^1(\Omega) + \hat{H}_{q,0}^1(\Omega)$ be a unique solution of the weak Dirichlet problem:

$$(\nabla K(\mathbf{u}, h), \nabla\varphi)_\Omega = (\text{Div}\,(\mu\mathbf{D}(\mathbf{u})) - \nabla\text{div}\,\mathbf{u}, \nabla\varphi)_\Omega \tag{3.182}$$

for any $\varphi \in \hat{H}_{q',0}^1(\Omega)$, subject to

$$K(\mathbf{u}, h) = \mu < \mathbf{D}(\mathbf{u})\mathbf{n}, \mathbf{n} > -(\mathcal{F}_2 h + \sigma\Delta_\Gamma h) - \text{div}\,\mathbf{u} \quad \text{on } \Gamma.$$

By Remark 3.2.6, we know the unique existence of $K_0(\mathbf{u})$ and $K(\mathbf{u}, h)$ satisfying the estimates:

$$\begin{aligned}
\|\nabla K_0(\mathbf{u})\|_{L_q(\Omega)} &\leq C\|\nabla\mathbf{u}\|_{H_q^1(\Omega)}, \\
\|\nabla K(\mathbf{u}, h)\|_{L_q(\Omega)} &\leq C(\|\nabla\mathbf{u}\|_{H_q^1(\Omega)} + \|h\|_{W_q^{3-1/q}(\Gamma)})
\end{aligned} \tag{3.183}$$

for some constant C depending on q. We consider the reduced Stokes equations:

$$\begin{cases}
\lambda\mathbf{u} - \text{Div}\,(\mu\mathbf{D}(\mathbf{u}) - K_0(\mathbf{u})\mathbf{I}) = \mathbf{f} & \text{in } \Omega, \\
(\mu\mathbf{D}(\mathbf{u}) - K_0(\mathbf{u})\mathbf{I})\mathbf{n} = \mathbf{h} & \text{on } \Gamma;
\end{cases} \tag{3.184}$$

$$\begin{cases}
\lambda\mathbf{u} - \text{Div}\,(\mu\mathbf{D}(\mathbf{u}, h) - K(\mathbf{u}, h)\mathbf{I}) = \mathbf{f} & \text{in } \Omega, \\
\lambda h + A_\kappa \cdot \nabla'_\Gamma h - \mathbf{n} \cdot \mathbf{u} + \mathcal{F}_1\mathbf{u} = d & \text{on } \Gamma, \\
(\mu\mathbf{D}(\mathbf{u}) - K(\mathbf{u}, h)\mathbf{I})\mathbf{n} - (\mathcal{F}_2 h + \sigma\Delta_\Gamma h)\mathbf{n} = \mathbf{h} & \text{on } \Gamma.
\end{cases} \tag{3.185}$$

Notice that both of the boundary conditions in Eq. (3.184) and (3.185) are equivalent to

$$(\mu \mathbf{D}(\mathbf{u})\mathbf{n})_\tau = \mathbf{h}_\tau \quad \text{and} \quad \operatorname{div} \mathbf{u} = \mathbf{n} \cdot \mathbf{h} \quad \text{on } \Gamma. \tag{3.186}$$

We now study the equivalence between Eqs. (3.176) and (3.184). The equivalence between Eqs. (3.178) and (3.185) are similarly studied. We first assume that Eq. (3.176) is uniquely solvable. Given $\mathbf{f} \in L_q(\Omega)^N$ in the right side of Eq. (3.184), let $g \in H_q^1(\Omega)$ be a unique solution of the variational equation:

$$\lambda(g, \varphi)_\Omega + (\nabla g, \nabla \varphi)_\Omega = (-\mathbf{f}, \nabla \varphi)_\Omega \quad \text{for any } \varphi \in H_{q',0}^1(\Omega) \tag{3.187}$$

subject to $g = \mathbf{n} \cdot \mathbf{h}$ on Γ. The unique existence of g is guaranteed for $\lambda \in \Sigma_{\epsilon_0, \lambda_0}$ with some large $\lambda_0 > 0$. From (3.187) it follows that

$$(g, \varphi)_\Omega = (-\lambda^{-1}(\mathbf{f} + \nabla g), \nabla \varphi)_\Omega, \tag{3.188}$$

and so $(g, \mathbf{g}) \in DI_q$ with $\mathbf{g} = \lambda^{-1}(\mathbf{f} + \nabla g)$. Thus, from the assumption we know that Eq. (3.176) admits unique solutions $\mathbf{u} \in H_q^2(\Omega)^N$ and $\mathfrak{p} \in H_q^1(\Omega) + \hat{H}_{q,0}^1(\Omega)$. In view of the first equation in Eq. (3.176) and Definition 3.4.1, for any $\varphi \in \hat{H}_{q',0}^1(\Omega)$ we have

$$\begin{aligned}
(\mathbf{f}, \nabla \varphi)_\Omega &= (\lambda \mathbf{u} - \operatorname{Div}(\mu \mathbf{D}(\mathbf{u}) - \mathfrak{p}\mathbf{I}), \nabla \varphi)_\Omega \\
&= (\lambda \mathbf{u}, \nabla \varphi)_\Omega - (\nabla \operatorname{div} \mathbf{u}, \nabla \varphi)_\Omega \\
&\quad - (\operatorname{Div}(\mu \mathbf{D}(\mathbf{u})) - \nabla \operatorname{div} \mathbf{u}, \nabla \varphi)_\Omega + (\nabla \mathfrak{p}, \nabla \varphi)_\Omega \\
&= \lambda(\lambda^{-1}(\mathbf{f} + \nabla g), \nabla \varphi)_\Omega - (\nabla g, \nabla \varphi)_\Omega + (\nabla(\mathfrak{p} - K_0(\mathbf{u})), \nabla \varphi)_\Omega,
\end{aligned}$$

and so,

$$(\nabla(\mathfrak{p} - K_0(\mathbf{u})), \nabla \varphi)_\Omega = 0 \quad \text{for any } \varphi \in \hat{H}_{q',0}^1(\Omega).$$

Moreover, by the second equation in Eq. (3.176) and (3.187), we have

$$\mathfrak{p} - K_0(\mathbf{u}) = -\mathbf{h} \cdot \mathbf{n} + \operatorname{div} \mathbf{u} = -g + g = 0 \quad \text{on } \Gamma.$$

Thus, the uniqueness implies that $\mathfrak{p} = K_0(\mathbf{u})$, which, combined with Eq. (3.176), shows that $\mathbf{u}$ satisfies Eq. (3.184).

Conversely, we assume that Eq. (3.184) is uniquely solvable. Let $\mathbf{f} \in L_q(\Omega)^N$ and $\mathbf{h} \in H_q^1(\Omega)$. Let $\theta \in H_q^1(\Omega) + \hat{H}_q^1(\Omega)$ be a unique solution of the weak Dirichlet problem:

$$(\nabla\theta, \nabla\varphi)_\Omega = (\mathbf{f}, \nabla\varphi)_\Omega \quad \text{for any } \varphi \in \hat{H}_{q',0}^1(\Omega),$$

subject to $\theta = \mathbf{n} \cdot \mathbf{h}$ on Γ. Setting $\mathsf{p} = \theta + \mathsf{q}$ in (3.176), we then have

$$\lambda\mathbf{u} - \mathrm{Div}\,(\mu\mathbf{D}(\mathbf{u}) - \mathsf{q}\mathbf{I}) = \mathbf{f} - \nabla\theta, \quad \mathrm{div}\,\mathbf{u} = g = \mathrm{div}\,\mathbf{g} \quad \text{in } \Omega,$$

$$(\mu\mathbf{D}(\mathbf{u}) - \mathsf{q}\mathbf{I})\mathbf{n} = \mathbf{h} - <\mathbf{h}, \mathbf{n}>\mathbf{n} \quad \text{on } \Gamma.$$

Let $\mathbf{f}' = \mathbf{f} - \nabla\theta$ and $\mathbf{h}' = \mathbf{h} - <\mathbf{h}, \mathbf{n}>\mathbf{n}$, and then

$$\mathbf{h}' \cdot \mathbf{n} = 0 \quad \text{on } \Gamma \quad \text{and} \quad (\mathbf{f}', \nabla\varphi)_\Omega = 0 \quad \text{for any } \varphi \in \hat{H}^1_{q',0}(\Omega). \tag{3.189}$$

Given $(g, \mathbf{g}) \in DI_q(\Omega)^N$, let $K \in H^1_q(\Omega) + \hat{H}^1_{q,0}(\Omega)$ be a solution to the weak Dirichlet problem:

$$(\nabla K, \nabla\varphi)_\Omega = (\lambda\mathbf{g} - \nabla g, \nabla\varphi)_\Omega \quad \text{for any } \varphi \in \hat{H}^1_{q',0}(\Omega), \tag{3.190}$$

subject to $K = -g$ on Γ. Let $\mathbf{u}$ be a solution of the equations:

$$\begin{cases} \lambda\mathbf{u} - \mathrm{Div}\,(\mu\mathbf{D}(\mathbf{u}) - K_0(\mathbf{u})\mathbf{I}) = \mathbf{f}' + \nabla K & \text{in } \Omega, \\ (\mu\mathbf{D}(\mathbf{u}) - K_0(\mathbf{u})\mathbf{I})\mathbf{n} = \mathbf{h}' + g\mathbf{n} & \text{on } \Gamma. \end{cases} \tag{3.191}$$

By (3.181), and (3.189)–(3.191), for any $\varphi \in \hat{H}^1_{q,0}(\Omega)$ we have

$$\begin{aligned} (\nabla K, \nabla\varphi)_\Omega &= (\mathbf{f}' + \nabla K, \nabla\varphi)_\Omega \\ &= (\lambda\mathbf{u} - \mathrm{Div}\,(\mu\mathbf{D}(\mathbf{u}) - K_0(\mathbf{u})\mathbf{I}), \nabla\varphi)_\Omega \\ &= (\lambda\mathbf{u}, \nabla\varphi)_\Omega - (\nabla\mathrm{div}\,\mathbf{u}, \nabla\varphi)_\Omega. \end{aligned} \tag{3.192}$$

Since $H^1_{q',0}(\Omega) \subset \hat{H}^1_{q',0}(\Omega)$, by (3.190) and (3.192) we have

$$\lambda(\mathrm{div}\,\mathbf{u} - g, \varphi)_\Omega + (\nabla(\mathrm{div}\,\mathbf{u} - g), \nabla\varphi)_\Omega = 0 \quad \text{for any } \varphi \in H^1_{q',0}(\Omega).$$

Here, we have used the fact that $\mathrm{div}\,\mathbf{g} = g \in H^1_q(\Omega)$. Recalling that $\mathbf{h}' \cdot \mathbf{n} = 0$ and putting (3.191) and the boundary condition of (3.181) together gives

$$g = (g\mathbf{n} + \mathbf{h}') \cdot \mathbf{n} = <\mu\mathbf{D}(\mathbf{u})\mathbf{n}, \mathbf{n}> - K_0(\mathbf{u}) = \mathrm{div}\,\mathbf{u}$$

on Γ. Thus, the uniqueness implies that $\mathrm{div}\,\mathbf{u} = g$ in $H^1_q(\Omega)$, which, combined with (3.190) and (3.192), leads to

$$(g, \nabla\varphi)_\Omega = (\mathbf{u}, \nabla\varphi)_\Omega \quad \text{for any } \varphi \in \hat{H}^1_{q',0}(\Omega),$$

because we may assume that $\lambda \neq 0$. Since $K = -g$ on Γ, by (3.192) we have

$$\lambda \mathbf{u} - \mathrm{Div}\,(\mu \mathbf{D}(\mathbf{u}) - (K_0(\mathbf{u}) - K)\mathbf{I}) = \mathbf{f}', \quad \mathrm{div}\,\mathbf{u} = g = \mathrm{div}\,\mathbf{g} \quad \text{in } \Omega,$$

$$(\mu \mathbf{D}(\mathbf{u}) - (K_0(\mathbf{u}) - K)\mathbf{I})\mathbf{n} = \mathbf{h}' \quad \text{on } \Gamma.$$

Thus, $\mathbf{u}$ and $\mathfrak{p} = K_0(\mathbf{u}) - K - \theta$ are required solutions of Eq. (3.176).

3.4.4 $\mathcal{R}$-Bounded Solution Operators for the Reduced Stokes Equations

In the following theorem, we state the existence of $\mathcal{R}$ bounded solution operators for the reduced Stokes equations (3.185).

Theorem 3.4.11 *Let $1 < q < \infty$ and $0 < \epsilon < \pi/2$. Let $\Lambda_{\kappa,\lambda_0}$ be the set defined in Theorem 3.4.8. Assume that Ω is a uniform C^3 domain and the weak Dirichlet problem is uniquely solvable for q. Let $A_0 = 0$ and A_κ ($\kappa \in (0, 1)$) be an $N - 1$ vector of real valued functions satisfying (3.164). Assume that $1 < q \leq r$ when $\kappa \in (0, 1)$. Set*

$$Y_q(\Omega) = \{(\mathbf{f}, d, \mathbf{h}) \mid \mathbf{f} \in L_q(\Omega)^N, \quad d \in W_q^{2-1/q}(\Gamma), \quad \mathbf{h} \in H_q^1(\Omega)^N\},$$

$$\mathcal{Y}_q(\Omega) = \{(F_1, \ldots, F_4) \mid F_1, F_3 \in L_q(\Omega)^N, \quad F_2 \in W_q^{2-1/q}(\Gamma)$$

$$F_4 \in H_q^1(\Omega)^N\}. \tag{3.193}$$

Then, there exist a constant $\lambda_* \geq 1$ and operator families:

$$\mathcal{A}_r(\lambda) \in Hol\,(\Lambda_{\kappa,\lambda_*\gamma_\kappa}, \mathcal{L}(\mathcal{Y}_q(\Omega), H_q^2(\Omega)^N)),$$

$$\mathcal{H}_r(\lambda) \in Hol\,(\Lambda_{\kappa,\lambda_*\gamma_\kappa}, \mathcal{L}(\mathcal{Y}_q(\Omega), H_q^3(\Omega)))$$

such that for any $\lambda \in \Lambda_{\kappa,\lambda_*\gamma_\kappa}$ and $(\mathbf{f}, d, \mathbf{h}) \in Y_q(\Omega)$,

$$\mathbf{u} = \mathcal{A}_r(\lambda)(\mathbf{f}, d, \lambda^{1/2}\mathbf{h}, \mathbf{h}), \quad h = \mathcal{H}_r(\lambda)(\mathbf{f}, d, \lambda^{1/2}\mathbf{h}, \mathbf{h}),$$

are solutions of (3.185), and

$$\mathcal{R}_{\mathcal{L}(\mathcal{Y}_q(\Omega), H_q^{2-j}(\Omega)^N)}(\{(\tau \partial_\tau)^\ell (\lambda^{j/2}\mathcal{A}_r(\lambda)) \mid \lambda \in \Lambda_{\kappa,\lambda_*\gamma_\kappa}\}) \leq r_b,$$

$$\mathcal{R}_{\mathcal{L}(\mathcal{Y}_q(\Omega), H_q^{3-k}(\Omega))}(\{(\tau \partial_\tau)^\ell (\lambda^k \mathcal{H}_r(\lambda)) \mid \lambda \in \Lambda_{\kappa,\lambda_*\gamma_\kappa}\}) \leq r_b$$

for $\ell = 0, 1$, $j = 0, 1, 2$ and $k = 0, 1$. Here, r_b is a constant depending on m_1, m_2, m_3, λ_*, p, q, and N, but independent of $\kappa \in (0, 1)$, and γ_κ is the number defined in Theorem 3.4.8.

If we assume that Ω is a uniformly C^3 domain whose inside has a finite covering and that the weak Dirichlet problem is uniquely solvable for $q' = q/(q - 1)$ in addition, then the uniqueness for Eq. (3.185) holds.

Remark 3.4.12 The norm of space $Y_q(\Omega)$ is defined by

$$\|(\mathbf{f}, d, \mathbf{h})\|_{Y_q(\Omega)} = \|\mathbf{f}\|_{L_q(\Omega)} + \|d\|_{W_q^{2-1/q}(\Gamma)} + \|\mathbf{h}\|_{H_q^1(\Omega)};$$

and the norm of space $\mathcal{Y}_q(\Omega)$ is defined by

$$\|(F_1, F_2, F_3, F_4)\|_{\mathcal{Y}_q(\Omega)} = \|(F_1, F_3)\|_{L_q(\Omega)} + \|F_2\|_{W_q^{2-1/q}(\Gamma)} + \|F_4\|_{W_q^1(\Omega)}.$$

Remark 3.4.13 By the equivalence of Eqs. (3.178) and (3.185) that was pointed out in Sect. 3.4.3, we see easily that Theorem 3.4.8 follows immediately from Theorem 3.4.11.

Concerning the existence of $\mathcal{R}$ bounded solution operator for Eq. (3.184), we have the following theorem.

Theorem 3.4.14 Let $1 < q < \infty$ and $0 < \epsilon < \pi/2$. Assume that Ω is a uniform C^2 domain and the weak Dirichlet problem is uniquely solvable for q. Set

$$\tilde{Y}_q(\Omega) = \{(\mathbf{f}, \mathbf{h}) \mid \mathbf{f} \in L_q(\Omega)^N, \ \mathbf{h} \in H_q^1(\Omega)^N\},$$

$$\tilde{\mathcal{Y}}_q(\Omega) = \{(F_1, F_3, F_4) \mid F_1, F_3 \in L_q(\Omega)^N, \ F_4 \in H_q^1(\Omega)^N\}. \tag{3.194}$$

Then, there exist a constant $\lambda_{**} \geq 1$ and operator family $\mathcal{A}_r^0(\lambda)$ with

$$\mathcal{A}_r^0(\lambda) \in Hol\,(\Sigma_{\sigma,\lambda_{**}}, \mathcal{L}(\tilde{\mathcal{Y}}_q(\Omega), H_q^2(\Omega)^N))$$

such that for any $\lambda \in \Sigma_{\sigma,\lambda_{**}}$ and $(\mathbf{f}, \mathbf{h}) \in \tilde{Y}_q(\Omega)$, $\mathbf{u} = \mathcal{A}_r^0(\lambda)(\mathbf{f}, \lambda^{1/2}\mathbf{h}, \mathbf{h})$ is a solution of (3.184), and

$$\mathcal{R}_{\mathcal{L}(\tilde{\mathcal{Y}}_q(\Omega), H_q^{2-j}(\Omega)^N)}(\{(\tau\partial_\tau)^\ell(\lambda^{j/2}\mathcal{A}_r^0(\lambda)) \mid \lambda \in \Sigma_{\sigma,\lambda_{**}}\}) \leq r_b$$

for $\ell = 0, 1$ and $j = 0, 1, 2$. Here, r_b is a constant depending on m_1, λ_0, p, q, and N.

If we assume that the weak Dirichlet problem is uniquely solvable for $q' = q/(q - 1)$ in addition, then the uniqueness for Eq. (3.184) holds.

Remark 3.4.15 Theorem 3.4.14 was proved by Shibata [43] and can be proved by the same argument as in the proof of Theorem 3.4.11. Thus, we may omit its proof.

3.4.5 Generation of C^0 Analytic Semigroup

In this subsection, we consider the following two initial-boundary value problems for the reduced Stokes operator:

$$\begin{cases} \partial_t \mathbf{u} - \text{Div}\,(\mu \mathbf{D}(\mathbf{u}) - K_0(\mathbf{u})\mathbf{I}) = 0 & \text{in } \Omega^\infty, \\ (\mu \mathbf{D}(\mathbf{u}) - K_0(\mathbf{u})\mathbf{I})\mathbf{n} = 0 & \text{on } \Gamma^\infty, \\ \mathbf{u}|_{t=0} = \mathbf{u}_0 & \text{in } \Omega; \end{cases} \tag{3.195}$$

$$\begin{cases} \partial_t \mathbf{v} - \text{Div}\,(\mu \mathbf{D}(\mathbf{v}) - K(\mathbf{v}, h)\mathbf{I}) = 0 & \text{in } \Omega^\infty, \\ \partial_t h - \mathbf{n} \cdot \mathbf{v} + \mathcal{F}_1 \mathbf{v} = 0 & \text{on } \Gamma^\infty, \\ (\mu \mathbf{D}(\mathbf{v}) - K(\mathbf{v}, h)\mathbf{I})\mathbf{n} - (\mathcal{F}_2 h + \sigma \Delta_\Gamma h)\mathbf{n} = 0 & \text{on } \Gamma^\infty, \\ \mathbf{v}|_{t=0} = \mathbf{v}_0 \text{ in } \Omega, \quad h|_{t=0} = h_0 \text{ on } \Gamma, \end{cases} \tag{3.196}$$

where we have set $\Omega^\infty = \Omega \times (0, \infty)$ and $\Gamma^\infty = \Gamma \times (0, \infty)$.

Let $\mathbf{u}$ and $(\mathbf{v}, h)$ be solutions of Eqs. (3.195) and (3.196), respectively. Roughly speaking, if $\mathbf{u} \in J_q(\Omega)$ for any $t > 0$, then, $\mathbf{u}$ and $\mathfrak{p} = K_0(\mathbf{u})$ are unique solutions of Eq. (3.158) with $g = \mathbf{g} = \mathbf{h} = 0$. And, if $\mathbf{v} \in J_q(\Omega)$ for any $t > 0$, then $\mathbf{v}$, $\mathfrak{q} = K(\mathbf{v}, h)$, and h are unique solutions of (3.159) with $d = 0$ and $(\mathbf{v}, h)|_{t=0} = (\mathbf{v}_0, h_0)$.

Let us introduce spaces and operators to describe (3.195) and (3.196) in the semigroup setting. Let

$$\mathcal{D}_q^1(\Omega) = \{\mathbf{u} \in J_q(\Omega) \cap H_q^2(\Omega)^N \mid (\mu \mathbf{D}(\mathbf{u})\mathbf{n})_\tau = 0 \text{ on } \Gamma\},$$

$$\mathcal{A}_q^1 \mathbf{u} = \text{Div}\,(\mu (\mathbf{D}(\mathbf{u}) - K_0(\mathbf{u})\mathbf{I}) \quad \text{for } \mathbf{u} \in \mathcal{D}_q^1(\Omega);$$

$$\mathcal{H}_q(\Omega) = \{(\mathbf{v}, h) \mid \mathbf{v} \in J_q(\Omega), \ h \in W_q^{2-1/q}(\Gamma)\},$$

$$\mathcal{D}_q^2(\Omega) = \{(\mathbf{v}, h) \in \mathcal{H}_q(\Omega) \mid \mathbf{v} \in H_q^2(\Omega)^N, \ h \in W_q^{3-1/q}(\Gamma), \tag{3.197}$$

$$(\mu \mathbf{D}(\mathbf{v})\mathbf{n})_\tau = 0 \quad \text{on } \Gamma\},$$

$$\mathcal{A}_q^2(\mathbf{v}, h) = (\text{Div}\,(\mu \mathbf{D}(\mathbf{v}) - K(\mathbf{v}, h)\mathbf{I}), \mathbf{n} \cdot \mathbf{v}|_\Gamma)$$

for $(\mathbf{v}, h) \in \mathcal{D}_q^2(\Omega)$. Since $\text{div}\,\mathbf{u} = 0$ for $\mathbf{u} \in \mathcal{D}_q^1(\Omega)$, by (3.186) $(\mu \mathbf{D}(\mathbf{u})\mathbf{n})_\tau = 0$ is equivalent to $(\mu \mathbf{D}(\mathbf{u}) - K_0(\mathbf{u})\mathbf{I})\mathbf{n} = 0$. And also, for $(\mathbf{v}, h) \in \mathcal{D}_q^2(\Omega)$, $(\mu \mathbf{D}(\mathbf{v})\mathbf{n})_\tau = 0$ is equivalent to

$$(\mu \mathbf{D}(\mathbf{v}) - K(\mathbf{v}, h)\mathbf{I})\mathbf{n} - (\mathcal{F}_2 h + \sigma \Delta_\Gamma h) = 0.$$

Using the symbols defined in (3.197), we see that Eqs. (3.195) and (3.196) are written as

$$\partial_t \mathbf{u} - \mathcal{A}_q^1 \mathbf{u} = 0 \quad (t > 0), \quad \mathbf{u}|_{t=0} = \mathbf{u}_0; \tag{3.198}$$

$$\partial_t U(t) - \mathcal{A}_q^2 U(t) = 0 \quad (t > 0), \quad U(t)|_{t=0} = U_0, \tag{3.199}$$

where $\mathbf{u}(t) \in \mathcal{D}_q^1(\Omega)$ for $t > 0$ and $\mathbf{u}_0 \in J_q(\Omega)$, and $U(t) = (\mathbf{v}, h) \in \mathcal{D}_q^2(\Omega)$ for $t > 0$ and $U_0 = (\mathbf{v}_0, h_0) \in \mathcal{H}_q(\Omega)$. The corresponding resolvent problem to (3.198) is that for any $\mathbf{f} \in J_q(\Omega)$ and $\lambda \in \Sigma_{\epsilon_0, \lambda_0}$ we find $\mathbf{u} \in \mathcal{D}_q^1(\Omega)$ uniquely solving the equation:

$$\lambda \mathbf{u} - \mathcal{A}_q^1 \mathbf{u} = \mathbf{f} \quad \text{in } \Omega \tag{3.200}$$

possessing the estimate:

$$|\lambda| \|\mathbf{u}\|_{L_q(\Omega)} + \|\mathbf{u}\|_{H_q^2(\Omega)} \leq C \|\mathbf{f}\|_{L_q(\Omega)}. \tag{3.201}$$

And also, the corresponding resolvent problem to (3.199) is that for any $F \in \mathcal{H}_q(\Omega)$ and $\lambda \in \Sigma_{\epsilon_0, \lambda_0}$ we find $U \in \mathcal{D}_q^2(\Omega)$ uniquely solving the equation:

$$\lambda U - \mathcal{A}_q^2 U = F \quad \text{in } \Omega \times \Gamma \tag{3.202}$$

possessing the estimate:

$$|\lambda| \|U\|_{\mathcal{H}_q(\Omega)} + \|U\|_{\mathcal{D}_q^2(\Omega)} \leq C \|F\|_{\mathcal{H}_q(\Omega)}, \tag{3.203}$$

where for $U = (\mathbf{v}, h)$ we have set

$$\|U\|_{\mathcal{H}_q(\Omega)} = \|\mathbf{v}\|_{L_q(\Omega)} + \|h\|_{W_q^{2-1/q}(\Gamma)}, \quad \|U\|_{\mathcal{D}_q^2(\Omega)} = \|\mathbf{v}\|_{H_q^2(\Omega)} + \|h\|_{W_q^{3-1/q}(\Gamma)}.$$

Since $\mathcal{R}$ boundedness implies boundedness as follows from Definition 3.4.7 with $n = 1$, by Theorem 3.4.14, we know the unique existence of $\mathbf{u} \in \mathcal{D}_q^1(\Omega)$ satisfying (3.200) and (3.201). And also, by Theorem 3.4.11, we know the unique existence of $U \in \mathcal{D}_q^2(\Omega)$ satisfying (3.202) and (3.203). Thus, using standard semigroup theory, we have the following theorem.

Theorem 3.4.16 *Let* $1 < q < \infty$. *Assume that* Ω *is a uniform* C^2 *domain in* $\mathbb{R}^N$ *and that the weak Dirichlet problem is uniquely solvable for* q *and* $q' = q/(q-1)$. *Then, problem* (3.198) *generates a* C^0 *analytic semigroup* $\{T_1(t)\}_{t \geq 0}$ *on*

$J_q(\Omega)$ *satisfying the estimates:*

$$\|T_1(t)\mathbf{u}_0\|_{L_q(\Omega)} + t(\|\partial_t T_1(t)\mathbf{u}_0\|_{L_q(\Omega)} + \|T_1(t)\mathbf{u}_0\|_{H_q^2(\Omega)}) \le Ce^{\gamma t}\|\mathbf{u}_0\|_{L_q(\Omega)},$$
$$(3.204)$$

$$\|\partial_t T_1(t)\mathbf{u}_0\|_{L_q(\Omega)} + \|T_1(t)\mathbf{u}_0\|_{H_q^2(\Omega)} \le Ce^{\gamma t}\|\mathbf{u}_0\|_{H_q^2(\Omega)} \qquad (3.205)$$

for any $t > 0$ *with some constants* $C > 0$ *and* $\gamma > 0$.

Theorem 3.4.17 *Let* $1 < q < \infty$. *Assume that* Ω *is a uniform* C^3 *domain in* $\mathbb{R}^N$ *and that the weak Dirichlet problem is uniquely solvable for* q *and* $q' = q/(q-1)$. *Then, problem* (3.199) *generates a* C^0 *analytic semigroup* $\{T_2(t)\}_{t \ge 0}$ *on* $\mathcal{H}_q(\Omega)$ *satisfying the estimates:*

$$\|T_2(t)U_0\|_{\mathcal{H}_q(\Omega)} + t(\|\partial_t T_2(t)U_0\|_{\mathcal{H}_q(\Omega)} + \|T_2(t)U_0\|_{\mathcal{D}_q^2(\Omega)}) \le Ce^{\gamma t}\|U_0\|_{\mathcal{H}_q(\Omega)},$$
$$(3.206)$$

$$\|\partial_t T_2(t)U_0\|_{\mathcal{H}_q(\Omega)} + \|T_2(t)U_0\|_{\mathcal{D}_q^2(\Omega)} \le Ce^{\gamma t}\|U_0\|_{\mathcal{D}_q^2(\Omega)} \qquad (3.207)$$

for any $t > 0$ *with some constants* $C > 0$ *and* $\gamma > 0$.

We now show the following maximal L_p–L_q regularity theorem for Eqs. (3.195) and (3.196).

Theorem 3.4.18 *Let* $1 < p, q < \infty$. *Assume that* Ω *is a uniform* C^2 *domain in* $\mathbb{R}^N$ *and that the weak Dirichlet problem is uniquely solvable for* q *and* $q' = q/(q-1)$. *Let* $\mathcal{D}_{q,p}^1(\Omega)$ *be a subspace of* $B_{q,p}^{2(1-1/p)}(\Omega)^N$ *defined by*

$$\mathcal{D}_{q,p}^1(\Omega) = (J_q(\Omega), \mathcal{D}_q^1(\Omega))_{1-1/p, p},$$

where $(\cdot, \cdot)_{1-1/p, p}$ *denotes a real interpolation functor. Then, there exists a* $\gamma > 0$ *such that for any initial data* $\mathbf{u}_0 \in \mathcal{D}_{q,p}^1(\Omega)$, *problem* (3.195) *admits a unique solution* $\mathbf{u}$ *with*

$$e^{-\gamma t}\mathbf{u} \in H_p^1((0, \infty), L_q(\Omega)^N) \cap L_p((0, \infty), H_q^2(\Omega)^N),$$

possessing the estimate:

$$\|e^{-\gamma t}\partial_t \mathbf{u}\|_{L_p((0,\infty), L_q(\Omega))} + \|e^{-\gamma t}\mathbf{u}\|_{L_p((0,\infty), H_q^2(\Omega))} \le C\|\mathbf{u}_0\|_{B_{q,p}^{2(1-1/p)}(\Omega)}.$$
$$(3.208)$$

Here, for any Banach space X *with norm* $\|\cdot\|_X$ *we have set*

$$\|e^{-\gamma t}f\|_{L_p((a,b), X)} = \left(\int_a^b (e^{-\gamma t}\|f(t)\|_X)^p \, dt\right)^{1/p}.$$

Remark 3.4.19 Since $\mathcal{D}_{q,p}^1(\Omega) \subset B_{q,p}^{2(1-1/p)}(\Omega)$, in view of a boundary trace theorem, we see that for $\mathbf{u}_0 \in \mathcal{D}_{q,p}^1(\Omega)$, we have

$$
\begin{cases}
\mathbf{u}_0 \in J_q(\Omega), \quad (\mu \mathbf{D}(\mathbf{u}_0)\mathbf{n})_\tau = 0 \quad \text{on } \Gamma & \text{for } \dfrac{2}{p} + \dfrac{1}{q} < 1, \\[3mm]
\mathbf{u}_0 \in J_q(\Omega) & \text{for } \dfrac{2}{p} + \dfrac{1}{q} > 1,
\end{cases}
$$

because $\mathbf{D}(\mathbf{u}_0) \in B_{q,p}^{1-2/p}(\Omega)$, and so $\mathbf{D}(\mathbf{u}_0)|_\Gamma$ exists for $\frac{2}{p} + \frac{1}{q} < 1$, but it does not exist for $\frac{2}{p} + \frac{1}{q} > 1$.

Theorem 3.4.20 *Let $1 < p, q < \infty$. Assume that Ω is a uniform C^3 domain in $\mathbb{R}^N$ and that the weak Dirichlet problem is uniquely solvable for q and $q' = q/(q-1)$. Let $\mathcal{D}_{q,p}^2(\Omega)$ be a subspace of $B_{q,p}^{2(1-1/p)}(\Omega)^N \times B_{q,p}^{3-1/p-1/q}(\Gamma)$ defined by*

$$
\mathcal{D}_{q,p}^2(\Omega) = (\mathcal{H}_q(\Omega), \mathcal{D}_q^2(\Omega))_{1-1/p,p}.
$$

Then, there exists a $\gamma > 0$ such that for any initial data $(\mathbf{u}_0, h_0) \in \mathcal{D}_{q,p}^2(\Omega)$, problem (3.196) admits a unique solution $U = (\mathbf{v}, h)$ with

$$
e^{-\gamma t} U \in H_p^1((0, \infty), \mathcal{H}_q(\Omega)) \cap L_p((0, \infty), \mathcal{D}_q^2(\Omega))
$$

possessing the estimate:

$$
\begin{aligned}
\|e^{-\gamma t} \partial_t U\|_{L_p((0,\infty),\mathcal{H}_q(\Omega))} + \|e^{-\gamma t} U\|_{L_p((0,\infty),\mathcal{D}_q^2(\Omega))} \\
\leq C(\|\mathbf{u}_0\|_{B_{q,p}^{2(1-1/p)}(\Omega)} + \|h_0\|_{B_{q,p}^{3-1/p-1/q}(\Gamma)}).
\end{aligned}
\tag{3.209}
$$

Proof of Theorem 3.4.20 We only prove Theorem 3.4.20, because Theorem 3.4.18 can be proved by the same argument. To prove Theorem 3.4.20, we observe that

$$
\left(\int_0^\infty (e^{-\gamma t} \|\partial_t T_2(t) U_0\|_{\mathcal{H}_q(\Omega)})^p \, dt \right)^{1/p}
$$

$$
= \left(\sum_{j=-\infty}^\infty \int_{2^j}^{2^{j+1}} (e^{-\gamma t} \|\partial_t T_2(t) U_0\|_{\mathcal{H}_q(\Omega)})^p \, dt \right)^{1/p}
$$

$$
\leq \left(\sum_{j=-\infty}^\infty (2^{j+1} - 2^j)(\sup_{t \in (2^j, 2^{j+1})} (e^{-\gamma t} \|\partial_t T_2(t) U_0\|_{\mathcal{H}_q(\Omega)})^p) \right)^{1/p}.
$$

We now introduce Banach spaces ℓ_p^s for $s \in \mathbb{R}$ and $1 \le p \le \infty$ which are sets of all sequences, $(a_j)_{j \in \mathbb{Z}}$, such that $\|(a_j)_{j \in \mathbb{Z}}\|_{\ell_p^s} < \infty$, where we have set

$$\|(a_j)_{j \in \mathbb{Z}}\|_{\ell_p^s} = \begin{cases} \left(\sum_{j=1}^{\infty} (2^{js}|a_j|)^p \right)^{1/p} & \text{for } 1 \le p < \infty, \\ \sup\{2^{js}|a_j| \mid j \in \mathbb{Z}\} & \text{for } p = \infty. \end{cases}$$

Let

$$a_j = \sup_{t \in (2^j, 2^{j+1})} e^{-\gamma t} \|\partial_t T_2(t) U_0\|_{\mathcal{H}_q(\Omega)},$$

and then

$$\left(\int_0^{\infty} (e^{-\gamma t} \|\partial_t T_2(t) \mathbf{u}_0\|_{\mathcal{H}_q(\Omega)})^p \, dt \right)^{1/p}$$

$$\le \left(\sum_{j=-\infty}^{\infty} (2^{j/p} a_j)^p \right)^{1/p} = \|(a_j)_{j \in \mathbb{Z}}\|_{\ell_p^{1/p}}. \tag{3.210}$$

By real interpolation theory (cf. Bergh and Löfström [8, 5.6.Theorem]), we have $\ell_p^{1/p} = (\ell_\infty^1, \ell_\infty^0)_{1-1/p,p}$. Moreover, by (3.206) and (3.207), we have

$$\|(a_j)_{j \in \mathbb{Z}}\|_{\ell_\infty^1} \le C\|U_0\|_{\mathcal{H}_q(\Omega)}, \qquad \|(a_j)_{j \in \mathbb{Z}}\|_{\ell_\infty^0} \le C\|U_0\|_{\mathcal{D}_q(\Omega)},$$

and therefore, by real interpolation

$$\|(a_j)_{j \in \mathbb{Z}}\|_{\ell_p^{1/p}} = \|(a_j)_{j \in \mathbb{Z}}\|_{(\ell_\infty^1, \ell_\infty^0)_{1-1/p,p}} \le C\|U_0\|_{(\mathcal{H}_q(\Omega), \mathcal{D}_q(\Omega))_{1-1/p,p}}.$$

Putting this and (3.210) together gives

$$\|e^{-\gamma t} \partial_t T_2(t) U_0\|_{L_p((0,\infty), \mathcal{H}_q(\Omega))} \le C\|U_0\|_{\mathcal{D}_{q,p}(\Omega)}.$$

Analogously, we have

$$\|e^{-\gamma t} T_2(t) U_0\|_{L_p((0,\infty), \mathcal{D}_q(\Omega))} \le C\|U_0\|_{\mathcal{D}_{q,p}(\Omega)}.$$

We now prove the uniqueness. Let U satisfy the homogeneous equation:

$$\partial_t U - \mathcal{A}_q^2 U = 0 \quad (t > 0), \qquad U|_{t=0} = 0, \tag{3.211}$$

and the condition:

$$e^{-\gamma t} U \in H_p^1((0, \infty), \mathcal{H}_q(\Omega)) \cap L_p((0, \infty), \mathcal{D}_q^2(\Omega)). \tag{3.212}$$

Let U_0 be the zero extension of U to $t < 0$. In particular, from (3.211) it follows that

$$\partial_t U_0 - \mathcal{A}_q^2 U_0 = 0 \quad (t \in \mathbb{R}). \tag{3.213}$$

For any $\lambda \in \mathbb{C}$ with $\operatorname{Re}\lambda > \gamma$, we set

$$\hat{U}(\lambda) = \int_{-\infty}^{\infty} e^{-\lambda t} U_0(t)\,dt = \int_0^{\infty} e^{-\lambda t} U(t)\,dt.$$

By (3.212) and Hölder's inequality, we have

$$\|\hat{U}(\lambda)\|_{\mathcal{D}_q(\Omega)} \le \left(\int_0^{\infty} e^{-(\operatorname{Re}\lambda-\gamma)tp'}\,dt \right)^{1/p'} \|e^{-\gamma t}U\|_{L_p((0,\infty),\mathcal{D}_q^2(\Omega))}$$

$$\le ((\operatorname{Re}\lambda - \gamma)p')^{-1/p'} \|e^{-\gamma t}U\|_{L_p((0,\infty),\mathcal{D}_q^2(\Omega))}.$$

Since $\lambda\hat{U}(\lambda) = \int_0^{\infty} e^{-\lambda t}\partial_t U(t)\,dt$, we also have

$$\|\lambda\hat{U}(\lambda)\|_{\mathcal{H}_q(\Omega)} \le ((\operatorname{Re}\lambda - \gamma)p')^{-1/p'} \|e^{-\gamma t}\partial_t U\|_{L_p((0,\infty),\mathcal{H}_q(\Omega))}.$$

Thus, by (3.213), we have $\hat{U}(\lambda) \in \mathcal{D}_q^2(\Omega)$ satisfies the homogeneous equation:

$$\lambda\hat{U}(\lambda) - \mathcal{A}_q^2\hat{U}(\lambda) = 0 \quad \text{in } \Omega \times \Gamma.$$

Since the uniqueness of the resolvent problem holds for $\lambda \in \Sigma_{\epsilon_0,\lambda_0}$, we have $\hat{U}(\lambda) = 0$ for any $\lambda \in \mathbb{C}$ with $\operatorname{Re}\lambda > \max(\lambda_0, \gamma)$. By the Laplace inverse transform, we have $U_0(t) = 0$ for $t \in \mathbb{R}$, that is $U(t) = 0$ for $t > 0$, which shows the uniqueness. This completes the proof of Theorem 3.4.20. □

3.4.6 Proof of Maximal Regularity Theorem

We first prove Theorem 3.4.2 with the help of Theorem 3.4.10.

Proof of Theorem 3.4.2 The key tool in the proof of Theorem 3.4.2 is the Weis operator valued Fourier multiplier theorem. To state it we need to make a few definitions. For a Banach space, X, $\mathcal{D}(\mathbb{R}, X)$ denotes the space of X-valued $C^{\infty}(\mathbb{R})$ functions with compact support $\mathcal{D}'(\mathbb{R}, X) = \mathcal{L}(\mathcal{D}(\mathbb{R}), X)$ the space of X-valued distributions. And also, $\mathcal{S}(\mathbb{R}, X)$ denotes the space of X-valued rapidly decreasing functions and $\mathcal{S}'(\mathbb{R}, X) = \mathcal{L}(\mathcal{S}(\mathbb{R}), X)$ the space of X-valued tempered distributions. Let Y be another Banach space. Then, given $m \in L_{1,\text{loc}}(\mathbb{R}, \mathcal{L}(X, Y))$, we define an operator $T_m : \mathcal{F}^{-1}\mathcal{D}(\mathbb{R}, X) \to \mathcal{S}'(\mathbb{R}, Y)$ by letting

$$T_m\phi = \mathcal{F}^{-1}[m\mathcal{F}[\phi]] \quad \text{for all } \mathcal{F}\phi \in \mathcal{D}(\mathbb{R}, X), \tag{3.214}$$

where $\mathcal{F}$ and $\mathcal{F}^{-1}$ denote the Fourier transform and its inversion formula, respectively.

Definition 3.4.21 A Banach space X is said to be a UMD Banach space, if the Hilbert transform is bounded on $L_p(\mathbb{R}, X)$ for some (and then all) $p \in (1, \infty)$. Here, the Hilbert transform H operating on $f \in \mathcal{S}(\mathbb{R}, X)$ is defined by

$$[Hf](t) = \frac{1}{\pi} \lim_{\epsilon \to 0+} \int_{|t-s|>\epsilon} \frac{f(s)}{t-s} \, ds \quad (t \in \mathbb{R}).$$

Theorem 3.4.22 (Weis [74]) *Let X and Y be two UMD Banach spaces and $1 < p < \infty$. Let m be a function in $C^1(\mathbb{R} \setminus \{0\}, \mathcal{L}(X, Y))$ such that*

$$\mathcal{R}_{\mathcal{L}(X,Y)}(\{m(\tau) \mid \tau \in \mathbb{R} \setminus \{0\}\}) = \kappa_0 < \infty,$$

$$\mathcal{R}_{\mathcal{L}(X,Y)}(\{\tau m'(\tau) \mid \tau \in \mathbb{R} \setminus \{0\}\}) = \kappa_1 < \infty.$$

Then, the operator T_m defined in (3.214) is extended to a bounded linear operator from $L_p(\mathbb{R}, X)$ into $L_p(\mathbb{R}, Y)$. Moreover, denoting this extension by T_m, we have

$$\|T_m f\|_{L_p(\mathbb{R},Y)} \leq C(\kappa_0 + \kappa_1)\|f\|_{L_p(\mathbb{R},X)} \quad \text{for all } f \in L_p(\mathbb{R}, X)$$

with some positive constant C depending on p.

We now construct a solution of Eq. (3.158). Let $\mathbf{f}_0$ be the zero extension of $\mathbf{f}$ outside of $(0, T)$, that is $\mathbf{f}_0(t) = \mathbf{f}(t)$ for $t \in (0, T)$ and $\mathbf{f}_0(t) = 0$ for $t \notin (0, T)$. Notice that $\mathbf{f}_0$, g, $\mathbf{g}$, and $\mathbf{h}$ are defined on the whole line $\mathbb{R}$. Thus, we first consider the equations:

$$\begin{cases} \partial_t \mathbf{u}_1 - \text{Div}\,(\mu \mathbf{D}(\mathbf{u}_1) - \mathfrak{q}_1\mathbf{I}) = \mathbf{f}_0 & \text{in } \Omega \times \mathbb{R}, \\ \text{div}\,\mathbf{u}_1 = g = \text{div}\,\mathbf{g} & \text{in } \Omega \times \mathbb{R}, \\ (\mu \mathbf{D}(\mathbf{u}_1) - \mathfrak{q}_1\mathbf{I})\mathbf{n} = \mathbf{h} & \text{on } \Gamma \times \mathbb{R}. \end{cases} \tag{3.215}$$

Let $\mathcal{F}_L$ be the Laplace transform with respect to the time variable t defined by

$$\hat{f}(\lambda) = \mathcal{F}_L[f](\lambda) = \int_{\mathbb{R}} e^{-\lambda t} f(t) \, dt$$

for $\lambda = \gamma + i\tau \in \mathbb{C}$. Obviously,

$$\mathcal{F}_L[f](\lambda) = \int_{\mathbb{R}} e^{-i\tau t} e^{-\gamma t} f(t)\, dt = \mathcal{F}[e^{-\gamma t} f](\tau).$$

Applying the Laplace transform to Eq. (3.215) gives

$$\begin{cases} \lambda \hat{\mathbf{u}}_1 - \mathrm{Div}\,(\mu \mathbf{D}(\hat{\mathbf{u}}_1) - \hat{\mathfrak{q}}_1 \mathbf{I}) = \hat{\mathbf{f}}_0 & \text{in } \Omega, \\[4pt] \mathrm{div}\,\hat{\mathbf{u}}_1 = \hat{g} = \mathrm{div}\,\hat{\mathbf{g}} & \text{in } \Omega, \\[4pt] (\mu \mathbf{D}(\hat{\mathbf{u}}_1) - \hat{\mathfrak{q}}_1 \mathbf{I})\mathbf{n} = \hat{\mathbf{h}} & \text{on } \Gamma. \end{cases} \qquad (3.216)$$

Applying Theorem 3.4.10, we have $\hat{\mathbf{u}}_1 = \mathcal{A}_2(\lambda)\mathbf{F}'_\lambda$ and $\hat{\mathfrak{q}}_1 = \mathcal{P}_2(\lambda)\mathbf{F}'_\lambda$ for $\lambda \in \Sigma_{\epsilon_0,\lambda_0}$, where

$$\mathbf{F}'_\lambda = (\hat{\mathbf{f}}_0(\lambda), \lambda^{1/2}\hat{\mathbf{h}}(\lambda), \hat{\mathbf{h}}(\lambda)\lambda^{1/2}\hat{g}(\lambda), \hat{g}(\lambda), \lambda\hat{\mathbf{g}}(\lambda)).$$

Let $\mathcal{F}_L^{-1}$ be the inverse Laplace transform defined by

$$\mathcal{F}_L^{-1}[g](t) = \frac{1}{2\pi} \int_{\mathbb{R}} e^{\lambda t} g(\tau)\, d\tau = e^{\gamma t}\frac{1}{2\pi} \int_{\mathbb{R}} e^{i\tau t} g(\tau)\, d\tau$$

for $\lambda = \gamma + i\tau \in \mathbb{C}$. Obviously,

$$\mathcal{F}_L^{-1}[g](t) = e^{\gamma t}\mathcal{F}^{-1}[g](t), \quad \mathcal{F}_L\mathcal{F}_L^{-1} = \mathcal{F}_L^{-1}\mathcal{F}_L = \mathbf{I}.$$

Setting

$$\Lambda_\gamma^{1/2} f = \mathcal{F}_L^{-1}[\lambda^{1/2}\mathcal{F}_L[f]] = e^{\gamma t}\mathcal{F}^{-1}[\lambda^{1/2}\mathcal{F}[e^{-\gamma t} f]],$$

and using the facts that

$$\lambda\hat{g}(\lambda) = \mathcal{F}_L[\partial_t \mathbf{g}](\lambda), \quad \lambda^{1/2}\hat{f}(\lambda) = \mathcal{F}_L[\Lambda_\gamma^{1/2} f] = \mathcal{F}[e^{-\gamma t}\Lambda_\gamma^{1/2} f]$$

for $f \in \{g, \mathbf{h}\}$, we define $\mathbf{u}_1$ and $\mathfrak{q}_1$ by

$$\mathbf{u}_1(\cdot, t) = \mathcal{F}_L[\mathcal{A}_2(\lambda)\mathbf{F}'_\lambda] = e^{\gamma t}\mathcal{F}^{-1}[\mathcal{A}_2(\lambda)\mathcal{F}[e^{-\gamma t} F'(t)](\tau)],$$

$$\mathfrak{q}_1(\cdot, t) = \mathcal{F}_L[\mathcal{P}_2(\lambda)\mathbf{F}'_\lambda] = e^{\gamma t}\mathcal{F}^{-1}[\mathcal{P}_2(\lambda)\mathcal{F}[e^{-\gamma t} F'(t)](\tau)],$$

with $F'(t) = (\mathbf{f}_0, \Lambda_\gamma^{1/2}\mathbf{h}, \mathbf{h}, \Lambda_\gamma^{1/2} g, g, \partial_t \mathbf{g})$, where γ is chosen as $\gamma > \lambda_0$, and so $\gamma + i\tau \in \Sigma_{\epsilon_0,\lambda_0}$ for any $\tau \in \mathbb{R}$. By Cauchy's theorem in the theory of one complex variable, $\mathbf{u}_1$ and $\mathfrak{q}_1$ are independent of choice of γ whenever $\gamma > \lambda_0$ and the condition (3.169) is satisfied for $\gamma > \lambda_0$. Noting that

$$\partial_t \mathbf{u}_1 = \mathcal{F}_L^{-1}[\lambda\mathcal{A}_1(\lambda)\mathbf{F}'_\lambda] = e^{\gamma t}\mathcal{F}^{-1}[\lambda\mathcal{A}_1(\lambda)\mathcal{F}[e^{-\gamma t} F'(t)](\tau)],$$

and applying Theorem 3.4.22, we have

$$\|e^{-\gamma t}\partial_t \mathbf{u}_1\|_{L_p(\mathbb{R}, L_q(\Omega))} + \|e^{-\gamma t}\mathbf{u}_1\|_{L_p(\mathbb{R}, H_q^2(\Omega))} + \|e^{-\gamma t}\nabla \mathfrak{q}_1\|_{L_p((\mathbb{R}, L_q(\Omega))}$$

$$\leq Cr_b\|e^{-\gamma t}F'\|_{L_p(\mathbb{R}, \mathcal{H}_q(\Omega))}$$

$$\leq Cr_b\{\|e^{-\gamma t}\mathbf{f}\|_{L_p((0,T), L_q(\Omega))} + \|e^{-\gamma t}\mathbf{g}\|_{L_p(\mathbb{R}, H_q^1(\Omega))} + \|e^{-\gamma t}\Lambda_\gamma^{1/2}\mathbf{g}\|_{L_p(\mathbb{R}, L_q(\Omega))}$$

$$+ \|e^{-\gamma t}\partial_t \mathbf{g}\|_{L_p(\mathbb{R}, L_q(\Omega))} + \|e^{-\gamma t}\mathbf{h}\|_{L_p(\mathbb{R}, H_q^1(\Omega))} + \|e^{-\gamma t}\Lambda_\gamma^{1/2}\mathbf{h}\|_{L_p(\mathbb{R}, L_q(\Omega))}\}.$$

$$(3.217)$$

We now write solutions $\mathbf{u}$ and $\mathfrak{q}$ of Eq. (3.158) by $\mathbf{u} = \mathbf{u}_1 + \mathbf{u}_2$ and $\mathfrak{q} = \mathfrak{q}_1 + \mathfrak{q}_2$, where $\mathbf{u}_2$ and $\mathfrak{q}_2$ are solutions of the following equations:

$$\begin{cases} \partial_t \mathbf{u}_2 - \mathrm{Div}\,(\mu\mathbf{D}(\mathbf{u}_2) - \mathfrak{q}_2\mathbf{I}) = 0, \quad \mathrm{div}\,\mathbf{u}_2 = 0 & \text{in } \Omega \times (0, \infty), \\ \qquad\qquad (\mu\mathbf{D}(\mathbf{u}_2) - \mathfrak{q}_2\mathbf{I})\mathbf{n} = 0 & \text{on } \Gamma \times (0, \infty), \quad (3.218) \\ \qquad\qquad \mathbf{u}_2 = \mathbf{u}_0 - \mathbf{u}_1|_{t=0} & \text{in } \Omega. \end{cases}$$

Notice that $\mathrm{div}\,\mathbf{u}_2 = 0$ in $\Omega \times (0, \infty)$ means that $\mathbf{u}_2 \in J_q(\Omega)$ for any $t > 0$. By real interpolation theory, we know that

$$\sup_{t\in(0,\infty)} e^{-\gamma t}\|\mathbf{u}_1(t)\|_{B_{q,p}^{2(1-1/p)}(\Omega)}$$

$$(3.219)$$

$$\leq C(\|e^{-\gamma t}\mathbf{u}_1\|_{L_p((0,\infty), H_q^2(\Omega))} + \|e^{-\gamma t}\partial_t \mathbf{u}_1\|_{L_p((0,\infty), L_q(\Omega))}).$$

In fact, this inequality follows from the following theory (cf. Tanabe[73, p.1]): Let X_1 and X_2 be two Banach spaces such that X_2 is a dense subset of X_1, and then

$$L_p((0, \infty), X_2) \cap H_p^1((0, \infty), X_1) \subset C([0, \infty), (X_1, X_2)_{1-1/p,p}), \quad (3.220)$$

and

$$\sup_{t\in(0,\infty)} \|u(t)\|_{(X_1,X_2)_{1-1/p,p}} \leq C(\|u\|_{L_p((0,\infty), X_2)} + \|\partial_t u\|_{L_p((0,\infty), X_1)}). \quad (3.221)$$

Since $B_{q,p}^{2(1-1/p)}(\Omega) = (L_p(\Omega), H_q^2(\Omega))_{1-1/p,p}$, we have (3.219). Thus,

$$\mathbf{u}_0 - \mathbf{u}_1|_{t=0} \in B_{q,p}^{2(1-1/p)}(\Omega).$$

By the compatibility condition (3.170) and (3.215), we have

$$(\mathbf{u}_0 - \mathbf{u}_1|_{t=0}, \nabla\varphi) = (\mathbf{u}_0 - \mathbf{g}|_{t=0}, \nabla\varphi) = 0 \quad \text{for any } \varphi \in \hat{H}_{q',0}^1(\Omega).$$

Moreover, if $2/p+1/q < 1$, then by the compatibility condition (3.171) and (3.215), we have

$$(\mu \mathbf{D}(\mathbf{u}_0 - \mathbf{u}_1|_{t=0})\mathbf{n})_\tau = (\mu \mathbf{D}(\mathbf{u}_0)\mathbf{n})_\tau - (\mathbf{h}|_{t=0})_\tau = 0 \quad \text{on } \Gamma.$$

Thus, if $2/p + /q \neq 1$, then $\mathbf{u}_0 - \mathbf{u}_1|_{t=0} \in \mathcal{D}_{q,p}^1(\Omega)$. Applying Theorem 3.4.18, we see that there exists a $\gamma' > 0$ such that Eq. (3.218) admits unique solutions $\mathbf{u}_2$ with $\mathfrak{q}_2 = K_0(\mathbf{u}_2)$ and

$$e^{-\gamma't}\mathbf{u}_2 \in H_p^1((0, \infty), L_q(\Omega)^N) \cap L_p((0, \infty), H_q^2(\Omega)^N) \tag{3.222}$$

possessing the estimate:

$$\|e^{-\gamma't}\partial_t\mathbf{u}_2\|_{L_p((0,\infty),L_q(\Omega))} + \|e^{-\gamma't}\mathbf{u}_2\|_{L_p((0,\infty),H_q^2(\Omega))}$$
$$\leq C\|\mathbf{u}_0 - \mathbf{u}_1|_{t=0}\|_{B_{q,p}^{2(1-1/p)}(\Omega)}. \tag{3.223}$$

Thus, setting $\mathbf{u} = \mathbf{u}_1 + \mathbf{u}_2$ and $\mathfrak{q} = \mathfrak{q}_1 + K_0(\mathbf{u}_2)$ and choosing γ_0 in such a way that $\gamma_0 > \max(\lambda_0, \gamma')$, by (3.215), (3.217)–(3.219), (3.222) and (3.223), we see that $\mathbf{u}$ and $\mathfrak{q}$ are required solutions of Eq. (3.158). Employing the same argument as in the proof of the uniqueness of Theorem 3.4.20, we can show the uniqueness. This completes the proof of Theorem 3.4.2.   □

Employing the same argument as above, we can show Theorem 3.4.3, and so we may omit the proof of Theorem 3.4.3. Thus, we finally give a

Proof of Theorem 3.4.4   Let d_0 be the zero extension of d outside of $(0, T)$, that is $d_0(t) = d(t)$ for $t \in (0, T)$ and $d_0(t) = 0$. Employing the same argument as in the proof of Theorem 3.4.2 above, we can show the existence of solutions, $\mathbf{w}$, $\mathfrak{r}$ and ρ, of the equations:

$$\begin{cases} \partial_t\mathbf{w} - \text{Div}\,(\mu\mathbf{D}(\mathbf{w}) - \mathfrak{r}\mathbf{I}) = 0, \quad \text{div}\,\mathbf{u} = 0 & \text{in } \Omega \times \mathbb{R}, \\ \partial_t\rho + A_\kappa \cdot \nabla'_\Gamma\rho - \mathbf{w}\cdot\mathbf{n} + \mathcal{F}_1\mathbf{w} = d_0 & \text{on } \Gamma \times \mathbb{R}, \\ (\mu\mathbf{D}(\mathbf{w}) - \mathfrak{r}\mathbf{I})\mathbf{n} - (\mathcal{F}_2\rho + \sigma\Delta_\Gamma\rho)\mathbf{n} = 0 & \text{on } \Gamma \times \mathbb{R}, \end{cases} \tag{3.224}$$

possessing the estimate:

$$\|e^{-\gamma t}\mathbf{w}\|_{L_p(\mathbb{R}, H_q^2(\Omega))} + \|e^{-\gamma t}\partial_t\mathbf{w}\|_{L_p(\mathbb{R}, L_q(\Omega))}$$
$$+ \|e^{-\gamma t}\rho\|_{L_p(\mathbb{R}, W_q^{3-1/q}(\Gamma))} + \|e^{-\gamma t}\partial_t\rho\|_{L_p(\mathbb{R}, W_q^{2-1/q}(\Gamma))} \tag{3.225}$$
$$\leq C\|e^{-\gamma t}d_0\|_{L_p(\mathbb{R}, W_q^{2-1/q}(\Gamma))} \leq C\|d\|_{L_q((0,T), W_q^{2-1/q}(\Gamma))}$$

for any $\gamma \geq \lambda_0 \kappa^{-b}$, where we have used (3.220) and (3.221). In particular, for any $\epsilon > 0$ and $\gamma \geq \lambda_0 \kappa^{-b}$, we have

$$\|e^{-\gamma t}\mathbf{w}\|_{L_p((-\infty,-\epsilon),H_q^2(\Omega))} + \|e^{-\gamma t}\rho\|_{L_p((-\infty,-\epsilon),W_q^{3-1/q}(\Gamma))}$$
$$\leq \|e^{-\gamma t}\mathbf{w}\|_{L_p(\mathbb{R},H_q^2(\Omega))} + \|e^{-\gamma t}\rho\|_{L_p(\mathbb{R},W_q^{3-1/q}(\Gamma))} \leq C\|d\|_{L_p((0,T),W_q^{2-1/q}(\Gamma))}.$$

By the monotonicity of $e^{-\gamma t}$, we have

$$\|e^{-\gamma t}\mathbf{w}\|_{L_p((-\infty,-\epsilon),H_q^2(\Omega))} + \|e^{-\gamma t}\rho\|_{L_p((-\infty,-\epsilon),W_q^{3-1/q}(\Gamma))}$$
$$\geq e^{\gamma\epsilon}(\|\mathbf{w}\|_{L_p((-\infty,-\epsilon),H_q^2(\Omega))} + \|\rho\|_{L_p((-\infty,-\epsilon),W_q^{3-1/q}(\Gamma))}).$$

Putting these inequalities together gives

$$\|\mathbf{w}\|_{L_p((-\infty,-\epsilon),H_q^2(\Omega))} + \|\rho\|_{L_p((-\infty,-\epsilon),W_q^{3-1/q}(\Gamma))}$$
$$\leq Ce^{-\gamma\epsilon}\|d\|_{L_p((0,T),W_q^{2-1/q}(\Gamma))}$$

for any $\gamma \geq \lambda_0 \kappa^{-b}$. Thus, letting $\gamma \to \infty$, we have

$$\|\mathbf{w}\|_{L_p((-\infty,-\epsilon),H_q^2(\Omega))} + \|\rho\|_{L_p((-\infty,-\epsilon),W_q^{3-1/q}(\Gamma))} = 0.$$

Since $\epsilon > 0$ is chosen arbitrarily, we have

$$\|\mathbf{w}\|_{L_p((-\infty,0),H_q^2(\Omega))} + \|\rho\|_{L_p((-\infty,0),W_q^{3-1/q}(\Gamma))} = 0,$$

which shows that $\mathbf{w} = 0$ and $\rho = 0$ for $t < 0$, because

$$\mathbf{w} \in C(\mathbb{R}, B_{q,p}^{2(1-1/p)}(\Omega)^N), \quad \rho \in C(\mathbb{R}, W_{q,p}^{3-1/p-1/q}(\Gamma)).$$

By the monotonicity of $e^{-\gamma t}$, we have

$$\|e^{-\gamma t}\mathbf{w}\|_{L_p(\mathbb{R},H_q^2(\Omega))} + \|e^{-\gamma t}\partial_t\mathbf{w}\|_{L_p(\mathbb{R},L_q(\Omega))}$$
$$\geq \|e^{-\gamma t}\mathbf{w}\|_{L_p((0,T),H_q^2(\Omega))} + \|e^{-\gamma t}\partial_t\mathbf{w}\|_{L_p((0,T),L_q(\Omega))}$$
$$\geq e^{-\gamma T}(\|\mathbf{w}\|_{L_p((0,T),H_q^2(\Omega))} + \|\partial_t\mathbf{w}\|_{L_p((0,T),L_q(\Omega))}).$$

Similarly, we have

$$\|e^{-\gamma t}\rho\|_{L_p(\mathbb{R},W_q^{3-1/q}(\Gamma))} + \|e^{-\gamma t}\partial_t\rho\|_{L_p(\mathbb{R},W_q^{2-1/q}(\Gamma))}$$
$$\geq e^{-\gamma T}(\|\rho\|_{L_p((0,T),W_q^{3-1/q}(\Gamma))} + \|\partial_t\rho\|_{L_p((0,T),W_q^{2-1/q}(\Gamma))}).$$

Thus, by (3.225), we have

$$
\begin{aligned}
\|\mathbf{w}\|_{L_p((0,T),H_q^2(\Omega))} &+ \|\partial_t \mathbf{w}\|_{L_p((0,T),L_q(\Omega))} \\
&+ \|\rho\|_{L_p((0,T),W_q^{3-1/q}(\Gamma))} + \|\partial_t \rho\|_{L_p((0,T),W_q^{2-1/q}(\Gamma))} \\
&\leq Ce^{\gamma T} \|d\|_{L_p((0,T),W_q^{2-1/q}(\Gamma))}
\end{aligned}
$$

for any $\gamma \geq \lambda_0 \kappa^{-b}$. This completes the proof of the existence part of Theorem 3.4.4. Employing the same argument as in the proof of the uniqueness of Theorem 3.4.20, we can show the uniqueness. This completes the proof of Theorem 3.4.4. □

3.5 $\mathcal{R}$ Bounded Solution Operators

In this section, we mainly prove Theorem 3.4.11. The operators $\mathcal{F}_1$ and $\mathcal{F}_2$ can be treated by perturbation method, and so the Sects. 3.5.1–3.5.8 below devote to proving the existence part for Eq. (3.185) for the following equations:

$$
\begin{cases}
\lambda \mathbf{u} - \mathrm{Div}\,(\mu \mathbf{D}(\mathbf{u}, h) - K(\mathbf{u}, h)\mathbf{I}) = \mathbf{f} & \text{in } \Omega, \\
\lambda h + A_\kappa \cdot \nabla'_\Gamma h - \mathbf{n} \cdot \mathbf{u} = d & \text{on } \Gamma, \\
(\mu \mathbf{D}(\mathbf{u}) - K(\mathbf{u}, h)\mathbf{I})\mathbf{n} - (\sigma \Delta_\Gamma h)\mathbf{n} = \mathbf{h} & \text{on } \Gamma.
\end{cases}
\tag{3.226}
$$

And then, in Sect. 3.5.9, we prove the existence part of Theorem 3.4.11 for Eq. (3.185) by using a perturbation method.

The existence part of Theorem 3.4.14 can be proved in the same manner as in the proof of the existence part of Theorem 3.4.11 and also has been proved by Shibata [43], and so we may omit its proof.

Concerning the uniqueness part, we first prove Theorem 3.4.14 in Sect. 3.5.10. Finally, the uniqueness part of Theorem 3.4.11 will be proved in Sect. 3.5.11 by showing *apriori* estimates for Eq. (3.186) under the assumption that Ω is a uniform C^3 domain whose inside has a finite covering.

3.5.1 Model Problem in $\mathbb{R}^N$; Constant μ Case

In this subsection, we assume that μ is a constant satisfying the assumption (3.164), that is $m_0 \leq \mu \leq m_1$. Given $\mathbf{u} \in H_q^2(\mathbb{R}^N)^N$, let $u = K_0(\mathbf{u})$ be a unique solution of the weak Laplace problem:

$$
(\nabla u, \nabla \varphi)_{\mathbb{R}^N} = (\mathrm{Div}\,(\mu \mathbf{D}(\mathbf{u})) - \nabla \mathrm{div}\,\mathbf{u}, \nabla \varphi)_{\mathbb{R}^N}
\tag{3.227}
$$

for any $\varphi \in \hat{H}^1_{q'}(\mathbb{R}^N)$. In this subsection, we consider the resolvent problem:

$$\lambda \mathbf{u} - \text{Div}\,(\mu \mathbf{D}(\mathbf{u}) - K_0(\mathbf{u})\mathbf{I}) = \mathbf{f} \quad \text{in } \mathbb{R}^N, \tag{3.228}$$

and prove the following theorem.

Theorem 3.5.1 *Let* $1 < q < \infty$, $0 < \epsilon < \pi/2$, *and* $\lambda_0 > 0$. *Then, there exists an operator family* $\mathcal{A}_0(\lambda) \in Hol\,(\Sigma_{\epsilon,\lambda_0}, \mathcal{L}(L_q(\mathbb{R}^N)^N, H^2_q(\mathbb{R}^N)^N))$ *such that for any* $\lambda = \gamma + i\tau \in \Sigma_{\epsilon,\lambda_0}$ *and* $\mathbf{f} \in L_q(\mathbb{R}^N)^N$, $\mathbf{u} = \mathcal{A}_0(\lambda)\mathbf{f}$ *is a unique solution of Eq.* (3.228) *and*

$$\mathcal{R}_{\mathcal{L}(L_q(\mathbb{R}^N)^N, H^{2-j}_q(\mathbb{R}^N)^N)}(\{(\tau\partial_\tau)^\ell(\lambda^{j/2}\mathcal{A}_0(\lambda)) \mid \lambda \in \Sigma_{\epsilon,\lambda_0}\}) \le r_b(\lambda_0) \tag{3.229}$$

for $\ell = 0, 1$ *and* $j = 0, 1, 2$, *where* $r_b(\lambda_0)$ *is a constant depending on* ϵ, λ_0, m_0, m_1, q *and* N, *but independent of* $\mu \in [m_0, m_1]$.

Proof We first consider the Stokes equations:

$$\lambda \mathbf{u} - \text{Div}\,(\mu(\mathbf{D}(\mathbf{u}) - \mathfrak{q}\mathbf{I}) = \mathbf{f}, \quad \text{div}\,\mathbf{u} = g = \text{div}\,\mathbf{g} \quad \text{in } \mathbb{R}^N. \tag{3.230}$$

Since $\text{Div}\,(\mu \mathbf{D}(\mathbf{u}) - \mathfrak{q}\mathbf{I}) = \mu\Delta\mathbf{u} + \mu\nabla\text{div}\,\mathbf{u} - \nabla\mathfrak{q}$, applying div to (3.230), we have

$$\lambda\text{div}\,\mathbf{g} - 2\mu\Delta g + \Delta\mathfrak{q} = \text{div}\,\mathbf{f},$$

and so,

$$\mathfrak{q} = 2\mu g + \Delta^{-1}(\text{div}\,\mathbf{f} - \lambda\text{div}\,\mathbf{g}).$$

Combining this with (3.230) gives

$$\lambda\mathbf{u} - \mu\Delta\mathbf{u} = \mathbf{f} - \nabla\Delta^{-1}\text{div}\,\mathbf{f} - \mu\nabla g + \lambda\nabla\Delta^{-1}\text{div}\,\mathbf{g}. \tag{3.231}$$

We now look for a solution formula for Eq. (3.228). Let g be a solution of the variational problem:

$$(\lambda g, \varphi)_{\mathbb{R}^N} + (\nabla g, \nabla\varphi)_{\mathbb{R}^N} = (-\mathbf{f}, \nabla\varphi)_{\mathbb{R}^N} \quad \text{for any } \varphi \in \hat{H}^1_{q'}(\mathbb{R}^N),$$

and then this g is given by $g = (\lambda - \Delta)^{-1}\text{div}\,\mathbf{f}$. According to (3.188), we set $\mathbf{g} = \lambda^{-1}(\mathbf{f} + \nabla g)$. Inserting these formulas into (3.231) gives

$$\lambda\mathbf{u} - \mu\Delta\mathbf{u} = \mathbf{f} - (\mu - 1)\nabla g = \mathbf{f} - (\mu - 1)(\lambda - \Delta)^{-1}\nabla\text{div}\,\mathbf{f}.$$

Thus, we have

$$\mathbf{u} = \mathcal{F}_\xi^{-1}\Big[\frac{\mathcal{F}[\mathbf{f}](\xi)}{\lambda + \mu|\xi|^2}\Big] + (\mu - 1)\mathcal{F}_\xi^{-1}\Big[\frac{\xi\xi \cdot \mathcal{F}[\mathbf{f}](\xi)}{(\lambda + \mu|\xi|^2)(\lambda + |\xi|^2)}\Big],$$

where $\mathcal{F}$ and $\mathcal{F}_\xi^{-1}$ denote the Fourier transform and its inversion formula defined by

$$\mathcal{F}[f](\xi) = \int_{\mathbb{R}^N} e^{-ix\cdot\xi} f(x)\,dx, \quad \mathcal{F}_\xi^{-1}[g(\xi)](x) = \frac{1}{(2\pi)^N}\int_{\mathbb{R}^N} e^{ix\cdot\xi} g(\xi)\,d\xi.$$

Thus, we define an operator family $\mathcal{A}_0(\lambda)$ acting on $\mathbf{f} \in L_q(\mathbb{R}^N)^N$ by

$$\mathcal{A}_0(\lambda)\mathbf{f} = \mathcal{F}_\xi^{-1}\Big[\frac{\mathcal{F}[\mathbf{f}](\xi)}{\lambda + \mu|\xi|^2}\Big] + (\mu - 1)\mathcal{F}_\xi^{-1}\Big[\frac{\xi\xi \cdot \mathcal{F}[\mathbf{f}](\xi)}{(\lambda + \mu|\xi|^2)(\lambda + |\xi|^2)}\Big].$$

To prove the $\mathcal{R}$-boundedness of $\mathcal{A}_0(\lambda)$, we use the following lemma.

Lemma 3.5.2 *Let $0 < \epsilon < \pi/2$. Then, for any $\lambda \in \Sigma_\epsilon$ and $x \in [0, \infty)$, we have*

$$|\lambda + x| \geq (\sin\frac{\epsilon}{2})(|\lambda| + x). \tag{3.232}$$

Proof Representing $\lambda = |\lambda|e^{i\theta}$ and using $\cos\theta \geq \cos(\pi - \epsilon) = -\cos\epsilon$ for $\lambda \in \Sigma_\epsilon$, we have (3.232). $\qquad\square$

Lemma 3.5.3 *Let $1 < q < \infty$ and let U be a subset of $\mathbb{C}$. Let $m = m(\lambda, \xi)$ be a function defined on $U \times (\mathbb{R}^N \setminus \{0\})$ which is infinitely differentiable with respect to $\xi \in \mathbb{R}^N \setminus \{0\}$ for each $\lambda \in U$. Assume that for any multi-index $\alpha \in \mathbb{N}_0^N$ there exists a constant C_α depending on α such that*

$$|\partial_\xi^\alpha m(\lambda, \xi)| \leq C_\alpha|\xi|^{-|\alpha|} \tag{3.233}$$

for any $(\lambda, \xi) \in U \times (\mathbb{R}^N \setminus \{0\})$. Set

$$\mathbf{b}(m) = \max_{|\alpha| \leq N+1} C_\alpha.$$

Let K_λ be an operator defined by

$$K_\lambda f = \mathcal{F}_\xi^{-1}[m(\lambda, \xi)\mathcal{F}[f](\xi)].$$

Then, the operator family $\{K_\lambda \mid \lambda \in U\}$ is $\mathcal{R}$-bounded on $\mathcal{L}(L_q(\mathbb{R}^N))$ and

$$\mathcal{R}_{\mathcal{L}(L_q(\mathbb{R}^N))}(\{K_\lambda \mid \lambda \in U\}) \leq C_{N,q}\mathbf{b}(m)$$

for some constant $C_{q,N}$ depending solely on q and N.

Proof Lemma 3.5.3 was proved by Enomoto and Shibata [18, Theorem 3.3] and Denk and Schnaubelt [16, Lemma 2.1]. □

By Lemma 3.5.2, we have

$$\left| \partial_\xi^\alpha \frac{\lambda^{j/2} \xi^\beta}{\lambda + \mu |\xi|^2} \right| \leq C_\alpha |\xi|^{-|\alpha|} \lambda_0^{-k/2},$$

$$\left| \partial_\xi^\alpha \frac{\xi_\ell \xi_m \lambda^{j/2} \xi^\beta}{(\lambda + \mu |\xi|^2)(\lambda + |\xi|^2)} \right| \leq C_\alpha |\xi|^{-|\alpha|} \lambda_0^{-k/2} \quad (\ell, m = 1, \ldots, N)$$

for any $j \in \mathbb{N}_0$, $k \in \mathbb{N}_0$ and $\beta \in \mathbb{N}_0^N$ such that $j + k + |\beta| = 2$ and for any $\alpha \in \mathbb{N}_0^N$ and $(\lambda, \xi) \in \Sigma_{\epsilon, \lambda_0} \times (\mathbb{R}^N \setminus \{0\})$. Thus, by Lemma 3.5.3, we have (3.229), which completes the proof of Theorem 3.5.1. □

We conclude this subsection by introducing some fundamental properties of $\mathcal{R}$-bounded operators and Bourgain's results concerning Fourier multiplier theorems with scalar multiplieres.

Proposition 3.5.4

(a) *Let X and Y be Banach spaces, and let $\mathcal{T}$ and $\mathcal{S}$ be $\mathcal{R}$-bounded families in $\mathcal{L}(X, Y)$. Then, $\mathcal{T} + \mathcal{S} = \{T + S \mid T \in \mathcal{T}, \ S \in \mathcal{S}\}$ is also an $\mathcal{R}$-bounded family in $\mathcal{L}(X, Y)$ and*

$$\mathcal{R}_{\mathcal{L}(X,Y)}(\mathcal{T} + \mathcal{S}) \leq \mathcal{R}_{\mathcal{L}(X,Y)}(\mathcal{T}) + \mathcal{R}_{\mathcal{L}(X,Y)}(\mathcal{S}).$$

(b) *Let X, Y and Z be Banach spaces, and let $\mathcal{T}$ and $\mathcal{S}$ be $\mathcal{R}$-bounded families in $\mathcal{L}(X, Y)$ and $\mathcal{L}(Y, Z)$, respectively. Then, $\mathcal{S}\mathcal{T} = \{ST \mid T \in \mathcal{T}, \ S \in \mathcal{S}\}$ also an $\mathcal{R}$-bounded family in $\mathcal{L}(X, Z)$ and*

$$\mathcal{R}_{\mathcal{L}(X,Z)}(\mathcal{S}\mathcal{T}) \leq \mathcal{R}_{\mathcal{L}(X,Y)}(\mathcal{T}) \mathcal{R}_{\mathcal{L}(Y,Z)}(\mathcal{S}).$$

(c) *Let $1 < p, q < \infty$ and let D be a domain in $\mathbb{R}^N$. Let $m = m(\lambda)$ be a bounded function defined on a subset U of $\mathbb{C}$ and let $M_m(\lambda)$ be a map defined by $M_m(\lambda)f = m(\lambda)f$ for any $f \in L_q(D)$. Then, $\mathcal{R}_{\mathcal{L}(L_q(D))}(\{M_m(\lambda) \mid \lambda \in U\}) \leq C_{N,q,D} \|m\|_{L_\infty(U)}$.*

(d) *Let $n = n(\tau)$ be a C^1-function defined on $\mathbb{R} \setminus \{0\}$ that satisfies the conditions $|n(\tau)| \leq \gamma$ and $|\tau n'(\tau)| \leq \gamma$ with some constant $c > 0$ for any $\tau \in \mathbb{R} \setminus \{0\}$. Let T_n be an operator-valued Fourier multiplier defined by $T_n f = \mathcal{F}^{-1}[n\mathcal{F}[f]]$ for any f with $\mathcal{F}[f] \in \mathcal{D}(\mathbb{R}, L_q(D))$. Then, T_n is extended to a bounded linear operator from $L_p(\mathbb{R}, L_q(D))$ into itself. Moreover, denoting this extension also by T_n, we have*

$$\|T_n\|_{\mathcal{L}(L_p(\mathbb{R}, L_q(D)))} \leq C_{p,q,D}\gamma.$$

Proof The assertions (a) and (b) follow from [17, p.28, Proposition 3.4], and the assertions (c) and (d) follow from [17, p.27, Remarks 3.2] (see also Bourgain [11]).   □

3.5.2 Perturbed Problem in $\mathbb{R}^N$

In this subsection, we consider the case where $\mu(x)$ is a real valued function satisfying (3.164). Let x_0 be any point in $\mathbb{R}^N$ and let d_0 be a positive number such that $B_{d_0}(x_0) \subset \mathbb{R}^N$. In view of (3.164), we assume that

$$|\mu(x) - \mu(x_0)| \le m_1 M_1 \quad \text{for } x \in B_{d_0}(x_0), \tag{3.234}$$

where we have set $M_1 = d_0$. We assume that $M_1 \in (0, 1)$ below. Let φ be a function in $C_0^\infty(\mathbb{R}^N)$ which equals 1 for $x \in B_{d_0/2}(x_0)$ and 0 outside of $B_{d_0}(x_0)$. Let

$$\tilde{\mu}(x) = \varphi(x)\mu(x) + (1 - \varphi(x))\mu(x_0). \tag{3.235}$$

Let $\tilde{K}_0(\mathbf{u}) \in \hat{H}_q^1(\mathbb{R}^N)$ be a unique solution of the weak Laplace problem:

$$(\nabla u, \nabla\varphi)_{\mathbb{R}^N} = (\text{Div}\,(\tilde{\mu}\mathbf{D}(\mathbf{u})) - \nabla\text{div}\,\mathbf{u}, \nabla\varphi)_{\mathbb{R}^N} \quad \text{for any } \varphi \in \hat{H}_{q'}^1(\mathbb{R}^N). \tag{3.236}$$

We consider the resolvent problem:

$$\lambda\mathbf{u} - \text{Div}\,(\tilde{\mu}\mathbf{D}(\mathbf{u}) - \tilde{K}_0(\mathbf{u})\mathbf{I}) = \mathbf{f} \quad \text{in } \mathbb{R}^N. \tag{3.237}$$

We shall prove the following theorem.

Theorem 3.5.5 *Let $1 < q < \infty$ and $0 < \epsilon < \pi/2$. Then, there exist $M_1 \in (0, 1)$, $\lambda_0 \ge 1$ and an operator family $\tilde{\mathcal{A}}_0(\lambda)$ with*

$$\tilde{\mathcal{A}}_0(\lambda) \in Hol\,(\Sigma_{\epsilon,\lambda_0}, \mathcal{L}(L_q(\mathbb{R}^N)^N, H_q^2(\mathbb{R}^N)^N))$$

such that for any $\lambda \in \Sigma_{\epsilon,\lambda_0}$ and $\mathbf{f} \in L_q(\mathbb{R}^N)^N$, $\mathbf{u} = \tilde{\mathcal{A}}(\lambda)\mathbf{f}$ is a unique solution of Eq. (3.237), and

$$\mathcal{R}_{\mathcal{L}(L_q(\mathbb{R}^N)^N, H_q^{2-j}(\mathbb{R}^N)^N)}(\{(\tau\partial_\tau)^\ell(\lambda^{j/2}\tilde{\mathcal{A}}_0(\lambda)) \mid \lambda \in \Sigma_{\epsilon,\lambda_0}\}) \le \tilde{r}_b$$

for $\ell = 0, 1$ and $j = 0, 1, 2$. Here, $\tilde{r}_b$ is a constant independent of M_1 and λ_0.

Proof Let $u = K_{x_0}(\mathbf{u}) \in \hat{H}_q^1(\mathbb{R}^N)$ be a unique solution of the weak Laplace equation:

$$(\nabla u, \nabla\varphi)_{\mathbb{R}^N} = (\text{Div}\,(\mu(x_0)\mathbf{D}(\mathbf{u}) - \nabla\text{div}\,\mathbf{u}, \nabla\varphi)_{\mathbb{R}^N} \tag{3.238}$$

for any $\varphi \in \hat{H}^1_{q'}(\mathbb{R}^N)$. We consider the resolvent problem:

$$\lambda\mathbf{u} - \text{Div}\,(\mu(x_0)\mathbf{D}(\mathbf{u}) - K_{x_0}(\mathbf{u})\mathbf{I}) = \mathbf{f} \quad \text{in } \mathbb{R}^N. \tag{3.239}$$

Let $\mathcal{B}_{x_0}(\lambda) \in \text{Hol}\,(\Sigma_{\epsilon,1}, \mathcal{L}(L_p(\mathbb{R}^N)^N, H^2_q(\mathbb{R}^N)^N))$ be a solution operator of Eq. (3.239) such that for any $\lambda \in \Sigma_{\epsilon,1}$ and $\mathbf{f} \in L_q(\mathbb{R}^N)^N$, $\mathbf{u} = \mathcal{B}_{x_0}(\lambda)\mathbf{f}$ is a unique solution of Eq. (3.239) and

$$\mathcal{R}_{\mathcal{L}(L_q(\mathbb{R}^N)^N, H^{2-j}_q(\mathbb{R}^N)^N)}(\{(\tau\partial_\tau)^\ell(\lambda^{j/2}\mathcal{B}_{x_0}(\lambda)) \mid \lambda \in \Sigma_{\epsilon,1}\}) \leq \gamma_0 \tag{3.240}$$

for $\ell = 0, 1$ and $j = 0, 1, 2$, where γ_0 is a constant independent of M_1 and $\nabla\varphi$. Such an operator is given in Theorem 3.5.1 with $\mu = \mu(x_0)$ and $\lambda_0 = 1$. Inserting the formula: $\mathbf{u} = \mathcal{B}_{x_0}(\lambda)\mathbf{f}$ into (3.237) gives

$$\lambda\mathbf{u} - \text{Div}\,(\tilde{\mu}(x)\mathbf{D}(\mathbf{u}) - \tilde{K}_0(\mathbf{u})\mathbf{I}) = \mathbf{f} - \mathcal{R}(\lambda)\mathbf{f} \quad \text{in } \mathbb{R}^N, \tag{3.241}$$

where we have set

$$\mathcal{R}(\lambda)\mathbf{f} = \text{Div}\,(\tilde{\mu}(x)\mathbf{D}(\mathcal{B}_{x_0}(\lambda)\mathbf{f}) - \mu(x_0)\mathbf{D}(\mathcal{B}_{x_0}(\lambda)\mathbf{f})) \\ - \nabla(\tilde{K}_0(\mathcal{B}_{x_0}(\lambda)\mathbf{f}) - K_{x_0}(\mathcal{B}_{x_0}(\lambda)\mathbf{f})). \tag{3.242}$$

We shall estimate $\mathcal{R}(\lambda)\mathbf{f}$. For any $\varphi \in \hat{H}^1_{q'}(\mathbb{R}^N)$, by (3.236) and (3.238), we have

$$(\nabla(\tilde{K}_0(\mathcal{B}_{x_0}(\lambda)\mathbf{f}) - K_{x_0}(\mathcal{B}_{x_0}(\lambda)\mathbf{f})), \nabla\varphi)_{\mathbb{R}^N} \\ = ((\text{Div}\,((\tilde{\mu}(x) - \mu(x_0))\mathbf{D}(\mathcal{B}_{x_0}(\lambda)\mathbf{f})), \nabla\varphi)_{\mathbb{R}^N}.$$

Since $\tilde{\mu}(x) - \mu(x_0) = \varphi(x)(\mu(x) - \mu(x_0))$, by (3.234) and (3.164), we have

$$\|\text{Div}\,((\tilde{\mu}(x) - \mu(x_0))\mathbf{D}(\mathcal{B}_{x_0}(\lambda)\mathbf{f})\|_{L_q(\mathbb{R}^N)} \\ \leq M_1\|\nabla^2\mathcal{B}_{x_0}(\lambda)\mathbf{f}\|_{L_q(\mathbb{R}^N)} + C_{m_1,\nabla\varphi}\|\nabla\mathcal{B}_{x_0}(\lambda)\mathbf{f}\|_{L_q(\mathbb{R}^N)}.$$

Here and in the following, $C_{m_1,\nabla\varphi}$ denotes a generic constant depending on m_1 and $\|\nabla\varphi\|_{L_\infty(\mathbb{R}^N)}$. Thus, we have

$$\|\mathcal{R}(\lambda)\mathbf{f}\|_{L_q(\mathbb{R}^N)} \leq CM_1\|\nabla^2\mathcal{B}_{x_0}(\lambda)\mathbf{f}\|_{L_q(\mathbb{R}^N)} + C_{m_1,\nabla\varphi}\|\nabla\mathcal{B}_{x_0}(\lambda)\mathbf{f}\|_{L_q(\mathbb{R}^N)}. \tag{3.243}$$

Here and in the following, C denotes a generic constants independent of M_1, m_1, and $\|\nabla\varphi\|_{L_\infty(\mathbb{R}^N)}$. Let λ_0 be any number ≥ 1 and let $n \in \mathbb{N}$, $\{\lambda_k\}^n_{k=1} \subset (\Sigma_{\epsilon,\lambda_0})^n$, and $\{F_k\}_{k=1}^n \subset (L_q(\mathbb{R}^N)^N)^n$. By (3.243), (3.240) and Proposition 3.5.4, we have

$$\int_0^1 \| \sum_{k=1}^n r_k(u) \mathcal{R}(\lambda_k) \mathbf{f}_k \|_{L_q(\mathbb{R}^N)}^q \, du$$

$$\leq 2^{q-1} M_1^q \int_0^1 \| \sum_{k=1}^n r_k(u) \nabla^2 \mathcal{B}_{x_0}(\lambda_k) \mathbf{f}_k \|_{L_q(\mathbb{R}^N)}^q \, du$$

$$+ 2^{q-1} C_{m_1, \nabla\varphi}^q \int_0^1 \| \sum_{k=1}^n r_k(u) \nabla \mathcal{B}_{x_0}(\lambda_k) \mathbf{f}_k \|_{L_q(\mathbb{R}^N)}^q \, du$$

$$\leq 2^{q-1} M_1^q \int_0^1 \| \sum_{k=1}^n r_k(u) \nabla^2 \mathcal{B}_{x_0}(\lambda_k) \mathbf{f}_k \|_{L_q(\mathbb{R}^N)}^q \, du$$

$$+ 2^{q-1} C_{m_1, \nabla\varphi}^q \lambda_0^{-q/2} \int_0^1 \| \sum_{k=1}^n r_k(u) \lambda_k^{1/2} \nabla \mathcal{B}_{x_0}(\lambda_k) \mathbf{f}_k \|_{L_q(\mathbb{R}^N)}^q \, du$$

$$\leq 2^{q-1} (M_1^q + C_{m_1, \nabla\varphi}^q \lambda_0^{-q/2}) \gamma_0^q \int_0^1 \| \sum_{k=1}^n r_k(u) \mathbf{f}_k \|_{L_q(\mathbb{R}^N)}^q \, du.$$

Choosing M_1 so small that $2^{q-1} M_1^q \gamma_0^q \leq (1/4)(1/2)^{q-1}$ and $\lambda_0 \geq 1$ so large that $2^{q-1} C_{m_1, \nabla\varphi}^q \gamma_0^q \lambda_0^{-q/2} \leq (1/4)(1/2)^{q-1}$, we have

$$\mathcal{R}_{\mathcal{L}(L_q(\mathbb{R}^N))}(\{\mathcal{R}(\lambda) \mid \lambda \in \Sigma_{\epsilon, \lambda_0}\}) \leq 1/2.$$

Analogously, we have

$$\mathcal{R}_{\mathcal{L}(L_q(\mathbb{R}^N))}(\{\tau \partial_\tau \mathcal{R}(\lambda) \mid \lambda \in \Sigma_{\epsilon, \lambda_0}\}) \leq 1/2.$$

Thus, $(\mathbf{I} - \mathcal{R}(\lambda))^{-1} = \mathbf{I} + \sum_{j=1}^\infty \mathcal{R}(\lambda)^j$ exists and

$$\mathcal{R}_{\mathcal{L}(L_q(\mathbb{R}^N))}(\{(\tau \partial_\tau)^\ell (\mathbf{I} - \mathcal{R}(\lambda))^{-1} \mid \lambda \in \Sigma_{\epsilon, \lambda_0}\}) \leq 4 \quad \text{for } \ell = 0, 1. \qquad (3.244)$$

Setting $\tilde{\mathcal{A}}_0(\lambda) = \mathcal{B}_{x_0}(\lambda)(\mathbf{I} - \mathcal{R}(\lambda))^{-1}$, by (3.240), (3.244) and Proposition 3.5.4, we see that $\tilde{\mathcal{A}}_0(\lambda)$ is a solution operator satisfying the required properties with $\tilde{r}_b = 4\gamma_0$.

To prove the uniqueness of solutions of Eq. (3.237), let $\mathbf{u} \in H_q^2(\mathbb{R}^N)^N$ be a solution of the homogeneous equation:

$$\lambda \mathbf{u} - \text{Div}\,(\tilde{\mu}\mathbf{D}(\mathbf{u}) - \tilde{K}_0(\mathbf{u})\mathbf{I}) = 0 \quad \text{in } \mathbb{R}^N.$$

And then, **u** satisfies the non-homogeneous equation:

$$\lambda\mathbf{u} - \mathrm{Div}\,(\mu(x_0)\mathbf{D}(\mathbf{u}) - K_{x_0}(\mathbf{u})\mathbf{I}) = R\mathbf{u} \quad \text{in } \mathbb{R}^N, \tag{3.245}$$

where we have set

$$R\mathbf{u} = -\mathrm{Div}\,((\tilde{\mu}(x) - \mu(x_0))\mathbf{D}(\mathbf{u})) + \nabla(\tilde{K}_0(\mathbf{u}) - K_{x_0}(\mathbf{u})).$$

Analogously to the proof of (3.234), we have

$$\|R\mathbf{u}\|_{L_q(\mathbb{R}^N)} \leq CM_1\|\nabla^2\mathbf{u}\|_{L_q(\mathbb{R}^N)} + C_{m_1,\nabla\varphi}\|\nabla\mathbf{u}\|_{L_q(\mathbb{R}^N)}. \tag{3.246}$$

On the other hand, applying Theorem 3.5.1 to (3.245) for $\lambda \in \Sigma_{\epsilon,1}$, we have

$$|\lambda|\|\mathbf{u}\|_{L_q(\mathbb{R}^N)} + |\lambda|^{1/2}\|\mathbf{u}\|_{H_q^1(\mathbb{R}^N)} + \|\mathbf{u}\|_{H_q^2(\mathbb{R}^N)} \leq C\|R\mathbf{u}\|_{L_q(\mathbb{R}^N)}. \tag{3.247}$$

Combining (3.246) and (3.247) gives

$$(\lambda_0^{1/2} - CC_{m_1,\nabla\varphi})\|\mathbf{u}\|_{H_q^1(\mathbb{R}^N)} + (1 - CM_1)\|\mathbf{u}\|_{H_q^2(\mathbb{R}^N)} \leq 0.$$

Choosing $M_1 \in (0, 1)$ so small that $1 - CM_1 > 0$ and $\lambda_0 \geq 1$ so large that $\lambda_0^{1/2} - CC_{m_1,\nabla\varphi} > 0$, we have $\mathbf{u} = 0$. This proves the uniqueness, and therefore we have proved Theorem 3.5.5. □

3.5.3 Model Problem in $\mathbb{R}_+^N$

In this section, we assume that μ, δ, and A_κ ($\kappa \in [0, 1)$) are constants and an $N - 1$ constant vector satisfying the conditions:

$$m_0 \leq \mu, \sigma \leq m_1, \quad A_0 = 0, \quad |A_\kappa| \leq m_2 \ (\kappa \in (0, 1)). \tag{3.248}$$

Let

$$\mathbb{R}_+^N = \{(x_1, \ldots, x_N) \in \mathbb{R}^N \mid x_N > 0\},$$

$$\mathbb{R}_0^N = \{(x_1, \ldots, x_N) \in \mathbb{R}^N \mid x_N = 0\},$$

$$\mathbf{n}_0 = {}^\top(0, \ldots, 0, -1).$$

Given $\mathbf{u} \in H_q^2(\mathbb{R}_+^N)^N$ and $h \in W_q^{3-1/q}(\mathbb{R}_0^N)$, let $K(\mathbf{u}, h) \in H_q^1(\mathbb{R}_+^N) + \hat{H}_{q,0}^1(\mathbb{R}_+^N)$ be a unique solution of the weak Dirichlet problem:

$$(\nabla K(\mathbf{u}, h), \nabla \varphi)_{\mathbb{R}_+^N} = (\mathrm{Div}\,(\mu \mathbf{D}(\mathbf{u})) - \nabla \mathrm{div}\,\mathbf{u}, \nabla \varphi)_{\mathbb{R}_+^N} \qquad (3.249)$$

for any $\varphi \in \hat{H}_{q',0}^1(\mathbb{R}^N)$, subject to $K(\mathbf{u}, h) = <\mu \mathbf{D}(\mathbf{u})\mathbf{n}_0, \mathbf{n}_0 > -\sigma \Delta' h - \mathrm{div}\,\mathbf{u}$ on $\mathbb{R}_0^N$, where $\Delta' h = \sum_{j=1}^{N-1} \partial^2 h / \partial x_j^2$. In this section, we consider the half space problem:

$$\begin{cases} \lambda \mathbf{u} - \mathrm{Div}\,(\mu \mathbf{D}(\mathbf{u}) - K(\mathbf{u}, h)\mathbf{I}) = \mathbf{f} & \text{in } \mathbb{R}_+^N, \\ \lambda h + A_\kappa \cdot \nabla' h - \mathbf{u} \cdot \mathbf{n}_0 = d & \text{on } \mathbb{R}_0^N, \\ (\mu \mathbf{D}(\mathbf{u}) - K(\mathbf{u}, h)\mathbf{I})\mathbf{n}_0 - \sigma(\Delta' h)\mathbf{n}_0 = \mathbf{h} & \text{on } \mathbb{R}_0^N, \end{cases} \qquad (3.250)$$

where $\nabla' = (\partial_1, \dots, \partial_{N-1})$. The last equations in (3.250) are equivalent to

$$(\mu \mathbf{D}(\mathbf{u})\mathbf{n}_0)_\tau = \mathbf{h}_\tau \quad \text{and} \quad \mathrm{div}\,\mathbf{u} = \mathbf{h} \cdot \mathbf{n}_0 \quad \text{on } \mathbb{R}_0^N.$$

Here, we have set $\mathbf{h}_\tau = \mathbf{h} - <\mathbf{h}, \mathbf{n}_0 > \mathbf{n}_0$. We shall show the following theorem.

Theorem 3.5.6 *Let* $1 < q < \infty$, *let* μ, σ, *and* A_κ *are constants and an* $N - 1$ *constant vector satisfying the conditions in (3.248). Let* $\Lambda_{\kappa,\lambda_0}$ *be the set defined in Theorem 3.4.8. Let* $Y_q(\mathbb{R}_+^N)$ *and* $\mathcal{Y}_q(\mathbb{R}_+^N)$ *be spaces defined by replacing* Ω *and* Γ *by* $\mathbb{R}_+^N$ *and* $\mathbb{R}_0^N$ *in Theorem 3.4.11. Then, there exist a constant* $\lambda_0 \geq 1$ *and operator families:*

$$\begin{aligned} \mathcal{A}_0(\lambda) &\in Hol\,(\Lambda_{\kappa,\lambda_0}, \mathcal{L}(\mathcal{Y}_q(\mathbb{R}_+^N), H_q^2(\mathbb{R}_+^N)^N)), \\ \mathcal{H}_0(\lambda) &\in Hol\,(\Lambda_{\kappa,\lambda_0}, \mathcal{L}(\mathcal{Y}_q(\mathbb{R}_+^N), H_q^3(\mathbb{R}_+^N))) \end{aligned} \qquad (3.251)$$

such that for any $\lambda = \gamma + i\tau \in \Lambda_{\kappa,\lambda_0}$ *and* $(\mathbf{f}, d, \mathbf{h}) \in Y_q(\mathbb{R}_+^N)$,

$$\mathbf{u} = \mathcal{A}_0(\lambda)(\mathbf{f}, d, \lambda^{1/2}\mathbf{h}, \mathbf{h}), \quad h = \mathcal{H}_0(\lambda)(\mathbf{f}, d, \lambda^{1/2}\mathbf{h}, \mathbf{h}),$$

are unique solutions of (3.250), and

$$\begin{aligned} \mathcal{R}_{\mathcal{L}(\mathcal{Y}_q(\mathbb{R}_+^N), H_q^{2-j}(\mathbb{R}_+^N)^N)}(\{(\tau \partial_\tau)^\ell (\lambda^{j/2}\mathcal{A}_0(\lambda)) \mid \lambda \in \Lambda_{\kappa,\lambda_0}\}) &\leq r_b, \\ \mathcal{R}_{\mathcal{L}(\mathcal{Y}_q(\mathbb{R}_+^N), H_q^{3-k}(\mathbb{R}_+^N))}(\{(\tau \partial_\tau)^\ell (\lambda^k \mathcal{H}_0(\lambda)) \mid \lambda \in \Lambda_{\kappa,\lambda_0}\}) &\leq r_b, \end{aligned} \qquad (3.252)$$

for $\ell = 0, 1$, $j = 0, 1, 2$ *and* $k = 0, 1$. *Here,* r_b *is a constant depending on* m_0, m_1, m_2, λ_0, q, *and* N.

Remark 3.5.7 In this section, what the constant c depends on m_0, m_1, m_2 means that the constant c depends on m_0, m_1, m_2, but is independent of μ, σ and A_κ whenever $\mu \in [m_0, m_1], \sigma \in [m_0, m_1]$, and $|A_\kappa| \leq m_2$ for $\kappa \in (0, 1)$.

To prove Theorem 3.5.6, as an auxiliary problem, we first consider the following equations:

$$\begin{cases} \lambda\mathbf{v} - \mathrm{Div}\,(\mu\mathbf{D}(\mathbf{v}) - \theta\mathbf{I}) = 0, \quad \mathrm{div}\,\mathbf{v} = 0 \quad \text{in } \mathbb{R}_+^N, \\ (\mu\mathbf{D}(\mathbf{v}) - \theta\mathbf{I})\mathbf{n}_0 = \mathbf{h} \quad \text{on } \mathbb{R}_0^N, \end{cases} \tag{3.253}$$

and we shall prove the following theorem, which was essentially proved by Shibata and Shimizu [54].

Theorem 3.5.8 *Let* $1 < q < \infty$, $\epsilon \in (0, \pi/2)$, *and* $\lambda_0 > 0$. *Let*

$$\mathcal{Y}_q'(\mathbb{R}_+^N) = \{(G_1, G_2) \mid G_1 \in L_q(\mathbb{R}_+^N)^N, \quad G_2 \in H_q^1(\mathbb{R}_+^N)^N\},$$

$$\hat{H}_q^1(\mathbb{R}_+^N) = \{\theta \in L_{q,loc}(\mathbb{R}_+^N) \mid \nabla\theta \in L_q(\mathbb{R}_+^N)\}.$$

Then, there exists a solution operator $\mathcal{V}(\lambda)$ *with*

$$\mathcal{V}(\lambda) \in Hol\,(\Sigma_{\epsilon,\lambda_0}, \mathcal{L}(\mathcal{Y}'(\mathbb{R}_+^N), H_q^2(\mathbb{R}_+^N)^N))$$

such that for any $\lambda = \gamma + i\tau \in \Sigma_{\epsilon,\lambda_0}$ *and* $\mathbf{h} \in H_q^1(\mathbb{R}_+^N)^N$, $\mathbf{v} = \mathcal{V}(\lambda)(\lambda^{1/2}\mathbf{h}, \mathbf{h})$ *are unique solutions of Eq.* (3.253) *with some* $\theta \in \hat{H}_q^1(\mathbb{R}_+^N)$ *and*

$$\mathcal{R}_{\mathcal{L}(\mathcal{Y}_q'(\mathbb{R}_+^N), H_q^{2-j}(\mathbb{R}_+^N)^N)}(\{(\tau\partial_\tau)^\ell(\lambda^{j/2}\mathcal{V}(\lambda)) \mid \lambda \in \Sigma_{\epsilon,\lambda_0}\}) \leq r_b(\lambda_0)$$

for $\ell = 0, 1$, *and* $j = 0, 1, 2$. *Here,* $r_b(\lambda_0)$ *is a constant depending on* m_0, m_1, m_2, ϵ, λ_0, N, *and* q.

Proof To prove Theorem 3.5.8, we start with the solution formulas of Eq. (3.253), which were obtained in Shibata and Shimizu [54] essentially, but for the sake of the completeness of the paper as much as possible and also for the later use, we will derive them in the following. Applying the partial Fourier transform with respect to $x' = (x_1, \ldots, x_{N-1})$ to Eq. (3.253), we have

$$\begin{cases} \lambda\hat{v}_j + \mu|\xi'|^2 - \mu\partial_N^2\hat{v}_j + i\xi_j\hat{\theta} = 0, \\ \lambda\hat{v}_N + \mu|\xi'|^2 - \mu\partial_N^2\hat{v}_N + \partial_N\hat{\theta} = 0 \quad (x_N > 0), \\ \displaystyle\sum_{j=1}^{N-1} i\xi_j\hat{v}_j + \partial_N\hat{v}_N = 0 \quad (x_N > 0), \\ \mu(\partial_N\hat{v}_j + i\xi_j\hat{v}_N) = g_j, \quad 2\mu\partial_N\hat{v}_N - \hat{\theta} = g_N \quad \text{for } x_N = 0. \end{cases} \tag{3.254}$$

Here, for $f = f(x', x_N), x' = (x_1, \ldots, x_{N-1}) \in \mathbb{R}^{N-1}, x_N \in (a, b)$, $\hat{f}$ denotes the partial Fourier transform of f with respect to x' defined by

$$\hat{f}(\xi', x_N) = \mathcal{F}'[f(\cdot, x_N)](\xi') = \int_{\mathbb{R}^{N-1}} e^{-ix'\cdot\xi'} f(x', x_N) \, dx'$$

with $\xi' = (\xi_1, \ldots, \xi_{N-1}) \in \mathbb{R}^{N-1}$ and $x' \cdot \xi' = \sum_{j=1}^{N-1} x_j \xi_j$, and we have set $g_j = -\hat{h}_j(\xi', 0)$. To obtain solution formula, we set

$$\hat{v}_j = \alpha_j e^{-Ax_N} + \beta_j e^{-Bx_N}, \quad \hat{\theta} = \omega e^{-Ax_N}$$

with $A = |\xi'|$ and $B = \sqrt{\lambda\mu^{-1} + |\xi'|^2}$, and then from (3.254) we have

$$\mu\alpha_j(B^2 - A^2) + i\xi_j\omega = 0, \quad \mu\alpha_N(B^2 - A^2) - A\omega = 0, \tag{3.255}$$

$$\sum_{k=1}^{N-1} i\xi_k\alpha_k - A\alpha_N = 0, \quad \sum_{k=1}^{N-1} i\xi_k\beta_k - B\beta_N = 0, \tag{3.256}$$

$$\mu\{(A\alpha_j + B\beta_j) - i\xi_j(\alpha_N + \beta_N)\} = g_j, \tag{3.257}$$

$$2\mu(A\alpha_N + B\beta_N) + \omega = g_N. \tag{3.258}$$

The solution formula of Eq. (3.250) was given in Shibata and Shimizu [54], but there is an error in the formula in [54, (4.17)] such as

$$\mu\{(A\alpha_j + B\beta_j) + i\xi_j(\alpha_N + \beta_N)\} = \hat{h}_j(\xi', 0),$$

which should read

$$\mu\{(A\alpha_j + B\beta_j) - i\xi_j(\alpha_N + \beta_N)\} = -\hat{h}_j(\xi', 0)$$

as (3.257) above. The formulas obtained in [54] are correct, but we repeat here how to obtain α_j, β_j and ω, because this error confuses readers.

We first drive 2×2 system of equations with respect to α_N and β_N. Multiplying (3.257) with $i\xi_j$, summing up the resultant formulas from $j = 1$ through $N - 1$ and writing $i\xi' \cdot m' = \sum_{j=1}^{N-1} i\xi_j m_j$ for $m_j \in \{\alpha_j, \beta_j, g_j\}$ give

$$Ai\xi' \cdot \alpha' + Bi\xi' \cdot \beta' + A^2(\alpha_N + \beta_N) = \mu^{-1}i\xi' \cdot g'. \tag{3.259}$$

By (3.256),

$$i\xi' \cdot \alpha' = A\alpha_N, \quad i\xi' \cdot \beta' = B\beta_N, \tag{3.260}$$

which, combined with (3.259), leads to

$$2A^2\alpha_N + (A^2 + B^2)\beta_N = \mu^{-1}i\xi' \cdot g'. \tag{3.261}$$

By (3.255),

$$\omega = \frac{\mu(B^2 - A^2)}{A}\alpha_N, \tag{3.262}$$

which, combined with (3.258), leads to

$$(A^2 + B^2)\alpha_N + 2AB\beta_N = \mu^{-1}Ag_N. \tag{3.263}$$

Thus, setting

$$\mathcal{L} = \begin{pmatrix} A^2 + B^2 & 2A^2 \\ 2AB & A^2 + B^2 \end{pmatrix} \quad \text{(Lopatinski matrix)},$$

we have

$$\mathcal{L}\begin{pmatrix} \beta_N \\ \alpha_N \end{pmatrix} = \begin{pmatrix} \mu^{-1}i\xi' \cdot g' \\ \mu^{-1}Ag_N \end{pmatrix}.$$

Since

$$\det \mathcal{L} = (A^2 + B^2)^2 - 4A^3B = A^4 - 4A^3B + 2A^2B^2 + B^4 = (B - A)D(A, B)$$

with

$$D(A, B) = B^3 + AB^2 + 3A^2B - A^3,$$

we have

$$\mathcal{L}^{-1} = \frac{1}{(B - A)D(A, B)}\begin{pmatrix} A^2 + B^2 & -2A^2 \\ -2AB & A^2 + B^2 \end{pmatrix}.$$

Thus, we have

$$\beta_N = \frac{1}{\mu(B - A)D(A, B)}((A^2 + B^2)i\xi' \cdot g' - 2A^3g_N),$$

$$\alpha_N = \frac{-1}{\mu(B - A)D(A, B)}(2ABi\xi' \cdot g' - (A^2 + B^2)Ag_N). \tag{3.264}$$

In particular,

$$\hat{v}_N = \alpha_N e^{-Ax_N} + \beta_N e^{-Bx_N} = \alpha_N(e^{-Ax_N} - e^{-Bx_N}) + (\alpha_N + \beta_N)e^{-Bx_N}.$$

We have

$$
\begin{aligned}
&\alpha_N + \beta_N \\
&= \frac{1}{(B-A)D(A,B)}((A^2 + B^2 - 2AB)i\xi' \cdot g' + ((A^2 + B^2)A - 2A^3)g_N) \\
&= \frac{1}{\mu(B-A)D(A,B)}((B-A)^2 i\xi' \cdot g' + A(B^2 - A^2)g_N) \\
&= \frac{1}{\mu(B-A)D(A,B)}((B-A)^2 i\xi' \cdot g' + A(B-A)(A+B)g_N) \\
&= \frac{1}{\mu D(A,B)}((B-A)i\xi' \cdot g' + A(A+B)g_N).
\end{aligned}
$$

$$(3.265)$$

Setting

$$\mathcal{M}(x_N) = \frac{e^{-Bx_N} - e^{-Ax_N}}{B - A},$$

we have

$$
\begin{aligned}
\hat{v}_N &= \frac{A}{\mu D(A,B)}\mathcal{M}(x_N)(2Bi\xi' \cdot g' - (A^2 + B^2)g_N) \\
&\quad + \frac{e^{-Bx_N}}{\mu D(A,B)}((B-A)i\xi' \cdot g' + A(A+B)g_N).
\end{aligned}
$$

$$(3.266)$$

By (3.262) and (3.264),

$$
\begin{aligned}
\omega &= \frac{\mu(B^2 - A^2)}{A}\alpha_N \\
&= \frac{\mu(B^2 - A^2)}{A}\frac{-1}{\mu(B-A)D(A,B)}(2ABi\xi' \cdot g' - (A^2 + B^2)Ag_N) \\
&= -\frac{(A+B)}{D(A,B)}(2Bi\xi' \cdot g' - (A^2 + B^2)g_N)
\end{aligned}
$$

and so

$$\hat{\theta} = -\frac{(A+B)e^{-Ax_N}}{D(A,B)}(2Bi\xi' \cdot g' - (A^2 + B^2)g_N). \qquad (3.267)$$

By (3.255),

$$
\begin{aligned}
\alpha_j &= -\frac{i\xi_j}{\mu(B^2 - A^2)}\omega \\
&= \frac{i\xi_j}{\mu(B^2 - A^2)}\frac{A+B}{D(A,B)}(2Bi\xi' \cdot g' - (A^2 + B^2)g_N) \\
&= \frac{i\xi_j}{\mu(B-A)D(A,B)}(2Bi\xi' \cdot g' - (A^2 + B^2)g_N).
\end{aligned}
\tag{3.268}
$$

By (3.257)

$$
\beta_j = \frac{1}{\mu B}g_j + \frac{1}{B}(i\xi_j(\alpha_N + \beta_N) - A\alpha_j).
$$

By (3.265) and (3.268)

$$
\begin{aligned}
&i\xi_j(\alpha_N + \beta_N) - A\alpha_j \\
&= \frac{i\xi_j}{\mu(B-A)D(A,B)}\{(B-A)^2 i\xi' \cdot g' + A(B-A)(A+B)g_N \\
&\quad - A(2Bi\xi' \cdot g' - (A^2 + B^2)g_N)\} \\
&= \frac{i\xi_j}{\mu(B-A)D(A,B)}\{(A^2 - 4AB + B^2)i\xi' \cdot g' + 2AB^2 g_N)\},
\end{aligned}
$$

and therefore

$$
\beta_j = \frac{1}{\mu B}g_j + \frac{i\xi_j}{\mu(B-A)D(A,B)B}\{(A^2 - 4AB + B^2)i\xi' \cdot g' + 2AB^2 g_N)\}.
\tag{3.269}
$$

Combining (3.268) and (3.269) gives

$$
\begin{aligned}
\hat{v}_j &= \frac{e^{-Bx_N}}{\mu B}g_j + \frac{i\xi_j e^{-Ax_N}}{\mu(B-A)D(A,B)}\{2Bi\xi' \cdot g' - (A^2 + B^2)g_N\} \\
&\quad + \frac{i\xi_j e^{-Bx_N}}{\mu(B-A)D(A,B)B}\{(A^2 - 4AB + B^2)i\xi' \cdot g' + 2AB^2 g_N)\} \\
&= \frac{1}{\mu B}g_j e^{-Bx_N} + Ii\xi' \cdot g' + IIg_N,
\end{aligned}
$$

with

$$
I = \frac{i\xi_j e^{-Ax_N}}{\mu(B-A)D(A,B)}2B + \frac{i\xi_j e^{-Bx_N}}{\mu(B-A)D(A,B)B}(A^2 - 4AB + B^2),
$$

$$
II = -\frac{i\xi_j e^{-Ax_N}}{\mu(B-A)D(A,B)}(A^2 + B^2) + \frac{i\xi_j e^{-Bx_N}}{\mu(B-A)D(A,B)}2AB
$$

We proceed as follows:

$$
\begin{aligned}
I &= \frac{i\xi_j(e^{-Ax_N} - e^{-Bx_N})}{\mu(B-A)D(A,B)}2B + \frac{i\xi_j e^{-Bx_N}}{\mu(B-A)D(A,B)B}(A^2 - 4AB + 3B^2) \\
&= -\frac{2i\xi_j B\mathcal{M}(x_N)}{\mu D(A,B)} + \frac{i\xi_j(3B-A)e^{-Bx_N}}{\mu D(A,B)B}; \\
II &= -\frac{i\xi_j(e^{-Ax_N} - e^{-Bx_N})}{\mu(B-A)D(A,B)}(A^2 + B^2) - \frac{i\xi_j e^{-Bx_N}(A^2 - 2AB + B^2)}{\mu(B-A)D(A,B)} \\
&= \frac{i\xi_j(A^2 + B^2)\mathcal{M}(x_N)}{\mu D(A,B)} - \frac{i\xi_j e^{-Bx_N}(B-A)}{\mu D(A,B)}.
\end{aligned}
$$

Therefore, we have

$$
\begin{aligned}
\hat{v}_j &= \frac{e^{-Bx_N}}{\mu B}g_j - \frac{i\xi_j \mathcal{M}(x_N)}{\mu D(A,B)}(2Bi\xi' \cdot g' - (A^2 + B^2)g_N) \\
&\quad + \frac{i\xi_j e^{-Bx_N}}{\mu D(A,B)B}((3B-A)i\xi' \cdot g' - B(B-A)g_N).
\end{aligned}
\tag{3.270}
$$

To define solution operators for Eq. (3.250), we make preparations.

Lemma 3.5.9 *Let $s \in \mathbb{R}$ and $0 < \epsilon < \pi/2$. Then, there exists a positive constant c depending on ϵ, m_1 and m_2 such that*

$$
c(|\lambda|^{1/2} + A) \le \operatorname{Re} B \le |B| \le (\mu^{-1}|\lambda|)^{1/2} + A, \tag{3.271}
$$

$$
c(|\lambda|^{1/2} + A)^3 \le |D(A,B)| \le 6((\mu^{-1}|\lambda|)^{1/2} + A)^3 \tag{3.272}
$$

for any $\lambda \in \Sigma_\epsilon$ and $\mu \in [m_1, m_2]$.

Proof The inequality in the left side of (3.271) follows immediately from Lemma 3.5.2. Notice that

$$
\begin{aligned}
D(A,B) &= B^3 + 3A^2 B + AB^2 - A^3 = B(B^2 + 2A^2) + A(A^2 + \mu^{-1}\lambda) - A^3 \\
&= B(\mu^{-1}\lambda + 4A^2) + \mu^{-1}A\lambda.
\end{aligned}
$$

If we consider the angle of $B(\mu^{-1}\lambda + 4A^2)$ and $-\mu^{-1}A\lambda$, then we see easily that $D(A,B) \neq 0$. Thus, studying the following three cases: $R_1|\lambda|^{1/2} \le A$, $R_1 A \le |\lambda|^{1/2}$ and $R_1^{-1}A \le |\lambda|^{1/2} \le R_1 A$ for sufficient large $R_1 > 0$, we can prove the inequality in the left side of (3.272). The detailed proof was given in Shibata and Shimizu [50]. The independence of the constant c of $\lambda \in \Sigma_\epsilon$ and $\mu \in [m_0, m_1]$ follows from the homogeneity: $\sqrt{\mu^{-1}(m^2\lambda) + (mA)^2} = m\sqrt{\mu^{-1}\lambda + A^2}$ and $D(mA, mB) = m^3 D(A,B)$ for any $m > 0$ and the compactness of the interval $[m_0, m_1]$. □

To introduce the key tool of proving the $\mathcal{R}$ boundedness in the half space, we make a definition.

Definition 3.5.10 Let V be a domain in $\mathbb{C}$, let $\varXi = V \times (\mathbb{R}^{N-1} \setminus \{0\})$, and let $m : \varXi \to \mathbb{C}; (\lambda, \xi') \mapsto m(\lambda, \xi')$ be C^1 with respect to τ, where $\lambda = \gamma + i\tau \in V$, and C^∞ with respect to $\xi' \in \mathbb{R}^{N-1} \setminus \{0\}$.

(1) $m(\lambda, \xi')$ is called a multiplier of order s with type 1 on $\varXi$, if the estimates:

$$|\partial_{\xi'}^{\kappa'} m(\lambda, \xi')| \le C_{\kappa'}(|\lambda|^{1/2} + |\xi'|)^{s-|\kappa'|},$$

$$|\partial_{\xi'}^{\kappa'}(\tau \partial_\tau m(\lambda, \xi'))| \le C_{\kappa'}(|\lambda|^{1/2} + |\xi'|)^{s-|\kappa'|}$$

hold for any multi-index $\kappa \in \mathbb{N}_0^N$ and $(\lambda, \xi') \in \varXi$ with some constant $C_{\kappa'}$ depending solely on κ' and V.

(2) $m(\lambda, \xi')$ is called a multiplier of order s with type 2 on $\varXi$, if the estimates:

$$|\partial_{\xi'}^{\kappa'} m(\lambda, \xi')| \le C_{\kappa'}(|\lambda|^{1/2} + |\xi'|)^s |\xi'|^{-|\kappa'|},$$

$$|\partial_{\xi'}^{\kappa'}(\tau \partial_\tau m(\lambda, \xi'))| \le C_{\kappa'}(|\lambda|^{1/2} + |\xi'|)^s |\xi'|^{-|\kappa'|}$$

hold for any multi-index $\kappa \in \mathbb{N}_0^N$ and $(\lambda, \xi') \in \varXi$ with some constant $C_{\kappa'}$ depending solely on κ' and V.

Let $\mathbf{M}_{s,i}(V)$ be the set of all multipliers of order s with type i on $\varXi$ for $i = 1, 2$. For $m \in \mathbf{M}_{s,i}(V)$, we set $M(m, V) = \max_{|\kappa'| \le N} C_{\kappa'}$.

Let $\mathcal{F}_{\xi'}^{-1}$ be the inverse partial Fourier transform defined by

$$\mathcal{F}_{\xi'}^{-1}[f(\xi', x_N)](x') = \frac{1}{(2\pi)^{N-1}} \int_{\mathbb{R}^{N-1}} e^{ix' \cdot \xi'} f(\xi', x_N) \, d\xi'.$$

Then, we have the following two lemmata which have been proved essentially by Shibata and Shimizu [54, Lemma 5.4 and Lemma 5.6].

Lemma 3.5.11 *Let* $0 < \epsilon < \pi/2$, $1 < q < \infty$, *and* $\lambda_0 > 0$. *Given* $m \in \mathbf{M}_{-2,1}(\Lambda_{\kappa,\lambda_0})$, *we define an operator* $L(\lambda)$ *by*

$$[L(\lambda)g](x) = \int_0^\infty \mathcal{F}_{\xi'}^{-1}[m(\lambda, \xi')\lambda^{1/2} e^{-B(x_N + y_N)} \hat{g}(\xi', y_N)](x') \, dy_N.$$

Then, we have

$$\mathcal{R}_{\mathcal{L}(L_q(\mathbb{R}_+^N), H_q^{2-j}(\mathbb{R}_+^N)^N)}(\{(\tau \partial_\tau)^\ell (\lambda^{j/2} \partial_x^\alpha L(\lambda)) \mid \lambda \in \Lambda_{\kappa,\lambda_0}\}) \le r_b(\lambda_0)$$

for $\ell = 0, 1$ *and* $j = 0, 1, 2$, *where* τ *denotes the imaginary part of* λ, *and* $r_b(\lambda_0)$ *is a constant depending on* $M(m, \Lambda_{\kappa,\lambda_0})$, ϵ, λ_0, N, *and* q.

Lemma 3.5.12 *Let* $0 < \epsilon < \pi/2$, $1 < q < \infty$, *and* $\lambda_0 > 0$. *Given* $m \in$ $\mathbf{M}_{-2,2}(\Lambda_{\kappa,\lambda_0})$, *we define operators* $L_i(\lambda)$ $(i = 1, \dots, 4)$ *by*

$$[L_1(\lambda)g](x) = \int_0^\infty \mathcal{F}_{\xi'}^{-1}[m(\lambda, \xi')Ae^{-B(x_N+y_N)}\hat{g}(\xi', y_N)](x')\, dy_N,$$

$$[L_2(\lambda)g](x) = \int_0^\infty \mathcal{F}_{\xi'}^{-1}[m(\lambda, \xi')Ae^{-A(x_N+y_N)}\hat{g}(\xi', y_N)](x')\, dy_N,$$

$$[L_3(\lambda)g](x) = \int_0^\infty \mathcal{F}_{\xi'}^{-1}[m(\lambda, \xi')A^2\mathcal{M}(x_N + y_N)\hat{g}(\xi', y_N)](x')\, dy_N,$$

$$[L_4(\lambda)g](x) = \int_0^\infty \mathcal{F}_{\xi'}^{-1}[m(\lambda, \xi')\lambda^{1/2}A\mathcal{M}(x_N + y_N)\hat{g}(\xi', y_N)](x')\, dy_N.$$

Then, we have

$$\mathcal{R}_{\mathcal{L}(L_q(\mathbb{R}_+^N), H_q^{2-j}(\mathbb{R}_+^N)^N)}(\{(\tau\partial_\tau)^\ell(\lambda^{j/2}\partial_x^\alpha L_i(\lambda)) \mid \lambda \in \Lambda_{\kappa,\lambda_0}\}) \leq r_b(\lambda_0)$$

for $\ell = 0, 1$ *and* $j = 0, 1, 2$, *where* τ *denotes the imaginary part of* λ, *and* $r_b(\lambda_0)$ *is a constant depending on* $M(m, \Lambda_{\kappa,\lambda_0})$, ϵ, λ_0, N, *and* q.

To construct solution operators, we use the following lemma.

Lemma 3.5.13 *Let* $0 < \epsilon < \pi/2$, $1 < q < \infty$ *and* $\lambda_0 > 0$. *Given multipliers,* $n_1 \in$ $\mathbf{M}_{-2,1}(\Lambda_{\kappa,\lambda_0})$, $n_2 \in \mathbf{M}_{-2,2}(\Lambda_{\kappa,\lambda_0})$, *and* $n_3 \in \mathbf{M}_{-1,2}(\Lambda_{\kappa,\lambda_0})$, *we define operators* $T_i(\lambda)$ $(i = 1, 2, 3)$ *by*

$$T_1(\lambda)h = \mathcal{F}_{\xi'}^{-1}[\lambda^{1/2}e^{-Bx_N}n_1(\lambda, \xi')\hat{h}(\xi', 0)](x'),$$

$$T_2(\lambda)h = \mathcal{F}_{\xi'}^{-1}[Ae^{-Bx_N}n_2(\lambda, \xi')\hat{h}(\xi', 0)](x'),$$

$$T_3(\lambda)h = \mathcal{F}_{\xi'}^{-1}[A\mathcal{M}(x_N)n_3(\lambda, \xi')\hat{h}(\xi', 0)](x').$$

Let

$$\mathcal{Z}_q(\mathbb{R}_+^N) = \{(G_3, G_4) \mid G_3 \in L_q(\mathbb{R}_+^N), \ G_4 \in H_q^1(\mathbb{R}_+^N)\}.$$

Then, there exist operator families $\mathcal{T}_i(\lambda) \in Hol\,(\Lambda_{\kappa,\lambda_0}, \mathcal{L}(\mathcal{Y}_q'(\mathbb{R}_+^N), H_q^2(\mathbb{R}_+^N)))$ *such that for any* $\lambda = \gamma + i\tau \in \Lambda_{\kappa,\lambda_0}$ *and* $h \in H_q^1(\mathbb{R}_+^N)$, $T_i(\lambda)h = \mathcal{T}_i(\lambda)(\lambda^{1/2}h, h)$ *and*

$$\mathcal{R}_{\mathcal{L}(\mathcal{Y}_q'(\mathbb{R}_+^N), H_q^{2-j}(\mathbb{R}_+^N))}(\{(\tau\partial_\tau)^\ell(\lambda^{j/2}\mathcal{T}_i(\lambda)) \mid \lambda \in \Lambda_{\kappa,\lambda_0}\}) \leq r_b(\lambda_0) \qquad (3.273)$$

for $\ell = 0, 1$, $j = 0, 1, 2$, *where* $r_b(\lambda_0)$ *is a constant depending on* $M(n_i, \Lambda_{\kappa,\lambda_0})$ $(i = 1, 2, 3)$, ϵ, λ_0, N, *and* q.

Proof By Volevich's trick we write

$$T_1(\lambda)h$$

$$= -\int_0^\infty \mathcal{F}_{\xi'}^{-1}[\frac{\partial}{\partial y_N}(\lambda^{1/2}e^{-B(x_N+y_N)}n_1(\lambda,\xi')\hat{h}(\xi',y_N))](x')\,dy_N$$

$$= -\int_0^\infty \mathcal{F}_{\xi'}^{-1}[\lambda^{1/2}e^{-B(x_N+y_N)}n_1(\lambda,\xi')\partial_N\hat{h}(\xi',y_N)](x')\,dy_N$$

$$+ \int_0^\infty \mathcal{F}_{\xi'}^{-1}[\lambda^{1/2}e^{-B(x_N+y_N)}\frac{\lambda^{1/2}}{\mu B}n_1(\lambda,\xi')\lambda^{1/2}\hat{h}(\xi',y_N)](x')\,dy_N$$

$$- \sum_{j=1}^{N-1}\int_0^\infty \mathcal{F}_{\xi'}^{-1}[Ae^{-B(x_N+y_N)}\frac{\lambda^{1/2}}{B}\frac{i\xi_j}{A}n_1(\lambda,\xi')\mathcal{F}[\partial_j h(\cdot,y_N)]](x')\,dy_N,$$

where we have used the formula:

$$B = \frac{\mu^{-1}\lambda + A^2}{\mu B} = \frac{\lambda}{\mu B} - \sum_{j=1}^{N-1}\frac{A}{B}\frac{i\xi_j}{A}i\xi_j.$$

Let

$$\mathcal{T}_1(\lambda)(G_3, G_4)$$

$$= -\int_0^\infty \mathcal{F}_{\xi'}^{-1}[\lambda^{1/2}e^{-B(x_N+y_N)}n_1(\lambda,\xi')\mathcal{F}[\partial_N G_4(\cdot,y_N)]](x')\,dy_N$$

$$+ \int_0^\infty \mathcal{F}_{\xi'}^{-1}[\lambda^{1/2}e^{-B(x_N+y_N)}\frac{\lambda^{1/2}}{\mu B}n_1(\lambda,\xi')\mathcal{F}[G_3(\cdot,y_N)]](x')\,dy_N$$

$$- \sum_{j=1}^{N-1}\int_0^\infty \mathcal{F}_{\xi'}^{-1}[Ae^{-B(x_N+y_N)}\frac{\lambda^{1/2}}{B}\frac{i\xi_j}{A}n_1(\lambda,\xi')\mathcal{F}[\partial_j G_4(\cdot,y_N)]](x')\,dy_N,$$

and then, $T_1(\lambda)h = \mathcal{T}_1(\lambda)(\lambda^{1/2}h, h)$. Moreover, Lemmas 3.5.11 and 3.5.12 yield
(3.273) with $j = 1$, because

$$n_1(\lambda,\xi') \in \mathbf{M}_{-2,1}(\Lambda_{\kappa,\lambda_0}), \qquad \frac{\lambda^{1/2}}{\mu B}n_1(\lambda,\xi') \in \mathbf{M}_{-2,1}(\Lambda_{\kappa,\lambda_0}),$$

$$\frac{\lambda^{1/2}}{B}\frac{i\xi_j}{A}n_1(\lambda,\xi') \in \mathbf{M}_{-2,2}(\Lambda_{\kappa,\lambda_0}).$$

Analogously, we can prove the existence of $\mathcal{T}_2(\lambda)$.

To construct $T_3(\lambda)$, we use the formula:

$$\frac{\partial}{\partial x_N}\mathcal{M}(x_N) = -e^{-Bx_N} - A\mathcal{M}(x_N),$$

and then, by Volevich's trick we have

$T_3(\lambda)h$

$$= -\int_0^\infty \mathcal{F}_{\xi'}^{-1}[\frac{\partial}{\partial y_N}(A\mathcal{M}(x_N + y_N)n_3(\lambda, \xi')\hat{h}(\xi', y_N))](x')\,dy_N = -I + II$$

with

$$I = \int_0^\infty \mathcal{F}_{\xi'}^{-1}[A\mathcal{M}(x_N + y_N)n_3(\lambda, \xi')\partial_N\hat{h}(\xi', y_N)](x')\,dy_N;$$

$$II = \int_0^\infty \mathcal{F}_{\xi'}^{-1}[(Ae^{-B(x_N+y_N)} + A^2\mathcal{M}(x_N + y_N))n_3(\lambda, \xi')\hat{h}(\xi', y_N)](x')\,dy_N.$$

Using the formula:

$$1 = \frac{B^2}{B^2} = \frac{\lambda^{1/2}}{\mu B^2}\lambda^{1/2} + \frac{A}{B^2}A = \frac{\lambda^{1/2}}{\mu B^2}\lambda^{1/2} - \sum_{j=1}^{N-1}\frac{i\xi_j}{B^2}i\xi_j,$$

we have

$$I = \int_0^\infty \mathcal{F}_{\xi'}^{-1}[\lambda^{1/2}A\mathcal{M}(x_N + y_N)\frac{\lambda^{1/2}}{\mu B^2}n_3(\lambda, \xi')\partial_N\hat{h}(\xi', y_N)](x')\,dy_N$$

$$+ \int_0^\infty \mathcal{F}_{\xi'}^{-1}[A^2\mathcal{M}(x_N + y_N)\frac{A}{B^2}n_3(\lambda, \xi')\partial_N\hat{h}(\xi', y_N)](x')\,dy_N;$$

$$II = \int_0^\infty \mathcal{F}_{\xi'}^{-1}[(Ae^{-B(x_N+y_N)} + A^2\mathcal{M}(x_N + y_N))$$

$$\times \frac{\lambda^{1/2}}{\mu B^2}n_3(\lambda, \xi')\lambda^{1/2}\hat{h}(\xi', y_N)](x')\,dy_N$$

$$- \sum_{j=1}^{N-1}\int_0^\infty \mathcal{F}_{\xi'}^{-1}[(Ae^{-B(x_N+y_N)} + A^2\mathcal{M}(x_N + y_N))$$

$$\times \frac{i\xi_j}{B^2}n_3(\lambda, \xi')\mathcal{F}[\partial_j h(\cdot, y_N)]](x')\,dy_N.$$

Let

$$T_3(\lambda)(G_3, G_4)$$

$$= -\int_0^\infty \mathcal{F}_{\xi'}^{-1}[\lambda^{1/2} A \mathcal{M}(x_N + y_N) \frac{\lambda^{1/2}}{\mu B^2} n_3(\lambda, \xi') \mathcal{F}[\partial_N G_4(\cdot, y_N)]](x') \, dy_N$$

$$- \int_0^\infty \mathcal{F}_{\xi'}^{-1}[A^2 \mathcal{M}(x_N + y_N) \frac{A}{B^2} n_3(\lambda, \xi') \mathcal{F}[\partial_N G_4(\cdot, y_N)]](x') \, dy_N$$

$$+ \int_0^\infty \mathcal{F}_{\xi'}^{-1}[(Ae^{-B(x_N+y_N)} + A^2 \mathcal{M}(x_N + y_N))$$

$$\times \frac{\lambda^{1/2}}{\mu B^2} n_3(\lambda, \xi') \mathcal{F}[G_3(\cdot, y_N)]](x') \, dy_N$$

$$- \sum_{j=1}^{N-1} \int_0^\infty \mathcal{F}_{\xi'}^{-1}[(Ae^{-B(x_N+y_N)} + A^2 \mathcal{M}(x_N + y_N))$$

$$\times \frac{i\xi_j}{B^2} n_3(\lambda, \xi') \mathcal{F}[\partial_j G_4(\cdot, y_N)]](x') \, dy_N,$$

and then $T_3(\lambda)h = T_3(\lambda)(\lambda^{1/2}h, h)$. Moreover, Lemma 3.5.12 yields (3.273) for $j = 3$, because

$$n_3(\lambda) \in \mathbf{M}_{-2,2}(\Lambda_{\kappa,\lambda_0}), \qquad \frac{\lambda^{1/2}}{\mu B^2} n_3(\lambda, \xi') \in \mathbf{M}_{-2,2}(\Lambda_{\kappa,\lambda_0}),$$

$$\frac{i\xi_j}{B^2} n_3(\lambda, \xi') \in \mathbf{M}_{-2,2}(\Lambda_{\kappa,\lambda_0}).$$

This completes the proof of Lemma 3.5.13. $\qquad\square$

Continuation of Proof of Theorem 3.5.8 Let

$$v_j(x) = \mathcal{F}_{\xi'}^{-1}[\hat{v}_j(\xi', x_N)](x'),$$

and then by (3.266) and (3.270) we have

$$v_N = \mathcal{F}_{\xi'}^{-1}\left[\frac{A}{\mu D(A, B)} \mathcal{M}(x_N)((A^2 + B^2)\hat{h}_N(\xi', 0) - 2B \sum_{\ell=1}^{N-1} i\xi_\ell \hat{h}_\ell(\xi', 0))\right](x')$$

$$- \mathcal{F}_{\xi'}^{-1}\left[\frac{Ae^{-Bx_N}}{\mu D(A, B)}((B - A) \sum_{\ell=1}^{N-1} \frac{i\xi_\ell}{A} \hat{h}_\ell(\xi', 0) + (A + B)\hat{h}_N(\xi', 0))\right](x');$$

$$v_k = -\mathcal{F}_{\xi'}^{-1}\left[\frac{\lambda^{1/2}}{\mu^2 B^3} \lambda^{1/2} e^{-Bx_N} \hat{h}_k(\xi', 0)\right](x') - \mathcal{F}_{\xi'}^{-1}\left[\frac{A}{\mu B^3} Ae^{-Bx_N} \hat{h}_k(\xi', 0)\right](x')$$

$$+ \mathcal{F}_{\xi'}^{-1} \Big[A\mathcal{M}(x_N) \frac{i\xi_k}{A} \frac{1}{\mu D(A, B)} (2B \sum_{\ell=1}^{N-1} i\xi_\ell \hat{h}_\ell(\xi', 0)$$

$$- (A^2 + B^2) \hat{h}_N(\xi', 0)) \Big] (x')$$

$$- \mathcal{F}_{\xi'}^{-1} \Big[A e^{-Bx_N} \frac{i\xi_k}{A} \frac{1}{\mu D(A, B)B} ((3B - A) \sum_{\ell=1}^{N-1} i\xi_\ell \hat{h}_\ell(\xi', 0)$$

$$- B(B - A) \hat{h}_N(\xi', 0)) \Big] (x'),$$

for $k = 1, \ldots, N - 1$, where we have used the formula $\dfrac{1}{\mu B} = \dfrac{\lambda}{\mu^2 B^3} + \dfrac{A^2}{\mu B^3}$ to treat the first term of $\hat{v}_j$ in (3.270). Since

$$\frac{Bi\xi_\ell}{\mu D(A, B)}, \quad \frac{A^2 + B^2}{\mu D(A, B)}, \quad \frac{i\xi_k}{A} \frac{Bi\xi_\ell}{\mu D(A, B)}, \quad \frac{i\xi_k}{A} \frac{A^2 + B^2}{\mu D(A, B)} \in \mathbf{M}_{-1,2}(\Sigma_{\epsilon,\lambda_0}),$$

$$\frac{B - A}{\mu D(A, B)} \frac{i\xi_\ell}{A}, \quad \frac{A + B}{\mu D(A, B)}, \quad \frac{A}{\mu B^3} \in \mathbf{M}_{-2,2}(\Sigma_{\epsilon,\lambda_0})$$

$$\frac{i\xi_k}{A} \frac{(3B - A)i\xi_\ell}{\mu D(A, B)B}, \quad \frac{i\xi_k}{A} \frac{B(B - A)}{\mu D(A, B)B} \in \mathbf{M}_{-2,2}(\Sigma_{\epsilon,\lambda_0}),$$

and $\dfrac{\lambda^{1/2}}{\mu^2 B^3} \in \mathbf{M}_{-2,1}(\Sigma_{\epsilon,\lambda_0})$, by Lemma 3.5.13 we have Theorem 3.5.8. □

We next consider the equations:

$$(3.274) \quad \begin{cases} \lambda \mathbf{w} - \mathrm{Div}\,(\mu \mathbf{D}(\mathbf{w}) - q\mathbf{I}) = 0, \quad \mathrm{div}\,\mathbf{w} = 0 & \text{in } \mathbb{R}_+^N, \\ \lambda h + A_\kappa \cdot \nabla' h - \mathbf{w} \cdot \mathbf{n}_0 = d & \text{on } \mathbb{R}_0^N, \\ (\mu \mathbf{D}(\mathbf{w}) - q\mathbf{I})\mathbf{n}_0 - \sigma(\Delta' h)\mathbf{n}_0 = 0 & \text{on } \mathbb{R}_0^N. \end{cases}$$

We shall prove the following theorem.

Theorem 3.5.14 *Let* $1 < q < \infty$ *and* $\epsilon \in (0, \pi/2)$. *Then, there exist a* $\lambda_1 > 0$ *and solution operators* $\mathcal{W}_\kappa(\lambda)$ *and* $\mathcal{H}_\kappa(\lambda)$ *with*

$$\mathcal{W}_\kappa(\lambda) \in Hol\,(\Lambda_{\kappa,\lambda_1}, \mathcal{L}(H_q^2(\mathbb{R}_+^N), H_q^2(\mathbb{R}_+^N)^N)),$$

$$\mathcal{H}_\kappa(\lambda) \in Hol\,(\Lambda_{\kappa,\lambda_1}, \mathcal{L}(H_q^2(\mathbb{R}_+^N), H_q^3(\mathbb{R}_+^N))),$$

such that for any $\lambda = \gamma + i\tau \in \Lambda_{\kappa,\lambda_1}$ and $d \in H_q^2(\mathbb{R}_+^N)$, $\mathbf{w} = \mathcal{W}_\kappa(\lambda)d$ and $h = \mathcal{H}_\kappa(\lambda)d$ are unique solutions of Eq. (3.274) with some $\mathfrak{q} \in \hat{H}_q^1(\Omega)$, and

$$\mathcal{R}_{\mathcal{L}(H_q^2(\mathbb{R}_+^N),H_q^{2-k}(\mathbb{R}_+^N)^N)}(\{(\tau\partial_\tau)^\ell(\lambda^{k/2}\mathcal{W}_\kappa(\lambda)) \mid \lambda \in \Lambda_{\kappa,\lambda_1}\}) \le r_b(\lambda_1),$$

$$\mathcal{R}_{\mathcal{L}(H_q^2(\mathbb{R}_+^N),H_q^{3-m}(\mathbb{R}_+^N))}(\{(\tau\partial_\tau)^\ell(\lambda^m\mathcal{H}_\kappa(\lambda)) \mid \lambda \in \Lambda_{\kappa,\lambda_1}\}) \le r_b(\lambda_1)$$

for $\ell = 0, 1$, $k = 0, 1, 2$, and $m = 0, 1$, where $r_b(\lambda_1)$ is a constant depending on m_0, m_1, m_2, ϵ, λ_1, N, and q.

Proof We start with solution formulas. Applying the partial Fourier transform to Eq. (3.274), we have the following generalized resolvent problem:

$$\lambda\hat{w}_j + \mu|\xi'|^2\hat{w}_j - \mu\partial_N^2\hat{w}_j + i\xi_j\hat{\mathfrak{q}} = 0 \quad (x_N > 0),$$

$$\lambda\hat{w}_N + \mu|\xi'|^2\hat{w}_N - \mu\partial_N^2\hat{w}_N + \partial_N\hat{\mathfrak{q}} = 0 \quad (x_N > 0),$$

$$\sum_{j=1}^{N-1} i\xi_j\hat{w}_j + \partial_N\hat{w}_N = 0 \quad (x_N > 0),$$

$$\mu(\partial_N\hat{w}_j(0) + i\xi_j\hat{w}_N(0)) = 0, \quad 2\mu\partial_N\hat{w}_N - \mathfrak{q} = -\sigma A^2\hat{h} \quad \text{for } x_N = 0,$$

$$\lambda\hat{h} + \sum_{j=1}^{N-1} i\xi_j A_{\kappa j}\hat{h} + \hat{w}_N = \hat{d} \quad \text{for } x_N = 0.$$

$$(3.275)$$

Here, we have set $A_\kappa = (A_{\kappa 1}, \ldots, A_{\kappa N-1})$. Using the solution formulas given in (3.266), (3.267) and (3.270) with $g_j = 0$ $(j = 1, \ldots, N-1)$, and $g_N = \sigma A^2\hat{h}$, we have

$$\hat{w}_j = \frac{i\xi_j\mathcal{M}(x_N)}{\mu D(A, B)}\sigma A^2(A^2 + B^2)\hat{h} - \frac{i\xi_j e^{-Bx_N}}{\mu D(A, B)}\sigma A^2(B - A)\hat{h},$$

$$\hat{w}_N = -\frac{A\mathcal{M}(x_N)}{\mu D(A, B)}\sigma A^2(A^2 + B^2)\hat{h} + \frac{e^{-Bx_N}}{\mu D(A, B)}\sigma A^3(A + B)\hat{h}, \quad (3.276)$$

$$\hat{\mathfrak{q}} = -\frac{(A + B)A^2(A^2 + B^2)e^{-Ax_N}}{D(A, B)}\hat{h}.$$

Inserting the formula of $\hat{w}_N|_{x_N=0}$ into the last equation in (3.275), we have

$$(\lambda + i\xi' \cdot A_\kappa)\hat{h} + \frac{\sigma A^3(A + B)}{\mu D(A, B)}\hat{h} = \hat{d},$$

where we have set $i\xi' \cdot A_\kappa = \sum_{j=1}^{N-1} i\xi_j A_{\kappa j}$, which implies that

$$\hat{h} = \frac{\mu D(A, B)}{E_\kappa} \hat{d} \qquad (3.277)$$

with $E_\kappa = \mu(\lambda + i\xi' \cdot A_\kappa)D(A, B) + \sigma A^3(A + B)$. Thus, we have the following solution formulas:

$$\hat{w}_j = i\xi_j \mathcal{M}(x_N)\frac{\sigma A^2(A^2 + B^2)}{E_\kappa}\hat{d} - i\xi_j e^{-Bx_N}\frac{\sigma A^2(B - A)}{E_\kappa}\hat{d},$$

$$\hat{w}_N = -A\mathcal{M}(x_N)\frac{\sigma A^2(A^2 + B^2)}{E_\kappa}\hat{d} + e^{-Bx_N}\frac{\sigma A^3(A + B)}{E_\kappa}\hat{d}, \qquad (3.278)$$

$$\hat{q} = \frac{\sigma(A + B)A^2(A^2 + B^2)e^{-Ax_N}}{E_\kappa}\hat{d}.$$

Concerning the estimation for E_κ, we have the following lemma.

Lemma 3.5.15

(1) *Let* $0 < \epsilon < \pi/2$ *and let* E_0 *be the function defined in* (3.277) *with* $A_0 = 0$. *Then, there exists a* $\lambda_1 > 0$ *and* $c > 0$ *such that the estimate:*

$$|E_0| \geq c(|\lambda| + A)(|\lambda|^{1/2} + A)^3 \qquad (3.279)$$

holds for $(\lambda, \xi') \in \Sigma_{\epsilon,\lambda_1} \times (\mathbb{R}^{N-1} \setminus \{0\})$.

(2) *Let* $\kappa \in (0, 1)$ *and let* E_κ *be the function defined in* (3.277). *Then, there exists a* $\lambda_1 > 0$ *and* $c > 0$ *such that*

$$|E_\kappa| \geq c(|\lambda| + A)(|\lambda|^{1/2} + A)^3 \qquad (3.280)$$

holds for $(\lambda, \xi') \in \mathbb{C}_{+,\lambda_1} \times (\mathbb{R}^{N-1} \setminus \{0\})$.

Here, the constant c *in* (1) *and* (2) *depends on* λ_1, m_0, m_1, *and* m_2.

Proof We first study the case where $|\lambda| \geq R_1 A$ for large $R_1 > 0$. Note that $|\lambda| \geq \lambda_1$. Since $|B| \leq A + \mu^{-1/2}|\lambda|^{1/2}$ and since $\Lambda_{\kappa,\lambda_1} \subset \Sigma_\epsilon$, by Lemma 3.5.9 we have

$$|E_\kappa| \geq \mu|\lambda||D(A, B)| - \mu|A_\kappa||A||D(A, B)| - \sigma A^3(A + \mu^{-1/2}|\lambda|^{1/2})$$

$$\geq c\mu|\lambda|(|\lambda|^{1/2} + A)^3 - \mu m_2 C R_1^{-1}|\lambda|(|\lambda|^{1/2} + A)^3$$

$$- \sigma R_1^{-1}|\lambda|(|\lambda|^{1/2} + A)^3 - \mu^{-1/2}\sigma|\lambda|^{1/2}(|\lambda|^{1/2} + A)^3$$

$$\geq (c\mu/2)|\lambda|(|\lambda|^{1/2} + A)^3 + ((c\mu/2) - \mu m_2 C R_1^{-1}$$

$$- \sigma R_1^{-1} - \sigma/(\mu|\lambda|)^{1/2})|\lambda|(|\lambda|^{1/2} + A)^3.$$

Thus, choosing $R_1 > 0$ and $\lambda_1 > 0$ so large that $(c\mu/4) - \mu m_2 C R_1^{-1} - \sigma R_1^{-1} \geq 0$ and $(c\mu/4) - \sigma/(\mu\lambda_1)^{1/2} \geq 0$, we have

$$|\tilde{E}_\kappa| \geq (c\mu/2)|\lambda|(|\lambda|^{1/2} + A)^3 \geq (c\mu/4)(|\lambda| + R_1 A)(|\lambda|^{1/2} + A)^3 \qquad (3.281)$$

provided that $|\lambda| \geq R_1 A$ and $\lambda \in \Lambda_{\kappa,\lambda_1}$. When $\kappa = 0$, we may assume that $m_2 = 0$ above.

We now consider the case where $|\lambda| \leq R_1 A$. We assume that $\lambda \in \Sigma_{\epsilon,\lambda_1}$. In this case, we have $A \geq R_1^{-1}|\lambda|^{1/2}\lambda_1^{1/2}$, and so, choosing λ_1 large enough, we have $B = A(1 + O(\lambda_1^{-1/2}))$. In particular, $D(A, B) = 4A^3(1 + O(\lambda_1^{-1/2}))$. Thus, we have

$$E_\kappa = 4\mu(\lambda + i\xi' \cdot A_\kappa)A^3(1 + O(\lambda_1^{-1/2})) + 2\sigma A^4(1 + O(\lambda_1^{-1/2})).$$

We first consider the case where $\kappa = 0$. Using Lemma 3.5.2, we have

$$|E_0| \geq |4\mu\lambda A^3 + 2\sigma A^4| - 4\mu|\lambda|A^3 O(\lambda_1^{-1/2}) - 2\sigma A^4 O(\lambda_1^{-1/2})$$

$$\geq (\sin \epsilon)(4\mu|\lambda|A^3 + 2\sigma A^4) - O(\lambda_1^{-1/2})(4\mu|\lambda|A^3 + 2\sigma A^4).$$

Thus, choosing $\lambda_1 > 0$ so large that $(\sin \epsilon/2) - O(\lambda_1^{-1/2}) \geq 0$, we have

$$|E_0| \geq (\sin \epsilon/2)(4\mu|\lambda|A^3 + 2\sigma A^4) \geq c(|\lambda| + A)A^3$$

$$\geq c/2^3(|\lambda| + A)(A + R_1^{-1}\lambda_1^{1/2}|\lambda|^{1/2})^3.$$

This completes the proof of (1).

We next consider the case of $\kappa \in (0, 1)$. Taking the real part gives

$$\text{Re } E_\kappa = 4\mu(\text{Re }\lambda)A^3(1 + O(\lambda_1^{-1/2})) + O(\lambda_1^{-1/2})(\text{Im }\lambda + A_\kappa \cdot \xi')A^3$$

$$+ 2\sigma A^4(1 + O(\lambda_1^{-1/2})).$$

Since $\text{Re }\lambda \geq \lambda_1 > 0$ and $|\lambda| \leq R_1 A$, we have

$$\text{Re } E_\kappa \geq 2\sigma A^4 - (4\mu(m_2 + R_1) + 2\sigma)O(\lambda_1^{-1/2})A^4,$$

and so, choosing $\lambda_1 > 0$ so large that $\sigma - (4\mu(m_2 + R_1) + 2\sigma)O(\lambda_1^{-1/2}) \geq 0$, we have

$$|E_\kappa| \geq \text{Re } E_\kappa \geq \sigma A^4 \geq (\sigma/2^4)(A + R_1^{-1}|\lambda|)(A + R_1^{-1}\lambda_1^{1/2}|\lambda|^{1/2})^3.$$

This completes the proof of Lemma 3.5.15. $\square$

Continuation of Proof of Theorem 3.5.14 Let $w_j = \mathcal{F}_{\xi'}^{-1}[\hat{w}_j]$, $\mathfrak{q} = \mathcal{F}_{\xi'}^{-1}[\hat{\mathfrak{q}}]$ and $\eta = \varphi(x_N)\mathcal{F}_{\xi'}^{-1}[e^{-Ax_N}\hat{h}]$, where $\varphi(x_N) \in C_0^\infty(\mathbb{R})$ equals to 1 for $x_N \in (-1, 1)$ and 0 for $x_N \notin [-2, 2]$. Notice that $\eta|_{x_N=0} = h$.

In view of (3.278) and Volevich's trick, we define $\mathcal{W}_{\kappa j}(\lambda)$ by

$$
\mathcal{W}_{\kappa j}(\lambda)d = \int_0^\infty \mathcal{F}_{\xi'}^{-1}\Big[-(Ae^{-B(x_N+y_N)} + A^2\mathcal{M}(x_N + y_N))
$$
$$
\times \frac{i\xi_j}{A}\frac{\sigma(A^2 + B^2)}{E_\kappa}\mathcal{F}[\Delta'd](\xi', y_N)
$$
$$
+ Ae^{-B(x_N+y_N)}\frac{i\xi_j}{A}\frac{\sigma B(B - A)}{E_\kappa}\mathcal{F}'[\Delta'd](\xi', y_N)\Big](x')\,dy_N
$$
$$
+ \int_0^\infty \mathcal{F}_{\xi'}^{-1}\Big[-A^2\mathcal{M}(x_N + y_N)\frac{\sigma(A^2 + B^2)}{E_\kappa}\mathcal{F}'[\partial_j\partial_N d](\xi', y_N)
$$
$$
+ Ae^{-B(x_N+y_N)}\frac{\sigma A(B - A)}{E_\kappa}\mathcal{F}'[\partial_j\partial_N d](\xi', y_N)\Big](x')\,dy_N,
$$

where we have used $\mathcal{F}'[\Delta'd](\xi', y_N) = -A^2\hat{d}(\xi', y_N)$. We have $\mathcal{W}_{\kappa j}(\lambda)d = w_j$. By Lemma 3.5.15, we see that

$$
\frac{A^2 + B^2}{E_\kappa}, \quad \frac{A^2 + B^2}{E_\kappa}\frac{\xi_j}{A}, \quad \frac{A(B - A)}{E_\kappa}, \quad \frac{B(B - A)}{E_\kappa}\frac{\xi_j}{A}
$$

belong to $\mathbf{M}_{-2,2}(\Lambda_{\kappa,\lambda_1})$, and so by Lemma 3.5.12, we have

$$
\mathcal{R}_{\mathcal{L}(H_q^2(\mathbb{R}_+^N), H_q^{2-k}(\mathbb{R}_+^N))}(\{(\tau\partial_\tau)^\ell(\lambda^{k/2}\mathcal{W}_{\kappa j}(\lambda)) \mid \lambda \in \Lambda_{\kappa,\lambda_1}\}) \le r_b(\lambda_1)
$$

for $\ell = 0, 1$ and $k = 0, 1, 2$, where $r_b(\lambda_1)$ is a constant depending on m_0, m_1, m_2 and λ_1.

Analogously, we have

$$
\mathcal{R}_{\mathcal{L}(H_q^2(\mathbb{R}_+^N), H_q^{2-k}(\mathbb{R}_+^N))}(\{(\tau\partial_\tau)^\ell(\lambda^{k/2}\mathcal{W}_{\kappa N}(\lambda)) \mid \lambda \in \Lambda_{\kappa,\lambda_1}\}) \le r_b(\lambda_1)
$$

for $\ell = 0, 1$ and $k = 0, 1, 2$. Thus, our final task is to construct $\mathcal{H}_\kappa(\lambda)$. In view of (3.277), we define $\mathcal{H}_\kappa(\lambda)$ acting on $d \in H_q^2(\mathbb{R}_+^N)$ by

$$
\mathcal{H}_\kappa(\lambda)d = \varphi(x_N)\mathcal{F}_{\xi'}^{-1}\Big[e^{-Ax_N}\frac{\mu D(A, B)}{E_\kappa}\hat{d}(\xi', 0)\Big](x').
$$

Since $\varphi(x_N)$ equals one for $x_N \in (-1, 1)$, we have $\mathcal{H}_\kappa(\lambda)d|_{x_N=0} = h$. Recalling the definition of $\hat{h}$ given in (3.277) and using Volevich's trick, we have $\mathcal{H}_\kappa(\lambda)d =$

$\varphi(x_N)\{\Omega_\kappa(\lambda)d + \mathcal{H}_\kappa^2(\lambda)d\}$ with

$$\Omega_\kappa(\lambda)d = \int_0^\infty \mathcal{F}_{\xi'}^{-1}\Big[Ae^{-A(x_N+y_N)}\frac{\mu D(A, B)}{E_\kappa}\varphi(y_N)\hat{d}(\xi', y_N)\Big](x')\,dy_N,$$

$$\mathcal{H}_\kappa^2(\lambda)d = -\int_0^\infty \mathcal{F}_{\xi'}^{-1}\Big[e^{-A(x_N+y_N)}\frac{\mu D(A, B)}{E_\kappa}\partial_N(\varphi(y_N)\hat{d}(\xi', y_N))\Big](x')\,dy_N.$$

We use the following lemma.

Lemma 3.5.16 *Let Λ be a domain in $\mathbb{C}$ and let $1 < q < \infty$. Let φ and ψ be two $C_0^\infty((-2, 2))$ functions. Given $m \in \mathbf{M}_{0,2}(\Lambda)$, we define operators $L_6(\lambda)$ and $L_7(\lambda)$ acting on $g \in L_q(\mathbb{R}_+^N)$ by*

$$[L_6(\lambda)g](x) = \varphi(x_N)\int_0^\infty \mathcal{F}_{\xi'}^{-1}\Big[e^{-A(x_N+y_N)}m(\lambda, \xi')\hat{g}(\xi', y_N)\psi(y_N)\Big]dy_N,$$

$$[L_7(\lambda)g](x) = \varphi(x_N)\int_0^\infty \mathcal{F}_{\xi'}^{-1}\Big[Ae^{-A(x_N+y_N)}m(\lambda, \xi')\hat{g}(\xi', y_N)\psi(y_N)\Big]dy_N.$$

Then,

$$\mathcal{R}_{\mathcal{L}(L_q(\mathbb{R}_+^N))}(\{(\tau\partial_\tau)^\ell L_k(\lambda) \mid \lambda \in \Lambda\}) \leq r_b \tag{3.282}$$

for $\ell = 0, 1$ and $k = 6, 7$, where r_b is a constant depending on $M(m, \Lambda)$. Here, $M(m, \Lambda)$ is the number defined in Definition 3.5.10.

Proof Using the assertion for $L_2(\lambda)$ in Lemma 3.5.12, we can show (3.282) immediately for $k = 7$, and so we show (3.282) only in the case that $k = 6$ below. In view of Definition 3.4.7, for any $n \in \mathbb{N}$, we take $\{\lambda_j\}_{j=1}^n \subset \Lambda$, $\{g_j\}_{j=1}^n \subset L_q(\mathbb{R}_+^N)$, and $r_j(u)$ $(j = 1, \ldots, n)$ are Rademacher functions. For the notational simplicity, we set

$$|||L_6(\lambda)g||| = \|\sum_{j=1}^n r_j(u)L_6(\lambda_j)g_j\|_{L_q((0,1), L_q(\mathbb{R}_+^N))}$$

$$= \Big(\int_0^1 \|\sum_{j=1}^n r_j(u)L_6(\lambda_j)g_j\|_{L_q(\mathbb{R}_+^N)}^q\,du\Big)^{1/q}.$$

By the Fubini–Tonelli theorem, we have

$$|||L_6(\lambda)g|||^q = \int_0^1 \int_0^\infty \int_{\mathbb{R}^{N-1}} |\sum_{j=1}^n r_j(u)L_6(\lambda_j)g_j|^q\,dy'dx_N\,du$$

$$= \int_0^\infty \Big(\int_0^1 \|\sum_{j=1}^n r_j(u)L_6(\lambda_j)g_j\|_{L_q(\mathbb{R}^{N-1})}^q\,du\Big)dx_N.$$

Since

$$|\partial_{\xi'}^{\alpha'}(e^{-A(x_N+y_N)}m(\lambda,\xi'))| \leq C_{\alpha'}|\xi'|^{-|\alpha'|}$$

for any $x_N \geq 0$, $y_N \geq 0$, $(\lambda,\xi') \in \Lambda \times (\mathbb{R}^{N-1} \setminus \{0\})$, and $\alpha' \in \mathbb{N}^{N-1}$, by Theorem 3.5.1 we have

$$\int_0^1 \left\| \sum_{j=1}^n r_j(u)\mathcal{F}_{\xi'}^{-1}\left[e^{-A(x_N+y_N)}m(\lambda_j,\xi')\hat{g}_j(\xi',y_N)\right](y') \right\|_{L_q(\mathbb{R}^{N-1})}^q du$$

$$\leq C_{N,q}M(m,\Lambda)\int_0^1 \left\| \sum_{j=1}^n r_j(u)g_j(\cdot,y_N) \right\|_{L_q(\mathbb{R}^{N-1})}^q du. \tag{3.283}$$

For any $x_N \geq 0$, by Minkowski's integral inequality, Lemma 3.5.3, and Hölder's inequality, we have

$$\left(\int_0^1 \left\| \sum_{j=1}^n r_j(u)L_6(\lambda_j)g_j \right\|_{L_q(\mathbb{R}^{N-1})}^q du\right)^{1/q}$$

$$= |\varphi(x_N)|\left(\int_0^1 \left\| \int_0^\infty \mathcal{F}_{\xi'}^{-1}[\sum_{j=1}^n r_j(u)e^{-A(x_N+y_N)}m(\lambda_j,\xi')\hat{g}_j(\xi',y_N)](y')\right.\right.$$

$$\left.\left. \times \psi(y_N)\,dy_N \right\|_{L_q(\mathbb{R}^{N-1})}^q du\right)^{1/q}$$

$$\leq |\varphi(x_N)|\left(\int_0^1 \left(\int_0^\infty \left\| \mathcal{F}_{\xi'}^{-1}[\sum_{j=1}^n r_j(u)e^{-A(x_N+y_N)}m(\lambda_j,\xi')\hat{g}_j(\xi',y_N)](y')\right.\right.\right.$$

$$\left.\left.\left. \times \psi(y_N) \right\|_{L_q(\mathbb{R}^{N-1})}\,dy_N\right)^q du\right)^{1/q}$$

$$\leq |\varphi(x_N)|\int_0^\infty \left(\int_0^1 \left\| \mathcal{F}_{\xi'}^{-1}[\sum_{j=1}^n r_j(u)e^{-A(x_N+y_N)}m(\lambda_j,\xi')\right.\right.$$

$$\left.\left. \times \hat{g}_j(\xi',y_N)](y') \right\|_{L_q(\mathbb{R}^{N-1})}^q du\right)^{1/q}|\psi(y_N)|\,dy_N$$

$$\leq C_{N,q}M(m,\Lambda)|\varphi(x_N)|\int_0^\infty \left(\int_0^1 \left\| \sum_{j=1}^n r_j(u)g_j(\cdot,y_N) \right\|_{L_q(\mathbb{R}^{N-1})}^q du\right)^{1/q}$$

$$\times |\psi(y_N)|\,dy_N$$

$$\leq C_{N,q}M(m,\Lambda)|\varphi(x_N)|\left(\int_0^\infty \int_0^1 \left\| \sum_{j=1}^n r_j(u)g_j(\cdot,y_N) \right\|_{L_q(\mathbb{R}^{N-1})}^q du\,dy_N\right)^{1/q}$$

$$\times \left(\int_0^\infty |\psi(y_N)|^{q'} \, dy_N \right)^{1/q'}$$

$$= C_{N,q} M(m, \Lambda) |\varphi(x_N)| \left(\int_0^1 \| \sum_{j=1}^n r_j(u) g_j(\cdot, y_N) \|_{L_q(\mathbb{R}_+^N)}^q \, du \right)^{1/q}$$

$$\times \left(\int_0^\infty |\psi(y_N)|^{q'} \, dy_N \right)^{1/q'}$$

Putting these inequalities together and using Hölder's inequality gives

$$\int_0^1 \| \sum_{j=1}^n r_j(u) L_6(\lambda_j) g_j \|_{L_q(\mathbb{R}_+^N)}^q \, du$$

$$\leq (C_{N,q} M(m, \Lambda))^q \int_0^\infty |\varphi(x_N)|^q dx_N \int_0^1 \| \sum_{j=1}^n r_j(u) g_j \|_{L_q(\mathbb{R}_+^N)}^q \, du$$

$$\times \left(\int_0^\infty |\psi(y_N)|^{q'} \, dy_N \right)^{q/q'},$$

and so, we have

$$\| \sum_{j=1}^n r_j L_6(\lambda_j) g_j \|_{L_q((0,1), L_q(\mathbb{R}_+^N))}$$

$$\leq C_{N,q} M(m, \Lambda) \|\varphi\|_{L_q(\mathbb{R})} \|\psi\|_{L_{q'}(\mathbb{R})} \| \sum_{j=1}^n r_j g_j \|_{L_q((0,1), L_q(\mathbb{R}_+^N))}.$$

This shows Lemma 3.5.16. $\square$

Continuation of Proof of Theorem 3.5.14 For $(j, \alpha', k) \in \mathbb{N}_0 \times \mathbb{N}_0^{N-1} \times \mathbb{N}_0$ with $j + |\alpha'| + k \leq 3$ and $j = 0, 1$, we write

$$\lambda^j \partial_{x'}^{\alpha'} \partial_N^k \mathcal{H}_\kappa(\lambda) d = \sum_{n=0}^k {}_k C_n (\partial_N^{k-n} \varphi(x_N))[\lambda^j \partial_{x'}^{\alpha'} \partial_N^n \Omega_\kappa(\lambda) d + \lambda^j \partial_{x'}^{\alpha'} \partial_N^n \mathcal{H}_\kappa^2(\lambda) d],$$

and then

$$\lambda^j \partial_{x'}^{\alpha'} \partial_N^n \Omega_\kappa(\lambda) d$$

$$= \int_0^\infty \mathcal{F}_{\xi'}^{-1} \Big[A e^{-A(x_N + y_N)} \frac{\mu \lambda^j (i\xi')^{\alpha'} (-A)^n D(A, B)}{(1 + A^2) E_\kappa} \varphi(y_N)$$

$$\mathcal{F}'[(1 - \Delta') d](\xi', y_N) \Big](x') \, dy_N;$$

$$\lambda^j \mathcal{H}_\kappa^2(\lambda)d = \int_0^\infty \mathcal{F}_{\xi'}^{-1}\Big[e^{-A(x_N+y_N)}\frac{\mu\lambda^j D(A,B)}{E_\kappa}$$

$$\partial_N(\varphi(y_N)\hat{d}(\xi',y_N))\big](x')\,dy_N;$$

$$\lambda^j \partial_{x'}^{\alpha'}\partial_N^n \mathcal{H}_\kappa^2(\lambda)d = \int_0^\infty \mathcal{F}_{\xi'}^{-1}\Big[e^{-A(x_N+y_N)}\frac{\mu\lambda^j (i\xi')^{\alpha'}(-A)^n D(A,B)}{(1+A^2)E_\kappa}$$

$$\partial_N(\varphi(y_N)\hat{d}(\xi',y_N))\big](x')\,dy_N$$

$$-\sum_{k=1}^{N-1}\int_0^\infty \mathcal{F}_{\xi'}^{-1}\Big[Ae^{-A(x_N+y_N)}\frac{\mu\lambda^j (i\xi')^{\alpha'}(-A)^n D(A,B)}{(1+A^2)E_\kappa}\frac{i\xi_j}{A}$$

$$\partial_N(\varphi(y_N)\mathcal{F}[\partial_j d(\cdot,y_N)](\xi'))\big](x')\,dy_N$$

for $|\alpha'|+n \geq 1$. Here, we have used the formula:

$$1 = \frac{1+A^2}{1+A^2} = \frac{1}{1+A^2} - \sum_{j=1}^{N-1}\frac{A}{1+A^2}\frac{i\xi_j}{A}i\xi_j$$

in the third equality. By Lemmas 3.5.9 and 3.5.15, we see that multipliers:

$$\frac{\lambda^j (i\xi')^{\alpha'} A^n D(A,B)}{(1+A^2)E_\kappa}, \quad \frac{\lambda^j D(A,B)}{E_\kappa},$$

$$\frac{\lambda^j (i\xi')^{\alpha'} A^n D(A,B)}{(1+A^2)E_\kappa}, \quad \frac{\lambda^j (i\xi')^{\alpha'} A^n D(A,B)}{(1+A^2)E_\kappa}\frac{\xi_j}{A}$$

belong to $\mathbf{M}_{0,2}(\Lambda_{\kappa,\lambda_1})$, because $j+|\alpha'|+n \leq 3$ and $j = 0,1$. Thus, using Lemma 3.5.16, we see that for any $n \in \mathbb{N}$, $\{\lambda_j\}_{j=1}^n \subset \Lambda_{\kappa,\lambda_1}$, and $\{d_j\}_{j=1}^n \subset H_q^2(\mathbb{R}_+^N)$, the inequality:

$$\Big\|\sum_{\ell=1}^n r_\ell(\cdot)(\partial_N^{k-n}\varphi)(\lambda_\ell)^j \partial_{x'}^{\alpha'}\partial_N^n (\Omega_\kappa(\lambda_\ell),\mathcal{H}_\kappa^2(\lambda_\ell))d_\ell\Big\|_{L_q((0,1),L_q(\mathbb{R}^N))}$$

$$\leq C\Big\|\sum_{\ell=1}^n r_\ell(\cdot)d_\ell\Big\|_{L_q((0,1),H_q^2(\mathbb{R}_+^N))}$$

holds, which leads to

$$\| \sum_{\ell=1}^{n} r_\ell(\cdot)(\lambda_\ell)^j \partial_{x'}^{\alpha'} \partial_N^k \mathcal{H}_\kappa(\lambda_\ell) d_\ell \|_{L_q((0,1),L_q(\mathbb{R}_+^N))}$$

$$\leq C \| \sum_{\ell=1}^{n} r_\ell(\cdot) d_\ell \|_{L_q((0,1),\dot{H}_q^2(\mathbb{R}_+^N))}.$$

Here, C is a constant depending on N, q, m_0, m_1, and m_2. This shows that

$$\mathcal{R}_{\mathcal{L}(H_q^2(\mathbb{R}_+^N), H_q^{3-k}(\mathbb{R}_+^N))}(\{\lambda^k \mathcal{H}_\kappa(\lambda) \mid \lambda \in \Lambda_{\kappa,\lambda_1}\}) \leq r_b(\lambda_1)$$

for $k = 0, 1$. Here, $r_b(\lambda_1)$ is a constant depending on N, q, m_0, m_1, and m_2, but independent of $\mu, \sigma \in [m_0, m_1]$ and $|A_\kappa| \leq m_2$ for $\kappa \in [0, 1)$. Analogously, we have

$$\mathcal{R}_{\mathcal{L}(H_q^2(\mathbb{R}_+^N), H_q^{3-k}(\mathbb{R}_+^N))}(\{\tau \partial_\tau(\lambda^k \mathcal{H}_\kappa(\lambda)) \mid \lambda \in \Lambda_{\kappa,\lambda_1}\}) \leq r_b(\lambda_1)$$

for $k = 0, 1$. This completes the proof of Theorem 3.5.14. □

Proof of Theorem 3.5.6 To prove Theorem 3.5.6, in view of the consideration in Sect. 3.4.3, we first consider the equation:

$$\operatorname{div} \mathbf{v} = g \quad \text{in } \mathbb{R}_+^N. \tag{3.284}$$

Here, g is a solution of the variational equation:

$$\lambda(g, \varphi)_{\mathbb{R}_+^N} + (\nabla g, \nabla \varphi)_{\mathbb{R}_+^N} = (-\mathbf{f}, \nabla \varphi)_{\mathbb{R}_+^N} \quad \text{for any } \varphi \in H_{q',0}^1(\mathbb{R}_+^N), \tag{3.285}$$

subject to $g = \rho$ on Γ. We have the following theorem.

Lemma 3.5.17 *Let* $1 < q < \infty$, $0 < \epsilon < \pi/2$, *and* $\lambda_0 > 0$. *Let*

$$Y_q''(\mathbb{R}_+^N) = \{(\mathbf{f}, \rho) \mid \mathbf{f} \in L_q(\mathbb{R}_+^N)^N, \quad \rho \in H_q^1(\mathbb{R}_+^N)\},$$

$$\mathcal{Y}_q''(\mathbb{R}_+^N) = \{(F_1, G_3, G_4) \mid F_1 \in L_q(\mathbb{R}_+^N)^N, \quad G_3 \in L_q(\mathbb{R}_+^N), \quad G_4 \in H_q^1(\mathbb{R}_+^N)\}.$$

Let g *be a solution of the variational problem* (3.285). *Then, there exists an operator family* $\mathcal{B}_0(\lambda) \in \text{Hol}(\Sigma_{\epsilon,\lambda_0}, \mathcal{L}(\mathcal{Y}_q''(\mathbb{R}_+^N), H_q^2(\mathbb{R}_+^N)^N))$ *such that for any* $\lambda \in \Sigma_{\epsilon,\lambda_0}$ *and* $(\mathbf{f}, \rho) \in Y_q''(\mathbb{R}_+^N)$, *problem* (3.284) *admits a solution* $\mathbf{v} = \mathcal{B}_0(\lambda)(\mathbf{f}, \lambda^{1/2}\rho, \rho)$, *and*

$$\mathcal{R}_{\mathcal{L}(\mathcal{Y}_q''(\mathbb{R}_+^N), H_q^{2-j}(\mathbb{R}_+^N)^N)}(\{(\tau \partial_\tau)^\ell(\lambda^{j/2} \mathcal{B}_0(\lambda)) \mid \lambda \in \Sigma_{\epsilon,\lambda_0}\}) \leq r_b(\lambda_0)$$

for $\ell = 0, 1$ *and* $j = 0, 1, 2$, *where* $r_b(\lambda_0)$ *is a constant depending on* ϵ, λ_0, N, *and* q.

Proof This lemma was proved in Shibata [45, Lemma 9.3.10], but for the sake of completeness of this lecture note as much as possible, we give a proof. Let g_1 be a solution of the equation:

$$(\lambda - \Delta)g_1 = \mathrm{div}\,\mathbf{f} \quad \text{in } \mathbb{R}_+^N, \quad g_1|_{x_N=0} = 0,$$

and let g_2 be a solution of the equation:

$$(\lambda - \Delta)g_2 = 0 \quad \text{in } \mathbb{R}_+^N, \quad g_2|_{x_N=0} = \rho.$$

And then, $g = g_1 + g_2$ is a solution of Eq. (3.285). To construct g_1 and g_2, we use the even extension, f^e, and odd extension, f^o, of a function, f, which has been introduced in (3.46). Let $\mathbf{f} = {}^\top(f_1, \ldots, f_N)$. Notice that $(\mathrm{div}\,\mathbf{f})^o = \sum_{j=1}^{N-1} \partial_j f_j^o + \partial_N f_N^e$. We define g_1 by letting

$$g_1 = \mathcal{F}_\xi^{-1}\left[\frac{\mathcal{F}[(\mathrm{div}\,\mathbf{f})^o](\xi)}{\lambda + |\xi|^2}\right] = \mathcal{F}_\xi^{-1}\left[\frac{\sum_{k=1}^{N-1} i\xi_k \mathcal{F}[f_k^o](\xi) + i\xi_N \mathcal{F}[f_N^e](\xi)}{\lambda + |\xi|^2}\right].$$

And also, the g_2 is defined by

$$g_2(x) = \mathcal{F}_{\xi'}^{-1}[e^{-B_0 x_N}\hat{\rho}(\xi', 0)](x') = \frac{\partial h}{\partial x_N}, \tag{3.286}$$

where we have set $B_0 = \sqrt{\lambda + |\xi'|^2}$ and $h(x) = -\mathcal{F}_{\xi'}^{-1}[B_0^{-1}e^{-B_0 x_N}\hat{\rho}(\xi', 0)](x')$. By Volevich's trick, we have

$$h(x) = \int_0^\infty \mathcal{F}_{\xi'}^{-1}[B_0^{-1}e^{-B_0(x_N+y_N)}\widehat{(\partial_N \rho)}(\xi', y_N)](x')\,dy_N$$

$$- \int_0^\infty \mathcal{F}_{\xi'}^{-1}[e^{-B_0(x_N+y_N)}\hat{\rho}(\xi', y_N)](x')\,dy_N$$

$$= \int_0^\infty \mathcal{F}_{\xi'}^{-1}[\frac{\lambda^{1/2}}{B_0^3}\lambda^{1/2}e^{-B_0(x_N+y_N)}\widehat{(\partial_N \rho)}(\xi', y_N)](x')\,dy_N$$

$$+ \int_0^\infty \mathcal{F}_{\xi'}^{-1}[\frac{A}{B_0^3}Ae^{-B_0(x_N+y_N)}\widehat{(\partial_N \rho)}(\xi', y_N)](x')\,dy_N$$

$$- \int_0^\infty \mathcal{F}_{\xi'}^{-1}[\frac{1}{B_0^2}\lambda^{1/2}e^{-B_0(x_N+y_N)}\widehat{(\lambda^{1/2}\rho)}(\xi', y_N)](x')\,dy_N$$

$$+ \sum_{j=1}^{N-1} \int_0^\infty \mathcal{F}_{\xi'}^{-1}[\frac{1}{B_0^2}\frac{i\xi_j}{A}Ae^{-B_0(x_N+y_N)}\widehat{(\partial_j \rho)}(\xi', y_N)](x')\,dy_N.$$

Let $\mathcal{Z}_q(\mathbb{R}^N_+)$ be the same space as in Lemma 3.5.13. We then define an operator $H(\lambda)$ acting on $(G_3, G_4) \in \mathcal{Z}_q(\mathbb{R}^N_+)$ by setting

$$
\begin{aligned}
H(\lambda)(G_3, G_4) = {} & \int_0^\infty \mathcal{F}_{\xi'}^{-1}[\frac{\lambda^{1/2}}{B_0^3} \lambda^{1/2} e^{-B_0(x_N + y_N)} \widehat{\partial_N G_4}(\xi', y_N)](x')\, dy_N \\
& + \int_0^\infty \mathcal{F}_{\xi'}^{-1}[\frac{A}{B_0^3} A e^{-B_0(x_N + y_N)} \widehat{\partial_N G_4}(\xi', y_N)](x')\, dy_N \\
& - \int_0^\infty \mathcal{F}_{\xi'}^{-1}[\frac{1}{B_0^2} \lambda^{1/2} e^{-B_0(x_N + y_N)} \widehat{G_3}(\xi', y_N)](x')\, dy_N \\
& + \sum_{j=1}^{N-1} \int_0^\infty \mathcal{F}_{\xi'}^{-1}[\frac{1}{B_0^2} \frac{i\xi_j}{A} A e^{-B_0(x_N + y_N)} \widehat{\partial_j G_4}(\xi', y_N)](x')\, dy_N.
\end{aligned}
$$

By Lemmas 3.5.11 and 3.5.12, we see that

$$
\begin{aligned}
& H(\lambda) \in \mathrm{Hol}\,(\Sigma_\epsilon, \mathcal{L}(\mathcal{Y}_q'(\mathbb{R}^N_+), H_q^2(\mathbb{R}^N_+))), \quad h = H(\lambda)(\lambda^{1/2}\rho, \rho), \\
& \mathcal{R}_{\mathcal{L}(\mathcal{Y}_q'(\mathbb{R}^N_+), H_q^{2-j}(\mathbb{R}^N_+))}(\{(\tau\partial_\tau)^\ell(\lambda^{j/2}H(\lambda)) \mid \lambda \in \Sigma_{\epsilon,\lambda_0}\}) \le r_b(\lambda)
\end{aligned}
\tag{3.287}
$$

for $\ell = 0, 1$ and $j = 0, 1, 2$. Moreover, we have

$$
(\lambda - \Delta)H(\lambda)(G_3, G_4) = 0 \quad \text{in } \mathbb{R}^N_+. \tag{3.288}
$$

Let $\mathbf{v}_1$ be an N-vector of functions defined by

$$
\mathbf{v}_1 = -\mathcal{F}_\xi^{-1}\Big[\frac{i\xi \mathcal{F}[g_1](\xi)}{|\xi|^2}\Big] = \mathcal{F}_\xi^{-1}\Big[\frac{\xi(\sum_{k=1}^{N-1} \xi_k \mathcal{F}[f_k^o](\xi) + \xi_N \mathcal{F}[f_N^e](\xi))}{(\lambda + |\xi|^2)|\xi|^2}\Big].
$$

We see that $\mathrm{div}\,\mathbf{v}_1 = g_1$ in $\mathbb{R}^N_+$. Moreover, by Lemmas 3.5.2 and 3.5.3, there exists an operator family $\mathcal{B}_0^1(\lambda) \in \mathrm{Hol}\,(\Sigma_\epsilon, \mathcal{L}(L_q(\mathbb{R}^N_+), H_q^2(\mathbb{R}^N_+)^N))$ such that $\mathbf{v}_1 = \mathcal{B}_0^1(\lambda)\mathbf{f}$ and

$$
\mathcal{R}_{\mathcal{L}(L_q(\mathbb{R}^N_+)^N, H_q^{2-j}(\mathbb{R}^N_+)^N)}(\{(\tau\partial_\tau)^\ell(\lambda^{j/2}\mathcal{B}_0^1(\lambda)) \mid \lambda \in \Sigma_{\epsilon,\lambda_0}\}) \le r_b(\lambda_0)
$$

for $\ell = 0, 1$, $j = 0, 1, 2$, and $\lambda_0 > 0$, where $r_b(\lambda_0)$ is a constant depending on ϵ, λ_0, N and q.

Let

$$
v_{2j} = \mathcal{F}_\xi^{-1}\Big[\frac{\xi_j \xi_N \mathcal{F}[h^e](\xi)}{|\xi|^2}\Big] = -\mathcal{F}_\xi^{-1}\Big[\frac{i\xi_j \mathcal{F}[g_2^o](\xi)}{|\xi|^2}\Big] \quad (j = 1, \ldots, N)
$$

and let $\mathbf{v}_2 = {}^\top(v_{21}, \ldots, v_{2N})$, and then by (3.286) we have $\operatorname{div} \mathbf{v}_2 = g_2$ in $\mathbb{R}^N_+$. Thus, we define an operator $\mathcal{B}^2_0(\lambda) = (\mathcal{B}^2_{01}(\lambda), \ldots, \mathcal{B}^2_{0N}(\lambda))$ acting on $(G_3, G_4) \in \mathcal{Z}_q(\mathbb{R}^N_+)$ by

$$\mathcal{B}^2_{0j}(\lambda)(G_3, G_4) = \mathcal{F}^{-1}_\xi\left[\frac{\xi_j \xi_N \mathcal{F}[H(\lambda)(G_3, G_4)^e](\xi)}{|\xi|^2}\right].$$

By (3.288), we have $\mathbf{v}_2 = \mathcal{B}^2_0(\lambda)(\lambda^{1/2}\rho, \rho)$. Noting that $\partial_N f^e = (\partial_N f)^o$ and $\partial_k \partial_N f^e = (\partial_k \partial_N f)^0$ $(k = 1, \ldots, N-1)$, we have

$$\lambda \mathcal{B}^2_{0j}(\lambda)(G_3, G_4) = \mathcal{F}^{-1}_\xi\left[\frac{\xi_j \xi_N \mathcal{F}[\lambda H(\lambda)(G_3, G_4)^e](\xi)}{|\xi|^2}\right],$$

$$\lambda^{1/2}\nabla \mathcal{B}^2_{0j}(\lambda)(G_3, G_4) = \mathcal{F}^{-1}_\xi\left[\frac{\xi_j \xi \mathcal{F}[\lambda^{1/2}(\partial_N H(\lambda)(G_3, G_4))^o](\xi)}{|\xi|^2}\right],$$

$$\partial_k \nabla \mathcal{B}^2_{0j}(\lambda)(G_3, G_4) = \mathcal{F}^{-1}_\xi\left[\frac{\xi_j \xi \mathcal{F}[\lambda^{1/2}(\partial_k \partial_N H(\lambda)(G_3, G_4))^o](\xi)}{|\xi|^2}\right],$$

Moreover, since $\partial^2_N H(\lambda)(G_3, G_4) = -(\lambda - \sum^{N-1}_{j=1}\partial^2_j)H(\lambda)(G_3, G_4)$ as follows from (3.288), we have

$$\partial^2_N \mathcal{B}^2_0(\lambda)(G_3, G_4) = \mathcal{F}^{-1}_\xi\left[\frac{\xi^2_N}{|\xi|^2}\mathcal{F}[((\lambda - \Delta')H(\lambda)(G_3, G_4))^e](\xi)\right].$$

Thus, by (3.287) and the Fourier multiplier theorem, we see that

$$\mathcal{R}_{\mathcal{L}(\mathcal{Y}'_q(\mathbb{R}^N), H^{2-j}_q(\mathbb{R}^N_+)^N)}(\{(\tau\partial_\tau)^\ell(\lambda^{j/2}\mathcal{B}^2_0(\lambda)) \mid \lambda \in \Sigma_{\epsilon, \lambda_0}\}) \leq r_b(\lambda_0)$$

for $\ell = 0, 1$, $j = 0, 1, 2$, and $\lambda_0 > 0$, where $r_b(\lambda_0)$ is a constant depending on ϵ, λ_0, N, and q. Since $\mathbf{v} = \mathbf{v}_1 + \mathbf{v}_2$ is a solution of Eq. (3.284), setting $\mathcal{B}_0(\lambda)(F_1, G_3, G_4) = \mathcal{B}^1_0(\lambda)F_1 + \mathcal{B}^2_0(\lambda)(G_3, G_4)$, we see that $\mathcal{B}_0(\lambda)$ is the required operator, which completes the proof of Lemma 3.5.17. □

We now prove Theorem 3.5.6. Let $(\mathbf{f}, d, \mathbf{h}) \in Y_q(\mathbb{R}^N_+)$. Let g be a solution of Eq. (3.285) with $\rho = \mathbf{n}_0 \cdot \mathbf{h}$, and let $\mathbf{u}$, q and h be solutions of the equations:

$$\begin{cases} \lambda \mathbf{u} - \operatorname{Div}(\mu \mathbf{D}(\mathbf{u}) - q\mathbf{I}) = \mathbf{f}, \quad \operatorname{div} \mathbf{u} = g = \operatorname{div} \mathbf{g} & \text{in } \mathbb{R}^N_+, \\ \lambda h + A_\kappa \cdot \nabla'_\Gamma h - \mathbf{u} \cdot \mathbf{n}_0 = d & \text{on } \mathbb{R}^N_0, \\ (\mu \mathbf{D}(\mathbf{u}) - q\mathbf{I} - \sigma(\Delta' h)\mathbf{I})\mathbf{n}_0 = \mathbf{h} & \text{on } \mathbb{R}^N_0. \end{cases} \quad (3.289)$$

Then, according to what pointed out in Sect. 3.4.3, $\mathbf{u}$ and h are solutions of Eq. (3.250). Thus, we shall look for $\mathbf{u}$, q and h below. In view of (3.187) and (3.188), applying Lemma 3.5.17 with $\rho = \mathbf{n}_0 \cdot \mathbf{h}$, we define $\mathbf{u}_0$ by $\mathbf{u}_0 = \mathcal{B}_0(\mathbf{f}, \lambda^{1/2}\mathbf{n}_0 \cdot \mathbf{h}, \mathbf{n}_0 \cdot \mathbf{h})$.

Notice that $\operatorname{div} \mathbf{u}_0 = g = \operatorname{div} \mathbf{g}_0$ with $\mathbf{g} = \lambda^{-1}(\mathbf{f} + \nabla g)$. We then look for $\mathbf{w}_0$, $\mathfrak{q}$, and h satisfying the equations:

$$
\begin{cases}
\lambda \mathbf{w}_0 - \operatorname{Div}(\mu \mathbf{D}(\mathbf{w}_0) - \mathfrak{q}\mathbf{I}) = \mathbf{f} - \mathbf{f}_0, & \operatorname{div} \mathbf{w}_0 = 0 \quad \text{in } \mathbb{R}_+^N, \\
\lambda h + A_\kappa \cdot \nabla' h - \mathbf{w}_0 \cdot \mathbf{n}_0 = d + d_0 & \text{on } \mathbb{R}_0^N, \\
(\mu \mathbf{D}(\mathbf{w}_0) - \mathfrak{q}\mathbf{I})\mathbf{n}_0 - \sigma(\Delta' h)\mathbf{n}_0 = \mathbf{h} - \mathbf{h}_0 & \text{on } \mathbb{R}_0^N,
\end{cases} \tag{3.290}
$$

where we have set

$$
\mathbf{f}_0 = \lambda \mathbf{u}_0 - \operatorname{Div}(\mu \mathbf{D}(\mathbf{u}_0)), \quad d_0 = \mathbf{u}_0 \cdot \mathbf{n}_0, \quad \mathbf{h}_0 = \mu \mathbf{D}(\mathbf{u}_0)\mathbf{n}_0.
$$

To solve Eq. (3.290), we first consider the equations:

$$
\begin{cases}
\lambda \mathbf{U}_1 - \operatorname{Div}(\mu \mathbf{D}(\mathbf{U}_1) - P_1\mathbf{I}) = \mathbf{F}, & \operatorname{div} \mathbf{U}_1 = 0 \quad \text{in } \mathbb{R}_+^N, \\
\partial_N(\mathbf{U}_1 \cdot \mathbf{n}_0) = 0, \quad P_1 = 0 & \text{on } \mathbb{R}_0^N.
\end{cases} \tag{3.291}
$$

For $\mathbf{F} = {}^\top(F_1, \ldots, F_N) \in L_q(\mathbb{R}_+^N)^N$, let $\tilde{\mathbf{F}} = {}^\top(F_1^e, \ldots, F_{N-1}^e, F_N^o)$. Let $\mathcal{B}_1(\lambda)$ and $\mathcal{P}_1(\lambda)$ be operators acting on $\mathbf{F} \in L_q(\mathbb{R}_+^N)^N$ defined by

$$
\mathcal{B}_1(\lambda)\mathbf{F} = \mathcal{F}_\xi^{-1}\left[\frac{\mathcal{F}[\tilde{\mathbf{F}}](\xi) - \xi\xi \cdot \mathcal{F}[\tilde{\mathbf{F}}](\xi)|\xi|^{-2}}{\lambda + \mu|\xi|^2}\right],
$$

$$
\mathcal{P}_1(\lambda)\mathbf{F} = \mathcal{F}_\xi^{-1}\left[\frac{\xi \cdot \mathcal{F}[\tilde{\mathbf{F}}](\xi)}{|\xi|^2}\right].
$$

As was seen in Shibata and Shimizu [54, p.587] or [50, Proof of Theorem 4.3], $\mathbf{U}_1 = \mathcal{B}_1(\lambda)\mathbf{F}$ and $P_1 = \mathcal{P}_1(\lambda)\mathbf{F}$ satisfy Eq. (3.291). Moreover, employing the same argument as in Sect. 3.5.1, by Lemmas 3.5.2 and 3.5.3, we see that

$$
\mathcal{B}_1(\lambda) \in \operatorname{Hol}(\Sigma_{\epsilon,\lambda_0}, \mathcal{L}(L_q(\mathbb{R}_+^N)^N, H_q^2(\mathbb{R}_+^N)^N)),
$$

$$
\mathcal{P}_1(\lambda) \in \operatorname{Hol}(\Sigma_{\epsilon,\lambda_0}, \mathcal{L}(L_q(\mathbb{R}_+^N)^N, \hat{H}_q^1(\mathbb{R}_+^N)))
$$

for any $\epsilon \in (0, \pi/2)$ and $\lambda_0 > 0$, and moreover

$$
\mathcal{R}_{\mathcal{L}(L_q(\mathbb{R}_+^N)^N, H_q^{2-j}(\mathbb{R}_+^N)^N)}(\{(\tau \partial_\tau)^\ell(\lambda^{j/2}\mathcal{B}_1(\lambda)) \mid \lambda \in \Sigma_{\epsilon,\lambda_0}\}) \leq r_b(\lambda_0)
$$

for $\ell = 0, 1$ and $j = 0, 1, 2$, where $r_b(\lambda_0)$ is a constant depending on ϵ, λ_0, m_0, and m_1. In particular, we set

$$
\mathbf{u}_1 = \mathcal{B}_1(\lambda)(\mathbf{f} - \mathbf{f}_0), \quad \mathfrak{q}_1 = \mathcal{P}_1(\lambda)(\mathbf{f} - \mathbf{f}_0). \tag{3.292}
$$

We now let $\mathbf{u} = \mathbf{u}_0 + \mathbf{u}_1 + \mathbf{U}_2$ and $q = q_1 + P_2$, and then

$$
\begin{cases}
\lambda \mathbf{U}_2 - \mathrm{Div}\,(\mu \mathbf{D}(\mathbf{U}_2) - P_2 \mathbf{I}) = 0, & \mathrm{div}\,\mathbf{U}_2 = 0 & \text{in } \mathbb{R}^N_+, \\
\lambda h + A_\kappa \cdot \nabla' h - \mathbf{U}_2 \cdot \mathbf{n}_0 = d + d_2 & & \text{on } \mathbb{R}^N_0, \\
(\mu \mathbf{D}(\mathbf{U}_2) - P_2 \mathbf{I})\mathbf{n}_0 - \sigma(\Delta' h)\mathbf{n}_0 = \mathbf{h} - \mathbf{h}_2 & & \text{on } \mathbb{R}^N_0,
\end{cases}
\tag{3.293}
$$

where we have set

$$
d_2 = \mathbf{n}_0 \cdot (\mathbf{u}_0 + \mathbf{u}_1), \quad \mathbf{h}_2 = \mu \mathbf{D}(\mathbf{u}_0 + \mathbf{u}_1).
$$

Thus, for $\mathbf{H} \in H_q^1(\mathbb{R}^N_+)^N$ we consider the equations:

$$
\begin{cases}
\lambda \mathbf{U}_2 - \mathrm{Div}\,(\mu \mathbf{D}(\mathbf{U}_2) - P_2 \mathbf{I}) = 0, & \mathrm{div}\,\mathbf{U}_2 = 0 & \text{in } \mathbb{R}^N_+, \\
(\mu \mathbf{D}(\mathbf{U}_2) - P_2 \mathbf{I})\mathbf{n}_0 = \mathbf{H} & & \text{on } \mathbb{R}^N_0,
\end{cases}
\tag{3.294}
$$

and then by Theorem 3.5.8, we see that $\mathbf{U}_2 = \mathcal{V}(\lambda)(\lambda^{1/2}\mathbf{H}, \mathbf{H})$ is a unique solutions of Eq. (3.294) with some $P_2 \in \hat{H}_q^1(\mathbb{R}^N_+)$. In particular, we set $\mathbf{u}_2 = \mathcal{V}(\lambda)(\lambda^{1/2}(\mathbf{h} - \mathbf{h}_2), (\mathbf{h} - \mathbf{h}_2))$.

We finally let $\mathbf{u} = \mathbf{u}_0 + \mathbf{u}_1 + \mathbf{u}_2 + \mathbf{u}_3$ and $q = q_1 + q_2 + q_3$, and then $\mathbf{u}_3, q_3$ and h are solutions of the equations:

$$
\begin{cases}
\lambda \mathbf{u}_3 - \mathrm{Div}\,(\mu \mathbf{D}(\mathbf{u}_3) - q_3 \mathbf{I}) = 0, & \mathrm{div}\,\mathbf{u}_3 = 0 & \text{in } \mathbb{R}^N_+, \\
\lambda h + A_\kappa \cdot \nabla' h - \mathbf{u}_3 \cdot \mathbf{n}_0 = d + d_3 & & \text{on } \mathbb{R}^N_0, \\
(\mu \mathbf{D}(\mathbf{u}_3) - q_3 \mathbf{I})\mathbf{n}_0 - \sigma(\Delta' h)\mathbf{n}_0 = 0 & & \text{on } \mathbb{R}^N_0,
\end{cases}
\tag{3.295}
$$

where $d_3 = \mathbf{n}_0 \cdot (\mathbf{u}_0 + \mathbf{u}_1 + \mathbf{u}_2)$. Setting $\mathcal{W}(\lambda) = {}^\top(\mathcal{W}_1(\lambda), \ldots, \mathcal{W}_N(\lambda))$, by Theorem 3.5.14, we see that $\mathbf{u}_3 = \mathcal{W}(\lambda)(d + d_3)$ and $h = \mathcal{H}_\kappa(\lambda)(d + d_3)$ are unique solutions of Eq. (3.295) with some $q_3 \in \hat{H}_q^1(\mathbb{R}^N_+)$. Since the composition of two $\mathcal{R}$-bounded operators is also $\mathcal{R}$ bounded as follows from Proposition 3.5.4, we see easily that given $\epsilon \in (0, \pi/2)$, there exist $\lambda_1 > 0$ and operator families $\mathcal{A}_0(\lambda)$ and $\mathcal{H}_0(\lambda)$ satisfying (3.251) such that $\mathbf{u} = \mathcal{A}_0(\lambda)(\mathbf{f}, d, \lambda^{1/2}\mathbf{h}, \mathbf{h})$ and $h = \mathcal{H}_0(\lambda)(\mathbf{f}, d, \lambda^{1/2}\mathbf{h}, \mathbf{h})$ are unique solutions of Eq. (3.250), and moreover the estimate (3.252) holds. This completes the proof of Theorem 3.5.6.

3.5.4 Problem in a Bent Half Space

Let $\Phi : \mathbb{R}^N \to \mathbb{R}^N : x \to y = \Phi(x)$ be a bijection of C^1 class and let Φ^{-1} be its inverse map. We assume that $\nabla \Phi$ and $\nabla \Phi^{-1}$ have the forms: $\nabla \Phi = \mathcal{A} + B(x)$ and $\nabla \Phi^{-1} = \mathcal{A}_{-1} + B_{-1}(y)$, where $\mathcal{A}$ and $\mathcal{A}_{-1}$ are $N \times N$ orthogonal matrices with constant coefficients and $B(x)$ and $B_{-1}(y)$ are matrices of functions in $C^2(\mathbb{R}^N)$ such that

$$\|(B, B_{-1})\|_{L_\infty(\mathbb{R}^N)} \le M_1, \quad \|\nabla(B, B_{-1})\|_{L_\infty(\mathbb{R}^N)} \le C_A,$$
$$\|\nabla^2(B, B_{-1})\|_{L_\infty(\mathbb{R}^N)} \le M_2. \tag{3.296}$$

Here, C_A is a constant depending on constants A, a_1, a_2 appearing in Definition 3.2.1. We choose $M_1 > 0$ small enough and M_2 large enough eventually, and so we may assume that $0 < M_1 \le 1 \le C_A \le M_2$. Let $\Omega_+ = \Phi(\mathbb{R}_+^N)$ and $\Gamma_+ = \Phi(\mathbb{R}_0^N)$. In the sequel, a_{ij} and $b_{ij}(x)$ denote the (i, j)th element of $\mathcal{A}_{-1}$ and $(B_{-1} \circ \Phi)(x)$, respectively.

Let $\mathbf{n}_+$ be the unit outer normal to Γ_+. Setting $\Phi^{-1} = {}^\top(\Phi_{-1,1}, \dots, \Phi_{-1,N})$, we see that Γ_+ is represented by $\Phi_{-,N}(y) = 0$, which yields that

$$\mathbf{n}_+(x) = -\frac{(\nabla\Phi_{-1,N}) \circ \Phi(x)}{|\nabla\Phi_{-1,N}) \circ \Phi(x)|} = -\frac{{}^\top(a_{N1} + b_{N1}(x), \dots, a_{NN} + b_{NN}(x))}{(\sum_{j=1}^N(a_{Nj} + b_{Nj}(x))^2)^{1/2}}. \tag{3.297}$$

Obviously, $\mathbf{n}_+$ is defined on $\mathbb{R}^N$ and $\mathbf{n}_+$ denotes the unit outer normal to Γ_+ for $y = \Phi(x', 0) \in \Gamma_+$. By (3.296), writing

$$\mathbf{n}_+ = -{}^\top(a_{N1}, \dots, a_{NN}) + \mathbf{b}_+(x), \tag{3.298}$$

we see that $\mathbf{b}_+$ is an N-vector defined on $\mathbb{R}^N$, which satisfies the estimates:

$$\|\mathbf{b}_+\|_{L_\infty(\mathbb{R}^N)} \le C_N M_1, \quad \|\nabla\mathbf{b}_+\|_{L_\infty(\mathbb{R}^N)} \le C_N C_A, \quad \|\nabla^2\mathbf{b}_+\|_{L_\infty(\mathbb{R}^N)} \le C_{M_2}. \tag{3.299}$$

We next give the Laplace–Beltrami operator on Γ_+. Let

$$g_{+ij}(x) = \frac{\partial\Phi}{\partial x_i}(x) \cdot \frac{\partial\Phi}{\partial x_j}(x) = \sum_{k=1}^N (a_{ik} + b_{ik}(x))(a_{jk} + b_{jk}(x)) = \delta_{ij} + \tilde{g}_{+ij}(x)$$

with $\tilde{g}_{+ij} = \sum_{k=1}^N (a_{ik}b_{jk}(x) + a_{jk}b_{ik}(x) + b_{ik}(x)b_{jk}(x))$. Since Γ_+ is given by $y_N = \Phi(x', 0)$, letting $G(x)$ be an $N \times N$ matrix whose (i, j)th element are $g_{+ij}(x)$, we see that $G(x', 0)$ is the 1st fundamental matrix of Γ_+. Let $g_+ := \sqrt{\det G}$ and let $g_+^{ij}(x)$ denote the (i, j)th component of the inverse matrix, G^{-1}, of G. By (3.296), we can write

$$g_+ = 1 + \tilde{g}_+, \quad g_+^{ij}(x) = \delta_{ij} + \tilde{g}_+^{ij}(x)$$

with

$$\|(\tilde{g}_+, \tilde{g}_+^{ij})\|_{L_\infty(\mathbb{R}^N)} \le C_N M_1, \quad \|\nabla(\tilde{g}_+, \tilde{g}_+^{ij})\|_{L_\infty(\mathbb{R}^N)} \le C_N C_A,$$
$$\|\nabla^2(\tilde{g}_+, \tilde{g}_+^{ij})\|_{L_\infty(\mathbb{R}^N)} \le C_{M_2}. \tag{3.300}$$

The Laplace–Beltrami operator Δ_{Γ_+} is given by

$$(\Delta_{\Gamma_+} f)(y) = \sum_{i,j=1}^{N-1} \frac{1}{g_+(x',0)} \frac{\partial}{\partial x_i} \{ g_+(x',0) g_+^{ij}(x',0) \frac{\partial}{\partial x_j} f(\Phi(x',0)) \}$$

$$= \Delta' f(\Phi(x',0)) + \mathcal{D}_+ f$$

$$(3.301)$$

for $y = \Phi(x',0) \in \Gamma_+$. Here,

$$(\mathcal{D}_+ f)(y) = \sum_{i,j=1}^{N-1} \tilde{g}_+^{ij}(x) \frac{\partial^2 (f \circ \Phi)}{\partial x_i \partial x_j}(x) + \sum_{j=1}^{N-1} g_+^j(x) \frac{\partial (f \circ \Phi)}{\partial x_j}(x) \quad \text{for } y = \Phi(x)$$

with

$$g_+^j(x) = \frac{1}{g_+(x)} \sum_{i=1}^{N-1} \frac{\partial}{\partial x_i} (g_+(x) g_+^{ij}(x)).$$

By (3.300)

$$\|\mathcal{D}_+ f\|_{H_q^1(\mathbb{R}_+^N)} \le C_N M_1 \|\nabla^3 f\|_{L_q(\mathbb{R}_+^N)} + C_{M_2} \|f\|_{H_q^2(\mathbb{R}_+^N)}. \tag{3.302}$$

We now formulate problem treated in this section. Let y_0 be any point of Γ_+ and let d_0 be a positive number such that

$$|\mu(y) - \mu(y_0)|, |\sigma(y) - \sigma(y_0)| \le m_1 M_1, \quad \text{for any } y \in \overline{\Omega_+} \cap B_{d_0}(y_0);$$

$$|A_\kappa(y) - A_\kappa(y_0)| \le m_2 M_1 \quad \text{for any } y \in \Gamma_+ \cap B_{d_0}(y_0).$$

$$(3.303)$$

In addition, μ, σ, and A_κ satisfy the following conditions:

$$m_0 \le \mu(y), \sigma(y) \le m_1, \quad |\nabla\mu(y)|, |\nabla\sigma(y)| \le m_1 \quad \text{for any } y \in \overline{\Omega_+},$$

$$|A_\kappa(y)| \le m_2 \quad \text{for any } y \in \Gamma_+, \quad \|A_\kappa\|_{W_r^{2-1/q}(\Omega_+)} \le m_3 \kappa^{-b}$$

$$(3.304)$$

for any $\kappa \in (0,1)$. In view of (3.164) and (3.304), to have (3.303) for given $M_1 \in (0,1)$ it suffices to choose $d_0 > 0$ in such a way that $d_0 \le M_1$ and $d_0^a \le M_1$. We assume that $N < r < \infty$ according to (3.164) and let $A_0 = 0$. Let $\varphi(y)$ be a function in $C_0^\infty(\mathbb{R}^N)$ which equals 1 for $y \in B_{d_0/2}(y_0)$ and 0 in the outside of $B_{d_0}(y_0)$. We assume that $\|\nabla\varphi\|_{H_\infty^1(\mathbb{R}^N)} \leq M_2$. Let

$$\mu_{y_0}(y) = \varphi(y)\mu(y) + (1 - \varphi(y))\mu(y_0),$$

$$\sigma_{y_0}(y) = \varphi(y)\sigma(y) + (1 - \varphi(y))\sigma(y_0),$$

$$A_{\kappa,y_0}(y) = \varphi(y)A_\kappa(y) + (1 - \varphi(y))A_\kappa(y_0).$$

In the following, C denotes a generic constant depending on $m_0, m_1, m_2, m_3, N, \epsilon$ and q, and C_{M_2} denotes a generic constant depending on $M_2, m_0, m_1, m_2, m_3, N, \epsilon$ and q.

Given $\mathbf{v} \in H_q^2(\Omega_+)^N$ and $h \in H_q^3(\Omega_+)$, let $K_b(\mathbf{v}, h)$ be a unique solution of the weak Dirichlet problem:

$$(\nabla K_b(\mathbf{v}, h), \nabla\varphi)_{\Omega_+} = (\mathrm{Div}\,(\mu_{y_0}\mathbf{D}(\mathbf{v})) - \nabla\mathrm{div}\,\mathbf{v}, \nabla\varphi)_{\Omega_+} \tag{3.305}$$

for any $\varphi \in \hat{H}_{q',0}^1(\Omega_+)$ subject to

$$K_b(\mathbf{v}, h) = <\mu_{y_0}\mathbf{D}(\mathbf{v})\mathbf{n}_+, \mathbf{n}_+> -\sigma_{y_0}\Delta_{\Gamma_+}h - \mathrm{div}\,\mathbf{v} \quad \text{on } \Gamma_+.$$

We then consider the following equations:

$$\begin{cases} \lambda\mathbf{v} - \mathrm{Div}\,(\mu_{y_0}\mathbf{D}(\mathbf{v}) - K_b(\mathbf{v}, h)\mathbf{I}) = \mathbf{g} & \text{in } \Omega_+, \\ \lambda h + A_{\kappa,y_0} \cdot \nabla_{\Gamma_+}h - \mathbf{v} \cdot \mathbf{n}_+ = g_d & \text{on } \Gamma_+, \\ (\mu_{y_0}\mathbf{D}(\mathbf{v}) - K_b(\mathbf{v}, h)\mathbf{I})\mathbf{n}_+ - \sigma_{y_0}(\Delta_{\Gamma_+}h)\mathbf{n}_+ = \mathbf{g}_b & \text{on } \Gamma_+. \end{cases} \tag{3.306}$$

The following theorem is a main result in this section.

Theorem 3.5.18 *Let $1 < q < \infty$ and $0 < \epsilon < \pi/2$. Let γ_κ be the number defined in Theorem 3.4.8. Then, there exist $M_1 \in (0, 1)$, $\tilde{\lambda}_0 \geq 1$ and operator families $\mathcal{A}_b(\lambda)$ and $\mathcal{H}_b(\lambda)$ with*

$$\mathcal{A}_b(\lambda) \in Hol\,(\Lambda_{\kappa,\tilde{\lambda}_0\gamma_\kappa}, \mathcal{L}(\mathcal{Y}_q(\Omega_+), H_q^2(\Omega_+)^N)),$$

$$\mathcal{H}_b(\lambda) \in Hol\,(\Lambda_{\kappa,\tilde{\lambda}_0\gamma_\kappa}, \mathcal{L}(\mathcal{Y}_q(\Omega_+), H_q^3(\Omega_+)))$$

such that for any $\lambda = \gamma + i\tau \in \Lambda_{\kappa,\tilde{\lambda}_0\gamma_\kappa}$ and $(\mathbf{g}, g_d, \mathbf{g}_b) \in Y_q(\Omega_+)$,

$$\mathbf{u} = \mathcal{A}_b(\lambda)(\mathbf{g}, g_d, \lambda^{1/2}\mathbf{g}_b, \mathbf{g}_b), \quad h = \mathcal{H}_b(\lambda)(\mathbf{g}, g_d, \lambda^{1/2}\mathbf{g}_b, \mathbf{g}_b)$$

are unique solutions of Eq. (3.306), *and*

$$\mathcal{R}_{\mathcal{L}(\mathcal{Y}_q(\Omega_+), H_q^{2-j}(\Omega_+)^N)}(\{(\tau\partial_\tau)^\ell(\lambda^{j/2}\mathcal{A}_b(\lambda)) \mid \lambda \in \Lambda_{\kappa,\tilde{\lambda}_0\gamma_\kappa}\}) \leq r_b,$$

$$\mathcal{R}_{\mathcal{L}(\mathcal{Y}_q(\Omega_+), H_q^{3-k}(\Omega_+))}(\{(\tau\partial_\tau)^\ell(\lambda^k\mathcal{H}_b(\lambda)) \mid \lambda \in \Lambda_{\kappa,\tilde{\lambda}_0\gamma_\kappa}\}) \leq r_b,$$

$$(3.307)$$

for $\ell = 0, 1$, $j = 0, 1, 2$, *and* $k = 0, 1$. *Here,* r_b *is a constant depending on* m_0, m_1, m_2, N, q, *and* ϵ, *but independent of* M_1 *and* M_2, *and moreover,* $\tilde{\lambda}_0$ *is a constant depending on* M_2.

Below, we shall prove Theorem 3.5.18. By the change of variables $y = \Phi(x)$, we transform Eq. (3.306) to a problem in the half-space. Let

$$y_0 = \Phi(x_0), \quad \tilde{\mu}(x) = \varphi(\Phi(x))(\mu(\Phi(x)) - \mu(\Phi(x_0))),$$

$$\tilde{\sigma}(x) = \varphi(\Phi(x))(\sigma(\Phi(x)) - \sigma(\Phi(x_0))),$$

$$\tilde{A}_\kappa(x) = \varphi(\Phi(x))(A_\kappa(\Phi(x)) - A_\kappa(\Phi(x_0))).$$

Notice that

$$\mu_{y_0}(\Phi(x)) = \mu(y_0) + \tilde{\mu}(x), \quad \sigma_{y_0}(\Phi(x)) = \sigma(y_0) + \tilde{\sigma}(x),$$

$$A_\kappa(\Phi(x', 0)) = A_\kappa(y_0) + \tilde{A}_\kappa(x).$$

We may assume that m_1, m_2, $m_3 \leq M_2$. Recalling that $\|\nabla\varphi\|_{H^1_\infty(\mathbb{R}^N)} \leq M_2$, by (3.303) and (3.304) we have

$$\|(\tilde{\mu}, \tilde{\sigma})\|_{L_\infty(\mathbb{R}^N)} \leq m_1 M_1, \quad \|\nabla(\tilde{\mu}, \tilde{\sigma})\|_{L_\infty(\mathbb{R}^N)} \leq C_{M_2},$$

$$\|\tilde{A}_\kappa\|_{L_\infty(\mathbb{R}^N_0)} \leq m_2 M_1,$$

$$\|\nabla\tilde{A}_\kappa\|_{W_q^{1-1/q}(\mathbb{R}^N_0)} \leq C_{M_2}\kappa^{-b} \quad \text{for } \kappa \in (0, 1).$$

$$(3.308)$$

Since $x = \Phi^{-1}(y)$, we have

$$\frac{\partial}{\partial y_j} = \sum_{k=1}^N (a_{kj} + b_{kj}(x))\frac{\partial}{\partial x_k} \tag{3.309}$$

where $(\nabla\Phi^{-1})(\Phi(x)) = (a_{ij} + b_{ij}(x))$. Let

$$\mathfrak{g} := \det\nabla\Phi, \quad \tilde{\mathfrak{g}} = \mathfrak{g} - 1.$$

By (3.296),

$$\|\tilde{g}\|_{L_\infty(\mathbb{R}^N)} \le C_N M_1, \quad \|\nabla \tilde{g}\|_{L_\infty(\mathbb{R}^N)} \le C_N C_A, \quad \|\nabla^2 \tilde{g}\|_{L_\infty(\mathbb{R}^N)} \le C_{M_2}.$$
(3.310)

By the change of variables: $y = \Phi(x)$, the weak Dirichlet problem:

$$(\nabla u, \nabla \varphi)_{\Omega_+} = (\mathbf{k}, \nabla \varphi)_{\Omega_+} \quad \text{for any } \varphi \in \hat{H}^1_{q',0}(\Omega_+),$$

subject to $u = k$ on Γ_+, is transformed to the following variational problem:

$$(\nabla v, \nabla \psi)_{\mathbb{R}^N_+} + (\mathcal{B}^0 \nabla v, \nabla \psi)_{\mathbb{R}^N_+} = (\mathbf{h}, \nabla \psi)_{\mathbb{R}^N_+} \quad \text{for any } \psi \in \hat{H}^1_{q',0}(\mathbb{R}^N_+),$$
(3.311)

subject to $v = h$, where $\mathbf{h} = g(\mathcal{A}_{-1} + B_{-1} \circ \Phi)\mathbf{k} \circ \Phi$ and $h = k \circ \Phi$. Moreover, $\mathcal{B}^0$ is an $N \times N$ matrix whose (ℓ, m)th component, $\mathcal{B}^0_{\ell m}$, given by

$$\mathcal{B}^0_{\ell m} = \tilde{g}\delta_{\ell m} + g \sum_{j=1}^{N} (a_{\ell j}b_{mj}(x) + a_{mj}b_{\ell j}(x) + b_{\ell j}b_{mj}(x)).$$

By (3.296), we have

$$\|\mathcal{B}^0_{\ell m}\|_{L_\infty(\mathbb{R}^N)} \le C_N M_1, \quad \|\nabla \mathcal{B}^0_{\ell m}\|_{L_\infty(\mathbb{R}^N)} \le C_N C_A,$$
$$\|\nabla^2 \mathcal{B}^0_{\ell m}\|_{L_\infty(\mathbb{R}^N)} \le C_{M_2}.$$
(3.312)

Lemma 3.5.19 *Let $1 < q < \infty$. Then, there exist an $M_1 \in (0, 1)$ and an operator $\mathcal{K}_1$ with*

$$\mathcal{K}_1 \in \mathcal{L}(L_q(\mathbb{R}^N_+)^N, H^1_q(\mathbb{R}^N_+) + \hat{H}^1_{q,0}(\mathbb{R}^N_+))$$

such that for any $\mathbf{f} \in L_q(\mathbb{R}^N_+)^N$ and $f \in H^1_q(\mathbb{R}^N_+)$, $v = \mathcal{K}_1(\mathbf{f}, f)$ is a unique solution of the variational problem:

$$(\nabla v, \nabla \psi)_{\mathbb{R}^N_+} + (\mathcal{B}^0 \nabla v, \nabla \psi)_{\mathbb{R}^N_+} = (\mathbf{f}, \nabla \psi)_{\mathbb{R}^N_+} \quad \text{for any } \psi \in \hat{H}^1_{q',0}(\mathbb{R}^N_+),$$
(3.313)

subject to $v = f$ on $\mathbb{R}^N_0$, which possesses the estimate:

$$\|\nabla v\|_{L_q(\mathbb{R}^N_+)} \le C_{M_2}(\|\mathbf{f}\|_{L_q(\mathbb{R}^N_+)} + \|f\|_{H^1_q(\mathbb{R}^N_+)}).$$
(3.314)

Proof We know the unique existence theorem of the variational problem:

$$(\nabla v, \nabla \psi)_{\mathbb{R}^N_+} = (\mathbf{f}, \nabla \psi)_{\mathbb{R}^N_+} \quad \text{for any } \psi \in \hat{H}^1_{q',0}(\mathbb{R}^N_+),$$

subject to $v = f$ on $\mathbb{R}_+^N$. Thus, choosing $M_1 > 0$ small enough in (3.312) and using the Banach fixed point theorem, we can easily prove the lemma. $\qquad\square$

Using the change of the unknown functions: $\mathbf{u} = \mathcal{A}_{-1}\mathbf{v} \circ \Phi$ as well as the change of variable: $y = \Phi(x)$, we will derive the problem in $\mathbb{R}_+^N$ from (3.306). Noting that $\mathcal{A} = {}^\top\mathcal{A}_{-1}$, by (3.309) we have

$$D_{ij}(\mathbf{v}) = \sum_{k,\ell=1}^N a_{ki}a_{\ell j}D_{k\ell}(\mathbf{u}) + b_{ij}^d : \nabla\mathbf{u} \tag{3.315}$$

with $b_{ij}^d : \nabla\mathbf{u} = \sum_{k,\ell=1}^N a_{kj}b_{\ell i}D_{k\ell}(\mathbf{u})$. Setting $\mathbf{b}_+(x) = {}^\top(b_{+1},\ldots,b_{+N})$ in (3.298), by (3.298) we have

$$< \mathbf{D}(\mathbf{v})\mathbf{n}_+, \mathbf{n}_+ > = < \mathbf{D}(\mathbf{u})\mathbf{n}_0, \mathbf{n}_0 > + \mathcal{B}^1 : \nabla\mathbf{u} \tag{3.316}$$

where we have set

$$\mathcal{B}^1 : \nabla\mathbf{u} = -2\sum_{i,j=1}^N a_{ji}b_{+i}D_{jN}(\mathbf{u}) + \sum_{i,j,k,\ell=1}^N a_{ki}a_{\ell j}b_{+i}b_{+j}D_{k\ell}(\mathbf{u})$$

$$+ \sum_{i,j=1}^N (b_{ij}^d : \nabla\mathbf{u})(a_{Ni} + b_{+i})(a_{Nj} + b_{+j}).$$

By (3.296), we have

$$\|\mathcal{B}^1 : \nabla\mathbf{u}\|_{L_q(\mathbb{R}_+^N)} \le C_N M_1 \|\nabla\mathbf{u}\|_{L_q(\mathbb{R}_+^N)},$$

$$\|\mathcal{B}^1 : \nabla\mathbf{u}\|_{H_q^1(\mathbb{R}_+^N)} \le C_N\{M_1\|\nabla^2\mathbf{u}\|_{L_q(\mathbb{R}_+^N)} + C_A\|\mathbf{u}\|_{H_q^1(\mathbb{R}_+^N)}\}. \tag{3.317}$$

And also,

$$\operatorname{div}\mathbf{v} = \operatorname{div}\mathbf{u} + \mathcal{B}^2 : \nabla\mathbf{u} \quad \text{with} \quad \mathcal{B}^2 : \nabla\mathbf{u} = \sum_{\ell,k=1}^M (\sum_{j=1}^N b_{kj}a_{\ell j})\frac{\partial u_\ell}{\partial x_k}. \tag{3.318}$$

By (3.296), we have

$$\|\mathcal{B}^2 : \nabla\mathbf{u}\|_{L_q(\mathbb{R}_+^N)} \le C_N M_1 \|\nabla\mathbf{u}\|_{L_q(\mathbb{R}_+^N)},$$

$$\|\mathcal{B}^2 : \nabla\mathbf{u}\|_{H_q^1(\mathbb{R}_+^N)} \le C_N\{M_1\|\nabla^2\mathbf{u}\|_{L_q(\mathbb{R}_+^N)} + C_A\|\mathbf{u}\|_{H_q^1(\mathbb{R}_+^N)}\}. \tag{3.319}$$

By (3.315), we have

$$\mathcal{A}_{-1} \mathrm{Div}\,(\mu_{y_0}\mathbf{D}(\mathbf{v})) = \mathrm{Div}\,(\mu(y_0)\mathbf{D}(\mathbf{u})) + \mathcal{R}^1 : \mathbf{u} \qquad (3.320)$$

with $\mathcal{R}^1 : \mathbf{u} = {}^{\top}(\mathcal{R}^1 : \mathbf{u}|_1, \ldots, \mathcal{R}^1 : \mathbf{u}|_N)$, and

$$\mathcal{R}^1 : \mathbf{u}|_s = \sum_{k=1}^{N} \frac{\partial}{\partial x_k}(\tilde{\mu}(x) D_{sk}(\mathbf{u})) + \sum_{i,j,k=1}^{N} a_{si} a_{kj} \frac{\partial}{\partial x_k}((\mu(y_0) + \tilde{\mu}(x))b_{ij}^d : \nabla \mathbf{u})$$

$$+ \sum_{j,k,\ell,m=1}^{N} a_{\ell j} b_{kj} \frac{\partial}{\partial x_k}\{(\mu(y_0) + \tilde{\mu}(x))D_{s\ell}(\mathbf{u})\} + \sum_{i,j,k=1}^{N} a_{si} b_{kj} \frac{\partial}{\partial x_k}\{(\mu(y_0) + \tilde{\mu}(x))b_{ij}^d : \nabla \mathbf{u}\}.$$

By (3.296) and (3.308),

$$\|\mathcal{R}^1 : \mathbf{u}\|_{L_q(\mathbb{R}^N_+)} \le C_N m_1 M_1 \|\nabla^2 \mathbf{u}\|_{L_q(\mathbb{R}^N_+)} + C_{M_2} \|\mathbf{u}\|_{H^1_q(\mathbb{R}^N_+)}. \qquad (3.321)$$

And also, by (3.315)

$$(\mathcal{A}_{-1} + \mathcal{B}_{-1} \circ \Phi^{-1}) \mathrm{Div}\,(\mu_{y_0}\mathbf{D}(\mathbf{v})) = \mathrm{Div}\,(\mu(y_0)\mathbf{D}(\mathbf{u})) + \mathcal{R}^2 : \mathbf{u}$$

with $\mathcal{R}^2 : \mathbf{u} = (\mathcal{R}^2 : \mathbf{u}|_1, \ldots, \mathcal{R}^2 : \mathbf{u}|_N)$ and

$$\mathcal{R}^2 : \mathbf{u}|_s - \mathcal{R}^1 : \mathbf{u}|_s$$

$$+ \sum_{i,j,k=1}^{N} b_{si}(a_{kj} + b_{kj}) \frac{\partial}{\partial x_k}[(\tilde{\mu}(x) + \mu(y_0))\{\sum_{\ell,m=1}^{N} a_{\ell i} a_{mj} D_{\ell m}(\mathbf{u}) + b_{ij}^d : \nabla \mathbf{u}\}].$$

By (3.296)

$$\|\mathcal{R}^2 : \mathbf{u}\|_{L_q(\mathbb{R}^N_+)} \le C_N m_1 M_1 \|\nabla^2 \mathbf{u}\|_{L_q(\mathbb{R}^N_+)} + C_{M_2} \|\mathbf{u}\|_{H^1_q(\mathbb{R}^N_+)}. \qquad (3.322)$$

And also, we have

$$(\mathcal{A}_{-1} + \mathcal{B}_{-1} \circ \Phi^{-1})(\nabla \mathrm{div}\,\mathbf{v}) \circ \Phi = \nabla \mathrm{div}\,\mathbf{u} + \mathcal{R}^3 : \mathbf{u}$$

with $\mathcal{R}^3 : \mathbf{u} = (\mathcal{R}^3 : \mathbf{u}|_1, \ldots, \mathcal{R}^3 : \mathbf{u}|_N)$ and

$$\mathcal{R}^3 : \mathbf{u}|_s = \frac{\partial}{\partial x_s}(\mathcal{B}^2 : \nabla \mathbf{u})$$

$$+ \sum_{k=1}^{N}\{\sum_{i=1}^{N}(a_{si} b_{ki} + b_{si}(a_{ki} + b_{ki}))\} \frac{\partial}{\partial x_k}(\mathrm{div}\,\mathbf{u} + \mathcal{B}^2 : \nabla \mathbf{u}).$$

By (3.296)

$$\|\mathcal{R}^3 : \mathbf{u}\|_{L_q(\mathbb{R}^N_+)} \le C_N m_1 (M_1 \|\nabla^2 \mathbf{u}\|_{L_q(\mathbb{R}^N_+)} + C_A \|\mathbf{u}\|_{H^1_q(\mathbb{R}^N_+)}). \qquad (3.323)$$

Let

$$\mathbf{f}(\mathbf{u}) := \mathfrak{g}(\mathcal{A}_{-1} + B_{-1} \circ \Phi)(\text{Div}\,(\mu_{y_0}\mathbf{D}(\mathbf{v})) - \nabla \text{div}\,\mathbf{v}) \circ \Phi,$$

and then

$$\mathbf{f}(\mathbf{u}) = \text{Div}\,(\mu(y_0)\mathbf{D}(\mathbf{u})) - \nabla \text{div}\,\mathbf{u} + \mathcal{R}^4 : \mathbf{u}$$

with $\mathcal{R}^4 : \mathbf{u} = (\mathcal{R}^4 : \mathbf{u}|_1, \dots \mathcal{R}^4 : \mathbf{u}|_N)$ and

$$\mathcal{R}^4 : \mathbf{u}|_s = \tilde{\mathfrak{g}}(\text{Div}\,(\mu(y_0)\mathbf{D}(\mathbf{u})) - \nabla \text{div}\,\mathbf{u}) + \mathfrak{g}\mathcal{R}^2 : \mathbf{u} - \mathfrak{g}\mathcal{R}^3 : \mathbf{u}.$$

By (3.296), (3.308), (3.310)–(3.322), and (3.323),

$$\|\mathcal{R}^4 : \mathbf{u}\|_{L_q(\mathbb{R}^N_+)} \le C_N (m_1 + 1) M_1 \|\nabla^2 \mathbf{u}\|_{L_q(\mathbb{R}^N_+)} + C_{M_2} \|\mathbf{u}\|_{H^1_q(\mathbb{R}^N_+)}. \qquad (3.324)$$

In view of (3.301), (3.316) and (3.318), setting

$$\rho = h \circ \Phi,$$

$$f(\mathbf{u}, \rho) = <\tilde{\mu}(x)\mathbf{D}(\mathbf{u})\mathbf{n}_0, \mathbf{n}_0 > +(\mu(y_0) + \tilde{\mu}(x))\mathcal{B}^1 : \nabla \mathbf{u}$$
$$- \tilde{\sigma}(x)\Delta'\rho - (\sigma(y_0) + \tilde{\sigma}(x))\mathcal{D}_+\rho - \mathcal{B}^2 : \nabla \mathbf{u},$$

we have

$$< \mu_{y_0}\mathbf{D}(\mathbf{v})\mathbf{n}_+, \mathbf{n}_+ > -\sigma_{y_0}\Delta_{\Gamma_+} h - \text{div}\,\mathbf{v}$$
$$=< \mu(y_0)\mathbf{D}(\mathbf{u})\mathbf{n}_0, \mathbf{n}_0 > -\sigma(y_0)\Delta'\rho - \text{div}\,\mathbf{u} + f(\mathbf{u}, \rho).$$

Thus, $K_1(\mathbf{u}, \rho) = K_b(\mathbf{v}, h) \circ \Phi$ satisfies the variational equation:

$$(\nabla K_1(\mathbf{u}, \rho), \nabla \psi)_{\mathbb{R}^N_+} + (\mathcal{B}^0 \nabla K_1(\mathbf{u}, \rho), \nabla \psi)_{\mathbb{R}^N_+}$$
$$= (\text{Div}\,(\mu(y_0)\mathbf{D}(\mathbf{u})) - \nabla \text{div}\,\mathbf{u} + \mathcal{R}^4 : \mathbf{u}, \nabla \psi)_{\mathbb{R}^N_+}$$

for any $\psi \in \hat{H}^1_{q',0}(\mathbb{R}^N_+)$, subject to

$$K_1(\mathbf{u}, \rho) =< \mu(y_0)\mathbf{D}(\mathbf{u})\mathbf{n}_0, \mathbf{n}_0 > -\sigma(y_0)\Delta'\rho - \text{div}\,\mathbf{u} + f(\mathbf{u}, \rho) \quad \text{on } \mathbb{R}^N_0.$$

Let $\tilde{K}(\mathbf{u}, \rho) \in H_q^1(\mathbb{R}_+^N) + \hat{H}_{q,0}^1(\mathbb{R}_+^N)$ be a unique solution of the weak Dirichlet problem:

$$(\nabla \tilde{K}(\mathbf{u}, \rho), \nabla \psi)_{\mathbb{R}_+^N} = (\mathrm{Div}\,(\mu(y_0)\mathbf{D}(\mathbf{u})) - \nabla \mathrm{div}\,\mathbf{u}, \nabla \psi)_{\mathbb{R}_+^N}$$

for any $\psi \in \hat{H}_{q',0}^1(\mathbb{R}_+^N)$ subject to

$$\tilde{K}(\mathbf{u}, \rho) = <\mu(y_0)\mathbf{D}(\mathbf{u})\mathbf{n}_0, \mathbf{n}_0> -\sigma(y_0)\Delta'\rho - \mathrm{div}\,\mathbf{u} \quad \text{on } \mathbb{R}_0^N.$$

Setting $K_1(\mathbf{u}, \rho) = \tilde{K}(\mathbf{u}, \rho) + K_2(\mathbf{u}, \rho)$, we then see that $K_2(\mathbf{u}, \rho)$ satisfies the variational equation:

$$(\nabla K_2(\mathbf{u}, \rho), \nabla \psi)_{\mathbb{R}_+^N} + (\mathcal{B}^0 \nabla K_2(\mathbf{u}, \rho), \nabla \psi)_{\mathbb{R}_+^N} = (\mathcal{R}^4 : \mathbf{u} - \mathcal{B}^0 \nabla \tilde{K}(\mathbf{u}, \rho), \nabla \psi)_{\mathbb{R}_+^N}$$

for any $\psi \in \hat{H}_{q',0}^1(\mathbb{R}_+^N)$, subject to $K_2(\mathbf{u}, \rho) = f(\mathbf{u}, \rho)$ on $\mathbb{R}_0^N$. In view of Lemma 3.5.19, we have

$$K_2(\mathbf{u}, \rho) = \mathcal{K}_1(\mathcal{R}^4 : \mathbf{u} - \mathcal{B}^0 \nabla \tilde{K}(\mathbf{u}, \rho), f(\mathbf{u}, \rho)).$$

By Lemma 3.5.19, (3.312), (3.317), (3.319), (3.302), and (3.324), we have

$$\|\nabla K_2(\mathbf{u}, \rho)\|_{L_q(\mathbb{R}_+^N)}$$

$$\leq C_N(1 + m_1)M_1(\|\nabla^2 \mathbf{u}\|_{L_q(\mathbb{R}_+^N)} + \|\nabla^3 \rho\|_{L_q(\mathbb{R}_+^N)}) \qquad (3.325)$$

$$+ C_{M_2}(\|\mathbf{u}\|_{H_q^1(\mathbb{R}_+^N)} + \|\rho\|_{H_q^2(\mathbb{R}_+^N)}).$$

Since

$$\mathcal{A}_{-1} \nabla K_b(\mathbf{v}, h)|_s = \sum_{i,k=1}^N a_{si}(a_{ki} + b_{ki})\frac{\partial}{\partial x_k} K_1(\mathbf{u}, \rho)$$

$$= \frac{\partial}{\partial x_s}\tilde{K}(\mathbf{u}, \rho) + \sum_{k=1}^N (\sum_{i=1}^N a_{si}b_{ki})\frac{\partial}{\partial x_k}\tilde{K}(\mathbf{u}, \rho)$$

$$+ \sum_{k=1}^N (\delta_{ks} + \sum_{i=1}^N a_{si}b_{ki})\frac{\partial}{\partial x_k} K_2(\mathbf{u}, \rho),$$

by (3.320) we see that the first equation of Eq. (3.306) is transformed to

$$\lambda\mathbf{u} - \mathrm{Div}\,(\mu(y_0)\mathbf{D}(\mathbf{u}) - \tilde{K}(\mathbf{u}, \rho)\mathbf{I}) + \mathcal{R}^5(\mathbf{u}, \rho) = \mathbf{h} \quad \text{in } \mathbb{R}_+^N,$$

where $\mathbf{h} = \mathcal{A}_{-1}\mathbf{g} \circ \Phi$, $R^5(\mathbf{u}, \rho) = (\mathcal{R}^5(\mathbf{u}, \rho)|_1, \dots, \mathcal{R}^5(\mathbf{u}, \rho)|_N)$, and

$$\mathcal{R}^5(\mathbf{u}, \rho)|_s = -\mathcal{R}^1 : \mathbf{u}|_s + \sum_{k=1}^{N}(\sum_{i=1}^{N} a_{si}b_{ki})\frac{\partial}{\partial x_k}\tilde{K}(\mathbf{u}, \rho)$$

$$+ \sum_{k=1}^{N}(\delta_{ks} + \sum_{i=1}^{N} a_{si}b_{ki})\frac{\partial}{\partial x_k}K_2(\mathbf{u}, \rho).$$

By (3.298), we have

$$\mathbf{v} \cdot \mathbf{n}_+ = -({}^{\top}\mathcal{A}_{-1}\mathbf{u}) \cdot {}^{\top}(a_{N1}, \dots, a_{NN}) + ({}^{\top}\mathcal{A}_{-1}\mathbf{u}) \cdot \mathbf{b}_+$$

$$= \mathbf{u} \cdot \mathbf{n}_0 + \mathbf{u} \cdot (\mathcal{A}_{-1}\mathbf{b}_+),$$

and so the second equation of Eq. (3.306) is transformed to

$$\lambda\rho + A_{\kappa}(y_0) \cdot \nabla'\rho - \mathbf{u} \cdot \mathbf{n}_0 + \mathcal{R}_{\kappa}^6(\mathbf{u}, \rho) = h_d$$

with $h_d = g_d \circ \Phi$ and

$$\mathcal{R}_0^6(\mathbf{u}, \rho) = -\mathbf{u} \cdot (\mathcal{A}_{-1}\mathbf{b}_+) \quad \text{for } \kappa = 0,$$

$$\mathcal{R}_{\kappa}^6(\mathbf{u}, \rho) = \tilde{A}_{\kappa}(x)\nabla'\rho - \mathbf{u} \cdot (\mathcal{A}_{-1}\mathbf{b}_+) \quad \text{for } \kappa \in (0, 1).$$

By (3.298) and (3.315), we have $\mathcal{A}_{-1}\mu_{y_0}\mathbf{D}(\mathbf{v})\mathbf{n}_+ = \mu(y_0)\mathbf{D}(\mathbf{u})\mathbf{n}_0 + \mathcal{R}_1^7(\mathbf{u})$, where $\mathcal{R}_1^7(\mathbf{u})$ is an N-vector of functions whose sth component, $\mathcal{R}_1^7(\mathbf{u})|_s$, is defined by

$$\mathcal{R}_1^7(\mathbf{u})|_s = -\tilde{\mu}(x)D_{sN}(\mathbf{u}) + (\mu(y_0) + \tilde{\mu}(x))$$

$$\times \sum_{i,j=1}^{N}(a_{ij}b_{+j}D_{si}(\mathbf{u}) + a_{si}b_{ij}^d : \nabla\mathbf{u}(-a_{Nj} + b_{+j})).$$

By (3.298),

$$\mathcal{A}_{-1}K_b(\mathbf{v}, h)\mathbf{n}_+ = \tilde{K}(\mathbf{u}, \rho)\mathbf{n}_0 + \tilde{K}(\mathbf{u}, \rho)\mathcal{A}_{-1}\mathbf{b}_+ + K_2(\mathbf{u}, \rho)(\mathbf{n}_0 + \mathcal{A}_{-1}\mathbf{b}_+).$$

By (3.301),

$$\mathcal{A}_{-1}\sigma_{y_0}(\Delta_{\Gamma_+}h)\mathbf{n}_+ = \sigma(y_0)(\Delta'\rho)\mathbf{n}_0 + \tilde{\sigma}(x)(\Delta'\rho)\mathbf{n}_0$$

$$+ (\tilde{\sigma}(x) + \sigma(y_0))\{(\Delta'\rho)(\mathcal{A}_{-1}\mathbf{b}_+) + (\mathcal{D}_+\rho)(\mathbf{n}_0 + \mathcal{A}_{-1}\mathbf{b}_+)\}.$$

Putting formulas above together yields that the third equation of Eq. (3.306) is transformed to the equation:

$$(\mu(y_0)\mathbf{D}(\mathbf{u}) - \tilde{K}(\mathbf{u}, \rho)\mathbf{I})\mathbf{n}_0 - \sigma(y_0)(\Delta'\rho)\mathbf{n}_0 + \mathcal{R}^7(\mathbf{u}, \rho) = \mathbf{h}_b \quad \text{on } \mathbb{R}_0^N,$$

where $\mathbf{h}_b = \mathcal{A}_{-1}\mathbf{g}_b \circ \Phi$, and

$$\mathcal{R}^7(\mathbf{u}, \rho) = \mathcal{R}_1^7(\mathbf{u}, \rho) - \tilde{K}(\mathbf{u}, \rho)(\mathcal{A}_{-1}\mathbf{b}_+) - K_2(\mathbf{u}, \rho)(\mathbf{n}_0 + \mathcal{A}_{-1}\mathbf{b}_+)$$
$$- \tilde{\sigma}(x)(\Delta'\rho)\mathbf{n}_0 - (\tilde{\sigma}(x) + \sigma(y_0))\{(\Delta'\rho)(\mathcal{A}_{-1}\mathbf{b}_+)$$
$$+ (\mathcal{D}_+\rho)(\mathbf{n}_0 + \mathcal{A}_{-1}\mathbf{b}_+)\}.$$

Summing up, we have seen that Eq. (3.306) is transformed to the following equations:

$$\begin{cases} \lambda\mathbf{u} - \text{Div}\,(\mu(y_0)\mathbf{D}(\mathbf{u}) - \tilde{K}(\mathbf{u}, \rho)\mathbf{I}) + \mathcal{R}^5(\mathbf{u}, \rho) = \mathbf{h} & \text{in } \mathbb{R}_+^N, \\ \lambda\rho + A_\kappa(y_0) \cdot \nabla'\rho - \mathbf{u} \cdot \mathbf{n}_0 + R_\kappa^6(\mathbf{u}, \rho) = h_d & \text{on } \mathbb{R}_0^N, \\ (\mu(y_0)\mathbf{D}(\mathbf{u}) - \tilde{K}(\mathbf{u}, \rho)\mathbf{I})\mathbf{n}_0 - \sigma(y_0)(\Delta'\rho)\mathbf{n}_0 + \mathcal{R}^7(\mathbf{u}, \rho) = \mathbf{h}_b & \text{on } \mathbb{R}_0^N, \end{cases}$$
$$(3.326)$$

where $\mathbf{h} = \mathcal{A}_{-1}\mathbf{g} \circ \Phi$, $h_d = g_d \circ \Phi$, $\mathbf{h}_b = \mathcal{A}_{-1}\mathbf{g}_d \circ \Phi$, and $\mathcal{R}^5(\mathbf{u}, \rho)$, $\mathcal{R}^6(\mathbf{u}, \rho)$ and $\mathcal{R}^7(\mathbf{u}, \rho)$ are linear in $\mathbf{u}$ and ρ and satisfy the estimates:

$$\|\mathcal{R}^5(\mathbf{u}, \rho)\|_{L_q(\mathbb{R}_+^N)} \leq CM_1(\|\nabla^2\mathbf{u}\|_{L_q(\mathbb{R}_+^N)} + \|\nabla^3\rho\|_{L_q(\mathbb{R}_+^N)})$$
$$+ C_{M_2}(\|\mathbf{u}\|_{H_q^1(\mathbb{R}_+^N)} + \|\rho\|_{H_q^2(\mathbb{R}_+^N)}),$$

$$\|\mathcal{R}_0^6(\mathbf{u}, \rho)\|_{W_q^{2-1/q}(\mathbb{R}_0^N)} \leq CM_1\|\nabla^2\mathbf{u}\|_{L_q(\mathbb{R}_+^N)} + C_{M_2}\|\mathbf{u}\|_{H_q^1(\mathbb{R}_+^N)},$$

$$\|\mathcal{R}_\kappa^6(\mathbf{u}, \rho)\|_{W_q^{2-1/q}(\mathbb{R}_0^N)} \leq CM_1(\|\nabla^2\mathbf{u}\|_{L_q(\mathbb{R}_+^N)} + \|\nabla^3\rho\|_{L_q(\mathbb{R}_+^N)})$$
$$+ C_{M_2}(\|\mathbf{u}\|_{H_q^1(\mathbb{R}_+^N)} + \kappa^{-b}\|\rho\|_{H_q^2(\mathbb{R}_+^N)}), \quad (3.327)$$

$$\|\mathcal{R}^7(\mathbf{u}, \rho)\|_{L_q(\mathbb{R}_+^N)} \leq CM_1(\|\nabla\mathbf{u}\|_{L_q(\mathbb{R}_+^N)} + \|\nabla^2\rho\|_{L_q(\mathbb{R}_+^N)})$$
$$+ C_{M_2}(\|\mathbf{u}\|_{L_q(\mathbb{R}_+^N)} + \|\rho\|_{H_q^1(\mathbb{R}_+^N)}),$$

$$\|\mathcal{R}^7(\mathbf{u}, \rho)\|_{H_q^1(\mathbb{R}_+^N)} \leq CM_1(\|\nabla^2\mathbf{u}\|_{L_q(\mathbb{R}_+^N)} + \|\nabla^3\rho\|_{L_q(\mathbb{R}_+^N)})$$
$$+ C_{M_2}(\|\mathbf{u}\|_{H_q^1(\mathbb{R}_+^N)} + \|\rho\|_{H_q^2(\mathbb{R}_+^N)}).$$

Here and in the following, C denotes a generic constant depending on N, q, m_1, and m_2 and C_{M_2} a generic constant depending on N, q, m_1, m_2, m_3 and M_2. By Theorem 3.5.6, there exists a large number λ_0 and operator families $\mathcal{A}_0(\lambda)$ and

$\mathcal{H}_0(\lambda)$ with

$$\mathcal{A}_0(\lambda) \in \mathrm{Hol}\,(\Lambda_{\kappa,\lambda_0},\,\mathcal{L}(\mathcal{Y}(\mathbb{R}^N_+),\,H^2_q(\mathbb{R}^N_+)^N)),$$

$$\mathcal{H}_0(\lambda) \in \mathrm{Hol}\,(\Lambda_{\kappa,\lambda_0},\,\mathcal{L}(\mathcal{Y}(\mathbb{R}^N_+),\,H^3_q(\mathbb{R}^N_+)))$$

such that for any $\lambda \in \Lambda_{\kappa,\lambda_0}$ and $(\mathbf{f}, d, \mathbf{h}) \in Y_q(\mathbb{R}^N_+)$, $\mathbf{u}$ and ρ with

$$\mathbf{u} = \mathcal{A}_0(\lambda)F_\lambda(\mathbf{f}, d, \mathbf{h}), \quad \rho = \mathcal{H}_0(\lambda)F_\lambda(\mathbf{f}, d, \mathbf{h}),$$

where $F_\lambda(\mathbf{f}, d, \mathbf{h}) = (\mathbf{f}, d, \lambda^{1/2}\mathbf{h}, \mathbf{h})$, are unique solutions of the equations:

$$\begin{cases} \lambda\mathbf{u} - \mathrm{Div}\,(\mu(y_0)\mathbf{D}(\mathbf{u}) - \tilde{K}(\mathbf{u}, \rho)\mathbf{I}) = \mathbf{f} & \text{in } \mathbb{R}^N_+, \\ \lambda\rho + A_\kappa(y_0) \cdot \nabla'\rho - \mathbf{u} \cdot \mathbf{n}_0 = d & \text{on } \mathbb{R}^N_0, \\ (\mu(y_0)\mathbf{D}(\mathbf{u}) - \tilde{K}(\mathbf{u}, \rho)\mathbf{I})\mathbf{n}_0 - \sigma(y_0)(\Delta'\rho)\mathbf{n}_0 = \mathbf{h}, & \text{on } \mathbb{R}^N_0, \end{cases}$$

and

$$\mathcal{R}_{\mathcal{L}(\mathcal{Y}(\mathbb{R}^N_+),\,H^{2-j}_q(\mathbb{R}^N_+)^N)}(\{(\tau\partial_\tau)^s(\lambda^{j/2}\mathcal{A}_0(\lambda)) \mid \lambda \in \Lambda_{\kappa,\lambda_0}\}) \leq r_b,$$

$$\mathcal{R}_{\mathcal{L}(\mathcal{Y}(\mathbb{R}^N_+),\,H^{3-k}_q(\mathbb{R}^N_+))}(\{(\tau\partial_\tau)^s(\lambda^k\mathcal{H}_0(\lambda)) \mid \lambda \in \Lambda_{\kappa,\lambda_0}\}) \leq r_b$$

for $s = 0, 1$, $j = 0, 1, 2$, and $k = 0, 1$. Here, r_b is a constant depending on ϵ, N, m_1, and m_2.

Let $\mathbf{u} = \mathcal{A}_0(\lambda)F_\lambda(\mathbf{h}, h_d, \mathbf{h}_b)$ and $\rho = \mathcal{H}_0(\lambda)F_\lambda(\mathbf{h}, h_d, \mathbf{h}_b)$ in (3.326). Then, Eq. (3.326) is rewritten as

$$\begin{cases} \lambda\mathbf{u} - \mathrm{Div}\,(\mu(y_0)\mathbf{D}(\mathbf{u}) - \tilde{K}(\mathbf{u}, \rho)\mathbf{I}) + \mathcal{R}^5(\mathbf{u}, \rho) \\ \quad = \mathbf{h} + \mathcal{R}^8(\lambda)F_\lambda(\mathbf{h}, h_d, \mathbf{h}_b) & \text{in } \mathbb{R}^N_+, \\ \lambda\rho + A_\kappa \cdot \nabla'\rho - \mathbf{u} \cdot \mathbf{n}_0 + R^6_\kappa(\mathbf{u}, \rho) \\ \quad = h_d + \mathcal{R}^8_d(\lambda)F_\lambda(\mathbf{h}, h_d, \mathbf{h}_b) & \text{in } \mathbb{R}^N_+, \\ (\mu(y_0)\mathbf{D}(\mathbf{u}) - \tilde{K}(\mathbf{u}, \rho)\mathbf{I})\mathbf{n}_0 - \sigma(y_0)(\Delta'\rho)\mathbf{n}_0 + \mathcal{R}^7(\mathbf{u}, \rho) \\ \quad = \mathbf{h}_b + \mathcal{R}^8_b(\lambda)F_\lambda(\mathbf{h}, h_d, \mathbf{h}_b) & \text{on } \mathbb{R}^N_0, \end{cases}$$
$$(3.328)$$

where we have set

$$\mathcal{R}^8(\lambda)(F_1, F_2, F_3, F_4)$$
$$= \mathcal{R}^5(\mathcal{A}_0(\lambda)(F_1, F_2, F_3, F_4), \mathcal{H}_0(\lambda)(F_1, F_2, F_3, F_4)),$$

$$\mathcal{R}_d^8(\lambda)(F_1, F_2, F_3, F_4)$$

$$= \mathcal{R}_\kappa^6(\mathcal{A}_0(\lambda)(F_1, F_2, F_3, F_4), \mathcal{H}_0(\lambda)(F_1, F_2, F_3, F_4)),$$

$$\mathcal{R}_b^8(\lambda)(F_1, F_2, F_3, F_4)$$

$$= \mathcal{R}^7(\mathcal{A}_0(\lambda)(F_1, F_2, F_3, F_4), \mathcal{H}_0(\lambda)(F_1, F_2, F_3, F_4)).$$

Let

$$\mathcal{R}^9(\lambda)F = (\mathcal{R}^8(\lambda)F, \mathcal{R}_d^8(\lambda)F, \mathcal{R}_b^8(\lambda)F)$$

for $F = (F_1, F_2, F_3, F_4) \in \mathcal{Y}_q(\mathbb{R}_+^N)$. Notice that

$$\mathcal{R}^9(\lambda)F = (\mathcal{R}^8(\lambda)F, \mathcal{R}_d^8(\lambda)F, \lambda^{1/2}\mathcal{R}_b^8(\lambda)F, \mathcal{R}_b^8(\lambda)F) \in \mathcal{Y}_q(\mathbb{R}_+^N),$$

for $F = (F_1, F_2, F_3, F_4) \in \mathcal{Y}_q(\mathbb{R}_+^N)$ and that the right side of Eq. (3.328) is written as $(\mathbf{h}, h_d, \mathbf{h}_b) + F^9(\lambda)F_\lambda(\mathbf{h}, h_d, \mathbf{h}_b)$. By (3.327), (3.299), Proposition 3.5.4, and Theorem 3.5.6, we have

$$\mathcal{R}_{\mathcal{L}(\mathcal{Y}_q(\mathbb{R}_+^N))}(\{(\tau\partial_\tau)^\ell(F_\lambda\mathcal{R}^9(\lambda)) \mid \lambda \in \Lambda_{\kappa,\lambda_1}\}) \leq CM_1 + C_{M_2}(\lambda_1^{-1/2} + \lambda_1^{-1}\gamma_\kappa) \tag{3.329}$$

for any $\lambda_1 \geq \lambda_0$. Here and in the following, C denotes a generic constant depending on N, ϵ, m_1, m_2, and C_A, and C_{M_2} denotes a generic constant depending on N, ϵ, m_1, m_2, m_3, C_A, and M_2. Choosing M_1 so small that $CM_1 \leq 1/4$ and choosing $\lambda_1 > 0$ so large that $C_{M_2}\lambda_1^{-1/2} \leq 1/8$ and $C_{M_2}\lambda_1^{-1}\gamma_\kappa \leq 1/8$, by (3.329) we have

$$\mathcal{R}_{\mathcal{L}(\mathcal{Y}_q(\mathbb{R}_+^N))}(\{(\tau\partial_\tau)^\ell(F_\lambda\mathcal{R}^9(\lambda)) \mid \lambda \in \Lambda_{\kappa,\lambda_1}\}) \leq 1/2 \tag{3.330}$$

for $\ell = 0, 1$. Since $\gamma_\kappa \geq 1$ and we may assume that $C_{M_2} \geq 1$, if $\lambda_1 \geq 64C_{M_2}^2\gamma_\kappa$, then $C_{M_2}\lambda_1^{-1/2} \leq 1/8$ and $C_{M_2}\lambda_1^{-1}\gamma_\kappa \leq 1/8$.

Recall that for $F = (F_1, F_2, F_3, F_4) \in \mathcal{Y}_q(\mathbb{R}_+^N)$ and $(\mathbf{h}, h_d, \mathbf{h}_b) \in Y_q(\mathbb{R}_+^N)$,

$$\|(F_1, F_2, F_3, F_4)\|_{\mathcal{Y}_q(\mathbb{R}_+^N)} = \|(F_1, F_3)\|_{L_q(\mathbb{R}_+^N)} + \|F_2\|_{W_q^{2-1/q}(\mathbb{R}_0^N)} + \|F_4\|_{H_q^1(\mathbb{R}_+^N)},$$

$$\|(\mathbf{h}, h_d, \mathbf{h}_b)\|_{X_q(\mathbb{R}_+^N)} = \|\mathbf{h}\|_{L_q(\mathbb{R}_+^N)} + \|h_d\|_{W_q^{2-1/q}(\mathbb{R}_0^N)} + \|\mathbf{h}_b\|_{H_q^1(\mathbb{R}_+^N)} \tag{3.331}$$

(cf. Remark 3.4.12, where Ω should be replaced by $\mathbb{R}_+^N$). By (3.330) we have

$$\|F_\lambda(\mathcal{R}^9(\lambda)F_\lambda(\mathbf{h}, h_d, \mathbf{h}_b))\|_{\mathcal{Y}_q(\mathbb{R}_+^N)} \leq (1/2)\|F_\lambda(\mathbf{h}, h_d, \mathbf{h}_b)\|_{\mathcal{Y}_q(\mathbb{R}_+^N)}. \tag{3.332}$$

In view of (3.331), when $\lambda \neq 0$, $\|\mathcal{F}_\lambda(\mathbf{h}, h_d, \mathbf{h}_b)\|_{\mathcal{Y}_q(\mathbb{R}_+^N)}$ is an equivalent norm to $\|(\mathbf{h}, h_d, \mathbf{h}_b)\|_{X_q(\mathbb{R}_+^N)}$. Thus, by (3.332) $(\mathbf{I} + \mathcal{R}^9(\lambda)F_\lambda)^{-1} = \sum_{j=1}^{\infty}(-\mathcal{R}^9(\lambda)F_\lambda)^j$ exists in $\mathcal{L}(X_q(\mathbb{R}_+^N))$. Setting

$$
\begin{aligned}
\mathbf{u} &= \mathcal{A}_0(\lambda)F_\lambda(\mathbf{I} + \mathcal{R}^9(\lambda)F_\lambda)^{-1}(\mathbf{h}, h_d, \mathbf{h}_b), \\
\rho &= \mathcal{H}_0(\lambda)F_\lambda(\mathbf{I} + \mathcal{R}^9(\lambda)F_\lambda)^{-1}(\mathbf{h}, h_d, \mathbf{h}_b),
\end{aligned}
\tag{3.333}
$$

by (3.328) we see that $\mathbf{u}$ and ρ are solutions of Eq. (3.326). In view of (3.328), $(\mathbf{I} + F_\lambda\mathcal{R}^9(\lambda))^{-1} = \sum_{j=0}^{\infty}(-F_\lambda\mathcal{R}^9(\lambda))^j$ exists in $\mathcal{L}(\mathcal{Y}_q(\mathbb{R}_+^N))$, and

$$
\mathcal{R}_{\mathcal{L}(\mathcal{Y}_q(\mathbb{R}_+^N))}(\{(\tau\partial_\tau)^\ell(\mathbf{I} + F_\lambda\mathcal{R}^9(\lambda))^{-1} \mid \lambda \in \Lambda_{\kappa,\lambda_1}\}) \leq 4
\tag{3.334}
$$

for $\ell = 0, 1$. Since

$$
\begin{aligned}
F_\lambda(\mathbf{I} + \mathcal{R}^9(\lambda)F_\lambda)^{-1} &= F_\lambda\sum_{j=0}^{\infty}(-\mathcal{R}^9(\lambda)F_\lambda)^j = (\sum_{j=0}^{\infty}(-F_\lambda\mathcal{R}^9(\lambda))^j)F_\lambda \\
&= (\mathbf{I} + F_\lambda\mathcal{R}^9(\lambda))^{-1}F_\lambda,
\end{aligned}
$$

defining operators $\mathcal{A}_1(\lambda)$ and $\mathcal{H}_1(\lambda)$ acting on $F = (F_1, F_2, F_3, F_4) \in \mathcal{Y}_q(\mathbb{R}_+^N)$ by

$$
\mathcal{A}_1(\lambda)F = \mathcal{A}_0(\lambda)(\mathbf{I} + F_\lambda\mathcal{R}^9(\lambda))^{-1}F, \quad \mathcal{H}_1(\lambda)F_1 = \mathcal{H}_0(\lambda)(\mathbf{I} + F_\lambda\mathcal{R}^9(\lambda))^{-1}F,
$$

by (3.333) $\mathbf{u} = \mathcal{A}_1(\lambda)F_\lambda(\mathbf{h}, h_d, \mathbf{h}_b)$ and $\rho = \mathcal{H}_1(\lambda)F_\lambda(\mathbf{h}, h_d, \mathbf{h}_b)$ are solutions of Eq. (3.326). Moreover, by (3.334) and Theorem 3.5.6

$$
\begin{aligned}
\mathcal{R}_{\mathcal{L}(\mathcal{Y}_q(\mathbb{R}_+^N), H_q^{2-j}(\mathbb{R}_+^N)^N)}(\{(\tau\partial_\tau)^\ell(\lambda^{j/2}\mathcal{A}_1(\lambda)) \mid \lambda \in \Lambda_{\kappa,\lambda_1\gamma_\kappa}\}) &\leq 4r_b, \\
\mathcal{R}_{\mathcal{L}(\mathcal{Y}_q(\mathbb{R}_+^N), H_q^{3-k}(\mathbb{R}_+^N))}(\{(\tau\partial_\tau)^\ell(\lambda^k\mathcal{H}_1(\lambda)) \mid \lambda \in \Lambda_{\kappa,\lambda_1\gamma_\kappa}\}) &\leq 4r_b,
\end{aligned}
\tag{3.335}
$$

for $\ell = 0, 1$, $j = 0, 1, 2$ and $k = 0, 1$. Recalling that

$$
\begin{aligned}
\mathbf{v} &= ({}^\top\mathcal{A}_{-1}\mathbf{u}) \circ \Phi^{-1}, \quad h = \rho \circ \Phi^{-1}, \\
\mathbf{h} &= \mathcal{A}_{-1}\mathbf{g} \circ \Phi, \quad h_d = g_d \circ \Phi, \quad \mathbf{h}_d = \mathcal{A}_{-1}\mathbf{g}_d \circ \Phi,
\end{aligned}
$$

we define operators $\mathcal{A}_b(\lambda)$ and $\mathcal{H}_b(\lambda)$ acting on $F = (F_1, F_2, F_3, F_4) \in \mathcal{Y}_q(\Omega_+)$ by

$$
\begin{aligned}
&\mathcal{A}_b(F_1, F_2, F_3, F_4) \\
&\quad = {}^\top\mathcal{A}_{-1}[\mathcal{A}_1(\lambda)(\mathcal{A}_{-1}F_1 \circ \Phi, F_2 \circ \Phi, \mathcal{A}_{-1}F_3 \circ \Phi, F_4 \circ \Phi)] \circ \Phi^{-1}, \\
&\mathcal{H}_b(F_1, F_2, F_3, F_4) = [\mathcal{H}_1(\lambda)(\mathcal{A}_{-1}F_1 \circ \Phi, F_2 \circ \Phi, \mathcal{A}_{-1}F_3 \circ \Phi, F_4 \circ \Phi)] \circ \Phi^{-1}.
\end{aligned}
$$

Obviously, given any $(\mathbf{g}, g_d, \mathbf{g}_b) \in Y_q(\Omega_+)$, $\mathbf{u} = \mathcal{A}_b(\lambda)F_\lambda(\mathbf{g}, g_d, \mathbf{g}_b)$ and $h = \mathcal{H}_b(\lambda)F_\lambda(\mathbf{g}, g_d, \mathbf{g}_b)$ are solutions of Eq. (3.306). From (3.296) we have

$$\|g \circ \Phi^{-1}\|_{H_q^\ell(\Omega_+)} \leq C_A \|g\|_{H_q^\ell(\mathbb{R}_+^N)} \quad \text{for } \ell = 0, 1, 2,$$

$$\|\nabla^3(g \circ \Phi^{-1})\|_{L_q(\Omega_+)} \leq C_A \|\nabla^2 g\|_{H_q^1(\mathbb{R}_+^N)} + C_{M_2} \|\nabla g\|_{L_q(\mathbb{R}_+^N)},$$

$$\|h \circ \Phi\|_{H_q^\ell(\mathbb{R}_+^N)} \leq C_A \|h\|_{H_q^\ell(\Omega_+)} \quad \text{for } \ell = 0, 1, 2,$$

and so, in view of (3.335) we can choose $\tilde{\lambda}_0 \geq \lambda_1$ suitably large such that $\mathcal{A}_b(\lambda)$ and $\mathcal{H}_b(\lambda)$ satisfy the estimates:

$$\mathcal{R}_{\mathcal{L}(Y_q(\Omega_+), H_q^{2-j}(\Omega_+)^N)}(\{(\tau\partial_\tau)^\ell(\lambda^{j/2}\mathcal{A}_b(\lambda)) \mid \lambda \in \Lambda_{\kappa, \lambda_1\gamma_\kappa}\}) \leq Cr_b,$$

$$\mathcal{R}_{\mathcal{L}(Y_q(\mathbb{R}_+^N), H_q^{3-k}(\mathbb{R}_+^N))}(\{(\tau\partial_\tau)^\ell(\lambda^k\mathcal{H}_b(\lambda)) \mid \lambda \in \Lambda_{\kappa, \lambda_1\gamma_\kappa}\}) \leq Cr_b,$$

for $\ell = 0, 1, j = 0, 1, 2$ and $k = 0, 1$, where C and r_b are constants independent of M_2. This completes the existence part of Theorem 3.5.18.

The uniqueness can be proved by showing a priori estimates of solutions of Eq. (3.326) with $(\mathbf{h}, h_d, \mathbf{h}_b) = (0, 0, 0)$ in the same manner as in the proof of Theorem 3.5.5. This completes the proof of Theorem 3.5.18 without the operators $\mathcal{L}_1$ and $\mathcal{L}_2$.

3.5.5 Some Preparation for the Proof of Theorem 3.4.11

In the following, we use the symbols given in Proposition 3.2.2 in Sect. 3.2.1 and we write $\Omega_j = \Phi_j(\mathbb{R}_+^N)$, and $\Gamma_j = \Phi_j(\mathbb{R}_0^N)$ for the sake of simplicity. Recall that $B_j^i = B_{r_0}(x_j^i)$. In view of the assumptions (3.164)–(3.166), we may assume that

$$|\mu(x) - \mu(x_j^i)| \leq M_1 \quad \text{for any } x \in B_j^i;$$

$$|\sigma(x) - \sigma(x_j^1)| \leq M_1 \quad \text{for any } x \in \Gamma_j \cap B_j^i;$$

$$|A_\kappa(x) - A_\kappa(x_j^1)| \leq M_1 \quad \text{for any } x \in \Gamma_j \cap B_j^i; \tag{3.336}$$

$$m_0 \leq \mu(x), \sigma(x) \leq m_1, \quad |\nabla\mu(x)|, |\nabla\sigma(x)| \leq m_1 \quad \text{for any } x \in \overline{\Omega},$$

$$|A_\kappa(x)| \leq m_2 \quad \text{for any } x \in \Gamma, \quad \|A_\kappa\|_{W_r^{2-1/q}(\Gamma)} \leq m_3\kappa^{-b} \tag{3.337}$$

for any $\kappa \in (0, 1)$. Here, m_0, m_1, m_2, m_3, b and r are constants given in (3.164).

We next prepare some propositions used to construct a parametrix.

Proposition 3.5.20 *Let X be a Banach space and X^* its dual space, while $\| \cdot \|_X$, $\| \cdot \|_{X^*}$, and $< \cdot, \cdot >$ denote the norm of X, the norm of X^*, and the duality pairing* *between of X and X^*, respectively. Let $n \in \mathbb{N}$, $l = 1, \ldots, n$, and $\{a_l\}_{l=1}^n \subset \mathbb{C}$, and let $\{f_j^l\}_{j=1}^\infty$ be sequences in X^* and $\{g_j^l\}_{j=1}^\infty$, $\{h_j\}_{j=1}^\infty$ be sequences of positive numbers. Assume that there exist maps $\mathcal{N}_j : X \to [0, \infty)$ such that*

$$| < f_j^l, \varphi > | \leq M_3 g_j^l \mathcal{N}_j(\varphi) \quad (l = 1, \ldots, n), \quad \left| \left\langle \sum_{l=1}^n a_l f_j^l, \varphi \right\rangle \right| \leq M_3 h_j \mathcal{N}_j(\varphi)$$

for any $\varphi \in X$ with some positive constant M_3 independent of $j \in \mathbb{N}$ and $l = 1, \ldots, n$. If

$$\sum_{j=1}^\infty \left(g_j^l \right)^q < \infty, \quad \sum_{j=1}^\infty (h_j)^q < \infty, \quad \sum_{j=1}^\infty (\mathcal{N}_j(\varphi))^{q'} \leq (M_4 \|\varphi\|_X)^{q'}$$

with $1 < q < \infty$ and $q' = q/(q-1)$ for some positive constant M_4, then the infinite sum $f^l = \sum_{j=1}^\infty f_j^l$ exists in the strong topology of X^* and

$$\|f^l\|_{X^*} \leq M_3 M_4 \left(\sum_{j=1}^\infty (g_j^l)^q \right)^{1/q}, \quad \left\| \sum_{l=1}^n a_l f^l \right\|_{X^*} \leq M_3 M_4 \left(\sum_{j=1}^\infty (h_j)^q \right)^{1/q}.$$

$$(3.338)$$

Proof For a proof, see Proposition 9.5.2 in Shibata [45]. $\square$

The following propositions are used to define the infinite sum of $\mathcal{R}$-bounded operator families defined on $\mathbb{R}^N$ and Ω_j.

Proposition 3.5.21 *Let $1 < q < \infty$, $i = 0, 1$, and $n \in \mathbb{N}_0$. Set $\mathcal{H}_j^0 = \mathbb{R}^N$ and $\mathcal{H}_j^1 = \Omega_j$. Let η_j^i be a function in $C_0^\infty(B_j^i)$ such that $\|\eta_j^i\|_{H_\infty^n(\mathbb{R}^N)} \leq c_1$ for any $j \in \mathbb{N}$ with some constant c_1 independent of $j \in \mathbb{N}$. Let f_j $(j \in \mathbb{N})$ be elements in $H_q^n(\mathcal{H}_j^i)$ such that $\sum_{j=1}^\infty \|f_j\|_{H_q^n(\mathcal{H}_j^i)}^q < \infty$. Then, $\sum_{j=1}^\infty \eta_j^i f_j$ converges some $f \in H_q^n(\Omega)$ strongly in $H_q^n(\Omega)$, and*

$$\|f\|_{H_q^n(\Omega)} \leq C_q \{ \sum_{j=1}^\infty \|f_j\|_{H_q^n(\mathcal{H}_j^i)}^q \}^{1/q}.$$

Proof For a proof, see Proposition 9.5.3 in Shibata [45]. $\square$

Proposition 3.5.22 *Let $1 < q < \infty$ and $n = 2, 3$. Then we have the following assertions.*

(1) *There exist extension maps $\mathbf{T}_j^n : W_q^{n-1/q}(\Gamma_j) \to H_q^n(\Omega_j)$ such that for any $h \in W_q^{n-1/q}(\Gamma_j)$, $\mathbf{T}_j^n h = h$ on Γ_j and $\|\mathbf{T}_j^n h\|_{H_q^n(\Omega_j)} \leq C\|h\|_{W_q^{n-1/q}(\Gamma_j)}$ with some constant $C > 0$ independent of $j \in \mathbb{N}$.*

(2) *There exists an extension map* $\mathbf{T}_\Gamma^n : W_q^{n-1/q}(\Gamma) \to H_q^n(\Omega)$ *such that for* $h \in W_q^{n-1/q}(\Gamma)$, $\mathbf{T}_\Gamma^n h = h$ *on* Γ *and* $\|\mathbf{T}_\Gamma^n h\|_{H_q^n(\Omega)} \le C\|h\|_{W_q^{n-1/q}(\Gamma)}$ *with some constant* $C > 0$.

Proof For a proof, see Proposition 9.5.4 in Shibata [45]. □

Proposition 3.5.23 *Let* $1 < q < \infty$ *and* $n = 2, 3$ *and let* $\eta_j \in C_0^\infty(B_j^1)$ $(j \in \mathbb{N})$ *with* $\|\eta_j\|_{H_\infty^n(\mathbb{R}^N)} \le c_2$ *for some constant* c_2 *independent of* $j \in \mathbb{N}$. *Then, we have the following two assertions:*

(1) *Let* f_j $(j \in \mathbb{N})$ *be functions in* $W_q^{n-1/q}(\Gamma_j)$ *satisfying the condition:* $\sum_{j=1}^\infty \|f_j\|_{W_q^{n-1/q}(\Gamma_j)}^q < \infty$, *and then the infinite sum* $\sum_{j=1}^\infty \eta_j f_j$ *converges to some* $f \in W_q^{n-1/q}(\Gamma)$ *strongly in* $W_q^{n-1/q}(\Gamma)$ *and*

$$\|f\|_{W_q^{n-1/q}(\Gamma)} \le C_q \{\sum_{j=1}^\infty \|f_j\|_{W_q^{n-1/q}(\Gamma_j)}^q\}^{1/q}.$$

(2) *For any* $h \in W_q^{n-1/q}(\Gamma)$,

$$\sum_{j=1}^\infty \|\eta_j h\|_{W_q^{n-1/q}(\Gamma_j)}^q \le C\|h\|_{W_q^{n-1/q}(\Gamma)}^q.$$

Proof For a proof, see Proposition 9.5.5 in Shibata [45]. □

3.5.6 Parametrix of Solutions of Eq. (3.184)

In this subsection, we construct a parametrix for Eq. (3.226). Let $\{\zeta_j^i\}_{j\in\mathbb{N}}$ and $\{\tilde{\zeta}_j^i\}_{j\in\mathbb{N}}$ $(i = 0, 1)$ be sequences of C_0^∞ functions given in Proposition 3.2.2, and let $(\mathbf{f}, d, \mathbf{h}) \in Y_q(\Omega)$ (cf. (3.193)). Recall that $\Omega_j = \Phi_j(\mathbb{R}_+^N)$ and $\Gamma_j = \Phi_j(\mathbb{R}_0^N)$. Let

$$\mu_j^i(x) = \tilde{\zeta}_j^i(x)\mu(x) + (1 - \tilde{\zeta}_j^i(x))\mu(x_j^i),$$
$$\sigma_j(x) = \tilde{\zeta}_j^1(x)\sigma(x) + (1 - \tilde{\zeta}_j^1(x))\sigma(x_j^1),$$
$$A_{\kappa,j}(x) = \tilde{\zeta}_j^1(x)A_\kappa(x) + (1 - \tilde{\zeta}_j^1(x))A_\kappa(x_j^1).$$

Notice that

$$\zeta_j^i\mu = \zeta_j^i\mu_j^i, \quad \zeta_j^1\sigma = \zeta_j^1\sigma_j, \quad \zeta_j^1 A_\kappa = \zeta_j^1 A_{\kappa,j},$$

because $\tilde{\zeta}_j^1 = 1$ on supp ζ_j^1. We consider the equations:

$$\lambda \mathbf{u}_j^0 - \text{Div}\,(\mu_j^0 \mathbf{D}(\mathbf{u}_j^0) - K_{0j}(\mathbf{u}_j^0)\mathbf{I}) = \tilde{\zeta}_j^0 \mathbf{f} \qquad \text{in } \mathbb{R}^N; \qquad (3.339)$$

$$\begin{cases} \lambda \mathbf{u}_j^1 - \text{Div}\,(\mu_j^1 \mathbf{D}(\mathbf{u}_j^1) - K_{1j}(\mathbf{u}_j^1, h_j)\mathbf{I}) = \tilde{\zeta}_j^1 \mathbf{f} & \text{in } \Omega_j, \\[4pt] \lambda h_j + A_{\kappa,j} \cdot \nabla_{\Gamma_j} h_j - \mathbf{n}_j \cdot \mathbf{u}_j = \tilde{\zeta}_j^1 d & \text{on } \Gamma_j, \qquad (3.340) \\[4pt] (\mu_j^1 \mathbf{D}(\mathbf{u}_j^1) - K_{1j}(\mathbf{u}_j^1, h_j)\mathbf{I})\mathbf{n}_j - \sigma_j(\Delta_{\Gamma_j} h_j)\mathbf{n}_j = \tilde{\zeta}_j^1 \mathbf{h} & \text{on } \Gamma_j. \end{cases}$$

Here, for $\mathbf{u} \in H_q^2(\mathbb{R}^N)^N$, $K_{0j}(\mathbf{u}) \in \hat{H}_q^1(\mathbb{R}^N)$ denotes a unique solution of the weak Laplace equation:

$$(\nabla K_{0j}(\mathbf{u}), \nabla \varphi)_{\mathbb{R}^N} = (\text{Div}\,(\mu_j^0 \mathbf{D}(\mathbf{u})) - \nabla \text{div}\,\mathbf{u}, \nabla \varphi)_{\mathbb{R}^N} \qquad (3.341)$$

for any $\varphi \in \hat{H}_{q'}^1(\mathbb{R}^N)$. And, for $\mathbf{u} \in H_q^2(\Omega_j)$ and $h \in H_q^3(\Omega_j)$, $K_{1j}(\mathbf{u}, h_j) \in H_q^1(\Omega_j) + \hat{H}_{q,0}^1(\Omega_j)$ denotes a unique solution of the weak Dirichlet problem:

$$(\nabla K_{1j}(\mathbf{u}, h), \nabla \varphi)_{\Omega_j} = (\text{Div}\,(\mu_j^1 \mathbf{D}(\mathbf{u})) - \nabla \text{div}\,\mathbf{u}, \nabla \varphi)_{\Omega_j} \qquad (3.342)$$

for any $\varphi \in \hat{H}_{q',0}^1(\Omega_j)$, subject to

$$K_{1j}(\mathbf{u}, h) = <\mu_j^1 \mathbf{D}(\mathbf{u})\mathbf{n}_j, \mathbf{n}_j> -\text{div}\,\mathbf{u} - \sigma_j \Delta_{\Gamma_j} h \qquad \text{on } \Gamma_j.$$

Moreover, we denote the unit outer normal to Γ_j by $\mathbf{n}_j$, which is defined on $\mathbb{R}^N$ and satisfies the estimate:

$$\|\mathbf{n}_j\|_{L_\infty(\mathbb{R}^N)} \leq C, \quad \|\nabla \mathbf{n}_j\|_{L_\infty(\mathbb{R}^N)} \leq C_A, \quad \|\nabla^2 \mathbf{n}_j\|_{L_\infty(\mathbb{R}^N)} \leq C_{M_2}.$$

Let $\nabla_{\Gamma_j} = (\partial_1, \ldots, \partial_{N-1})$ with $\partial_j = \partial/\partial x_j$ for $y = \Phi_j(x', 0) \in \Gamma_j$ and let Δ_{Γ_j} be the Laplace–Beltrami operator on Γ_j, which have the form:

$$\Delta_{\Gamma_j} f = \Delta' f + \mathcal{D}_{\Gamma_j} f \qquad \text{on } \Phi_j^{-1}(\Gamma_j),$$

where $\Delta' f = \sum_{j=1}^{N-1} \partial_j^2 f$ and $\mathcal{D}_{\Gamma_j} f = \sum_{k,\ell=1}^{N-1} a_{k\ell}^j \partial_k \partial_\ell f + \sum_{k=1}^{N-1} a_k^j \partial_k f$, and $a_{k\ell}^j$ and a_k^j satisfy the following estimates:

$$\|a_{k\ell}^j\|_{L_\infty(\mathbb{R}^N)} \leq C M_1, \quad \|(\partial_1 a_{k\ell}^j, \ldots, \partial_{N-1} a_{k\ell}^j, a_k^j)\|_{L_\infty(\mathbb{R}^N)} \leq C_A,$$

$$\|(\partial_1 a_{k\ell}^j, \ldots, \partial_{N-1} a_{k\ell}^j, a_k^j)\|_{H_\infty^1(\mathbb{R}^N)} \leq C_{M_2}.$$

Notice that $\mathbf{n}_j = \mathbf{n}$ and $\Delta_{\Gamma_j} = \Delta_\Gamma$ on $\Gamma_j \cap B_j^1 = \Gamma \cap B_j^1$. We know the existence of $K_{0j}(\mathbf{u}_j^0) \in \hat{H}_q^1(\mathbb{R}^N)$ possessing the estimate:

$$\|\nabla K_{0j}(\mathbf{u}_j^0)\|_{L_q(\mathbb{R}^N)} \le C\|\nabla \mathbf{u}_j^0\|_{H_q^1(\mathbb{R}^N)}. \tag{3.343}$$

Let ρ be a function in $C_0^\infty(B_{r_0})$ such that $\int_{\mathbb{R}^N} \rho \, dx = 1$. Below, this ρ is fixed. Since $K_{0j}(\mathbf{u}_j^0) + c$ also satisfy the variational Eq. (3.341) for any constant c, adjusting constants we may assume that

$$\int_{B_j^0} K_{0j}(\mathbf{u}_j^0)\rho(x - x_j^0) \, dx = 0. \tag{3.344}$$

Moreover, choosing $M_1 \in (0, 1)$ suitably small, we have the unique existence of solutions $K_{1j}(\mathbf{u}_j^1, h_j) \in H_q^1(\Omega_j) + \hat{H}_{q,0}^1(\Omega_j)$ of Eq. (3.342) possessing the estimates:

$$\|\nabla K_{1j}(\mathbf{u}_j^1, h_j)\|_{L_q(\Omega_j)} \le C(\|\nabla \mathbf{u}_j^1\|_{H_q^1(\Omega_j)} + \|h_j\|_{W_q^{3-1/q}(\Gamma_j)}). \tag{3.345}$$

Let $Y_q(\Omega_j)$ and $\mathcal{Y}_q(\Omega_j)$ be the spaces defined in (3.193) replacing Ω by Ω_j. By Theorem 3.5.1 and Theorem 3.5.18, there exist constants $M_1 \in (0, 1)$ and $\lambda_0 \ge 1$, which are independent of $j \in \mathbb{N}$, and operator families

$$\mathcal{S}_{0j}(\lambda) \in \text{Hol}\,(\Sigma_{\epsilon,\lambda_0}, \mathcal{L}(L_q(\mathbb{R}^N)^N, H_q^2(\mathbb{R}^N)^N)),$$

$$\mathcal{S}_{1j}(\lambda) \in \text{Hol}\,(\Lambda_{\kappa,\lambda_0\gamma_\kappa}, \mathcal{L}(\mathcal{Y}_q(\Omega_j), H_q^2(\Omega_j)^N)),$$

$$\mathcal{H}_j(\lambda) \in \text{Hol}\,(\Lambda_{\kappa,\lambda_0\gamma_\kappa}, \mathcal{L}(\mathcal{Y}_q(\Omega_j), H_q^3(\Omega_j)))$$

such that for each $j \in \mathbb{N}$, Eq. (3.339) admits a unique solution $\mathbf{u}_j^0 = \mathcal{S}_{0j}(\lambda)\tilde{\zeta}_j^0 \mathbf{f}$ and Eq. (3.340) admits unique solutions $\mathbf{u}_j^1 = \mathcal{S}_{1j}(\lambda)\tilde{\zeta}_j^1 \mathbf{F}_\lambda(\mathbf{f}, d, \mathbf{h})$ and $h_j = \mathcal{H}_j(\lambda)\tilde{\zeta}_j^1 \mathbf{F}_\lambda(\mathbf{f}, d, \mathbf{h})$, where $\mathbf{F}_\lambda(\mathbf{f}, d, \mathbf{h}) = (\mathbf{f}, d, \lambda^{1/2}\mathbf{h}, \mathbf{h})$, and $\tilde{\zeta}_j^1 \mathbf{F}_\lambda(\mathbf{f}, d, \mathbf{h}) = (\tilde{\zeta}_j^1\mathbf{f}, \tilde{\zeta}_j^1 d, \lambda^{1/2}\tilde{\zeta}_j^1\mathbf{h}, \tilde{\zeta}_j^1\mathbf{h})$. Moreover, there exists a number $r_b > 0$ independent of M_1, M_2, and $j \in \mathbb{N}$ such that

$$\mathcal{R}_{\mathcal{L}(L_q(\mathbb{R}^N)^N, H_q^{2-k}(\mathbb{R}^N)^N)}(\{(\tau\partial_\tau)^\ell(\lambda^{k/2}\mathcal{S}_{0j}(\lambda)) \mid \lambda \in \Sigma_{\epsilon,\lambda_0}\}) \le r_b,$$

$$\mathcal{R}_{\mathcal{L}(\mathcal{Y}_q(\Omega_j), H_q^{2-k}(\Omega_j)^N)}(\{(\tau\partial_\tau)^\ell(\lambda^{k/2}\mathcal{S}_{1j}(\lambda)) \mid \lambda \in \Lambda_{\epsilon,\lambda_0\gamma_\kappa}\}) \le r_b, \tag{3.346}$$

$$\mathcal{R}_{\mathcal{L}(\mathcal{Y}_q(\Omega_j), H_q^{3-n}(\Omega_j))}(\{(\tau\partial_\tau)^\ell(\lambda^n\mathcal{H}_j(\lambda)) \mid \lambda \in \Lambda_{\kappa,\lambda_0\gamma_\kappa}\}) \le r_b,$$

for $\ell = 0, 1$, $j \in \mathbb{N}$, $k = 0, 1, 2$, and $n = 0, 1$. Notice that $\lambda_0\gamma_\kappa \ge \lambda_0$.

By (3.346), we have

$$|\lambda| \|\mathbf{u}_j^0\|_{L_q(\mathbb{R}^N)} + |\lambda|^{1/2} \|\mathbf{u}_j^0\|_{H_q^1(\mathbb{R}^N)} + \|\mathbf{u}_j^0\|_{H_q^2(\mathbb{R}^N)} \le r_b \|\zeta_j^0 \mathbf{f}\|_{L_q(\mathbb{R}^N)},$$

$$|\lambda| \|\mathbf{u}_j^1\|_{L_q(\Omega_j)} + |\lambda|^{1/2} \|\mathbf{u}_j^1\|_{H_q^1(\Omega_j)} + \|\mathbf{u}_j^1\|_{H_q^2(\Omega_j)}$$

$$+ |\lambda| \|h_j\|_{H_q^2(\Omega_j)} + \|h_j\|_{H_q^3(\Omega_j)}$$

$$\le r_b(\|\tilde{\zeta}_j^1 \mathbf{f}\|_{L_q(\Omega_j)} + \|\tilde{\zeta}_j^1 d\|_{W_q^{2-1/q}(\Gamma_j)} + |\lambda|^{1/2} \|\mathbf{h}\|_{L_q(\Omega_j)} + \|\mathbf{h}\|_{H_q^1(\Omega_j)}) \tag{3.347}$$

for $\lambda \in \Sigma_{\kappa, \lambda_0 \gamma_\kappa}$. Let

$$\mathbf{u} = \sum_{i=0}^{1} \sum_{j=1}^{\infty} \zeta_j^i \mathbf{u}_j^i, \quad h = \sum_{j=1}^{\infty} \zeta_j^1 h_j. \tag{3.348}$$

Then, by (3.339), (3.340), (3.347), Proposition 3.5.21, and Proposition 3.5.23, we have $\mathbf{u} \in H_q^2(\Omega)^N$, $h \in H_q^3(\Omega)$, and

$$|\lambda| \|\mathbf{u}\|_{L_q(\Omega)} + |\lambda|^{1/2} \|\mathbf{u}\|_{H_q^1(\Omega)} + \|\mathbf{u}\|_{H_q^2(\Omega)} + |\lambda| \|h\|_{H_q^2(\Omega)} + \|h\|_{H_q^3(\Omega)}$$

$$\le C_q r_b(\|\mathbf{f}\|_{L_q(\Omega)} + \|d\|_{W_q^{2-1/q}(\Gamma)} + |\lambda|^{1/2} \|\mathbf{h}\|_{L_q(\Omega)} + \|\mathbf{h}\|_{H_q^1(\Omega)})$$

for $\lambda \in \Lambda_{\kappa, \lambda_0 \gamma_\kappa}$.

Moreover, we have

$$\begin{cases} \lambda \mathbf{u} - \operatorname{Div}(\mu \mathbf{D}(\mathbf{u}) - K(\mathbf{u}, h)\mathbf{I}) = \mathbf{f} - V^1(\lambda)(\mathbf{f}, d, \mathbf{h}) & \text{in } \Omega, \\ \lambda h + A_\kappa \cdot \nabla_\Gamma' h - \mathbf{u} \cdot \mathbf{n} = d - V_\kappa^2(\lambda)(\mathbf{f}, d, \mathbf{h}) & \text{on } \Gamma, \\ (\mu \mathbf{D}(\mathbf{u}) - K(\mathbf{u}, h)\mathbf{I} - (\sigma \Delta_\Gamma h)\mathbf{I})\mathbf{n} = \mathbf{h} - V^3(\lambda)(\mathbf{f}, d, \mathbf{h}) & \text{on } \Gamma, \end{cases} \tag{3.349}$$

where we have set

$$V^1(\lambda)(\mathbf{f}, d, \mathbf{h}) = V_1^1(\lambda)(\mathbf{f}, d, \mathbf{h}) + V_2^1(\lambda)(\mathbf{f}, d, \mathbf{h}),$$

$$V_1^1(\lambda)(\mathbf{f}, d, \mathbf{h}) = \sum_{i=0}^{1} \sum_{j=1}^{\infty} [\operatorname{Div}(\mu(\mathbf{D}(\zeta_j^i \mathbf{u}_j^i) - \zeta_j^i \mathbf{D}(\mathbf{u}_j^i)))$$

$$+ \operatorname{Div}(\zeta_j^i \mu_j^i \mathbf{D}(\mathbf{u}_j^i)) - \zeta_j^i \operatorname{Div}(\mu_j^i \mathbf{D}(\mathbf{u}_j^i))],$$

$$V_2^1(\lambda)(\mathbf{f}, d, \mathbf{h}) = \nabla K(\mathbf{u}, h) - \sum_{j=1}^{\infty} \zeta_j^0 \nabla K_{0j}(\mathbf{u}_j^0) - \sum_{j=1}^{\infty} \zeta_j^1 \nabla K_{1j}(\mathbf{u}_j^1, h_j),$$

$$V_\kappa^2(\lambda)(\mathbf{f}, d, \mathbf{h}) = \sum_{j=1}^\infty A_\kappa(x) \cdot ((\nabla_\Gamma' \zeta_j^1) h_j),$$

$$V^3(\lambda)(\mathbf{f}, d, \mathbf{h}) = V_1^3(\lambda)(\mathbf{f}, d, \mathbf{h}) - V_2^3(\lambda)(\mathbf{f}, d, \mathbf{h}) - V_3^3(\lambda)(\mathbf{f}, d, \mathbf{h}),$$

$$V_1^3(\lambda)(\mathbf{f}, d, \mathbf{h}) = \sum_{j=1}^\infty \mu(\mathbf{D}(\zeta_j^1 \mathbf{u}_j^1) - \zeta_j^1 \mathbf{D}(\mathbf{u}_j^1))\mathbf{n},$$

$$V_2^3(\lambda)(\mathbf{f}, d, \mathbf{h}) = \{\sum_{j=1}^\infty \zeta_j^1 K_{1j}(\mathbf{u}_j^1, h_j) - K(\mathbf{u}, h)\}\mathbf{n},$$

$$V_3^3(\lambda)(\mathbf{f}, d, \mathbf{h}) = \sum_{j=1}^\infty \sigma(\Delta_{\Gamma_j}(\zeta_j^1 h_j^1) - \zeta_j^1 \Delta_{\Gamma_j} h_j^1).$$

For $F = (F_1, F_2, F_3, F_4) \in \mathcal{Y}_q(\Omega)$, we define operators $\mathcal{A}_p(\lambda)$ and $\mathcal{B}_p(\lambda)$ acting on F by

$$\mathcal{A}_p(\lambda)F = \sum_{j=1}^\infty \zeta_j^0 \mathcal{S}_j^0(\lambda) \tilde{\zeta}_j^0 F_1 + \sum_{j=1}^\infty \zeta_j^1 \mathcal{S}_{1j}(\lambda) \tilde{\zeta}_j^1 F,$$

$$\mathcal{B}_p(\lambda)F = \sum_{j=1}^\infty \zeta_j^1 \mathcal{H}_j(\lambda) \tilde{\zeta}_j^1 F. \tag{3.350}$$

Then, by Proposition 3.5.21 and (3.346), we have $\mathbf{u} = \mathcal{A}_p(\lambda)\mathbf{F}_\lambda(\mathbf{f}, d, \mathbf{h})$, $h = \mathcal{H}_p(\lambda)\mathbf{F}_\lambda(\mathbf{f}, d, \mathbf{h})$, and

$$\mathcal{A}_p(\lambda) \in \mathrm{Hol}\,(\Lambda_{\kappa,\lambda_1}, \mathcal{L}(\mathcal{Y}_q(\Omega), H_q^2(\Omega)^N)),$$

$$\mathcal{B}_p(\lambda) \in \mathrm{Hol}\,(\Lambda_{\kappa,\lambda_1}, \mathcal{L}(\mathcal{Y}_q(\Omega), H_q^3(\Omega))),$$

$$\mathcal{R}_{\mathcal{L}(\mathcal{Y}_q(\Omega), H_q^{2-j}(\Omega)^N)}(\{(\tau \partial_\tau)^\ell (\lambda^{j/2} \mathcal{A}_p(\lambda)) \mid \lambda \in \Lambda_{\kappa,\lambda_1}\})$$

$$\leq (C + C_{M_2} \lambda_1^{-1/2}) r_b, \tag{3.351}$$

$$\mathcal{R}_{\mathcal{L}(\mathcal{Y}_q(\Omega), H_q^{3-k}(\Omega))}(\{(\tau \partial_\tau)^\ell (\lambda^k \mathcal{B}_p(\lambda)) \mid \lambda \in \Lambda_{\kappa,\lambda_1}\})$$

$$\leq (C + C_{M_2} \lambda_1^{-1}) r_b$$

for $\ell = 0, 1$, $j = 0, 1, 2$, and $k = 0, 1$ for any $\lambda_1 \geq \lambda_0 \gamma_\kappa$.

3.5.7   Estimates of the Remainder Terms

For $F = (F_1, F_2, F_3, F_4) \in \mathcal{Y}_q(\Omega)$, let

$$\mathcal{V}^1(\lambda)F = \mathcal{V}_1^1(\lambda)F + \mathcal{V}_2^1(\lambda)F,$$

$$\mathcal{V}_1^1(\lambda)F = \sum_{j=1}^{\infty}[\mathrm{Div}\,(\mu(\mathbf{D}(\zeta_j^0 \mathcal{S}_{0j}(\lambda)\tilde{\zeta}_j^0 F_1) - \zeta_j^0 \mathbf{D}(\mathcal{S}_{0j}(\lambda)\tilde{\zeta}_j^0 F_1)))$$

$$+ \mathrm{Div}\,(\zeta_j^0 \mu_j^0 \mathbf{D}(\mathcal{S}_{0j}(\lambda)\tilde{\zeta}_j^0 F_1)) - \zeta_j^0 \mathrm{Div}\,(\mu_j^0 \mathbf{D}(\mathcal{S}_{0j}(\lambda)\tilde{\zeta}_j^0 F_1))]$$

$$+ \sum_{j=1}^{\infty}[\mathrm{Div}\,(\mu(\mathbf{D}(\zeta_j^1 \mathcal{S}_{1j}(\lambda)\tilde{\zeta}_j^1 F) - \zeta_j^1 \mathbf{D}(\mathcal{S}_{1j}(\lambda)\tilde{\zeta}_j^1 F)))$$

$$+ \mathrm{Div}\,(\zeta_j^1 \mu_j^1 \mathbf{D}(\mathcal{S}_{1j}(\lambda)\tilde{\zeta}_j^1 F)) - \zeta_j^1 \mathrm{Div}\,(\mu_j^1 \mathbf{D}(\mathcal{S}_{1j}(\lambda)\tilde{\zeta}_j^1 F))],$$

$$\mathcal{V}_2^1(\lambda)F = \nabla K(\mathcal{A}_p(\lambda)F, \mathcal{B}_p(\lambda)F) - \sum_{j=1}^{\infty} \zeta_j^0 \nabla K_{0j}(\mathcal{S}_{0j}(\lambda)\tilde{\zeta}_j^0 F_1)$$

$$- \sum_{j=1}^{\infty} \zeta_j^1 \nabla K_{1j}(\mathcal{S}_{1j}(\lambda)\tilde{\zeta}_j^1 F, \mathcal{H}_j(\lambda)\tilde{\zeta}_j^1 F),$$

$$\mathcal{V}_\kappa^2(\lambda)F = \sum_{j=1}^{\infty} A_\kappa(x) \cdot ((\nabla_\Gamma' \zeta_j^1)\mathcal{H}_j(\lambda)\tilde{\zeta}_j^1 F),$$

$$\mathcal{V}^3(\lambda)F = \mathcal{V}_1^3(\lambda)F + \mathcal{V}_2^3(\lambda)F + \mathcal{V}_3^3(\lambda)F,$$

$$\mathcal{V}_1^3(\lambda)F = \sum_{j=1}^{\infty} \mu(\mathbf{D}(\zeta_j^1 \mathcal{S}_{1j}(\lambda)\tilde{\zeta}_j^1 F) - \zeta_j^1 \mathbf{D}(\zeta_j^1 \mathcal{S}_{1j}(\lambda)\tilde{\zeta}_j^1 F))\mathbf{n}$$

$$\mathcal{V}_2^3(\lambda)F = \{\sum_{j=1}^{\infty} \zeta_j^1 K_{1j}(\mathcal{S}_{1j}(\lambda)\tilde{\zeta}_j^1 F, \mathcal{H}_j(\lambda)\tilde{\zeta}_j^1 F) - K(\mathcal{A}_p(\lambda)F, \mathcal{B}_p(\lambda)F)\}\mathbf{n},$$

$$\mathcal{V}_3^3(\lambda)F = \sum_{j=1}^{\infty} \sigma(\Delta_{\Gamma_j}(\zeta_j^1 \mathcal{H}_j(\lambda)\tilde{\zeta}_j^1 F) - \zeta_j^1 \Delta_{\Gamma_j}(\mathcal{H}_j(\lambda)\tilde{\zeta}_j^1 F)).$$

Notice that $\mathcal{V}_\kappa^2(\lambda)F = 0$ for $\kappa = 0$.
    Let

$$V(\lambda)(\mathbf{f}, d, \mathbf{h}) = (V^1(\lambda)(\mathbf{f}, d, \mathbf{h}), V^2(\lambda)(\mathbf{f}, d, \mathbf{h}), V^3(\lambda)(\mathbf{f}, d, \mathbf{h})),$$

$$\mathcal{V}(\lambda)F = (\mathcal{V}^1(\lambda)F, \mathcal{V}_\kappa^2(\lambda)F, \mathcal{V}^3(\lambda)F).$$

Since $\mathbf{u}_j^0 = \mathcal{S}_{0j}(\lambda)\tilde{\xi}_j^0\mathbf{f}$, $\mathbf{u}_j^1 = \mathcal{S}_{1j}(\lambda)\tilde{\xi}_j^1\mathbf{F}_\lambda(\mathbf{f}, d, \mathbf{h})$, and $h_j = \mathcal{H}_j(\lambda)\zeta_j^1\mathbf{F}_{\tilde{\lambda}}(\mathbf{f}, d, \mathbf{h})$, we have

$$V(\lambda)(\mathbf{f}, d, \mathbf{h}) = \mathcal{V}(\lambda)\mathbf{F}_\lambda(\mathbf{f}, d, \mathbf{h}). \tag{3.352}$$

In what follows, we shall prove that

$$\mathcal{R}_{\mathcal{L}(\mathcal{Y}_q(\Omega))}(\{(\tau\partial_\tau)^\ell(\mathbf{F}_\lambda\mathcal{V}(\lambda)) \mid \lambda \in \Lambda_{\kappa,\tilde{\lambda}_0}\})$$
$$\leq C_q r_b(\epsilon + C_{M_2,\epsilon}(\tilde{\lambda}_0^{-1}\gamma_\kappa + \tilde{\lambda}_0^{-1/2})) \tag{3.353}$$

for $\ell = 0, 1$ and $\tilde{\lambda}_0 \geq \lambda_0\gamma_\kappa$, where γ_κ is the number given in Theorem 3.4.8.

To prove (3.353), we use Propositions 3.2.2, 3.5.4, and 3.5.20–3.5.23, (3.301), (3.302), (3.336), (3.337), (3.28) and (3.346). In the following, $\tilde{\lambda}_0$ is any number such that $\tilde{\lambda}_0 \geq \lambda_0\gamma_\kappa$. We start with the following estimate of $\mathcal{V}_1^1(\lambda)$:

$$\mathcal{R}_{\mathcal{L}(\mathcal{Y}_q(\Omega), L_q(\Omega)^N)}(\{(\tau\partial_\tau)^\ell\mathcal{V}_1^1(\lambda) \mid \lambda \in \Lambda_{\kappa,\tilde{\lambda}_0}\}) \leq C_{M_2}r_b\tilde{\lambda}_0^{-1/2} \tag{3.354}$$

for $\ell = 0, 1$. In fact, since $D_{\ell,m}(\zeta_j^i\mathbf{u}) - \zeta_j^i D_{\ell m}(\mathbf{u}) = (\partial_\ell\zeta_j^i)u_m + (\partial_m\zeta_j^i)u_\ell$, and $\operatorname{div}(\zeta_j^i\mathbf{u}) - \zeta_j^i\operatorname{div}\mathbf{u} = \sum_{k=1}^N(\partial_k\zeta_j^i)u_k$, for any $n \in \mathbb{N}$, $\{\lambda_\ell\}_{\ell=1}^n \subset (\Lambda_{\kappa,\tilde{\lambda}_0})^n$, and $\{F_\ell = (F_{1\ell}, F_{2\ell}, F_{3\ell}, F_{4\ell})\}_{\ell=1}^n \subset \mathcal{Y}_q(\Omega)^n$, we have

$$\int_0^1 \left\| \sum_{\ell=1}^n r_\ell(u)\mathcal{V}_1^1(\lambda_\ell)F_\ell \right\|_{L_q(\Omega)}^q du$$

$$\leq C_q^q M_2^q \sum_{j=1}^\infty \left\{ \int_0^1 \left\| \sum_{\ell=1}^n r_\ell(u)\mathcal{S}_{0j}(\lambda_\ell)\tilde{\xi}_j^0 F_{1\ell} \right\|_{H_q^1(\mathbb{R}^N)}^q du \right.$$

$$\left. + \int_0^1 \left\| \sum_{\ell=1}^n r_\ell(u)\mathcal{S}_{1j}(\lambda_\ell)\tilde{\xi}_j^1 F_\ell \right\|_{H_q^1(\Omega_j)}^q du \right\}$$

$$\leq C_q^q M_2^q \tilde{\lambda}_0^{-q/2} \sum_{j=1}^\infty \left\{ \int_0^1 \left\| \sum_{\ell=1}^n r_\ell(u)\lambda_\ell^{1/2}\mathcal{S}_{0j}(\lambda_\ell)\tilde{\xi}_j^0 F_{1\ell} \right\|_{H_q^1(\mathbb{R}^N)}^q du \right.$$

$$\left. + \int_0^1 \left\| \sum_{\ell=1}^n r_\ell(u)\lambda_\ell^{1/2}\mathcal{S}_{1j}(\lambda_\ell)\tilde{\xi}_j^1 F_\ell \right\|_{H_q^1(\Omega_j)}^q du \right\}$$

$$\leq C_q^q M_2^q \tilde{\lambda}_0^{-q/2} r_b^q \sum_{j=1}^\infty \left\{ \int_0^1 \left\| \sum_{\ell=1}^n r_\ell(u)\tilde{\xi}_j^0 F_{1\ell} \right\|_{L_q(\mathbb{R}^N)}^q du \right.$$

$$\left. + \int_0^1 \left\| \sum_{\ell=1}^n r_\ell(u)\tilde{\xi}_j^1 F_\ell \right\|_{\mathcal{Y}_q(\Omega_j)}^q du \right\}$$

$$\leq C_q^{2q} M_2^q \tilde{\lambda}_0^{-q/2} r_b^q \int_0^1 \left\| \sum_{\ell=1}^n r_\ell(u)F_\ell \right\|_{L_q(\Omega)}^q du.$$

This shows that

$$\mathcal{R}_{\mathcal{L}(\mathcal{Y}_q(\Omega), L_q(\Omega)^N)}(\{\mathcal{V}_1^1(\lambda) \mid \lambda \in \Lambda_{\kappa, \tilde{\lambda}_0}\}) \leq C_{M_2} r b \tilde{\lambda}_0^{-1/2}.$$

Analogously, we can show that

$$\mathcal{R}_{\mathcal{L}(\mathcal{Y}_q(\Omega), L_q(\Omega)^N)}(\{\tau \partial_\tau \mathcal{V}_1^1(\lambda) \mid \lambda \in \Lambda_{\kappa, \tilde{\lambda}_0}\}) \leq C_{M_2} r b \tilde{\lambda}_0^{-1/2},$$

and therefore we have (3.354).

For $r \in (N, \infty)$ and $q \in (1, r]$, by the extension of functions defined on Γ_j to Ω_j and Sobolev's imbedding theorem, we have

$$\|ab\|_{W_q^{2-1/q}(\Gamma_j)} \leq C_{q,r,K} \|a\|_{H_q^2(\Omega_j)} \|b\|_{W_q^{2-1/q}(\Gamma_j)}$$

for any $a \in H_r^2(\Omega)$ and $b \in W_q^{2-1/q}(\Gamma_j)$. Applying this inequality, we have

$$\|A_\kappa \cdot ((\nabla_\Gamma' \zeta_j^1) \mathcal{H}_j(\lambda) \tilde{\zeta}_j^1 F)\|_{W_q^{2-1/q}(\Gamma_j)} \leq C_{q,r} M_2 m_3 \kappa^{-b} \|\mathcal{H}_j(\lambda) \tilde{\zeta}_j^1 F\|_{H_q^2(\Omega_j)}$$

for $\kappa \in (0, 1)$. Thus, employing the same argument as in the proof of (3.354), we have

$$\mathcal{R}_{\mathcal{L}(\mathcal{Y}_q(\Omega), W_q^{2-1/q}(\Gamma))}(\{(\tau \partial_\tau)^\ell \mathcal{V}_\kappa^2(\lambda) \mid \lambda \in \Lambda_{\kappa, \tilde{\lambda}_0}\}) \leq C_{M_2} r b \tilde{\lambda}_0^{-1} \kappa^{-b}$$

for $\ell = 0, 1$ and $\kappa \in (0, 1)$.

Employing the same argument as in the proof of (3.354), we also have

$$\mathcal{R}_{\mathcal{L}(\mathcal{Y}_q(\Omega), L_q(\Gamma)^N)}(\{(\tau \partial_\tau)^\ell (\lambda^{1/2} \mathcal{V}_m^3(\lambda)) \mid \lambda \in \Lambda_{\kappa, \tilde{\lambda}_0}\}) \leq C_{M_2} r b \tilde{\lambda}_0^{-1/2},$$

$$\mathcal{R}_{\mathcal{L}(\mathcal{Y}_q(\Omega), H_q^1(\Gamma)^N)}(\{(\tau \partial_\tau)^\ell \mathcal{V}_m^3(\lambda) \mid \lambda \in \Lambda_{\kappa, \tilde{\lambda}_0}\}) \leq C_{M_2} r b \tilde{\lambda}_0^{-1/2},$$

for $\ell = 0, 1$, $m = 1$ and 3. Noting that $\mu \zeta_j^1 = \mu_j^1 \zeta_j^1$ and $\sigma \zeta_j^1 = \sigma_j^1 \zeta_j^1$, we have

$$\sum_{j=1}^{\infty} \zeta_j^1 K_{1j}(\mathcal{S}_{1j}(\lambda) \tilde{\zeta}_j^1 F, \mathcal{H}_j(\lambda) \tilde{\zeta}_j^1 F) - K(\mathcal{A}_p(\lambda) F, \mathcal{B}_p(\lambda) F)$$

$$= \sum_{j=1}^{\infty} \mu < \zeta_j^1 \mathbf{D}(\mathcal{S}_{1j}(\lambda) \tilde{\zeta}_j^1 F) - \mathbf{D}(\zeta_j^1 \mathcal{S}_{1j}(\lambda) \tilde{\zeta}_j^1 F), \mathbf{n} >$$

$$- \sum_{j=1}^{\infty} \{\zeta_j^1 \operatorname{div} \mathcal{S}_{1j}(\lambda) \tilde{\zeta}_j^1 F - \operatorname{div}(\zeta_j^1 \mathcal{S}_{1j}(\lambda) \tilde{\zeta}_j^1 F)\}$$

$$- \sum_{j=1}^{\infty} \sigma \{\zeta_j^1 \Delta_{\Gamma_j}(\mathcal{H}_j(\lambda) \tilde{\zeta}_j^1 F) - \Delta_{\Gamma_j}(\zeta_j^1 \mathcal{H}_j(\lambda) \tilde{\zeta}_j^1 F)\}$$

(3.355)

on Γ, where we have used $\Delta_\Gamma = \Delta_{\Gamma_j}$ and $\mathbf{n} = \mathbf{n}_j$ on $\Gamma_j \cap B_j^1$. Employing the same argument as in the proof of (3.354), we have

$$\mathcal{R}_{\mathcal{L}(\mathcal{Y}_q(\Omega), L_q(\Gamma)^N)}(\{(\tau\partial_\tau)^\ell(\lambda^{1/2}\mathcal{V}_2^3(\lambda)) \mid \lambda \in \Lambda_{\kappa,\tilde{\lambda}_0}\}) \le C_{M_2} r_b \tilde{\lambda}_0^{-1/2};$$

$$\mathcal{R}_{\mathcal{L}(\mathcal{Y}_q(\Omega), H_q^1(\Gamma)^N)}(\{(\tau\partial_\tau)^\ell \mathcal{V}_2^3(\lambda) \mid \lambda \in \Lambda_{\kappa,\tilde{\lambda}_0}\}) \le C_{M_2} r_b \tilde{\lambda}_0^{-1/2}$$

for $\ell = 0, 1$.

The final task is to prove that

$$\mathcal{R}_{\mathcal{L}(\mathcal{Y}_q(\Omega), L_q(\Gamma)^N)}(\{(\tau\partial_\tau)^\ell \mathcal{V}_2^1(\lambda) \mid \lambda \in \Lambda_{\kappa,\tilde{\lambda}_0}\}) \le C_{q,r}(\epsilon + C_{M_2,\epsilon}\tilde{\lambda}_0^{-1/2})r_b \tag{3.356}$$

for $\ell = 0, 1$. For this purpose, we use Lemma 3.2.13 and the following lemma.

Lemma 3.5.24 *Let $1 < q < \infty$. For $\mathbf{u} \in H_q^2(\mathbb{R}^N)$, let $K_{0j}(\mathbf{u})$ be a unique solution of the weak Laplace equation (3.341) satisfying (3.344). Then, we have*

$$\|K_{0j}(\mathbf{u})\|_{L_q(B_j^0)} \le C\|\nabla\mathbf{u}\|_{L_q(\mathbb{R}^N)}. \tag{3.357}$$

Proof Let ρ be the same function in (3.344). Let ψ be any function in $C_0^\infty(B_j^0)$ and we set

$$\tilde{\psi}(x) = \psi(x) - \rho(x - x_j^0)\int_{\mathbb{R}^N} \psi(y)\,dy.$$

Then,

$$\tilde{\psi} \in C_0^\infty(B_j^0), \quad \int_{\mathbb{R}^N} \tilde{\psi}\,dx = 0, \quad \|\tilde{\psi}\|_{L_{q'}(B_j^0)} \le C_{q'}\|\psi\|_{q'(B_j^0)}. \tag{3.358}$$

Moreover,

$$\tilde{\psi} \in \hat{H}_q^1(\mathbb{R}^N)^* = \hat{H}_{q'}^{-1}(\mathbb{R}^N), \quad \|\tilde{\psi}\|_{\hat{H}_{q'}^{-1}(\mathbb{R}^N)} \le C_{q'}\|\psi\|_{L_{q'}(\mathbb{R}^N)}. \tag{3.359}$$

In fact, by Lemma 3.2.13, for any $\varphi \in \hat{H}_{q'}^1(\mathbb{R}^N)$, there exists a constant e_j for which

$$\|\varphi - e_j\|_{L_q(B_j^0)} \le c_q\|\nabla\varphi\|_{L_q(B_j^0)}.$$

Thus, by (3.358), we have

$$|(\tilde{\psi}, \varphi)_{\mathbb{R}^N}| = |(\tilde{\psi}, \varphi - e_j)_{\mathbb{R}^N}| \le \|\tilde{\psi}\|_{L_{q'}(B_j^0)}\|\varphi - e_j\|_{L_q(B_j^0)}$$

$$\le C_q\|\tilde{\psi}\|_{q'(B_j^0)}\|\nabla\varphi\|_{L_q(B_j^0)},$$

which yields (3.359). Let Ψ be a function in $\hat{H}^1_{q'}(\mathbb{R}^N)$ such that $\nabla\Psi \in H^1_{q'}(\mathbb{R}^N)^N$,

$$(\nabla\Psi, \nabla\theta)_{\mathbb{R}^N} = (\tilde{\psi}, \theta)_{\mathbb{R}^N} \quad \text{for any } \theta \in \hat{H}^1_q(\mathbb{R}^N),$$

$$\|\nabla\Psi\|_{H^1_{q'}(\mathbb{R}^N)} \leq C(\|\tilde{\psi}\|_{L_{q'}(\mathbb{R}^N)} + \|\tilde{\psi}\|_{\hat{H}^{-1}_{q'}(\mathbb{R}^N)}). \tag{3.360}$$

By (3.358) and (3.359), we have

$$\|\nabla\Psi\|_{H^1_{q'}(\mathbb{R}^N)} \leq C_{q'}\|\psi\|_{L_{q'}(\mathbb{R}^N)}. \tag{3.361}$$

By (3.344), (3.360), and the divergence theorem of Gauß, we have

$$(K_{0j}(\mathbf{u}), \psi)_{\mathbb{R}^N} = (K_{0j}(\mathbf{u}), \tilde{\psi})_{\mathbb{R}^N} = (\nabla K_{0j}(\mathbf{u}), \nabla\Psi)_{\mathbb{R}^N}$$

$$= (\mathrm{Div}\,(\mu^0_j \mathbf{D}(\mathbf{u})) - \nabla\,\mathrm{div}\,\mathbf{u}, \nabla\Psi)_{\mathbb{R}^N}$$

$$= -(\mu^0_j \mathbf{D}(\mathbf{u}), \nabla^2\Psi)_{\mathbb{R}^N} + (\mathrm{div}\,\mathbf{u}, \Delta\Psi)_{\mathbb{R}^N},$$

and therefore by (3.361)

$$|(K_{0j}(\mathbf{u}), \psi)_{\mathbb{R}^N}| \leq C\|\nabla\mathbf{u}\|_{L_q(\mathbb{R}^N)}\|\psi\|_{L_{q'}(\mathbb{R}^N)},$$

which proves (3.357). This completes the proof of Lemma 3.5.24. □

Lemma 3.5.25 *Let* $1 < q < \infty$. *For* $\mathbf{u} \in H^2_q(\Omega_j)$ *and* $h \in H^3_q(\Omega_j)$, *let* $K_{1j}(\mathbf{u}, h) \in H^1_q(\Omega_q) + \hat{H}^1_{q,0}(\Omega_j)$ *be a unique solution of the weak Dirichlet problem* (3.342). *Then, we have*

$$\|K_{1j}(\mathbf{u}, h)\|_{L_q(\Omega_j \cap B^1_j)} \leq C(\|\nabla\mathbf{u}\|_{L_q(\Omega_j)} + \|h\|_{H^2_q(\Omega_j)}$$

$$+ \|\nabla^2\mathbf{u}\|^{1/q}_{L_q(\Omega_j)}\|\nabla\mathbf{u}\|^{1-1/q}_{L_q(\Omega_j)} + \|h\|^{1/q}_{H^3_q(\Omega_j)}\|h\|^{1-1/q}_{H^2(\Omega_j)}).$$

Here, the constant C *depends on* q *and* C_A.

Remark 3.5.26 By Young's inequality, we have

$$\|K_{1j}(\mathbf{u}, h)\|_{L_q(\Omega_j \cap B^1_j)} \leq \epsilon(\|\nabla^2\mathbf{u}\|_{L_q(\Omega_j)} + \|h\|_{H^3_q(\Omega_j)})$$

$$+ C_\epsilon(\|\nabla\mathbf{u}\|_{L_q(\Omega_j)} + \|h\|_{H^2_q(\Omega_j)}) \tag{3.362}$$

for any $\epsilon \in (0, 1)$ with some constant $C_{\epsilon,q}$ depending on ϵ and q.

Proof For a proof, see Lemma 3.4 in Shibata [42]. □

To prove (3.356), we divide $\mathcal{V}_2^1(\lambda)$ into two parts as $\mathcal{V}_2^1(\lambda) = \nabla\mathcal{V}_{21}^1(\lambda) + \mathcal{V}_{22}^1(\lambda)$, where

$$\mathcal{V}_{21}^1(\lambda)F = K(\mathcal{A}_p(\lambda)F, \mathcal{B}_p(\lambda)F) - \sum_{j=1}^{\infty} \zeta_j^0 K_{0j}(\mathcal{S}_{0j}(\lambda)\tilde{\zeta}_j^0 F_1)$$

$$- \sum_{j=1}^{\infty} \zeta_j^1 K_{1j}(\mathcal{S}_{1j}(\lambda)\tilde{\zeta}_j^0 F, \mathcal{H}_j(\lambda)\tilde{\zeta}_j^1 F),$$

$$\mathcal{V}_{22}^1(\lambda)F = \sum_{j=1}^{\infty} (\nabla\zeta_j^0) K_{0j}(\mathcal{S}_{0j}(\lambda)\tilde{\zeta}_j^0 F_1)$$

$$+ \sum_{j=1}^{\infty} (\nabla\zeta_j^1) K_{1j}(\mathcal{S}_{1j}(\lambda)\tilde{\zeta}_j^0 F, \mathcal{H}_j(\lambda)\tilde{\zeta}_j^1 F).$$

By (3.182), (3.341), and (3.342), for any $\varphi \in \hat{H}_{q'0}^1(\Omega)$ we have $(\nabla\mathcal{V}_{21}^1 F, \nabla\varphi)_\Omega = I - II$, where

$$I = (\text{Div}\,(\mu\mathbf{D}(\mathcal{A}_p(\lambda)F)) - \nabla\text{div}\,(\mathcal{A}_p(\lambda)F), \nabla\varphi)_\Omega,$$

$$II = \sum_{j=1}^{\infty} ((\nabla\zeta_j^0) K_{0j}(\mathcal{S}_{0j}(\lambda)\tilde{\zeta}_j^0 F_1), \nabla(\varphi - e_j))_\Omega$$

$$+ \sum_{j=1}^{\infty} ((\nabla\zeta_j^1) K_{1j}(\mathcal{S}_{1j}(\lambda)\tilde{\zeta}_j^1 F, \mathcal{H}_j(\lambda)\tilde{\zeta}_j^1 F), \nabla\varphi)_\Omega$$

$$+ \sum_{j=1}^{\infty} (\nabla K_{0j}(\mathcal{S}_{0j}(\lambda)\tilde{\zeta}_j^0 F_1), \nabla(\zeta_j^0(\varphi - e_j)))_\Omega$$

$$+ \sum_{j=1}^{\infty} (\nabla K_{1j}(\mathcal{S}_{1j}(\lambda)\tilde{\zeta}_j^1 F, \mathcal{H}_j(\lambda)\tilde{\zeta}_j^1 F), \nabla(\zeta_j^1\varphi))_\Omega$$

$$- \sum_{j=1}^{\infty} ((\nabla\zeta_j^0)\nabla K_{0j}(\mathcal{S}_{0j}(\lambda)\tilde{\zeta}_j^0 F_1), \varphi - e_j)_\Omega$$

$$- \sum_{j=1}^{\infty} ((\nabla\zeta_j^1)\nabla K_{1j}(\mathcal{S}_{1j}(\lambda)\tilde{\zeta}_j^1 F, \mathcal{H}_j(\lambda)\tilde{\zeta}_j^1 F), \varphi)_\Omega.$$

Here and in the following, $e_j = c_j^0(\varphi)$ are constants given in Lemma 3.2.13. By the definition (3.350), we have

$$I = \sum_{j=1}^{\infty}(\mathrm{Div}\,(\mu\mathbf{D}(\zeta_j^0\mathcal{S}_{0j}(\lambda)\tilde{\zeta}_j^0 F_1)) - \nabla\mathrm{div}\,(\zeta_j^0\mathcal{S}_{0j}(\lambda)\tilde{\zeta}_j^0 F_1), \nabla\varphi)_{\mathbb{R}^N}$$

$$+ \sum_{j=1}^{\infty}(\mathrm{Div}\,(\mu\mathbf{D}(\zeta_j^1\mathcal{S}_{1j}(\lambda)\tilde{\zeta}_j^1 F)) - \nabla\mathrm{div}\,(\zeta_j^1\mathcal{S}_{1j}(\lambda)\tilde{\zeta}_j^1 F), \nabla\varphi)_{\Omega}$$

$$= \sum_{j=1}^{\infty}(\zeta_j^0\mathrm{Div}\,(\mu_j^0\mathbf{D}(\mathcal{S}_{0j}(\lambda)\tilde{\zeta}_j^0 F_1)) - \zeta_j^0\nabla\mathrm{div}\,(\mathcal{S}_{0j}(\lambda)\tilde{\zeta}_j^0 F_1), \nabla\varphi)_{\mathbb{R}^N}$$

$$+ \sum_{j=1}^{\infty}(\zeta_j^1\mathrm{Div}\,(\mu_j^1\mathbf{D}(\mathcal{S}_{1j}(\lambda)\tilde{\zeta}_j^1 F)) - \zeta_j^1\nabla\mathrm{div}\,(\mathcal{S}_{1j}(\lambda)\tilde{\zeta}_j^1 F), \nabla\varphi)_{\Omega} + III,$$

where

$$III = \sum_{j=1}^{\infty}(\mathrm{Div}\,(\mu\mathbf{D}(\zeta_j^0\mathcal{S}_{0j}(\lambda)\tilde{\zeta}_j^0 F_1)) - \zeta_j^0\mathrm{Div}\,(\mu\mathbf{D}(\mathcal{S}_{0j}(\lambda)\tilde{\zeta}_j^0 F_1)), \nabla\varphi)_{\mathbb{R}^N}$$

$$- \sum_{j=1}^{\infty}(\nabla\mathrm{div}\,(\zeta_j^0\mathcal{S}_{0j}(\lambda)\tilde{\zeta}_j^0 F_1) - \zeta_j^0\nabla\mathrm{div}\,(\mathcal{S}_{0j}(\lambda)\tilde{\zeta}_j^0 F_1), \nabla\varphi)_{\Omega}$$

$$+ \sum_{j=1}^{\infty}(\mathrm{Div}\,(\mu\mathbf{D}(\zeta_j^1\mathcal{S}_{1j}(\lambda)\tilde{\zeta}_j^1 F)) - \zeta_j^1\mathrm{Div}\,(\mu\mathbf{D}(\mathcal{S}_{1j}(\lambda)\tilde{\zeta}_j^1 F)), \nabla\varphi)_{\mathbb{R}^N}$$

$$- \sum_{j=1}^{\infty}(\nabla\mathrm{div}\,(\zeta_j^1\mathcal{S}_{1j}(\lambda)\tilde{\zeta}_j^1 F) - \zeta_j^1\nabla\mathrm{div}\,(\mathcal{S}_{1j}(\lambda)\tilde{\zeta}_j^1 F), \nabla\varphi)_{\Omega}. \qquad (3.363)$$

Since $\zeta_j^0(\varphi - e_j) \in \hat{H}_{q',0}^1(\mathbb{R}^N)$, and $\zeta_j^1\varphi \in \hat{H}_{q',0}^1(\Omega_j)$, by (3.341) and (3.342), we have

$$II = \sum_{j=1}^{\infty}(\zeta_j^0\mathrm{Div}\,(\mu_j^0\mathbf{D}(\mathcal{S}_{0j}(\lambda)\tilde{\zeta}_j^0 F_1)) - \zeta_j^0\nabla\mathrm{div}\,(\mathcal{S}_{0j}(\lambda)\tilde{\zeta}_j^0 F_1), \nabla\varphi)_{\mathbb{R}^N}$$

$$+ \sum_{j=1}^{\infty}(\zeta_j^1\mathrm{Div}\,(\mu_j^1\mathbf{D}(\mathcal{S}_{1j}(\lambda)\tilde{\zeta}_j^1 F)) - \zeta_j^1\nabla\mathrm{div}\,(\mathcal{S}_{1j}(\lambda)\tilde{\zeta}_j^1 F), \nabla\varphi)_{\Omega} + IV,$$

where we have set

IV

$$= \sum_{j=1}^{\infty} \Big\{ 2(K_{0j}(\mathcal{S}_{0j}(\lambda)\tilde{\zeta}_j^0 F_1)(\nabla \zeta_j^0), \nabla \varphi)_{\Omega} + (K_{0j}(\mathcal{S}_{0j}(\lambda)\tilde{\zeta}_j^0 F_1)(\Delta \zeta_j^0), \varphi - e_j)_{\Omega}$$

$$- (\mu_j^0 \mathbf{D}(\mathcal{S}_{0j}(\lambda)\tilde{\zeta}_j^0 F_1) : (\nabla^2 \zeta_j^0), \varphi - e_j)_{\Omega} - (\mu_j^0 \mathbf{D}(\mathcal{S}_{0j}(\lambda)\tilde{\zeta}_j^0 F_1)(\nabla \zeta_j^0), \nabla \varphi)_{\Omega}$$

$$+ (\text{div}\,(\mathcal{S}_{0j}(\lambda)\tilde{\zeta}_j^0 F_1)(\Delta \zeta_j^0), \varphi - e_j)_{\Omega} + (\text{div}\,(\mathcal{S}_{0j}(\lambda)\tilde{\zeta}_j^0 F_1)(\nabla \zeta_j^0), \nabla \varphi)_{\Omega}$$

$$+ 2(K_{1j}(\mathcal{S}_{1j}(\lambda)\tilde{\zeta}_j^1 F, \mathcal{H}_j(\lambda)\tilde{\zeta}_j^1 F)(\nabla \zeta_j^1), \nabla \varphi)_{\Omega}$$

$$+ (K_{1j}(\mathcal{S}_{1j}(\lambda)\tilde{\zeta}_j^1 F, \mathcal{H}_j(\lambda)\tilde{\zeta}_j^1 F)(\Delta \zeta_j^1), \varphi)_{\Omega}$$

$$- (\mu_j^1 \mathbf{D}(\mathcal{S}_{1j}(\lambda)\tilde{\zeta}_j^1 F_1) : (\nabla^2 \zeta_j^1), \varphi)_{\Omega} - (\mu_j^1 \mathbf{D}(\mathcal{S}_{1j}(\lambda)\tilde{\zeta}_j^1 F_1)(\nabla \zeta_j^1), \nabla \varphi)_{\Omega}$$

$$+ (\text{div}\,(\mathcal{S}_{1j}(\lambda)\tilde{\zeta}_j^1 F_1)(\Delta \zeta_j^1), \varphi)_{\Omega} + (\text{div}\,(\mathcal{S}_{1j}(\lambda)\tilde{\zeta}_j^1 F_1)(\nabla \zeta_j^1), \nabla \varphi)_{\Omega} \Big\}.$$

Thus, we have

$$(\nabla \mathcal{V}_{21}^1(\lambda) F, \nabla \varphi)_{\Omega} = III + IV. \tag{3.364}$$

We let define operators $\mathcal{L}(\lambda)$ and $\mathcal{M}(\lambda)$ acting on $F \in \mathcal{Y}_q(\Omega)$ by the following formulas:

$$\mathcal{L}(\lambda) F = \sum_{j=1}^{\infty} (\text{Div}\,(\mu \mathbf{D}(\zeta_j^0 \mathcal{S}_{0j}(\lambda)\tilde{\zeta}_j^0 F_1)) - \zeta_j^0 \text{Div}\,(\mu \mathbf{D}(\mathcal{S}_{0j}(\lambda)\tilde{\zeta}_j^0 F_1))$$

$$- \sum_{j=1}^{\infty} (\nabla \text{div}\,(\zeta_j^0 \mathcal{S}_{0j}(\lambda)\tilde{\zeta}_j^0 F_1) - \zeta_j^0 \nabla \text{div}\,(\mathcal{S}_{0j}(\lambda)\tilde{\zeta}_j^0 F_1))$$

$$+ \sum_{j=1}^{\infty} (\text{Div}\,(\mu \mathbf{D}(\zeta_j^1 \mathcal{S}_{1j}(\lambda)\tilde{\zeta}_j^1 F)) - \zeta_j^1 \text{Div}\,(\mu \mathbf{D}(\mathcal{S}_{1j}(\lambda)\tilde{\zeta}_j^1 F))$$

$$- \sum_{j=1}^{\infty} (\nabla \text{div}\,(\zeta_j^1 \mathcal{S}_{1j}(\lambda)\tilde{\zeta}_j^1 F) - \zeta_j^1 \nabla \text{div}\,(\mathcal{S}_{1j}(\lambda)\tilde{\zeta}_j^1 F))$$

$$+ 2 \sum_{j=1}^{\infty} (\nabla \zeta_j^0) K_{0j}(\mathcal{S}_{0j}(\lambda)\tilde{\zeta}_j^0 F_1)$$

$$+ 2 \sum_{j=1}^{\infty} (\nabla \zeta_j^1) K_{1j}(\mathcal{S}_{1j}(\lambda)\tilde{\zeta}_j^1 F, \mathcal{H}_j(\lambda)\tilde{\zeta}_j^1 F)$$

$$- \sum_{j=1}^{\infty} \mu_j^0 \mathbf{D}(\mathcal{S}_{0j}(\lambda)\tilde{\zeta}_j^0 F_1)(\nabla \zeta_j^0) + \sum_{j=1}^{\infty} \operatorname{div}(\mathcal{S}_{0j}(\lambda)\tilde{\zeta}_j^0 F_1)(\nabla \zeta_j^0)$$

$$- \sum_{j=1}^{\infty} \mu_j^1 \mathbf{D}(\mathcal{S}_{1j}(\lambda)\tilde{\zeta}_j^1 F)(\nabla \zeta_j^1) + \sum_{j=1}^{\infty} \operatorname{div}(\mathcal{S}_{1j}(\lambda)\tilde{\zeta}_j^1 F_1)(\nabla \zeta_j^1);$$

$$<< \mathcal{M}(\lambda)F, \varphi >> = - \sum_{j=1}^{\infty} (\mu_j^0 \mathbf{D}(\mathcal{S}_{0j}(\lambda)\tilde{\zeta}_j^0 F_1) : (\nabla^2 \zeta_j^0), \varphi - e_j)_{\Omega}$$

$$+ \sum_{j=1}^{\infty} (\operatorname{div}(\mathcal{S}_{0j}(\lambda)\tilde{\zeta}_j^0 F_1)(\Delta \zeta_j^0), \varphi - e_j)_{\Omega}$$

$$+ \sum_{j=1}^{\infty} (K_{0j}(\mathcal{S}_{0j}(\lambda)\tilde{\zeta}_j^0 F_1)(\Delta \zeta_j^0), \varphi - e_j)_{\Omega}$$

$$- \sum_{j=1}^{\infty} (\mu_j^1 \mathbf{D}(\mathcal{S}_{1j}(\lambda)\tilde{\zeta}_j^1 F) : (\nabla^2 \zeta_j^1), \varphi)_{\Omega}$$

$$+ \sum_{j=1}^{\infty} (\operatorname{div}(\mathcal{S}_{1j}(\lambda)\tilde{\zeta}_j^1 F)(\Delta \zeta_j^1), \varphi)_{\Omega}$$

$$+ \sum_{j=1}^{\infty} (K_{1j}(\mathcal{S}_{1j}(\lambda)\tilde{\zeta}_j^1 F, \mathcal{H}_j(\lambda)\tilde{\zeta}_j^1 F)(\Delta \zeta_j^1), \varphi)_{\Omega}.$$

Here and in the following, $\hat{W}_{q,0}^{-1}(\Omega)$ denotes the dual space of $\hat{H}_{q',0}^1(\Omega)$, and $<< \cdot, \cdot >>$ denotes the duality between $\hat{W}_{q,0}^{-1}(\Omega)$ and $\hat{H}_{q',0}^1(\Omega)$.

Moreover, by (3.182) and (3.342), for $x \in \Gamma$ we have

$$\mathcal{V}_{21}^1(\lambda)F = < \mu \mathbf{D}(\mathcal{A}_p(\lambda)F)\mathbf{n}, \mathbf{n} > -\sigma \Delta_{\Gamma} \mathcal{B}_p(\lambda)F - \operatorname{div} \mathcal{A}_p(\lambda)F$$

$$- \sum_{j=1}^{\infty} \zeta_j^1 \{ < (\mu(x_j^1)\mathbf{D}(\mathcal{S}_{1j}(\lambda)\tilde{\zeta}_j^1 F)\mathbf{n}_j, \mathbf{n}_j > -\sigma(x_j^1)\Delta_{\Gamma_j} \mathcal{H}_j(\lambda)\tilde{\zeta}_j^1 F$$

$$- \operatorname{div} \mathcal{S}_{1j}(\lambda)\tilde{\zeta}_j^1 F \}$$

$$= \sum_{j=1}^{\infty} < \mu \mathbf{D}(\zeta_j^1 \mathcal{S}_{1j}(\lambda)\tilde{\zeta}_j^1 F)\mathbf{n}, \mathbf{n} > - \sum_{j=1}^{\infty} \sigma \Delta_{\Gamma}(\zeta_j^1 \mathcal{H}_j(\lambda)\tilde{\zeta}_j^1 F)$$

$$-\sum_{j=1}^{\infty} \text{div}\,(\zeta_j^1 \mathcal{S}_{1j}(\lambda)\tilde{\zeta}_j^1 F)$$

$$-\sum_{j=1}^{\infty} \zeta_j^1 \{< \mu(x_j^1)\mathbf{D}(\mathcal{S}_{1j}(\lambda)\tilde{\zeta}_j^1 F)\mathbf{n}_j, \mathbf{n}_j > -\sigma(x_j^1)\Delta_{\Gamma_j}\mathcal{H}_j(\lambda)\tilde{\zeta}_j^1 F$$

$$-\text{div}\,\mathcal{S}_{1j}(\lambda)\tilde{\zeta}_j^1 F\}.$$

Thus, we define an operator $\mathcal{L}_b(\lambda)$ acting on $F \in \mathcal{Y}_q(\Omega)$ by letting

$$\mathcal{L}_b(\lambda)F = \sum_{j=1}^{\infty}[< \mu(x)(\mathbf{D}(\zeta_j^1 \mathcal{S}_{1j}(\lambda)\tilde{\zeta}_j^1 F) - \zeta_j^1 \mathbf{D}(\mathcal{S}_{1j}(\lambda)\tilde{\zeta}_j^1 F))\mathbf{n}, \mathbf{n} >$$

$$-\sigma(\Delta_\Gamma(\zeta_j^1 \mathcal{H}_j(\lambda)\tilde{\zeta}_j^1 F) - \zeta_j^1 \Delta_\Gamma \mathcal{H}_j(\lambda)\tilde{\zeta}_j^1 F) - (\nabla\zeta_j^1)\mathcal{S}_{1j}(\lambda)\tilde{\zeta}_j^1 F],$$

and then, $\mathcal{V}_{21}^1(\lambda)F = \mathcal{L}_b(\lambda)F$ on Γ.

We now prove the $\mathcal{R}$ boundedness of operator families $\mathcal{L}(\lambda)$, $\mathcal{M}(\lambda)$ and $\mathcal{L}_b(\lambda)$. We first prove that

$$\mathcal{R}_{\mathcal{L}(\mathcal{Y}_q(\Omega), \hat{W}_{q,0}^{-1}(\Omega))}(\{(\tau\partial_\tau)^\ell \mathcal{M}(\lambda) \mid \lambda \in \Lambda_{\kappa,\tilde{\lambda}_0}\}) \le (\epsilon + C_{q,\epsilon}\tilde{\lambda}_0^{-1/2})r_b \qquad (3.365)$$

for $\ell = 0, 1$. In fact, if we set

$$<< \mathcal{M}_j^0(\lambda)F, \varphi >> = -(\mu(x_j^0)\mathbf{D}(\mathcal{S}_{0j}(\lambda)\tilde{\zeta}_j^0 F_1) : (\nabla^2 \zeta_j^0), \varphi - e_j)_\Omega$$

$$+ (\text{div}\,(\mathcal{S}_{0j}(\lambda)\tilde{\zeta}_j^0 F_1)(\Delta\zeta_j^0), \varphi - e_j)_\Omega$$

$$+ (K_{0j}(\mathcal{S}_{0j}(\lambda)\tilde{\zeta}_j^0 F_1)(\Delta\zeta_j^0), \varphi - e_j)_\Omega,$$

$$<< \mathcal{M}_j^1(\lambda)F, \varphi >> = -(\mu(x_j^1)\mathbf{D}(\mathcal{S}_{1j}(\lambda)\tilde{\zeta}_j^1 F) : (\nabla^2 \zeta_j^1), \varphi)_\Omega$$

$$+ (\text{div}\,(\mathcal{S}_{1j}(\lambda)\tilde{\zeta}_j^1 F)(\Delta\zeta_j^1), \varphi)_\Omega$$

$$+ (K_{1j}(\mathcal{S}_{1j}(\lambda)\tilde{\zeta}_j^1 F, \mathcal{H}_j(\lambda)\tilde{\zeta}_j^1 F)(\Delta\zeta_j^1), \varphi)_\Omega,$$

then, by Lemmas 3.2.13, 3.5.24 and (3.362), we have

$$| << \mathcal{M}_j^0(\lambda)F, \varphi >> | \le C_{M_2}\|\nabla\mathcal{S}_{0j}(\lambda)\tilde{\zeta}_j^0 F\|_{L_q(\mathbb{R}^N)}\|\nabla\varphi\|_{L_{q'}(B_j^0)},$$

$$| << \mathcal{M}_j^1(\lambda)F, \varphi >> | \le \epsilon(\|\mathcal{S}_{1j}\tilde{\zeta}_j^1 F\|_{H_q^2(\Omega_j)} + \|\mathcal{H}_j(\lambda)\tilde{\zeta}_j^1 F\|_{H_q^3(\Omega_j)})$$

$$+ C_{\epsilon,M_2}(\|\mathcal{S}_{1j}\tilde{\zeta}_j^1 F\|_{H_q^1(\Omega_j)} + \|\mathcal{H}_j(\lambda)\tilde{\zeta}_j^1 F\|_{H_q^2(\Omega_j)})\|\nabla\varphi\|_{L_q(B_j^1 \cap \Omega)}.$$

By (3.27), we have

$$\sum_{j=1}^{\infty} \|\nabla \varphi\|_{L_{q'}(B_j^0)}^{q'} + \sum_{j=1}^{\infty} \|\nabla \varphi\|_{L_{q'}(B_j^1 \cap \Omega)}^{q'} \leq C_{q'} \|\nabla \varphi\|_{L_{q'}(\Omega)}^{q'}$$

for any $\varphi \in \hat{H}_{q',0}^1(\Omega)$. By (3.346), (3.27), and Proposition 3.5.23, we have

$$\sum_{j=1}^{\infty} \|\nabla \mathcal{S}_{0j}(\lambda) \tilde{\zeta}_j^0 F_1\|_{L_q(\mathbb{R}^N)}^q + \sum_{j=1}^{\infty} (\|\mathcal{S}_{1j}(\lambda) \tilde{\zeta}_j^1 F\|_{H_q^2(\Omega_j)}^q + \|\mathcal{H}_j(\lambda) \tilde{\zeta}_j^1 F\|_{H_q^3(\Omega_j)}^q)$$

$$\leq r_b^q (\sum_{j=1}^{\infty} \|\tilde{\zeta}_j^0 F_1\|_{L_q(\mathbb{R}^N)}^q + \sum_{j=1}^{\infty} \|\tilde{\zeta}_j^1 F\|_{\mathcal{Y}_q(\Omega_j)}^q) \leq r_b^q C_q \|F\|_{\mathcal{Y}_q(\Omega)}^q < \infty.$$

Thus, by Proposition 3.5.20, $\mathcal{M}(\lambda)F = \sum_{j=1}^{\infty} \mathcal{M}_j^0(\lambda)F + \sum_{j=1}^{\infty} \mathcal{M}_j^1(\lambda)F$ exists in $\hat{W}_{q,0}^{-1}(\Omega)$ for any $F \in \mathcal{Y}_q(\Omega)$ and

$$\|\mathcal{M}(\lambda)F\|_{\hat{W}_{q,0}^{-1}(\Omega)}^q$$

$$\leq C_{M_2}^q \sum_{j=1}^{\infty} \|\nabla \mathcal{S}_{0j}(\lambda) \tilde{\zeta}_j^0 F_1\|_{L_q(\mathbb{R}^N)}^q + \epsilon^q \sum_{j=1}^{\infty} \|\mathcal{S}_{1j}(\lambda) \tilde{\zeta}_j^1 F\|_{H_q^2(\Omega_j)}^q$$

$$+ \epsilon^q \sum_{j=1}^{\infty} \|\mathcal{H}_j(\lambda) \tilde{\zeta}_j^1 F\|_{H_q^3(\Omega_j)}^q + C_{\epsilon,M_2}^q \sum_{j=1}^{\infty} \|\mathcal{S}_{1j}(\lambda) \tilde{\zeta}_j^1 F\|_{H_q^1(\Omega_j)}^q$$

$$+ C_{\epsilon,M_2}^q \sum_{j=1}^{\infty} \|\mathcal{H}_j(\lambda) \tilde{\zeta}_j^1 F\|_{H_q^1(\Omega_j)}^q.$$

Analogously, by Proposition 3.5.20 we have

$$\|\sum_{\ell=1}^{n} r_\ell(u) \mathcal{M}(\lambda_\ell) F_\ell\|_{\hat{W}_{q,0}^{-1}(\Omega)}^q \leq C_{M_2}^q \sum_{j=1}^{\infty} \|\sum_{\ell=1}^{n} r_\ell(u) \nabla \mathcal{S}_{0j}(\lambda_\ell) \tilde{\zeta}_j^0 F_{1\ell}\|_{L_q(\mathbb{R}^N)}^q$$

$$+ \epsilon^q \sum_{j=1}^{\infty} \{\|\sum_{\ell=1}^{n} r_\ell(u) \mathcal{S}_{1j}(\lambda_\ell) \tilde{\zeta}_j^1 F_\ell\|_{H_q^2(\Omega_j)}^q$$

$$+ \sum_{\ell=1}^{n} r_\ell(u) \mathcal{H}_j(\lambda_\ell) \tilde{\zeta}_j^1 F_\ell\|_{H_q^3(\Omega_j)}^q\}$$

$$+ C_{\epsilon,M_2}^q \sum_{j=1}^{\infty} \{ \| \sum_{\ell=1}^{n} r_\ell(u) \mathcal{S}_{1j}(\lambda_\ell) \tilde{\zeta}_j^1 F_\ell \|_{H_q^1(\Omega_j)}^q$$

$$+ \| \sum_{\ell=1}^{n} r_\ell(u) \mathcal{H}_j(\lambda_\ell) \tilde{\zeta}_j^1 F_\ell \|_{H_q^1(\Omega_j)}^q \}.$$

Noting that $\Omega \cap B_j^1 = \Omega_j \cap B_j^1$, by (3.346), (3.28), Proposition 3.5.23, and Proposition 3.5.4, we have

$$\int_0^1 \| \sum_{\ell=1}^{n} r_\ell(u) \mathcal{M}(\lambda_\ell) F_\ell \|_{\hat{W}_{q,0}^{-1}(\Omega)}^q \, du$$

$$\leq C_{M_2}^q \tilde{\lambda}_0^{-q/2} r_b^q \int_0^1 \sum_{j=1}^{\infty} \| \sum_{\ell=1}^{n} r_\ell(u) \tilde{\zeta}_j^0 F_{1\ell} \|_{L_q(\mathbb{R}^N)}^q \, du$$

$$+ \epsilon^q r_b^q \int_0^1 \sum_{j=1}^{\infty} \| \sum_{\ell=1}^{n} r_\ell(u) \tilde{\zeta}_j^1 F_\ell \|_{\mathcal{Y}_q(\Omega_j)}^q \, du$$

$$+ C_{\epsilon,M_2}^q \tilde{\lambda}_0^{-q/2} r_b^q \int_0^1 \sum_{j=1}^{\infty} \| \sum_{\ell=1}^{n} r_\ell(u) \tilde{\zeta}_j^1 F_\ell \|_{\mathcal{Y}_q(\Omega_j)}^q \, du$$

$$\leq C_q(\epsilon^q + C_{\epsilon,M_2}^q \tilde{\lambda}_0^{-q/2}) r_b^q \int_0^1 \| \sum_{\ell=1}^{n} r_\ell(u) F_\ell \|_{\mathcal{Y}_q(\Omega)}^q \, du,$$

which shows (3.365). Analogously, we can prove

$$\mathcal{R}_{\mathcal{L}(\mathcal{Y}_q(\Omega), L_q(\Omega)^N)}(\{(\tau \partial_\tau)^\ell \mathcal{L}(\lambda) \mid \lambda \in \Lambda_{\kappa,\tilde{\lambda}_0}\}) \leq C_{M_2} r_b \tilde{\lambda}_0^{-1/2},$$

$$\mathcal{R}_{\mathcal{L}(\mathcal{Y}_q(\Omega), H_q^1(\Omega)^N)}(\{(\tau \partial_\tau)^\ell \mathcal{L}_b(\lambda) \mid \lambda \in \Lambda_{\kappa,\tilde{\lambda}_0}\}) \leq C_{M_2} r_b \tilde{\lambda}_0^{-1/2} \tag{3.366}$$

for $\ell = 0, 1$.

We now use the following lemma.

Lemma 3.5.27 *Let* $1 < q < \infty$. *Then, there exists a linear map* $\mathcal{E}$ *from* $\hat{W}_{q,0}^{-1}(\Omega)$ *into* $L_q(\Omega)^N$ *such that for any* $F \in \hat{W}_{q,0}^{-1}(\Omega)$, $\|\mathcal{E}(F)\|_{L_q(\Omega)} \leq C \|F\|_{\hat{W}_{q,0}^{-1}(\Omega)}$ *and*

$$< F, \varphi >= (\mathcal{E}(F), \nabla\varphi)_\Omega \quad \text{for all } \varphi \in \hat{H}_{q',0}^1(\Omega).$$

Proof The lemma follows from the Hahn-Banach theorem by indentifying $\hat{H}_{q',0}^1(\Omega)$ with a closed subspace of $L_{q'}(\Omega)^N$ via the mapping: $\varphi \mapsto \nabla\varphi$. $\square$

Applying Lemma 3.5.27 and using (3.364) and (3.365), we have

$$(\nabla \mathcal{V}_{21}^1(\lambda)F, \nabla\varphi)_\Omega = (\mathcal{L}(\lambda)F + \mathcal{E}(\mathcal{M}(\lambda)F), \nabla\varphi)_\Omega \quad \text{for all } \varphi \in \hat{H}_{q',0}^1(\Omega), \tag{3.367}$$

subject to $\mathcal{V}_{21}^1(\lambda)F = \mathcal{L}_b(\lambda)F$ on Γ, and

$$\mathcal{R}_{\mathcal{L}(\mathcal{Y}_q(\Omega), L_q(\Omega)^N)}(\{(\tau\partial_\tau)^\ell \mathcal{E} \circ \mathcal{M}(\lambda) \mid \lambda \in \Sigma_{\sigma,\tilde{\lambda}_0}\}) \leq C(\epsilon + C_{q,\epsilon}\tilde{\lambda}_0^{-1/2})r_b, \tag{3.368}$$

where $\mathcal{E} \circ \mathcal{M}(\lambda)$ denotes a bounded linear operator family acting on F by $\mathcal{E} \circ \mathcal{M}(\lambda)F = \mathcal{E}(\mathcal{M}(\lambda)F)$. By Remark 3.2.6, we have $\mathcal{V}_{21}^1(\lambda)F = \mathcal{L}_b(\lambda)F + K(\mathcal{L}(\lambda)F + \mathcal{E}(\mathcal{M}(\lambda)F) - \nabla\mathcal{L}_b(\lambda)F)$, and so by (3.366) and (3.368), we see that $\nabla\mathcal{V}_{21}^1(\lambda) \in \text{Hol}(\Sigma_{\sigma,\tilde{\lambda}_0}, \mathcal{L}(\mathcal{Y}_q(\Omega), L_q(\Omega)^N))$ and

$$\begin{aligned} &\mathcal{R}_{\mathcal{L}(\mathcal{Y}_q(\Omega), L_q(\Omega)^N)}(\{(\tau\partial_\tau)^\ell \nabla\mathcal{V}_{21}^1(\lambda) \mid \lambda \in \Sigma_{\sigma,\tilde{\lambda}_0}\}) \\ &\leq C_q(\epsilon + C_{M2,\epsilon}\tilde{\lambda}_0^{-1/2})r_b \quad \text{for } \ell = 0, 1. \end{aligned} \tag{3.369}$$

Finally, by Lemma 3.5.24, (3.362), (3.28), and Proposition 3.5.20, we have

$$\mathcal{V}_{22}^1(\lambda) \in \text{Hol}(\Sigma_{\sigma,\tilde{\lambda}_0}, \mathcal{L}(\mathcal{Y}_q(\Omega), L_q(\Omega)^N)),$$

$$\mathcal{R}_{\mathcal{L}(\mathcal{Y}_q(\Omega), L_q(\Omega)^N)}(\{(\tau\partial_\tau)^\ell \mathcal{V}_{21}^1(\lambda) \mid \lambda \in \Sigma_{\sigma,\tilde{\lambda}_0}\}) \leq C_q(\epsilon + C_\epsilon\tilde{\lambda}_0^{-1/2})r_b$$

for $\ell = 0, 1$, which, combined with (3.369) and the formula: $\mathcal{V}_2^1(\lambda) = \nabla\mathcal{V}_{21}^1(\lambda) + \mathcal{V}_{22}^2(\lambda)$, leads to (3.356).

3.5.8 Proof of Theorem 3.4.11, Existence Part for Eq. (3.226)

Choosing ϵ so small that $C_{q,r}r_b\epsilon \leq 1/4$, and $\tilde{\lambda}_0$ so large that

$$C_q r_b C_{M2,\epsilon}(\tilde{\lambda}_0^{-1}\gamma_k + \tilde{\lambda}_0^{-1/2}) \leq 1/4 \tag{3.370}$$

in (3.353), we have

$$\mathcal{R}_{\mathcal{L}(\mathcal{Y}_q(\Omega))}(\{(\tau\partial_\tau)^\ell \mathbf{F}_\lambda \mathcal{V}(\lambda) \mid \lambda \in \Lambda_{\kappa,\tilde{\lambda}_0}\}) \leq 1/2 \tag{3.371}$$

for $\ell = 0, 1$. Let λ_* be a large number for which $\lambda_* \geq (8C_q r_b C_{M2,\epsilon})^2$, and then setting $\tilde{\lambda}_0 = \lambda_* \gamma_k$, we have (3.370). By (3.370), $(\mathbf{I} - \mathbf{F}_\lambda \mathcal{V}(\lambda))^{-1} = \sum_{j=1}^\infty (\mathbf{F}_\lambda \mathcal{V}(\lambda))^j$ exists in $\text{Hol}(\Lambda_{\kappa,\tilde{\lambda}_0}, \mathcal{L}(\mathcal{Y}_q(\Omega)))$ and

$$\mathcal{R}_{\mathcal{L}(\mathcal{Y}_q(\Omega))}(\{(\tau\partial_\tau)^\ell (\mathbf{I} - \mathbf{F}_\lambda \mathcal{V}(\lambda))^{-1} \mid \lambda \in \Lambda_{\kappa,\tilde{\lambda}_0}\}) \leq 4 \tag{3.372}$$

for $\ell = 0, 1$. Moreover, by (3.352) and (3.371)

$$\|\mathbf{F}_\lambda V(\lambda)(\mathbf{f}, d, \mathbf{h})\|_{\mathcal{Y}_q(\Omega)} \leq (1/2)\|\mathbf{F}_\lambda(\mathbf{f}, d, \mathbf{h})\|_{\mathcal{Y}_q(\Omega)}. \tag{3.373}$$

Since $\|\mathbf{F}_\lambda(\mathbf{f}, d, \mathbf{h})\|_{\mathcal{Y}_q(\Omega)}$ gives an equivalent norm in $Y_q(\Omega)$ for $\lambda \neq 0$, by (3.373) $(\mathbf{I} - V(\lambda))^{-1} = \sum_{j=0}^\infty V(\lambda)^j$ exists in $\mathcal{L}(Y_q(\Omega))$. Since $\mathbf{u} = \mathcal{A}_p(\lambda)\mathbf{F}_\lambda(\mathbf{f}, d, \mathbf{h})$ and $h = \mathcal{B}_p(\lambda)\mathbf{F}_\lambda(\mathbf{f}, d, \mathbf{h})$ satisfy Eq. (3.349), setting

$$\mathbf{v} = \mathcal{A}_p(\lambda)\mathbf{F}_\lambda(\mathbf{I} - V(\lambda))^{-1}(\mathbf{f}, d, \mathbf{h}), \quad \rho = \mathcal{B}_p(\lambda)(\lambda)\mathbf{F}_\lambda(\mathbf{I} - V(\lambda))^{-1}(\mathbf{f}, d, \mathbf{h}),$$

we see that $\mathbf{v} \in H_q^2(\Omega)^N$, $\rho \in H_q^3(\Omega)$ and $\mathbf{v}$ and ρ satisfy the equations:

$$\begin{cases} \lambda\mathbf{v} - \mathrm{Div}\,(\mu\mathbf{D}(\mathbf{v}) - K(\mathbf{v}, \rho)\mathbf{I}) = \mathbf{f} & \text{in } \Omega, \\ \lambda\rho + A_\kappa \cdot \nabla'_\Gamma \rho - \mathbf{v} \cdot \mathbf{n} = d & \text{on } \Gamma, \\ (\mu\mathbf{D}(\mathbf{v}) - K(\mathbf{v}, \rho)\mathbf{I} - (\sigma\Delta_\Gamma\rho)\mathbf{I})\mathbf{n} = \mathbf{h} & \text{on } \Gamma. \end{cases} \tag{3.374}$$

Moreover, by (3.352) we have $\mathbf{F}_\lambda(\mathbf{I} - V(\lambda))^{-1} = (\mathbf{I} - \mathbf{F}_\lambda V(\lambda))^{-1}\mathbf{F}_\lambda$. Thus, setting

$$\mathcal{A}_r(\lambda) = \mathcal{A}_p(\lambda)(\mathbf{I} - \mathbf{F}_\lambda V(\lambda))^{-1}, \quad \mathcal{H}_r(\lambda) = \mathcal{B}_p(\lambda)(\mathbf{I} - \mathbf{F}_\lambda V(\lambda))^{-1}$$

we see that $\mathbf{v} = \mathcal{A}_r(\lambda)\mathbf{F}_\lambda(\mathbf{f}, d, \mathbf{h})$ and $\rho = \mathcal{H}_r(\lambda)\mathbf{F}_\lambda(\mathbf{f}, d, \mathbf{h})$ are solutions of Eq. (3.374). Since we may assume that $\lambda_*\gamma_\kappa \geq \lambda_1$ in (3.351), by (3.351) and (3.372), we have

$$\begin{aligned} \mathcal{R}_{\mathcal{L}(\mathcal{Y}_q(\Omega), H_q^{2-j}(\Omega)^N)}(\{(\tau\partial_\tau)^\ell(\lambda^{j/2}\mathcal{A}_r(\lambda)) \mid \lambda \in \Lambda_{\kappa, \lambda_*\gamma_\kappa}\}) &\leq C_q r b, \\ \mathcal{R}_{\mathcal{L}(\mathcal{Y}_q(\Omega), H_q^{3-k}(\Omega))}(\{(\tau\partial_\tau)^\ell(\lambda^k\mathcal{H}_r(\lambda)) \mid \lambda \in \Lambda_{\kappa, \lambda_*\gamma_\kappa}\}) &\leq C_q r b, \end{aligned} \tag{3.375}$$

for $\ell = 0, 1$, $j = 0, 1, 2$ and $k = 0, 1$. This completes the proof of the existence part of Theorem 3.4.11 for Eq. (3.226).

3.5.9 A Proof of the Existence Part of Theorem 3.4.11

We now prove the existence of $\mathcal{R}$-bounded solution operators of Eq. (3.185). Let $\mathcal{A}_r(\lambda)$ and $\mathcal{H}_r(\lambda)$ be the operators constructed in the previous subsection. Let $(\mathbf{f}, d, \mathbf{h}) \in Y_q(\Omega)$. Let $\mathbf{u} = \mathcal{A}_r(\lambda)\mathbf{F}_\lambda(\mathbf{f}, d, \mathbf{h})$ and $h = \mathcal{H}_r(\lambda)\mathbf{F}_\lambda(\mathbf{f}, d, \mathbf{h})$, where,

$\mathbf{F}(\mathbf{f}, d, \mathbf{h}) = (\mathbf{f}, d, \lambda^{1/2}\mathbf{h}, \mathbf{h}) \in \mathcal{Y}_q(\Omega)$ for $(\mathbf{f}, d, \mathbf{h}) \in Y_q(\Omega)$. Then $\mathbf{u}$ and h satisfy the equations:

$$\begin{cases} \lambda\mathbf{u} - \mathrm{Div}\,(\mu\mathbf{D}(\mathbf{u}, h) - K(\mathbf{u}, h)\mathbf{I}) = \mathbf{f} & \text{in } \Omega, \\ \lambda h + A_\kappa \cdot \nabla'_\Gamma h - \mathbf{n} \cdot \mathbf{u} + \mathcal{F}_1\mathbf{u} = d + \mathcal{F}_1\mathbf{u} & \text{on } \Gamma, \\ (\mu\mathbf{D}(\mathbf{u}) - K(\mathbf{u}, h)\mathbf{I})\mathbf{n} - (\mathcal{F}_2 h + \sigma\Delta_\Gamma h)\mathbf{n} = \mathbf{h} - (\mathcal{F}_2 h)\mathbf{n} & \text{on } \Gamma. \end{cases}$$

$$(3.376)$$

Let

$$V_{R1}(\lambda)(\mathbf{f}, d, \mathbf{h}) = -\mathcal{F}_1\mathcal{A}_r(\lambda)\mathbf{F}_\lambda(\mathbf{f}, d, \mathbf{h}), \quad \mathcal{V}_{R1}(\lambda)F = -\mathcal{F}_1\mathcal{A}_r(\lambda)F,$$

$$V_{R2}(\lambda)(\mathbf{f}, d, \mathbf{h}) = \mathcal{F}_2\mathcal{H}_r(\lambda)\mathbf{F}_\lambda(\mathbf{f}, d, \mathbf{h}), \quad \mathcal{V}_{R2}(\lambda)F = \mathcal{F}_2\mathcal{H}_r(\lambda)F$$

with $\mathbf{F}_\lambda(\mathbf{f}, d, \mathbf{h}) = (\mathbf{f}, d, \lambda^{1/2}\mathbf{h}, \mathbf{h})$ and $F = (F_1, F_2, F_3, F_4) \in \mathcal{Y}_q(\Omega)$. Notice that

$$\mathcal{F}_1\mathbf{u} = -V_{R1}(\lambda)(\mathbf{f}, d, \mathbf{h}) = -\mathcal{V}_{R1}(\lambda)\mathbf{F}_\lambda(\mathbf{f}, d, \mathbf{h}),$$

$$\mathcal{F}_2 h = V_{R2}(\lambda)(\mathbf{f}, d, \mathbf{h}) = \mathcal{V}_{R2}(\lambda)\mathbf{F}_\lambda(\mathbf{f}, d, \mathbf{h}).$$

Let $V_R(\lambda) = (0, V_{R1}(\lambda), V_{R2}(\lambda))$ and $\mathcal{V}_R(\lambda) = (0, \mathcal{V}_{R1}(\lambda), \mathcal{V}_{R2}(\lambda))$, and then

$$V_R(\lambda) = \mathcal{V}_R(\lambda)\mathbf{F}_\lambda. \qquad (3.377)$$

By (3.377) and (3.161), we have

$$\mathcal{R}_{\mathcal{L}(\mathcal{Y}_q(\Omega))}(\{(\tau\partial_\tau)^\ell \mathbf{F}_\lambda\mathcal{V}_R(\lambda) \mid \lambda \in \Lambda_{\kappa,\lambda_2}\}) \le M_0(\lambda_2^{-1/2} + \lambda_2^{-1})r_b \quad (\ell = 0, 1)$$

for any $\lambda_2 > \lambda_*\gamma_\kappa$. Choosing λ_1 so large, we have

$$\mathcal{R}_{\mathcal{L}(\mathcal{Y}_q(\Omega))}(\{(\tau\partial_\tau)^\ell \mathbf{F}_\lambda\mathcal{V}_R(\lambda) \mid \lambda \in \Lambda_{\kappa,\lambda_2}\}) \le 1/2 \quad (\ell = 0, 1). \qquad (3.378)$$

By (3.377) and (3.378),

$$\|\mathbf{F}_\lambda V_R(\lambda)(\mathbf{f}, d, \mathbf{h})\|_{\mathcal{Y}_q(\Omega)} \le (1/2)\|\mathbf{F}_\lambda(\mathbf{f}, d, \mathbf{h})\|_{\mathcal{Y}_q(\Omega)}.$$

Since $\|\mathbf{F}_\lambda(\mathbf{f}, d, \mathbf{h})\|_{\mathcal{Y}_q(\Omega)}$ gives an equivalent norm in $Y_q(\Omega)$ for $\lambda \neq 0$, by (3.373) $(\mathbf{I} - V_R(\lambda))^{-1} = \sum_{j=0}^\infty V_R(\lambda)^j$ exists in $\mathcal{L}(Y_q(\Omega))$. Since $\mathbf{u} = \mathcal{A}_r(\lambda)\mathbf{F}_\lambda(\mathbf{f}, d, \mathbf{h})$ and $h = \mathcal{H}_r(\lambda)\mathbf{F}_\lambda(\mathbf{f}, d, \mathbf{h})$ satisfy Eq. (3.376), setting

$$\mathbf{v} = \mathcal{A}_r(\lambda)\mathbf{F}_\lambda(\mathbf{I} - V_R(\lambda))^{-1}(\mathbf{f}, d, \mathbf{h}), \quad \rho = \mathcal{H}_r(\lambda)\mathbf{F}_\lambda(\mathbf{I} - V_R(\lambda))^{-1}(\mathbf{f}, d, \mathbf{h}),$$

we see that $\mathbf{v} \in H_q^2(\Omega)^N$, $\rho \in H_q^3(\Omega)$ and $\mathbf{v}$ and ρ satisfy the equations:

$$\begin{cases} \lambda\mathbf{v} - \text{Div}\,(\mu\mathbf{D}(\mathbf{v}) - K(\mathbf{v}, \rho)\mathbf{I}) = \mathbf{f} & \text{in } \Omega, \\ \lambda\rho + A_\kappa \cdot \nabla'_\Gamma\rho - \mathbf{v}\cdot\mathbf{n} + \mathcal{F}_1\mathbf{v} = d & \text{on } \Gamma, \\ (\mu\mathbf{D}(\mathbf{v}) - K(\mathbf{v}, \rho)\mathbf{I} - ((\mathcal{F}_2\rho + \sigma\Delta_\Gamma)\rho)\mathbf{I})\mathbf{n} = \mathbf{h} & \text{on } \Gamma. \end{cases} \tag{3.379}$$

Moreover, by (3.377) we have $\mathbf{F}_\lambda(\mathbf{I} - \mathcal{V}_R(\lambda))^{-1} = (\mathbf{I} - \mathbf{F}_\lambda\mathcal{V}_R(\lambda))^{-1}\mathbf{F}_\lambda$. Thus, setting

$$\tilde{\mathcal{A}}_r(\lambda) = \mathcal{A}_r(\lambda)(\mathbf{I} - \mathbf{F}_\lambda\mathcal{V}(\lambda))^{-1}, \quad \tilde{\mathcal{H}}_r(\lambda) = \mathcal{H}_r(\lambda)(\mathbf{I} - \mathbf{F}_\lambda\mathcal{V}(\lambda))^{-1}$$

we see that $\mathbf{v} = \tilde{\mathcal{A}}_r(\lambda)\mathbf{F}_\lambda(\mathbf{f}, d, \mathbf{h})$ and $\rho = \tilde{\mathcal{H}}_r(\lambda)\mathbf{F}_\lambda(\mathbf{f}, d, \mathbf{h})$ are solutions of Eq. (3.379). And, by (3.377) and (3.378), there exists a large number $\lambda_{**} \geq \lambda_*$ for which

$$\mathcal{R}_{\mathcal{L}(\mathcal{Y}_q(\Omega), H_q^{2-j}(\Omega)^N)}(\{(\tau\partial_\tau)^\ell(\lambda^{j/2}\tilde{\mathcal{A}}_r(\lambda)) \mid \lambda \in \Lambda_{\kappa, \lambda_{**}\gamma_\kappa}\}) \leq C_q r_b,$$

$$\mathcal{R}_{\mathcal{L}(\mathcal{Y}_q(\Omega), H_q^{3-k}(\Omega))}(\{(\tau\partial_\tau)^\ell(\lambda^k\tilde{\mathcal{H}}_r(\lambda)) \mid \lambda \in \Lambda_{\kappa, \lambda_{**}\gamma_\kappa}\}) \leq C_q r_b,$$

for $\ell = 0, 1$, $j = 0, 1, 2$ and $k = 0, 1$. This completes the proof of the existence part of Theorem 3.4.11.

3.5.10 Uniqueness

In this subsection, we shall prove the uniqueness part of Theorem 3.4.14. The uniqueness part of Theorem 3.4.11 will be proved in the next subsection.

Let $\mathbf{u} \in H_q^2(\Omega)^N$ satisfy the homogeneous equations:

$$\lambda\mathbf{u} - \text{Div}\,(\mu\mathbf{D}(\mathbf{u}) - K_0(\mathbf{u})\mathbf{I}) = 0 \text{ in } \Omega, \quad (\mu\mathbf{D}(\mathbf{u}) - K_0(\mathbf{u})\mathbf{I})\mathbf{n} = 0 \text{ on } \Gamma. \tag{3.380}$$

We shall prove that $\mathbf{u} = 0$ below. Let λ_0 be a large positive number such that for any $\lambda \in \Sigma_{\epsilon, \lambda_0}$ the existence part of Theorem 3.4.14 holds for $q' = q/(q-1)$. Let $J_q(\Omega)$ be a solenoidal space defined in (3.163) and let $\mathbf{g}$ be any element in $J_{q'}(\Omega)$. Let $\mathbf{v} \in H_{q'}^2(\Omega)^N$ be a solution of the equations:

$$\bar{\lambda}\mathbf{v} - \text{Div}\,(\mu\mathbf{D}(\mathbf{v}) - K_0(\mathbf{v})\mathbf{I}) = \mathbf{g} \text{ in } \Omega, \quad (\mu\mathbf{D}(\mathbf{v}) - K_0(\mathbf{v})\mathbf{I})\mathbf{n} = 0 \text{ on } \Gamma. \tag{3.381}$$

We first observe that $\mathbf{v} \in J_{q'}(\Omega)$. In fact, for any $\varphi \in \hat{H}_{q,0}^1(\Omega)$, we have

$$0 = (\mathbf{g}, \nabla\varphi)_\Omega = \bar{\lambda}(\mathbf{v}, \nabla\varphi)_\Omega - (\text{Div}\,(\mu\mathbf{D}(\mathbf{v})), \nabla\varphi)_\Omega + (\nabla K_0(\mathbf{v}), \nabla\varphi)_\Omega$$

$$= \bar{\lambda}(\mathbf{v}, \nabla\varphi)_\Omega - (\nabla\text{div}\,\mathbf{v}, \nabla\varphi)_\Omega. \tag{3.382}$$

Since $H^1_{q,0}(\Omega) \subset \hat{H}^1_{q,0}(\Omega)$, for any $\varphi \in H^1_{q,0}(\Omega)$, we have

$$0 = \bar{\lambda}(\operatorname{div} \mathbf{v}, \varphi)_\Omega + (\nabla \operatorname{div} \mathbf{v}, \nabla\varphi)_\Omega. \qquad (3.383)$$

Choosing $\lambda_0 > 0$ larger if necessary, we have the uniquness of the resolvent problem (3.383) for the weak Dirichlet operator, and so $\operatorname{div} \mathbf{v} = 0$. Putting this and (3.382) together gives $(\mathbf{v}, \nabla\varphi)_\Omega = 0$ for any $\varphi \in \hat{H}^1_{q,0}(\Omega)$, that is $\mathbf{v} \in J_{q'}(\Omega)$. Analogously, we see that $\mathbf{u} \in J_q(\Omega)$, because $\mathbf{u}$ satisifies Eq. (3.380).

Since $K_0(\mathbf{v}) \in H^1_{q'}(\Omega) + \hat{H}^1_{q',0}(\Omega)$, we write $K_0(\mathbf{v}) = A_1 + A_2$ with $A_1 \in H^1_{q'}(\Omega)$ and $A_2 \in \hat{H}^1_{q',0}(\Omega)$. Since $\mathbf{u} \in J_q(\Omega)$, we see that $\operatorname{div} \mathbf{u} = 0$ in Ω. Thus, by the definition of the solenoidal space $J_q(\Omega)$ and the divergence theorem of Gauss

$$(\mathbf{u}, \nabla K_0(\mathbf{v}))_\Omega = (\mathbf{u}, \nabla A_1)_\Omega + (\mathbf{u}, \nabla A_2)_\Omega = (\mathbf{u} \cdot \mathbf{n}, A_1)_\Gamma - (\operatorname{div} \mathbf{u}, A_1)_\Omega$$

$$= (\mathbf{u} \cdot \mathbf{n}, K_0(\mathbf{v}))_\Gamma,$$

where we have used $K_0(\mathbf{v}) = A_1$ on Γ. Analogously, we have

$$(\nabla K_0(\mathbf{u}), \mathbf{v})_\Omega = (K_0(\mathbf{u}), \mathbf{v} \cdot \mathbf{n})_\Gamma.$$

Thus, by the divergence theorem of Gauss we have

$$(\mathbf{u}, \mathbf{g})_\Omega = (\mathbf{u}, \bar{\lambda}\mathbf{v}) - (\mathbf{u}, \operatorname{Div}(\mu \mathbf{D}(\mathbf{v}) - K_0(\mathbf{v})\mathbf{I}))_\Omega$$

$$= \lambda(\mathbf{u}, \mathbf{v}) - (\mathbf{u}, (\mu \mathbf{D}(\mathbf{v}) - K_0(\mathbf{v}))\mathbf{n})_\Gamma + (\frac{\mu}{2}\mathbf{D}(\mathbf{u}), \mathbf{D}(\mathbf{v}))_\Omega$$

$$= \lambda(\mathbf{u}, \mathbf{v}) + \frac{1}{2}(\mu \mathbf{D}(\mathbf{u}), \mathbf{D}(\mathbf{v}))_\Omega.$$

Analogously, we have

$$0 = (\lambda\mathbf{u} - \operatorname{Div}(\mu \mathbf{D}(\mathbf{u}) - K_0(\mathbf{u})\mathbf{I}), \mathbf{v})_\Omega - (\mathbf{u}, \operatorname{Div}(\mu \mathbf{D}(\mathbf{v}) - K_0(\mathbf{v})\mathbf{I}))_\Omega$$

$$= \lambda(\mathbf{u}, \mathbf{v}) + \frac{1}{2}(\mu \mathbf{D}(\mathbf{u}), \mathbf{D}(\mathbf{v}))_\Omega.$$

Combining these two equalities yields that

$$(\mathbf{u}, \mathbf{g})_\Omega = 0 \quad \text{for any } \mathbf{g} \in J_{q'}(\Omega). \qquad (3.384)$$

For any $\mathbf{f} \in C_0^\infty(\Omega)^N$, let $\psi \in \hat{H}^1_{q',0}(\Omega)$ be a solution to the variational equation $(\mathbf{f}, \nabla\varphi)_\Omega = (\nabla\psi, \nabla\varphi)_\Omega$ for any $\varphi \in \hat{H}^1_{q,0}(\Omega)$. Let $\mathbf{g} = \mathbf{f} - \nabla\psi$, and then $\mathbf{g} \in J_{q'}(\Omega)$ and $(\mathbf{u}, \nabla\psi)_\Omega = 0$, because $\mathbf{u} \in J_q(\Omega)$. Thus, by (3.384), $(\mathbf{u}, \mathbf{f})_\Omega = (\mathbf{u}, \mathbf{g})_\Omega = 0$, which, combined with the arbitrariness of the choice of $\mathbf{f}$, leads to $\mathbf{u} = 0$. This completes the proof of the uniqueness part of Theorem 3.4.14.

3.5.11 A Priori Estimate

In this subsection, we prove a priori estimates of solutions of Eq. (3.185), from which the uniqueness part of Theorem 3.4.11 follows immediately. In the previous subsection, we used the dual problem for Eq. (3.184) to prove the uniqueness, but in the present case, it is not clear what is a suitable dual problem for Eq. (3.185) to prove the uniqueness. This is the reason why we consider the a priori estimates.

To prove the a priori estimates, we need a slight restriction of the domain Ω. Namely, in this subsection, we assume that Ω is a uniform C^3 domain whose inside has a finite covering (cf. Definition 3.2.3), which is used to estimate the local norm of $K(\mathbf{u}, h)$.

The following theorem is the main result of this section.

Theorem 3.5.28 *Let* $1 < q < \infty$. *Let* Ω *be a uniformly* C^3 *domain whose inside has a finite covering. Then, there exists a* $\lambda_0 > 0$ *such that for any* $\lambda \in \Lambda_{\sigma, \lambda_0 \gamma_\kappa}$ *and* $(\mathbf{u}, h) \in H_q^2(\Omega)^N \times H_q^3(\Omega)$ *satisfying Eq. (3.185), we have*

$$|\lambda| \|\mathbf{u}\|_{L_q(\Omega)} + |\lambda|^{1/2} \|\mathbf{u}\|_{H_q^1(\Omega)} + \|\mathbf{u}\|_{H_q^2(\Omega)} + |\lambda| \|h\|_{H_q^2(\Omega)} + \|h\|_{H_q^3(\Omega)}$$

$$\leq C\{\|\mathbf{f}\|_{L_q(\Omega)} + \|d\|_{W_q^{2-1/q}(\Gamma)} + |\lambda|^{1/2} \|\mathbf{h}\|_{L_q(\Omega)} + \|\mathbf{h}\|_{H_q^1(\Omega)}\}. \tag{3.385}$$

Corollary 3.5.29 *Let* $1 < q < \infty$. *Let* Ω *be a uniformly* C^3 *domain whose inside has a finite covering. Then, there exists a* $\lambda_0 > 0$ *such that the uniqueness holds for Eq. (3.185) for any* $\lambda \in \Lambda_{\sigma, \lambda_0 \gamma_\kappa}$.

In what follows, we shall prove Theorem 3.5.28. We first consider Eq. (3.226). Let $\mathbf{u} \in H_q^2(\Omega)^N$ and $h \in H_q^3(\Omega)$ satisfy Eq. (3.226). We use the same notation as in Sect. 3.5.6, Proposition 3.2.2 and Definition 3.2.3. Let ψ^0 be the function given in Definition 3.2.3, that is $\psi_0 = \sum_{j=1}^{\infty} \zeta_j^0$, where ζ_j^0 are the functions given in Proposition 3.2.2. Notice that supp $\psi^0 \subset \Omega$. Let $\mathbf{u}^0 = \psi^0 \mathbf{u}$, and then $\mathbf{u}^0$ satisfies the equations:

$$\lambda \mathbf{u}^0 - \mathrm{Div}\,(\mu \mathbf{D}(\mathbf{u}^0) - K_0(\mathbf{u}^0)\mathbf{I}) = \mathbf{f}^0 \qquad \text{in } \Omega,$$
$$(\mu \mathbf{D}(\mathbf{u}^0) - K_0(\mathbf{u}_0)\mathbf{I})\mathbf{n} = 0 \qquad \text{on } \Gamma. \tag{3.386}$$

Here, we have set

$$\mathbf{f}^0 = \psi^0 \mathbf{f} + \psi^0 \mathrm{Div}\,(\mu \mathbf{D}(\mathbf{u})) - \mathrm{Div}\,(\mu \mathbf{D}(\psi^0 \mathbf{u})) - (\psi^0 \nabla K(\mathbf{u}, h) - \nabla K_0(\psi^0 \mathbf{u})),$$

and we have used the fact:

$$K_0(\mathbf{u}_0) = <\mu \mathbf{D}(\mathbf{u}^0)\mathbf{n}, \mathbf{n}> - \mathrm{div}\,\mathbf{u}^0 = 0 \quad \text{on } \Gamma.$$

We also let $\mathbf{u}_j^1 = \zeta_j^1 \mathbf{u}$ and $h_j = \zeta_j^1 h$, and then $\mathbf{u}_j^1$ and h_j satisfy the equations:

$$\begin{cases} \lambda \mathbf{u}_j^1 - \text{Div}\,(\mu(x_j^1)\mathbf{D}(\mathbf{u}_j^1) - K_{1j}(\mathbf{u}_j^1, h_j^1)\mathbf{I}) = \mathbf{f}_j^1 & \text{in } \Omega_j, \\ \lambda h_j + A_\kappa(x_j^1) \cdot \nabla_{\Gamma_j} h_j - \mathbf{n}_j \cdot \mathbf{u}_j = d_j & \text{on } \Gamma_j, \\ (\mu(x_j^1)\mathbf{D}(\mathbf{u}_j^1) - K_{1j}(\mathbf{u}_j^1, h_j^1)\mathbf{I})\mathbf{n}_j - \sigma(x_j^1)(\Delta_{\Gamma_j} h_j)\mathbf{n}_j = \mathbf{h}_j & \text{on } \Gamma_j. \end{cases}$$
$$(3.387)$$

Here, we have set

$$\mathbf{f}_j^1 = \zeta_j^1 \mathbf{f} + \zeta_j^1 \text{Div}\,(\mu(x)\mathbf{D}(\mathbf{u})) - \text{Div}\,(\mu(x)\mathbf{D}(\zeta_j^1 \mathbf{u}))$$
$$+ \text{Div}\,((\mu(x) - \mu(x_j^1))\mathbf{D}(\zeta_j^1 \mathbf{u})) - (\zeta_j^1 \nabla K(\mathbf{u}, h) - \nabla K_{1j}(\zeta_j^1 \mathbf{u}, \zeta_j^1 h));$$
$$d_j = \zeta_j^1 d - \zeta_j^1 (A_\kappa(x) - A_\kappa(x_j^1)) \cdot \nabla_{\Gamma_j} h_j$$
$$- A_\kappa(x_j^1) \cdot (\zeta_j^1 \nabla_{\Gamma_j} h - \nabla_{\Gamma_j}(\zeta_j^1 h));$$
$$\mathbf{h}_j = \zeta_j^1 \mathbf{h} - \{\zeta_j^1 (\mu(x) - \mu(x_j^1))\mathbf{D}(\mathbf{u}) + \mu(x_j^1)(\zeta_j^1 \mathbf{D}(\mathbf{u}) - \mathbf{D}(\zeta_j^1 \mathbf{u}))\}\mathbf{n}$$
$$+ (\zeta_j^1 K(\mathbf{u}, h) - K_{1j}(\zeta_j^1 \mathbf{u}, \zeta_j^1 h))\mathbf{n}$$
$$+ \zeta_j^1 (\sigma(x) - \sigma(x_j^1))(\Delta_\Gamma h)\mathbf{n} + \sigma(x_j^1)(\zeta_j^1 \Delta_\Gamma h - \Delta_\Gamma(\zeta_j^1 h))\mathbf{n}.$$

Set

$$E_\lambda(\mathbf{u}, h) = |\lambda|^q \|\mathbf{u}\|_{L_q(\Omega)}^q + |\lambda|^{q/2} \|\mathbf{u}\|_{H_q^1(\Omega)}^q + \|\mathbf{u}\|_{H_q^2(\Omega)}^q$$
$$+ |\lambda|^q \|h\|_{H_q^2(\Omega)}^q + \|h\|_{H_q^3(\Omega)}^q.$$

Employing the similar argument to that in Sect. 3.5.7 and using Theorem 3.4.14 to estimate $\mathbf{u}^0$ in addition, for any positive number ω we have

$$E_\lambda(\mathbf{u}, h) \le C\{\|\mathbf{f}^0\|_{L_q(\Omega)}^q + \sum_{j=1}^{\infty} (\|\mathbf{f}_j^1\|_{L_q(\Omega_j)}^q + \|d_j\|_{W_q^{2-1/q}(\Gamma_j)}^q$$
$$+ |\lambda|^{q/2} \|\mathbf{h}_j\|_{L_q(\Omega_j)}^q + \|\mathbf{h}_j\|_{H_q^1(\Omega_j)}^q)\}$$
$$\le C\{\|\mathbf{f}\|_{L_q(\Omega)}^q + \|d\|_{W_q^{2-1/q}(\Gamma)}^q + |\lambda|^{q/2} \|\mathbf{h}\|_{L_q(\Omega)}$$
$$+ \|\mathbf{h}\|_{H_q^1(\Omega)}^q + \gamma_\kappa^q \|h\|_{H_q^2(\Omega)}^q$$
$$+ (\omega^q + M_1^q)(\|\mathbf{u}\|_{H_q^2(\Omega)}^q + \|h\|_{H_q^3(\Omega)}^q + |\lambda|^{q/2} \|h\|_{H_q^2(\Omega)})$$

$$+ C_{\omega, M_2} (\|\mathbf{u}\|_{H_q^1(\Omega)} + |\lambda|^{q/2} \|\mathbf{u}\|_{L_q(\Omega)} + \|h\|_{H_q^2(\Omega)}^q$$

$$+ |\lambda|^{q/2} \|h\|_{H_q^1(\Omega)}^q) + \|K(\mathbf{u}, h)\|_{L_q(\mathcal{O})}^q\}. \tag{3.388}$$

Here, C_{ω, M_2} is a constant depending on ω and M_2, and we have used the assumption that

$$\text{supp } \nabla \psi^0 \cup \left(\bigcup_{j=1}^{\infty} \text{supp } \nabla \zeta_j^1 \right) = \mathcal{O} \tag{3.389}$$

in Definition 3.2.3.

To estimate $\|K(\mathbf{u}, h)\|_{L_q(\mathcal{O})}$, we need the following Poincarés' type lemma.

Lemma 3.5.30 *Let* $1 < q < \infty$ *and let* Ω *be a uniformly* C^2 *domain whose inside has a finite covering. Let* $\mathcal{O}$ *be a set given in* (3.389). *Then, we have*

$$\|\varphi\|_{L_q(\mathcal{O})} \le C_{q,\mathcal{O}} \|\nabla \varphi\|_{L_q(\Omega)} \quad \text{for any } \varphi \in \hat{H}_{q,0}^1(\Omega)$$

with some constant $C_{q,\mathcal{O}}$ *depending solely on* $\mathcal{O}$ *and* q.

Proof Let $\mathcal{O}_i$ ($i = 1, \ldots, \iota$) be the sub-domains given in Definition 3.2.3, and then it is sufficient to prove that

$$\|\varphi\|_{L_q(\mathcal{O}_i)} \le C \|\nabla \varphi\|_{L_q(\Omega)} \quad \text{for any } \varphi \in \hat{H}_{q,0}^1(\Omega) \text{ and } i = 1, \ldots, \iota.$$

If $\mathcal{O}_i \subset \Omega_R$ for some $R > 0$, since $\varphi|_\Gamma = 0$, by the usual Poincarés' inequality we have

$$\|\varphi\|_{L_q(\mathcal{O}_i)} \le \|\varphi\|_{L_q(\Omega_R)} \le C \|\nabla \varphi\|_{L_q(\Omega_R)} \quad \text{for any } \varphi \in \hat{H}_{q,0}^1(\Omega).$$

Let $\mathcal{O}_i$ be a subdomain for which the condition (b) in Definition 3.2.3 holds. Since the norms for $\varphi(\mathcal{A} \circ \tau(y))$ and $\varphi(y)$ are equivalent, without loss of generality we may assume that

$$\mathcal{O}_i \subset \{x = (x', x_N) \in \mathbb{R}^N \mid a(x') < x_N < b, \quad x' \in D\} \subset \Omega,$$

$$\{x = (x', x_N) \in \mathbb{R}^N \mid x_N = a(x') \quad x' \in D\} \subset \Gamma.$$

Since $\varphi \in \hat{H}_{q,0}^1(\Omega)$, we can write

$$\varphi(x', x_N) = \int_{a(x')}^{x_N} (\partial_s \varphi)(x', s) \, ds.$$

because $\varphi(x', a(x')) = 0$. By Hardy inequality (3.48), we have

$$\left(\int_{a(x')}^b |\varphi(x', x_N)|^q x_N^{-q} \, dx_N \right)^{1/q}$$

$$\leq \left(\int_{a(x')}^b \left(\int_{a(x')}^{x_N} |(\partial_s \varphi)(x', s)| \, ds \right)^q x_N^{-q} \, dx_N \right)^{1/q}$$

$$\leq \frac{q}{q-1} \left(\int_{a(x')}^b |s \partial_s \varphi(x', s)|^q \, s^{-q} \, ds \right)^{1/q},$$

and so, by Fubini's theorem we have

$$\left(\int_{\mathcal{O}_i} |\varphi(x)|^q \, dx \right)^{1/q} \leq \left(\int_D dx' \int_{a(x')}^b |\varphi(x', x_N)|^q \, dx_N \right)^{1/q}$$

$$\leq \left(\int_D dx' \int_{a(x')}^b |\varphi(x', x_N)|^q \, x_N^{-q} b^q \, dx_N \right)^{1/q}$$

$$\leq \frac{qb}{q-1} \left(\int_D dx' \int_{a(x')}^b |\partial_N \varphi(x)|^q \, dx_N \right)^{1/q}$$

$$\leq bq' \|\nabla \varphi\|_{L_q(\Omega)}.$$

This completes the proof of Lemma 3.5.30.   □

We now prove that for any $\omega > 0$ there exists a constant C_{ω, M_2} depending on ω and M_2 such that

$$\|K(\mathbf{u}, h)\|_{L_q(\mathcal{O})}$$
$$\leq \omega(\|\mathbf{u}\|_{H_q^2(\Omega)} + \|h\|_{H_q^3(\Omega)}) + C_{\omega, M_2}(\|\mathbf{u}\|_{H_q^1(\Omega)} + \|h\|_{H_q^2(\Omega)}). \tag{3.390}$$

For this purpose, we estimate $|(K(\mathbf{u}, h), \psi)_\Omega|$ for any $\psi \in C_0^\infty(\mathcal{O})$. By Lemma 3.5.30, for any $\varphi \in \hat{H}_{q,0}^1(\Omega)$ we have

$$|(\varphi, \psi)_\Omega| \leq \|\varphi\|_{L_q(\mathcal{O})} \|\psi\|_{L_{q'}(\mathcal{O})} \leq C_{q, \mathcal{O}} \|\nabla \varphi\|_{L_q(\mathcal{O})} \|\psi\|_{L_{q'}(\mathcal{O})}.$$

Thus, by the Hahn-Banach theorem, there exists a $\mathbf{g} \in L_{q'}(\Omega)^N$ such that $\|\mathbf{g}\|_{L_{q'}(\Omega)} \leq C_{q', \mathcal{O}} \|\psi\|_{L_{q'}(\mathcal{O})}$ and

$$(\varphi, \psi)_\Omega = (\nabla \varphi, \mathbf{g})_\Omega \tag{3.391}$$

for any $\varphi \in \hat{H}_{q,0}^1(\Omega)$. In particular, $\operatorname{div} \mathbf{g} = -\psi$, and therefore $\|\operatorname{div} \mathbf{g}\|_{L_{q'}(\Omega)} \leq \|\psi\|_{L_{q'}(\mathcal{O})}$. By the assumption of the unique existence of solutions of the weak Dirichlet problem and its regularity theorem, Theorem 3.2.8 in Sect. 3.2.6 below, there exists a $\Psi \in \hat{H}^1_{q',0}(\Omega)$ such that $\nabla^2 \Psi \in L_q(\Omega)^N$, Ψ satisfies the weak Dirichlet problem:

$$(\nabla \Psi, \nabla \varphi)_\Omega = (\mathbf{g}, \nabla \varphi)_\Omega \quad \text{for any } \varphi \in \hat{H}^1_{q,0}(\Omega) \tag{3.392}$$

and the estimate:

$$\|\nabla \Psi\|_{H^1_{q'}(\Omega)} \leq C_{q,\mathcal{O}} \|\psi\|_{L_{q'}(\mathcal{O})}. \tag{3.393}$$

Let $L = K(\mathbf{u}, h) - \{< \mu \mathbf{D}(\mathbf{u})\mathbf{n}, \mathbf{n} > -\sigma \Delta_\Gamma h - \operatorname{div} \mathbf{u}\}$, where the Laplace–Beltrami operator Δ_Γ is suitably extended in Ω, and then $L \in \hat{H}^1_{q,0}(\Omega)$. Thus, by (3.391), (3.392) with $\varphi = L$ and the divergence theorem of Gauß,

$$|(L, \psi)_\Omega| = |(\nabla L, \mathbf{g})_\Omega| = |(\nabla \Psi, \nabla L)_\Omega|$$

$$\leq |(\operatorname{Div}(\mu \mathbf{D}(\mathbf{u})) - \nabla \operatorname{div} \mathbf{u}, \nabla \Psi)_\Omega|$$

$$+ |(\nabla\{< \mu \mathbf{D}(\mathbf{u})\mathbf{n}, \mathbf{n} > -\sigma \Delta_\Gamma h - \operatorname{div} \mathbf{u}\}, \nabla \Psi)_\Omega|$$

$$\leq C_{M_2}\{(\|\nabla \mathbf{u}\|_{L_q(\Gamma)} + \|(h, \nabla h, \nabla^2 h)\|_{L_q(\Gamma)})\|\nabla \Psi\|_{L_q(\Gamma)}$$

$$+ (\|\nabla \mathbf{u}\|_{L_q(\Omega)} + \|h\|_{H^2_q(\Omega)})\|\nabla^2 \Psi\|_{L_q(\Omega)}\}.$$

Using the interpolation inequality: $\|v\|_{L_q(\Gamma)} \leq C\|\nabla v\|^{1/q}_{L_q(\Omega)}\|v\|^{1-1/q}_{L_q(\Omega)}$ and (3.393), we have

$$|(L, \psi)_\Omega| \leq \{\omega(\|\nabla^2 \mathbf{u}\|_{L_q(\Omega)} + \|h\|_{H^3_q(\Omega)})$$

$$+ C_{\omega, M_2}(\|\nabla \mathbf{u}\|_{L_q(\Omega)} + \|h\|_{H^2_q(\Omega)})\}\|\psi\|_{L_{q'}(\mathcal{O})},$$

which leads to

$$\|L\|_{L_q(\Omega)} \leq \omega(\|\nabla^2 \mathbf{u}\|_{L_q(\Omega)} + \|h\|_{H^3_q(\Omega)}) + C_{\omega, M_2}(\|\nabla \mathbf{u}\|_{L_q(\Omega)} + \|h\|_{H^2_q(\Omega)}).$$

Thus, we have (3.390).

Putting (3.388) and (3.390) together and choosing ω and M_1 small enough and λ_0 large enough, we have (3.385). This completes the proof of Theorem 3.5.28 for Eq. (3.226).

For Eq. (3.185), using the result we have proved just now, we see that

$$E_\lambda(\mathbf{u}, h) \leq C(\|\mathbf{f}\|_{L_q(\Omega)} + \|d\|_{W^{2-1/q}_q(\Gamma)} + |\lambda|^{1/2}\|\mathbf{h}\|_{L_q(\Omega)} + \|\mathbf{h}\|_{H^1_q(\Omega)} + R)$$

with

$$R = \|\mathcal{F}_1 \mathbf{u}\|_{W^{2-1/q}_q(\Gamma)} + |\lambda|^{1/2}\|\mathcal{F}_2 h\|_{L_q(\Omega)} + \|\mathcal{F}_2 h\|_{H^1_q(\Omega)}).$$

Since $R \leq C(\|\mathbf{u}\|_{H_q^1(\Omega)} + |\lambda|^{1/2}\|h\|_{H_q^2(\Omega)})$ as follows from (3.161), choosing $|\lambda|$ suitably large, R can be absorbed by $E_\lambda(\mathbf{u}, h)$, which completes the proof of Theorem 3.5.28.

3.6 Local Well-Posedness for Arbitrary Initial Velocity Fields in a Uniform C^3 Domain

In this section, we shall prove the unique existence of local in time solutions of Eq. (3.157). Namely, we shall prove the following theorem.

Theorem 3.6.1 *Let* $2 < p < \infty$, $N < q < \infty$, $B > 0$ *and let* Ω *be a uniform* C^3 *domain whose inside has a finite covering. Assume that* $2/p + N/q < 1$ *and that the weak Dirichlet problem is uniquely solvable for indices* q *and* $q' = q/(q - 1)$. *Let* $\mathbf{u}_0 \in B_{q,p}^{2(1-1/p)}(\Omega)^N$ *and* $\rho_0 \in W_{q,p}^{3-1/p-/q}(\Gamma)$ *be initial data such that* $\|\mathbf{u}_0\|_{B_{q,p}^{2(1-1/p)}(\Omega)} \leq B$ *holds and the compatibility conditions:*

$$\mathbf{u}_0 - \mathbf{g}(\mathbf{u}_0, \Psi_\rho|_{t=0}) \in J_q(\Omega), \quad div\, \mathbf{u}_0 = g(\mathbf{u}_0, \Psi_\rho|_{t=0}) \quad in\ \Omega,$$

$$(\mu \mathbf{D}(\mathbf{u}_0)\mathbf{n})_\tau = \mathbf{h}'(\mathbf{u}_0, \Psi_\rho|_{t=0}) \quad on\ \Gamma$$

are satisfied. Then, there exist a time $T > 0$ *and a small number* $\epsilon > 0$ *depending on* B *such that if* $\|\rho_0\|_{B_{q,p}^{3-1/p-1/q}(\Gamma)} \leq \epsilon$, *then problem* (3.157) *admits unique solutions* $\mathbf{u}$, q *and* ρ *with*

$$\mathbf{u} \in L_p((0, T), H_q^2(\Omega)^N) \cap H_p^1((0, T), L_q(\Omega)^N),$$

$$\mathsf{q} \in L_p((0, T), H_q^1(\Omega) + \hat{H}_{q,0}^1(\Omega)),$$

$$\rho \in L_p((0, T), W_q^{3-1/q}(\Gamma)) \cap H_p^1((0, T), W_q^{2-1/q}(\Gamma))$$

possessing the estimate (3.100) *and*

$$E_{p,q,T}(\mathbf{u}, \rho) \leq CB$$

for some constant C *independent of* B. *Here, we have set*

$$E_{p,q,T}(\mathbf{u}, \rho) = \|\mathbf{u}\|_{L_p((0,T),H_q^2(\Omega))} + \|\partial_t \mathbf{u}\|_{L_p((0,T),L_q(\Omega))}$$

$$+ \|\partial_t \rho\|_{L_\infty((0,T),W_q^{1-1/q}(\Gamma))} + \|\partial_t \rho\|_{L_p((0,T),W_q^{2-1/q}(\Omega))}$$

$$+ \|\rho\|_{L_p((0,T),W_q^{3-1/p}(\Omega))}.$$

In what follows, we shall prove Theorem 3.6.1 by the Banach fixed point theorem. Since $N < q < \infty$, by Sobolev's inequality we have the following estimates:

$$\|f\|_{L_\infty(\Omega)} \leq C\|f\|_{H_q^1(\Omega)},$$

$$\|fg\|_{H_q^1(\Omega)} \leq C\|f\|_{H_q^1(\Omega)}\|g\|_{H_q^1(\Omega)},$$

$$\|fg\|_{W_q^{1-1/q}(\Gamma)} \leq C\|f\|_{W_q^{1-1/q}(\Gamma)}\|g\|_{W_q^{1-1/q}(\Gamma)}. \tag{3.394}$$

Moreover, since $2/p + N/q < 1$, we have

$$\|f\|_{H_\infty^1(\Omega)} \leq C\|f\|_{B_{q,p}^{2(1-1/p)}(\Omega)}, \quad \|f\|_{H_\infty^2(\Gamma)} \leq C\|f\|_{B_{q,p}^{3-1/p-1/q}(\Gamma)}. \tag{3.395}$$

We define an underlying space $\mathbf{U}_T$ by letting

$$\mathbf{U}_T = \{(\mathbf{u}, \rho) \mid \mathbf{u} \in H_p^1((0, T), L_q(\Omega)^N) \cap L_p((0, T), H_q^2(\Omega)^N),$$

$$\rho \in H_p^1((0, T), W_q^{2-1/q}(\Gamma)) \cap L_p((0, T), W_q^{3-1/q}(\Gamma)),$$

$$\mathbf{u}|_{t=0} = \mathbf{u}_0 \quad \text{in } \Omega, \quad \rho|_{t=0} = \rho_0 \quad \text{on } \Gamma,$$

$$E_{p,q,T}(\mathbf{u}, \rho) \leq L, \quad \sup_{t \in (0,T)} \|\rho(\cdot, t)\|_{H_\infty^1(\Gamma)} \leq \delta\},$$

where L is a number determined later. Since L is chosen large and ϵ small eventually, we may assume that $0 < \epsilon < 1 < L$ in the following. Thus, for example, we will use the inequality: $1 + \epsilon + L < 3L$ below.

Let $(\mathbf{v}, h) \in \mathbf{U}_T$ and let $\mathbf{u}$, q and ρ be solutions of the linear equations:

$$\begin{cases} \partial_t \mathbf{u} - \text{Div}\,(\mu\mathbf{D}(\mathbf{u}) - q\mathbf{I}) = \mathbf{f}(\mathbf{v}, \Psi_h) & \text{in } \Omega^T, \\ \text{div}\,\mathbf{u} = g(\mathbf{u}, \Psi_h) = \text{div}\,\mathbf{g}(\mathbf{v}, \Psi_h) & \text{in } \Omega^T, \\ \partial_t \rho + <\mathbf{u}_\kappa \mid \nabla'_\Gamma \rho> - \mathbf{u} \cdot \mathbf{n} = d_\kappa(\mathbf{v}, \Psi_h) & \text{on } \Gamma^T, \\ (\mu\mathbf{D}(\mathbf{u})\mathbf{n})_\tau = \mathbf{h}'(\mathbf{v}, \Psi_h) & \text{on } \Gamma^T, \\ <\mu\mathbf{D}(\mathbf{u})\mathbf{n}, \mathbf{n}> -q - \sigma(\Delta_\Gamma \rho + \mathbb{B}\rho) = h_N(\mathbf{v}, \Psi_h) & \text{on } \Gamma^T, \\ (\mathbf{u}, \rho)|_{t=0} = (\mathbf{u}_0, \rho_0) & \text{in } \Omega \times \Gamma. \end{cases} \tag{3.396}$$

Here, the operator $\mathbb{B}$ is defined in (3.149) and we have set

$$d_\kappa(\mathbf{v}, \Psi_h) = d(\mathbf{v}, \Psi_h) + <\mathbf{u}_\kappa - \mathbf{v} \mid \nabla'_\Gamma h>.$$

Let $\Psi_h = \omega H_h \mathbf{n}$, where H_h is an extension of h to Ω such that

$$\|H_h(\cdot,t)\|_{H_q^k(\Omega)} \le C\|h(\cdot,t)\|_{W_q^{k-1/q}(\Gamma)},$$
$$\|\partial_t H_h(\cdot,t)\|_{H_q^\ell(\Omega)} \le C\|\partial_t h(\cdot,t)\|_{W_q^{\ell-1/q}(\Gamma)} \tag{3.397}$$

for $k = 1,2,3$ and $\ell = 1,2$ with some constants C. In particular, for $(\mathbf{v}, h) \in \mathbf{U}_T$, we may assume that

$$\sup_{t \in (0,T)} \|\Psi_h(\cdot,t)\|_{H_\infty^1(\mathbb{R}^N)} \le \delta. \tag{3.398}$$

In fact, we can assume that $\sup_{t \in (0,T)} \|h(\cdot,t)\|_{H_\infty^1(\Gamma)} \le \delta$ with smaller δ.

Since $(\mathbf{v}, h) \in \mathbf{U}_T$, we have

$$E_{p,q,T}(\mathbf{v}, h) \le L, \tag{3.399}$$

that is

$$\|\partial_t h\|_{L_\infty((0,T),W_q^{1-1/q}(\Gamma))} + \|\partial_t h\|_{L_p((0,T),W_q^{2-1/q}(\Gamma))} + \|h\|_{L_p((0,T),W_q^{3-1/q}(\Gamma))}$$
$$+ \|\partial_t \mathbf{v}\|_{L_p((0,T),L_q(\Omega))} + \|\mathbf{v}\|_{L_p((0,T),H_q^2(\Omega))} \le L. \tag{3.400}$$

Moreover, we use the assumption:

$$\|\mathbf{u}_0\|_{B_{q,p}^{2(1-1/p)}(\Omega)} \le B, \quad \|\rho_0\|_{B_{q,p}^{3-1/p-1/q}(\Gamma)} \le \epsilon, \tag{3.401}$$

where ϵ is a small constant determined later.

To obtain the estimates of $\mathbf{u}$ and ρ, we shall use Corollary 3.4.6. Thus, we shall estimate the nonlinear functions appearing in the right hand side of Eq. (3.396). We start with proving that

$$\|\mathbf{f}(\mathbf{v}, \Psi_h)\|_{L_p((0,T),L_q(\Omega))} \le C\{T^{1/p}(L + B)^2 + (\epsilon + T^{1/p'}L)L\}. \tag{3.402}$$

The definition of $\mathbf{f}(\mathbf{v}, \Psi_h)$ is given by replacing $\xi(t)$, ρ and $\mathbf{u}$ by 0, h and $\mathbf{v}$ in (3.113). Since $|\mathbf{V}_0(\mathbf{k})| \le C|\mathbf{k}|$ when $|\mathbf{k}| \le \delta$, by (3.398) we have

$$\|\mathbf{f}(\mathbf{v}, \Psi_h)\|_{L_q(\Omega)} \le C\{\|\mathbf{v}\|_{L_\infty(\Omega)}\|\nabla\mathbf{v}\|_{L_q(\Omega)} + \|\partial_t \Psi_h\|_{L_\infty(\Omega)}\|\nabla\mathbf{v}\|_{L_q(\Omega)}$$
$$+ \|\nabla\Psi_h\|_{L_\infty(\Omega)}\|\partial_t\mathbf{v}\|_{L_q(\Omega)} + \|\nabla\Psi_h\|_{L_\infty(\Omega)}\|\nabla^2\mathbf{v}\|_{L_q(\Omega)}$$
$$+ \|\nabla^2\Psi_h\|_{L_q(\Omega)}\|\nabla\mathbf{v}\|_{L_\infty(\Omega)}\}.$$

By (3.397) and (3.394), we have

$$\|\mathbf{v}(\cdot,t)\|_{L_\infty(\Omega)} \leq C\|\mathbf{v}(\cdot,t)\|_{H_q^1(\Omega)},$$

$$\|\partial_t \Psi_h(\cdot,t)\|_{L_\infty(\Omega)} \leq C\|\partial_t h(\cdot,t)\|_{W_q^{1-1/q}(\Gamma)},$$

$$\|\nabla \Psi_h(\cdot,t)\|_{L_\infty(\Omega)} \leq C\|h(\cdot,t)\|_{W_q^{2-1/q}(\Gamma)}, \tag{3.403}$$

$$\|\nabla \mathbf{v}(\cdot,t)\|_{L_\infty(\Omega)} \leq C\|\mathbf{v}(\cdot,t)\|_{H_q^2(\Omega)},$$

and so,

$$\|\mathbf{f}(\mathbf{v},\Psi_h)\|_{L_p((0,T),L_q(\Omega))} \leq C\{\|\mathbf{v}\|^2_{L_\infty((0,T),H_q^1(\Omega))} T^{1/p}$$

$$+ \|\mathbf{v}\|_{L_\infty((0,T),H_q^1(\Omega))}\|\partial_t h\|_{L_\infty((0,T),W_q^{1-1/q}(\Gamma))} T^{1/p} \tag{3.404}$$

$$+ \|h\|_{L_\infty((0,T),W_q^{2-1/q}(\Gamma))}(\|\partial_t \mathbf{v}\|_{L_p((0,T),L_q(\Omega))} + \|\mathbf{v}\|_{L_p((0,T),H_q^2(\Omega))})\}.$$

Here and in the following, we use the estimate:

$$\|f\|_{L_p((0,T),X)} \leq T^{1/p}\|f\|_{L_\infty((0,T),X)}$$

for the lower order term. In what follows, we shall use the following inequalities:

$$\|\mathbf{v}\|_{L_\infty((0,T),B_{q,p}^{2(1-1/p)}(\Omega))} \leq C\{\|\mathbf{u}_0\|_{B_{q,p}^{2(1-1/p)}(\Omega)}$$

$$+ \|\mathbf{v}\|_{L_p((0,T),H_q^2(\Omega))} + \|\partial_t \mathbf{v}\|_{L_p((0,T),L_q(\Omega))}\}, \tag{3.405}$$

$$\|h\|_{L_\infty((0,T),B_{q,p}^{3-1/p-1/q}(\Gamma))} \leq C\{\|\rho_0\|_{B_{q,p}^{3-1/p-1/q}(\Gamma)}$$

$$+ \|h\|_{L_p((0,T),W_q^{3-1/q}(\Gamma))} + \|\partial_t h\|_{L_p((0,T),W_q^{2-1/q}(\Gamma))}\}, \tag{3.406}$$

$$\|h(\cdot,t)\|_{L_\infty((0,T),W_q^{2-1/q}(\Gamma))} \leq \|\rho_0\|_{W_q^{2-1/q}(\Gamma)} + \int_0^T \|\partial_s h(\cdot,s)\|_{W_q^{2-1/q}(\Gamma)}\, ds$$

$$\leq \|\rho_0\|_{W_q^{2-1/q}(\Gamma)} + T^{1/p'}L, \tag{3.407}$$

for some constant C independent of T. The inequalities: (3.405) and (3.406) will be proved later. Combining (3.404) with (3.405), (3.407), (3.400) and (3.401) gives (3.402).

We next consider $d_\kappa(\mathbf{u},\Psi_h)$. Since $\xi(t) = 0$, by (3.133) we have

$$d(\mathbf{v},\Psi_h) = \mathbf{v}\cdot\mathbf{V}_\Gamma(\bar\nabla\Psi_h)\bar\nabla\Psi_h \otimes \bar\nabla\Psi_h - \partial_t h < \mathbf{n}, \mathbf{V}_\Gamma(\bar\nabla\Psi_h)\bar\nabla\Psi_h \otimes \bar\nabla\Psi_h > .$$

Applying (3.126), (3.398) and (3.394), we have

$$\|\mathbf{V}_\Gamma(\bar{\nabla}\Psi_h)\bar{\nabla}\Psi_h \otimes \bar{\nabla}\Psi_h\|_{W_q^{1-1/q}(\Gamma)} \leq C\|h(\cdot,t)\|_{W_q^{2-1/q}(\Gamma)},$$

$$\|\mathbf{V}_\Gamma(\bar{\nabla}\Psi_h)\bar{\nabla}\Psi_h \otimes \bar{\nabla}\Psi_h\|_{W_q^{2-1/q}(\Gamma)} \leq C\|h(\cdot,t)\|_{W_q^{2-1/q}(\Gamma)}\|h(\cdot,t)\|_{W_q^{3-1/q}(\Gamma)},$$

and so, by (3.403) and (3.394)

$$\|d(\mathbf{v},\Psi_h)\|_{W_q^{1-1/q}(\Gamma)} \leq C(\|\partial_t h(\cdot,t)\|_{W_q^{1-1/q}(\Gamma)} + \|\mathbf{v}(\cdot,t)\|_{H_q^1(\Omega)})\|h(\cdot,t)\|_{W_q^{2-1/q}(\Gamma)},$$

$$\|d(\mathbf{v},\Psi_h)\|_{W_q^{2-1/q}(\Gamma)} \leq C\{(\|\partial_t h(\cdot,t)\|_{W_q^{2-1/q}(\Gamma)} + \|\mathbf{v}(\cdot,t)\|_{H_q^2(\Omega)})\|h(\cdot,t)\|_{W_q^{2-1/q}(\Gamma)}$$

$$+ (\|\partial_t h(\cdot,t)\|_{W_q^{1-1/q}(\Gamma)}$$

$$+ \|\mathbf{v}(\cdot,t)\|_{H_q^1(\Omega)})\|h(\cdot,t)\|_{W_q^{2-1/q}(\Gamma)}\|h(\cdot,t)\|_{W_q^{3-1/q}(\Gamma)}\}.$$

Thus, by (3.400), (3.401), (3.405), and (3.407), we have

$$\sup_{t\in(0,T)} \|d(\mathbf{v},\Psi_h)\|_{W_q^{1-1/q}(\Gamma)} \leq C(L+B)(\epsilon + T^{1/p'}L),$$

$$\|d(\mathbf{v},\Psi_h)\|_{L_p((0,T),W_q^{2-1/q}(\Gamma))} \leq C(L+B)L(\epsilon + T^{1/p'}L). \tag{3.408}$$

By (3.156) and (3.394), we have

$$\| <\mathbf{u}_\kappa - \mathbf{v} \mid \nabla_\Gamma' h> \|_{W_q^{1-1/q}(\Gamma)}$$

$$\leq C(\|\mathbf{u}_0\|_{B_{q,p}^{2(1-1/p)}(\Omega)} + \|\mathbf{v}(\cdot,t)\|_{H_q^1(\Omega)})\|h(\cdot,t)\|_{W_q^{2-1/q}(\Gamma)},$$

$$\| <\mathbf{u}_\kappa - \mathbf{v} \mid \nabla_\Gamma' h> \|_{W_q^{2-1/q}(\Gamma)} \tag{3.409}$$

$$\leq C\{(\|\mathbf{u}_\kappa\|_{H_q^2(\Omega)} + \|\mathbf{v}(\cdot,t)\|_{H_q^2(\Omega)})\|h(\cdot,t)\|_{W_q^{2-1/q}(\Gamma)}$$

$$+ \|\mathbf{u}_\kappa - \mathbf{v}(\cdot,t)\|_{H_q^1(\Omega)}\|h(\cdot,t)\|_{W_q^{3-1/q}(\Gamma)}\}.$$

By (3.405), (3.407), (3.400), and (3.401), we have

$$\| <\mathbf{u}_\kappa - \mathbf{v} \mid \nabla_\Gamma' h> \|_{L_\infty((0,T),W_q^{1-1/q}(\Gamma))}$$

$$\leq C(B+L)(\epsilon + T^{1/p'}L). \tag{3.410}$$

By the definition of $\mathbf{u}_\kappa$, we have

$$\mathbf{u}_\kappa - \mathbf{u}_0 = \frac{1}{\kappa}\int_0^\kappa (T(s)\tilde{\mathbf{u}}_0 - \mathbf{u}_0)\,ds,$$

and so, we have

$$\|\mathbf{u}_\kappa - \mathbf{u}_0\|_{L_q(\Omega)} \leq \frac{1}{\kappa} \int_0^\kappa \left(\int_0^s \|\partial_s T(r)\tilde{\mathbf{u}}_0\|_{L_q(\Omega)} \, dr \right) ds$$

$$\leq C\kappa^{1/p'} \|\mathbf{u}_0\|_{B_{q,p}^{2(1-1/p)}(\Omega)}.$$

Let s be a positive number such that $1 < s < 2(1 - 1/p)$, and then by interpolation theory and (3.156) we have

$$\|\mathbf{u}_\kappa - \mathbf{u}_0\|_{H_q^1(\Omega)} \leq C\|\mathbf{u}_\kappa - \mathbf{u}_0\|_{L_q(\Omega)}^{1-1/s}\|\mathbf{u}_\kappa - \mathbf{u}_0\|_{W_q^s(\Omega)}^{1/s}$$

$$\leq C\kappa^{(1-1/s)/p'}\|\mathbf{u}_0\|_{B_{q,p}^{2(1-1/p)}(\Omega)}. \tag{3.411}$$

Writing $\mathbf{v}(\cdot, t) - \mathbf{u}_0 = \int_0^t \partial_r \mathbf{v}(\cdot, r) \, dr$, we have

$$\|\mathbf{v}(\cdot, t) - \mathbf{u}_0\|_{L_q(\Omega)} \leq LT^{1/p'}.$$

On the other hand, by (3.405)

$$\|\mathbf{v}(\cdot, t) - \mathbf{u}_0\|_{B_{q,p}^{2(1-1/p)}(\Omega)} \leq L + \|\mathbf{u}_0\|_{B_{q,p}^{2(1-1/p)}(\Omega)},$$

and so,

$$\|\mathbf{v}(\cdot, t) - \mathbf{u}_0\|_{H_q^1(\Omega)} \leq (L + B)T^{(1-1/s)/p'}. \tag{3.412}$$

Putting (3.411) and (3.412) together gives

$$\sup_{t \in (0,T)} \|\mathbf{v}(\cdot, t) - \mathbf{u}_\kappa\|_{H_q^1(\Omega)} \leq C(\kappa^{(1-1/s)/p'} + T^{(1-1/s)/p'})(L + B). \tag{3.413}$$

By (3.409)

$$\| <\mathbf{u}_\kappa - \mathbf{v} \mid \nabla_\Gamma' h> \|_{L_p((0,T),W_q^{2-1/q}(\Gamma))}$$

$$\leq C\{(\|\mathbf{u}_\kappa\|_{H_q^2(\Omega)}T^{1/p} + \|\mathbf{v}\|_{L_p((0,T),H_q^2(\Omega))})\|h\|_{L_\infty((0,T),W_q^{2-1/q}(\Gamma))}$$

$$+ \|\mathbf{u}_\kappa - \mathbf{v}\|_{L_\infty((0,T),H_q^1(\Omega))}\|h\|_{L_p((0,T),W_q^{3-1/q}(\Gamma))}\}.$$

Thus, by (3.156), (3.400), (3.401), (3.407) and (3.413), we have

$$\| < \mathbf{u}_\kappa - \mathbf{v} \mid \nabla'_\Gamma h > \|_{L_p((0,T), W_q^{2-1/q}(\Gamma))}$$

$$\leq C\{(B\kappa^{-1/p}T^{1/p} + L)(\epsilon + T^{1/p'}L) \tag{3.414}$$

$$+ L(L + B)(\kappa^{(1-1/s)/p'} + T^{(1-1/s)/p'})\}$$

for some constant $s \in (1, 2(1 - 1/p))$. Putting (3.408), (3.410), and (3.414) gives

$$\|\tilde{d}_\kappa(\mathbf{v}, \Psi_h)\|_{L_\infty((0,T), W_q^{1-1/q}(\Gamma))} \leq C(L + B)(\epsilon + T^{1/p'}L),$$

$$\|\tilde{d}_\kappa(\mathbf{v}, \Psi_h)\|_{L_p((0,T), W_q^{2-1/q}(\Gamma))} \leq C\{(L + B\kappa^{-1/p}T^{1/p})(\epsilon + T^{1/p'}L)$$

$$+ L(L + B)(\epsilon + T^{1/p'}L + \kappa^{(1-1/s)/p'} + T^{(1-1/s)/p'})\}, \tag{3.415}$$

where s is a constant $\in (1, 2(1 - 1/p))$.

We now estimate $g(\mathbf{v}, \Psi_h)$ and $\mathbf{g}(\mathbf{v}, \Psi_h)$, which are defined in (3.108). In view of Corollary 3.4.6, we have to extend $g(\mathbf{v}, \Psi_h)$ and $\mathbf{g}(\mathbf{v}, \Psi_h)$ to the whole time interval $\mathbb{R}$. Before turning to the extension of these functions, we make a few definitions. Let $\tilde{\mathbf{u}}_0 \in B_{q,p}^{2(1-1/p)}(\mathbb{R}^N)^N$ be an extension of $\mathbf{u}_0 \in B_{q,p}^{2(1-1/p)}(\Omega)^N$ to $\mathbb{R}^N$ such that

$$\mathbf{u}_0 = \tilde{\mathbf{u}}_0 \quad \text{in } \Omega, \quad \|\tilde{\mathbf{u}}_0\|_{X(\mathbb{R}^N)} \leq C\|\mathbf{u}_0\|_{X(\Omega)}$$

for $X \in \{H_q^k, B_{q,p}^{2(1-1/p)}\}$. Let

$$T_v(t)\mathbf{u}_0 = e^{-(2-\Delta)t}\tilde{\mathbf{u}}_0 = \mathcal{F}^{-1}[e^{-(|\xi|^2+2)t}\mathcal{F}[\tilde{\mathbf{u}}_0](\xi)], \tag{3.416}$$

and then $T_v(0)\mathbf{u}_0 = \mathbf{u}_0$ in Ω and

$$\|e^t T_v(t)\mathbf{u}_0\|_{X(\Omega)} \leq C\|\mathbf{u}_0\|_{X(\Omega)},$$

$$\|e^t T_v(\cdot)\mathbf{u}_0\|_{H_p^1((0,\infty), L_q(\Omega))} + \|e^t T_v(\cdot)\mathbf{u}_0\|_{L_p((0,\infty), H_q^2(\Omega))}$$

$$\leq C\|\mathbf{u}_0\|_{B_{q,p}^{2(1-1/p)}(\Omega)}. \tag{3.417}$$

Let $\mathbf{W}$, P, and Ξ be solutions of the equations:

$$\begin{cases} \partial_t \mathbf{W} + \lambda_0 \mathbf{W} - \text{Div}\,(\mu \mathbf{D}(\mathbf{W}) - P\mathbf{I}) = 0, \quad \text{div}\,\mathbf{W} = 0 & \text{in } \Omega \times (0,\infty), \\ \partial_t \Xi + \lambda_0 \Xi - \mathbf{W} \cdot \mathbf{n} = 0 & \text{on } \Gamma \times (0,\infty), \\ (\mu \mathbf{D}(\mathbf{W})\mathbf{n} - P\mathbf{I})\mathbf{n} - (\sigma \Delta_\Gamma \Xi)\mathbf{n} = 0 & \text{on } \Gamma \times (0,\infty), \\ (\mathbf{W}, \Xi)|_{t=0} = (0, \rho_0) & \text{in } \Omega \times \Gamma. \end{cases}$$

Since

$$\|\mathbf{W}\|_{L_\infty((0,\infty),B_{q,p}^{2(1-1/p)}(\Omega))}$$

$$\leq C(\|\mathbf{W}\|_{L_p((0,\infty),H_q^2(\Omega))} + \|\partial_t \mathbf{W}\|_{L_p((0,\infty),L_q(\Omega))}),$$

$$\|\varXi\|_{L_\infty((0,\infty),B_{q,p}^{3-1/p-1/q}(\Gamma))}$$

$$\leq C(\|\varXi\|_{L_p((0,\infty),W_q^{3-1/q}(\Gamma))} + \|\partial_t \varXi\|_{L_p((0,\infty),W_q^{2-1/q}(\Gamma))}),$$

as follow from real interpolation results (3.220) and (3.221), choosing λ_0 large enough, by Theorem 3.4.3, we know the existence of $\mathbf{W}$ and $\varXi$ with

$$\mathbf{W} \in L_p((0,\infty), H_q^2(\Omega)^N) \cap H_p^1((0,\infty), L_q(\Omega)^N),$$

$$\varXi \in L_p((0,\infty), W_q^{3-1/q}(\Gamma)) \cap H_p^1((0,\infty), W_q^{2-1/q}(\Gamma))$$

possessing the estimates:

$$\|e^t \mathbf{W}\|_{L_\infty((0,\infty),B_{q,p}^{2(1-1/p)}(\Omega))} + \|e^t \varXi\|_{L_\infty((0,\infty),B_{q,p}^{3-1/p-1/q}(\Gamma))}$$

$$+ \|e^t \mathbf{W}\|_{L_p((0,\infty),H_q^2(\Omega))} + \|e^t \partial_t \mathbf{W}\|_{L_p((0,\infty),L_q(\Omega))}$$

$$+ \|e^t \varXi\|_{L_p((0,\infty),W_q^{3-1/q}(\Gamma))} + \|e^t \partial_t \varXi\|_{L_p((0,\infty),W_q^{2-1/q}(\Gamma))}$$

$$\leq C\|\rho_0\|_{B_{q,p}^{3-1/p-1/q}(\Gamma)}.$$

Moreover, by the trace theorem and the kinematic equation, we have

$$\|e^t \partial_t \varXi\|_{L_\infty((0,\infty),W_q^{1-1/q}(\Gamma))}$$

$$\leq \lambda_0 \|e^t \varXi\|_{L_\infty((0,\infty),W_q^{1-1/q}(\Gamma))} + \|e^t \mathbf{n} \cdot \mathbf{W}\|_{L_\infty((0,\infty),W_q^{1-1/q}(\Gamma))}$$

$$\leq C\|\rho_0\|_{W_{q,p}^{3-1/p-1/q}(\Gamma)}.$$

Setting $T_h(t)\rho_0 = \Psi_\varXi$, we have $T_h(0)\rho_0 = \Psi_{\rho_0}$ in Ω, and

$$\|e^t T_h(\cdot)\rho_0\|_{L_\infty(0,\infty),B_{q,p}^{3-1/p}(\Omega))} + \|e^t \partial_t T_h(\cdot)\rho_0\|_{L_\infty((0,\infty),H_q^1(\Omega))}$$

$$+ \|e^t T_h(\cdot)\rho_0\|_{L_p((0,\infty),H_q^3(\Omega))} + \|e^t \partial_t T_h(\cdot)\rho_0\|_{L_p((0,\infty),H_q^2(\Omega))} \qquad (3.418)$$

$$\leq C\|\rho_0\|_{B_{q,p}^{3-1/p-1/q}(\Gamma)}.$$

Let $\psi(t)$ be a function of $C^\infty(\mathbb{R})$ which equals one for $t > -1$ and zero for $t < -2$. Given a function, $f(t)$, defined on $(0, T)$, the extension, $e_T[f]$, of f is defined by letting

$$e_T[f](t) = \begin{cases} 0 & \text{for } t < 0, \\ f(t) & \text{for } 0 < t < T, \\ f(2T - t) & \text{for } T < t < 2T, \\ 0 & \text{for } t > 2T. \end{cases} \tag{3.419}$$

Obviously, $e_T[f](t) = f(t)$ for $t \in (0, T)$ and $e_T[f](t) = 0$ for $t \notin (0, 2T)$. Moreover, if $f|_{t=0} = 0$, then

$$\partial_t e_T[f](t) = \begin{cases} 0 & \text{for } t < 0, \\ (\partial_t f)(t) & \text{for } 0 < t < T, \\ -(\partial_t f)(2T - t) & \text{for } T < t < 2T, \\ 0 & \text{for } t > 2T. \end{cases} \tag{3.420}$$

We now define the extensions, $\mathcal{E}_1[\mathbf{v}]$, and $\mathcal{E}_2[\Psi_h]$, of $\mathbf{v}$ and Ψ_h to $\mathbb{R}$ by letting

$$\mathcal{E}_1[\mathbf{v}] = e_T[\mathbf{v} - T_v(t)\mathbf{u}_0] + \psi(t)T_v(|t|)\mathbf{u}_0,$$
$$\mathcal{E}_2[\Psi_h] = e_T[\Psi_h - T_h(t)\rho_0] + \psi(t)T_h(|t|)\rho_0. \tag{3.421}$$

Since $\mathbf{v}|_{t=0} = T_v(0)\mathbf{u}_0 = \mathbf{u}_0$ and $\Psi_h|_{t=0} = T_h(0)\rho_0$, we can differentiate $\mathcal{E}_1[\mathbf{v}]$ and $\mathcal{E}_2[\Psi_h]$ once with respect to t and we can use the formula (3.420). Obviously, we have

$$\mathcal{E}_1[\mathbf{v}] = \mathbf{v}, \quad \mathcal{E}_2[\Psi_h] = \Psi_h \quad \text{in } \Omega^T. \tag{3.422}$$

Let

$$\tilde{g}(\mathbf{v}, \Psi_h) = -\{J_0(\nabla\mathcal{E}_2[\Psi_h])\text{div}\,\mathcal{E}_1[\mathbf{v}]$$
$$+ (1 + J_0(\nabla\mathcal{E}_2[\Psi_h]))\mathbf{V}_0(\nabla\mathcal{E}_2[\Psi_h]) : \nabla\mathcal{E}_1[\mathbf{v}]\}, \tag{3.423}$$

$$\tilde{\mathbf{g}}(\mathbf{v}, \Psi_h) = -(1 + J_0(\nabla\mathcal{E}_2[\Psi_h]))^\top\mathbf{V}_0(\nabla\mathcal{E}_2[\Psi_h])\mathcal{E}_1[\mathbf{v}],$$

and then, applying the Hanzawa transform: $x = y + \mathcal{E}_2[\Psi_h]$ instead of $x = y + \Psi_h$, by (3.108), (3.109), and (3.422), we have

$$\tilde{g}(\mathbf{v}, \Psi_h) = g(\mathbf{v}, \Psi_h),$$
$$\tilde{\mathbf{g}}(\mathbf{v}, \Psi_h) = \mathbf{g}(\mathbf{v}, \Psi_h) \quad \text{in } \Omega^T, \tag{3.424}$$
$$\text{div}\,\tilde{\mathbf{g}}(\mathbf{v}, \Psi_h) = \tilde{g}(\mathbf{v}, \Psi_h) \quad \text{in } \Omega \times \mathbb{R}.$$

By (3.418) and (3.407), we have

$$\sup_{t\in\mathbb{R}}\|\mathcal{E}_2[\Psi_h]\|_{H^1_\infty(\Omega)} \le C(\sup_{t\in(0,T)}\|\Psi_h\|_{H^2_q(\Omega)} + \|T(\cdot)\rho_0\|_{L_\infty,H^2_{\tilde{q}}(\Omega)})$$

$$\le C(\|\rho_0\|_{W^{3-1/p1/q}_q(\Gamma)} + T^{1/p'}L).$$

Thus, we choose T and $\|\rho\|_{W^{3-1/p-1/q}_{q,p}(\Omega)}$ so small that

$$\sup_{t\in\mathbb{R}}\|\mathcal{E}_2[\Psi_h]\|_{H^1_\infty(\Omega)} \le \delta. \tag{3.425}$$

Since $\mathbf{V}_0(0) = 0$, we may write $\tilde{\mathbf{g}}(\mathbf{v}, \Psi_h)$ as

$$\tilde{\mathbf{g}}(\mathbf{v}, \Psi_h) = \mathbf{V_g}(\nabla\mathcal{E}_2[\Psi_h])\nabla\mathcal{E}_2[\Psi_h] \otimes \mathcal{E}_1(\mathbf{v}) \tag{3.426}$$

with some matrix of C^1 functions, $\mathbf{V_g}(\mathbf{k})$, defined on $|\mathbf{k}| < \delta$ such that $\|(\mathbf{V}_g, \mathbf{V}'_g)\|_{L_\infty(|\mathbf{k}|\le\delta)} \le C$, where $\mathbf{V}'_g$ denotes the derivative of $\mathbf{V_g}$ with respect to $\mathbf{k}$. We may write the time derivative of $\tilde{\mathbf{g}}(\mathbf{v}, \Psi_h)$ as

$$\partial_t\tilde{\mathbf{g}}(\mathbf{v}, \Psi_h) = \mathbf{V_g}(\nabla\mathcal{E}_2[\Psi_h])\nabla\mathcal{E}_2[\Psi_h] \otimes (\partial_t\mathcal{E}_1(\mathbf{v}))$$
$$+ \mathbf{V_g}(\nabla\mathcal{E}_2[\Psi_h])(\partial_t\nabla\mathcal{E}_2[\Psi_h]) \otimes \mathcal{E}_1(\mathbf{v})$$
$$+ (\mathbf{V}'_\mathbf{g}(\nabla\mathcal{E}_2[\Psi_h])\partial_t\nabla\mathcal{E}_2[\Psi_h])\nabla\mathcal{E}_2[\Psi_h] \otimes \mathcal{E}_1(\mathbf{v}),$$

and so by (3.425) we have

$$\|\partial_t\tilde{\mathbf{g}}(\mathbf{v}, \Psi_h)\|_{L_q(\Omega)} \le C\{\|\nabla\mathcal{E}_2[\Psi_h]\|_{H^1_q(\Omega)}\|\partial_t\mathcal{E}_1[\mathbf{v}]\|_{L_q(\Omega)}$$
$$+ \|\partial_t\nabla\mathcal{E}_2[\Psi_h]\|_{L_q(\Omega)}\|\mathcal{E}_1[\mathbf{v}]\|_{H^1_q(\Omega)}\}. \tag{3.427}$$

Thus, using (3.420), (3.394), (3.407), (3.404), (3.400), (3.401), (3.417), and (3.418), for any $\gamma \ge 0$, we have

$$\|e^{-\gamma t}\partial_t\tilde{\mathbf{g}}(\mathbf{v}, \Psi_h)\|_{L_p(\mathbb{R},L_q(\Omega))}$$
$$\le C(\|\Psi_h\|_{L_\infty((0,T),H^2_q(\Omega))} + \|T_h(\cdot)\Psi_{\rho_0}\|_{L_\infty((0,\infty),H^2_{\tilde{q}}(\Omega))})$$
$$\times (\|\partial_t\mathbf{v}\|_{L_p((0,T),L_q(\Omega))} + \|T_v(\cdot)\mathbf{u}_0\|_{H^1_p((0,\infty),L_q(\Omega))})$$
$$+ T^{1/p}(\|\partial_t\Psi_h\|_{L_\infty((0,T),H^1_q(\Omega))} + \|\partial_t T_h(\cdot)\rho_0\|_{L_\infty((0,\infty),H^1_{\tilde{q}}(\Omega))})$$
$$\times (\|\mathbf{v}\|_{L_\infty((0,T),H^1_q(\Omega))} + \|T_v(\cdot)\mathbf{u}_0\|_{L_\infty((0,\infty),H^1_{\tilde{q}}(\Omega))})$$
$$\le C(LT^{1/p'} + \epsilon + (L + \epsilon)T^{1/p})(L + B)$$
$$\le C(\epsilon + L(T^{1/p'} + T^{1/p}))(L + B). \tag{3.428}$$

We next prove that

$$\|e^{-\gamma t}\tilde{g}(\mathbf{v}, \Psi_h)\|_{H_p^{1/2}(\mathbb{R}, L_q(\Omega))} + \|e^{-\gamma t}\tilde{g}(\mathbf{v}, \Psi_h)\|_{L_p(\mathbb{R}, H_q^1(\Omega))}$$

$$\leq C((\epsilon + T^{1/p'}L)^{1/2}L^{1/2} + \epsilon + T^{1/p'}L)(L + B) \tag{3.429}$$

for any $\gamma \geq 0$. In the sequel, to estimate $\|fg\|_{H_p^{1/2}(\mathbb{R}, L_q(\Omega))}$, we use the following two lemmas.

Lemma 3.6.2 *Let* $1 < p < \infty$ *and* $N < q < \infty$. *Let*

$$f \in L_\infty(\mathbb{R}, H_q^1(\Omega)) \cap H_\infty^1(\mathbb{R}, L_q(\Omega)),$$

$$g \in H_p^{1/2}(\mathbb{R}, L_q(\Omega)) \cap L_p(\mathbb{R}, H_q^1(\Omega)).$$

Then, we have

$$\|fg\|_{H_p^{1/2}(\mathbb{R}, L_q(\Omega))} + \|fg\|_{L_p(\mathbb{R}, H_q^1(\Omega))}$$

$$\leq C(\|f\|_{L_\infty(\mathbb{R}, H_q^1(\Omega))}^{1/2}\|\partial_t f\|_{L_\infty(\mathbb{R}, L_q(\Omega))}^{1/2} + \|f\|_{L_\infty(\mathbb{R}, H_q^1(\Omega))})$$

$$\times (\|g\|_{H_p^{1/2}(\mathbb{R}, L_q(\Omega))} + \|g\|_{L_p(\mathbb{R}, H_q^1(\Omega))}).$$

Proof We can prove the lemma by using the complex interpolation:

$$H_p^{1/2}(\mathbb{R}, L_q(\Omega)) \cap L_p(\mathbb{R}, H_q^{1/2}(\Omega))$$

$$= (L_p(\mathbb{R}, L_q(\Omega)), H_p^1(\mathbb{R}, L_q(\Omega)) \cap L_p(\mathbb{R}, H_q^1(\Omega)))_{[1/2]}.$$

$$\square$$

To estimate $\|\nabla v\|_{H_p^{1/2}(\mathbb{R}, L_q(\Omega))}$, we use the following lemma.

Lemma 3.6.3 *Let* $1 < p, q < \infty$ *and let* Ω *be a uniform* C^2 *domain. Then,*

$$H_p^1(\mathbb{R}, L_q(\Omega)) \cap L_p(\mathbb{R}, H_q^2(\Omega)) \subset H_p^{1/2}(\mathbb{R}, H_q^1(\Omega))$$

and

$$\|u\|_{H_p^{1/2}(\mathbb{R}, H_q^1(\Omega))} \leq C\{\|u\|_{L_p(\mathbb{R}, H_q^2(\Omega))} + \|\partial_t u\|_{L_p(\mathbb{R}, L_q(\Omega))}\}.$$

Remark 3.6.4 This lemma was mentioned in Shibata–Shimizu [53] in the case that Ω is bounded and was proved by Shibata [48] in the case that Ω is a uniform C^2 domain.

Since $J_0(0) = 0$ and $\mathbf{V}_0(0) = 0$, by (3.425) we may write

$$\tilde{g}(\mathbf{v}, \Psi_h) = V_g(\nabla \mathcal{E}_2[\Psi_h])\nabla \mathcal{E}_2[\Psi_h] \otimes \nabla \mathcal{E}_1[\mathbf{v}] \tag{3.430}$$

with some matrix of C^1 functions, $V_g(\mathbf{k})$, defined on $|\mathbf{k}| < \delta$ such that $\|(V_g, V_g')\|_{L_\infty(|\mathbf{k}| \leq \delta)} \leq C$, where V_g' denotes the derivative of V_g with respect to $\mathbf{k}$. By (3.394), (3.397), (3.425), (3.421), (3.418), (3.407), and (3.401), we have

$$\|V_g(\nabla \mathcal{E}_2[\Psi_h])\nabla \mathcal{E}_2[\Psi_h]\|_{L_\infty(\mathbb{R}, H_q^1(\Omega))} \leq C\|\mathcal{E}_2[\Psi_h]\|_{L_\infty(\mathbb{R}, H_q^2(\Omega))}$$

$$\leq C(\|T_h(\cdot)\rho_0\|_{L_\infty(\mathbb{R}, H_q^2(\Omega))} + \|\Psi_h\|_{L_\infty(\mathbb{R}, H_q^2(\Omega))}) \leq C(\epsilon + T^{1/p'}L);$$

$$\|\partial_t(V_g(\nabla \mathcal{E}_2[\Psi_h])\nabla \mathcal{E}_2[\Psi_h])\|_{L_\infty(\mathbb{R}, L_q(\Omega))} \leq C\|\partial_t \mathcal{E}_2[\Psi_h]\|_{L_\infty(\mathbb{R}, H_q^1(\Omega))}$$

Thus, by Lemmas 3.6.2 and 3.6.3 we have

$$\|e^{-\gamma t}\tilde{g}(\mathbf{v}, \Psi_h)\|_{H_p^{1/2}(\mathbb{R}, L_q(\Omega))} + \|e^{-\gamma t}\tilde{g}(\mathbf{v}, \Psi_h)\|_{L_p(\mathbb{R}, H_q^1(\Omega))}$$

$$\leq C((\epsilon + T^{1/p'}L)^{1/2}L^{1/2} + \epsilon + T^{1/p'}L) \tag{3.431}$$

$$\times (\|e^{-\gamma t}\mathcal{E}_1[\mathbf{v}]\|_{H_p^1(\mathbb{R}, L_q(\Omega))} + \|e^{-\gamma t}\mathcal{E}_1[\mathbf{v}]\|_{L_p(\mathbb{R}, H_q^2(\Omega))})$$

for any $\gamma \geq 0$, which, combined with (3.400), (3.401), and (3.417), leads to (3.429).

We finally estimate $\mathbf{h}'(\mathbf{v}, \Psi_h)$ and $h_N(\mathbf{v}, h)$ given in (3.139) and (3.152). In view of (3.139), we may write $\mathbf{h}'(\mathbf{v}, \Psi_h)$ as

$$\mathbf{h}'(\mathbf{v}, \Psi_h) = \mathbf{V}_{\mathbf{h}}'(\bar{\nabla}\Psi_h)\bar{\nabla}\Psi_h \otimes \nabla \mathbf{v} \tag{3.432}$$

with some matrix of C^1 functions, $\mathbf{V}_{\mathbf{h}}'(\bar{\mathbf{k}}) = \mathbf{V}_{\mathbf{h}}'(y, \bar{\mathbf{k}})$, defined on $\overline{\Omega} \times \{\bar{\mathbf{k}} \mid |\bar{\mathbf{k}}| \leq \delta\}$ possessing the estimate :

$$\sup_{|\bar{\mathbf{k}}| \leq \delta} \|(\mathbf{V}_{\mathbf{h}}'(\cdot, \bar{\mathbf{k}}), \partial_{\bar{\mathbf{k}}}\mathbf{V}_{\mathbf{h}}'(\cdot, \bar{\mathbf{k}}))\|_{H_\infty^1(\Omega)} \leq C$$

with some constant C. Here, $\partial_{\bar{\mathbf{k}}}\mathbf{V}_{\mathbf{h}}'$ denotes the derivative of $\mathbf{V}_{\mathbf{h}}'$ with respect to $\bar{\mathbf{k}}$. We extend $\mathbf{h}'(\mathbf{v}, \Psi_h)$ to the whole time interval $\mathbb{R}$ by letting

$$\tilde{\mathbf{h}}'(\mathbf{v}, \Psi_h) = \mathbf{V}_{\mathbf{h}}'(\bar{\nabla}\mathcal{E}_2[\Psi_h]))\bar{\nabla}\mathcal{E}_2[\Psi_h] \otimes \nabla \mathcal{E}_1[\mathbf{v}]. \tag{3.433}$$

Employing the same argument as in the proof of (3.429), for any $\gamma > 0$ we have

$$\|e^{-\gamma t}\tilde{\mathbf{h}}'(\mathbf{v}, \Psi_h)\|_{H_p^{1/2}(\mathbb{R}, L_q(\Omega))} + \|e^{-\gamma t}\tilde{\mathbf{h}}'(\mathbf{v}, \Psi_\rho)\|_{L_p(\mathbb{R}, H_q^1(\Omega))}$$

$$\leq C((\epsilon + T^{1/p'}L)^{1/2}L^{1/2} + \epsilon + T^{1/p'}L)(L + B). \tag{3.434}$$

We finally consider $h_N(\mathbf{v}, \Psi_h)$. In (3.152) we may write

$$- < \mathbf{n}, \mu \mathbf{D}(\mathbf{v}) \mathbf{V}_\Gamma(\bar{\mathbf{k}}) \bar{\mathbf{k}} > - < \mathbf{n}, \mu(\mathcal{D}_\mathbf{D}(\mathbf{k}) \nabla \mathbf{v})(\mathbf{n} + \mathbf{V}_\Gamma(\bar{\mathbf{k}}) \bar{\mathbf{k}}) >$$

$$= \mathbf{V}_{\mathbf{h},N}(\bar{\mathbf{k}}) \nabla \Psi_h \otimes \nabla \mathbf{v}$$

with some matrix of C^1 functions, $\mathbf{V}_{\mathbf{h},N}(\bar{\mathbf{k}}) = \mathbf{V}_{\mathbf{h},N}(y, \bar{\mathbf{k}})$, defined on $\overline{\Omega} \times \{\mathbf{k} \mid |\mathbf{k}| \leq \delta\}$ such that

$$\sup_{|\bar{\mathbf{k}}| \leq \delta} \|(\mathbf{V}_{\mathbf{h},N}(\cdot, \bar{\mathbf{k}}), \partial_{\bar{\mathbf{k}}} \mathbf{V}_{\mathbf{h},N}(\cdot, \bar{\mathbf{k}}))\|_{H^1_\infty(\Omega)} \leq C.$$

Thus, we may write $h_N(\mathbf{v}, \Psi_h)$ as

$$h_N(\mathbf{v}, \Psi_h) = \mathbf{V}_{\mathbf{h},N}(\bar{\nabla} \Psi_h) \nabla \Psi_h \otimes \nabla \mathbf{v} + \sigma \mathbf{V}'_\Gamma(\bar{\nabla} \Psi_h) \bar{\nabla} \Psi_h \otimes \bar{\nabla}^2 \Psi_h, \tag{3.435}$$

and so, we can define the extension of $h_N(\mathbf{v}, \Psi_h)$ by letting

$$\begin{aligned} \tilde{h}_N(\mathbf{v}, \Psi_h) &= \mathbf{V}_{\mathbf{h},N}(\bar{\nabla} \mathcal{E}_2[\Psi_h]) \bar{\nabla} \mathcal{E}_2[\Psi_h] \otimes \nabla \mathcal{E}_1[\mathbf{v}] \\ &\quad + \sigma V^1_\Gamma(\bar{\nabla} \mathcal{E}_2[\Psi_h]) \bar{\nabla} \mathcal{E}_2[\Psi_h] \otimes \bar{\nabla}^2 \mathcal{E}_2[\Psi_h]. \end{aligned} \tag{3.436}$$

Using Lemmas 3.6.2, 3.6.3, (3.417), and (3.418), we have

$$\|\tilde{h}_N(\mathbf{v}, \Psi_h)\|_{L_p(\mathbb{R}, H^1_q(\Omega))} + \|\tilde{h}_N(\mathbf{v}, \Psi_h)\|_{H^{1/2}_p(\mathbb{R}, L_q(\Omega))}$$

$$\leq C((\epsilon + T^{1/p'} L)^{1/2} L^{1/2} + \epsilon + T^{1/p'} L)$$

$$\times (\|e^{-\gamma t} \mathcal{E}_1[\mathbf{v}]\|_{L_p(\mathbb{R}, H^2_q(\Omega))} + \|e^{-\gamma t} \mathcal{E}_1[\mathbf{v}]\|_{H^1_p(\mathbb{R}, L_q(\Omega))}$$

$$+ \|e^{-\gamma t} \bar{\nabla}^2 \mathcal{E}_2[\Psi_h]\|_{L_p(\mathbb{R}, H^1_q(\Omega))} + \|e^{-\gamma t} \bar{\nabla}^2 \mathcal{E}_2[\Psi_h]\|_{H^{1/2}_p(\mathbb{R}, L_q(\Omega))}).$$

By the fact that $H^1_p(\mathbb{R}, L_q(\Omega)) \subset H^{1/2}_p(\mathbb{R}, L_q(\Omega))$, we have

$$\|e^{-\gamma t} \bar{\nabla}^2 \mathcal{E}_2[\Psi_h]\|_{H^{1/2}_p(\mathbb{R}, L_q(\Omega))} \leq \|e^{-\gamma t} \bar{\nabla}^2 \mathcal{E}_2[\Psi_h]\|_{H^1_p(\mathbb{R}, L_q(\Omega))} \leq C(\epsilon + L).$$

Therefore, by (3.417), (3.418), (3.400), and (3.401), we have

$$\|\tilde{h}_N(\mathbf{v}, \Psi_h)\|_{L_p(\mathbb{R}, H^1_q(\Omega))} + \|\tilde{h}_N(\mathbf{v}, \Psi_h)\|_{H^{1/2}_p(\mathbb{R}, L_q(\Omega))}$$

$$\leq C((\epsilon + T^{1/p'} L)^{1/2} L^{1/2} + \epsilon + T^{1/p'} L)(B + L). \tag{3.437}$$

Let

$$\tilde{E}_{p,q,T}(\mathbf{u}, \rho) = \|\mathbf{u}\|_{L_p((0,T), H_q^2(\Omega))} + \|\partial_t \mathbf{u}\|_{L_p((0,T), L_q(\Omega))}$$

$$+ \|\rho\|_{L_p((0,T), W_q^{3-1/q}(\Gamma))} + \|\partial_t \rho\|_{L_p((0,T), W_q^{2-1/q}(\Gamma))}.$$

Notice that

$$E_{p,q,T}(\mathbf{u}, \rho) = \tilde{E}_{p,q,T}(\mathbf{u}, \rho) + \|\partial_t \rho\|_{L_\infty((0,T), H_q^1(\Omega))}.$$

Applying Corollary 3.4.6 and using the estimates (3.402), (3.415), (3.428), (3.434), (3.437), and (3.401), we have

$$\tilde{E}_{p,q,T}(\mathbf{u}, \rho) \le C e^{\gamma \kappa^{-b} T} D_\epsilon \tag{3.438}$$

with

$$D_\epsilon = B + \kappa^{-b}\epsilon + T^{1/p}(L + B)^2$$

$$+ (\epsilon + T^{1/p'}L)L + (L + B\kappa^{-1/p}T^{1/p})(\epsilon + T^{1/p'}L)$$

$$+ L(L + B)(\epsilon + T^{1/p'}L + \kappa^{(1-1/s)/p'} + T^{(1-1/s)/p'})$$

$$+ (\epsilon + L(T^{1/p'} + T^{1/p}) + (\epsilon + T^{1/p'}L)^{1/2}L^{1/2})(L + B)\}$$

for some positive constants γ and C independent of B, ϵ and T. Combining (3.405) and (3.406) with (3.438) yields that

$$\|\mathbf{u}\|_{L_\infty((0,T), B_{q,p}^{2(1-1/p)}(\Omega))} \le C(B + \tilde{E}_{p,q,T}(\mathbf{u}, \rho)),$$

$$\|\rho\|_{L_\infty((0,T), B_{q,p}^{3-1/p-1/q}(\Gamma))} \le C(\epsilon + \tilde{E}_{p,q,T}(\mathbf{u}, \rho)). \tag{3.439}$$

Noting that $3 - 1/p - 1/q > 2 - 1/q$, by the kinematic equation in Eq. (3.396) and (3.439) we have

$$\sup_{t \in (0,T)} \|\partial_t \rho(\cdot, t)\|_{W_q^{1-1/q}(\Gamma)}$$

$$\le \|\mathbf{u}_\kappa\|_{H_q^1(\Omega)} \sup_{t \in (0,T)} \|\rho(\cdot, t)\|_{W_q^{2-1/q}(\Gamma)}$$

$$+ C \sup_{t \in (0,T)} \|\mathbf{u}(\cdot, t)\|_{H_q^1(\Omega)} + \sup_{t \in (0,T)} \|d(\cdot, t)\|_{W_q^{1-1/q}(\Gamma)}$$

$$\le C B(\epsilon + \tilde{E}_{p,q,T}) + C(B + \tilde{E}_{p,q,T}(\mathbf{u}, \rho)) + C(L + B)(\epsilon + T^{1/p'}L). \tag{3.440}$$

Since we may assume that $0 < \epsilon < 1$ and $B \geq 1$, combining (3.438) and (3.440) yields that

$$E_{p,q,T}(\mathbf{u}, \rho) \leq A_1 B + A_2 (L + B)(\epsilon + T^{1/p'} L) + A_3 B e^{\gamma \kappa^{-b} T} D_\epsilon \qquad (3.441)$$

for some constants A_1, A_2 and A_3 independent of B, ϵ, κ and T. Let $\kappa = \epsilon = T$, and then we have $D_\epsilon = B + \epsilon^{1-b} + D'_\epsilon$ with

$$D'_\epsilon = \epsilon^{1/p}(L + B)^2(\epsilon + \epsilon^{1/p'} L)L + (L + B)(\epsilon + \epsilon^{1/p'} L)$$

$$+ L(L + B)(\epsilon + \epsilon^{1/p'} L + 2\epsilon^{(1-1/s)/p'})$$

$$+ (\epsilon + L(\epsilon^{1/p'} + \epsilon^{1/p}) + (\epsilon + \epsilon^{1/p'} L)^{1/2} L^{1/2})(L + B).$$

Choosing $\epsilon \in (0, 1)$ so small that

$$D'_\epsilon \leq 1, \quad \epsilon^{1-b} \leq B, \quad \gamma \kappa^{-b} T = \gamma \epsilon^{1-b} \leq 1,$$

$$(L + B)(\epsilon + T^{1/p'} L) = (L + B)(\epsilon + \epsilon^{1/p'} L) \leq 2B$$

by (3.441), we have $E_{p,q,T}(\mathbf{u}, \rho) \leq A_1 B + 2A_2 B + 3A_3 e B$. Thus, setting $L = A_1 B + 2A_2 B + 3A_3 e B$, we finally obtain

$$E_{p,q,T}(\mathbf{u}, \rho) \leq L. \qquad (3.442)$$

Let $\mathcal{M}$ be a map defined by letting $\mathcal{M}(\mathbf{v}, h) = (\mathbf{u}, \rho)$, and then by (3.442) $\mathcal{M}$ maps $\mathbf{U}_T$ into itself. We can also prove that $\mathcal{M}$ is a contraction map. Namely, choosing $\kappa = \epsilon = T$ smaller if necessary, we can show that for any $(\mathbf{v}_i, h_i) \in \mathbf{U}_T$ ($i = 1, 2$),

$$E_{p,q,T}(\mathcal{M}(\mathbf{v}_1, h_1) - \mathcal{M}(\mathbf{v}_2, h_2)) \leq (1/2) E_{p,q,T}((\mathbf{v}_1, \rho_1) - (\mathbf{v}_2, \rho_2)).$$

Thus, by the contraction mapping principle, we have Theorem 3.6.1, which completes the proof of Theorem 3.6.1.

We finally prove the inequalities (3.405) and (3.406). Let $\mathcal{E}_1[\mathbf{v}]$ be the function given in (3.421). By (3.422) and (3.221), we have

$$\|\mathbf{v}\|_{L_\infty((0,T), B_{q,p}^{2(1-1/p)}(\Omega))} \leq \|\mathcal{E}_1[\mathbf{v}]\|_{L_\infty((0,T), B_{q,p}^{2(1-1/p)}(\Omega))}$$

$$\leq C\{\|\mathcal{E}_1[\mathbf{v}]\|_{L_p((0,\infty), H_q^2(\Omega))} + \|\partial_t \mathcal{E}_1[\mathbf{v}]\|_{L_p((0,\infty), L_q(\Omega))}\},$$

which, combined with (3.417), leads to the inequality (3.405). Analogously, using $\mathcal{E}_2[\Psi_h]$ given in (3.421) and (3.221), we have

$$\|\Psi_h\|_{L_\infty((0,T),B_{q,p}^{3-1/p-1/q}(\Gamma))}$$

$$\leq C\{\|\Psi_h\|_{L_p((0,T),H_q^3(\Omega))} + \|\partial_t\Psi_h\|_{L_p((0,T),H_q^2(\Omega))}\},$$

which, combined with (3.418) and (3.98), leads to the inequality in (3.406).

3.7 Global Well-Posedness in a Bounded Domain Closed to a Ball

In this section, we study the global well-posedness of Eq. (3.1). As was stated in Sect. 3.3.4, we consider the problem in the following setting. Let B_R be the ball of radius R centered at the origin and let S_R be a sphere of radius R centered at the origin. We assume that

(A.1) $|\Omega| = |B_R| = \dfrac{R^N \omega_N}{N}$, where $|D|$ denotes the Lebesgue measure of a
    Lebesgue measurable set D in $\mathbb{R}^N$ and ω_N is the area of S_1

(A.2) $\displaystyle\int_\Omega x\,dx = 0$.

(A.3) Γ is a normal perturbation of S_R given by

$$\Gamma = \{x = y + \rho_0(y)\mathbf{n}(y) \mid y \in S_R\}$$

with a given small function $\rho_0(y)$ defined on S_R.

Here, $\mathbf{n} = y/|y|$ is the unit outer normal to S_R. $\mathbf{n}$ is extended to $\mathbb{R}^N$ by $\mathbf{n} = R^{-1}y$. Let Γ_t be given by

$$\Gamma_t = \{x = y + \rho(y,t)\mathbf{n} + \xi(t) \mid y \in S_R\}$$
$$= \{x = y + R^{-1}\rho(y,t)y + \xi(t) \mid y \in S_R\},$$
(3.443)

where $\rho(y,t)$ is an unknown function with $\rho(y,0) = \rho_0(y)$ for $y \in S_R$. Let $\xi(t)$ be the barycenter point of the domain Ω_t defined by

$$\xi(t) = \frac{1}{|\Omega|}\int_{\Omega_t} x\,dx.$$

Here, by (3.9), we have $|\Omega_t| = |\Omega|$. Notice that $\xi(t)$ is also unknown. It follows from the assumption (A.2) that $\xi(0) = 0$. Moreover, by (3.18) we have

$$\xi'(t) = \frac{1}{|\Omega|} \int_{\Omega_t} \mathbf{v}(x, t) \, dx. \tag{3.444}$$

Given any function ρ defined on S_R, let H_ρ be a suitable extension of ρ such that $H_\rho = \rho$ on S_R and (3.98) holds. We define the Hanzawa transform centered at $\xi(t)$ by letting

$$x = \mathbf{h}_\rho(y, t) = (1 + R^{-1}H_\rho(y, t))y + \xi(t) \quad \text{for } y \in B_R. \tag{3.445}$$

In the following, we set $\Psi_\rho = R^{-1}H_\rho y$ and we assume that

$$\sup_{t \in (0,T)} \|\Psi_\rho(\cdot, t)\|_{H^1_\infty(B_R)} \leq \delta. \tag{3.446}$$

We will choose δ so small that several conditions hold. For example, since $|\Psi_\rho(y, t) - \Psi_\rho(y', t)| \leq \|\nabla\Psi_\rho(\cdot, t)\|_{L_\infty(B_R)}|y - y'|$, if $0 < \delta < 1/2$, then

$$|\mathbf{h}_\rho(y, t) - \mathbf{h}_\rho(y', t)| \geq (1 - \delta)|y - y'| \geq (1/2)|y - y'| \quad \text{for any } y, y' \in B_R,$$

and so the Hanzawa transform is a bijective map from B_R onto Ω_t with

$$\Omega_t = \{x = \mathbf{h}_\rho(y, t) \mid y \in B_R\} \quad \text{for } t \in (0, T). \tag{3.447}$$

Let $\mathbf{v}_0(x)$ be an initial data for Eq. (3.1), and then we set $\mathbf{u}_0(y) = \mathbf{v}_0(\mathbf{h}_{\rho_0}(y))$, where $\mathbf{h}_{\rho_0}(y) = y + R^{-1}H_{\rho_0} y$. Notice that $\mathbf{h}_{\rho_0}(y) = \mathbf{h}_\rho(y, 0)$ if $\rho|_{t=0} = \rho_0$. Let $\mathbf{v}$ and $\mathfrak{p}$ satisfy Eq. (3.1) and we set

$$\mathbf{u}(y, t) = \mathbf{v}(\mathbf{h}_\rho(y, t), t), \quad \mathfrak{q}(y, t) = \mathfrak{p}(\mathbf{h}_\rho(y, t), t), \quad p_0 = \frac{\sigma(N - 1)}{R}. \tag{3.448}$$

We then see from the consideration in Sect. 3.3 that $\mathbf{u}$, $\mathfrak{q}$ and ρ satisfy the following equations:

$$\begin{cases}
\partial_t \mathbf{u} - \text{Div}\,(\mu\mathbf{D}(\mathbf{u}) - \mathfrak{q}\mathbf{I}) = \mathbf{f}(\mathbf{u}, \Psi_\rho) & \text{in } B_R^T, \\
\text{div}\,\mathbf{u} = g(\mathbf{u}, H_\rho) = \text{div}\,\mathbf{g}(\mathbf{u}, \Psi_\rho) & \text{in } B_R^T, \\
\partial_t \rho - \mathbf{n} \cdot P\mathbf{u} = \tilde{d}(\mathbf{u}, \Psi_\rho) & \text{on } S_R^T, \\
\Pi_0(\mu\mathbf{D}(\mathbf{u})\mathbf{n}) = \mathbf{h}'(\mathbf{u}, \Psi_\rho) & \text{on } S_R^T, \\
< \mu\mathbf{D}(\mathbf{u})\mathbf{n}, \mathbf{n} > -\mathfrak{q} - \sigma\mathcal{B}\rho = h_N(\mathbf{u}, \Psi_\rho) & \text{on } S_R^T, \\
(\mathbf{u}, \rho)|_{t=0} = (\mathbf{u}_0, \rho_0) & \text{in } B_R \times S_R,
\end{cases} \tag{3.449}$$

where $B_R^T = B_R \times (0, T)$, $S_R^T = S_R \times (0, T)$, $\mathbf{n} = y/|y|$ for $y \in S_R$, $P\mathbf{u} = \mathbf{u} - \dfrac{1}{|B_R|} \displaystyle\int_{B_R} \mathbf{u}\, dy$, $\mathcal{B} = \Delta_{S_R} + \dfrac{N-1}{R^2}$, and Δ_{S_R} is the Laplace–Beltrami operator on S_R. Here, the functions in the right side of (3.449), $\mathbf{f}(\mathbf{u}, \Psi_\rho)$, $g(\mathbf{u}, \Psi_\rho)$, $\mathbf{g}(\mathbf{u}, \Psi_\rho)$, and $\mathbf{h}'(\mathbf{u}, \Psi_\rho)$, are nonlinear terms given in (3.113), (3.108), and (3.139), respectively. And, $h_N(\mathbf{u}, \Psi_\rho)$ is given in the formula (3.459) of Sect. 3.7.1 below. Since $dx = dy + J_0(\mathbf{k})dy$ with

$$J_0(\mathbf{k}) = \det \begin{pmatrix} \dfrac{\partial \Psi_{\rho 1}(y,t)}{\partial y_1} & \cdots & \dfrac{\partial \Psi_{\rho 1}(y,t)}{\partial y_N} \\ \vdots & \ddots & \vdots \\ \dfrac{\partial \Psi_{\rho N}(y,t)}{\partial y_1} & \cdots & \dfrac{\partial \Psi_{\rho N}(y,t)}{\partial y_N} \end{pmatrix}$$

for $\Psi_\rho = {}^T(\Psi_{\rho 1}, \ldots, \Psi_{\rho N})$, by (3.444) we have

$$\xi'(t) = \frac{1}{|B_R|} \int_{B_R} \mathbf{u}(y, t)\, dy + \frac{1}{|B_R|} \int_{B_R} \mathbf{u}(y, t) J_0(\mathbf{k})\, dy, \tag{3.450}$$

where $\mathbf{k} = \nabla \Psi_\rho$. Thus, by (3.132) and (3.133), we have $\partial_t \rho - \mathbf{n} \cdot P\mathbf{u} = \tilde{d}(\mathbf{u}, \Psi_\rho)$, where $\tilde{d}(\mathbf{u}, \Psi_\rho)$ is given by letting

$$\tilde{d}(\mathbf{u}, \Psi_\rho) = d(\mathbf{u}, \Psi_\rho) - <\mathbf{u} \mid \nabla_\Gamma' \rho> + \frac{1}{|\Omega|} \int_{B_R} \mathbf{u}(y, t) J_0(\mathbf{k})\, dy, \tag{3.451}$$

with $d(\mathbf{u}, \Psi_\rho)$ given in (3.133).

We now state our main result of this section. For this purpose, we make several definitions. From the assumptions (A.1) and (A.2) it follows that ρ_0 should satisfy the following conditions:

$$\frac{R^N \omega_N}{N} = \int_\Omega dx = \int_{|\omega|=1} \int_0^{R + \rho_0(R\omega)} r^{N-1}\, dr\, d\omega = \int_{|\omega|=1} \frac{(R + \rho_0(R\omega))^N}{N}\, d\omega,$$

$$0 = \int_\Omega x_i\, dx = \int_{|\omega|=1} \int_0^{R + \rho_0(R\omega)} \omega_i r^N\, dr\, d\omega$$

$$= \int_{|\omega|} \frac{(R + \rho_0(R\omega))^{N+1}}{N+1} \omega_i\, d\omega$$

for $i = 1, \ldots, N$, and so, we have the compatibility conditions for ρ_0 as follows:

$$\sum_{k=1}^{N} {}_N C_k \int_{S_R} (R^{-1} \rho_0(y))^k\, d\omega = 0, \quad \sum_{k=1}^{N+1} {}_{N+1} C_k \int_{S_R} y_i (R^{-1} \rho_0(y))^k\, d\omega = 0$$

$$\tag{3.452}$$

where $_NC_k = \frac{N!}{k!(N-k)!}$ and $d\omega$ denotes the surface element of S_R, because $\int_{S_1} d\omega = \omega_N$ and $\int_{S_1} \omega_i \, d\omega = 0$. Let $\mathcal{R}_d = \{\mathbf{u} \mid \mathbf{D}(\mathbf{u}) = 0\}$, and then $\mathcal{R}_d$ is a finite dimensional vector space spanned by constant N-vectors and first order polynomials $x_i \mathbf{e}_j - x_j \mathbf{e}_i$ $(i, j = 1, \ldots, N)$, where $\mathbf{e}_i$ are N-vector whose ith component is 1 and other components are zero. Let M_d be the dimension of $\mathcal{R}_d$ and let $\mathbf{p}_\ell = |B_R|^{-1/2}\mathbf{e}_\ell$ $(\ell = 1, \ldots, N)$, and $\mathbf{p}_\ell$ $(\ell = N + 1, \ldots, M_d)$ be one of $c_R^1(x_i \mathbf{e}_j - x_j \mathbf{e}_i)$ with $c_R^1 = \sqrt{(N+2)/(2R^2|B_R|)}$. Since $(\mathbf{p}_\ell, \mathbf{p}_m)_{B_R} = \delta_{\ell, m}$ for $\ell, m = 1, \ldots, M_d$, the set $\{\mathbf{p}_\ell\}_{\ell=1}^{M_d}$ forms a orthogonal basis of $\mathcal{R}_d$ with respect to the $L_2(B_R)$ inner-product $(\cdot, \cdot)_{B_R}$.

We know that $\mathcal{B}x_i = 0$ on S_R for $i = 1, \ldots, N$. Setting $\varphi_1 = |S_R|^{-1/2}$ and $\varphi_\ell = c_R^2 x_{\ell-1}$ $(\ell = 2, \ldots, N+1)$ with $c_R^2 = \sqrt{N/(R^{N+1}\omega_N)}$, we see that $(\varphi_i, \varphi_j)_{S_R} = \delta_{ij}$ $(i, j = 1, \ldots, N+1)$, and therefore the set $\{\varphi_i\}_{i=1}^{N+1}$ forms an orthogonal basis of the space $\{\psi \mid \mathcal{B}\psi = 0 \text{ on } S_R\} \cup \mathbb{C}$ with respect to the $L_2(S_R)$ inner-product $(\cdot, \cdot)_{S_R}$. In this section, we set

$$\mathcal{S}_{p,q}((a,b)) = \{(\mathbf{u}, \mathsf{q}, \rho) \mid \mathbf{u} \in H_p^1((a,b), L_q(B_R)^N) \cap L_p((a,b), H_q^2(B_R)^N),$$

$$\mathsf{q} \in L_p((a,b), H_q^1(S_R)),$$

$$\rho \in H_p^1((a,b), W_q^{2-1/q}(S_R)) \cap L_p((a,b), W_q^{3-1/q}(S_R))\}. \tag{3.453}$$

Our main result of this section is the following.

Theorem 3.7.1 *Let p and q be real numbers such that $2 < p < \infty$, $N < q < \infty$ and $2/p + N/q < 1$. Assume that (A.1), (A.2) and (A.3) hold. Then, there exists a small number $\epsilon > 0$ such that if initial data $\mathbf{u}_0 \in B_{q,p}^{2(1-1/p)}(B_R)$ and $\rho_0 \in B_{q,p}^{3-1/p-1/q}(S_R)$ satisfy the smallness condition:*

$$\|\mathbf{u}_0\|_{B_{q,p}^{2(1-1/p)}(B_R)} + \|\rho_0\|_{B_{q,p}^{3-1/p-1/q}(S_R)} \le \epsilon, \tag{3.454}$$

the compatibility conditions (3.452) and

$$\begin{aligned} div\,\mathbf{u}_0 &= g(\mathbf{u}_0, \rho_0) = div\,\mathbf{g}(\mathbf{u}_0, \rho_0) \quad in\ B_R, \\ (\mu\mathbf{D}(\mathbf{u}_0)\mathbf{n})_\tau &= \mathbf{g}'(\mathbf{u}_0, \rho_0) \quad on\ S_R, \end{aligned} \tag{3.455}$$

and the orthogonal condition:

$$(\mathbf{v}_0, \mathbf{e}_i)_\Omega = 0, \quad (\mathbf{v}_0, x_i \mathbf{e}_j - x_j \mathbf{e}_i)_\Omega = 0, \tag{3.456}$$

for $i, j = 1, \ldots, N$, *then problem* (3.449) *with* $T = \infty$ *admits a unique solution* $(\mathbf{u}, \mathfrak{q}, \rho) \in \mathcal{S}_{p,q}((0, \infty))$ *possessing the estimate:*

$$\|e^{\eta t} \partial_t \mathbf{u}\|_{L_p((0,\infty), L_q(B_R))} + \|e^{\eta t} \mathbf{u}\|_{L_p((0,\infty), H_q^2(B_R))} + \|e^{\eta t} \nabla \mathfrak{q}\|_{L_p((0,\infty), L_q(B_R))}$$

$$+ \|e^{\eta t} \partial_t \rho\|_{L_p((0,\infty), W_q^{2-1/q}(S_R))} + \|e^{\eta t} \rho\|_{L_p((0,\infty), W_q^{3-1/q}(S_R))} \leq C\epsilon$$

for some positive constants C *and* η *independent of* ϵ.

3.7.1 Derivation of Nonlinear Term $h_N(\mathbf{u}, \Psi_\rho)$ in Eq. (3.449)

In this subsection, we derive the nonlinear term $h_N(\mathbf{u}, \Psi_\rho)$ in (3.449). Let $\omega \in S_1$ be represented by $\omega = \omega(p_1, \ldots, p_{N-1})$ with a local coordinate $(p_1, \ldots, p_{N-1})$, and then for $x = (R + \rho)\omega + \xi(t) \in \Gamma_t$, we have

$$\frac{\partial x}{\partial p_j} = (R + \rho)\tau_j + \frac{\partial \rho}{\partial p_j}\omega$$

where $\tau_j = \frac{\partial \omega}{\partial p_j}$. Since $\tau_j \cdot \omega = 0$, the (i, j)th component of the first fundamental form G_t of Γ_t is given by

$$g_{tij} = \frac{\partial x}{\partial p_i} \cdot \frac{\partial x}{\partial p_j} = (R + \rho)^2 g_{ij} + \frac{\partial \rho}{\partial p_i}\frac{\partial \rho}{\partial p_j},$$

where $g_{ij} = \tau_i \cdot \tau_j$ are the (i, j)th elements of the first fundamental form, G, of S_1, and so

$$G_t = (R + \rho)^2 (G + (R + \rho)^{-2}\nabla_p\rho \otimes \nabla_p\rho)$$

$$= (R + \rho)^2 G(\mathbf{I} + (R + \rho)^{-2}(G^{-1}\nabla_p\rho) \otimes \nabla_p\rho),$$

where $\nabla_p\rho = {}^\top(\partial\rho/\partial p_1, \ldots, \partial\rho/\partial p_{N-1})$. Since

$$\det(\mathbf{I} + \mathbf{a}' \otimes \mathbf{b}') = 1 + \mathbf{a}' \cdot \mathbf{b}', \quad (\mathbf{I} + \mathbf{a}' \otimes \mathbf{b}')^{-1} = \mathbf{I} - \frac{\mathbf{a}' \otimes \mathbf{b}'}{1 + \mathbf{a}' \cdot \mathbf{b}'} \quad (3.457)$$

for any $N - 1$ vectors $\mathbf{a}'$ and $\mathbf{b}' \in \mathbb{R}^{N-1}$, we have

$$G_t^{-1} = (R + \rho)^{-2}\Big(\mathbf{I} - \frac{(R + \rho)^{-2}(G^{-1}\nabla_p\rho) \otimes \nabla_p\rho}{1 + (R + \rho)^{-2} < G^{-1}\nabla_p\rho, \nabla_p\rho >}\Big)G^{-1}$$

$$= (R + \rho)^{-2}G^{-1} + O_2.$$

Here and in the following, O_2 denotes a symbol of the form:

$$O_2 = a_0 \Psi_\rho^2 + \sum_{j=1}^N b_j \Psi_\rho \frac{\partial \Psi_\rho}{\partial y_j} + \sum_{i,j=1}^N c_{ij} \frac{\partial \Psi_\rho}{\partial y_i} \frac{\partial \Psi_\rho}{\partial y_j}$$

with some coefficients a_0, b_j and c_{ij} satisfying the estimate:

$$|(a_0, b_j, c_{ij})(y, t)| \leq C, \quad |\nabla(a, b_j, a_{ij})(y, t)| \leq C(|\nabla \Psi_\rho(y, t)| + |\nabla^2 \Psi_\rho(y, t)|)$$

provided that (3.446) holds. In particular, we have

$$g_t^{ij} = (R + \rho)^{-2} g^{ij} + O_2,$$

componentwise.

We next calculate the Christoffel symbols of Γ_t. Since

$$\tau_{ti} = (R + \rho)\tau_i + \frac{\partial \rho}{\partial p_i} \omega,$$

$$\tau_{tij} = (R + \rho)\tau_{ij} + \frac{\partial \rho}{\partial p_j} \tau_i + \frac{\partial \rho}{\partial p_i} \tau_j + \frac{\partial^2 \rho}{\partial p_i \partial p_j} \omega,$$

we have

$$< \tau_{tij}, \tau_{tm} > = (R + \rho)^2 < \tau_{ij}, \tau_m > + (R + \rho)\left(\frac{\partial \rho}{\partial p_m} \ell_{ij} + g_{i\ell} \frac{\partial \rho}{\partial p_j} + g_{j\ell} \frac{\partial \rho}{\partial p_i}\right)$$

$$+ \frac{\partial^2 \rho}{\partial p_i \partial p_j} \frac{\partial \rho}{\partial p_m},$$

and so

$$\Lambda_{tij}^k = g_t^{km} < \tau_{tij}, \tau_{tm} >$$

$$= < (R + \rho)^{-2} g^{km} + O_2, (R + \rho)^2 < \tau_{ij}, \tau_m >$$

$$+ (R + \rho)\left(\frac{\partial \rho}{\partial p_m} \ell_{ij} + g_{im} \frac{\partial \rho}{\partial p_j} + g_{jm} \frac{\partial \rho}{\partial p_i}\right) + \frac{\partial^2 \rho}{\partial p_i \partial p_j} \frac{\partial \rho}{\partial p_m} >$$

$$= \Lambda_{ij}^k + (R + \rho)^{-1} g^{km}\left(\frac{\partial \rho}{\partial p_m} \ell_{ij} + \delta_i^k \frac{\partial \rho}{\partial p_j} + \delta_j^k \frac{\partial \rho}{\partial p_i}\right)$$

$$+ ((R + \rho)^{-2} g^{km} \frac{\partial \rho}{\partial p_m} + O_2) \frac{\partial^2 \rho}{\partial p_i \partial p_j} + O_2.$$

Thus,

$$\Delta_{\Gamma_t} f = g_t^{ij} (\partial_i \partial_j f - \Lambda_{tij}^k \partial_k f)$$

$$= (R + \rho)^{-2} g^{ij} (\partial_i \partial_j f - \Lambda_{ij}^k \partial_k f) + (A^k \nabla_p^2 \rho) \partial_k f + O_2 \otimes (\bar{\nabla}'^2 f)$$

where $\bar{\nabla}'^2 f$ is an $(N-1)^2 + N$ vector of the form: $\bar{\nabla}'^2 f = (\partial_i \partial_j f, \partial_i f, f \mid i, j = 1, \ldots, N-1)$, $\partial_i = \partial / \partial p_i$, and

$$A^k \nabla_p^2 \rho = ((R + \rho)^{-4} g^{k\ell} g^{ij} \frac{\partial \rho}{\partial p_\ell} + O_2) \frac{\partial^2 \rho}{\partial p_i \partial p_j},$$

and so

$$H(\Gamma_t) \mathbf{n}_t = \Delta_{\Gamma_t} [(R + \rho) \omega + \xi(t)]$$

$$= (R + \rho)^{-2} g^{ij} (\partial_i \partial_j - \Lambda_{ij}^k \partial_k)((R + \rho)\omega) + (A^k \nabla_p^2 \rho) \partial_k ((R + \rho)\omega)$$

$$+ O_2 \otimes \bar{\nabla}'^2 ((R + \rho)\omega)$$

$$= (R + \rho)^{-1} g^{ij} (\partial_i \partial_j \omega - \Lambda_{ij}^k \partial_k \omega) + (R + \rho)^{-2} g^{ij} (\partial_i \rho \partial_j \omega + \partial_j \rho \partial_i \omega)$$

$$+ (R + \rho)^{-2} g^{ij} (\partial_i \partial_j \rho - \Lambda_{ij}^k \partial_k \rho)\omega + (A^k \nabla_p^2 \rho)(\partial_k \rho)\omega$$

$$+ (A^k \nabla_p^2 \rho)(R + \rho) \partial_k \omega + O_2 \otimes \bar{\nabla}'^2 \rho$$

Combining this formula with (3.122), recalling $\mathbf{n} = \omega$ in this case and using $< \partial_i \omega, \omega \geq 0$, $< \omega \tau_\ell \geq 0$, $\Delta_{S_1} \omega = -(N-1)\omega$, and (3.115) gives

$$< H(\Gamma_t) \mathbf{n}_t, \mathbf{n}_t >$$

$$= -(R + \rho)^{-1} (N-1) + (R + \rho)^{-2} \Delta_{S_1} \rho + (O_1 + O_2) \otimes \nabla_p^2 \rho + O_2$$

where O_1 is a symbol of the form:

$$O_1 = a_0 \rho + \sum_{j=1}^{N-1} b_j \frac{\partial \rho}{\partial p_j}$$

with coefficients a_0 and b_j satisfying the estimates:

$$|(a_0, b_j)(y, t)| \leq C, \quad |\nabla(a_0, b_j)(y, t)| \leq C(|\nabla \Psi_\rho(y, t)| + |\nabla^2 \Psi_\rho(y, t)|)$$

provided that (3.446) holds. Since

$$(R + \rho)^{-1} = R^{-1} - \rho R^{-2} + O(\rho^2),$$

$$(R + \rho)^{-2} \Delta_{S_1} \rho = R^{-2} \Delta_{S_1} \rho + 2R^{-3} \rho \Delta_{S_1} \rho + O_2 \otimes \nabla_p^2 \rho,$$

we have

$$< H(\Gamma_t) \mathbf{n}_t, \mathbf{n}_t >$$

$$= -\frac{N-1}{R} + \mathcal{B}\rho + (O_1 + O_2) \otimes \nabla_p^2 \rho + O_2. \tag{3.458}$$

Setting $p_0 = -(N-1)/R$, we have

$$< \mu \mathbf{D}(\mathbf{u})\mathbf{n}, \mathbf{n} > - \mathfrak{q} - \sigma \mathcal{B}\rho = h_N(\mathbf{u}, \Psi_\rho)$$

on S_R^T. In view of (3.147), (3.150), and (3.435), $h_N(\mathbf{u}, \Psi_\rho)$ may be defined by letting

$$h_N(\mathbf{u}, \Psi_\rho) = \mathbf{V}_{h,N}(\bar{\nabla}\Psi_\rho)\bar{\nabla}\Psi_\rho \otimes \nabla \mathbf{u} + \sigma \tilde{\mathbf{V}}'_\Gamma(\bar{\nabla}\Psi_\rho)\bar{\nabla}\Psi_\rho \otimes \bar{\nabla}^2 \Psi_\rho, \tag{3.459}$$

where $\mathbf{V}_{h,N}(\bar{\mathbf{k}})$ and $\tilde{\mathbf{V}}'_\Gamma(\bar{\mathbf{k}})$ are functions defined on $\overline{B_R} \times \{\bar{\mathbf{k}} \mid |\bar{\mathbf{k}}| \leq \delta\}$ possessing the estimate:

$$\sup_{|\bar{\mathbf{k}}| \leq \delta} \|(\mathbf{V}_{h,N}(\cdot, \bar{\mathbf{k}}), \partial_{\bar{\mathbf{k}}} \mathbf{V}_{h,N}(\cdot, \bar{\mathbf{k}}))\|_{H^1_\infty(B_R)} \leq C,$$

$$\sup_{|\bar{\mathbf{k}}| \leq \delta} \|(\tilde{\mathbf{V}}'_\Gamma(\cdot, \bar{\mathbf{k}}), \partial_{\bar{\mathbf{k}}} \tilde{\mathbf{V}}'_\Gamma(\cdot, \bar{\mathbf{k}}))\|_{H^1_\infty(B_R)} \leq C,$$

for some constant C.

3.7.2   Local Well-Posedness

In this subsection, we prove the local well-posedness of Eq. (3.449).

Theorem 3.7.2 *Let* $N < q < \infty$, $2 < p < \infty$ *and* $T > 0$. *Assume that* $2/p + N/q < 1$. *Then, there exists a constant* $\epsilon > 0$ *depending on* T *such that if initial data* $\mathbf{u}_0 \in B^{2(1-1/p)}_{q,p}(B_R)$ *and* $\rho_0 \in B^{3-1/p-1q}_{q,p}(S_R)$ *satisfy the smallness condition:*

$$\|\mathbf{u}_0\|_{B^{2(1-1/p)}_{q,p}(B_R)} + \|\rho_0\|_{B^{3-1/p-1/q}_{q,p}(S_R)} \leq \epsilon^2 \tag{3.460}$$

and the compatibility condition:

$$div\, \mathbf{u}_0 = g(\mathbf{u}_0, \rho_0) \quad in\ B_R, \quad \Pi_0(\mu \mathbf{D}(\mathbf{u}_0)\mathbf{n}) = \mu \mathbf{h}'(\mathbf{u}_0, \rho_0) \quad on\ S_R, \qquad (3.461)$$

then problem (3.449) admits unique solutions $(\mathbf{u}, \mathsf{q}, \rho) \in \mathcal{S}_{p,q}((0, T))$ *possessing the estimates:*

$$\sup_{0<t<T} \|\Psi_\rho(\cdot, t)\|_{H^1_\infty(B_R)} \le \delta, \quad E_{p,q,T}(\mathbf{u}, \rho) \le \epsilon.$$

Here and in the following, for $\eta \in \mathbb{R}$ *and* $0 \le a < b \le \infty$, *we set*

$$E_{p,q,T}(\mathbf{u}, \rho) = \|\partial_t \mathbf{u}\|_{L_p((0,T),L_q(B_R))} + \|\mathbf{u}\|_{L_p((0,T),H^2_q(B_R))}$$

$$+ \|\partial_t \rho\|_{L_p((0,T),W^{2-1/q}_q(S_R))} + \|\rho\|_{L_p((0,T),W^{3-1/q}_q(S_R))}$$

$$+ \|\partial_t \rho\|_{L_\infty((0,T),W^{1-1/q}_q(S_R))}.$$

To prove Theorem 3.7.2 the key tool is the maximal L_p–L_q regularity for the following linear equations:

$$\begin{cases} \partial_t \mathbf{u} - \mathrm{Div}\,(\mu \mathbf{D}(\mathbf{u}) - \mathsf{q}\mathbf{I}) = \mathbf{f} & \text{in } B^T_R, \\ \mathrm{div}\,\mathbf{u} = g = \mathrm{div}\,\mathbf{g} & \text{in } B^T_R, \\ \partial_t \rho - \mathbf{n} \cdot \mathbf{P}\mathbf{u} = d & \text{on } S^T_R, \qquad (3.462) \\ \mu \mathbf{D}(\mathbf{u}) - \mathsf{q}\mathbf{I})\mathbf{n} - \sigma(\mathcal{B}\rho)\mathbf{n} = \mathbf{h} & \text{on } S^T_R, \\ (\mathbf{u}, \rho)|_{t=0} = (\mathbf{u}_0, \rho_0) & \text{in } B_R \times S_R. \end{cases}$$

Setting $\mathcal{F}_1 \mathbf{u} = \dfrac{1}{|B_R|} \displaystyle\int_{B_R} \mathbf{u}\, dx$, and $\mathcal{F}_2 \rho = \sigma R^{-2}(N - 1)\rho$, by Theorems 3.4.2 and 3.4.3, we have the following theorem.

Theorem 3.7.3 *let* $1 < p, q < \infty$ *and* $2/p + 1/q \ne 0$. *Let* $T > 0$. *Then, there exists a* $\gamma_0 > 0$ *such that the following assertion holds: Let* $\mathbf{u}_0 \in B^{2(1-1/p)}_{q,p}(B_R)^N$ *and* $\rho_0 \in B^{3-1/p-1/q}_{q,p}(S_R)$ *be initial data for Eq. (3.462) and let* $\mathbf{f}, g, \mathbf{g}, d, \mathbf{h}$ *be given functions in the right side of Eq. (3.462) with*

$$\mathbf{f} \in L_p((0, T), L_q(B_R)^N), \quad e^{-\gamma t} g \in H^1_p(\mathbb{R}, H^1_q(B_R)) \cap H^{1/2}_p(\mathbb{R}, L_q(B_R)),$$

$$e^{-\gamma t}\mathbf{g} \in H^1_p(\mathbb{R}, L_q(B_R)), \quad d \in L_p((0, T), W^{2-1/q}_q(S_R)),$$

$$e^{-\gamma t}\mathbf{h} \in H^1_p(\mathbb{R}, H^1_q(B_R)^N) \cap H^{1/2}_p(\mathbb{R}, L_q(B_R)^N)$$

for all $\gamma \geq \gamma_0$. *Assume that the compatibility condition:* $\operatorname{div} \mathbf{u}_0 = g|_{t=0}$ *in* B_R *holds. In addition, the compatibility condition:* $\boldsymbol{\Pi}_0(\mu \mathbf{D}(\mathbf{u}_0)) = \boldsymbol{\Pi}_0(\mathbf{h}|_{t=0})$ *on* Γ *holds provided* $2/p + 1/q < 1$. *Then, problem* (3.462) *admits unique solutions* $(\mathbf{u}, \mathsf{q}, \rho) \in \mathcal{S}_{p,q}((0, \infty))$ *possessing the estimate:*

$$
\|\partial_t \mathbf{u}\|_{L_p((0,T)L_q(B_R))} + \|\mathbf{u}\|_{L_p((0,T),H_q^2(B_R))} + \|\partial_t \rho\|_{L_p((0,T),W_q^{2-1/q}(S_R))}
$$

$$
+ \|\rho\|_{L_p((0,T),W_q^{3-1/q}(S_R))} \leq Ce^{\gamma T} (\|\mathbf{u}_0\|_{B_{q,p}^{2(1-1/p)}(B_R)} + \|\rho_0\|_{B_{q,p}^{3-1/p-1/q}(S_R)}
$$

$$
+ \|\mathbf{f}\|_{L_p((0,T),L_q(B_R))} + \|e^{-\gamma t}(g, \mathbf{h})\|_{L_p(\mathbb{R}, H_q^1(B_R))} + \|e^{-\gamma t}(g, \mathbf{h})\|_{H_p^{1/2}(\mathbb{R}, L_q(B_R))}
$$

$$
+ \|e^{-\gamma t}\partial_t g\|_{L_p(\mathbb{R}, L_q(B_R))} + \|d\|_{L_p((0,T),W_q^{2-1/q}(S_R))})
\tag{3.463}
$$

for any $\gamma \geq \gamma_0$ *with some constant* C *independent of* γ.

Proof of Theorem 3.7.2 In what follows, using the Banach fixed point argument, we prove Theorem 3.7.2. Let $\mathcal{U}_\epsilon$ be the underlying space defined by

$$
\mathcal{U}_\epsilon = \{(\mathbf{u}, \rho) \mid \mathbf{u} \in H_p^1((0, T), L_q(B_R)^N) \cap L_p((0, T), H_q^2(B_R)^N),
$$

$$
\rho \in H_q^1((0, T), W_q^{2-1/q}(S_R)) \cap L_p((0, T), W_q^{3-1/q}(S_R)),
$$

$$
\mathbf{u}|_{t=0} = \mathbf{u}_0 \quad \text{in } B_R, \quad \rho|_{t=0} = \rho_0 \quad \text{on } S_R,
$$

$$
\sup_{0<t<T} \|\Psi_\rho(\cdot, t)\|_{H_\infty^1(B_R)} \leq \delta, \quad E_{p,q,T}(\mathbf{u}, \rho) \leq \epsilon\}.
\tag{3.464}
$$

We recall that $\Psi_\rho = H_\rho \mathbf{n}$ and H_ρ is a suitable extension of ρ to B_R such that $H_\rho = \rho$ on S_R and (3.98) holds. Since $N < q < \infty$ and $2/p + N/q < 1$, we have

$$
\sup_{t\in(0,T)} \|f(\cdot, t)\|_{H_\infty^1(B_R)} \leq C_{p,q} \sup_{t\in(0,T)} \|f(\cdot, t)\|_{B_{q,p}^{2(1-1/p)}(B_R)},
$$

$$
\sup_{t\in(0,T)} \|f(\cdot, t)\|_{H_\infty^2(B_R)} \leq C_{p,q} \sup_{t\in(0,T)} \|f(\cdot, t)\|_{B_{q,p}^{3-1/p}(B_R)}.
\tag{3.465}
$$

Let $(\mathbf{v}, h) \in \mathcal{U}_\epsilon$. We then consider the linear equations:

$$
\begin{cases}
\partial_t \mathbf{u} - \operatorname{Div}(\mu \mathbf{D}(\mathbf{u}) - \mathsf{q}\mathbf{I}) = \mathbf{f}(\mathbf{v}, \Psi_h) & \text{in } B_R^T, \\
\operatorname{div} \mathbf{u} = g(\mathbf{v}, H_h) = \operatorname{div} \mathbf{g}(\mathbf{v}, \Psi_h) & \text{in } B_R^T, \\
\partial_t \rho - \mathbf{n} \cdot P\mathbf{u} = \tilde{d}(\mathbf{v}, \Psi_h) & \text{on } S_R^T, \\
(\mu \mathbf{D}(\mathbf{u}) - \mathsf{q}\mathbf{I})\mathbf{n} - \sigma(\mathcal{B}\rho)\mathbf{n} = \mathbf{h}(\mathbf{v}, \Psi_h) & \text{on } S_R^T, \\
(\mathbf{u}, \rho)|_{t=0} = (\mathbf{u}_0, \rho_0) & \text{in } B_R \times S_R.
\end{cases}
\tag{3.466}
$$

Noting (3.405) and (3.406), we have

$$\|\mathbf{v}\|_{L_\infty((0,T),B_{q,p}^{2(1-1/p)}(B_R))} \le C\epsilon,$$

$$\|\rho\|_{L_\infty((0,T),B_{q,p}^{3-1/p-1/q}(S_R))} \le C\epsilon,$$

$$\sup_{t\in(0,T)} \|\Psi_h(\cdot,t)\|_{H_\infty^1(B_R)} \le \delta,$$

$$\|\partial_t\mathbf{v}\|_{L_p((0,T),L_q(B_R))} + \|\mathbf{v}\|_{L_p((0,T),H_q^2(B_R))}$$

$$+ \|\partial_t h\|_{L_p((0,T),W_q^{2-1/q}(S_R))} + \|h\|_{L_p((0,T),W_q^{3-1/q}(S_R))}$$

$$+ \|\partial_t h\|_{L_\infty((0,T),W_q^{1-1/q}(S_R))} \le \epsilon.$$

(3.467)

Employing the same argument as in the proof of Theorem 3.6.1, we have

$$\|\mathbf{f}(\mathbf{v},\Psi_h)\|_{L_q(B_R)} \le C\{(\|\mathbf{v}(\cdot,t)\|_{H_q^1(B_R)} + \|\partial_t\Psi_h(\cdot,t)\|_{H_q^1(B_R)})\|\mathbf{v}(\cdot,t)\|_{H_q^1(B_R)}$$

$$+ \|\Psi_h(\cdot,t)\|_{H_q^2(B_R)}(\|\partial_t\mathbf{v}(\cdot,t)\|_{L_q(B_R)} + \|\mathbf{v}(\cdot,t)\|_{H_q^2(B_R)})\}.$$

(3.468)

Putting (3.468), (3.405), (3.406), (3.464), and (3.467), we have

$$\|\mathbf{f}(\mathbf{v},\Psi_h)\|_{L_p((0,mT),L_q(B_R))} \le C\epsilon^2.$$

(3.469)

Here and in the following, C denotes a generic constant independent of ϵ. Moreover, since we choose $\epsilon > 0$ small enough eventually, we may assume that $0 < \epsilon < 1$, and so $\epsilon^s \le \epsilon^2$ for any $s \ge 2$.

We next consider $\tilde{d}(\mathbf{v},\Psi_h)$. In view of (3.450), for $(\mathbf{v},h) \in \mathcal{U}_\epsilon$

$$\xi'(t) = \frac{1}{|B_R|}\int_{B_R} \mathbf{v}(y,t)\,dy + \frac{1}{|B_R|}\int_{B_R} \mathbf{v}(y,t)\mathbf{J}_0(\mathbf{k})\,dy.$$

Here and in the following, $\mathbf{k} = \nabla\Psi_h$. Thus, noting (3.467), we have

$$\sup_{t\in(0,T)} |\xi'(t)| \le \sup_{t\in(0,T)} \|\mathbf{v}(\cdot,t)\|_{L_q(B_R)} \le C\epsilon,$$

and so by (3.467) and (3.98) we have

$$\| <\xi'(t) \mid \nabla_\Gamma' h> \|_{L_\infty((0,T),W_q^{1-1/q}(S_R))}$$

$$\le C\epsilon\|h\|_{L_\infty((0,T),W_q^{2-1/q}(S_R))} \le C\epsilon^2,$$

$$\| <\xi'(t) \mid V_\Gamma(\bar{\mathbf{k}})\bar{\nabla}\Psi_h \otimes \bar{\nabla}\Psi_h> \|_{L_\infty((0,T),W_q^{1-1/q}(S_R))}$$

$$\leq C\epsilon \|h\|^2_{L_\infty((0,T),W_q^{2-1/q}(S_R))} \leq C\epsilon^3,$$

$$\| < \xi'(t) \mid \nabla'_\Gamma h > \|_{L_p((0,T),W_q^{2-1/q}(S_R))}$$

$$\leq C\epsilon \|h\|_{L_p((0,T),W_q^{3-1/q}(S_R))} \leq C\epsilon^2,$$

$$\| < \xi'(t) \mid V_\Gamma(\bar{\mathbf{k}})\bar{\nabla}\Psi_h \otimes \bar{\nabla}\Psi_h > \|_{L_p((0,T),W_q^{2-1/q}(S_R))}$$

$$\leq C\epsilon(1 + \|h\|_{L_\infty((0,T),W_q^{2-1/q}(S_R))})\|h\|_{L_p((0,T),W_q^{3-1/q}(S_R))} \leq C\epsilon^2(1 + \epsilon).$$

Estimating the rest of $d(\mathbf{v}, h)$ given in (3.133) defined by replacing $\mathbf{u}$ and ρ by $\mathbf{v}$ and h in the similar manner to the proof of (3.408), we have

$$\|d(\mathbf{v}, \Psi_h)\|_{L_\infty((0,T),W_q^{1-1/q}(S_R))} \leq C\epsilon^2, \quad \|d(\mathbf{v}, \Psi_h)\|_{L_p((0,T),W_q^{2-1/q}(S_R))} \leq C\epsilon^2.$$

Moreover, by (3.394), (3.405), (3.406), and (3.467), we have

$$\| < \mathbf{v} \mid \nabla'_\Gamma h > \|_{L_\infty((0,T),W_q^{1-1/q}(S_R))}$$

$$\leq C\|\mathbf{v}\|_{L_\infty((0,T),H_q^1(B_R))}\|h\|_{L_\infty((0,T),W_q^{2-1/q}(S_R))} \leq C\epsilon^2;$$

$$\| < \mathbf{v} \mid \nabla'_\Gamma h > \|_{L_p((0,T),W_q^{2-1/q}(S_R))}$$

$$\leq C\{\|\mathbf{v}\|_{L_p((0,T),H_q^2(B_R))}\|h\|_{L_\infty((0,T),W_q^{2-1/q}(S_R))}$$

$$+ \|\mathbf{v}\|_{L_\infty((0,T),H_q^1(B_R))}\|h\|_{L_p((0,T),W_q^{3-1/q}(S_R))}\} \leq C\epsilon^2.$$

And also,

$$\int_{B_R} |\mathbf{v}(y,t)J_0(\mathbf{k})| \, dy \leq |B_R|^{1/q'}\|\mathbf{v}(\cdot,t)\|_{L_q(B_R)}\|\nabla\Psi_h(\cdot,t)\|_{L_\infty(B_R)},$$

and so by (3.467) we have

$$\left\| \int_{B_R} \mathbf{v}J_0 \, dy \right\|_{L_\infty((0,T),W_q^{1-1/q}(S_R))} \leq C\epsilon^2,$$

$$\left\| \int_{B_R} \mathbf{v}J_0 \, dy \right\|_{L_p((0,T),W_q^{2-1/q}(S_R))} \leq C\epsilon^2.$$

Combining estimates obtained above gives

$$\|\tilde{d}(\mathbf{v}, \Psi_h)\|_{L_\infty((0,T),W_q^{1-1/q}(S_R))} \leq C\epsilon^2,$$

$$\|\tilde{d}(\mathbf{v}, \Psi_h)\|_{L_p((0,T),W_q^{2-1/q}(S_R))} \leq C\epsilon^2. \tag{3.470}$$

We now consider $g(\mathbf{v}, \Psi_h)$ and $\mathbf{g}(\mathbf{v}, \Psi_h)$. Let $\tilde{g}(\mathbf{v}, \Psi_h)$ and $\tilde{\mathbf{g}}(\mathbf{v}, \Psi_h)$ be the extension of $g(\mathbf{v}, \Psi_h)$ and $\mathbf{g}(\mathbf{v}, \Psi_h)$ to the whole time line $\mathbb{R}$ given in (3.426) and (3.430), respectively. By (3.427) we have

$$\|e^{-\gamma t}\partial_t\tilde{\mathbf{g}}(\mathbf{v}, \Psi_h)\|_{L_p(\mathbb{R}, L_q(B_R))}$$

$$\leq C\{\|\mathcal{E}_2[\Psi_h]\|_{L_\infty(\mathbb{R}, H_q^2(B_R))}\|e^{-\gamma t}\partial_t\mathcal{E}_1[\mathbf{v}]\|_{L_p(\mathbb{R}, L_q(B_R))}$$

$$+ \|e^{-\gamma t}\mathcal{E}_2[\Psi_h]\|_{L_p(\mathbb{R}, H_q^2(B_R))}\|\mathcal{E}_1[\mathbf{v}]\|_{L_\infty(\mathbb{R}, H_q^1(B_R))}\},$$

and so by (3.417), (3.418), and (3.467), we have

$$\|e^{-\gamma t}\partial_t\tilde{\mathbf{g}}(\mathbf{v}, \Psi_h)\|_{L_p(\mathbb{R}, L_q(B_R))} \leq C\epsilon^2. \tag{3.471}$$

Analogously, by (3.417), (3.418), Lemmas 3.6.2, 3.6.3, (3.464), (3.405), and (3.406), we have

$$\|e^{-\gamma t}\tilde{g}(\mathbf{v}, \Psi_h)\|_{H_p^{1/2}(\mathbb{R}, L_q(B_R))} + \|e^{-\gamma t}\tilde{g}(\mathbf{v}, \Psi_h)\|_{L_p(\mathbb{R}, H_q^1(B_R))} \leq C\epsilon^2. \tag{3.472}$$

We next consider $\mathbf{h}'(\mathbf{v}, \Psi_h)$. Let $\tilde{\mathbf{h}}'(\mathbf{v}, \Psi_h)$ be the extension of $\mathbf{h}'(\mathbf{v}, \Psi_h)$ to the whole time line $\mathbb{R}^N$ given by (3.433). By Lemmas 3.6.2 and 3.6.3, we have

$$\|e^{-\gamma t}\mathbf{h}'(\mathbf{v}, \Psi_h)\|_{H_p^{1/2}(\mathbb{R}, L_q(B_R))} + \|e^{-\gamma t}\tilde{\mathbf{h}}'(\mathbf{v}, \Psi_h)\|_{L_p(\mathbb{R}, H_q^1(\Omega))}$$

$$\leq C(\|\bar{\nabla}\mathcal{E}_2[\Psi_h]\|_{L_\infty(\mathbb{R}, H_q^1(B_R))} + \|\partial_t\bar{\nabla}\mathcal{E}_2[\Psi_h]\|_{L_\infty(\mathbb{R}, L_q(B_R))})$$

$$\times (\|e^{-\gamma t}\mathcal{E}_1[\mathbf{v}]\|_{L_p(\mathbb{R}, H_q^2(B_R))} + \|e^{-\gamma t}\partial_t\mathcal{E}_1[\mathbf{v}]\|_{L_p(\mathbb{R}, L_q(B_R))}).$$

Thus, by (3.417), (3.418), and (3.467), we have

$$\|e^{-\gamma t}\tilde{\mathbf{h}}'(\mathbf{v}, \Psi_h))\|_{H_p^{1/2}(\mathbb{R}, L_q(B_R))} + \|e^{-\gamma t}\tilde{\mathbf{h}}'(\mathbf{v}, \Psi_h)\|_{L_p(\mathbb{R}, H_q^1(B_R))} \leq C\epsilon^2. \tag{3.473}$$

We finally consider $h_N(\mathbf{v}, \Psi_h)$ given in (3.459). Let $\tilde{h}_N(\mathbf{v}, \Psi_h)$ be the extension of $h_N(\mathbf{v}, \Psi_h)$ to the whole time interval $\mathbb{R}$ given by

$$\tilde{h}_N(\mathbf{v}, \Psi_h) = \mathbf{V}_{h,N}(\bar{\nabla}\mathcal{E}_2[\Psi_h])(\bar{\nabla}\mathcal{E}_2[\Psi_h], \nabla\mathcal{E}_1[\mathbf{v}])$$
$$+ \sigma\tilde{\mathbf{V}}'_\Gamma(\bar{\nabla}\mathcal{E}_2[\Psi_h])(\bar{\nabla}\mathcal{E}_2[\Psi_h], \bar{\nabla}^2\mathcal{E}_2[\Psi_h]). \tag{3.474}$$

By Lemmas 3.6.2 and 3.6.3, we have

$$\|e^{-\gamma t}\tilde{h}_N(\mathbf{v}, \Psi_h)\|_{H_p^{1/2}(\mathbb{R}, L_q(B_R))} + \|e^{-\gamma t}\tilde{h}_N(\mathbf{v}, \Psi_h)\|_{L_p(\mathbb{R}, H_q^1(B_R))}$$

$$\leq C(\|\partial_t\bar{\nabla}\mathcal{E}_2[\Psi_h]\|_{L_\infty(\mathbb{R}, L_q(B_R))} + \|\bar{\nabla}\mathcal{E}_2[\Psi_h]\|_{L_\infty(\mathbb{R}, H_q^1(B_R))})$$

$$\times (\|e^{-\gamma t}\partial_t \mathcal{E}_1[\mathbf{v}]\|_{L_p(\mathbb{R},L_q(B_R))} + \|e^{-\gamma t}\mathcal{E}_1[\mathbf{v}]\|_{L_p(\mathbb{R},H_q^2(B_R))}$$

$$+ \|e^{-\gamma t}\partial_t \mathcal{E}_2[\Psi_h]\|_{L_p(\mathbb{R},H_q^2(B_R))} + \|e^{-\gamma t}\mathcal{E}_2[\Psi_h]\|_{L_p(\mathbb{R},H_q^3(B_R))}).$$

Thus, by (3.417), (3.418), and (3.467), we have

$$\|e^{-\gamma t}\tilde{h}(\mathbf{v},\Psi_h)\|_{H_p^{1/2}((\mathbb{R},L_q(B_R))} + \|e^{-\gamma t}\tilde{h}(\mathbf{v},\Psi_h)\|_{L_p((\mathbb{R},H_q^1(B_R))} \le C\epsilon^2. \tag{3.475}$$

Applying Theorem 3.7.3 to Eq. (3.466) and using (3.469)–(3.473), and (3.475), we have

$$\|\mathbf{u}\|_{L_p((0,T),H_q^2(B_R))} + \|\partial_t \mathbf{u}\|_{L_p((0,T),L_q(B_R))}$$

$$+ \|\rho\|_{L_p((0,T),W_q^{3-1/q}(S_R))} + \|\partial_t \rho\|_{L_p((0,T),W_q^{2-1/q}(S_R))} \le Ce^{\gamma T}\epsilon^2 \tag{3.476}$$

for some positive constants C and γ. Moreover, by the third equation in (3.466), we have

$$\|\partial_t \rho\|_{L_\infty((0,T),W_q^{1-1/q}(S_R))}$$
$$\le C(\|\mathbf{u}\|_{L_\infty((0,T),H_q^1(B_R))} + \|\tilde{d}(\mathbf{v},\Psi_h)\|_{L_\infty((0,T),W_q^{1-1/q}(S_R))}), \tag{3.477}$$

which, combined with (3.405), (3.406), and (3.470), yields

$$\|\partial_t \rho\|_{L_\infty((0,T),W_q^{1-1/q}(S_R))} \le C(\|\mathbf{u}_0\|_{B_{q,p}^{2(1-1/p)}(B_R)}$$

$$+ \|\mathbf{u}\|_{L_p((0,T),H_q^2(B_R))} + \|\partial_t \mathbf{u}\|_{L_p((0,T),L_q(B_R))} + C\epsilon^2).$$

Combining this inequality with (3.476) gives

$$E_{p,q,T}(\mathbf{u},\rho) \le Ce^{\gamma T}\epsilon^2 \tag{3.478}$$

for some positive constants C and γ.

Let $\mathcal{P}$ be a map defined by $\mathcal{P}(\mathbf{v},h) = (\mathbf{u},\rho)$, and then choosing ϵ so small that $Ce^{\gamma T}\epsilon \le 1$, by (3.478) we see that $E_{p,q,T}(\mathcal{P}(\mathbf{v},h)) \le \epsilon$, which shows that $\mathcal{P}$ maps $\mathcal{U}_\epsilon$ into itself. Let $(\mathbf{v}_i,h_i)$ $(i=1,2)$ be any two elements of $\mathcal{U}_\epsilon$, and then we have

$$E_{p,q,T}(\mathcal{P}(\mathbf{v}_1,h_1) - \mathcal{P}(\mathbf{v}_2,h_2)) \le Ce^{\gamma T}\epsilon E_{p,q,T}((\mathbf{v}_1,h_1) - (\mathbf{v}_2,h_2))$$

for some positive constants C and γ, where we have used $E_{p,q,T}((\mathbf{v}_i,\rho_i)) \le \epsilon$ for $i=1,2$. Choosing ϵ so small that $Ce^{\gamma T}\epsilon \le 1/2$, we see that $\mathcal{P}$ is a contraction map from $\mathcal{U}_\epsilon$ into itself, and so by the Banach fixed point theorem, there exists a unique $(\mathbf{u},\rho) \in \mathcal{U}_\epsilon$ satisfying $\mathcal{P}(\mathbf{u},\rho) = (\mathbf{u},\rho)$. Thus, $(\mathbf{u},\rho)$ is a required unique solution of Eq. (3.449), which completes the proof of Theorem 3.7.2.

3.7.3 Decay Estimates of Solutions for the Linearized Equations

To prove Theorem 3.7.1 the key tool is decay properties of solutions of the Stokes equations:

$$
\begin{cases}
\partial_t \mathbf{u} - \text{Div}\,(\mu \mathbf{D}(\mathbf{u}) - \mathfrak{p}\mathbf{I}) = \mathbf{f} & \text{in } B_R^T, \\
\text{div}\,\mathbf{u} = g = \text{div}\,\mathbf{g} & \text{in } B_R^T, \\
\partial_t \rho - \mathbf{n} \cdot P\mathbf{u} = d & \text{on } S_R^T, \\
(\mu \mathbf{D}(\mathbf{u}) - \mathfrak{p}\mathbf{I})\mathbf{n} - \sigma(\mathcal{B}\rho)\mathbf{n} = \mathbf{h} & \text{on } S_R^T, \\
(\mathbf{u}, \rho)|_{t=0} = (\mathbf{u}_0, \rho_0) & \text{in } B_R \times S_R.
\end{cases}
\tag{3.479}
$$

Here and in the following, $\mathbf{n} = y/|y| \in S_1$. We will prove the following theorem.

Theorem 3.7.4 *Let $1 < p, q < \infty$ and $2/p + 1/q \neq 0$. Let $\{\mathbf{p}_\ell\}_{\ell=1}^M$ and $\{\varphi_j\}_{j=1}^{N+1}$ be orthogonal basis of $\mathcal{R}_d = \{\mathbf{u} \mid \mathbf{D}(\mathbf{u}) = 0\}$ and $\{\psi \mid \mathcal{B}\psi = 0 \quad \text{on } S_R\} \cup \mathbb{C}$, which are given before Theorem 3.7.1. Then, there exists an $\eta > 0$ such that the following assertion holds: Let $\mathbf{u}_0 \in B_{q,p}^{2(1-1/p)}(B_R)^N$ and $\rho_0 \in B_{q,p}^{3-1/p-1/q}(S_R)$ be initial data for Eq. (3.479) and let $\mathbf{f}, g, \mathbf{g}, d, \mathbf{h}$ be given functions in the right side of Eq. (3.479). Assume that*

$$
\mathbf{f} \in L_p((0, T), L_q(B_R)^N), \quad d \in L_p((0, T), W_q^{2-1/q}(S_R)),
$$

and that there exist $\tilde{g}_\eta, \tilde{\mathbf{g}}_\eta$ and $\tilde{\mathbf{h}}_\eta$ such that $e^{\eta t} g = \tilde{g}_\eta$, $e^{\eta t} \mathbf{g} = \tilde{\mathbf{g}}_\eta$, $e^{\eta t} \mathbf{h} = \tilde{\mathbf{h}}_\eta$ for $t \in (0, T)$, $\text{div}\,\tilde{\mathbf{g}}_\eta = \tilde{g}_\eta$ for $t \in \mathbb{R}$, and

$$
\tilde{g}_\eta \in L_p(\mathbb{R}, H_q^1(B_R)) \cap H_p^{1/2}(\mathbb{R}, L_q(B_R)), \quad \tilde{\mathbf{g}}_\eta \in H_p^1(\mathbb{R}, L_q(B_R)),
$$

$$
\tilde{\mathbf{h}}_\eta \in L_p(\mathbb{R}, H_q^1(B_R)^N) \cap H_p^{1/2}(\mathbb{R}, L_q(B_R)^N).
$$

Assume that the compatibility condition: $\text{div}\,\mathbf{u}_0 = g|_{t=0}$ in B_R holds. In addition, the compatibility condition: $(\mu \mathbf{D}(\mathbf{u}_0)\mathbf{n})_\tau = \mathbf{h}_\tau|_{t=0}$ on Γ holds provided $2/p + 1/q < 1$. Then, problem (3.479) admits unique solutions $(\mathbf{u}, \mathfrak{p}, \rho) \in \mathcal{S}_{p,q}((0, T))$ possessing the estimate:

$$
\mathcal{I}_{p,q,T}(\mathbf{u}, \rho; \eta) \leq C \Big\{ \mathcal{J}_{p,q,T}(\mathbf{u}_0, \rho_0, \mathbf{f}, g, \mathbf{g}, d, \mathbf{h}; \eta)
$$

$$
+ \sum_{\ell=1}^M \Big(\int_0^T (e^{\eta s} |(\mathbf{u}(\cdot, s), \mathbf{p}_\ell)_{B_R}|)^p \, ds \Big)^{1/p}
$$

$$
+ \sum_{j=1}^{N+1} \Big(\int_0^T (e^{\eta s} |(\rho(\cdot, s), \varphi_j)_{S_R}|)^p \, ds \Big)^{1/p} \Big\}
\tag{3.480}
$$

for some constant C independent of η. Here and in the following, we set

$$\mathcal{I}_{p,q,T}(\mathbf{u}, \rho; \eta) = \|e^{\eta t}\mathbf{u}\|_{L_p((0,T),H_q^2(B_R))} + \|e^{\eta t}\partial_t\mathbf{u}\|_{L_p((0,T),L_q(B_R))}$$

$$+ \|e^{\eta t}\rho\|_{L_p((0,T),W_q^{3-1/q}(S_R))} + \|e^{\eta t}\partial_t\rho\|_{L_p((0,T),W_q^{2-1/q}(S_R))};$$

$$\mathcal{J}_{p,q,T}(\mathbf{u}_0, \rho_0, \mathbf{f}, g, \mathbf{g}, d, \mathbf{h}; \eta) = \|\mathbf{u}_0\|_{B_{q,p}^{2(1-1/p)}(B_R)} + \|\rho_0\|_{B_{q,p}^{3-1/p-1/q}(S_R)}$$

$$+ \|e^{\eta t}\mathbf{f}\|_{L_p((0,T),L_q(B_R))} + \|e^{\eta t}d\|_{L_p((0,T),W_q^{2-1/q}(S_R))} + \|\partial_t\mathbf{g}_\eta\|_{L_p(\mathbb{R},L_q(B_R))}$$

$$+ \|(g_\eta, \mathbf{h}_\eta)\|_{H_p^{1/2}(\mathbb{R},L_q(B_R))} + \|(g_\eta, \mathbf{h}_\eta)\|_{L_p(\mathbb{R},H_q^1(B_R))}.$$

To prove Theorem 3.7.3, we first consider the following shifted equations:

$$
\begin{cases}
\partial_t\mathbf{u}_1 + \lambda_1\mathbf{u}_1 - \mathrm{Div}\,(\mu\mathbf{D}(\mathbf{u}_1) - \mathfrak{p}_1\mathbf{I}) = \mathbf{f} & \text{in } B_R^T, \\
\mathrm{div}\,\mathbf{u}_1 = g = \mathrm{div}\,\mathbf{g} & \text{in } B_R^T, \\
\partial_t\rho_1 + \lambda_1\rho_1 - \mathbf{n}\cdot P\mathbf{u}_1 = d & \text{on } S_R^T, \\
(\mu\mathbf{D}(\mathbf{u}_1) - \mathfrak{p}_1\mathbf{I})\mathbf{n} - \sigma(\mathcal{B}\rho_1)\mathbf{n} = \mathbf{h} & \text{on } S_R^T, \\
(\mathbf{u}_1, \rho_1)|_{t=0} = (\mathbf{u}_0, \rho_0) & \text{on } B_R \times S_R.
\end{cases}
\tag{3.481}
$$

For the shifted equation (3.481), we have

Theorem 3.7.5 *Let* $1 < p, q < \infty$ *and* $T > 0$. *Assume that* $2/p + 1/q \neq 1$. *Let* λ_0 *be a constant given in Theorem 3.4.8. Let* $\mathbf{u}_0 \in B_{q,p}^{2(1-1/p)}(B_R)^N$ *and* $\rho_0 \in B_{q,p}^{3-1/p-1/q}(S_R)$ *be initial data for Eq. (3.481) and let* $\mathbf{f}, g, \mathbf{g}, d, \mathbf{h}$ *be given functions in the right side of Eq. (3.481) satisfying the same condition as in Theorem 3.7.4. Moreover, there exist* $\tilde{g}_0$, $\tilde{\mathbf{g}}_0$ *and* $\tilde{\mathbf{h}}_0$ *such that* $g = \tilde{g}_0$, $\mathbf{g} = \tilde{\mathbf{g}}_0$ *and* $\mathbf{h} = \tilde{\mathbf{h}}_0$ *for* $t \in (0, T)$, $\mathrm{div}\,\tilde{\mathbf{g}}_0 = \tilde{g}_0$ *for* $t \in \mathbb{R}$, *and*

$$\tilde{g}_0 \in L_p(\mathbb{R}, H_q^1(B_R)) \cap H_p^{1/2}(\mathbb{R}, L_q(B_R)), \quad \tilde{\mathbf{g}}_0 \in H_p^1(\mathbb{R}, L_q(B_R)),$$

$$\tilde{\mathbf{h}}_0 \in L_p(\mathbb{R}, H_q^1(B_R)^N) \cap H_p^{1/2}(\mathbb{R}, L_q(B_R)^N).$$

Assume that the compatibility condition: $\mathrm{div}\,\mathbf{u}_0 = g|_{t=0}$ *in* B_R *holds. In addition, the compatibility condition:* $(\mu\mathbf{D}(\mathbf{u}_0)\mathbf{n})_\tau = \mathbf{h}_\tau|_{t=0}$ *on* Γ *holds provided* $2/p + 1/q < 1$. *Then, for any* $\lambda_1 > \lambda_0$, *problem (3.481) admits unique solutions* $(\mathbf{u}_1, \mathfrak{p}_1, \rho_1) \in S_{p,q}((0, T))$ *possessing the estimate:*

$$\mathcal{I}_{p,q,T}(\mathbf{u}_1, \rho_1; 0) \leq C\mathcal{J}_{p,q,T}(\mathbf{u}_0, \rho_0, \mathbf{f}, g, \mathbf{g}, d, \mathbf{h}; 0)$$

for some constant C independent of T.

Proof Let $\mathcal{A}_2(\lambda)$, $\mathcal{P}_2(\lambda)$ and $\mathcal{H}_2(\lambda)$ be operators given in Theorem 3.4.8. Let λ_1 and λ_2 be numbers for which $\lambda_1 - \lambda_0 > \lambda_2 > 0$. Then, for any $\lambda \in -\lambda_2 + \Sigma_\epsilon$ we have $\lambda + \lambda_1 \in \lambda_0 + \Sigma_\epsilon$, and so by Theorem 3.4.8 we have

$$\mathcal{R}_{\mathcal{L}(\mathcal{X}_q(\Omega), H_q^{2-j}(\Omega)^N)}(\{(\tau \partial_\tau)^\ell (\lambda^{j/2} \mathcal{A}_2(\lambda + \lambda_1)) \mid \lambda \in -\lambda_2 + \Sigma_{\epsilon_0}\}) \leq r_b;$$

$$\mathcal{R}_{\mathcal{L}(\mathcal{X}_q(\Omega), L_q(\Omega)^N)}(\{(\tau \partial_\tau)^\ell (\nabla \mathcal{P}_2(\lambda + \lambda_1)) \mid \lambda \in -\lambda_2 + \Sigma_{\epsilon_0}\}) \leq r_b;$$

$$\mathcal{R}_{\mathcal{L}(\mathcal{X}_q(\Omega), W_q^{3-k}(\Gamma))}(\{(\tau \partial_\tau)^\ell (\lambda^k \mathcal{H}_2(\lambda + \lambda_1)) \mid \lambda \in -\lambda_2 + \Sigma_{\epsilon_0}\}) \leq r_b.$$

for $\ell = 0, 1$, $j = 0, 1, 2$, and $k = 0, 1$. Thus, we can choose $\gamma = 0$ in the argument on Sect. 3.4.6, and so we have the theorem. □

For any $\eta > 0$, $e^{\eta t}\mathbf{u}_1$, $e^{\eta t}\mathfrak{p}_1$ and $e^{\eta t}\rho_1$ satisfy the equations:

$$\left.\begin{array}{r}
\partial_t(e^{\eta t}\mathbf{u}_1) + (\lambda_1 - \eta)e^{\eta t}\mathbf{u}_1 - \mathrm{Div}\,(\mu\mathbf{D}(e^{\eta t}\mathbf{u}_1) - e^{\eta t}\mathfrak{p}_1\mathbf{I}) = e^{\eta t}\mathbf{f} \\
\mathrm{div}\,e^{\eta t}\mathbf{u}_1 = e^{\eta t}g = \mathrm{div}\,e^{\eta t}\mathbf{g}
\end{array}\right\} \quad \text{in } B_R^T,$$

$$\left.\begin{array}{r}
\partial_t(e^{\eta t}\rho_1) + (\lambda_1 - \eta)e^{\eta t}\rho_1 - \mathbf{n} \cdot P(e^{\eta t}\mathbf{u}_1) = e^{\eta t}d \\
(\mu\mathbf{D}(e^{\eta t}\mathbf{u}_1) - e^{\eta t}\mathfrak{p}_1\mathbf{I})\mathbf{n}) - \sigma(\mathcal{B}(e^{\eta t}\rho_1))\mathbf{n} = e^{\eta t}\mathbf{h}
\end{array}\right\} \quad \text{on } S_R^T,$$

$$(e^{\eta t}\mathbf{u}_1, e^{\eta t}\rho_1)|_{t=0} = (\mathbf{u}_0, \rho_0) \quad \text{on } B_R \times S_R.$$

Given $\eta > 0$, we choose $\lambda_1 > 0$ in such a way that $\lambda_1 - \lambda_0 > \eta > 0$, and then by Theorem 3.7.5, we have the following corollary.

Corollary 3.7.6 *Let $1 < p, q < \infty$, $T > 0$ and $\eta > 0$. Assume that $2/p + 1/q \neq 1$. Let $\mathbf{u}_0 \in B_{q,p}^{2(1-1/p)}(B_R)^N$ and $\rho_0 \in B_{q,p}^{3-1/p-1/q}(S_R)$ be initial data for Eq. (3.481) and let $\mathbf{f}$, g, $\mathbf{g}$, d, $\mathbf{h}$ be given functions in the right side of Eq. (3.481) satisfying the same conditions as in Theorem 3.7.4. Assume that the compatibility condition: $\mathrm{div}\,\mathbf{u}_0 = g|_{t=0}$ in B_R holds. In addition, the compatibility condition: $(\mu\mathbf{D}(\mathbf{u}_0)\mathbf{n})_\tau = \mathbf{h}_\tau|_{t=0}$ on Γ holds provided $2/p + 1/q < 1$. Then, there exists a $\lambda_1 > 0$ such that problem (3.481) admits unique solutions $(\mathbf{u}_1, \mathfrak{p}_1, \rho_1) \in S_{p,q}((0, T))$ possessing the estimate:*

$$\mathcal{I}_{p,q,T}(\mathbf{u}_1, \rho_1; \eta) \leq C\mathcal{J}_{p,q,T}(\mathbf{u}_0, \rho_0, \mathbf{f}, g, \mathbf{g}, d, \mathbf{h}; \eta) \tag{3.482}$$

for some constant C.

We consider solutions $\mathbf{u}$, $\mathfrak{p}$ and ρ of problem (3.479) of the form: $\mathbf{u} = \mathbf{u}_1 + \mathbf{v}$, $\mathfrak{p} = \mathfrak{p}_1 + \mathfrak{q}$ and $\rho = \rho_1 + h$, where $\mathbf{u}_1$, $\mathfrak{p}_1$ and ρ_1 are solutions of the shifted equations (3.481), and then $\mathbf{v}$, $\mathfrak{q}$ and h should satisfy the equations:

$$\left\{\begin{array}{ll}
\partial_t\mathbf{v} - \mathrm{Div}\,(\mu\mathbf{D}(\mathbf{v}) - \mathfrak{q}\mathbf{I}) = -\lambda_1\mathbf{u}_1, \quad \mathrm{div}\,\mathbf{v} = 0 & \text{in } B_R^T, \\
\partial_t h - \mathbf{n} \cdot P\mathbf{v} = -\lambda_1\rho_1 & \text{on } S_R^T, \\
(\mu\mathbf{D}(\mathbf{v}) - \mathfrak{q}\mathbf{I})\mathbf{n} - \sigma(\mathcal{B}h)\mathbf{n} = 0 & \text{on } S_R^T, \\
(\mathbf{v}, h)|_{t=0} = (0, 0) & \text{on } B_R \times S_R.
\end{array}\right. \tag{3.483}$$

Recall the definition of $J_q(\Omega)$ given in (3.163) in Sect. 3.4.5, that is

$$J_q(B_R) = \{\mathbf{f} \in L_q(B_R)^N \mid (\mathbf{f}, \nabla\varphi)_{B_R} = 0 \quad \text{for any } \varphi \in \hat{H}^1_{q',0}(B_R)\}.$$

Recall that

$$H^1_{q,0}(B_R) = \{\varphi \in H^1_q(B_R) \mid \varphi|_{S_R} = 0\},$$

$$\hat{H}^1_{q,0}(B_R) = \{\varphi \in L_{q,\mathrm{loc}}(B_R) \mid \nabla\varphi \in L_q(B_R)^N, \ \varphi|_{S_R} = 0\}.$$

Since $C_0^\infty(B_R)$ is dense in $\hat{H}^1_{q',0}(B_R)$, the necessary and sufficient condition in order that $\mathbf{u} \in J_q(B_R)$ is that $\mathrm{div}\,\mathbf{u} = 0$ in B_R. Let $\psi \in H^1_{q,0}(B_R)$ be a solution of the variational equation:

$$(\nabla\psi, \nabla\varphi)_{B_R} = (\mathbf{u}_1, \nabla\varphi)_{B_R} \quad \text{for any } \varphi \in H^1_{q',0}(B_R), \tag{3.484}$$

and let $\mathbf{w} = \mathbf{u}_1 - \nabla\psi$. Then, $\mathbf{w} \in J_q(B_R)$ and

$$\|\mathbf{w}\|_{L_q(B_R)} + \|\psi\|_{H^1_q(B_R)} \le C\|\mathbf{u}_1\|_{L_q(B_R)}. \tag{3.485}$$

Using $\mathbf{w}$ and ψ, we can rewrite the first equation in (3.483) as follows:

$$\partial_t \mathbf{v} - \mathrm{Div}\,(\mu\mathbf{D}(\mathbf{v}) - (\mathfrak{q} + \lambda_1\psi)\mathbf{I}) = -\lambda_1\mathbf{w}, \quad \mathrm{div}\,\mathbf{v} = 0 \quad \text{in } B_R^T.$$

Thus, in what follows we may assume that

$$\mathbf{u}_1 \in H^1_p((0, T), J_q(B_R)) \cap L_p((0, T), H^2_q(B_R)^N). \tag{3.486}$$

According to the argument in Sect. 3.4.3, we introduce a functional $P(\mathbf{v}, h) \in H^1_q(B_R) + \hat{H}^1_{q,0}(B_R)$ that is a unique solution of the weak Dirichlet problem

$$(\nabla P(\mathbf{v}, h), \nabla\varphi)_{B_R} = (\mathrm{Div}\,(\mu\mathbf{D}(\mathbf{v})) - \nabla\mathrm{div}\,\mathbf{v}, \nabla\varphi)_{B_R} \tag{3.487}$$

for any $\varphi \in \hat{H}^1_{q',0}(B_R)$, subject to

$$P(\mathbf{v}, h) = \mu < \mathbf{D}(\mathbf{v})\mathbf{n}, \mathbf{n} > -\sigma(\mathcal{B}h) - \mathrm{div}\,\mathbf{v} \quad \text{on } S_R. \tag{3.488}$$

And then, to handle problem (3.483) in the semigroup setting, we consider the initial value problem:

$$\begin{cases} \partial_t \mathbf{v} - \mathrm{Div}\,(\mu\mathbf{D}(\mathbf{v}) - P(\mathbf{v}, h)\mathbf{I}) = 0 & \text{in } B_R \times (0, \infty), \\ \partial_t h - \mathbf{n} \cdot P\mathbf{v} = 0 & \text{on } S_R \times (0, \infty), \\ (\mu\mathbf{D}(\mathbf{v}) - \mathfrak{q}\mathbf{I})\mathbf{n} - \sigma(\mathcal{B}h)\mathbf{n} = 0 & \text{on } S_R \times (0, \infty), \\ (\mathbf{v}, h)|_{t=0} = (\mathbf{v}_0, \rho_0) & \text{on } B_R \times S_R. \end{cases} \tag{3.489}$$

Note that $(\mu\mathbf{D}(\mathbf{v}) - P(\mathbf{v}, h)\mathbf{I})\mathbf{n} - \sigma(\mathcal{B}h)\mathbf{n} = 0$ on $S_R \times (0, \infty)$ is equivalent to

$$(\mathbf{D}(\mathbf{v})\mathbf{n})_\tau = 0, \quad \text{div}\,\mathbf{v} = 0 \quad \text{on } S_R \times (0, \infty). \tag{3.490}$$

Defining $\mathcal{H}_q(B_R)$, $\mathcal{D}_q(B_R)$ and $\mathcal{A}_q(\mathbf{v}, h)$ by

$$\mathcal{H}_q(B_R) = \{(\mathbf{v}, h) \mid \mathbf{v} \in J_q(B_R), \quad h \in W_q^{2-1/q}(S_R)\},$$

$$\mathcal{D}_q(B_R) = \{(\mathbf{v}, h) \in \mathcal{H}_q(B_R) \mid \mathbf{v} \in H_q^2(B_R)^N, \quad h \in W_q^{3-1/q}(S_R),$$

$$(\mathbf{D}(\mathbf{v})\mathbf{n})_\tau = 0 \text{ on } S_R\},$$

$$\mathcal{A}_q(\mathbf{v}, h) = (\text{Div}\,(\mathbf{D}(\mathbf{v}) - P(\mathbf{v}, h)\mathbf{I}), -\mathbf{n} \cdot P\mathbf{v}) \quad \text{for } (\mathbf{v}, h) \in \mathcal{D}_q(B_R),$$
$$\tag{3.491}$$

we see that Eq. (3.489) is formulated by

$$\partial_t U = \mathcal{A}_q U \quad (t > 0), \quad U|_{t=0} = U_0 \tag{3.492}$$

with $U = (\mathbf{v}, h) \in \mathcal{D}_q(B_R)$ for $t > 0$ and $U_0 = (\mathbf{u}_0, \rho_0) \in \mathcal{H}_q(B_R)$. According to Theorem 3.4.17, we see that $\mathcal{A}_q$ generates a C^0 semigroup $\{T(t)\}_{t\geq 0}$ on $\mathcal{H}_q(B_R)$. Moreover, if we define

$$\dot{J}_q(B_R) = \{\mathbf{f} \in J_q(B_R) \mid (\mathbf{f}, \mathbf{p}_\ell)_{B_R} = 0 \ (\ell = 1, \dots, M)\};$$

$$\dot{W}_q^\ell(S_R) = \{g \in W_q^\ell(S_R) \mid (g, \varphi_j)_{S_R} = 0 \ (j = 1, \dots, N+1)\};$$

$$\dot{\mathcal{H}}_q(B_R) = \{(\mathbf{f}, g) \mid \mathbf{f} \in \dot{J}_q(B_R), \quad g \in \dot{W}_q^{2-1/q}(S_R)\}; \tag{3.493}$$

$$\|(\mathbf{f}, g)\|_{\mathcal{H}_q} = \|\mathbf{f}\|_{L_q(B_R)} + \|g\|_{W_q^{2-1/q}(S_R)};$$

$$\|(\mathbf{v}, h)\|_{\mathcal{D}_q} = \|\mathbf{v}\|_{H_q^2(B_R)} + \|h\|_{W_q^{3-1/q}(S_R)},$$

then we have

Theorem 3.7.7 *Let $1 < q < \infty$. Then, $\{T(t)\}_{t\geq 0}$ is exponentially stable on $\dot{\mathcal{H}}_q$, that is*

$$\|T(t)(\mathbf{f}, g)\|_{\mathcal{H}_q} \leq Ce^{-\eta_1 t}\|(\mathbf{f}, g)\|_{\mathcal{H}_q} \tag{3.494}$$

for any $t > 0$ and $(\mathbf{f}, g) \in \dot{\mathcal{H}}_q(B_R)$ with some positive constants C and η_1.

Postponing the proof of Theorem 3.7.7 to the next section, we continue to prove Theorem 3.7.3. Let

$$\tilde{\mathbf{u}}_1 = \mathbf{u}_1 - \sum_{\ell-1}^{M}(\mathbf{u}_1(\cdot, t), \mathbf{p}_\ell)_{B_R}\mathbf{p}_\ell, \quad \tilde{\rho}_1 = \rho_1 - \sum_{j=1}^{N+1}(\rho_1(\cdot, t), \varphi_j)_{S_R}\varphi_j,$$

and then $(\tilde{\mathbf{u}}_1, \mathbf{p}_\ell)_{B_R} = 0 \ (\ell = 1, \dots, M)$ and $(\tilde{\rho}_1, \varphi_j)_{S_R} = 0 \ (j = 1, \dots, N+1)$. Moreover, since div $\mathbf{p}_\ell = 0$, by (3.486) we have $\tilde{\mathbf{u}}_1 \in L_p((0, T), \dot{J}_q(B_R))$. Let

$$(\tilde{\mathbf{v}}, \tilde{h})(\cdot, s) = \int_0^s T(s - r)(-\lambda_1 \tilde{\mathbf{u}}_1(\cdot, r), -\lambda_1 \tilde{\rho}_1(\cdot, r)) \, dr,$$

and then by the Duhamel principle, $\tilde{\mathbf{v}}$ and $\tilde{h}$ satisfy the equations:

$$\begin{cases} \partial_t \tilde{\mathbf{v}} - \mathrm{Div} \, (\mu \mathbf{D}(\tilde{\mathbf{v}}) - P(\tilde{\mathbf{v}}, \tilde{h})\mathbf{I}) = -\lambda_1 \tilde{\mathbf{u}}_1, & \mathrm{div} \, \tilde{\mathbf{v}} = 0 & \text{in } B_R^T, \\ \partial_t \tilde{h} - \mathbf{n} \cdot P\tilde{\mathbf{v}} = -\lambda_1 \tilde{\rho}_1 & & \text{on } S_R^T, \\ (\mu \mathbf{D}(\tilde{\mathbf{v}}) - P(\tilde{\mathbf{v}}, \tilde{h})\mathbf{I})\mathbf{n} - \sigma(\mathcal{B}\tilde{h})\mathbf{n} = 0 & & \text{on } S_R^T, \\ (\tilde{\mathbf{v}}, \tilde{h})|_{t=0} = (0, 0) & \text{on } B_R \times S_R. \end{cases} \tag{3.495}$$

By (3.494),

$$\|(\tilde{\mathbf{v}}, \tilde{h})(\cdot, s)\|_{\mathcal{H}_q} \le C \int_0^s e^{-\eta_1(s-r)} \|(\tilde{\mathbf{u}}_1(\cdot, r), \tilde{\rho}_1(\cdot, r))\|_{\mathcal{H}_q} \, dr$$

$$\le C \Big(\int_0^s e^{-\eta_1(s-r)} \, dr \Big)^{1/p'} \Big(\int_0^s e^{-\eta_1(s-r)} \|(\tilde{\mathbf{u}}_1(\cdot, r), \tilde{\rho}_1(\cdot, r))\|_{\mathcal{H}_q}^p \, dr \Big)^{1/p}.$$

Choosing $\eta > 0$ smaller if necessary, we may assume that $0 < \eta p < \eta_1$ without loss of generality. Thus, by the inequality above we have

$$\int_0^t (e^{\eta s} \|(\tilde{\mathbf{v}}, \tilde{h})(\cdot, s)\|_{\mathcal{H}_q})^p \, ds$$

$$\le C \int_0^t \Big(\int_0^s e^{\eta s p} e^{-\eta_1(s-r)} \|(\mathbf{u}_1(\cdot, r), \rho_1(\cdot, r))\|_{\mathcal{H}_q}^p \, dr \Big) \, ds$$

$$= \int_0^t \Big(\int_0^s e^{-(\eta_1 - p\eta)(s-r)} (e^{\eta r} \|(\mathbf{u}_1(\cdot, r), \rho_1(\cdot, r))\|_{\mathcal{H}_q})^p \, dr \Big) \, ds$$

$$= \int_0^t (e^{\eta r} \|(\mathbf{u}_1(\cdot, r), \rho_1(\cdot, r))\|_{\mathcal{H}_q})^p \Big(\int_r^t e^{-(\eta_1 - p\eta)(s-r)} \, ds \Big) \, dr$$

$$\le (\eta_1 - p\eta)^{-1} \int_0^T (e^{\eta r} \|(\mathbf{u}_1(\cdot, r), \rho_1(\cdot, r))\|_{\mathcal{H}_q})^p \, dr,$$

which, combined with (3.482), leads to

$$\|e^{\eta s}(\tilde{\mathbf{v}}, \tilde{h})\|_{L_p((0,t), \mathcal{H}_q)} \le C \mathcal{J}_{p,q,T}(\mathbf{u}_0, \rho_0, \mathbf{f}, g, \mathbf{g}, d, \mathbf{h}; \eta) \tag{3.496}$$

for any $t \in (0, T)$. If $\tilde{\mathbf{w}}$ and $\tilde{k}$ satisfy the shifted equations:

$$
\begin{cases}
\partial_t \tilde{\mathbf{w}} + \lambda_1 \tilde{\mathbf{w}} - \mathrm{Div}\,(\mu \mathbf{D}(\tilde{\mathbf{w}}) - P(\tilde{\mathbf{w}}, \tilde{k})\mathbf{I}) = \mathbf{f}, & \mathrm{div}\,\tilde{\mathbf{w}} = 0 \quad \text{in } B_R^T, \\
\partial_t \tilde{k} + \lambda_1 \tilde{k} - \mathbf{n} \cdot P\tilde{\mathbf{w}} = d & \text{on } S_R^T, \\
(\mu \mathbf{D}(\tilde{\mathbf{w}}) - P(\tilde{\mathbf{w}}, \tilde{k})\mathbf{I})\mathbf{n} - \sigma(\mathcal{B}\tilde{k})\mathbf{n} = 0 & \text{on } S_R^T, \\
(\tilde{\mathbf{w}}, \tilde{k})|_{t=0} = (0, 0) \quad \text{on } B_R \times S_R,
\end{cases}
$$

where we have set

$$
\mathbf{f} = -\lambda_1 \tilde{\mathbf{u}}_1 + \lambda_1 \tilde{\mathbf{v}}, \quad d = -\lambda_1 \tilde{\rho}_1 + \lambda_1 \tilde{h},
$$

by (3.482) and (3.496), we have

$$
\mathcal{I}_{p,q,T}(\tilde{\mathbf{w}}, \tilde{k}; \eta) \le C \mathcal{J}_{p,q,T}(\mathbf{u}_0, \rho_0, \mathbf{f}, f_d, \mathbf{f}_d, g, \mathbf{h}; \eta).
$$

But, noting that $\tilde{\mathbf{v}}$ and $\tilde{h}$ satisfy Eq. (3.495), by the uniqueness we see that $\tilde{\mathbf{w}} = \tilde{\mathbf{v}}$ and $\tilde{k} = \tilde{h}$ for $t \in (0, T)$, and so we have

$$
\mathcal{I}_{p,q,T}(\tilde{\mathbf{v}}, \tilde{h}; \eta) \le C \mathcal{J}_{p,q,T}(\mathbf{u}_0, \rho_0, \mathbf{f}, f_d, \mathbf{f}_d, g, \mathbf{h}; \eta). \tag{3.497}
$$

Let

$$
\mathbf{v} = \tilde{\mathbf{v}} - \lambda_1 \sum_{\ell=1}^{M} \int_0^t (\mathbf{u}_1(\cdot, s), \mathbf{p}_\ell)_{B_R}\, ds\, \mathbf{p}_\ell,
$$

$$
h = \tilde{h} - \lambda_1 \sum_{j=1}^{N+1} \int_0^t (\rho_1(\cdot, s), \varphi_j)_{S_R}\, ds\, \varphi_j.
$$

In this case,

$$
P(\mathbf{v}, h) = P(\tilde{\mathbf{v}}, \tilde{h}) + \lambda_1 (N-1) R^{-2} \sigma \int_0^t (\rho_1(\cdot, s), \varphi_1)_{S_R}\, ds\, \varphi_1.
$$

In fact, letting

$$
\mathcal{C} = \lambda_1 (N-1) R^{-2} \sigma \int_0^t (\rho_1(\cdot, s), \varphi_1)_{S_R}\, ds\, \varphi_1
$$

for notational simplicity, we have

$$
(\nabla(P(\mathbf{v}, h) - (P(\tilde{\mathbf{v}}, \tilde{h}) + \mathcal{C})), \nabla \psi)_{B_R} = 0
$$

for any $\psi \in \hat{H}^1_{q',0}(B_R)$, because $\nabla \varphi_1 = 0$. Moreover, on S_R we have

$$P(\mathbf{v}, h) - (P(\tilde{\mathbf{v}}, \tilde{h}) + C) = 0$$

because $\mathbf{D}(\mathbf{p}_\ell) = 0$, $\operatorname{div} \mathbf{p}_\ell = 0$, and

$$\mathcal{B}h = \mathcal{B}\tilde{h} - (N-1)R^{-2}\lambda_1 \int_0^t (\rho_1(\cdot, s), \varphi_1)_{S_R} \, ds \, \varphi_1.$$

Thus, we have $P(\mathbf{v}, h) = P(\tilde{\mathbf{v}}, \tilde{h}) + C$.

By (3.495), we have

$$\partial_t \mathbf{v} - \operatorname{Div}(\mu \mathbf{D}(\mathbf{v}) - P(\mathbf{v}, h)\mathbf{I}) = -\lambda_1 \mathbf{u}_1, \quad \operatorname{div} \mathbf{v} = 0 \quad \text{in } B_R^T,$$

$$(\mu \mathbf{D}(\mathbf{v}) - P(\mathbf{v}, h)\mathbf{I})\mathbf{n} - \sigma(\mathcal{B}h)\mathbf{n} = 0 \qquad \qquad \text{on } S_R^T.$$

Recall that $\mathbf{p}_\ell \cdot \mathbf{n}|_{S_R} = 0$ for $\ell = N+1, \ldots, M$. Moreover, recalling that $\mathbf{p}_\ell = |B_R|^{-1}\mathbf{e}_\ell$ ($\ell = 1, \ldots, N$), we have

$$P\mathbf{p}_\ell = |B_R|^{-1}\left(\mathbf{e}_\ell - |B_R|^{-1}\int_{B_R} \mathbf{e}_\ell \, dy\right) = 0,$$

and therefore,

$$\partial_t h - \mathbf{n} \cdot P\mathbf{v} = \partial_t \tilde{h} - \mathbf{n} \cdot P\tilde{\mathbf{v}} - \lambda_1 \sum_{j=1}^{N+1}(\rho_1(\cdot, t), \varphi_j)_{S_R} \varphi_j = -\lambda_1\rho_1 \quad \text{on } S_R^T.$$

Summing up, we have proved that $\mathbf{v}$, $q = P(\mathbf{v}, h)$, and h satisfy the Eq. (3.483).

By (3.497), we have

$$\|e^{\eta t}\partial_t(\mathbf{v}, h)\|_{L_p((0,T),\mathcal{H}_q)} \leq C\mathcal{J}_{p,q,T}(\mathbf{u}_0, \rho_0, \mathbf{f}, f_d, \mathbf{f}_d, g, \mathbf{h}; \eta) \qquad (3.498)$$

for any $t \in (0, T)$.

To estimate $\|e^{\eta t}(\mathbf{v}, h)\|_{L_p((0,T),\mathcal{D}_q)}$, we use the following lemma.

Lemma 3.7.8 *Let $1 < q < \infty$. Let $\mathbf{u} \in H_q^2(B_R)^N \cap J_q(B_R)$ and $\rho \in W_q^{3-1/q}(S_R)$ satisfy the equations:*

$$\begin{cases} -\operatorname{Div}(\mu \mathbf{D}(\mathbf{u}) - P(\mathbf{u}, \rho)\mathbf{I}) = \mathbf{f} & \text{in } B_R, \\ \mathbf{n} \cdot P\mathbf{u} = g & \text{on } S_R, \\ (\mu \mathbf{D}(\mathbf{u}) - P(\mathbf{u}, \rho)\mathbf{I})\mathbf{n} - \sigma(\mathcal{B}\rho)\mathbf{n} = 0 & \text{on } S_R. \end{cases} \qquad (3.499)$$

Then, there exists a constant $C > 0$ such that

$$\|(\mathbf{u}, \rho)\|_{\mathcal{D}_q} \leq C \Big\{ \|(\mathbf{f}, g)\|_{\mathcal{H}_q} + \sum_{\ell=1}^{M} |(\mathbf{u}, \mathbf{p}_\ell)_{B_R}| + \sum_{j=1}^{N+1} |(\rho, \varphi_j)_{S_R}| \Big\}. \tag{3.500}$$

Postponing the proof of Lemma 3.7.8 to the next section, we continue to prove Theorem 3.7.3. By (3.483), $\mathbf{v}$ and h satisfy the elliptic equations:

$$-\operatorname{Div}(\mu \mathbf{D}(\mathbf{v}) - P(\mathbf{v}, h)\mathbf{I}) = -\lambda_1 \mathbf{u}_1 - \partial_t \mathbf{v}, \quad \operatorname{div} \mathbf{v} = 0 \qquad \text{in } B_R,$$

$$\mathbf{n} \cdot P\mathbf{v} = \lambda_1 \rho_1 + \partial_t h \qquad \text{on } S_R,$$

$$(\mu \mathbf{D}(\mathbf{v} - P(\mathbf{v}, h)\mathbf{I})\mathbf{n} - \sigma(\mathcal{B}h)\mathbf{n} = 0 \qquad \text{on } S_R,$$

and therefore, applying Lemma 3.7.8 and using (3.498) yield that

$$\|e^{\eta t}\mathbf{v}\|_{L_p((0,T), H_q^2(B_R))} + \|e^{\eta t} h\|_{L_p((0,T), W_q^{3-1/q}(S_R))}$$

$$\leq C \Big\{ \mathcal{J}_{p,q,T}(\mathbf{u}_0, \rho_0, \mathbf{f}, f_d, \mathbf{f}_d, g, \mathbf{h}; \eta)$$

$$+ \sum_{\ell=1}^{M} \Big(\int_0^T (e^{\eta s} |(\mathbf{v}(\cdot, s), \mathbf{p}_\ell)_{B_R}|)^p \, ds \Big)^{1/p} \tag{3.501}$$

$$+ \sum_{j=1}^{M} \Big(\int_0^T (e^{\eta s} |(h(\cdot, s), \varphi_j)_{S_R}|)^p \, ds \Big)^{1/p} \Big\}.$$

Let $\mathbf{u} = \mathbf{u}_1 + \mathbf{v}$, $\mathfrak{p} = \mathfrak{p}_1 + \mathfrak{q}$ and $\rho = \rho_1 + h$. By (3.481) and (3.483), $\mathbf{u}$, $\mathfrak{p}$ and ρ satisfy the equations (3.479). Since

$$\Big(\int_0^T (e^{\eta s} |(\mathbf{v}(\cdot, s), \mathbf{p}_\ell)_{B_R}|)^p \, ds \Big)^{1/p}$$

$$\leq \Big(\int_0^T (e^{\eta s} |(\mathbf{u}(\cdot, s), \mathbf{p}_\ell)_{B_R}|)^p \, ds \Big)^{1/p} + C J_{\eta, T},$$

$$\Big(\int_0^T (e^{\eta s} |(h(\cdot, s), \varphi_j)_{S_R}|)^p \, ds \Big)^{1/p} \tag{3.502}$$

$$\leq \Big(\int_0^T (e^{\eta s} |(\rho(\cdot, s), \varphi_j)_{S_R}|)^p \, ds \Big)^{1/p} + C J_{\eta, T},$$

where we have set $J_{\eta, T} = \mathcal{J}_{p,q,T}(\mathbf{u}_0, \rho_0, \mathbf{f}, f_d, \mathbf{f}_d, g, \mathbf{h}; \eta)$, as follows from (3.482), by (3.498), (3.501), and (3.502), we see that $\mathbf{u}$, $\mathfrak{p}$ and ρ satisfy the inequality (3.480).

3.7.4 Exponential Stability of Continuous Analytic Semigroup Associated with Eq. (3.489)

In this subsection, we shall prove Theorem 3.7.7 stated in the previous subsection. For this purpose, we consider the equations:

$$(\lambda \mathbf{I} - \mathcal{A}_q)U = F \tag{3.503}$$

for $F = (\mathbf{f}, g) \in \dot{J}_q(B_R)$ and $U = (\mathbf{v}, h) \in \mathcal{D}_q(B_R) \cap \dot{J}_q(B_R)$, which is the resolvent problem corresponding to Eq. (3.492). Here, $\mathbf{I}$ is the identity operator, $\dot{J}_q(B_R)$ is the space defined in (3.493), and $\mathcal{D}_q(B_R)$ and $\mathcal{A}_q$ are the domain and the operator defined in (3.491). Since $\mathcal{R}$ boundedness implies the usual boundedness of operator families, by Theorem 3.4.8 and the observation in Sect. 3.4.3, we have the following theorem.

Theorem 3.7.9 *Let* $1 < q < \infty$ *and* $0 < \epsilon_0 < \pi/2$. *Then, there exists a* $\lambda_0 > 0$ *such that for any* $\lambda \in \Sigma_{\epsilon_0, \lambda_0}$ *and* $F \in \mathcal{H}_q(B_R)$, *Eq.* (3.503) *admits a unique solution* $U \in \mathcal{D}_q(B_R)$ *possessing the estimate:*

$$|\lambda| \|U\|_{\mathcal{H}_q} + \|U\|_{\mathcal{D}_q} \le C \|F\|_{\mathcal{H}_q} \tag{3.504}$$

for some constant $C > 0$. *Here,* $\mathcal{H}_q(B_R)$ *is the space defined in* (3.491).

Our task in this subsection is to prove the following theorem.

Theorem 3.7.10 *Let* $1 < q < \infty$ *and let* $\mathbb{C}_+ = \{\lambda \in \mathbb{C} \mid Re\, \lambda \ge 0\}$. *Then, for any* $\lambda \in \mathbb{C}_+$ *and* $F \in \dot{\mathcal{H}}_q(B_R)$, *Eq.* (3.503) *admits a unique solution* $U \in \mathcal{D}_q(B_R) \cap \dot{\mathcal{H}}_q(B_R)$ *possessing the estimate* (3.504).

In the following, we shall prove Theorem 3.7.10. We start with the following lemma.

Lemma 3.7.11 *Let* $1 < q < \infty$, *let* $\lambda \in \mathbb{C} \setminus (-\infty, 0)$ *and* $F = (\mathbf{f}, g) \in \dot{\mathcal{H}}_q(B_R)$. *If* $U = (\mathbf{v}, h) \in \mathcal{D}_q(B_R)$ *satisfies Eq.* (3.503), *then* $U = (\mathbf{v}, h)$ *belongs to* $\dot{\mathcal{H}}_q(B_R)$.

Proof First we prove that $\mathbf{v} \in \dot{J}_q(B_R)$. By (3.503), we know that $\mathbf{v} \in J_q(B_R) \cap H_q^2(B_R)^N$ satisfies the equations:

$$\lambda \mathbf{v} - \mathrm{Div}\,(\mu \mathcal{D}(\mathbf{v}) - P(\mathbf{v}, h)\mathbf{I}) = \mathbf{f}, \quad \mathrm{div}\,\mathbf{v} = 0 \quad \text{in } B_R,$$

$$(\mu \mathcal{D}(\mathbf{v}) - P(\mathbf{v}, \mathfrak{q})\mathbf{I})\mathbf{n} - \sigma(\mathcal{B}h)\mathbf{n} = 0 \quad \text{on } S_R.$$

Since $F = (\mathbf{f}, g) \in \dot{\mathcal{H}}_q(B_R)$, we know that $\mathbf{f} \in J_q(B_R)$ and $(\mathbf{f}, \mathbf{p}_\ell)_{B_R} = 0$ for $\ell = 1, \ldots, M$, and so by the divergence theorem of Gauß we have

$$0 = (\mathbf{f}, \mathbf{p}_\ell)_{B_R} = (\lambda \mathbf{v} - \mathrm{Div}\,(\mu(\mathbf{D}(\mathbf{v}) - P(\mathbf{v}, h)\mathbf{I}), \mathbf{p}_\ell)_{B_R}$$

$$= \lambda(\mathbf{v}, \mathbf{p}_\ell)_{B_R} - \sigma(\mathcal{B}h, \mathbf{n} \cdot \mathbf{p}_\ell)_{S_R} + \frac{\mu}{2}(\mathbf{D}(\mathbf{v}), \mathbf{D}(\mathbf{p}_\ell))_{B_R} - (P(\mathbf{v}, h), \mathrm{div}\,\mathbf{p}_\ell)_{B_R}.$$

We see that

$$(\mathcal{B}h, \mathbf{n} \cdot \mathbf{p}_\ell)_{S_R} = 0 \quad (\ell = 1, \ldots, M). \tag{3.505}$$

In fact, recalling that $\mathbf{p}_\ell = |B_R|^{-1}\mathbf{e}_\ell$ $(\ell = 1, \ldots, N)$ and $\mathbf{n} = y/|y| \in S_1$, we have

$$
\begin{aligned}
(\mathcal{B}h, \mathbf{n} \cdot \mathbf{p}_\ell)_{S_R} &= R^{-1}|B_R|^{-1}(h, \mathcal{B}y_\ell)_{S_R} \\
&= R^{-3}|B_R|^{-1}(h, (N-1+\Delta_{S_1})y_\ell)_{S_R} = 0
\end{aligned}
$$

for $\ell = 1, \ldots, N$. Moreover, $\mathbf{p}_\ell \cdot \mathbf{n} = 0$ for $\ell = N+1, \ldots, M$ because $\mathbf{p}_\ell$ $(\ell = N+1, \ldots, M)$ are equal to $c_{ij}(x_i\mathbf{e}_j - x_j\mathbf{e}_i)$ for some i and j and constant c_{ij}, and therefore $(\mathcal{B}h, \mathbf{n} \cdot \mathbf{p}_\ell)_{S_R} = 0$ for $\ell = N+1, \ldots, M$.

Since $\mathbf{D}(\mathbf{p}_\ell) = 0$ and $\operatorname{div} \mathbf{p}_\ell = 0$, we have $\lambda(\mathbf{v}, \mathbf{p}_\ell)_{B_R} = 0$, which, combined with $\lambda \neq 0$, leads to $(\mathbf{v}, \mathbf{p}_\ell)_{B_R} = 0$, that is, $\mathbf{v} \in \dot{J}_q(B_R) \cap \mathcal{D}_q(B_R)$.

Next, we prove that $h \in \dot{W}_q^{3-1/q}(S_R)$. We know that $h \in W_q^{3-1/q}(S_R)$ and g satisfies the equation:

$$\lambda h - \mathbf{n} \cdot P\mathbf{v} = g \quad \text{on } S_R.$$

Since $(g, \varphi_j)_{S_R} = 0$, by the divergence theorem of Gauß we have

$$
\begin{aligned}
0 - (g, \varphi_j)_{S_R} &= \lambda(h, \varphi_j)_{S_R} - (P\mathbf{v} \cdot \mathbf{n}, \varphi_j)_{S_R} \\
&= \lambda(h, \varphi_j)_{S_R} - \int_{B_R} \operatorname{div}((P\mathbf{v})\varphi_j) \, dx.
\end{aligned}
$$

Since $\operatorname{div} P\mathbf{v} = \operatorname{div} \mathbf{v} = 0$ and since $\partial_\ell\varphi_j$ are constants, we have

$$
\begin{aligned}
&\int_{B_R} \operatorname{div}((P\mathbf{v})\varphi_j) \, dx \\
&= \int_{B_R} (\operatorname{div}(P\mathbf{v}))\varphi_j \, dx + \sum_{\ell=1}^{N+1} \int_{B_R} \left(v_\ell - |B_R|^{-1}\int_{B_R} v_\ell \, dy\right)(\partial_\ell\varphi_j) \, dx \\
&= 0,
\end{aligned}
\tag{3.506}
$$

Thus, $\lambda(h, \varphi_j)_{S_R} = 0$, which, combined with $\lambda \neq 0$, leads to $(h, \varphi_j)_{S_R} = 0$. Therefore, we have $h \in \dot{W}_q^{3-1/q}(S_R)$. This completes the proof of Lemma 3.7.11.

□

Combining Theorem 3.7.9 and Lemma 3.7.11, we have the following Corollary.

Corollary 3.7.12 *Let* $1 < q < \infty$ *and* $0 < \epsilon_0 < \pi/2$. *Then, there exists a positive constant* λ_0 *such that for any* $\lambda \in \Sigma_{\epsilon_0, \lambda_0}$ *and* $(\mathbf{f}, g) \in \mathcal{H}_q(B_R)$, *Eq.* (3.503) *admits a unique solution* $(\mathbf{v}, h) \in \mathcal{D}_q \cap \mathcal{H}_q(B_R)$ *possessing the estimates* (3.504).

In view of Corollary 3.7.12, in order to prove Theorem 3.7.10 it suffices to prove the following theorem.

Theorem 3.7.13 *Let* $1 < q < \infty$ *and let* λ_0 *be the same positive number as in Corollary 3.7.12. Let*

$$Q_{\lambda_0} = \{\lambda \in \mathbb{C} \mid Re\, \lambda \geq 0,\ |\lambda| \leq \lambda_0\}.$$

Then, for any $\lambda \in Q_{\lambda_0}$ *and* $(\mathbf{f}, g) \in \mathcal{H}_q(B_R)$, *Eq.* (3.503) *admits a unique solution* $(\mathbf{v}, h) \in \mathcal{D}_q(B_R) \cap \mathcal{H}_q(B_R)$ *possessing the estimate:*

$$\|(\mathbf{v}, h)\|_{\mathcal{D}_q} \leq C \|(\mathbf{f}, g)\|_{\mathcal{H}_q} \tag{3.507}$$

with some constant C *independent of* $\lambda \in Q_{\lambda_0}$.

Proof We write $\dot{\mathcal{D}}_q = \mathcal{D}_q(B_R) \cap \dot{\mathcal{H}}_q(B_R)$ for the sake of simplicity. We first observe that

$$\mathcal{A}_q \dot{\mathcal{D}}_q \subset \dot{\mathcal{H}}_q(B_R). \tag{3.508}$$

In fact, for $(\mathbf{v}, h) \in \dot{\mathcal{D}}_q$, we set $\mathcal{A}_q(\mathbf{v}, h) = (\mathbf{f}, g)$, that is $\mathrm{Div}\,(\mu(\mathbf{D}(\mathbf{v}) - P(\mathbf{v}, h)\mathbf{I}) = \mathbf{f}$ in B_R and $\mathbf{n} \cdot P\mathbf{v} = g$ on S_R. For any $\varphi \in \hat{H}^1_{q',0}(B_R)$, by (3.487) and the fact that $\mathrm{div}\,\mathbf{v} = 0$, we have

$$(\mathbf{f}, \nabla\varphi)_{B_R} = (\mathrm{Div}\,(\mu\mathbf{D}(\mathbf{v}) - P(\mathbf{v}, h)\mathbf{I}), \nabla\varphi)_{B_R} = (\nabla\mathrm{div}\,\mathbf{v}, \nabla\varphi)_{B_R} = 0,$$

which implies $\mathbf{f} \in J_q(B_R)$.

Next, we observe that

$$(\mathbf{f}, \mathbf{p}_\ell)_{B_R} = (\mathrm{Div}\,(\mu\mathbf{D}(\mathbf{v}) - P(\mathbf{v}, h)\mathbf{I}), \mathbf{p}_\ell)_{B_R}$$

$$= \sigma(\mathcal{B}h, \mathbf{n} \cdot \mathbf{p}_\ell)_{S_R} - \frac{\mu}{2}(\mathbf{D}(\mathbf{v}), \mathbf{D}(\mathbf{p}_\ell))_{B_R} + (P(\mathbf{v}, h), \mathrm{div}\,\mathbf{p}_\ell)_{B_R}.$$

Thus, by (3.505) and the facts that $\mathbf{D}(\mathbf{p}_\ell) = 0$ and $\mathrm{div}\,\mathbf{p}_\ell = 0$, we have $(\mathbf{f}, \mathbf{p}_\ell)_{B_R} = 0$, and so, $\mathbf{f} \in \dot{J}_q(B_R)$.

Finally, by (3.506) we have

$$(g, \varphi_j)_{S_R} = (\mathbf{n} \cdot P\mathbf{v}, \varphi_j)_{S_R} = \int_{B_R} \mathrm{div}\,((P\mathbf{v})\varphi_j)\, dx = 0,$$

which leads to $g \in \dot{W}_q^{3-1/q}(S_R)$. This completes the proof of (3.508).

In view of Corollary 3.7.12, $(\lambda_0 I - \mathcal{A}_q)^{-1}$ exists as a bounded linear operator from $\dot{\mathcal{H}}_q(B_R)$ onto $\dot{\mathcal{D}}_q$, and then, the equation (3.503) is rewritten as

$$(\mathbf{f}, g) = (\lambda I - \mathcal{A}_q)(\mathbf{v}, h) = (\lambda - \lambda_0)(\mathbf{v}, h) + (\lambda_0 I - \mathcal{A}_q)(\mathbf{v}, h)$$

$$= (I + (\lambda - \lambda_0)(\lambda_0 I - \mathcal{A}_q)^{-1})(\lambda_0 I - \mathcal{A}_q)(\mathbf{v}, h).$$

If $(I + (\lambda - \lambda_0)(\lambda_0 I - \mathcal{A}_q)^{-1})^{-1}$ exists as a bounded linear operator from $\dot{\mathcal{H}}_q(B_R)$ into itself, then we have

$$(\mathbf{v}, h) = (\lambda_0 I - \mathcal{A}_q)^{-1}(I + (\lambda - \lambda_0)(\lambda_0 I - \mathcal{A}_q)^{-1})^{-1}(\mathbf{f}, g). \tag{3.509}$$

Thus, our task is to prove the existence of the inverse operator $(I + (\lambda - \lambda_0)(\lambda_0 I - \mathcal{A}_q)^{-1})^{-1}$. Since $H_q^2(B_R)^N$ and $W_q^{3-1/q}(S_R)$ are compactly embedded into $L_q(B_R)^N$ and $W_q^{2-1/q}(S_R)$, respectively, as follows from the Rellich compact embedding theorem, $(\lambda_0 I - \mathcal{A}_q)^{-1}$ is a compact operator from $\dot{\mathcal{H}}_q$ into itself. Thus, in view of Riesz-Schauder theory, in order to prove the existence of the inverse operator $(I + (\lambda - \lambda_0)(\lambda_0 I - \mathcal{A}_q)^{-1})^{-1}$ it suffices to prove that the kernel of the map $I + (\lambda - \lambda_0)(\lambda_0 I - \mathcal{A}_q)^{-1}$ is trivial. Thus, let $(\mathbf{f}, g)$ be an element in $\dot{\mathcal{H}}_q(B_R)$ such that

$$(I + (\lambda - \lambda_0)(\lambda_0 I - \mathcal{A}_q)^{-1})(\mathbf{f}, g) = (0, 0). \tag{3.510}$$

Our task is to prove that $(\mathbf{f}, g) = (0, 0)$. Since $(\mathbf{f}, g) = -(\lambda - \lambda_0)(\lambda_0 I - \mathcal{A}_q)^{-1}(\mathbf{f}, g) \in \dot{\mathcal{D}}_q$, we have $(\lambda_0 I - \mathcal{A}_q)(\mathbf{f}, g) = -(\lambda - \lambda_0)(\mathbf{f}, g)$, and so, $(\mathbf{f}, g) \in \dot{\mathcal{D}}_q$ satisfies the homogeneous equation:

$$(\lambda I - \mathcal{A}_q)(\mathbf{f}, g) = (0, 0). \tag{3.511}$$

Namely, $(\mathbf{f}, g) \in \dot{\mathcal{D}}_q$ satisfies the homogeneous equations:

$$\begin{cases} \lambda \mathbf{f} - \mathrm{Div}\,(\mu \mathbf{D}(\mathbf{f}) - P(\mathbf{f}, g)I) = 0 & \text{in } B_R, \\ \lambda g - \mathbf{n} \cdot P\mathbf{f} = 0 & \text{on } S_R, \\ (\mu \mathbf{D}(\mathbf{f}) - P(\mathbf{f}, g)I)\mathbf{n} - \sigma(\mathcal{B}g)\mathbf{n} = 0 & \text{on } S_R. \end{cases} \tag{3.512}$$

First we consider the case where $2 \leq q < \infty$. Since $(\mathbf{f}, g) \in \dot{\mathcal{D}}_q \subset \dot{\mathcal{D}}_2$, by (3.512) and the divergence theorem of Gauß, we have

$$0 = (\lambda \mathbf{f} - \mathrm{Div}\,(\mathbf{D}(\mathbf{f}) - P(\mathbf{f}, g)I), \mathbf{f})_{B_R}$$

$$= \lambda \|\mathbf{f}\|_{L_2(B_R)}^2 - \sigma(\mathcal{B}g, \mathbf{n} \cdot \mathbf{f})_{S_R} + \frac{\mu}{2}\|\mathbf{D}(\mathbf{f})\|_{L_2(B_R)}^2 - (P(\mathbf{f}, g), \mathrm{div}\,\mathbf{f})_{B_R}.$$

For $h \in H_q^2(S_R)$ and $\mathbf{g} = {}^\top(g_1, \ldots, g_N)$, we have

$$(\mathcal{B}h, P\mathbf{g} \cdot \mathbf{n})_{S_R} = (\mathcal{B}h, \mathbf{g} \cdot \mathbf{n})_{S_R}, \tag{3.513}$$

because recalling $\mathbf{n} = y/|y| \in S_1$, we have

$$\sum_{j=1}^N |B_R|^{-1} \int_{B_R} g_\ell \, dy \, R^{-1}(\mathcal{B}h, y_\ell)_{S_R}$$

$$= \sum_{j=1}^N |B_R|^{-1} \int_{B_R} g_\ell \, dy \, R^{-1}(h, R^{-2}(N - 1 + \Delta_{S_1})y_\ell)_{S_R} = 0.$$

Moreover, $\operatorname{div} \mathbf{f} = 0$, because $\mathbf{f} \in \dot{J}_q(B_R)$. Thus, noting that $\lambda g = P\mathbf{f} \cdot \mathbf{n}$ on S_R, we have

$$\lambda \|\mathbf{f}\|_{L_2(B_R)}^2 - \sigma\bar{\lambda}(\mathcal{B}g, g)_{S_R} + \frac{\mu}{2}\|\mathbf{D}(\mathbf{f})\|_{L_2(B_R)}^2 = 0. \tag{3.514}$$

To treat $(\mathcal{B}g, g)_{S_R}$, we use the following lemma.

Lemma 3.7.14 *Let*

$$\dot{H}_2^2(S_R) = \{h \in H_2^2(S_R) \mid (h, 1)_{S_R} = 0, \quad (h, x_j)_{S_R} = 0 \ (j = 1, \ldots, N)\}.$$

Then,

$$-(\mathcal{B}h, h) \geq c\|h\|_{L_2(S_R)}^2 \tag{3.515}$$

for any $h \in \dot{H}_2^2(S_R)$ with some constant $c > 0$.

Postponing the proof of Lemma 3.7.14, we continue the proof of Theorem 3.7.10. Since $g \in \dot{W}_q^{3-1/q}(S_R) \subset \dot{H}_2^2(S_R)$, taking the real part of (3.514) we have

$$0 = \operatorname{Re} \lambda(\|\mathbf{f}\|_{L_2(B_R)}^2 - \sigma(\mathcal{B}g, g)_{S_R}) + \frac{\mu}{2}\|\mathbf{D}(\mathbf{f})\|_{L_2(B_R)}^2$$

$$\geq \operatorname{Re} \lambda(\|\mathbf{f}\|_{L_2(B_R)}^2 + c\sigma\|g\|_{L_2(S_R)}^2) + \frac{\mu}{2}\|\mathbf{D}(\mathbf{f})\|_{L_2(B_R)}^2,$$

which, combined with $\operatorname{Re} \lambda \geq 0$, leads to $\mathbf{D}(\mathbf{f}) = 0$. But, $(\mathbf{f}, \mathbf{p}_\ell)_{B_R} = 0$ for $\ell = 1, \ldots, M$, and so, $\mathbf{f} = 0$. Thus, by the first equation in (3.512), $\nabla P(\mathbf{f}, g) = 0$, and so, $P(\mathbf{f}, g) = f_0$ with some constant f_0, which, combined with the third equation in (3.512), leads to $\mathcal{B}g = -\sigma^{-1}f_0$ on S_R. Since $(g, 1)_{S_R} = |S_R|(g, \varphi_1)_{S_R} = 0$, we have

$$-\sigma^{-1}|S_R|f_0 = (\mathcal{B}g, 1)_{S_R} = (g, \Delta_{S_R}1)_{S_R} + R^{-2}(N - 1)(g, 1)_{S_R} = 0,$$

and so, $f_0 = 0$, which implies that $\mathcal{B}g = 0$ on S_R. Recalling that $g \in \dot{W}_q^{3-1/q}(S_R) \subset \dot{H}_2^2(S_R)$, by Lemma 3.7.14 $g = 0$. Thus, we have $(\mathbf{f}, g) = (0, 0)$, and so, the formula (3.509) holds. Namely, problem (3.503) admits a unique solution $(\mathbf{v}, h) \in \dot{\mathcal{D}}_q$ possessing the estimate:

$$\|(\mathbf{v}, h)\|_{\mathcal{D}_q} \leq C_\lambda \|(\mathbf{f}, g)\|_{\mathcal{H}_q(B_R)} \tag{3.516}$$

with some constant C_λ depending on λ, when $2 \leq q < \infty$ and $\lambda \in Q_{\lambda_0}$.

Before considering the case where $1 < q < 2$, at this point we give a

Proof of Lemma 3.7.14 Let $\{\lambda_j\}_{j=1}^\infty$ be the set of all eigen-values of the Laplace–Beltrami operator Δ_{S_R} on S_R. We may assume that $\lambda_1 > \lambda_2 > \lambda_3 > \cdots > \lambda_j > \cdots \to -\infty$, and then $\lambda_1 = 0$ and $\lambda_2 = -(N-1)R^{-2}$. Let E_j be the eigen-space corresponding to λ_j, and then the dimension of E_j is finite (cf. Neri [31, Chapter III, Spherical Harmonics]). Let $d_j = \dim E_j$, and then $d_1 = 1$ and $d_2 = N$. Especially, $E_1 = \{a \mid a \in \mathbb{C}\}$ and $E_2 = \{a_1 x_1 + \cdots + a_N x_N \mid a_i \in \mathbb{C} \ (i = 1, \ldots, N)\}$. Let $\{\varphi_{ij}\}_{j=1}^{d_i}$ be the orthogonal basis of E_i in $L_2(S_R)$, and then for any $h \in \dot{H}_2^2(S_R)$ we have

$$h = \sum_{i=3}^\infty \sum_{j=1}^{d_i} a_{ij} \varphi_{ij} \quad (a_{ij} = (h, \varphi_{ij})_{S_R}),$$

because $(h, \varphi_{ij})_{S_R} = 0$ for $i = 1, 2$. Thus, we have

$$-(\mathcal{B}h, h)_{S_R} = \sum_{i=3}^\infty \sum_{j=1}^{d_i} |a_{ij}|^2 (-\lambda_i - (N-1)R^{-1}) \|\varphi_{ij}\|_{L_2(S_R)}^2.$$

Since $-\lambda_i - (N-1)R^{-1} \geq c$ with some positive constant c for any $i \geq 3$, we have (3.515), which completes the proof of Lemma 3.7.14. □

Next, we consider the case where $1 < q < 2$. Let $(\mathbf{f}, g) \in \dot{\mathcal{D}}_q$ satisfy the homogeneous equations (3.512). First, we prove that

$$(\mathbf{f}, \mathbf{g})_{B_R} = 0 \quad \text{for any } \mathbf{g} \in \dot{J}_{q'}(B_R). \tag{3.517}$$

Let $(\mathbf{u}, \rho) \in \dot{\mathcal{D}}_{q'}$ be a solution of the equations:

$$\begin{cases} \bar{\lambda}\mathbf{u} - \mathrm{Div}\,(\mu \mathbf{D}(\mathbf{u}) - P(\mathbf{u}, \rho)\mathbf{I}) = \mathbf{g} & \text{in } B_R, \\ \bar{\lambda}\rho - \mathbf{n} \cdot P\mathbf{u} = 0 & \text{on } S_R, \\ (\mu \mathbf{D}(\mathbf{u}) - P(\mathbf{u}, \rho)\mathbf{I})\mathbf{n} - \sigma(\mathcal{B}\rho)\mathbf{n} = 0 & \text{on } S_R. \end{cases} \tag{3.518}$$

Since $\lambda \in Q_{\lambda_0}$, $\bar{\lambda} \in Q_{\lambda_0}$, and moreover $2 < q' < \infty$, and so, by the fact proved above we know the unique existence of $(\mathbf{u}, \rho) \in \mathcal{D}_{q'}$. By (3.512), (3.518) and the divergence theorem of Gauß, we have

$$(\mathbf{f}, \mathbf{g})_{B_R} = (\mathbf{f}, \bar{\lambda}\mathbf{u} - \mathrm{Div}\,(\mu\mathbf{D}(\mathbf{u}) - P(\mathbf{u}, \rho)\mathbf{I}))_{B_R}$$

$$= \lambda(\mathbf{f}, \mathbf{u})_{B_R} - (\mathbf{f} \cdot \mathbf{n}, \sigma\mathcal{B}\rho)_{S_R} + \frac{\mu}{2}(\mathbf{D}(\mathbf{f}), \mathbf{D}(\mathbf{u}))_{B_R} - (\mathrm{div}\,\mathbf{f}, P(\mathbf{u}, \rho))_{B_R}.$$

Noting that $P\mathbf{f} \cdot \mathbf{n} = \lambda g$ and $\mathrm{div}\,\mathbf{f} = 0$ and using (3.513), we have

$$(\mathbf{f}, \mathbf{g})_{B_R} = \lambda(\mathbf{f}, \mathbf{u})_{B_R} + \sigma\lambda\{(\nabla_{S_R}g, \nabla_{S_R}\rho)_{S_R}$$

$$- R^{-2}(N-1)(g, \rho)_{S_R}\} + \frac{\mu}{2}(\mathbf{D}(\mathbf{f}), \mathbf{D}(\mathbf{u}))_{B_R}. \tag{3.519}$$

On the other hand, we have

$$0 = (\lambda\mathbf{f} - \mathrm{Div}\,(\mu\mathbf{D}(\mathbf{f}) - P(\mathbf{f}, g)\mathbf{I}), \mathbf{u})_{B_R}$$

$$= \lambda(\mathbf{f}, \mathbf{u})_{B_R} - \sigma(\mathcal{B}g, \mathbf{n} \cdot \mathbf{u})_{S_R} + \frac{\mu}{2}(\mathbf{D}(\mathbf{f}), \mathbf{D}(\mathbf{u}))_{B_R} - (P(\mathbf{f}, g), \mathrm{div}\,\mathbf{u})_{B_R}.$$

Noting that $P\mathbf{u} \cdot \mathbf{n} = \bar{\lambda}\rho$ and $\mathrm{div}\,\mathbf{u} = 0$ and using (3.513), we have

$$0 = \lambda(\mathbf{f}, \mathbf{u})_{B_R} + \sigma\lambda\{(\nabla_{S_R}g, \nabla_{S_R}\rho)_{S_R}$$

$$- R^{-2}(N-1)(g, \rho)_{S_R}\} + \frac{\mu}{2}(\mathbf{D}(\mathbf{f}), \mathbf{D}(\mathbf{u}))_{B_R},$$

which, combined with (3.519), leads to (3.517).

Next, we prove that $(\mathbf{f}, \mathbf{g})_{B_R} = 0$ for any $\mathbf{g} \in L_{q'}(B_R)^N$. Given $\mathbf{g} \in L_{q'}(B_R)^N$, let $\psi \in \hat{H}^1_{q',0}(B_R)$ be a solution to the variational equation:

$$(\nabla\psi, \nabla\varphi)_{B_R} = (\mathbf{g}, \nabla\varphi)_{B_R} \quad \text{for any } \varphi \in \hat{H}^1_{q,0}(B_R).$$

Let $\mathbf{h} = \mathbf{g} - \nabla\psi$ and we decompose $\mathbf{g}$ as

$$\mathbf{g} = \nabla\psi + \mathbf{h} - \sum_{j=1}^{M}(\mathbf{h}, \mathbf{p}_\ell)_{B_R}\mathbf{p}_\ell + \sum_{j=1}^{M}(\mathbf{h}, \mathbf{p}_\ell)_{B_R}\mathbf{p}_\ell.$$

Since $\mathbf{f} \in \dot{J}_q(B_R)$, we have $(\mathbf{f}, \mathbf{g})_{B_R} = (\mathbf{f}, \mathbf{h} - \sum_{j=1}^{M}(\mathbf{h}, \mathbf{p}_\ell)\mathbf{p}_\ell)_{B_R}$. Since $\mathrm{div}\,\mathbf{p}_\ell = 0$, $\mathbf{h} - \sum_{j=1}^{M}(\mathbf{h}, \mathbf{p}_\ell)\mathbf{p}_\ell \in \dot{J}_{q'}(B_R)$, and so, by (3.517) $(\mathbf{f}, \mathbf{h} - \sum_{j=1}^{M}(\mathbf{h}, \mathbf{p}_\ell)\mathbf{p}_\ell)_{B_R} = 0$, which implies that $(\mathbf{f}, \mathbf{g})_{B_R} = 0$ for any $\mathbf{g} \in L_{q'}(B_R)$. Thus, we have $\mathbf{f} = 0$. By the first equation in (3.512), $\nabla P(\mathbf{f}, g) = 0$ in B_R, which leads to $P(\mathbf{f}, g) = f_0$ with some constant f_0. Thus, by the third equation in (3.512), we have $\mathcal{B}g = $

$-\sigma^{-1}f_0$ on S_R. Since $(g, 1)_{S_R} = 0$, we have $-\sigma^{-1}f_0|S_R| = (\mathcal{B}g, 1)_{S_R} = R^{-2}(N-1)(g, 1)_{S_R} = 0$, which leads to $f_0 = 0$. Thus, we have $\mathcal{B}g = 0$ on S_R. By the hypoellipticity of the operator Δ_{S_R}, we see that $g \in H_2^2(S_R)$, and so, $g \in \dot{H}_2^2(S_R)$, which, combined with Lemma 3.7.14, leads to $g = 0$. Thus, the formula (3.509) holds, and therefore problem (3.503) admits a unique solution $(\mathbf{v}, h) \in \dot{\mathcal{D}}_q$ possessing the estimate (3.516) when $1 < q < 2$ and $\lambda \in Q_{\lambda_0}$.

Finally, we prove that the constant in the estimate (3.516) is independent of $\lambda \in Q_{\lambda_0}$. Let $\lambda \in Q_{\lambda_0}$ and $\mu \in \mathbb{C}$, and we consider the equation:

$$(\mu I - \mathcal{A}_q)(\mathbf{v}, h) = (\mathbf{f}, g). \tag{3.520}$$

We write this equation as follows:

$$(\mathbf{f}, g) = ((\mu - \lambda)I + (\lambda I - \mathcal{A}_q))(\mathbf{v}, h) = (I + (\mu - \lambda)(\lambda I - \mathcal{A}_q)^{-1})(\lambda I - \mathcal{A}_q)(\mathbf{v}, h).$$

Since $\|(\mu - \lambda)(\lambda I - \mathcal{A}_q)^{-1}\|_{\mathcal{L}(\dot{\mathcal{H}}_q(B_R))} \le |\mu - \lambda|C_\lambda$ as follows from (3.516), where $\|\cdot\|_{\mathcal{L}(\dot{\mathcal{H}}_q(B_R))}$ denotes the operator norm of the bounded linear operator from $\dot{\mathcal{H}}_q(B_R)$ into itself, choosing $\mu \in \mathbb{C}$ in such a way that $|\mu - \lambda|C_\lambda \le 1/2$, we see that the inverse operator $(I + (\mu - \lambda)(\lambda I - \mathcal{A}_q)^{-1})^{-1}$ exists as a bounded linear operator from $\dot{\mathcal{H}}_q(B_R)$ into itself and

$$\|(I + (\mu - \lambda)(\lambda I - \mathcal{A}_q)^{-1})^{-1}\|_{\mathcal{L}(\dot{\mathcal{H}}_q(B_R))} \le 2.$$

Thus, $(\mathbf{v}, h) = (\lambda I - \mathcal{A}_q)^{-1}(I + (\mu - \lambda)(\lambda I - \mathcal{A}_q)^{-1})^{-1}(\mathbf{f}, g)$ belongs to $\dot{\mathcal{D}}_q$ and solves the Eq. (3.520). Moreover,

$$\|(\mathbf{v}, h)\|_{\mathcal{D}_q}$$
$$\le \|(\lambda I - \mathcal{A}_q)^{-1}\|_{\mathcal{L}(\mathcal{H}_q, \mathcal{D}_q)}\|(I + (\mu - \lambda)(\lambda I - \mathcal{A}_q)^{-1}\|_{\mathcal{L}(\dot{\mathcal{H}}_q(B_R))}\|(\mathbf{f}, g)\|_{\mathcal{H}_q}$$
$$\le 2C_\lambda\|(\mathbf{f}, g)\|_{\mathcal{H}_q},$$

provided that $|\mu - \lambda| \le (2C_\lambda)^{-1}$, where $\|\cdot\|_{\mathcal{L}(\dot{\mathcal{H}}_q, \mathcal{D}_q)}$ denotes the operator norm of bounded linear operators from $\dot{\mathcal{H}}_q(B_R)$ into $\mathcal{D}_q(B_R)$. Since Q_{λ_0} is a compact set, we have (3.507), which completes the proof of Theorem 3.7.10. $\qquad\square$

Proof of Lemma 3.7.8 Finally we prove Lemma 3.7.8. Let

$$\tilde{\mathbf{u}} = \mathbf{u} - \sum_{\ell=1}^{M}(\mathbf{u}, \mathbf{p}_\ell)_{B_R}\,\mathbf{p}_\ell, \quad \tilde{\rho} = \rho - \sum_{j=1}^{N+1}(\rho, \varphi_j)_{S_R}\,\varphi_j.$$

Since $\mathbf{D}(\mathbf{p}_\ell) = 0$, div $\mathbf{p}_\ell = 0$, and $\mathcal{B}\varphi_j = 0$ $(j = 2, \ldots, N+1)$, we have

$$(\nabla P(\tilde{\mathbf{u}}, \tilde{\rho}), \nabla \psi)_{B_R} = (\mu \mathrm{Div}\, \mathbf{D}(\mathbf{u}) - \nabla \mathrm{div}\, \mathbf{u}, \nabla \psi)_{B_R} = (\nabla P(\mathbf{u}, \rho), \nabla \psi)_{B_R}$$

for any $\psi \in \hat{H}^1_{q',0}(B_R)$, subject to

$$P(\tilde{\mathbf{u}}, \tilde{\rho}) = <\mu \mathbf{D}(\mathbf{u})\mathbf{n}, \mathbf{n}> -\sigma \mathcal{B}\rho + \frac{\sigma(N-1)}{R^2}(\rho, \varphi_1)_{S_R}\varphi_1 - \mathrm{div}\, \mathbf{u} \quad \text{on } \Gamma,$$

and so, $P(\tilde{\mathbf{u}}, \tilde{\varphi}) = P(\mathbf{u}, \rho) + \frac{\sigma(N-1)}{R^2}(\rho, \varphi_1)_{S_R}\varphi_1$. Since $\nabla \varphi_1 = 0$, $\tilde{\mathbf{u}}$ and $\tilde{\rho}$ satisfy
Eq. (3.499). Moreover, $(\tilde{\mathbf{u}}, \tilde{\rho}) \in \mathcal{H}_q(B_R) \cap \mathcal{D}_q(B_R)$, and therefore by (3.508) and
Theorem 3.7.10 with $\lambda = 0$, we have $\|(\tilde{\mathbf{u}}, \tilde{\rho})\|_{\mathcal{D}_q(B_R)} \leq C\|(\mathbf{f}, g)\|_{\mathcal{H}_q(B_R)}$, which,
combined with the estimate:

$$\|(\mathbf{u}, \rho)\|_{\mathcal{D}_q(B_R)} \leq \|(\tilde{\mathbf{u}}, \tilde{\rho})\|_{\mathcal{D}_q(B_R)} + C\left\{ \sum_{\ell=1}^{M} |(\mathbf{u}, \mathbf{p}_\ell)_{B_R}| + \sum_{j=1}^{N+1} |(\rho, \varphi_j)_{S_R}| \right\},$$

leads to (3.500). This completes the proof of Lemma 3.7.8.

3.7.5   Global Wellposedness, a Proof of Theorem 3.7.1

In this section, we prove Theorem 3.7.1. Assume that the initial data $\mathbf{u}_0 \in B^{2-2/p}_{q,p}(B_R)^N$ and $\rho_0 \in W^{3-1/p-1/q}_{q,p}(S_R)$ satisfy the smallness condition:

$$\|\mathbf{u}_0\|_{H^2_q(B_R)} + \|\rho_0\|_{W^{3-1/p-1/q}_{q,p}(S_R)} \leq \epsilon \tag{3.521}$$

with small constant $\epsilon > 0$ as well as the compatibility condition (3.171). For the
notational simplicity, we write

$$\mathcal{I} = \|\mathbf{u}_0\|_{H^2_q(B_R)} + \|\rho_0\|_{W^{3-1/p-1/q}_{q,p}(S_R)},$$

$$\tilde{E}_{p,q,T}(\mathbf{u}, \rho; \eta) = \|e^{\eta t}\mathbf{u}\|_{L_p((0,T),H^2_q(B_R))} + \|e^{\eta t}\partial_t \mathbf{u}\|_{L_p((0,T),L_q(B_R))}$$

$$+ \|e^{\eta t}\rho\|_{L_p((0,T),W^{3-1/q}_q(S_R))} + \|e^{\eta t}\partial_t \rho\|_{L_p((0,T),W^{2-1/q}_q(S_R))}$$

$$E_{p,q,T}(\mathbf{u}, \rho; \eta) = \tilde{E}_{p,q,T}(\mathbf{u}, \rho; \eta) + \|e^{\eta t}\partial_t \rho\|_{L_\infty((0,T),W^{1-1/q}_q(S_R))}.$$

Notice that

$$\|e^{\eta t}\mathbf{u}\|_{L_\infty((0,T),B^{2(1-1/p)}_{q,p}(B_R))} + \|e^{\eta t}\rho\|_{L_\infty((0,T),B^{3-1/p-1/q}_{q,p}(S_R))}$$

$$\leq C(\mathcal{I} + \tilde{E}_{p,q,T}(\mathbf{u}, \rho; \eta)),$$

which follows from (3.405) and (3.406). Since we choose ϵ small enough eventually, we may assume that $0 < \mathcal{I} \leq \epsilon < 1$. Let T_0 be a positive number > 2. In view of Theorem 3.7.2, there exists a constant $\epsilon_1 > 0$ depending on T_0 such that if $\mathcal{I} \leq \epsilon_1$, then for any $T \in (0, T_0]$ problem (3.479) admits a unique solution $(\mathbf{u}, \mathfrak{q}, \rho) \in S_{p,q}((0, T))$ satisfying the condition:

$$\sup_{0 < t < T} \|\Psi_\rho(\cdot, t)\|_{H^1_\infty(B_R)} \leq \delta \tag{3.522}$$

where $\delta \in (0, 1/4)$ is the same constant as in (3.446). We shall prove that $\mathbf{u}$, $\mathfrak{q}$ and ρ can be prolonged beyond T_0 provided that $\epsilon > 0$ is small enough. Note that if we write solutions of Eq. (3.449) by $U_T = (\mathbf{u}_T, \mathfrak{q}_T, \rho_T) \in S_{p,q}((0, T))$, then by the uniqueness of local in time solutions, we see that $U_T = U_{T'}$ in $(0, T)$ for any T and $T' \in (0, T_0]$ with $T < T'$, and so in what follows we write $(\mathbf{u}_T, \mathfrak{q}_T, \rho_T)$ simply by $(\mathbf{u}, \mathfrak{q}, \rho)$. Below, it is assumed that $0 < \epsilon \leq \epsilon_1$.

To prove Theorem 3.7.1, it suffices to prove that the inequality:

$$E_{p,q,T}(\mathbf{u}, \rho; \eta) \leq M_3(\mathcal{I} + E_{p,q,T}(\mathbf{u}, \rho; \eta)^2 + E_{p,q,T}(\mathbf{u}, \rho; \eta)^3) \tag{3.523}$$

holds for any $T \in (0, T_0]$ with some constant $M_3 > 0$ independent of ϵ, T, and T_0, where η is the same positive constant as in Theorem 3.7.3.

In fact, let $r_0(\epsilon)$ and $r_\pm(\epsilon)$ be three different solutions of the algebraic equation: $x^3 + x^2 + \epsilon - M_3^{-1}x = 0$ with $r_0(\epsilon) = M_3\epsilon + O(\epsilon^2)$, $r_+(\epsilon) = M_3^{-1} + O(M_3^{-2}) + O(\epsilon)$, and $r_-(\epsilon) = -1 - M_3^{-1} + O(M_3^{-2}) + O(\epsilon)$ as $M_3 \to \infty$ and $\epsilon \to 0$. Since $E_{p,q,T}(\mathbf{u}, \rho; \eta) \geq 0 > r_-(\epsilon)$, by (3.523) one of the following cases holds:

$$E_{p,q,T}(\mathbf{u}, \rho; \eta) \leq r_0(\epsilon), \quad E_{p,q,T}(\mathbf{u}, \rho; \eta) \geq r_+(\epsilon).$$

Since we may change M_3 larger for (3.523) to hold if necessary, and since we choose $\epsilon_1 > 0$ relatively small, we may assume that

$$e\epsilon_1 < M_3/2. \tag{3.524}$$

By (3.524), $E_{p,q,1/\eta}(\mathbf{u}, \rho; \eta) \leq e\epsilon_1 < M_3/2 < r_+(\epsilon)$, and therefore,

$$E_{p,q,T}(\mathbf{u}, \rho; \eta) \leq r_0(\epsilon) \quad \text{for } T \in (0, 1/\eta).$$

But, $E_{p,q,T}(\mathbf{u}, \rho; \eta)$ is a continuous function with respect to $T \in (0, T_0)$, which yields that

$$E_{p,q,T}(\mathbf{u}, \rho; \eta) \leq r_0(\epsilon) \quad \text{for any } T \in (0, T_0). \tag{3.525}$$

Let $T_0' = T_0 - 1/2$. Thus, choosing $\epsilon > 0$ small enough and employing the same argument as that in proving Theorem 3.7.2, we see that there exist unique solutions $(\mathbf{v}, \mathfrak{p}, h) \in \mathcal{S}_{p,q}((T_0', T_0' + 1))$ of the equations:

$$
\begin{cases}
\partial_t \mathbf{v} - \mathrm{Div}\,(\mu \mathbf{D}(\mathbf{v}) - \mathfrak{p}\mathbf{I}) = \mathbf{f}(\mathbf{v}, \Psi_h) & \text{in } B_R \times (T_0', T_0' + 1), \\
\mathrm{div}\,\mathbf{v} = g(\mathbf{v}, \Psi_h) = \mathrm{div}\,\mathbf{g}(\mathbf{v}, \Psi_h) & \text{in } B_R \times (T_0', T_0' + 1), \\
\partial_t h - \omega \cdot \mathbf{P}\mathbf{v} = \tilde{d}(\mathbf{v}, \Psi_h) & \text{on } S_R \times (T_0', T_0' + 1), \\
(\mu \mathbf{D}(\mathbf{v})\omega)_\tau = \mathbf{h}'(\mathbf{v}, \Psi_h) & \text{on } S_R \times (T_0', T_0' + 1), \\
< \mu \mathbf{D}(\mathbf{v})\omega, \omega > -\mathfrak{p} - \sigma \mathcal{B}\rho = h_N(\mathbf{v}, \Psi_h) & \text{on } S_R \times (T_0', T_0' + 1), \\
(\mathbf{v}, h)|_{t=T_0'} = (\mathbf{u}(\cdot, T_0'), \rho(\cdot, T_0')) & \text{in } B_R \times S_R,
\end{cases}
\tag{3.526}
$$

which satisfies the condition:

$$
\sup_{T_0' < t < T_0' + 1} \|\Psi_h(\cdot, t)\|_{H^1_\infty(B_R)} \leq \delta.
$$

Let

$$
\mathbf{u}_1 = \begin{cases} \mathbf{u} & (0 < t \leq T_0'), \\ \mathbf{v} & (T_0' < t < T_0' + 1), \end{cases} \qquad
\mathfrak{q}_1 = \begin{cases} \mathfrak{q} & (0 < t \leq T_0'), \\ \mathfrak{p} & (T_0' < t < T_0' + 1), \end{cases}
$$

$$
\rho_1 = \begin{cases} \rho & (0 < t \leq T_0'), \\ h & (T_0' < t < T_0' + 1), \end{cases}
$$

and then $(\mathbf{u}_1, \mathfrak{q}_1, \rho_1)$ belongs to $\mathcal{S}_{p,q}((0, T_0' + 1))$ and satisfies the condition:

$$
\sup_{0 < t < T_0' + 1} \|\Psi_{\rho_1}(\cdot, t)\|_{H^1_\infty(B_R)} \leq \delta,
$$

and Eq. (3.479) in $(0, T_0' + 1)$. Since $T_0' + 1 = T_0 + 1/2$, we can prolong the solutions. Repeating this argument, we can prolong $\mathbf{u}$, $\mathfrak{q}$ and ρ to the time interval $(0, \infty)$.

Below, we prove (3.523). Notice that $(\mathbf{u}, \mathfrak{q}, \rho) \in \mathcal{S}_{p,q}((0, T))$ satisfies Eq. (3.449). We shall estimate the right side of Eq. (3.449). In view of (3.468), we have

$$
\|e^{\eta t} \mathbf{f}(\mathbf{u}, \Psi_\rho)\|_{L_p((0,T), L_q(B_R))} \leq C(\mathcal{I} + E_{p,q,T}(\mathbf{u}, \rho, \eta)) E_{p,q,T}(\mathbf{u}, \rho, \eta).
\tag{3.527}
$$

Here and in the following, we use the inequality:

$$
E_{p,q,T}(\mathbf{u}, \rho, 0) \leq E_{p,q,T}(\mathbf{u}, \rho, \eta).
$$

Employing the same argument as in the proof of (3.470), we have

$$\|e^{\eta t}\tilde{d}(\mathbf{u}, \Psi_\rho)\|_{L_\infty((0,T),W_q^{1-1/q}(S_R))} \leq C(\mathcal{I} + E_{p,q,T}(\mathbf{u}, \rho, \eta))^2,$$

$$\|e^{\eta t}\tilde{d}(\mathbf{u}, \Psi_\rho)\|_{L_p((0,T),W_q^{2-1/q}(S_R))} \leq C(\mathcal{I} + E_{p,q,T}(\mathbf{u}, \rho, \eta))^2 E_{p,q,T}(\mathbf{u}, \rho, \eta).$$
$$(3.528)$$

We now consider $g(\mathbf{u}, \Psi_\rho)$ and $\mathbf{g}(\mathbf{u}, \Psi_\rho)$. Let $\tilde{g}_\eta = \tilde{g}(e^{\eta t}\mathbf{u}, \Psi_\rho)$ and $\tilde{\mathbf{g}}_\eta = \tilde{\mathbf{g}}(e^{\eta t}\mathbf{u}, \Psi_\rho)$, where $\tilde{g}(\mathbf{v}, \psi_h)$ and $\tilde{\mathbf{g}}(\mathbf{v}, \Psi_h)$ are the functions defined in (3.423). Since $e^{\eta t}\mathbf{u}|_{t=0} = \mathbf{u}_0$, we note that in (3.421)

$$\mathcal{E}_1[e^{\eta t}\mathbf{u}] = e_T[e^{\eta t}\mathbf{u} - T_v(t)\mathbf{u}_0] + \psi(t)T_v(|t|)\mathbf{u}_0.$$

Moreover, we may assume that $0 < \eta < 1$, and so the way of estimating $\tilde{g}_\eta$ and $\tilde{\mathbf{g}}_\eta$ is the same as in Sect. 3.6. Moreover, by the same reason as in (3.424), we have

$$\tilde{g}_\eta = e^{\eta t}g(\mathbf{u}, \Psi_\rho), \quad \tilde{\mathbf{g}}_\eta = e^{\eta t}\mathbf{g}(\mathbf{u}, \Psi_\rho) \quad \text{for } t \in (0, T),$$
$$\operatorname{div}\tilde{\mathbf{g}}_\eta = \tilde{g}_\eta \quad \text{for } t \in \mathbb{R}. \tag{3.529}$$

Using (3.427), (3.417), and (3.418), we have

$$\|\tilde{\mathbf{g}}_\eta\|_{L_p(\mathbb{R},L_q(B_R))} \leq C(\mathcal{I} + E_{p,q,T}(\mathbf{u}, \rho, \eta))E_{p,q,T}(\mathbf{u}, \rho, \eta). \tag{3.530}$$

Employing the same argument as in proving (3.431) and using Lemmas 3.6.2, 3.6.3, (3.417), and (3.418), we have

$$\|\tilde{g}_\eta\|_{H_p^{1/2}(\mathbb{R},L_q(B_R))} + \|\tilde{g}_\eta\|_{L_p(\mathbb{R},H_q^1(B_R))}$$
$$\leq C(\mathcal{I} + E_{p,q,T}(\mathbf{u}, \rho, \eta))E_{p,q,T}(\mathbf{u}, \rho, \eta). \tag{3.531}$$

We now consider $\mathbf{h}'(\mathbf{u}, \Psi_\rho)$. Since $\mathbf{h}'(\mathbf{v}, \Psi_h)$ is written like (3.432), following (3.433) we define $\tilde{\mathbf{h}}'_\eta$ by setting

$$\mathbf{h}'_\eta = \mathbf{v}'_\mathbf{h}(\bar{\nabla}\mathcal{E}_2[\Psi_\rho])\bar{\nabla}\mathcal{E}_2[\Psi_\rho] \otimes \nabla\mathcal{E}_1[e^{\eta t}\mathbf{u}].$$

And then, we have $e^{\eta t}\mathbf{h}'(\mathbf{u}, \Psi_\rho) = \tilde{\mathbf{h}}'_\eta$ for $t \in (0, T)$. Moreover, using Lemmas 3.6.2, 3.6.3, (3.417), and (3.418), we have

$$\|\tilde{\mathbf{h}}'_\eta\|_{H_p^{1/2}(\mathbb{R},L_q(B_R))} + \|\tilde{\mathbf{h}}'_\eta\|_{L_p(\mathbb{R},H_q^1(B_R))}$$
$$\leq C(\mathcal{I} + E_{p,q,T}(\mathbf{u}, \rho, \eta))E_{p,q,T}(\mathbf{u}, \rho, \eta). \tag{3.532}$$

We finally consider $h_N(\mathbf{u}, \Psi_\rho)$, which is given in (3.459). According to the formula (3.459), we define $\tilde{h}_{N,\eta}$ by setting

$$\tilde{h}_{N,\eta} = \mathbf{V}_{h,N}(\bar{\nabla}\mathcal{E}_2[\Psi_\rho])\bar{\nabla}\mathcal{E}_2[\Psi_\rho] \otimes \nabla\mathcal{E}_1[e^{\eta t}\mathbf{v}]$$
$$+ \sigma\tilde{\mathbf{V}}'_\Gamma(\bar{\nabla}\mathcal{E}_2[\Psi_\rho])\bar{\nabla}\mathcal{E}_2[\Psi_\rho] \otimes \bar{\nabla}^2\mathcal{E}_2[e^{\eta t}\Psi_\rho].$$

We have $e^{\eta t}h_N(\mathbf{u}, \Psi_\rho) = \tilde{h}_{N,\eta}$ for $t \in (0, T)$. Since $T_h(t)\rho_0|_{t=0} = \Psi_{\rho_0} = e^{\eta t}\Psi_\rho|_{t=0}$, we note that in (3.421)

$$\mathcal{E}_2[e^{\eta t}\Psi_\rho] = e_T[e^{\eta t}\Psi_\rho - T_h(t)\rho_0] + \psi(t)T_h(|t|)\rho_0.$$

The way of estimating $\tilde{h}_N$ is the same as in Sect. 6, and so, noting that

$$\|\bar{\nabla}^2\mathcal{E}_2[e^{\eta t}\Psi_\rho]\|_{H_p^{1/2}(\mathbb{R}, L_q(B_R))} \le \|\bar{\nabla}^2\mathcal{E}_2[e^{\eta t}\Psi_\rho]\|_{H_p^1(\mathbb{R}, L_q(B_R))},$$

by Lemmas 3.6.2, 3.6.3, (3.417), and (3.418), we have

$$\|\tilde{h}_{N,\eta}\|_{H_p^{1/2}(\mathbb{R}, L_q(B_R))} + \|\tilde{h}_{N,\eta}\|_{L_p(\mathbb{R}, H_q^1(B_R))} \tag{3.533}$$
$$\le C(\mathcal{I} + E_{p,q,T}(\mathbf{u}, \rho, \eta))E_{p,q,T}(\mathbf{u}, \rho, \eta).$$

Applying Theorem 3.7.4 and using (3.527), (3.528), and (3.530)–(3.533), we have

$$\tilde{E}_{p,q,T}(\mathbf{u}, \rho; \eta) \le C\{\mathcal{I} + E_{p,q,T}(\mathbf{u}, \rho; \eta)^2 + E_{p,q,T}(\mathbf{u}, \rho; \eta)^3$$
$$+ \sum_{\ell=1}^M \left(\int_0^T (e^{\eta s}|(\mathbf{u}(\cdot, s), \mathbf{p}_\ell)_{B_R}|)^p \, ds\right)^{1/p} \tag{3.534}$$
$$+ \sum_{j=1}^{N+1} \left(\int_0^T (e^{\eta s}|(\rho(\cdot, s), \varphi_j)_{S_R}|)^p \, ds\right)^{1/p}\}.$$

Here, we have used the inequalities:

$$(\mathcal{I} + x)^2 \le \mathcal{I} + x^2, \quad (\mathcal{I} + x)^2 x \le \mathcal{I} + 3x^2 + x^3$$

for any $0 < \mathcal{I} < 1$ and $x > 0$.

By (3.477), we have

$$\|e^{\eta t}\partial_t\rho\|_{L_\infty((0,T), W_q^{1-1/q}(S_R))}$$
$$\le C(\|e^{\eta t}\mathbf{u}\|_{L_\infty((0,T), H_q^1(B_R))} + \|e^{\eta t}\tilde{d}(\mathbf{v}, \Psi_h)\|_{L_\infty((0,T), W_q^{1-1/q}(S_R))}),$$

and so, by (3.528) and (3.405), we have

$$\|e^{\eta t}\partial_t\rho\|_{L_\infty((0,T),W_q^{1-1/q}(S_R))} \le C(\mathcal{I} + \tilde{E}_{p,q,T}(\mathbf{u},\rho;\eta) + E_{p,q,T}(\mathbf{u},\rho;\eta)^2),$$

which, combined with (3.534), leads to

$$\begin{aligned}
E_{p,q,T}(\mathbf{u},\rho;\eta) &\le C\{\mathcal{I} + E_{p,q,T}(\mathbf{u},\rho;\eta)^2 + E_{p,q,T}(\mathbf{u},\rho;\eta)^3 \\
&\quad + \sum_{\ell=1}^{M}\left(\int_0^T (e^{\eta s}|(\mathbf{u}(\cdot,s),\mathbf{p}_\ell)_{B_R}|)^p\,ds\right)^{1/p} \\
&\quad + \sum_{j=1}^{N+1}\left(\int_0^T (e^{\eta s}|(\rho(\cdot,s),\varphi_j)_{S_R}|)^p\,ds\right)^{1/p}\}.
\end{aligned} \tag{3.535}$$

Our final task is to prove that

$$\sum_{\ell=1}^{M}\left(\int_0^t (e^{\eta s}|(\mathbf{u}(\cdot,s),\mathbf{p}_\ell)_{B_R}|)^p\,ds\right)^{1/p} + \sum_{j=1}^{N+1}\left(\int_0^t (e^{\eta s}|(\rho(\cdot,s),\varphi_j)_{S_R}|)^p\,ds\right)^{1/p}$$

$$\le C(\mathcal{I} + E_{p,q,T}(\mathbf{u},\rho;\eta))E_{p,q,T}(\mathbf{u},\rho;\eta) \tag{3.536}$$

with some constant $C > 0$ independent of ϵ and $T > 0$. If we have (3.536), then putting (3.535) and (3.536) together gives (3.523).

From now on, we prove (3.536). Let $\mathbf{h}_z^{-1}(x,t)$ be the inverse map of the Hanzawa transform $x = \mathbf{h}_z(y,t)$ for each $t \in [0,T]$, which is the diffeomorphism from B_R onto Ω_t of H_q^3 class for each $t \in [0,T]$. Let $\mathbf{v} = \mathbf{u} \circ \mathbf{h}_z^{-1}$ and $\mathfrak{p} = (\mathfrak{q} + \sigma(N-1)R^{-1})\circ\mathbf{h}_z^{-1}$, and then $\mathbf{v}$ and $\mathfrak{p}$ satisfy the equations (3.1). Let J be the Jacobian of the Hanzawa transform. By (3.9), we have $|\Omega_t| = |\Omega|$, which, combined with (A.1), leads to $|\Omega_t| = |B_R|$. Since Γ_t is given by (3.443), using the polar coordinates: $x - \xi(t) = s\omega$ for $\omega \in S_1$ and $s \in (0, R + \rho(R\omega,t))$, we have

$$|B_R| = |\Omega_t| = \int_{\Omega_t} dx = \int_{|\omega|=1} d\omega \int_0^{R+\rho(R\omega,t)} s^{N-1}\,ds$$

$$= \frac{1}{N}\int_{|\omega|=1}(R + \rho(R\omega,t))^N\,d\omega = |B_R| + \frac{1}{N}\sum_{j=1}^{N} {}_NC_j R^{1-j}\int_{S_R}\rho(y,t)^j\,d\tau,$$

which leads to

$$(\rho,1)_{S_R} = -\sum_{j=2}^{N} {}_NC_j R^{1-j}\int_{S_R}\rho(y,t)^j\,d\tau. \tag{3.537}$$

Since $\|\rho(\cdot, t)\|_{L_\infty(S_R)} \leq \|H_\rho(\cdot, t)\|_{L_\infty(B_R)} \leq \delta$ as follows from $H_\rho|_{S_R} = \rho$ on S_R, by (3.537) and (3.406) we have

$$\left(\int_0^T (e^{\eta t} |(\rho(\cdot, t), \varphi_1)_{S_R}|)^p \, dt\right)^{1/p}$$

$$\leq C \sup_{0<t<T} \|\rho(\cdot, t)\|_{W_q^{2-1/q}(S_R)} \|e^{\eta t} \rho\|_{L_p((0,T), L_q(S_R))} \tag{3.538}$$

$$\leq C(\mathcal{I} + E_{p,q,T}(\mathbf{u}, \rho; \eta)) E_{p,q,T}(\mathbf{u}, \rho; \eta).$$

Since $\xi(t) = |\Omega|^{-1} \int_{\Omega_t} x \, dx$ and since $|\Omega_t| = |\Omega|$, we have

$$0 = \int_{\Omega_t} x \, dx - |\Omega| \xi(t) = \int_{\Omega_t} (x - \xi(t)) \, dx.$$

Using the polar coordinates again, we have

$$0 = \int_{\Omega_t} (x_i - \xi_i(t)) \, dx = \int_{|\omega|=1} d\omega \int_0^{R+\rho(R\omega, t)} (s\omega_i) s^{N-1} \, ds$$

$$= \frac{1}{N+1} \int_{|\omega|=1} \omega_i (R + \rho(R\omega, t))^{N+1} \, d\omega$$

$$= \frac{R^{N+1}}{N+1} \int_{|\omega|=1} \omega_i \, d\omega + R^N \int_{|\omega|=1} \rho(R\omega, t) \omega_i \, d\omega$$

$$+ \frac{1}{N+1} \sum_{k=2}^{N+1} {}_{N+1}C_k R^{N+1-k} \int_{|\omega|=1} \rho(R\omega, t)^k \omega_i \, d\omega$$

$$= (\rho, x_i)_{S_R} + \frac{1}{N+1} \sum_{k=2}^{N+1} {}_{N+1}C_k R^{1-k} (\rho^k, x_i)_{S_R}.$$

Recalling that φ_j equals constant $\times x_j$ ($j = 2, \ldots, N+1$), we have

$$\left(\int_0^T (e^{\eta t} |(\rho(\cdot, t), \varphi_i)_{S_R}|)^p \, dt\right)^{1/p}$$

$$\leq C \sup_{0<s<T} \|\rho(\cdot, t)\|_{W_q^{2-1/q}(S_R)} \|e^{\eta t} \rho\|_{L_p((0,T), L_q(S_R))} \tag{3.539}$$

$$\leq C(\mathcal{I} + E_{p,q,T}(\mathbf{u}, \rho; \eta)) E_{p,q,T}(\mathbf{u}, \rho; \eta).$$

By (3.14) and (3.17), we have

$$\int_{\Omega_t} \mathbf{v}(x, t) \, dx = \int_{\Omega} \mathbf{v}_0(x) \, dx, \tag{3.540}$$

$$\int_{\Omega_t} (x_i v_j(x, t) - x_j v_i(x, t)) \, dx = \int_{\Omega} (x_i v_{0j}(x) - x_j v_{0i}(x)) \, dx. \tag{3.541}$$

Putting (3.540), (3.541) and (3.456) together gives

$$(\mathbf{v}, \mathbf{e}_i)_{\Omega_t} = 0, \quad (\mathbf{v}, x_i \mathbf{e}_j - x_j \mathbf{e}_i)_{\Omega_t} = 0 \tag{3.542}$$

for $i, j = 1, \ldots, N$. Since the Jacobian J of the Hanzawa transform has the form: $J = 1 + J_0(\mathbf{k})$, it then follows from (3.542) that

$$0 = \int_{B_R} \mathbf{u}(y, t) \cdot \mathbf{p}_\ell(y + H_\rho(y, t)\mathbf{n}(y) + \xi(t)) J \, dy$$

$$= (\mathbf{u}, \mathbf{p}_\ell)_{B_R} + \int_{B_R} \mathbf{u}(y, t) \cdot \mathbf{p}_\ell(y) J_0(\mathbf{k}) \, dy$$

$$+ \int_{B_R} \mathbf{u}(y, t) \tilde{\mathbf{p}}_\ell(H_\rho(y, t)y + \xi(t))(1 + J_0(\mathbf{k})) \, dy$$

where $\tilde{\mathbf{p}}_\ell = 0$ for $\ell = 1, \ldots, N$ and for $\ell = N + 1, \ldots, M$

$$\mathbf{p}_\ell(H_\rho(y, t)y + \xi(t)) = c_R^1((H_\rho(y, t)y_i + \xi_i(t))\mathbf{e}_j - (H_\rho(y, t)y_j + \xi_j(t))\mathbf{e}_i)$$

for some (i, j) with $1 \le i < j \le N$. Since $\xi(0) = 0$ as follows from (A.2) and since $|J_0(\mathbf{k})| \le C|\nabla\Psi_\rho(\cdot, t)|$, using (3.444) and (3.446), we have

$$|\xi(t)| \le \frac{1}{|B_R|} \int_0^t \int_{B_R} |\mathbf{u}(y, t)|(1 + |J_0(\mathbf{k})|) \, dy$$

$$\le C\left(\int_0^T e^{-\eta p's} \, ds\right)^{1/p'} \|e^{\eta t} \mathbf{u}\|_{L_p((0,T), L_q(B_R))},$$

and so, we have

$$\left(\int_0^T (e^{\eta t}|(\mathbf{u}(\cdot, t), \mathbf{p}_\ell)_{B_R}|)^p \, dt\right)^{1/p}$$

$$\le \left(\int_0^T \left(e^{\eta t}\left|\int_{B_R} \mathbf{u}(y, t) \cdot \mathbf{p}_\ell(y) J_0(\mathbf{k}) \, dy\right|\right)^p \, dt\right)^{1/p}$$

$$+ \left(\int_0^T \left(e^{\eta t}\left|\int_{B_R} \mathbf{u}(y, t) \tilde{\mathbf{p}}_\ell(\Psi_\rho(y, t) + \xi(t)) \, dy\right|\right)^p \, dt\right)^{1/p}$$

$$\leq C \Big(\int_0^T (e^{\eta t} \| \mathbf{u}(\cdot, t) \|_{L_q(B_R)})^p \, dt \Big)^{1/p} (\| \Psi_\rho \|_{L_\infty((0,T), L_\infty(B_R))} + \sup_{t \in (0,T)} |\xi(t)|)$$

$$\leq C(\mathcal{I} + E_{p,q,T}(\mathbf{u}, \rho; \eta)) E_{p,q,T}(\mathbf{u}, \rho; \eta).$$

Putting this and (3.539) together gives (3.536). This completes the proof of Theorem 3.7.1.

3.8 Global Well-Posedness in an Exterior Domain

In this section, we consider Eq. (3.1) in the case where $\sigma = 0$ and Ω is an exterior domain in $\mathbb{R}^N$ whose boundary Γ is a compact C^2 hypersurface. The problem we consider in this section is the following:

$$
\begin{cases}
\partial_t \mathbf{v} + (\mathbf{v} \cdot \nabla) \mathbf{v} - \mathrm{Div}\,(\mu \mathbf{D}(\mathbf{v}) - \mathfrak{p}\mathbf{I}) = 0 & \text{in } \bigcup_{0<t<T} \Omega_t \times \{t\}, \\[2ex]
\mathrm{div}\,\mathbf{v} = 0 & \text{in } \bigcup_{0<t<T} \Omega_t \times \{t\}, \\[2ex]
(\mu \mathbf{D}(\mathbf{v}) - \mathfrak{p}\mathbf{I})\mathbf{n}_t = 0 & \text{on } \bigcup_{0<t<T} \Gamma_t \times \{t\}, \qquad (3.543)\\[2ex]
V_n = \mathbf{v} \cdot \mathbf{n}_t & \text{on } \bigcup_{0<t<T} \Gamma_t \times \{t\}, \\[2ex]
\mathbf{v}|_{t=0} = \mathbf{v}_0 \quad \text{in } \Omega_0, \quad \Omega_t|_{t=0} = \Omega_0 = \Omega
\end{cases}
$$

In this section, we assume that μ is a positive constant. If we use the Hanazawa transform to transform Ω_t to some fixed domain, as was discussed in Sect. 3.6, we have to require $W_q^{3-1/q}$ regularity of the height function ρ representing Γ_t. However, this regularity is obtained by surface tension, that is the Laplace–Beltrami operator on Γ_t. We now consider the case where the surface tension is not taken into account. Thus, we can not obtain $W_q^{3-1/q}$ regularity of the height function. Thus, we can not use the Hanazawa transform in the present case. Another method is to use the Lagrange transform. However, we can not expect the exponential decay unlike Sect. 3.7 for the solutions of the Stokes equations with free boundary condition, because Ω is unbounded domain. The decay of the solutions of the Stokes equations with free boundary condition is only polynomial order, which is not sufficient to controle the term: first derivatives of $\int_0^t \mathbf{v}(y, s) \, ds$ times the second derivatives of $\mathbf{v}$.

To overcome this difficulty, the idea here is to use the Lagrange transform only near the boundary. Let R be a positive number for which $\mathcal{O} = \mathbb{R}^N \setminus \Omega \subset B_{R/2}$. Let $\kappa \in C_0^\infty(B_{2R})$ equal one in B_R. Let $\mathbf{u}(y, t)$ be the velocity field in the Lagrange coordinates $\{\xi\}$. We consider the partial Lagrange transform:

$$x = X_{\mathbf{u}}(y, t) = y + \int_0^t \kappa(y)\mathbf{u}(y, s)\, ds. \tag{3.544}$$

Assume that

$$\int_0^T \|\kappa(\cdot)\mathbf{u}(\cdot, s)\|_{H^1_\infty(\Omega)}\, ds \le \delta, \tag{3.545}$$

where $\delta > 0$ is a small number that will be chosen in such a way that several conditions hold. For example, if $\delta < 1/2$, then the map $x = X_{\mathbf{u}}(y, t)$ is injective for each $t \in (0, T)$. Let

$$\Psi(y, t) = \int_0^t \kappa(y)\mathbf{u}(y, s)\, ds,$$

and so $X_{\mathbf{u}}(y, t) = y + \Psi(y, t)$. Let

$$\Omega_t = \{x = X_{\mathbf{u}}(y, t) \mid y \in \Omega\}, \quad \Gamma_t = \{x = X_{\mathbf{u}}(y, t) \mid y \in \Gamma\}.$$

Let $y = X_{\mathbf{u}}^{-1}(x, t)$ be the inverse of the transformation: $x = X_{\mathbf{u}}(y, t)$ given in (3.544) and set

$$\mathbf{v}(x, t) = \mathbf{u}(X_{\mathbf{u}}^{-1}(x, t), t), \quad \mathfrak{p}(x, t) = \mathfrak{q}(X_{\mathbf{u}}^{-1}(x, t), t).$$

We observe that

$$V_n = \frac{\partial x}{\partial t} \cdot \mathbf{n}_t = \mathbf{u}(\xi, t) \cdot \mathbf{n}_t = \mathbf{v}(x, t) \cdot \mathbf{n}_t$$

on Γ_t, because $\kappa = 1$ on Γ_t, and so the kinematic equation is automatically satisfied. If $\mathbf{v}$ and $\mathfrak{p}$ satisfy Eq. (3.543), then employing the same argument as in Sect. 3.3.2 and using the formula:

$$\mathbf{n}_t = \frac{^\top A\mathbf{n}}{|^\top A\mathbf{n}|} \quad \text{with } A = \left(\frac{\partial x}{\partial y}\right)^{-1} = \mathbf{I} + {}^\top V_0(\nabla \Psi(y, t)), \tag{3.546}$$

which will be given in the beginning of Sect. 3.8.3 below, and (3.112), we see that $\mathbf{u}(y, t)$ and $\mathfrak{q}(y, t)$ satisfy the following equations:

$$\begin{cases} \rho \partial_t \mathbf{u} - \mathrm{Div}\,(\mu \mathbf{D}(\mathbf{u}) - \mathfrak{q}\mathbf{I}) = \mathbf{f}(\mathbf{u}), & \text{in } \Omega^T, \\[2mm] \mathrm{div}\,\mathbf{u} = g(\mathbf{u}) = \mathrm{div}\,\mathbf{g}(\mathbf{u}) & \text{in } \Omega^T, \\[2mm] (\mu \mathbf{D}(\mathbf{u}) - \mathfrak{q}\mathbf{I})\mathbf{n} = \mathbf{h}(\mathbf{u}) & \text{on } \Gamma^T, \\[2mm] \mathbf{u}|_{t=0} = \mathbf{u}_0 & \text{in } \Omega. \end{cases} \tag{3.547}$$

Here, $\mathbf{u}_0 = \mathbf{v}_0$, $\mathbf{n}$ is the unit outer normal to Γ, and the nonlinear terms in the right side of Eq. (3.547) are given as follows:

$$\mathbf{f}(\mathbf{u})|_i = -\rho \sum_{j,k=1}^{N} (1-\kappa) u_j (\delta_{jk} + V_{0jk}(\mathbf{k})) \frac{\partial u_i}{\partial y_k}$$

$$-\rho \sum_{\ell=1}^{N} \frac{\partial \Psi_\ell}{\partial y_i} \left(\frac{\partial u_\ell}{\partial t} + \sum_{j,k=1}^{N} (1-\kappa) u_j (\delta_{jk} + V_{0jk}(\mathbf{k})) \frac{\partial u_\ell}{\partial y_k} \right)$$

$$+ \mu \left(\sum_{j=1}^{N} \frac{\partial}{\partial y_j} (\mathcal{D}_{\mathbf{D}}(\mathbf{k}) \nabla \mathbf{u})_{ij} \right.$$

$$+ \sum_{j,k=1}^{N} V_{0jk}(\mathbf{k}) \frac{\partial}{\partial y_k} (\mathbf{D}(\mathbf{u})_{ij} + (\mathcal{D}_{\mathbf{D}}(\mathbf{k}) \nabla \mathbf{u})_{ij})$$

$$+ \sum_{j,k,\ell=1}^{N} \frac{\partial \Psi_\ell}{\partial y_i} (\delta_{jk} + V_{0jk}(\mathbf{k})) \frac{\partial}{\partial y_k} (\mathbf{D}(\mathbf{u})_{\ell j} + (\mathcal{D}_{\mathbf{D}}(\mathbf{k}) \nabla \mathbf{u})_{\ell j}) \right), \qquad (3.548)$$

$$g(\mathbf{u}) = -(J_0(\mathbf{k}) \operatorname{div} \mathbf{u} + (1 + J_0(\mathbf{k})) V_0(\mathbf{k}) : \nabla \mathbf{u}),$$

$$\mathbf{g}(\mathbf{u}) = -(1 + J_0(\mathbf{k}))^\top V_0(\mathbf{k}) \mathbf{u}, \qquad\qquad (3.549)$$

$$\mathbf{h}(\mathbf{u}) = -\mu \{ \mathbf{D}(\mathbf{u})^\top V_0(\mathbf{k}) \mathbf{n} + (\mathcal{D}_{\mathbf{D}}(\mathbf{k}) \nabla \mathbf{u})(\mathbf{I} + {}^\top V_0(\mathbf{k})) \mathbf{n}$$

$$+ {}^\top (\nabla \Psi)(\mathbf{D}(\mathbf{u}) + \mathcal{D}_{\mathbf{D}}(\mathbf{k}) \nabla \mathbf{u})(\mathbf{I} + {}^\top V_0(\mathbf{k})) \mathbf{n} \}, \qquad (3.550)$$

where $\mathbf{k} = \nabla \int_0^t \kappa(y) \mathbf{u}(y, s) \, ds$, $\Psi = \int_0^t \kappa(y) \mathbf{u}(y, s) \, ds$, and we have used the fact that $\partial_t \Psi_j = \kappa u_j$.

The main result of this section is the following theorem that shows the unique existence theorem of global in time solutions of Eq. (3.547) and asymptotics as $t \to \infty$.

Theorem 3.8.1 *Let $N \geq 3$ and let q_1 and q_2 be exponents such that*

$$max\left(N, \frac{2N}{N-2}\right) < q_2 < \infty, \quad 1/q_1 = 1/q_2 + 1/N.$$

Let b and p be numbers defined by

$$b = \frac{3N}{2q_2} + \frac{1}{2}, \quad p = \frac{2q_2(1+\sigma)}{q_2 - N} \qquad (3.551)$$

with some very small positive number σ. Then, there exists an $\epsilon > 0$ such that if initial data $\mathbf{u}_0 \in B_{q_2,p}^{2(1-1/p)}(\Omega)^N \cap B_{q_1/2,p}^{2(1-1/p)}(\Omega)^N$ satisfies the compatibility *condition:*

$$div\, \mathbf{u}_0 = 0 \quad in\ \Omega, \quad \mathbf{D}(\mathbf{u}_0)\mathbf{n} - <\mathbf{D}(\mathbf{u}_0)\mathbf{n}, \mathbf{n}> \mathbf{n} = 0 \quad on\ \Gamma,$$

and the smallness condition:

$$\|\mathbf{u}_0\|_{B^{2(1-1/p)}_{q_2,p}} + \|\mathbf{u}_0\|_{B^{2(1-1/p)}_{q_1/2,p}} \le \epsilon, \tag{3.552}$$

then Eq. (3.547) admits unique solutions $\mathbf{u}$ *and* $\mathfrak{q}$ *with*

$$\mathbf{u} \in L_p((0,\infty), H^2_{q_2}(\Omega)^N) \cap H^1_p((0,\infty), L_{q_2}(\Omega)^N),$$
$$\mathfrak{q} \in L_p((0,\infty), H^1_{q_2}(\Omega) + \hat{H}^1_{q_2,0}(\Omega)), \tag{3.553}$$

possessing the estimate $[\mathbf{u}]_\infty \le C\epsilon$ *with*

$$[\mathbf{u}]_T = \Big\{ \int_0^T (<s>^b \|\mathbf{u}(\cdot,s)\|_{H^1_\infty(\Omega)})^p\, ds$$
$$+ \int_0^T (<s>^{(b-\frac{N}{2q_1})} \|\mathbf{u}(\cdot,s)\|_{H^1_{q_1}(\Omega)})^p\, ds$$
$$+ (\sup_{0<s<T} <s>^{\frac{N}{2q_1}} \|\mathbf{u}(\cdot,s)\|_{L_{q_1}(\Omega)})^p$$
$$+ \int_0^T (<s>^{(b-\frac{N}{2q_2})} (\|\mathbf{u}(\cdot,s)\|_{H^2_{q_2}(\Omega)} + \|\partial_t \mathbf{u}(\cdot,s)\|_{L_{q_2}(\Omega)}))^p\, ds \Big\}^{1/p}. \tag{3.554}$$

Here, $<s> = (1+s^2)^{\frac{1}{2}}$ *and* C *is a constant that is independent of* ϵ.

Remark 3.8.2 Let $p' = p/(p-1)$, that is $1/p' = 1 - 1/p$. And then,

$$\frac{1}{p'} = \frac{(1+2\sigma)q_2 + N}{2q_2(1+\sigma)}.$$

We choose $\sigma > 0$ small enough in such a way that the following relations hold:

$$1 < q_1 < 2, \quad \frac{N}{q_1} > b > \frac{1}{p'}, \quad \Big(\frac{N}{q_1} - b\Big)p > 1, \quad \Big(b - \frac{N}{2q_2}\Big)p > 1, \quad b \ge \frac{N}{2q_1},$$
$$b \ge \frac{N}{q_2}, \quad \Big(\frac{N}{2q_2} + \frac{1}{2}\Big)p' < 1, \quad bp' > 1, \quad \Big(b - \frac{N}{2q_2}\Big)p' > 1, \quad \frac{N}{q_2} + \frac{2}{p} < 1. \tag{3.555}$$

Remark 3.8.3 The exponent q_2 is used to control the nonlinear terms, and so q_2 is chosen in such a way that $N < q_2 < \infty$. Let

$$\frac{1}{q_1} = \frac{1}{N} + \frac{1}{q_2}, \qquad \frac{1}{q_3} = \frac{1}{q_1} + \frac{1}{q_2}. \tag{3.556}$$

Since we have to take $q_3 > 1$, we have

$$1 > \frac{1}{q_1} + \frac{1}{q_2} = \frac{1}{N} + \frac{2}{q_2},$$

which leads to

$$q_2 > \frac{2N}{N-1}.$$

Thus, we assume that

$$\max(N, \frac{2N}{N-2}) < q_2 < \infty.$$

Remark 3.8.4 If we choose $\delta > 0$ small enough in (3.545), then $x = X_{\mathbf{u}}(y, t)$ becomes a diffeomorphism with suitable regularity from Ω onto Ω_t, and so the original problem (3.543) is globally well-posed.

3.8.1 Maximal L_p–L_q Regularity in an Exterior Domain

In this subsection, we study the maximal L_p–L_q regularity of solutions to the Stokes equations with free boundary condition:

$$\begin{cases} \partial_t \mathbf{u} - \mathrm{Div}\,(\mu \mathbf{D}(\mathbf{u}) - q\mathbf{I}) = \mathbf{f}, \quad \mathrm{div}\,\mathbf{u} = g = \mathrm{div}\,\mathbf{g} & \text{in } \Omega^T, \\ (\mu \mathbf{D}(\mathbf{u}) - q\mathbf{I})\mathbf{n} = \mathbf{h} & \text{on } \Gamma^T, \\ \mathbf{u}|_{t=0} = \mathbf{u}_0 & \text{in } \Omega. \end{cases} \tag{3.557}$$

Let

$$\hat{H}^1_{q,0}(\Omega) = \{\varphi \in L_{q,\mathrm{loc}}(\Omega) \mid \nabla \varphi \in L_q(\Omega)^N, \quad \varphi|_\Gamma = 0\},$$

$$J_q(\Omega) = \{\mathbf{u} \in L_q(\Omega)^N \mid (\mathbf{u}, \nabla \varphi)_\Omega = 0 \quad \text{for any } \varphi \in \hat{H}^1_{q',0}(\Omega)\}.$$

We start with the following proposition.

Proposition 3.8.5 *Let* $1 < q < \infty$. *If* $\mathbf{u} \in H_q^1(\Omega)$ *satisfies* $div\,\mathbf{u} = 0$ *in* Ω, *then* $\mathbf{u} \in J_q(\Omega)$.

To prove Proposition 3.8.5, we need the following lemma.

Lemma 3.8.6 *Let* $1 < q < \infty$, $m \in \mathbb{N}_0$ *and let* G *be a bounded domain whose boundary* ∂G *is a hypersurface of* C^{m+1} *class. Let* $H_{q,0}^0(G) = L_q(G)$ *and for* $m \geq 1$ *let*

$$H_{q,0}^m(G) = \{f \in H^m(G) \mid \partial_x^\alpha f|_{\partial G} = 0 \quad for\ |\alpha| \leq m - 1\}.$$

Then, there exists a linear operator $\mathbb{B} : H_{q,0}^m(G) \to H_{q,0}^{m+1}(G)^N$ *having the following properties:*

(1) *There exists a* $\rho \in C_0^\infty(G)$ *such that* $\rho \geq 0$, $\int_G \rho\,dx = 1$, *and*

$$div\,\mathbb{B}[f] = f - \rho \int_\Omega f\,dx.\ In\ particular,\ if\ \int_G f\,dx = 0,\ then\ div\,\mathbb{B}[f] = f.$$

(2) *We have the estimate:*

$$\|\mathbb{B}[f]\|_{H_q^{k+1}(G)} \leq C_{q,k,G}\|f\|_{H_q^k(G)} \quad (k = 0, \ldots, m).$$

(3) *If* $f = \partial g/\partial x_i$ *with some* $g \subset H_{q,0}^{m+1}(G)$, *then*

$$\|\mathbb{B}[f]\|_{H_q^k(G)} \leq C_{q,k,G}\|g\|_{H_q^k(G)} \quad (k = 0, \ldots, m).$$

Remark 3.8.7

(1) Since $\mathbb{B}[f] \in H_{q,0}^{m+1}(G)$, if the 0 extension of $\mathbb{B}[f]$ to $\mathbb{R}^N$ is written simply by $\mathbb{B}[f]$, then $\mathbb{B}[f] \in H_q^{m+1}(\mathbb{R}^N)^N$ and supp $\mathbb{B}[f] \subset G$.
(2) Lemma 3.8.6 was proved by Bogovski [9, 10] (cf. also Galdi [19]).

To apply Lemma 3.8.6, we use the following lemma.

Lemma 3.8.8 *Let* $1 < q < \infty$ *and let* $2R < L_1 < L_2 < L_3 < L_4 < 5R$. *Let* χ *be a function in* $C^\infty(\mathbb{R}^N)$ *such that* $\chi(x) = 1$ *for* $x \in B_{L_2}$ *and* $\chi(x) = 0$ *for* $x \notin B_{L_3}$. *If* $\mathbf{v} \in H_q^2(G)^N$, $G \in \{\mathbb{R}^N, \Omega, \Omega_{5R}\}$, *satisfies* $div\,\mathbf{v} = 0$ *in* D_{L_1,L_4}, *then* $(\nabla\chi) \cdot \mathbf{v} \in H_{q,0,a}^3(D_{L_2,L_3})$. *Here, we have set*

$$H_{q,0,a}^3(D_{L_2,L_3}) = \{f \in H_{q,0}^3(D_{L_2,L_3}) \mid \int_{D_{L_2,L_3}} f(x)\,dx = 0\}.$$

Here and in the following, we set $D_{L,M} = \{x \in \mathbb{R}^N \mid L < |x| < M\}$ *for* $0 < L < M$.

To prove Lemma 3.8.8, we need the following lemma.

Lemma 3.8.9 *Let* $1 < q < \infty$ *and let* $2R < L_1 < L_2 < L_3 < L_4 < L_5 < L_6 <$ $5R$. *Let* χ *be a function in* $C^\infty(\mathbb{R}^N)$ *such that* $\operatorname{supp} \nabla \chi \subset D_{L_3, L_4}$. *If* $\mathbf{u} \in H_q^2(G)$, $G \in \{\mathbb{R}^N, \Omega, \Omega_{5R}\}$, *satisfies* $\operatorname{div} \mathbf{u} = 0$ *in* D_{L_1, L_6}, *then there exists a* $\mathbf{v} \in H_q^2(\mathbb{R}^N)^N$ *such that* $\operatorname{supp} \mathbf{v} \subset D_{L_2, L_5}$, $\operatorname{div} \mathbf{v} = 0$ *in* $\mathbb{R}^N$ *and* $(\nabla \chi) \cdot \mathbf{v} = (\nabla \chi) \cdot \mathbf{u}$ *in* $\mathbb{R}^N$.

Proof Let $A_0, A_1, \ldots, A_5$ and $B_0, B_1, \ldots, B_5$ be numbers such that $L_2 = A_5 <$ $A_4 < A_3 < A_2 < A_1 < A_0 < L_3 < L_4 < B_0 < B_1 < B_2 < B_3 < B_4 < B_5 =$ L_5. Let $\varphi \in C_0^\infty(\mathbb{R}^N)$ such that $\varphi(x) = 1$ for $A_2 < |x| < B_2$ and $\varphi(x) = 0$ for $|x| < A_3$ or $|x| > B_3$. Note that $\varphi(x) = 1$ on $\operatorname{supp} \nabla \chi$. Let $E = D_{A_4, A_1} \cup D_{B_1, B_4}$. Since $\operatorname{div} \mathbf{u} = 0$ in D_{L_1, L_6} and $\operatorname{supp} \varphi \subset D_{A_3, B_3}$, we have $\operatorname{div} (\varphi \mathbf{u}) = (\nabla \varphi) \cdot \mathbf{u}$ in D_{A_4, B_4} and $\operatorname{div} (\varphi \mathbf{u}) = (\nabla \varphi) \cdot \mathbf{u} \in H_{q,0}^2(E)$. Moreover, we have

$$\int_E (\nabla \varphi) \cdot \mathbf{u} \, dx = \int_{D_{A_4, B_4}} (\nabla \varphi) \cdot \mathbf{u} \, dx = \int_{D_{A_4, B_4}} \operatorname{div} (\varphi \mathbf{u}) \, dx$$

$$= -\int_{S_{A_4}} \frac{x}{|x|} \cdot (\varphi \mathbf{u}) \, d\tau + \int_{S_{B_4}} \frac{x}{|x|} \cdot (\varphi \mathbf{u}) \, d\tau = 0,$$

which leads to $(\nabla \varphi) \cdot \mathbf{u} \in H_{q,0,a}^2(E)$. By Lemma 3.8.6, $\mathbf{v} = \varphi \mathbf{u} - \mathbb{B}[(\nabla \varphi) \cdot \mathbf{u}]$ has the properties: $\mathbf{v} \in H_q^2(\mathbb{R}^N)$, $\operatorname{supp} \mathbf{v} \subset D_{A_3, B_3}$, and $\operatorname{div} \mathbf{v} = 0$ in $\mathbb{R}^N$. Moreover, $(\nabla \chi) \cdot \mathbf{u} = (\nabla \chi) \cdot \mathbf{v}$ in $\mathbb{R}^N$, because $\varphi = 1$ on $\operatorname{supp} \nabla \chi$ and $\mathbb{B}[(\nabla \varphi) \cdot \mathbf{u}] = 0$ on $\operatorname{supp} \nabla \chi$. $\square$

Proof of Lemma 3.8.8 By Lemma 3.8.9, there exists a $\mathbf{w} \in H_q^2(\mathbb{R}^N)^N$ such that $(\nabla \chi) \cdot \mathbf{v} = (\nabla \chi) \cdot \mathbf{w}$ in $\mathbb{R}^N$, $\operatorname{supp} \mathbf{w} \subset B_{L_4}$ and $\operatorname{div} \mathbf{w} = 0$ in $\mathbb{R}^N$. Then, the assertion follows from the following observation:

$$\int_{D_{L_2, L_3}} (\nabla \chi) \cdot \mathbf{v} \, dx = \int_{D_{L_2, L_3}} (\nabla \chi) \cdot \mathbf{w} \, dx = \int_{B_{L_4}} (\nabla \chi) \cdot \mathbf{w} \, dx$$

$$= \int_{B_{L_4}} \operatorname{div} (\chi \mathbf{w}) \, dx = \int_{S_{L_4}} \frac{x}{|x|} \cdot (\chi \mathbf{w}) \, d\tau = 0,$$

because $\chi|_{S_{L_4}} = 0$. This completes the proof of Lemma 3.8.8. $\square$

Proof of Proposition 3.8.5 We use the Sobolev cut off function. Let $\chi(t) \in$ $C_0^\infty((-1, 1))$ equal one for $|t| \le 1/2$, and set

$$\chi_L(x) = \chi\left(\frac{\ln \ln |x|}{\ln \ln L}\right).$$

Notice that

$$|\nabla \chi_L(x)| \le \frac{c}{\ln \ln L} \frac{1}{|x| \ln |x|} \tag{3.558}$$

and $\nabla \chi_L(x)$ vanishes outside of $\dot{\Omega}_L$, where $\dot{\Omega}_L = \{x \in \Omega \mid \exp \sqrt{\ln L} < |x| < L\}$. Since $\mathbf{u} \in H^1_q(\Omega)$ and div $\mathbf{u} = 0$ in Ω, for any $\varphi \in \hat{H}^1_{q',0}(\Omega)$ we have

$$(\mathbf{u}, \nabla \varphi)_\Omega = \lim_{L \to \infty} (\chi_L \mathbf{u}, \nabla \varphi)_\Omega = - \lim_{L \to \infty} (\nabla \chi_L \cdot \mathbf{u}, \varphi)_\Omega.$$

Let φ_0 be the zero extension of φ to $\mathbb{R}^N$, that is $\varphi_0(x) = \varphi(x)$ for $x \in \Omega$ and $\varphi_0(x) = 0$ for $x \notin \Omega$. Since $\varphi \in \hat{H}^1_{q',0}(\Omega)$, $\varphi_0 \in \hat{H}^1_{q'}(\mathbb{R}^N)$, where

$$\hat{H}^1_{q'}(\mathbb{R}^N) = \{\psi \in L_{q',\mathrm{loc}}(\mathbb{R}^N) \mid \nabla \psi \in L_{q'}(\mathbb{R}^N)^N\}.$$

Then, we know (cf. [19]) that there exist constants $c \neq 0$ and C for which

$$\left\| \frac{\varphi_0 - c}{d} \right\|_{L_{q'}(\mathbb{R}^N)} \leq C \|\nabla \varphi_0\|_{L_{q'}(\mathbb{R}^N)} = C \|\nabla \varphi\|_{L_{q'}(\Omega)},$$

where $d(x) = (1 + |x|) \log(2 + |x|)$. Noting that supp $\nabla \chi_L \subset \Omega_R$, we have

$$(\nabla \chi_L \cdot \mathbf{u}, \varphi)_\Omega = c \int_{\dot{\Omega}_L} (\nabla \chi_L(x)) \cdot \mathbf{u}(x) \, dx$$

$$+ \int_{\dot{\Omega}_L} (d(x)(\nabla \chi_L)(x) \cdot \mathbf{u}(x)) \frac{\varphi_0(x) - c}{d(x)} \, dx.$$

By Lemma 3.8.8, we have

$$\int_{\dot{\Omega}_L} (\nabla \chi_L(x)) \cdot \mathbf{u}(x) \, dx = 0.$$

Moreover, by (3.558) and Hölder's inequality, we have

$$\left| \int_{\dot{\Omega}_L} (d(x)(\nabla \chi_L)(x) \cdot \mathbf{u}(x)) \frac{\varphi_0(x) - c}{d(x)} \, dx \right|$$

$$\leq \left(\int_{\dot{\Omega}_L} (|d(x)(\nabla \chi_L)(x)||\mathbf{u}(x)|)^q \, dx \right)^{1/q} \left\| \frac{\varphi_0 - c}{d} \right\|_{L_{q'}(\mathbb{R}^N)}$$

$$\leq \frac{C}{\ln \ln L} \left(\int_{\dot{\Omega}_L} |\mathbf{u}(x)|^q \, dx \right)^{1/q} \|\nabla \varphi\|_{L_q(\Omega)} \to 0 \quad \text{as } L \to \infty.$$

Therefore, we have $(\mathbf{u}, \nabla \varphi)_\Omega = 0$ for any $\varphi \in \hat{H}^1_{q',0}(\Omega)$, that is $\mathbf{u} \in J_q(\Omega)$. This completes the proof of Proposition 3.8.5. □

We next consider the weak Dirichlet problem:

$$(\nabla u, \nabla \varphi)_\Omega = (\mathbf{f}, \nabla \varphi)_\Omega \quad \text{for any } \varphi \in \hat{H}^1_{q',0}(\Omega). \tag{3.559}$$

Then, we know the following fact.

Proposition 3.8.10 *Let $1 < q < \infty$ and let Ω be an exterior domain in $\mathbb{R}^N$ ($N \geq 2$). Then, the weak Dirichlet problem is uniquely solvable. Namely, for any $\mathbf{f} \in L_q(\Omega)^N$, problem (3.559) admits a unique solution $u \in \hat{H}^1_{q,0}(\Omega)$ possessing the estimate: $\|\nabla u\|_{L_q(\Omega)} \leq C\|\mathbf{f}\|_{L_q(\Omega)}$.*

Remark 3.8.11

(1) This proposition was proved by Pruess and Simonett [37, Section 7.4] and by Shibata [48, Theorem 18] independently.
(2) Let $\Omega = \mathbb{R}^N \setminus S_1$ and $\Gamma = S_1$, where S_1 denotes the unit sphere in $\mathbb{R}^N$. Let

$$f(x) = \begin{cases} \ln|x| & N = 2, \\ |x|^{-(N-2)} - 1 & N \geq 3. \end{cases}$$

Then, $f(x)$ satisfies the strong Dirichlet problem: $\Delta f = 0$ in Ω and $f|_\Gamma = 0$. Moreover, $f \in \hat{H}^1_{q,0}(\Omega)$ provided that $q > N/(N-1)$. However, f does not satisfy the weak Dirichlet problem:

$$(\nabla f, \nabla \varphi)_\Omega = 0 \quad \text{for any } \varphi \in \hat{H}^1_{q',0}(\Omega).$$

In fact, $C_0^\infty(\Omega)$ is not dense in $\hat{H}^1_{q',0}(\Omega)$ when $1 < q' < N$. The detailed has been studied in Shibata [48, Appendix A].

By Theorem 3.4.2 in Sect. 3.4, we have the following theorem.

Theorem 3.8.12 *Let $1 < p, q < \infty$ with $2/p + 1/q \neq 1$ and $0 < T < \infty$. Let*

$$\mathbf{u}_0 \in B^{2(1-1/p)}_{q,p}(\Omega)^N, \quad \mathbf{f} \in L_p((0,T), L_q(\Omega)^N),$$

$$g \in L_p(\mathbb{R}, H^1_q(\Omega)) \cap H^{1/2}_p(\mathbb{R}, L_q(\Omega)), \quad \mathbf{g} \in H^1_p(\mathbb{R}, L_q(\Omega)^N), \tag{3.560}$$

$$\mathbf{h} \in H^{1/2}_p(\mathbb{R}, L_q(\Omega)^N) \cap L_p(\mathbb{R}, H^1_q(\Omega)^N)$$

which satisfy the compatibility condition:

$$\operatorname{div} \mathbf{u}_0 = g|_{t=0} \quad \text{in } \Omega$$

and, in addition,

$$(\mu \mathbf{D}(\mathbf{u}_0)\mathbf{n} - \mathbf{h}|_{t=0})_\tau = 0 \quad \text{on } \Gamma \tag{3.561}$$

provided that $2/p + 1/q < 1$. Then, problem (3.557) admits unique solutions $\mathbf{u}$ and $\mathfrak{p}$ with

$$\mathbf{u} \in L_p((0, T), H_q^2(\Omega)^N) \cap H_p^1((0, T), L_q(\Omega)^N),$$

$$\mathfrak{p} \in L_p((0, T), H_q^1(\Omega) + \hat{H}_{q,0}^1(\Omega)) \tag{3.562}$$

satisfying the estimates

$$\|\mathbf{u}\|_{L_p((0,T),H_q^2(\Omega))} + \|\partial_t \mathbf{u}\|_{L_p((0,T),L_q(\Omega))}$$

$$\leq C_\gamma e^{\gamma T} \big[\|\mathbf{u}_0\|_{B_{q,p}^{2(1-1/p)}(\Omega)} + \|\mathbf{f}\|_{L_p((0,T),L_q(\Omega))} + \|(g, \mathbf{h})\|_{L_p(\mathbb{R}, H_q^1(\Omega))}$$

$$+ \|(g, \mathbf{h})\|_{H_p^{1/2}(\mathbb{R}, L_q(\Omega))} + \|g\|_{H_p^1(\mathbb{R}, L_q(\Omega))} \big] \tag{3.563}$$

for some positive constants C and γ.

Remark 3.8.13 In the case where $2/p + 1/q < 1$, $\mu\mathbf{D}(\mathbf{u}_0) \in B_{q,p}^{1-2/p}(\Omega)$ and $1 - 2/p > 1/q$, and so $\mu\mathbf{D}(\mathbf{u}_0)|_\Gamma$ exists. Since $\mathbf{h} \in H_p^{1/2}(\mathbb{R}, L_q(\Omega)^N) \cap L_p(\mathbb{R}, H_q^1(\Omega)^N)$, by complex interpolation theory, $\mathbf{h} \in H_p^{\theta/2}(\mathbb{R}, H_q^{1-\theta}(\Omega)^N)$ for any $\theta \in (0, 1)$. Since $2/p + 1/q < 1$, we can choose θ in such a way that $1 - \theta > 1/q$ and $1/p < \theta/2$. Thus, $\mathbf{h}|_{t=0} \in H_q^{1-\theta}(\Omega)^N$, and so the trace of $\mathbf{h}|_{t=0}$ to Γ exists.

3.8.2 Local Well-Posedness of Eq. (3.547)

In this subsection, we prove the local well-posedness of Eq. (3.547). The following theorem is the main result of this subsection.

Theorem 3.8.14 *Let $2 < p < \infty$, $N < q < \infty$ and $S > 0$. Let Ω be an exterior domain in $\mathbb{R}^N$ ($N \geq 2$) whose boundary Γ is a C^2 compact hypersurface. Assume that $2/p + N/q < 1$. Then, there exists a time $T > 0$ depending on S such that if initial data $\mathbf{u}_0 \in B_{q,p}^{2(1-1/p)}(\Omega)^N$ satisfies $\|\mathbf{u}_0\|_{B_{q,p}^{2(1-1/p)}(\Omega)} \leq S$ and the compatibility condition:*

$$\text{div}\,\mathbf{u}_0 = 0 \ \text{in}\ \Omega, \quad (\mathbf{D}(\mathbf{u}_0)\mathbf{n})_\tau = 0 \ \text{on}\ \Gamma, \tag{3.564}$$

then problem (3.543) admits a unique solution $(\mathbf{u}, \mathfrak{q})$ with

$$\mathbf{u} \in L_p((0, T), H_q^2(\Omega)^N) \cap H_p^1((0, T), L_q(\Omega)^N),$$

$$\mathfrak{q} \in L_p((0, T), H_q^1(\Omega) + \hat{H}_{q,0}^1(\Omega))$$

possessing the estimate:

$$\|\mathbf{u}\|_{L_p((0,T),H_q^2(\Omega))} + \|\partial_t\mathbf{u}\|_{L_p((0,T),L_q(\Omega))} \le CS,$$

$$\int_0^T \|\kappa(\cdot)\mathbf{u}(\cdot,s)\|_{H_\infty^1(\Omega)}\,ds \le \delta$$

for some constant $C > 0$ independent of T and S. Here, δ is the constant appearing in (3.545).

Proof Let T and L be a positive numbers determined late and let

$$\mathcal{I}_T = \{\mathbf{u} \in L_p((0,T), H_q^2(\Omega)^N) \cap H_p^1((0,T), L_q(\Omega)^N) \mid \mathbf{u}|_{t=0} = \mathbf{u}_0,$$

$$<\mathbf{u}>_T := \|\mathbf{u}\|_{L_p((0,T),H_q^2(\Omega))} + \|\partial_t\mathbf{u}\|_{L_p((0,T),L_q(\Omega))} \le L,$$

$$\int_0^T \|\kappa(\cdot)\mathbf{u}(\cdot,s)\|_{H_\infty^1(\Omega)}\,ds \le \delta\}.$$

Since $T > 0$ is chosen small enough eventually, we may assume that $0 < T \le 1$. Given $\mathbf{v} \in \mathcal{I}_T$, let $\mathbf{u}$ be a solution of linear equations:

$$\begin{cases} \partial_t\mathbf{u} - \mathrm{Div}\,(\mu\mathbf{D}(\mathbf{u}) - q\mathbf{I}) = \mathbf{f}(\mathbf{v}) & \text{in } \Omega^T, \\ \mathrm{div}\,\mathbf{u} = g(\mathbf{v}) = \mathrm{div}\,\mathbf{g}(\mathbf{v}) & \text{in } \Omega^T, \\ (\mu\mathbf{D}(\mathbf{u}) - q\mathbf{I})\mathbf{n} = \mathbf{h}(\mathbf{v}) & \text{on } \Gamma^T, \\ \mathbf{u}|_{t=0} = \mathbf{u}_0 & \text{in } \Omega. \end{cases} \tag{3.565}$$

Notice that

$$<\mathbf{v}>_T \le L, \quad \int_0^T \|\kappa(\cdot)\mathbf{v}(\cdot,s)\|_{H_\infty^1(\Omega)}\,ds \le \delta. \tag{3.566}$$

To solve (3.565), we use Theorem 3.8.12. To this end, we introduce $\mathcal{E}[\mathbf{v}] = \mathcal{E}_1[\mathbf{v}]$, which is the functions defined in (3.421) of Sect. 3.6, and e_T, which is the extension map defined by (3.419). Recall that the following formulas hold:

$$<\mathcal{E}_1[\mathbf{v}]>_\infty \le C(\|\mathbf{u}_0\|_{B_{q,p}^{2(1-1/p)}(\Omega)} + <\mathbf{v}>_T) \le C(S + L). \tag{3.567}$$

For the sake of simplicity, we may write $g(\mathbf{v})$, $\mathbf{g}(\mathbf{v})$ and $\mathbf{h}(\mathbf{v})$ given in Eqs. (3.549) and (3.550) as

$$g(\mathbf{v}) = v_1\left(\int_0^t \nabla(\kappa\mathbf{v})\,ds\right)\int_0^t \nabla(\kappa\mathbf{v})\,ds \otimes \nabla\mathbf{v},$$

$$\mathbf{g}(\mathbf{v}) = v_2(\int_0^t \nabla(\kappa\mathbf{v})\,ds) \int_0^t \nabla(\kappa\mathbf{v})\,ds \otimes \mathbf{v},$$

$$\mathbf{h}(\mathbf{v}) = v_3(\int_0^t \nabla(\kappa\mathbf{v})\,ds) \int_0^t \nabla(\kappa\mathbf{v})\,ds \otimes \nabla\mathbf{v},$$

with some matrices of C^1 functions $v_1(\mathbf{k})$, $v_2(\mathbf{k})$, and $v_3(\mathbf{k})$ defined for $|\mathbf{k}| < \delta$. Note that $v_3(\mathbf{k})$ depends also on $x \in \mathbb{R}^N$ with

$$\sup_{|\mathbf{k}| \leq \delta} \|(v_3(\cdot, \mathbf{k}), \partial_{\mathbf{k}} v_3(\cdot, \mathbf{k})\|_{H^1_\infty(\mathbb{R}^N)} \leq C$$

with some constant C, because $\mathbf{n}$ is defined on $\mathbb{R}^N$ with $\|\mathbf{n}\|_{H^1_\infty(\mathbb{R}^N)} < \infty$.

Let e_T be the operator defined in (3.419), and set

$$g_0 = e_T[g(\mathbf{v})], \quad \mathbf{g}_0 = e_T[\mathbf{g}(\mathbf{v})], \quad \mathbf{h}_0 = e_T[\mathbf{h}(\mathbf{v})].$$

Since

$$g_0(\cdot, t) = \begin{cases} 0 & t < 0, \\ [g(\mathbf{v})](\cdot, t) & 0 < t < T, \\ [g(\mathbf{v})](\cdot, 2T - t) & T < t < 2T, \\ 0 & t > 2T, \end{cases}$$

$$\mathbf{g}_0(\cdot, t) = \begin{cases} 0 & t < 0, \\ [\mathbf{g}(\mathbf{v})](\cdot, t) & 0 < t < T, \\ [\mathbf{g}(\mathbf{v})](\cdot, 2T - t) & T < t < 2T, \\ 0 & t > 2T, \end{cases}$$

as follows from $g(\mathbf{v}) = 0$ and $\mathbf{g}(\mathbf{v}) = 0$ at $t = 0$, and since $\operatorname{div} \mathbf{g}(\mathbf{v}) = g(\mathbf{v})$ for $0 < t < T$, we have $\operatorname{div} \mathbf{g}_0 = g_0$ for any $t \in \mathbb{R}$. Moreover, we have

$$g_0 = e_T[v_1(\int_0^t \nabla(\kappa\mathbf{v})\,ds) \int_0^t \nabla(\kappa\mathbf{v})\,ds \otimes \nabla\mathcal{E}_1[\mathbf{v}]],$$

$$\mathbf{g}_0 = e_T[v_2(\int_0^t \nabla(\kappa\mathbf{v})\,ds) \int_0^t \nabla(\kappa\mathbf{v})\,ds \otimes \mathcal{E}_1[\mathbf{v}]], \quad (3.568)$$

$$\mathbf{h}_0 = e_T[v_3(\int_0^t \nabla(\kappa\mathbf{v})\,ds) \int_0^t \nabla(\kappa\mathbf{v})\,ds \otimes \nabla\mathcal{E}_1[\mathbf{v}]],$$

because $\mathcal{E}_1[\mathbf{v}] = \mathbf{v}$ in $(0, T)$. Here, $\mathcal{E}_1[\mathbf{v}]$ has been defined in (3.421).

Let $\mathbf{u}$ and q be solutions of the linear equations:

$$\begin{cases} \partial_t \mathbf{u} - \text{Div}\,(\mu \mathbf{D}(\mathbf{u}) - q\mathbf{I}) = \mathbf{f} & \text{in } \Omega^T, \\ \text{div}\,\mathbf{u} = g_0 = \text{div}\,\mathbf{g}_0 & \text{in } \Omega^T, \\ (\mu \mathbf{D}(\mathbf{u}) - q\mathbf{I})\mathbf{n} = \mathbf{h}_0 & \text{on } \Gamma^T, \\ \mathbf{u}|_{t=0} = \mathbf{u}_0 & \text{in } \Omega, \end{cases} \tag{3.569}$$

and then $\mathbf{u}$ and q are also solutions of the Eq. (3.565), because $g_0 = g(\mathbf{v})$, $g_0 = \mathbf{g}(\mathbf{v})$ and $\mathbf{h}_0 = \mathbf{h}(\mathbf{v})$ for $t \in (0, T)$. In the following, using the Banach fixed point theorem, we prove that there exists a unique $\mathbf{u} \in \mathcal{I}_T$ such that $\mathbf{u} = \mathbf{v}$, which is a required solution of Eq. (3.547).

Applying Theorem 3.8.12 gives that

$$< \mathbf{u} >_T \leq C\{\|\mathbf{u}_0\|_{B_{q,p}^{2(1-1/p)}(\Omega)} + \|\mathbf{f}_0\|_{L_p(\mathbb{R}, L_q(\Omega))} + \|(g_0, \mathbf{h}_0)\|_{L_p(\mathbb{R}, H_q^1(\Omega))}$$

$$+ \|(g_0, \mathbf{h}_0)\|_{H_p^{1/2}(\mathbb{R}, L_q(\Omega))} + \|\partial_t g_0\|_{L_p(\mathbb{R}, L_q(\Omega))}\}, \tag{3.570}$$

provided that the right hand side in (3.570) is finite. In the following, C denotes generic constants independent of T and L. Recalling $\mathbf{k} = \nabla \int_0^t (\kappa \mathbf{v})\,ds$ and the definition of $\mathbf{f}(\mathbf{u})|_i$ given in (3.548), we have

$$\|\mathbf{f}(\mathbf{v})\|_{L_q(\Omega)} \leq C\left\{ \|\mathbf{v}(\cdot, t)\|_{L_\infty(\Omega)} \|\nabla \mathbf{v}(\cdot, t)\|_{L_q(\Omega)} \right.$$

$$+ \int_0^T \|\mathbf{v}(\cdot, s)\|_{H_\infty^1(\Omega)}\,ds \|(\nabla^2 \mathbf{v}, \partial_t \mathbf{v})\|_{L_q(\Omega)}$$

$$\left. + \int_0^T \|\mathbf{v}(\cdot, s)\|_{H_q^2(\Omega)}\,ds \|\nabla \mathbf{v}(\cdot, t)\|_{L_\infty(\Omega)} \right\}. \tag{3.571}$$

By Hölder's inequality, the Sobolev inequality and the assumption: $2/p + N/q < 1$,

$$\int_0^T \|\mathbf{v}(\cdot, s)\|_{H_\infty^1(\Omega)}\,ds \leq C\left(\int_0^T \|\mathbf{v}(\cdot, s)\|_{H_q^2(\Omega)}^p\,ds \right)^{1/p} T^{1/p'} \leq CT^{1/p'} L,$$

$$\int_0^T \|\mathbf{v}(\cdot, s)\|_{H_q^2(\Omega)}\,ds \leq C\left(\int_0^T \|\mathbf{v}(\cdot, s)\|_{H_q^2(\Omega)}^p\,ds \right)^{1/p} T^{1/p'} \leq CT^{1/p'} L,$$

$$\|\mathbf{v}(\cdot, t)\|_{L_\infty(\Omega)} \leq C\|\mathbf{v}(\cdot, t)\|_{H_q^1(\Omega)},$$

$$\|\nabla \mathbf{v}(\cdot, t)\|_{L_\infty(\Omega)} \leq C\|\mathbf{v}(\cdot, t)\|_{B_{q,p}^{2(1-1/p)}(\Omega)}. \tag{3.572}$$

Moreover, by (3.220) and (3.460), we have

$$\sup_{0\leq t\leq T} \|\mathbf{v}(\cdot,t)\|_{B_{q,p}^{2(1-1/p)}(\Omega)} \leq \sup_{t\in(0,\infty)} \|\mathcal{E}_1[\mathbf{v}]\|_{B_{q,p}^{2(1-1/p)}(\Omega)} \leq C(S+L).$$

(3.573)

By (3.572) and (3.573)

$$\sup_{0\leq t\leq T} (\|\mathbf{v}(\cdot,t)\|_{L_\infty(\Omega)}\|\mathbf{v}(\cdot,t)\|_{H_q^1(\Omega)}) \leq C \sup_{0<t<T} \|\mathbf{v}(\cdot,t)\|_{H_q^1(\Omega)}^2$$

$$\leq C \sup_{0<t<T} \|\mathbf{v}(\cdot,t)\|_{B_{q,p}^{2(1-1/p)}(\Omega)}^2 \leq C(S+L)^2,$$

(3.574)

because $2(1-1/p) > 1$ as follows from $p > 2$ and $N < q < \infty$. Combining (3.571), (3.572) and (3.574), we have

$$\|\mathbf{f}(\mathbf{v})\|_{L_p((0,T),L_q(\Omega))} \leq C(S+L)^2(T^{1/p'} + T^{1/p}).$$

(3.575)

We next consider the estimate of g_0 and $\mathbf{h}_0$. By (3.419), (3.566), (3.567), the Sobolev inequality and the assumption: $N < q < \infty$, we have

$$\|g_0(\cdot,t)\|_{H_q^1(\Omega)} \leq C \int_0^T \|\mathbf{v}(\cdot,s)\|_{H_q^2(\Omega)} ds \|\mathbf{v}(\cdot,t)\|_{H_q^2(\Omega)} \quad \text{for } t \in (0,T),$$

$$\|g_0(\cdot,t)\|_{H_q^1(\Omega)} \leq C \int_0^{2T-t} \|\mathbf{v}(\cdot,s)\|_{H_q^2(\Omega)} ds \|\mathbf{v}(\cdot,2T-t)\|_{H_q^2(\Omega)}$$

for $t \in (T, 2T)$, and $\|g_0(\cdot,t)\|_{H_q^1(\Omega)} = 0$ for $t \notin [0, 2T]$. Thus,

$$\|g_0(\cdot,t)\|_{H_q^1(\Omega)} \leq C \begin{cases} 0 & t < 0, \\ T^{1/p'} <\mathbf{v}>_T \|\mathbf{v}(\cdot,t)\|_{H_q^2(\Omega)} & 0 < t < T, \\ T^{1/p'} <\mathbf{v}>_T \|\mathbf{v}(\cdot,2T-t)\|_{H_q^2(\Omega)} & T < t < 2T, \\ 0 & t > 2T, \end{cases}$$

which, combined with (3.566), gives that

$$\|g_0\|_{L_p(\mathbb{R},H_q^1(\Omega))} \leq CL^2 T^{1/p'}.$$

(3.576)

Analogously,

$$\|\mathbf{h}_0\|_{L_p(\mathbb{R},H_q^1(\Omega))} \leq CL^2 T^{1/p'}.$$

(3.577)

Since

$$\| \int_0^t \nabla(\kappa \mathbf{v}) \, ds \|_{L_\infty((0,T),H_q^1(\Omega))} \leq CT^{1/p'} < \mathbf{v} >_T \leq CT^{1/p'}L,$$

$$\| \partial_t \int_0^t \nabla(\kappa \mathbf{v}) \, ds \|_{L_\infty((0,T),L_q(\Omega))} \leq C\|\mathbf{v}\|_{L_\infty((0,T),H_q^1(\Omega))} \leq C(S+L)$$

as follows from (3.573), applying Lemmas 3.6.2 and 3.6.3 to g_0 and $\mathbf{h}_0$, and using the formula of g_0 and $\mathbf{h}_0$ given in (3.568) and the estimates (3.566), we have

$$\|g_0\|_{H_p^{1/2}(\mathbb{R},L_q(\Omega))} \leq C(L+S)LT^{1/2p'}, \tag{3.578}$$

$$\|\mathbf{h}_0\|_{H_p^{1/2}(\mathbb{R},L_q(\Omega))} \leq C(L+S)LT^{1/2p'}. \tag{3.579}$$

We finally estimate $\partial_t g_0$. To this end, we write

$$\partial_t g_0 = \mathbf{v}_2(\int_0^t \nabla(\kappa \mathbf{v}) \, ds) \int_0^t \nabla(\kappa \mathbf{v}) \, ds \otimes \partial_t \mathcal{E}_1[\mathbf{v}](\cdot, t)$$

$$+ \mathbf{v}_2(\int_0^t \nabla(\kappa \mathbf{v}) \, ds)\nabla(\kappa \mathbf{v}) \otimes \mathcal{E}_1[\mathbf{v}](\cdot, t)$$

$$+ \mathbf{v}_2'(\int_0^t \nabla(\kappa \mathbf{v}) \, ds)\nabla(\kappa \mathbf{v}) \int_0^t \nabla(\kappa \mathbf{v}) \, ds \otimes \mathcal{E}_1[\mathbf{v}](\cdot, t) \quad \text{for } t \in (0, T),$$

$$\partial_t g_0 = \mathbf{v}_2(\int_0^{2T-t} \nabla(\kappa \mathbf{v}) \, ds) \int_0^{2T-t} \nabla(\kappa \mathbf{v}) \, ds \otimes \partial_t \mathcal{E}_1[\mathbf{v}](\cdot, t)$$

$$+ \mathbf{v}_2(\int_0^{2T-t} \nabla(\kappa \mathbf{v}) \, ds)\nabla(\kappa \mathbf{v}) \otimes \mathcal{E}_1[\mathbf{v}](\cdot, t)$$

$$+ \mathbf{v}_2'(\int_0^{2T-t} \nabla(\kappa \mathbf{v}) \, ds)\nabla(\kappa \mathbf{v}) \int_0^{2T-t} \nabla(\kappa \mathbf{v}) \, ds \otimes \mathcal{E}_1[\mathbf{v}](\cdot, t)$$

for $t \in (T, 2T)$, and $\partial_t g_0 = 0$ for $t \notin [0, 2T]$, where $\mathbf{v}_2'(\mathbf{k}) = \nabla_\mathbf{k} \mathbf{v}_2(\mathbf{k})$. By (3.573),

$$\|\partial_t g_0\|_{L_p(\mathbb{R},L_q(\Omega))} \leq C(T^{\frac{1}{p'}} + T^{\frac{1}{p}})(L+S)^2,$$

which, combined with (3.570), (3.575)–(3.579), leads to

$$< \mathbf{u} >_T \leq C(L+S)^2(T^{\frac{1}{p'}} + T^{\frac{1}{p}} + T^{\frac{1}{2p'}}).$$

Choosing $T > 0$ so small that

$$C(L+S)(T^{\frac{1}{p'}} + T^{\frac{1}{p}} + T^{\frac{1}{2p'}}) \leq 1/2,$$

we have $< \mathbf{u} >_T \leq (C + 1/2)S + L/2$. Thus, choosing $L = (2C + 1)S$, we have

$$< \mathbf{u} >_T \leq L. \tag{3.580}$$

Moreover, we have

$$\int_0^T \|\kappa(\cdot)\mathbf{u}(\cdot, s)\|_{H_\infty^1(\Omega)} \, ds \leq C_q \|\kappa\|_{H_\infty^1(\Omega)} T^{1/p'} \left(\int_0^T \|\mathbf{u}(\cdot, s)\|_{H_q^2(\Omega)}^p \, ds \right)^{1/p}$$

$$\leq C_q \|\kappa\|_{H_\infty^1(\Omega)} L T^{1/p'},$$

and so choosing $T > 0$ in such a way that $C_q \|\kappa\|_{H_\infty^1(\Omega)} L T^{1/p'} \leq \delta$, we have

$$\int_0^T \|\kappa(\cdot)\mathbf{u}(\cdot, s)\|_{H_\infty^1(\Omega)} \, ds \leq \delta.$$

Thus, $\mathbf{u} \in \mathcal{I}_T$. Let $\mathbf{Q}$ be a map defined by $\mathbf{Qv} = \mathbf{u}$, and then $\mathbf{Q}$ maps $\mathcal{I}_T$ into itself.

Given $\mathbf{v}_i \in \mathcal{I}_T$ ($i = 1, 2$), considering the equations satisfied by $\mathbf{u}_2 - \mathbf{u}_1 = \mathbf{Qv}_2 - \mathbf{Qv}_1$ and employing the same argument as that in proving (3.580), we can show that

$$< \mathbf{Qv}_1 - \mathbf{Qv}_2 >_T \leq C(L + S)(T^{\frac{1}{p'}} + T^{\frac{1}{p}} + T^{\frac{1}{2p'}}) < \mathbf{v}_1 - \mathbf{v}_2 >_T$$

holds. Choosing T smaller if necessary, we may assume that $C(L + S)(T^{\frac{1}{p'}} + T^{\frac{1}{p}} + T^{\frac{1}{2p'}}) \leq 1/2$, and so $\mathbf{Q}$ is a contration map on $\mathcal{I}_T$. By the Banach fixed point theorem, there exists a unique $\mathbf{u} \in \mathcal{I}_T$ such that $\mathbf{Qu} = \mathbf{u}$, which is a required unique solution of Eq. (3.543). This completes the proof of Theorem 3.8.14. $\qquad\square$

Employing the similar argument to that in Sect. 3.7.2, we can prove the following theorem, which is used to prove the global well-posedness.

Theorem 3.8.15 *Let $2 < p < \infty$, $N < q < \infty$ and $T > 0$. Let Ω be an exterior domain in $\mathbb{R}^N$ ($N \geq 2$), whose boundary Γ is a C^2 compact hypersurface. Assume that $2/p + N/q < 1$. Then, there exists an $\epsilon_1 > 0$ depending on T such that if initial data $\mathbf{u}_0 \in B_{q,p}^{2(1-1/p)}(\Omega)^N$ satisfies $\|\mathbf{u}_0\|_{B_{q,p}^{2(1-1/p)}(\Omega)} \leq \epsilon_1$ and the compatibility condition (3.564), then problem (3.543) admits a unique solution $(\mathbf{u}, \mathfrak{q})$ with*

$$\mathbf{u} \in L_p((0, T), H_q^2(\Omega)^N) \cap H_p^1((0, T), L_q(\Omega)^N),$$

$$\mathfrak{q} \in L_p((0, T), H_q^1(\Omega) + \hat{H}_{q,0}^1(\Omega))$$

possessing the estimate:

$$\|\mathbf{u}\|_{L_p((0,T),H_q^2(\Omega))} + \|\partial_t \mathbf{u}\|_{L_p((0,T),L_q(\Omega))} \leq \epsilon_1, \qquad \int_0^T \|\kappa(\cdot)\mathbf{u}(\cdot,s)\|_{H_\infty^1(\Omega)} \leq \delta.$$

3.8.3 A New Formulation of Eq. (3.547)

Let $T > 0$ and let

$$\mathbf{u} \in H_p^1((0,T), L_q(\Omega)^N) \cap L_p((0,T), H_q^2(\Omega)^N),$$

$$\mathfrak{q} \in L_p((0,T), H_q^1(\Omega) + \hat{H}_{q,0}^1(\Omega)) \tag{3.581}$$

be solutions of Eq. (3.547) satisfying the condition (3.545). In what follows, we rewrite Eq. (3.547) in order that the nonlinear terms have suitable decay properties.

In the following, we repeat the argument in Sects. 3.3.2 and 3.3.3. Let $\mathbf{V}_0(\mathbf{k}) = (V_{0ij}(\mathbf{k}))$ be the $N \times N$ matrix defined in (3.103) in Sect. 3.3.2 and set $\frac{\partial y}{\partial x} = \mathbf{I} + \mathbf{V}_0(\mathbf{k}) := (a_{ij}(t))$, where $\mathbf{k} = \{k_{ij} \mid i, j = 1, \ldots, N\}$ are the variables corresponding to $\int_0^t \nabla(\kappa(y)\mathbf{v}(y,s))\,ds$. And also, let $\mathbf{n}_t = {}^\top(n_{t1}, \ldots, n_{tN})$ and $\mathbf{n} = {}^\top(n_1, \ldots, n_N)$ be respective the unit outer normals to Γ_t and Γ. Since

$$0 = \mathbf{n}_t \cdot dx = \sum_{j=1}^N n_{tj} dx_j = \sum_{j,k=1}^N n_{tj} \frac{\partial x_j}{\partial y_k} dy_k,$$

we see that ${}^\top \frac{\partial x}{\partial y} \mathbf{n}_t$ is parallel to $\mathbf{n}$, that is ${}^\top \frac{\partial x}{\partial y} \mathbf{n}_t = c\mathbf{n}$ for some $c \in \mathbb{R} \setminus \{0\}$, and so we have $\mathbf{n}_t = c\, {}^\top \frac{\partial y}{\partial x} \mathbf{n}$. Since $|\mathbf{n}_t| = 1$, we have (3.546). Thus, by (3.103) and (3.309) we have

$$\frac{\partial}{\partial x_i} = \sum_{j=1}^N a_{ji}(t) \frac{\partial}{\partial y_j}, \quad n_{ti} = d(t) \sum_{j=1}^N a_{ji}(t) n_j \tag{3.582}$$

where $d(t) = |(\mathbf{I} + {}^\top\mathbf{V}_0(\mathbf{k}))\mathbf{n}|$. Let J be the Jacobian of the partial Lagrange transform (3.544) and set

$$\ell_{ij} = \delta_{ij} + \int_0^t \frac{\partial}{\partial y_j}(\kappa(y)u_i(y,s))\,ds$$

where $\mathbf{u} = {}^\top(u_1, \ldots, u_N)$. Let

$$a_{ij}(t) = \delta_{ij} + \tilde{a}_{ij}(t), \quad J(t) = 1 + \tilde{J}(t), \quad \ell_{ij}(t) = \delta_{ij} + \tilde{\ell}_{ij}(t) \tag{3.583}$$

with

$$\tilde{a}_{ij}(t) = b_{ij}\Big(\int_0^t \nabla(\kappa(y)\mathbf{v}(y, s))\, ds\Big), \quad \tilde{J}(t) = K\Big(\int_0^t \nabla(\kappa(y)\mathbf{v}(y, s))\, ds\Big),$$

$$\tilde{\ell}_{ij}(t) = m_{ij}\Big(\int_0^t \nabla(\kappa(y)\mathbf{v}(y, s))\, ds\Big) := \int_0^t \frac{\partial}{\partial y_j}(\kappa(y)u_i(y, t))\, ds. \tag{3.584}$$

Here, if we use the symbols defined in Sect. 3.3.2, then $b_{ij} = V_{0ij}$ and $K = J_0$, and so b_{ij} and K are smooth functions defined on $\{\mathbf{k} \mid |\mathbf{k}| \leq \delta\}$ such that $b_{ij}(0) = K(0) = 0$.

Let $\mathbf{v}(x, t) = \mathbf{u}(y, t)$ and $\mathfrak{p}(x, t) = \mathfrak{q}(y, t)$, and then $\mathbf{v}$ and $\mathfrak{p}$ are solutions of Eq. (3.1) with

$$\Omega_t = \{x = y + \int_0^t \kappa(y)\mathbf{u}(y, s)\, ds \mid y \in \Omega\},$$

$$\Gamma_t = \{x = y + \int_0^t \kappa(y)\mathbf{u}(y, s)\, ds \mid y \in \Gamma\}.$$

By (3.582),

$$\frac{\partial v_i}{\partial x_j} + \frac{\partial v_j}{\partial x_i} = D_{ij,t}(\mathbf{u}) := D_{ij}(\mathbf{u}) + \tilde{D}_{ij}(t)\nabla \mathbf{u}$$

with

$$D_{ij}(\mathbf{u}) = \frac{\partial u_i}{\partial y_j} + \frac{\partial u_j}{\partial y_j}, \quad \tilde{D}_{ij}(t)\nabla \mathbf{u} = \sum_{k=1}^N \Big(\tilde{a}_{kj}(t)\frac{\partial u_i}{\partial y_k} + \tilde{a}_{ki}(t)\frac{\partial u_j}{\partial y_k}\Big). \tag{3.585}$$

By (3.107) in Sect. 3.3.2 we also have an important formula:

$$\operatorname{div} \mathbf{v} = \sum_{j=1}^N \frac{\partial v_j}{\partial x_j} = \sum_{j,k=1}^N J(t)a_{kj}(t)\frac{\partial u_j}{\partial y_k} = \sum_{j,k=1}^N \frac{\partial}{\partial y_k}(J(t)a_{kj}(t)u_j), \tag{3.586}$$

and so, the divergence free condition: $\operatorname{div} \mathbf{v} = 0$ leads to

$$\sum_{j,k=1}^N (\tilde{a}_{kj}(t) + \tilde{J}(t)a_{kj}(t))\frac{\partial u_j}{\partial y_k} = \sum_{j,k=1}^N \frac{\partial}{\partial y_k}\{(\tilde{a}_{kj}(t) + \tilde{J}(t)a_{kj}(t))u_j\} = 0. \tag{3.587}$$

And then, Eq. (3.547) is written as follows:

$$\sum_{i=1}^{N} \ell_{is}(t)(\partial_t u_i + (1-\kappa) \sum_{j,k=1}^{N} u_j a_{kj}(t)\frac{\partial u_i}{\partial y_k})$$

$$-\mu \sum_{i,j,k=1}^{N} \ell_{is}(t)a_{kj}(t)\frac{\partial}{\partial y_k} D_{ij,t}(\mathbf{u}) - \frac{\partial \mathfrak{q}}{\partial y_s} = 0 \qquad \text{in } \Omega^T,$$

$$\sum_{j,k=1}^{N} J(t)a_{kj}(t)\frac{\partial u_j}{\partial y_k} = \sum_{j,k=1}^{N} \frac{\partial}{\partial y_k}(J(t)a_{kj}(t)u_j) = 0 \qquad \text{in } \Omega^T,$$

$$\mu \sum_{i,j,k=1}^{N} \ell_{is}(t)a_{kj}(t)D_{ij,t}(\mathbf{u})n_k - \mathfrak{q}n_s = 0 \qquad \text{on } \Gamma^T,$$

$$\mathbf{u}|_{t=0} = \mathbf{u}_0 \qquad \text{in } \Omega,$$

where s runs from 1 through N. Here, we have used the fact that $(\ell_{ij}) = \mathbf{A}^{-1}$.
In order to get some decay properties of the nonlinear terms, we write

$$\int_0^t \nabla(\kappa(y)\mathbf{u}(y,s))\,ds = \int_0^T \nabla(\kappa(y)\mathbf{u}(y,s))\,ds - \int_t^T \nabla(\kappa(y)\mathbf{u}(y,s))\,ds.$$

In (3.584), by the Taylor formula we write

$$a_{ij}(t) = a_{ij}(T) + \mathcal{A}_{ij}(t), \qquad \ell_{ij}(t) = \ell_{ij}(T) + \mathcal{L}_{ij}(t),$$

$$D_{ij,t}(\mathbf{u}) = D_{ij,T}(\mathbf{u}) + \mathcal{D}_{ij}(t)\nabla\mathbf{u}, \qquad J(t) = J(T) + \mathcal{J}(t) \qquad (3.588)$$

with

$$\mathcal{A}_{ij}(t) = -\int_0^1 b'_{ij}\left(\int_0^T \nabla(\kappa(y)\mathbf{u}(y,s))\,ds - \theta \int_t^T \nabla(\kappa(y)\mathbf{u}(y,s))\,ds\right)d\theta$$

$$\times \int_t^T \nabla(\kappa(y)\mathbf{u}(y,s))\,ds$$

$$\mathcal{L}_{ij}(t) = -\int_t^T \frac{\partial}{\partial y_j}(\kappa(y)u_i(y,s))\,ds,$$

$$\mathcal{D}_{ij}(t)\nabla\mathbf{u} = \sum_{k=1}^{N}(\mathcal{A}_{kj}(t)\frac{\partial u_i}{\partial y_k} + \mathcal{A}_{ki}(t)\frac{\partial u_j}{\partial y_k}),$$

$$\mathcal{J}(t) = -\int_0^1 K'\left(\int_0^T \nabla(\kappa(y)\mathbf{u}(y,s))\,ds - \theta \int_t^T \nabla(\kappa(y)\mathbf{u}(y,s))\,ds\right)d\theta$$

$$\times \int_t^T \nabla(\kappa(y)\mathbf{u}(y,s))\,ds,$$

where b'_{ij} and K' are derivatives of b_{ij} and K with respect to $\mathbf{k}$. By the relation:

$$\sum_{s=1}^{N} \ell_{is}(T) a_{sm}(T) = \delta_{si}, \tag{3.589}$$

the first equation in (3.588) is rewritten as follows:

$$\partial_t u_m - \sum_{j,k=1}^{N} a_{kj}(T) \frac{\partial}{\partial y_k} (\mu D_{mj,T}(\mathbf{u}) - \delta_{mj} \mathsf{q}) = \tilde{f}_m(\mathbf{u})$$

with

$$\tilde{f}_m(\mathbf{u}) = -\sum_{s=1}^{N} a_{sm}(T)\{\sum_{i=1}^{N} \mathcal{L}_{is}(t) \partial_t u_i + \sum_{i,j,k=1}^{N} (1-\kappa)\ell_{is}(t) a_{kj}(t) u_j \frac{\partial u_i}{\partial y_k}\}$$

$$+ \mu \sum_{s=1}^{N} a_{sm}(T) \Big\{ \sum_{i,j,k=1}^{N} \ell_{is}(T) a_{kj}(T) \frac{\partial}{\partial y_k} (\mathcal{D}_{ij}(t) \nabla \mathbf{u}) \tag{3.590}$$

$$+ \sum_{i,j,k=1}^{N} \ell_{is}(T) \mathcal{A}_{kj}(t) \frac{\partial}{\partial y_k} D_{ij,t}(\mathbf{u}) + \sum_{i,j,k=1}^{N} \mathcal{L}_{is}(t) a_{kj}(t) \frac{\partial}{\partial y_k} D_{ij,t}(\mathbf{u}) \Big\}.$$

Next, by (3.586)

$$\widetilde{\mathrm{div}}\, \mathbf{u} = \tilde{g}(\mathbf{u}) = \mathrm{div}\, \tilde{\mathbf{g}}(\mathbf{u})$$

with

$$\widetilde{\mathrm{div}}\, \mathbf{u} = \sum_{j,k=1}^{N} J(T) a_{kj}(T) \frac{\partial u_j}{\partial y_k} = \sum_{j,k=1}^{N} \frac{\partial}{\partial y_k} (J(T) a_{kj}(T) u_j),$$

$$\tilde{g}(\mathbf{u}) = \sum_{j,k=1}^{N} (J(T) \mathcal{A}_{kj}(t) + \mathcal{J}(t) a_{kj}(t)) \frac{\partial u_j}{\partial y_k}, \tag{3.591}$$

$$\tilde{g}_k(\mathbf{u}) = \sum_{j=1}^{N} (J(T) \mathcal{A}_{kj}(t) + \mathcal{J}(t) a_{kj}(t)) u_j, \quad \tilde{\mathbf{g}}(\mathbf{u}) = {}^{\top}(\tilde{g}_1(\mathbf{u}), \dots, \tilde{g}_N(\mathbf{u})).$$

Finally, we consider the boundary condition. Let $\tilde{\mathbf{n}}$ be an N-vector defined on $\mathbb{R}^N$ such that $\tilde{\mathbf{n}} = \mathbf{n}$ on Γ and $\|\tilde{\mathbf{n}}\|_{H^2_\infty(\mathbb{R}^N)} \leq C$. In what follows, $\tilde{\mathbf{n}}$ is simply written by $\mathbf{n} = {}^\top(n_1, \ldots, n_N)$. By (3.582) and (3.589)

$$\sum_{j,k=1}^N a_{kj}(T)(\mu D_{mj,T}(\mathbf{u}) - \delta_{mj}\mathsf{q})n_k = \tilde{h}_m(\mathbf{u})$$

with

$$\tilde{h}_m(\mathbf{u}) = -\mu \sum_{j,k=1}^N (a_{kj}(T)\mathcal{D}_{mj}(t)\nabla\mathbf{u} + \mathcal{A}_{kj}(t)D_{mj,t}(\mathbf{u}))n_k$$

$$- \mu \sum_{i,j,k,s=1}^N a_{sm}(T)\mathcal{L}_{is}(t)a_{kj}(t)D_{ij,t}(\mathbf{u})n_k.$$

$$(3.592)$$

By (3.586),

$$\sum_{j,k=1}^N a_{kj}(T)\frac{\partial}{\partial y_k}(\mu D_{mj,T}(\mathbf{u}) - \delta_{mj}\mathsf{q})$$

$$= J(T)^{-1} \sum_{k=1}^N \frac{\partial}{\partial y_k}[\sum_{j=1}^N \{J(T)a_{kj}(T)(\mu D_{mj,T}(\mathbf{u}) - \delta_{mj}\mathsf{q})\}].$$

Thus, letting

$$S_{mk}(\mathbf{u}, \mathsf{q}) = \sum_{j=1}^N J(T)a_{kj}(T)(D_{mj,T}(\mathbf{u}) - \delta_{mj}\mathsf{q}), \quad \tilde{\mathbf{S}}(\mathbf{u}, \mathsf{q}) = (S_{ij}(\mathbf{u}, \mathsf{q})),$$

$$\tilde{\mathbf{f}}(\mathbf{u}) = {}^\top(\tilde{f}_1(\mathbf{u}), \ldots, \tilde{f}_N(\mathbf{u})), \quad \tilde{\mathbf{h}}(\mathbf{u}) = {}^\top(\tilde{h}_1(\mathbf{u}), \ldots, \tilde{h}_N(\mathbf{u})),$$

and using (3.586), we see that $\mathbf{u}$ and q satisfy the following equations:

$$\begin{cases} \partial_t\mathbf{u} - J(T)^{-1}\text{Div}\,\tilde{\mathbf{S}}(\mathbf{u}, \mathsf{q}) = \tilde{\mathbf{f}}(\mathbf{u}) & \text{in } \Omega^T, \\ \widetilde{\text{div}}\,\mathbf{u} = \tilde{g}(\mathbf{u}) = \text{div}\,\tilde{\mathbf{g}}(\mathbf{u}) & \text{in } \Omega^T, \\ \tilde{\mathbf{S}}(\mathbf{u}, \mathsf{q})\mathbf{n} = J(T)\tilde{\mathbf{h}}(\mathbf{u}) & \text{on } \Gamma^T, \\ \mathbf{u}|_{t=0} = \mathbf{u}_0 & \text{in } \Omega. \end{cases}$$

$$(3.593)$$

This is a new formula of Eq. (3.547) which is satisfied by local in time solutions **u** and q of Eq. (3.547). The corresponding linear equations to Eq. (3.593) is the followings:

$$
\begin{cases}
\partial_t \mathbf{u} - J(T)^{-1}\mathrm{Div}\,\tilde{\mathbf{S}}(\mathbf{u},\, \mathsf{q}) = \mathbf{f} & \text{in } \Omega^T, \\[4pt]
\widetilde{\mathrm{div}}\,\mathbf{u} = g = \mathrm{div}\,\mathbf{g} & \text{in } \Omega^T, \\[4pt]
\tilde{\mathbf{S}}(\mathbf{u},\, \mathsf{q})\mathbf{n} = \mathbf{h} & \text{on } \Gamma^T, \\[4pt]
\mathbf{u}|_{t=0} = \mathbf{u}_0 & \text{in } \Omega.
\end{cases}
\tag{3.594}
$$

We call Eq. (3.594) the slightly perturbed Stokes equations.

3.8.4 Slightly Perturbed Stokes Equations

In this subsection we summarize some results obtained by Shibata [49] concerning the slightly perturbed Stokes equations. Let r be an exponent such that $N < r < \infty$. Let $a_{ij}(T)$, $\tilde{a}_{ij}(T)$, $J(T)$ and $\tilde{J}(T)$ be functions defined in (3.583) with $t = T$. We assume that

$$
\|(\tilde{a}_{ij}(T),\, \tilde{J}(T))\|_{L_\infty(\Omega)} + \|\nabla(\tilde{a}_{ij}(T),\, \tilde{J}(T))\|_{L_r(\Omega)} \le \sigma
\tag{3.595}
$$

with some small constant $\sigma > 0$. In view of Theorem 3.8.15, we can choose $\sigma > 0$ small as much as we want if we choose the initial data small. Since $\mathrm{supp}\,\kappa \subset B_{2R}$, $\tilde{a}_{ij}(T)$ and $\tilde{J}(T)$ vanish for $x \notin B_{2R}$. In the following, we write $a_{ij}(T)$, $\tilde{a}_{ij}(T)$, $J(T)$ and $\tilde{J}(T)$ simply by a_{ij}, $\tilde{a}_{ij}$, J and $\tilde{J}$, respectively. And also, we write $\mathbf{A} = (a_{ij}(T))$.

To state the compatibility condition for Eq. (3.594), we modify the equation slightly. Notice that it follows from (3.586) with $u_j = \delta_{mj}\mathsf{q}$ that

$$
\sum_{j,k=1}^N J^{-1}\frac{\partial}{\partial y_k}(Ja_{kj}\delta_{mj}\mathsf{q}) = \sum_{j,k=1}^N a_{kj}\frac{\partial}{\partial y_k}(\delta_{mj}\mathsf{q}) = \sum_{k=1}^N a_{km}\frac{\partial \mathsf{q}}{\partial y_k}
$$

$$
= {}^\top \mathbf{A}\nabla \mathsf{q}|_m.
$$

In the following, we set

$$
\tilde{\nabla}\mathsf{q} = {}^\top \mathbf{A}\nabla \mathsf{q}, \quad \tilde{\mathbf{D}}(\mathbf{u}) = (D_{ij,T}(\mathbf{u})), \quad \tilde{\mathbf{n}} = d_{\mathbf{n}}^{-1}\,{}^\top \mathbf{A}\mathbf{n}, \quad \tilde{n}_i = d_{\mathbf{n}}^{-1}\sum_{j=1}^N a_{ji}(T)n_j,
$$

with $d_{\mathbf{n}} = (\sum_{i,j=1}^N a_{ij}(T) n_i n_j)^{1/2}$, where we have set $\tilde{\mathbf{n}} = {}^\top(\tilde{n}_1, \ldots, \tilde{n}_N)$ and $\mathbf{n} = {}^\top(n_1, \ldots, n_N)$. Note that $|\tilde{\mathbf{n}}| = 1$. And then, we can write

$$J^{-1} \operatorname{Div} \tilde{\mathbf{S}}(\mathbf{u}, q) = J^{-1} \operatorname{Div}(J A \mu \tilde{\mathbf{D}}(\mathbf{u})) - \tilde{\nabla} q,$$
$$\tilde{\mathbf{S}}(\mathbf{u}, q) \mathbf{n} = J d_{\mathbf{n}} (\mu \tilde{\mathbf{D}}(\mathbf{u}) - q \mathbf{I}) \tilde{\mathbf{n}}. \tag{3.596}$$

And then, Eq. (3.594) is rewritten as

$$\begin{cases} \partial_t \mathbf{u} - J^{-1} \operatorname{Div}(J A \mu \tilde{\mathbf{D}}(\mathbf{u})) + \tilde{\nabla} q = \mathbf{f} & \text{in } \Omega^T, \\ \widetilde{\operatorname{div}} \mathbf{u} = g = \operatorname{div} \mathbf{g} & \text{in } \Omega^T, \\ \mu \tilde{\mathbf{D}}(\mathbf{u}) \tilde{\mathbf{n}} - q \tilde{\mathbf{n}} = (J d_{\mathbf{n}})^{-1} \mathbf{h} & \text{on } \Gamma^T, \\ \mathbf{u}|_{t=0} = \mathbf{u}_0 & \text{in } \Omega. \end{cases} \tag{3.597}$$

To obtain the maximal L_p–L_q regularity and some decay properties of solutions of Eq. (3.597), we first consider the following generalized resolvent problem corresponding to Eq. (3.593):

$$\begin{cases} \lambda \mathbf{u} - J^{-1} \operatorname{Div}(J A \mu \tilde{\mathbf{D}}(\mathbf{u})) + \tilde{\nabla} q = \mathbf{f} & \text{in } \Omega^T, \\ \widetilde{\operatorname{div}} \mathbf{u} = g = \operatorname{div} \mathbf{g} & \text{in } \Omega^T, \\ \mu \tilde{\mathbf{D}}(\mathbf{u}) \tilde{\mathbf{n}} - q \tilde{\mathbf{n}} = (J d_{\mathbf{n}})^{-1} \mathbf{h} & \text{on } \Gamma^T, \\ \mathbf{u}|_{t=0} = \mathbf{u}_0 & \text{in } \Omega. \end{cases} \tag{3.598}$$

The following theorem was proved by Shibata [49].

Theorem 3.8.16 *Let $1 < q \leq r$ and $0 < \epsilon_0 < \pi/2$. Assume that Ω is an exterior domain whose boundary Γ is a compact C^2 hypersurface. Let*

$$X_q''(\Omega) = \{(\mathbf{f}, g, \mathbf{g}, \mathbf{h}) \mid \mathbf{f}, \mathbf{g} \in L_q(\Omega)^N, \quad g \in H_q^1(\Omega), \quad \mathbf{h} \in H_q^1(\Omega)^N\},$$

$$\mathcal{X}_q''(\Omega) = \{(F_1, \ldots, F_N) \mid F_1, F_4, F_6 \in L_q(\Omega)^N, \quad F_2 \in H_q^1(\Omega),$$

$$F_3 \in L_q(\Omega), \quad F_5 \in H_q^1(\Omega)^N\}.$$

Then, there exist a constant $\lambda_0 > 0$ and operator families $\mathcal{A}_s(\lambda)$ and $\mathcal{P}_s(\lambda)$ with

$$\mathcal{A}_s(\lambda) \in Hol(\Sigma_{\epsilon_0, \lambda_0}, \mathcal{L}(\mathcal{X}_q''(\Omega), H_q^2(\Omega)^N)),$$

$$\mathcal{P}_s(\lambda) \in Hol(\Sigma_{\epsilon_0, \lambda_0}, \mathcal{L}(\mathcal{X}_q''(\Omega), H_q^1(\Omega) + \hat{H}_{q,0}^1(\Omega)))$$

such that for any $\lambda \in \Sigma_{\epsilon_0, \lambda_0}$ *and* $(\mathbf{f}, g, \mathbf{g}, \mathbf{h}) \in X''_q(\Omega)$, $\mathbf{u} = \mathcal{A}_s(\lambda)\mathbf{F}_\lambda$ *and* $\mathfrak{q} = \mathcal{P}_s(\lambda)\mathbf{F}_\lambda$ *are unique solutions of Eq.* (3.598), *where*

$$\mathbf{F}_\lambda = (\mathbf{f}, g, \lambda^{1/2}g, \lambda\mathbf{g}, \mathbf{h}, \lambda^{1/2}\mathbf{h}).$$

Moreover, we have

$$\mathcal{R}_{\mathcal{L}(X''_q(\Omega), H_q^{2-j}(\Omega)^N)}(\{(\tau\partial_\tau)^\ell(\lambda^{j/2}\mathcal{A}_s(\lambda)) \mid \lambda \in \Sigma_{\epsilon_0, \lambda_0}\}) \leq r_b,$$

$$\mathcal{R}_{\mathcal{L}(X''_q(\Omega), L_q(\Omega)^N)}(\{(\tau\partial_\tau)^\ell(\nabla\mathcal{P}_s(\lambda)) \mid \lambda \in \Sigma_{\epsilon_0, \lambda_0}\}) \leq r_b$$

for $\ell = 0, 1$ *and* $j = 0, 1, 2$ *with some constant* r_b.

To obtain the decay properties of solutions to Eq. (3.597), we first consider the time shifted equations:

$$\begin{cases} \partial_t\mathbf{u} + \lambda_0\mathbf{u} - J^{-1}\mathrm{Div}\,(JA\mu\tilde{\mathbf{D}}(\mathbf{u})) + \tilde{\nabla}\mathfrak{q} = \mathbf{f} & \text{in } \Omega^T, \\ \widetilde{\mathrm{div}}\,\mathbf{u} = g = \mathrm{div}\,\mathbf{g} & \text{in } \Omega^T, \\ \mu\tilde{\mathbf{D}}(\mathbf{u})\tilde{\mathbf{n}} - \mathfrak{q}\tilde{\mathbf{n}} = (Jd_\mathbf{n})^{-1}\mathbf{h} & \text{on } \Gamma^T, \\ \mathbf{u}|_{t=0} = \mathbf{u}_0 & \text{in } \Omega. \end{cases} \qquad (3.599)$$

Employing the same argument as in Sect. 3.4.6, we have the following maximal L_p–L_q regularity theorem for Eq. (3.599) with large $\lambda_0 > 0$.

Theorem 3.8.17 *Let* $1 < p, q < \infty$ *and assume that* $2/p + N/q \neq 1$. *Then, there exist constants* $\sigma > 0$ *and* $\lambda_0 > 0$ *such that if* (3.595) *holds, then the following assertion holds: Let* $\mathbf{u}_0 \in B_{q,p}^{2(1-1/p)}(\Omega)^N$ *be initial data for Eq.* (3.599) *and let* $\mathbf{f}, g, \mathbf{g}, d, \mathbf{h}$ *be given functions for Eq.* (3.599) *with*

$$\mathbf{f} \in L_p(\mathbb{R}, L_q(\Omega)^N), \quad g \in H_p^1(\mathbb{R}, H_q^1(\Omega)) \cap H_p^{1/2}(\mathbb{R}, L_q(B_R)),$$

$$\mathbf{g} \in H_p^1(\mathbb{R}, L_q(\Omega)^N), \quad \mathbf{h} \in H_p^1(\mathbb{R}, H_q^1(\Omega)^N) \cap H_p^{1/2}(\mathbb{R}, L_q(\Omega)^N).$$

Assume that the compatibility conditions:

$$\widetilde{\mathrm{div}}\,\mathbf{u}_0 = g|_{t=0} \quad \text{in } \Omega.$$

When $2/p + 1/q < 1$, *in addition we assume that*

$$Jd_\mathbf{n}\tilde{\mathbf{\Pi}}_0(\mu\tilde{\mathbf{D}}(\mathbf{u}_0)\tilde{\mathbf{n}}) = \tilde{\mathbf{\Pi}}_0(\mathbf{h}|_{t=0}) \quad \text{on } \Gamma,$$

where for any N-vector $\mathbf{d}$ *we set* $\tilde{\Pi}_0 \mathbf{d} = \mathbf{d} - <\mathbf{d}, \tilde{\mathbf{n}}> \tilde{\mathbf{n}}$. *Then, problem* (3.599) *admits unique solutions* $\mathbf{u}$ *and* $\mathfrak{q}$ *with*

$$\mathbf{u} \in H_p^1((0, T), L_q(\Omega)^N) \cap L_p((0, T), H_q^2(\Omega)^N),$$

$$\mathfrak{q} \in L_p((0, T), H_q^1(\Omega) + \hat{H}_{q,0}^1(\Omega))$$

possessing the estimate:

$$\|\mathbf{u}\|_{L_p((0,T),H_q^2(\Omega))} + \|\partial_t \mathbf{u}\|_{L_p((0,T),L_q(\Omega))}$$

$$\leq C(\|\mathbf{u}_0\|_{B_{q,p}^{2(1-1/p)}(\Omega)} + \|\mathbf{f}\|_{L_p(\mathbb{R}, L_q(\Omega))} + \|(g, \mathbf{h})\|_{H_p^{1/2}(\mathbb{R}, L_q(\Omega))}$$

$$+ \|(g, \mathbf{h})\|_{L_p(\mathbb{R}, H_q^1(\Omega))} + \|\partial_t g\|_{L_p(\mathbb{R}, L_q(\Omega))})$$

for some constant C.

Since $\partial_t(<t>^b \mathbf{u}) = <t>^b \partial_t \mathbf{u} + b <t>^{b-1} \mathbf{u}$, if $\mathbf{u}$ and $\mathfrak{q}$ satisfy Eq. (3.599), then $<t>^b \mathbf{u}$ and $<t>^b \mathfrak{q}$ satisfy the equations:

$$\partial_t(<t>^b \mathbf{u}) + \lambda_0(<t>^b \mathbf{u}) - J(T)^{-1}\text{Div}\,(JA\mu\tilde{\mathbf{D}}(<t>^b \mathbf{u}))$$

$$+\tilde{\nabla}(<t>^b \mathfrak{q}) = <t>^b \tilde{\mathbf{f}} + b <t>^{b-1} \mathbf{u} \quad \text{in } \Omega^T,$$

$$\widetilde{\text{div}} <t>^b \mathbf{u} = <t>^b \tilde{g} = \text{div}\,(<t>^b \tilde{\mathbf{g}}) \quad \text{in } \Omega^T,$$

$$\mu\tilde{\mathbf{D}}(<t>^b \mathbf{u})\tilde{\mathbf{n}} - <t>^b \mathfrak{q}\tilde{\mathbf{n}} = <t>^b (Jd_{\mathbf{n}})^{-1}\tilde{\mathbf{h}} \quad \text{on } \Gamma^T,$$

$$<t>^b \mathbf{u}|_{t=0} = \mathbf{u}_0 \quad \text{in } \Omega.$$

Thus, repeated use of Theorem 3.8.17 yields that

$$\| <t>^b \mathbf{u}\|_{L_p((0,T),H_q^2(\Omega))} + \| <t>^b \partial_t \mathbf{u}\|_{L_p((0,T),L_q(\Omega))}$$

$$\leq C(\|\mathbf{u}_0\|_{B_{q,p}^{2(1-1/p)}(\Omega)} + \| <t>^b \mathbf{f}\|_{L_p(\mathbb{R}, L_q(\Omega))}$$

$$+ \|(g_b, \mathbf{h}_b)\|_{H_p^{1/2}(\mathbb{R}, L_q(\Omega))} + \|(g_b, \mathbf{h}_b)\|_{L_p(\mathbb{R}, H_q^1(\Omega))}$$

$$+ \|\partial_t g_b\|_{L_p(\mathbb{R}, L_q(\Omega))}), \tag{3.600}$$

provided that the right hand side is finite. Here, g_b, $\mathbf{g}_b$ and $\mathbf{h}_b$ are suitable extension of $<t>^b g$, $<t>^b \mathbf{g}$, and $<t>^b \mathbf{h}$ to the whole time interval $\mathbb{R}$ such that $<t> g = g_b$, $<t>^b \mathbf{g} = \mathbf{g}_b$, $<t>^b \mathbf{h} = \mathbf{h}_b$ for $t \in (0, T)$ and div $\mathbf{g}_b = g_b$ for $t \in \mathbb{R}$.

We next consider so called L_p–L_q decay estimate of semigroup associated with Eq. (3.597). Thus, we formulate Eq. (3.597) in the semigroup setting. For this purpose, we have to eliminate q. We start with the weak Dirichlet problem:

$$(\tilde{\nabla}u, J\tilde{\nabla}\varphi)_\Omega = (\mathbf{f}, J\tilde{\nabla}\varphi)_\Omega \quad \text{for any } \varphi \in \hat{H}^1_{q',0}(\Omega). \tag{3.601}$$

Here,

$$\tilde{\nabla}\varphi = {}^\top(\sum_{k-1}^N a_{k1}\frac{\partial\varphi}{\partial x_k}, \dots, \sum_{k-1}^N a_{kN}\frac{\partial\varphi}{\partial x_k}) = {}^\top\mathbf{A}\nabla\varphi.$$

Since $\tilde{a}_{ij}$ and $\tilde{J}$ vanish outside of B_{2R}, $\widetilde{\text{div}}\,\mathbf{u} = \text{div}\,\mathbf{u}$ and $\tilde{\nabla}\varphi = \nabla\varphi$ in $\mathbb{R}^N \setminus B_{2R}$. Thus, by Proposition 3.8.10, we have the following result.

Proposition 3.8.18 *Let* $1 < q \leq r$. *Then, for any* $\mathbf{f} \in L_q(\Omega)^N$ *problem* (3.601) *admits a unique solution* $u \in \hat{H}^1_{q,0}(\Omega)$ *possessing the estimate:*

$$\|\nabla u\|_{L_q(\Omega)} \leq C\|\mathbf{f}\|_{L_q(\Omega)}.$$

We next consider the following slightly perturbed Dirichlet problem. Given $\mathbf{u} \in H^2_q(\Omega)^N$, let $K(\mathbf{u})$ be a unique solution of the weak Dirichlet problem:

$$(\tilde{\nabla}K(\mathbf{u}), J\tilde{\nabla}\varphi)_\Omega = (\mu J^{-1}\text{Div}\,(J\mathbf{A}\tilde{\mathbf{D}}(\mathbf{u})) - \tilde{\nabla}\text{div}\,\mathbf{u}, J\tilde{\nabla}\varphi)_\Omega \tag{3.602}$$

for any $\varphi \in \hat{H}^1_{q',0}(\Omega)$, subject to

$$K(\mathbf{u}) = <\tilde{\mathbf{D}}(\mathbf{u})\tilde{\mathbf{n}}, \tilde{\mathbf{n}}> -\widetilde{\text{div}}\,\mathbf{u}.$$

This problem also can be treated by small perturbation of the weak Dirichlet problem.

We now consider the evolution equations:

$$\begin{cases} \partial_t\mathbf{u} - J^{-1}\text{Div}\,(J\mathbf{A}\mu\tilde{\mathbf{D}}(\mathbf{u})) + \tilde{\nabla}K(\mathbf{u}) = 0 & \text{in } \Omega \times (0, \infty), \\ \mu\tilde{\mathbf{D}}(\mathbf{u})\tilde{\mathbf{n}} - K(\mathbf{u})\tilde{\mathbf{n}} = 0 & \text{on } \Gamma \times (0, \infty), \\ \mathbf{u}|_{t=0} = \mathbf{u}_0, \end{cases} \tag{3.603}$$

which is corresponding to Eq. (3.594) with $\lambda_0 = 0$, $\mathbf{f} = g = \mathbf{g} = \mathbf{h} = 0$, and $\mathbf{u}_0 \in \tilde{J}_q(\Omega)$. Here, we have set

$$\tilde{J}_q(\Omega) = \{\mathbf{f} \in L_q(\Omega) \mid (\mathbf{f}, J\tilde{\nabla}\varphi)_\Omega = 0 \quad \text{for any } \varphi \in \hat{H}^1_{q',0}(\Omega)\}.$$

Given $\mathbf{f} \in \tilde{J}_q(\Omega)$, let $\mathbf{u} \in H_q^2(\Omega)^N$ and $\mathfrak{q} \in H_q^1(\Omega) + \hat{H}_{q,0}^1(\Omega)$ be solutions of the resolvent equations:

$$\begin{cases} \lambda \mathbf{u} - J^{-1}\mathrm{Div}\,(JA\mu\tilde{\mathbf{D}}(\mathbf{u})) + \tilde{\nabla}K(\mathbf{u}) = \mathbf{f}, \quad \widetilde{\mathrm{div}}\,\mathbf{u} = 0 \quad \text{in } \Omega, \\ \mu\tilde{\mathbf{D}}(\mathbf{u})\tilde{\mathbf{n}} - K(\mathbf{u})\tilde{\mathbf{n}} = 0 \quad \text{on } \Gamma. \end{cases} \tag{3.604}$$

Employing the same argument as in Sect. 3.4.3, we have $\mathfrak{q} = K(\mathbf{u})$, and so by Theorem 3.8.16 we have $\mathbf{u} = \mathcal{A}_s(\lambda)\mathbf{f}$ and

$$|\lambda|\|\mathbf{u}\|_{L_q(\Omega)} + \|\mathbf{u}\|_{H_q^2(\Omega)} \leq r_b\|\mathbf{f}\|_{L_q(\Omega)}$$

for any $\lambda \in \Sigma_{\epsilon_0,\lambda_0}$. Let

$$\mathcal{D}_{s,q}(\Omega) = \{\mathbf{u} \in \tilde{J}_q(\Omega) \cap H_q^2(\Omega)^N \mid \mu\tilde{\mathbf{D}}(\mathbf{u})\tilde{\mathbf{n}} - K(\mathbf{u})\tilde{\mathbf{n}}|_\Gamma = 0\},$$

$$\mathcal{A}_s\mathbf{u} = J^{-1}\mathrm{Div}\,(JA\mu\tilde{\mathbf{D}}(\mathbf{u})) - \tilde{\nabla}K(\mathbf{u}) \quad \text{for } \mathbf{u} \in \mathcal{D}_{s,q}(\Omega).$$

Then, Eq. (3.603) is written by

$$\dot{\mathbf{u}} - \mathcal{A}_s\mathbf{u} = 0 \quad t > 0, \quad \mathbf{u}|_{t=0} = \mathbf{u}_0$$

for $\mathbf{u}_0 \in \tilde{J}_q(\Omega)$. Notice that $\mu\tilde{\mathbf{D}}(\mathbf{u})\tilde{\mathbf{n}} - K(\mathbf{u})\tilde{\mathbf{n}}|_\Gamma = 0$ for $\mathbf{u} \in \mathcal{D}_{s,q}(\Omega)$ is equivalent to

$$\tilde{\Pi}_0[\mu\tilde{\mathbf{D}}(\mathbf{u})\tilde{\mathbf{n}}] = 0 \quad \text{on } \Gamma. \tag{3.605}$$

We then see that the operator $\mathcal{A}_s$ generates a C^0 analytic semigroup $\{T_s(t)\}_{t\geq 0}$ on $\tilde{J}_q(\Omega)$. Moreover, we have the following theorem.

Theorem 3.8.19 *Assume that $N \geq 3$ and that μ is a positive constant. Then, there exists a $\sigma > 0$ such that if the assumption (3.595) holds, then for any $q \in (1, r]$, there exists a C^0 analytic semigroup $\{T_s(t)\}_{t\geq 0}$ associated with Eq. (3.603) such that for any $\mathbf{u}_0 \in \tilde{J}_q(\Omega)$, $\mathbf{u}$ is a unique solution of Eq. (3.603) with*

$$\mathbf{u} = T_s(t)\mathbf{u}_0$$

$$\in C^0([0,\infty), \tilde{J}_q(\Omega)) \cap C^1((0,\infty), L_q(\Omega)^N) \cap C^0((0,\infty), \mathcal{D}_{s,q}(\Omega)).$$

Moreover, for any $p \in [q, \infty)$ or $p = \infty$ and for any $\mathbf{f} \in \tilde{J}_q(\Omega)$ and $t > 0$ we have the following estimates:

$$\|T_s(t)\mathbf{f}\|_{L_p(\Omega)} \leq C_{q,p}t^{-\frac{1}{2}\left(\frac{1}{q}-\frac{1}{p}\right)}\|\mathbf{f}\|_{L_q(\Omega)},$$

$$\|\nabla T_s(t)\mathbf{f}\|_{L_p(\Omega)} \leq C_{q,p}t^{-\frac{1}{2}-\frac{1}{2}\left(\frac{1}{q}-\frac{1}{p}\right)}\|\mathbf{f}\|_{L_q(\Omega)}. \tag{3.606}$$

Remark 3.8.20 This theorem was proved by Shibata [49].

We finally consider the following equations:

$$\begin{cases} \partial_t \mathbf{u} - J^{-1}\mathrm{Div}\,(J\mathbf{A}\mu\tilde{\mathbf{D}}(\mathbf{u})) + \tilde{\nabla}\mathfrak{q} = \mathbf{f}, & \widetilde{\mathrm{div}}\,\mathbf{u} = 0 & \text{in } \Omega^T, \\ \mu\tilde{\mathbf{D}}(\mathbf{u})\tilde{\mathbf{n}} - \mathfrak{q}\tilde{\mathbf{n}} = 0 & \text{on } \Gamma^T, & (3.607) \\ \mathbf{u}|_{t=0} = 0 & \text{in } \Omega. \end{cases}$$

Let $\psi \in \hat{H}^1_{q,0}(\Omega)$ be a solution of the weak Dirichlet problem:

$$(\tilde{\nabla}\psi, J\tilde{\nabla}\varphi)_\Omega = (\mathbf{f}, J\tilde{\nabla}\varphi)_\Omega \quad \text{for any } \varphi \in \hat{H}^1_{q',0}(\Omega).$$

Let $\mathbf{g} = \mathbf{f} - \tilde{\nabla}\psi$, and then $\mathbf{g} \in \tilde{J}_q(\Omega)$ and

$$\|\mathbf{g}\|_{L_q(\Omega)} + \|\nabla\psi\|_{L_q(\Omega)} \le C\|\mathbf{f}\|_{L_q(\Omega)}.$$

Using this decomposition, we can rewrite Eq. (3.607) as

$$\begin{cases} \partial_t \mathbf{u} - J^{-1}\mathrm{Div}\,(J\mathbf{A}\mu\tilde{\mathbf{D}}(\mathbf{u})) + \tilde{\nabla}(\mathfrak{q} - \psi) = \mathbf{g}, & \widetilde{\mathrm{div}}\,\mathbf{u} = 0 & \text{in } \Omega^T, \\ \mu\tilde{\mathbf{D}}(\mathbf{u})\tilde{\mathbf{n}} - (\mathfrak{q} - \psi)\tilde{\mathbf{n}} = 0 & \text{on } \Gamma^T, \\ \mathbf{u}|_{t=0} = 0 & \text{in } \Omega, \end{cases}$$

where we have used the formula (3.596) and $\psi|_\Gamma = 0$. Since $\mathbf{g} \in \tilde{J}_q(\Omega)$ for any $t \in (0, T)$, we see that by Duhamel's principle, we have

$$\mathbf{u} = \int_0^t T(t - s)\mathbf{g}(s)\,ds = \int_0^t T(t - s)(\mathbf{f}(s) - \nabla\psi(s))\,ds. \quad (3.608)$$

Moreover, we have $\mathfrak{q} = K(\mathbf{u}) + \psi$. This is a solution formula of Eq. (3.607).

3.8.5 Estimates for the Nonlinear Terms

Let $\tilde{\mathbf{f}}(\mathbf{u})$, $\tilde{g}(\mathbf{u})$, $\tilde{\mathbf{g}}(\mathbf{u})$ and $\tilde{\mathbf{h}}(\mathbf{u})$ are functions defined in Sect. 3.8.3. In this subsection, we estimate these functions. In the following we write

$$\| < t >^\alpha \mathbf{w}\|_{L_p((0,T),X)} = \left\{ \int_0^T (< t >^\alpha \|\mathbf{w}(\cdot, t)\|_X)^p\,dt \right\}^{1/p} \quad 1 \le p < \infty,$$

$$\| < t >^\alpha \mathbf{w}\|_{L_\infty((0,T),X)} = \operatorname*{esssup}_{0<t<T} < t >^\alpha \|\mathbf{w}(\cdot, t)\|_X \quad p = \infty.$$

Let $\tilde{\mathbf{f}} = {}^{\top}(\tilde{f}_1(\mathbf{u}), \ldots, f_N(\mathbf{u}))$ be the vector of functions given in (3.590). We first prove that

$$\| <t>^b \tilde{\mathbf{f}}\|_{L_p((0,T),L_{q_1/2}(\Omega))} + \| <t>^b \tilde{\mathbf{f}}\|_{L_p((0,T),L_{q_2}(\Omega))}$$

$$\leq C(\mathcal{I} + [\mathbf{u}]_T^2), \tag{3.609}$$

where $[\mathbf{u}]_T$ is the norm defined in Theorem 3.8.1 and $\mathcal{I} = \|\mathbf{u}_0\|_{B^{2(1-1/p)}_{q_2,p}(\Omega)} + \|\mathbf{u}_0\|_{B^{2(1-1/p)}_{q_1/2,p}(\Omega)}$. Here and in the following, C denotes generic constants independent of $\mathcal{I}$, $[\mathbf{u}]_T$, δ, and T. The value of C may change from line to line. Since we choose $\mathcal{I}$ small enough eventually, we may assume that $0 < \mathcal{I} \leq 1$. Especially, we use the estimates:

$$\mathcal{I}^2 \leq \mathcal{I}, \quad \mathcal{I}[\mathbf{u}]_T \leq \frac{1}{2}(\mathcal{I}^2 + [\mathbf{u}]_T^2) \leq \mathcal{I} + [\mathbf{u}]_T^2$$

below.

Since

$$\int_{\alpha}^{\beta} \|\nabla(\kappa\mathbf{u}(\cdot,s))\|_{L_\infty(\Omega)}\, ds$$

$$\leq C(1+\alpha)^{-b+\frac{1}{p'}}\left(\int_{\alpha}^{\beta} (<s>^b \|\mathbf{u}(\cdot,s)\|_{H^1_\infty(\Omega)})^p\, ds\right)^{1/p},$$

$$\int_{\alpha}^{\beta} \|\nabla^2(\kappa\mathbf{u}(\cdot,s))\|_{L_q(\Omega)}\, ds$$

$$\leq C(1+\alpha)^{-b+\frac{N}{2q_2}+\frac{1}{p'}}\left(\int_{\alpha}^{\beta} (<s>^{b-\frac{N}{2q_2}} \|\mathbf{u}(\cdot,s)\|_{H^2_{q_2}(\Omega)})^p\, ds\right)^{1/p}$$

for any $0 \leq \alpha < \beta \leq T$, where $q \in [1, q_2]$, we have

$$\int_{\alpha}^{\beta} \|\nabla(\kappa\mathbf{u}(\cdot,s))\|_{L_\infty(\Omega)}\, ds \leq C[\mathbf{u}]_T(1+\alpha)^{-b+\frac{1}{p'}},$$

$$\int_{\alpha}^{\beta} \|\nabla^2(\kappa\mathbf{u}(\cdot,s))\|_{L_q(\Omega)}\, ds \leq C[\mathbf{u}]_T \tag{3.610}$$

for any $0 \leq \alpha < \beta \leq T$, where $q \in [1, q_2]$, because $b > \frac{N}{2q_2} + \frac{1}{p'}$ as follows from (3.555). By (3.405), we have

$$\sup_{t\in(0,T)} <t>^{b-\frac{N}{2q_2}} \|\mathbf{u}(\cdot,t)\|_{B^{2(1-1/p)}_{q_2,p}(\Omega)} \leq C(\|\mathbf{u}_0\|_{B^{2(1-1/p)}_{q_2,p}(\Omega)}$$

$$+ \| <t>^{b-\frac{N}{2q_2}} \mathbf{u}\|_{L_p((0,T),H^2_{q_2}(\Omega))} + \| <t>^{b-\frac{N}{2q_2}} \partial_t\mathbf{u}\|_{L_p((0,T),L_{q_2}(\Omega))}\}. \tag{3.611}$$

Since $2/p + N/q_2 < 1$, $B_{q_2,p}^{2(1-1/p)}(\Omega)$ is continuously imbedded into $H_\infty^1(\Omega)$, and so by (3.611)

$$\| <t>^{b-\frac{N}{q_2}} \mathbf{u}\|_{L_\infty((0,T),H_\infty^1(\Omega))} \le C(\mathcal{I} + [\mathbf{u}]_T). \qquad (3.612)$$

Applying (3.545), (3.610) and (3.611) to the formulas in (3.583) and (3.584) and using the fact that $-b + \frac{1}{p'} < -\frac{N}{2q_2}$ and $-b + \frac{N}{2q_2} \le -\frac{N}{2q_2}$, which follows from (3.555), give

$$\|(a_{ij}(t), J(t), \ell_{ij}(t), \mathcal{A}_{ij}(t), \mathcal{J}(t), \mathcal{L}_{ij}(t))\|_{L_\infty(\Omega)} \le C,$$

$$\|(\mathcal{A}_{ij}(t), \mathcal{J}(t), \mathcal{L}_{ij}(t))\|_{L_\infty(\Omega)}$$

$$\le C \int_t^T \|\nabla(\kappa\mathbf{u}(\cdot, s))\|_{L_\infty(\Omega)}\, ds \le C[\mathbf{u}]_T <t>^{-b+\frac{1}{p'}} \le C[\mathbf{u}]_T <t>^{-\frac{N}{2q_2}},$$

$$\|\nabla(a_{ij}(t), J(t), \ell_{ij}(t), \mathcal{A}_{ij}(t), \mathcal{J}(t), \mathcal{L}_{ij}(t))\|_{L_q(\Omega)}$$

$$\le C \int_0^T \|\nabla^2(\kappa\mathbf{u}(\cdot, s))\|_{L_q(\Omega)} \le C[\mathbf{u}]_T,$$

$$\|\partial_t(a_{ij}(t), J(t), \ell_{ij}(t), \mathcal{A}_{ij}(t), \mathcal{J}(t), \mathcal{L}_{ij}(t))\|_{L_\infty(\Omega)} \le C\|\nabla(\kappa\mathbf{u}(\cdot, t))\|_{L_\infty(\Omega)}$$

$$\le C(\mathcal{I} + [\mathbf{u}]_T) <t>^{-b+\frac{N}{2q_2}} \le C(\mathcal{I} + [\mathbf{u}]_T) <t>^{-\frac{N}{2q_2}} \qquad (3.613)$$

for any $t \in (0, T]$, where $q \in [1, q_2]$. Moreover, we have

$$(\tilde{a}_{ij}, \tilde{J}, \tilde{\ell}_{ij}, \mathcal{A}_{ij}, \mathcal{J}, \mathcal{L}_{ij})(x, t) = 0 \quad \text{for } x \notin B_{2R} \text{ and } t \in [0, T]. \qquad (3.614)$$

By (3.613) and (3.614),

$$\|a_{sm}(T)\mathcal{L}_{is}(t)\partial_t u_i(t)\|_{L_q(\Omega)} \le C[\mathbf{u}]_T <t>^{-b+\frac{1}{p'}} \|\partial_t u_i(t)\|_{L_{q_2}(\Omega)}$$

for any $q \in [1, q_2]$. Since $\frac{1}{p'} < b - \frac{N}{2q_2}$ as follows from (3.555), we have

$$\| <t>^b a_{sm}(T)\mathcal{L}_{is}\partial_t u_i\|_{L_p((0,T),L_q(\Omega))} \le C(\mathcal{I} + [\mathbf{u}]_T^2)$$

for any $q \in [1, q_2]$.

Next, by Hölder's inequality,

$$<t>^b \|\mathbf{u}(\cdot, t) \cdot \nabla\mathbf{u}(\cdot, t)\|_{L_{q_1/2}(\Omega)}$$

$$\le <t>^{\frac{N}{2q_1}} \|\mathbf{u}(\cdot, t)\|_{L_{q_1}(\Omega)} <t>^{b-\frac{N}{2q_1}} \|\nabla\mathbf{u}(\cdot, t)\|_{L_{q_1}(\Omega)}$$

and so, by (3.613), we have

$$\| <t>^b a_{sm}(T)\ell_{is}a_{kj}u_j \frac{\partial u_i}{\partial \xi_k} \|_{L_p((0,T),L_{q_1/2}(\Omega))} \le C[\mathbf{u}]_T^2.$$

Since

$$<t>^b \|\mathbf{u}\cdot\nabla\mathbf{u}(\cdot,t)\|_{L_{q_2}(\Omega)}$$

$$\le <t>^{\frac{N}{2q_2}} \|\mathbf{u}(\cdot,t)\|_{L_\infty(\Omega)} <t>^{b-\frac{N}{2q_2}} \|\nabla\mathbf{u}(\cdot,t)\|_{L_{q_2}(\Omega)},$$

by (3.613)

$$\| <t>^b a_{sm}(T)\ell_{is}a_{kj}u_j \frac{\partial u_i}{\partial \xi_k} \|_{L_p((0,T),L_{q_2}(\Omega))} \le C(\mathcal{I}+[\mathbf{u}]_T)[\mathbf{u}]_T \le C(\mathcal{I}+[\mathbf{u}]_T^2).$$

Since

$$\frac{\partial}{\partial \xi_k}(\mathcal{D}_{ij}(t)\nabla\mathbf{u}) = \sum_{m=1}^{N}(\mathcal{A}_{mj}(t)\frac{\partial^2 u_i}{\partial \xi_k \partial \xi_m} + \mathcal{A}_{mi}(t)\frac{\partial^2 u_j}{\partial \xi_k \partial \xi_m})$$

$$+ \sum_{m=1}^{N}((\frac{\partial}{\partial \xi_m}\mathcal{A}_{mj}(t))\frac{\partial u_i}{\partial \xi_m} + (\frac{\partial}{\partial \xi_k}\mathcal{A}_{mi}(t))\frac{\partial u_j}{\partial \xi_m}),$$

by (3.613) and (3.614),

$$<t>^b \|\frac{\partial}{\partial \xi_k}(\mathcal{D}_{ij}(\cdot)\nabla\mathbf{u})\|_{L_q(\Omega)}$$

$$\le C[\mathbf{u}]_T\{<t>^{b-\frac{N}{2q_2}} \|\nabla^2\mathbf{u}(\cdot t)\|_{L_{q_2}(\Omega)} + <t>^b \|\nabla\mathbf{u}(\cdot,t)\|_{L_\infty(\Omega)}\}, \tag{3.615}$$

for any $q \in [1, q_2]$, and therefore

$$\| <t>^b a_{sm}(T)\ell_{is}(T)a_{kj}(T)\frac{\partial}{\partial \xi_k}(\mathcal{D}_{ij}(\cdot)\nabla\mathbf{u})\|_{L_p((0,T),L_q(\Omega))} \le C[\mathbf{u}]_T^2$$

for any $q \in [1, q_2]$. Since

$$\frac{\partial}{\partial \xi_k}D_{ij,T}(\mathbf{u}) = \sum_{m=1}^{N}(a_{mj}(T)\frac{\partial^2 u_i}{\partial \xi_k \partial \xi_m} + a_{mi}(T)\frac{\partial^2 u_j}{\partial \xi_k \partial \xi_m})$$

$$+ \sum_{m=1}^{N}((\frac{\partial}{\partial \xi_m}a_{mj}(T))\frac{\partial u_i}{\partial \xi_m} + (\frac{\partial}{\partial \xi_k}a_{mi}(T))\frac{\partial u_j}{\partial \xi_m}),$$

by (3.613) and (3.614),

$$< t >^b \|a_{sm}(T)\ell_{is}(T)\mathcal{A}_{kj}(t)\frac{\partial}{\partial \xi_k} D_{ij,T}(\mathbf{u})\|_{L_q(\Omega)}$$

$$\leq C[\mathbf{u}]_T\{< t >^{b-\frac{N}{2q_2}} \|\nabla^2\mathbf{u}(\cdot t)\|_{L_{q_2}(\Omega)} + < t >^b \|\nabla\mathbf{u}(\cdot, t)\|_{L_\infty(\Omega)}\},$$

and so

$$\| < t >^b a_{sm}(T)\ell_{is}(T)\mathcal{A}_{kj}\frac{\partial}{\partial \xi_k} D_{ij,T}(\mathbf{u})\|_{L_p((0,T),L_q(\Omega))} \leq C[\mathbf{u}]_T^2$$

for any $q \in [1, q_2]$. Analogously, we have

$$\| < t >^b a_{sm}(T)\mathcal{L}_{is}a_{kj}\frac{\partial}{\partial \xi_k} D_{ij,T}(\mathbf{u})\|_{L_p((0,T),L_q(\Omega))} \leq C[\mathbf{u}]_T^2$$

for any $q \in [1, q_2]$. Summing up, we have obtained (3.609).

We now consider $\tilde{g}$ and $\mathbf{h} = {}^\top(\tilde{h}_1(\mathbf{u}), \dots, \tilde{h}_N(\mathbf{u}))$, which have been defined in (3.591) and (3.592), respectively. To estimate the $H_p^{\frac{1}{2}}$ norm, we use the following lemma.

Lemma 3.8.21 *Let* $f \in H_\infty^1(\mathbb{R}, L_\infty(\Omega))$ *and* $g \in H_p^{\frac{1}{2}}(\mathbb{R}, L_{q_2}(\Omega))$. *Assume that* $f(x, t) = 0$ *for* $(x, t) \notin B_R \times \mathbb{R}$. *Then,*

$$\|fg\|_{H_p^{\frac{1}{2}}(\mathbb{R},L_q(\Omega))} \leq C_q \|f\|_{H_\infty^1(\mathbb{R},L_\infty(\Omega))}\|g\|_{H_p^{\frac{1}{2}}(\mathbb{R},L_{q_2}(\Omega))}. \tag{3.616}$$

for any $q \in [1, q_2]$ *with some constant* C_q *depending on* q *and* q_2.

Proof To prove the lemma, we use the fact that

$$H_p^{\frac{1}{2}}(\mathbb{R}, L_q(\Omega)) = (L_p(\mathbb{R}, L_q(\Omega)), H_p^1(\mathbb{R}, L_q(\Omega)))_{[\frac{1}{2}]}, \tag{3.617}$$

where $(\cdot, \cdot)_{[1/2]}$ denotes a complex interpolation functor. Let $q \in [1, q_2]$. Noting that $f(x, t) = 0$ for $(x, t) \notin B_R \times \mathbb{R}$, we have

$$\|\partial_t(fg)\|_{L_q(\Omega)} \leq \|\partial_t f\|_{L_\infty(\Omega)}\|g\|_{L_{q_2}(\Omega)} + \|f\|_{L_\infty(\Omega)}\|\partial_t g\|_{L_{q_2}(\Omega)},$$

and therefore

$$\|\partial_t(fg)\|_{L_p(\mathbb{R},L_q(\Omega))} \leq C\|f\|_{H_\infty^1(\mathbb{R},L_\infty(\Omega))}\|g\|_{H_p^1(\mathbb{R},L_{q_2}(\Omega))}$$

for any $q \in [1, q_2]$. Moreover, we easily see that

$$\|fg\|_{L_p(\mathbb{R}, L_q(\Omega))} \leq C\|f\|_{L_\infty(\mathbb{R}, L_\infty(\Omega))}\|g\|_{L_p(\mathbb{R}, L_{q_2}(\Omega))}.$$

Thus, by (3.617), we have (3.616), which completes the proof of Lemma 3.8.21. □

To use the maximal L_p–L_q estimate, we have to extend $\tilde{g}$, $\tilde{\mathbf{g}}$ and $\tilde{\mathbf{h}}$ to $\mathbb{R}$ with respect to time variable. For this purpose, we introduce an extension operator $\tilde{e}_T$. Let f be a function defined on $(0, T)$ such that $f|_{t=T} = 0$, and then $\tilde{e}_T$ is an operator acting on f defined by

$$[\tilde{e}_T f](t) = \begin{cases} 0 & (t > T), \\ f(t) & (0 < t < T), \\ f(-t) & (-T < t < 0), \\ 0 & (t < -T). \end{cases} \tag{3.618}$$

Lemma 3.8.22 Let $1 < p < \infty$, $1 \leq q \leq q_2$, and $0 \leq a \leq b$. Let $f \in H_\infty^1((0, T), L_\infty(\Omega))$ and $g \in H_p^1((0, T), L_{q_2}(\Omega)) \cap L_p((0, T), H_{q_2}^2(\Omega))$. Assume that $f|_{t=T} = 0$ and $f = 0$ for $(x, t) \notin B_R^T$. Let $< t >= (1 + t^2)^{1/2}$. Then, we have

$$\|\tilde{e}_T(< t >^a f\nabla g)\|_{H_p^{\frac{1}{2}}(\mathbb{R}, L_q(\Omega))}$$

$$\leq C\| < t >^{\frac{N}{2q_2}} f\|_{H_\infty^1((0,T),L_\infty(\Omega))}(\| < t >^{b-\frac{N}{2q_2}} g\|_{L_p((0,T),H_{q_2}^2(\Omega))}$$

$$+ \| < t >^{b-\frac{N}{2q_2}} \partial_t g\|_{L_p((0,T),L_{q_2}(\Omega))} + \|g|_{t=0}\|_{B_{q_2,p}^{2(1-1/p)}(\Omega)}). \tag{3.619}$$

Proof Let $f_0(t) =< t >^{a-b+\frac{N}{2q_2}} f(t)$ and $g_0(t) =< t >^{b-\frac{N}{2q_2}} g(t)$, and then $< t >^a f\nabla g = f_0\nabla g_0$. Let h be a function in $B_{q_2,p}^{2(1-1/p)}(\mathbb{R}^N)$ such that $h = g|_{t=0}$ in Ω and $\|h\|_{B_{q_2,p}^{2(1-1/p)}(\Omega)} \leq C\|g|_{t=0}\|_{B_{q_2,p}^{2(1-1/p)}(\Omega)}$. Similarly to (3.416), we define $T_v(t)h$ by letting $T_v(t)h = e^{-(2-\Delta)}h$.

Recall the operator e_T defined in (3.419) and note that $g_0|_{t=0} = g|_{t=0} = T_v(t)h|_{t=0}$ in Ω. Let

$$G(t) = e_T[g_0 - T_v(\cdot)h](t) + T_v(t)h$$

for $t > 0$ and let

$$[\iota g](t) = \begin{cases} G(t) & (t > 0), \\ G(-t) & (t < 0), \end{cases} \qquad [\iota f](t) = \begin{cases} 0 & (t > T), \\ f_0(t) & (0 < t < T), \\ f_0(-t) & (-T < t < 0), \\ 0 & (t < -T). \end{cases}$$

Since $G(t) = g_0(t)$ for $0 < t < T$, we have

$$\tilde{e}_T[< t >^a f\nabla g](t) = \begin{cases} 0 & (t > T) \\ f_0(t)\nabla g_0(t) & (0 < t < T) \\ f_0(-t)\nabla g_0(-t) & (-T < t < 0) \\ 0 & (t < -T) \end{cases}$$

$$= \begin{cases} 0 & (t > T) \\ f_0(t)\nabla G(t) & (0 < t < T) \\ f_0(-t)\nabla G(-t) & (-T < t < 0) \\ 0 & (t < -T) \end{cases} = [\iota f](t)\nabla[\iota g](t).$$

By Lemma 3.8.21,

$$\|\tilde{e}_T[< t >^a f\nabla g]\|_{H_p^{\frac{1}{2}}(\mathbb{R}, L_q(\Omega))} = \|[\iota f]\nabla[\iota g]\|_{H_p^{\frac{1}{2}}(\mathbb{R}, L_q(\Omega))}$$

$$\le C\|\iota f\|_{H_\infty^1(\mathbb{R}, L_q(\Omega))}\|\nabla(\iota g)\|_{H_p^{\frac{1}{2}}(\mathbb{R}, L_{q_2}(\Omega))}.$$

Since $f_0(t)|_{t=T} = 0$, we have

$$\|\iota f\|_{H_\infty^1(\mathbb{R}, L_\infty(\Omega))} = 2\|f_0\|_{H_\infty^1((0,T), L_\infty(\Omega))} \le \| < t >^{\frac{N}{2q_2}} f\|_{H_\infty^1((0,T), L_\infty(\Omega))},$$

because $a - b \le 0$.

To estimate $\|\nabla(\iota g)\|_{H_p^{\frac{1}{2}}(\mathbb{R}, L_{q_2}(\Omega))}$, we use Lemma 3.6.3. And then, by Lemma 3.6.3 and (3.417), we have

$$\|\nabla(\iota g)\|_{H_p^{\frac{1}{2}}(\mathbb{R}, L_{q_2}(\Omega))}$$

$$\le C(\|\iota g\|_{H_p^1(\mathbb{R}, L_{q_2}(\Omega))} + \|\iota g\|_{L_p(\mathbb{R}, H_{q_2}^2(\Omega))})$$

$$\le C(\|G\|_{H_p^1((0,\infty), L_{q_2}(\Omega))} + \|G\|_{L_p((0,\infty), H_{q_2}^2(\Omega))})$$

$$\le C(\|g_0 - T_v(\cdot)h\|_{H_p^1((0,T), L_{q_2}(\Omega))} + \|g_0 - T_v(\cdot)h\|_{L_p((0,T), H_{q_2}^2(\Omega))})$$

$$+ \|T_v(\cdot)h\|_{H^1_p((0,T),L_{q_2}(\Omega))} + \|T_v(\cdot)h\|_{L_p((0,\infty),H^2_{q_2}(\Omega))})$$

$$\leq C(\| < t >^{b-\frac{N}{2q_2}} \partial_t g\|_{L_p((0,T),L_{q_2}(\Omega))} + \| < t >^{b-\frac{N}{2q_2}} g\|_{L_p((0,T),H^2_{q_2}(\Omega))}$$

$$+ \|g|_{t=0}\|_{B^{2(1-1/p)}_{q_2,p}(\Omega)}).$$

This completes the proof of Lemma 3.8.22. $\qquad\qquad\square$

Recall the definitions of $\tilde{g}(\mathbf{u})$ and $\tilde{h}_m(\mathbf{u})$ given in (3.591) and (3.592). By Lemma 3.8.22 and (3.613)

$$\|\tilde{e}_T[< t >^a \tilde{g}(\mathbf{u})]\|_{H^{\frac{1}{2}}_p(\mathbb{R},L_q(\Omega))}$$

$$\leq \sum_{j,k=1}^N \| < t >^{\frac{N}{2q_2}} (J(T)\mathcal{A}_{kj}(\cdot) + T_v(\cdot)a_{kj}(\cdot))\|_{H^1_\infty((0,T),L_\infty(\Omega))}$$

$$\times (\| < t >^{b-\frac{N}{2q_2}} \mathbf{u}\|_{L_p((0,T),H^2_{q_2}(\Omega))} + \| < t >^{b-\frac{N}{2q_2}} \partial_t\mathbf{u}\|_{L_p((0,T),L_{q_2}(\Omega))}$$

$$+ \|\mathbf{u}_0\|_{B^{2(1-1/p)}_{q_2,p}(\Omega)})$$

$$\leq C(\mathcal{I} + [\mathbf{u}]^2_T) \tag{3.620}$$

for any $a \in [0, b]$ and $q \in [1, q_2]$. Analogously, we have

$$\|\tilde{e}_T[< t >^a \tilde{\mathbf{h}}(\mathbf{u})]\|_{H^{\frac{1}{2}}_p(\mathbb{R},L_q(\Omega))} \leq C(\mathcal{I} + [\mathbf{u}]^2_T) \tag{3.621}$$

for any $a \in [0, b]$ and $q \in [1, q_2]$.

Next, by (3.613), (3.614) and (3.618),

$$\|\tilde{e}_T[< t >^a \tilde{g}(\mathbf{u})]\|_{L_p(\mathbb{R},H^1_q(\Omega))}$$

$$\leq \sum_{j,k=1}^N \| < t >^{\frac{N}{2q_2}} (J(T)\mathcal{A}_{kj}(\cdot) + \mathcal{J}(\cdot)a_{kj}(\cdot))\|_{L_\infty((0,T),L_\infty(\Omega))}$$

$$\times \| < t >^{b-\frac{N}{2q_2}} \mathbf{u}\|_{L_p((0,T),H^2_{q_2}(\Omega))}$$

$$+ \sum_{j,k=1}^N \|\nabla(J(T)\mathcal{A}_{kj}(\cdot) + \mathcal{J}(\cdot)a_{kj}(\cdot))\|_{L_\infty((0,T),L_q(\Omega))}$$

$$\times \| < t >^b \mathbf{u}\|_{L_p((0,T),H^1_\infty(\Omega))}$$

$$\leq C[\mathbf{u}]^2_T \tag{3.622}$$

for any $a \in [0, b]$ and $q \in [1, q_2]$. Analogously, we have

$$\|\tilde{e}_T[< t >^b \tilde{\mathbf{h}}(\mathbf{u})]\|_{L_p((0,T), H_q^1(\Omega))} \leq C[\mathbf{u}]_T^2 \tag{3.623}$$

for any $a \in [0, b]$ and $q \in [1, q_2]$.

We finally consider $\tilde{\mathbf{g}} = {}^\top(\tilde{g}_1(\mathbf{u}), \ldots, \tilde{g}_N(\mathbf{u}))$. Let $\tilde{g}_k(\mathbf{u})$ be functions given in (3.591). Since

$$\partial_t \tilde{g}_k(\mathbf{u}) = \sum_{j=1}^N (J(T)\partial_t \mathcal{A}_{kj}(t) + (\partial_t J(t))a_{kj}(t) + J(t)(\partial_t a_{kj}(t))u_j)$$

$$+ \sum_{j=1}^N (J(T)\mathcal{A}_{kj}(t) + J(t)a_{kj}(t))\partial_t u_j$$

and since $\|(J(T), a_{kj}(t), \mathcal{J}(t))\|_{L_\infty(\Omega)} \leq C$ as follows from (3.613), by (3.614) we have

$$\|\tilde{e}_T[< t >^a \partial_t g_k(\mathbf{u})]\|_{L_p(\mathbb{R}, L_q(\Omega))}$$

$$\leq \sum_{j=1}^N (\| < t >^{\frac{N}{2q_2}} \partial_t(\mathcal{A}_{kj}, \mathcal{J}, a_{kj})\|_{L_\infty((0,T), L_\infty(\Omega))}$$

$$\times \| < t >^{b - \frac{N}{2q_2}} \mathbf{u}\|_{L_p((0,T), L_{q_2}(\Omega))}$$

$$+ \| < t >^{\frac{N}{2q_2}} (\mathcal{A}_{kj}, \mathcal{J})\|_{L_\infty((0,T), L_\infty(\Omega))} \| < t >^{b - \frac{N}{2q_2}} \partial_t \mathbf{u}\|_{L_p((0,T), L_{q_2}(\Omega))}),$$

which, combined with (3.613), leads to

$$\|\tilde{e}_T[< t >^a \tilde{\mathbf{g}}(\mathbf{u})]\|_{L_p(\mathbb{R}, L_q(\Omega))} \leq C(\mathcal{I} + [\mathbf{u}]_T^2) \tag{3.624}$$

for any $a \in [0, b]$ and $q \in [1, q_2]$.

3.8.6 A Proof of Theorem 3.8.1

The strategy of proving Theorem 3.8.1 is to prolong local in time solutions to any time interval, which is the same idea as that in Sect. 3.7.5. Let T be a positive number >2. Let $\mathbf{u}$ and $\mathfrak{q}$ be solutions of Eq. (3.547), which satisfy the regularity condition (3.553) and the condition (3.545). Let $[\cdot]_T$ be the norm defined in (3.554). To prove Theorem 3.8.1 it suffices to prove that

$$[\mathbf{u}]_T \leq C(\mathcal{I} + [\mathbf{u}]_T^2) \tag{3.625}$$

for some constant $C > 0$, where

$$\mathcal{I} = \|\mathbf{u}_0\|_{B^{2(1-1/p)}_{q_2,p}(\Omega)} + \|\mathbf{u}_0\|_{B^{2(1-1/p)}_{q_1/2,p}(\Omega)}.$$

If we show (3.625), employing the same argument as that in Sect. 3.7.5 and using Theorem 3.8.15 concerning the almost global unique existence theorem, we can show that there exists a small constant $\epsilon > 0$ such that if $\mathcal{I} \leq \epsilon$ then $[\mathbf{u}]_T \leq C\epsilon$ for some constant $C > 0$ independent of ϵ, and so we can prolong $\mathbf{u}$ to any time interval beyond $(0, T)$. Thus, we have Theorem 3.8.1. In view of Theorem 3.8.15, there exists an $\epsilon_1 > 0$ such that if $\|\mathbf{u}_0\|_{B^{2(1-1/p)}_{q_2,p}(\Omega)} \leq \epsilon_1$, then $\mathbf{u}$ and $\mathfrak{q}$ mentioned above surely exist. We assume that $0 < \epsilon \leq \epsilon_1$. Thus, our task below is to prove (3.625). In the following, we use the results stated in Sect. 3.8.4 with $r = q_2$ and Sect. 3.8.5.

As was seen in Sect. 3.8.2, $\mathbf{u}$ and $\mathfrak{q}$ satisfy Eq. (3.593). To estimate $\mathbf{u}$, we divide $\mathbf{u}$ and $\mathfrak{q}$ into two parts as $\mathbf{u} = \mathbf{w} + \mathbf{v}$, and $\mathfrak{q} = \mathfrak{r} + \mathfrak{p}$, where $\mathbf{w}$ and $\mathfrak{r}$ are solutions of the equations:

$$\begin{cases} \partial_t \mathbf{w} + \lambda_0 \mathbf{w} - J(T)^{-1}\mathrm{Div}\,(J A \mu \tilde{\mathbf{D}}(\mathbf{w})) + \tilde{\nabla}\mathfrak{q} = \tilde{\mathbf{f}}(\mathbf{u}) & \text{in } \Omega^T, \\ \widetilde{\mathrm{div}}\,\mathbf{w} = \tilde{g}(\mathbf{u}) = \mathrm{div}\,\tilde{\mathbf{g}}(\mathbf{u}) & \text{in } \Omega^T, \\ J d_{\mathbf{n}}(\mu \tilde{\mathbf{D}}(\mathbf{w})\tilde{\mathbf{n}} - \mathfrak{q}\tilde{\mathbf{n}}) = \tilde{\mathbf{h}}(\mathbf{u}) & \text{on } \Gamma^T, \\ \mathbf{w}|_{t=0} = \mathbf{u}_0 & \text{in } \Omega, \end{cases} \tag{3.626}$$

and $\mathbf{v}$ and $\mathfrak{p}$ are solutions of the equations:

$$\begin{cases} \partial_t \mathbf{v} - J(T)^{-1}\mathrm{Div}\,(J A \mu \tilde{\mathbf{D}}(\mathbf{v})) + \tilde{\nabla}\mathfrak{r} = -\lambda_0 \mathbf{w} & \text{in } \Omega^T, \\ \widetilde{\mathrm{div}}\,\mathbf{v} = 0 & \text{in } \Omega^T, \\ \mu \tilde{\mathbf{D}}(\mathbf{v})\tilde{\mathbf{n}} - \mathfrak{r}\tilde{\mathbf{n}} = 0 & \text{on } \Gamma^T, \\ \mathbf{v}|_{t=0} = 0 & \text{in } \Omega. \end{cases} \tag{3.627}$$

Concerning the estimate of $\mathbf{w}$, applying (3.600) and using estimations (3.609), (3.620)–(3.624), we have

$$\| <t>^b \partial_t \mathbf{w}\|_{L_p((0,T),L_q(\Omega))} + \| <t>^b \mathbf{w}\|_{L_p((0,T),H^2_q(\Omega))} \leq C(\mathcal{I} + [\mathbf{u}]^2_T) \tag{3.628}$$

for $q = q_1/2, q_1, q_2$. By Sobolev's inequality, we have

$$\int_0^T (<s>^b \|\mathbf{w}(\cdot, s)\|_{H^1_\infty(\Omega)})^p \, ds \leq \int_0^T (<s>^b \|\mathbf{w}(\cdot, s)\|_{H^2_{q_2}(\Omega)})^p \, ds, \tag{3.629}$$

because $N < q_2 < \infty$. By (3.405),

$$\sup_{0<t<T} <s>^{\frac{N}{2q_1}} \|\mathbf{w}(\cdot,t)\|_{L_{q_1}(\Omega)} \le C(\|\mathbf{u}_0\|_{B^{2(1-1/q_1)}_{q_1,p}(\Omega)}$$

$$+ \| <t>^{\frac{N}{2q_1}} \partial_t \mathbf{w}\|_{L_p((0,T),L_{q_1}(\Omega))} + \| <t>^{\frac{N}{2q_1}} \mathbf{w}\|_{L_p((0,T),H^2_{q_1}(\Omega))}).$$

$$(3.630)$$

Since $q_1/2 < q_1 < q_2$, we have $\|\mathbf{u}_0\|_{B^{2(1-1/q_1)}_{q_1,p}(\Omega)} \le \mathcal{I}$. Thus, putting (3.628)–(3.630) together yields that

$$[\mathbf{w}]_T \le C(\mathcal{I} + [\mathbf{u}]^2_T). \tag{3.631}$$

We next consider $\mathbf{v}$. Let ψ be a solution of the weak Dirichlet problem

$$(\tilde{\nabla}\psi, J\tilde{\nabla}\varphi)_\Omega = (-\lambda_0 \mathbf{w}, J\tilde{\nabla}\varphi)_\Omega \quad \text{for any } \varphi \in \hat{H}^1_{q',0}(\Omega).$$

Let $\mathbf{Pw} = -\lambda_0 \mathbf{w} - \nabla\psi$. Since

$$\|\nabla\psi\|_{L_q(\Omega)} \le C_q \|\mathbf{w}\|_{L_q(\Omega)}$$

for any $q \in (1, q_2]$, we have

$$\|\mathbf{Pw}\|_{L_q(\Omega)} \le C_q \|\mathbf{w}\|_{L_q(\Omega)} \tag{3.632}$$

for any $q \in (1, q_2]$. Moreover, by (3.608), we have

$$\mathbf{v}(\cdot,t) = \int_0^t T(t-s)(\mathbf{Pw})(\cdot,s)\,ds. \tag{3.633}$$

Using the estimates (3.606) and (3.632) yields that

$$\|\nabla^j \mathbf{v}(\cdot,t)\|_{L_r(\Omega)} \le C_{r,\tilde{q}_1} \int_0^{t-1} (t-s)^{-\frac{j}{2}-\frac{N}{2}\left(\frac{1}{\tilde{q}_1}-\frac{1}{r}\right)} \|\mathbf{w}(\cdot,s)\|_{L_{\tilde{q}_1}(\Omega)}\,ds$$

$$+ C_{r,\tilde{q}_2} \int_{t-1}^t (t-s)^{-\frac{j}{2}-\frac{N}{2}\left(\frac{1}{\tilde{q}_2}-\frac{1}{r}\right)} \|\mathbf{w}(\cdot,s)\|_{L_{\tilde{q}_2}(\Omega)}\,ds$$

$$(3.634)$$

for $j = 0, 1$, for any $t > 1$ and for any indices r, $\tilde{q}_1$ and $\tilde{q}_2$ such that $1 < \tilde{q}_1, \tilde{q}_2 \le r \le \infty$ and $\tilde{q}_1, \tilde{q}_2 \le q_2$, where $\nabla^0 \mathbf{v} = \mathbf{v}$ and $\nabla^1 \mathbf{v} = \nabla \mathbf{v}$.

Recall that $T > 2$. In what follows, we prove that

$$\left(\int_2^T (<t>^b \|\mathbf{v}(\cdot,t)\|_{H^1_\infty(\Omega)})^p \, dt\right)^{1/p} \leq C(\mathcal{I} + [\mathbf{u}]_T^2), \tag{3.635}$$

$$\sup_{2 \leq t \leq T} (<t>^{\frac{N}{2q_1}} \|\mathbf{v}(\cdot,t)\|_{L_{q_1}(\Omega)}) \leq C(\mathcal{I} + [\mathbf{u}]_T^2), \tag{3.636}$$

$$\left(\int_2^T (<t>^{b-\frac{N}{2q_1}} \|\mathbf{v}(\cdot,t)\|_{H^1_{q_1}(\Omega)})^p \, dt\right)^{1/p} \leq C(\mathcal{I} + [\mathbf{u}]_T^2), \tag{3.637}$$

$$\left(\int_2^T (<t>^{b-\frac{N}{2q_2}} \|\mathbf{v}(\cdot,t)\|_{L_{q_2}(\Omega)})^p \, dt\right)^{1/p} \leq C(\mathcal{I} + [\mathbf{u}]_T^2). \tag{3.638}$$

By (3.634) with $r = \infty$, $\tilde{q}_1 = q_1/2$ and $\tilde{q}_2 = q_2$,

$$\|\mathbf{v}(\cdot,t)\|_{H^1_\infty(\Omega)} \leq C \int_0^t \|T(t-s)\mathbf{P}\mathbf{w}(\cdot,s)\|_{H^1_\infty(\Omega)} \, ds$$

$$= C(I_\infty(t) + II_\infty(t) + III_\infty(t))$$

with

$$I_\infty(t) = \int_0^{t/2} (t-s)^{-\frac{N}{q_1}} \|\mathbf{w}(\cdot,s)\|_{L_{q_1/2}(\Omega)} \, ds,$$

$$II_\infty(t) = \int_{t/2}^{t-1} (t-s)^{-\frac{N}{q_1}} \|\mathbf{w}(\cdot,s)\|_{L_{q_1/2}(\Omega)} \, ds,$$

$$III_\infty(t) = \int_{t-1}^t (t-s)^{-\frac{N}{2q_2}-\frac{1}{2}} \|\mathbf{w}(\cdot,s)\|_{L_{q_2}(\Omega)} \, ds.$$

Since

$$I_\infty(t)$$

$$\leq (t/2)^{-\frac{N}{q_1}} \left(\int_0^{t/2} <s>^{-bp'} ds\right)^{1/p'} \left(\int_0^{t/2} (<s>^b \|\mathbf{w}(\cdot,s)\|_{L_{q_1/2}(\Omega)})^p \, ds\right)^{1/p}$$

$$\leq C(bp'-1)^{-1/p'}(\mathcal{I} + [\mathbf{u}]_T^2) t^{-\frac{N}{q_1}}$$

as follows from the condition: $bp' > 1$ in (3.555), by the condition: $(\frac{N}{q_1} - b)p > 1$ in (3.555), we have

$$\int_2^T (<t>^b I_\infty(t))^p \, dt \leq C \int_2^T <t>^{-\left(\frac{N}{q_1}-b\right)p} \, dt (\mathcal{I} + [\mathbf{u}]_T^2)^p$$

$$\leq C((\frac{N}{q_1} - b)p - 1)^{-1}(\mathcal{I} + [\mathbf{u}]_T^2)^p.$$

By Hölder's inequality,

$$< t >^b II_\infty(t)$$

$$\leq C \int_{t/2}^{t-1} (t-s)^{-\frac{N}{q_1}} < s >^b \|\mathbf{w}(\cdot, s)\|_{L_{q_1/2}(\Omega)} ds$$

$$\leq C \Big(\int_{t/2}^{t-1} (t-s)^{-\frac{N}{q_1}} ds \Big)^{1/p'}$$

$$\times \Big(\int_{t/2}^{t-1} (t-s)^{-\frac{N}{q_1}} (< s >^b \|\mathbf{w}(\cdot, s)\|_{L_{q_1/2}(\Omega)})^p ds \Big)^{1/p}$$

$$\leq C \Big(\frac{N}{q_1} - 1 \Big)^{-1/p'} \Big(\int_{t/2}^{t-1} (t-s)^{-\frac{N}{q_1}} (< s >^b \|\mathbf{w}(\cdot, s)\|_{L_{q_1/2}(\Omega)})^p ds \Big)^{1/p},$$

because $N/q_1 = N/q_2 + 1 > 1$. By the Fubini–Tonelli theorem and (3.631),

$$\int_2^T (< t >^b II_\infty(t))^p dt$$

$$\leq C \Big(\frac{N}{q_1} - 1 \Big)^{-\frac{p}{p'}} \int_2^T dt \int_{t/2}^{t-1} (t-s)^{-\frac{N}{q_1}} (< s >^b \|\mathbf{w}(\cdot, s)\|_{L_{q_1/2}(\Omega)})^p ds$$

$$\leq C \Big(\frac{N}{q_1} - 1 \Big)^{-\frac{p}{p'}} \int_1^{T-1} (< s >^b \|\mathbf{w}(\cdot, s)\|_{L_{q_1/2}(\Omega)})^p ds \int_{s+1}^{2s} (t-s)^{-\frac{N}{q_1}} dt$$

$$\leq C \Big(\frac{N}{q_1} - 1 \Big)^{-p} (\mathcal{I} + [\mathbf{u}]_T^2)^p.$$

Since $\frac{N}{2q_2} + \frac{1}{2} < 1$ as follows from $q_2 > N$, by Hölder's inequality,

$$< t >^b III_\infty(t)$$

$$\leq C \int_{t-1}^t (t-s)^{-\frac{N}{2q_2} - \frac{1}{2}} < s >^b \|\mathbf{w}(\cdot, s)\|_{L_{q_2}(\Omega)} ds$$

$$\leq C \Big(\int_{t-1}^t (t-s)^{-\frac{N}{2q_2} - \frac{1}{2}} ds \Big)^{1/p'}$$

$$\times \Big(\int_{t-1}^t (t-s)^{-\frac{N}{2q_2} - \frac{1}{2}} (< s >^b \|\mathbf{w}(\cdot, s)\|_{L_{q_2}(\Omega)})^p ds \Big)^{1/p}$$

$$\leq C \Big(\frac{N}{2q_2} - \frac{1}{2} \Big)^{-1/p'} \Big(\int_{t-1}^t (t-s)^{-\frac{N}{2q_2} - \frac{1}{2}} (< s >^b \|\mathbf{w}(\cdot, s)\|_{L_{q_2}(\Omega)})^p ds \Big)^{1/p}.$$

By the Fubini–Tonelli theorem, we have

$$\int_2^T (< t >^b III_\infty(t))^p \, dt$$

$$\leq C\left(1 - \frac{N}{2q_2}\right)^{-\frac{p}{p'}}$$

$$\times \int_2^T dt \int_{t-1}^t (t-s)^{-\frac{N}{2q_2}-\frac{1}{2}} (< s >^b \|\mathbf{w}(\cdot,s)\|_{L_{q_2}(\Omega)})^p \, ds$$

$$\leq C\left(1 - \frac{N}{2q_2}\right)^{-\frac{p}{p'}}$$

$$\times \int_1^T (< s >^b \|\mathbf{w}(\cdot,s)\|_{L_{q_2}(\Omega)})^p \, ds \int_s^{s+1} (t-s)^{-\frac{N}{2q_2}-\frac{1}{2}} \, dt$$

$$= C\left(1 - \frac{N}{2q_2}\right)^{-p} (\mathcal{I} + [\mathbf{u}]_T^2)^p.$$

Summing up, we have obtained (3.635).

Next, we prove (3.636). By (3.634) with $r = q_1$, $\tilde{q}_1 = q_1/2$ and $\tilde{q}_2 = q_1$,

$$\|\mathbf{v}(\cdot,t)\|_{L_{q_1}(\Omega)} \leq C(I_{q_1,1}(t) + II_{q_1,1}(t) + III_{q_1,1}(t))$$

with

$$I_{q_1,1}(t) = \int_0^{t/2} (t-s)^{-\frac{N}{2q_1}} \|\mathbf{w}(\cdot,s)\|_{L_{q_1/2}(\Omega)} \, ds,$$

$$II_{q_1,1}(t) = \int_{t/2}^{t-1} (t-s)^{-\frac{N}{2q_1}} \|\mathbf{w}(\cdot,s)\|_{L_{q_1/2}(\Omega)} \, ds,$$

$$III_{q_1,1}(t) = \int_{t-1}^t \|\mathbf{w}(\cdot,s)\|_{L_{q_1}(\Omega)} \, ds.$$

By (3.631)

$$I_{q_1,1}(t) \leq (t/2)^{-\frac{N}{2q_1}} \left(\int_0^{t/2} < s >^{-bp'} ds\right)^{1/p'}$$

$$\times \left(\int_0^T (< s >^b \|\mathbf{w}(\cdot,s)\|_{L_{q_1/2}(\Omega)})^p \, ds\right)^{1/p}$$

$$\leq Ct^{-\frac{N}{2q_1}} (\mathcal{I} + [\mathbf{u}]_T^2).$$

Analogously, by Hölder's inequality and (3.631),

$$II_{q_1,1}(t) \le C \int_{t/2}^{t-1} (t-s)^{-\frac{N}{2q_1}} <s>^{-b} <s>^b \|\mathbf{w}(\cdot,s)\|_{L_{q_1/2}(\Omega)} \, ds$$

$$\le C <t>^{-b} \left(\int_{t/2}^{t-1} (t-s)^{-\frac{Np'}{2q_1}} \, ds \right)^{1/p'}$$

$$\times \left(\int_0^T (<s>^b \|\mathbf{w}(\cdot,s)\|_{L_{q_1/2}(\Omega)})^p \, ds \right)^{1/p}$$

$$= C \left(1 - \frac{Np'}{2q_1} \right)^{1/p'} <t>^{-b-\frac{N}{2q_1}+\frac{1}{p'}} (\mathcal{I} + [\mathbf{u}]_T^2)$$

$$\le C \left(1 - \frac{Np'}{2q_1} \right)^{1/p'} <t>^{-\frac{N}{2q_1}} (\mathcal{I} + [\mathbf{u}]_T^2),$$

because $b > \frac{1}{p'}$. Finally, by (3.631),

$$III_{q_1,1}(t) \le Ct^{-b} \int_{t-1}^t <s>^b \|\mathbf{w}(\cdot,s)\|_{L_{q_1/2}(\Omega)} \, ds$$

$$\le Ct^{-b} \left(\int_{t-1}^t \, ds \right)^{1/p'} \left(\int_0^T (<s>^b \|\mathbf{w}(\cdot,s)\|_{L_{q_1/2}(\Omega)})^p \, ds \right)^{1/p}$$

$$\le Ct^{-b} (\mathcal{I} + [\mathbf{u}]_T^2).$$

Summing up, we have obtained (3.636).

Next, we prove (3.637). By (3.634) with $r = q_1$, $\bar{q}_1 = q_1/2$ and $\bar{q}_2 = q_1$,

$$\|\mathbf{v}(\cdot,t)\|_{H_{q_1}^1(\Omega)} \le C(I_{q_1,2}(t) + II_{q_1,2}(t) + III_{q_1,2}(t))$$

with

$$I_{q_1,2}(t) = \int_0^{t/2} (t-s)^{-\frac{N}{2q_1}} \|\mathbf{w}(\cdot,s)\|_{L_{q_1/2}(\Omega)} \, ds,$$

$$II_{q_1,2}(t) = \int_{t/2}^{t-1} (t-s)^{-\frac{N}{2q_1}} \|\mathbf{w}(\cdot,s)\|_{L_{q_1/2}(\Omega)} \, ds,$$

$$III_{q_1,2}(t) = \int_{t-1}^t (t-s)^{-\frac{1}{2}} \|\mathbf{w}(\cdot,s)\|_{L_{q_1}(\Omega)} \, ds.$$

By (3.631),

$$I_{q_1,2}(t) \le (t/2)^{-\frac{N}{2q_1}} \left(\int_0^{t/2} <s>^{-bp'} ds \right)^{1/p'}$$

$$\times \left(\int_0^{t/2} (<s>^b \|\mathbf{w}(\cdot, s)\|_{L_{q_1/2}(\Omega)})^p ds \right)^{1/p}$$

$$\le Ct^{-\frac{N}{2q_1}} (\mathcal{I} + [\mathbf{u}]_T^2),$$

and so, by the condition: $(\frac{N}{q_1} - b)p > 1$ in (3.555)

$$\left(\int_2^T (<t>^{b-\frac{N}{2q_1}} I_{q_1,2}(t))^p dt \right)^{1/p} \le C \left((\frac{N}{q_1} - b)p - 1 \right)^{-1/p} (\mathcal{I} + [\mathbf{u}]_T^2).$$

By Hölder's inequality,

$$<t>^{b-\frac{N}{2q_1}} II_{q_1,2}(t)$$

$$\le C <t>^{-\frac{N}{2q_1}} \int_{t/2}^{t-1} (t-s)^{-\frac{N}{2q_1}} <s>^b \|\mathbf{w}(\cdot, s)\|_{L_{q_1/2}(\Omega)} ds$$

$$\le C <t>^{-\frac{N}{2q_1}} \left(\int_{t/2}^{t-1} (t-s)^{-\frac{Np'}{2q_1}} ds \right)^{1/p'}$$

$$\times \left(\int_0^T (<s>^b \|\mathbf{w}(\cdot, s)\|_{L_{q_1/2}(\Omega)})^p ds \right)^{1/p}$$

$$\le C(1+t)^{-\left(\frac{N}{q_1} - \frac{1}{p'}\right)} (\mathcal{I} + [\mathbf{u}]_T^2).$$

Since $(\frac{N}{q_1} - \frac{1}{p'})p > 1$ as follows from $\frac{N}{q_1} = 1 + \frac{N}{q_2} > 1 = \frac{1}{p} + \frac{1}{p'}$, we have

$$\left(\int_2^T (<t>^{b-\frac{N}{2q_1}} II_{q_1,2}(t))^p dt \right)^{1/p} \le C \left((\frac{N}{q_1} - b)p - 1 \right)^{-1/p} (\mathcal{I} + [\mathbf{u}]_T^2).$$

Since

$$<t>^{b-\frac{N}{2q_1}} III_{q_1,2}(t) \le \int_{t-1}^t (t-s)^{-\frac{1}{2}} <s>^{b-\frac{N}{2q_1}} \|\mathbf{w}(\cdot, s)\|_{L_{q_1}(\Omega)} ds$$

$$\le \left(\int_{t-1}^t (t-s)^{-\frac{1}{2}} ds \right)^{1/p'}$$

$$\times \left(\int_{t-1}^t (t-s)^{-\frac{1}{2}} (<s>^b \|\mathbf{w}(\cdot, s)\|_{L_{q_1}(\Omega)})^p ds \right)^{1/p},$$

by the Fubini–Tonelli theorem, we have

$$\int_2^T (<t>^{b-\frac{N}{2q_1}} III_{q_1,2}(t))^p \, dt$$

$$\leq 2^{\frac{p}{p'}} \int_2^T dt \int_{t-1}^t (t-s)^{-\frac{1}{2}} (<s>^b \|\mathbf{w}(\cdot,s)\|_{L_{q_1}(\Omega)})^p \, ds$$

$$\leq 2^{\frac{p}{p'}} \int_0^T (<s>^b \|\mathbf{w}(\cdot,s)\|_{L_{q_1}(\Omega)})^p \, ds$$

$$\times \int_s^{s+1} (t-s)^{-\frac{1}{2}} \, dt = 2^p \| <t>^b \mathbf{w}\|_{L_p((0,T),L_{q_1}(\Omega))},$$

which, combined with (3.628) with $q = q_1$, leads to

$$\left(\int_2^T (<t>^{b-\frac{N}{2q_1}} III_{q_1,2}(t))^p \, dt \right)^{1/p} \leq C(\mathcal{I} + [\mathbf{u}]_T^2).$$

Summing up, we have obtained (3.637).

Finally, we prove (3.638). By (3.634) with $r = q_2$, $\tilde{q}_1 = q_1/2$ and $\tilde{q}_2 = q_2$,

$$\|\mathbf{v}(\cdot,t)\|_{L_{q_2}(\Omega)} \leq C(I_{q_2}(t) + II_{q_2}(t) + III_{q_2}(t))$$

with

$$I_{q_2}(t) = \int_0^{t/2} (t-s)^{-\frac{N}{2}\left(\frac{2}{q_1}-\frac{1}{q_2}\right)} \|\mathbf{w}(\cdot,s)\|_{L_{q_1/2}(\Omega)} \, ds,$$

$$II_{q_2}(t) = \int_{t/2}^{t-1} (t-s)^{-\frac{N}{2}\left(\frac{2}{q_1}-\frac{1}{q_2}\right)} \|\mathbf{w}(\cdot,s)\|_{L_{q_1/2}(\Omega)} \, ds,$$

$$III_{q_2}(t) = \int_{t-1}^t \|\mathbf{w}(\cdot,s)\|_{L_{q_2}(\Omega)} \, ds.$$

By Hölder's inequality,

$$I_{q_2}(t) \leq (t/2)^{-\frac{N}{2}\left(\frac{2}{q_1}-\frac{1}{q_2}\right)} \left(\int_0^{t/2} <s>^{-bp'} \, ds \right)^{1/p'}$$

$$\times \left(\int_0^{t/2} (<s>^b \|\mathbf{w}(\cdot,s)\|_{L_{q_1/2}(\Omega)})^p \, ds \right)^{1/p}$$

$$\leq C <t>^{-\frac{N}{2}\left(\frac{2}{q_1}-\frac{1}{q_2}\right)} (\mathcal{I} + [\mathbf{u}]_T^2)$$

for $t \geq 2$. Since

$$\frac{N}{2}\left(\frac{2}{q_1} - \frac{1}{q_2}\right) - \left(b - \frac{N}{2q_2}\right) = \frac{N}{q_1} - b,$$

by the condition: $(\frac{N}{q_1} - b)p > 1$ in (3.555),

$$\left(\int_2^T (<t>^{b-\frac{N}{2q_2}} I_{q_2}(t))^p \, dt\right)^{1/p}$$

$$\leq C\left(\int_2^T t^{-\left(\frac{N}{q_1}-b\right)p} \, dt\right)^{1/p} (\mathcal{I} + [\mathbf{u}]_T^2)$$

$$\leq C\left((\frac{N}{q_1} - b)p - 1\right)^{-1/p} (\mathcal{I} + [\mathbf{u}]_T^2).$$

Since

$$\frac{N}{2}\left(\frac{2}{q_1} - \frac{1}{q_2}\right) = \frac{N}{2}\left(\frac{1}{q_2} + \frac{2}{N}\right) = \frac{N}{2q_2} + 1 > 1,$$

by Hölder's inequality

$$<t>^{b-\frac{N}{2q_2}} II_{q_2}(t)$$

$$\leq C \int_{t/2}^{t-1} (t-s)^{-\left(\frac{N}{2q_2}+1\right)} <s>^{b-\frac{N}{2q_2}} \|\mathbf{w}(\cdot, s)\|_{L_{q_1/2}(\Omega)} \, ds$$

$$\leq C\left(\int_{t/2}^{t-1} (t-s)^{-\left(\frac{N}{2q_2}+1\right)} \, ds\right)^{1/p'}$$

$$\times \left(\int_{t/2}^{t-1} (t-s)^{-\left(\frac{N}{2q_2}+1\right)} (<s>^b \|\mathbf{w}(\cdot, s)\|_{L_{q_1/2}(\Omega)})^p \, ds\right)^{1/p}$$

$$\leq C\left(\frac{N}{2q_2}\right)^{-1/p'} \left(\int_{t/2}^{t-1} (t-s)^{-\left(\frac{N}{2q_2}+1\right)} (<s>^b \|\mathbf{w}(\cdot, s)\|_{L_{q_1/2}(\Omega)})^p \, ds\right)^{1/p},$$

and so, by the Fubini–Tonelli theorem and (3.631)

$$\int_2^T (<t>^{b-\frac{N}{2q_2}} II_{q_2}(t))^p \, dt$$

$$\leq C\left(\frac{N}{2q_2}\right)^{-p/p'} \int_2^T dt \int_{t/2}^{t-1} (t-s)^{-\left(\frac{N}{2q_2}+1\right)} (<s>^b \|\mathbf{w}(\cdot, s)\|_{L_{q_1/2}(\Omega)})^p \, ds$$

$$\leq C\Big(\frac{N}{2q_2}\Big)^{-p/p'} \int_0^T (<s>^b \|\mathbf{w}(\cdot,s)\|_{L_{q_1/2}(\Omega)})^p \, ds$$

$$\times \int_{s+1}^{2s} (t-s)^{-\left(\frac{N}{2q_2}+1\right)} dt \leq C\Big(\frac{N}{2q_2}\Big)^{-p} (\mathcal{I}+[\mathbf{u}]_T^2)^p.$$

Analogously, by Hölder's inequality

$$<t>^{b-\frac{N}{2q_2}} III_{q_2}(t) \leq C \int_{t-1}^t <s>^{b-\frac{N}{2q_2}} \|\mathbf{w}(\cdot,s)\|_{L_{q_2}(\Omega)} \, ds,$$

$$\leq C\Big(\int_{t-1}^t ds\Big)^{1/p'} \Big(\int_{t-1}^t (<s>^b \|\mathbf{w}(\cdot,s)\|_{L_{q_2}(\Omega)})^p \, ds\Big)^{1/p}$$

$$= C\Big(\int_{t-1}^t (<s>^b \|\mathbf{w}(\cdot,s)\|_{L_{q_2}(\Omega)})^p \, ds\Big)^{1/p},$$

and so, by the Fubini–Tonelli theorem and (3.631)

$$\int_2^T (<t>^{b-\frac{N}{2q_2}} III_{q_2}(t))^p \, dt \leq C \int_2^T dt \int_{t-1}^t (<s>^b \|\mathbf{w}(\cdot,s)\|_{L_{q_2}(\Omega)})^p \, ds$$

$$\leq C \int_0^T (<s>^b \|\mathbf{w}(\cdot,s)\|_{L_{q_2}(\Omega)})^p \, ds \int_s^{s+1} dt \leq C(\mathcal{I}+[\mathbf{u}]_T^2)^p.$$

Summing up, we have obtained (3.638).

We next consider the case where $\iota \in (0,2)$. In view of Theorem 3.8.16, the usual maximal L_p–L_q theorem holds for Eq. (3.627), and so we have

$$\|\mathbf{v}\|_{L_p((0,2),H_q^2(\Omega))} + \|\partial_t \mathbf{v}\|_{L_p((0,2),L_q(\Omega))}$$

$$\leq C_q \|\lambda_0 \mathbf{w}\|_{L_p((0,2),L_q(\Omega))} \leq C(\mathcal{I}+[\mathbf{u}]_T^2) \tag{3.639}$$

for any $q \in [q_1/2, q_2]$. Since $2(1-1/p) > N/q_2+1$ as follows from $2/p+N/q_2 < 1$, by (3.405), Sobolev's inequality, and (3.639), we have

$$\Big(\int_0^2 \|\mathbf{v}(\cdot,t)\|_{H_\infty^1(\Omega)} \, dt\Big)^{1/p} \leq C \sup_{0<t<2} \|\mathbf{v}(\cdot,t)\|_{B_{q_2,p}^{2(1-1/p)}(\Omega)}$$

$$\leq C(\|\mathbf{u}_0\|_{B_{q_2,p}^{2(1-1/p)}(\Omega)} + \|\mathbf{v}\|_{L_p((0,2),H_{q_2}^2(\Omega))} + \|\partial_t \mathbf{v}\|_{L_p((0,2),L_{q_2}(\Omega))})$$

$$\leq C(\mathcal{I}+[\mathbf{u}]_T^2). \tag{3.640}$$

Moreover, by (3.405) and (3.639)

$$\sup_{0<t<2} \|\mathbf{v}(\cdot,t)\|_{L_{q_1}(\Omega)}$$

$$\leq C(\|\mathbf{u}_0\|_{B^{2(1-1/p)}_{q_1,p}(\Omega)} + \|\mathbf{v}\|_{L_p((0,2),H^2_{q_1}(\Omega))} + \|\partial_t\mathbf{v}\|_{L_p((0,2),L_{q_1}(\Omega))})$$

$$\leq C(\mathcal{I} + [\mathbf{u}]^2_T).$$

$$(3.641)$$

Combining (3.635)–(3.641), we have

$$\| < t >^b \mathbf{v}\|_{L_p((0,T),H^1_\infty(\Omega))} + \| < t >^{\frac{N}{2q_1}} \mathbf{v}\|_{L_\infty((0,T),L_{q_1}(\Omega))}$$

$$+ \| < t >^{b-\frac{N}{2q_1}} \mathbf{v}\|_{L_p((0,T),H^1_{q_1}(\Omega))} + \| < t >^{b-\frac{N}{2q_2}} \mathbf{v}\|_{L_p((0,T),L_{q_2}(\Omega))}$$

$$\leq C(\mathcal{I} + [\mathbf{u}]^2_T).$$

$$(3.642)$$

From (3.627), $\mathbf{v}$ satisfies the equations:

$$\partial_t\mathbf{v} + \lambda_0\mathbf{v} - J(T)^{-1}\,\mathrm{Div}\,(J\mathbf{A}\mu\tilde{\mathbf{D}}(\mathbf{v})) + \tilde{\nabla}\mathfrak{r} = -\lambda_0\mathbf{w} + \lambda_0\mathbf{v} \quad \text{in } \Omega^T,$$

$$\widetilde{\mathrm{div}}\,\mathbf{v} = 0 \quad \text{in } \Omega^T,$$

$$\mu\tilde{\mathbf{D}}(\mathbf{v})\tilde{\mathbf{n}} - \mathfrak{r}\tilde{\mathbf{n}} = 0 \quad \text{on } \Gamma^T,$$

$$\mathbf{v}|_{t=0} = 0 \quad \text{in } \Omega,$$

and so by (3.600) we have

$$\| < t >^{b-\frac{N}{2q_2}} \mathbf{v}\|_{L_p((0,T),H^2_{q_2}(\Omega))} + \| < t >^{b-\frac{N}{2q_2}} \partial_t\mathbf{v}\|_{L_p((0,T),L_{q_2}(\Omega))}$$

$$\leq C\| < t >^{b-\frac{N}{2q_2}} (\mathbf{v},\mathbf{w})\|_{L_p((0,T),L_{q_2}(\Omega))},$$

which, combined with (3.642), leads to

$$[\mathbf{v}]_T \leq C(\mathcal{I} + [\mathbf{u}]^2_T).$$

$$(3.643)$$

Since $\mathbf{u} = \mathbf{w} + \mathbf{v}$, by (3.631) and (3.643), we see that $\mathbf{u}$ satisfies the inequality (3.625), which completes the proof of Theorem 3.8.1.

Acknowledgements This research is partially supported by JSPS Grant-in-aid for Scientific Research (A) 17H01097, Toyota Central Research Institute Joint Research Fund, and Top Global University Project. Adjunct faculty member in the Department of Mechanical Engineering and Materials Science, University of Pittsburgh.

References

1. H. Abels, The initial-value problem for the Navier–Stokes equations with a free surface in L^q-Sobolev spaces. Adv. Differential Equ. **10**, 45–64 (2005)
2. H. Abels, On general solutions of two-phase flows for viscous incompressible fluids. Interfaces Free Boud. **9**, 31–65 (2007)
3. G. Allain, Small-time existence for Navier–Stokes equations with a free surface. Appl. Math. Optim. **16**, 37–50 (1987)
4. H. Amann, M. Hieber, G. Simonett, Bounded H^∞-calculus for elliptic operators. Differ. Integral Eq. **7**, 613–653 (1994)
5. J.T. Beale, The initial value problem for the Navier–Stokes equations with a free surface. Commun. Pure Appl. Math. **34**, 359–392 (1980)
6. J.T. Beale, Large-time regularity of viscous surface waves. Arch. Ration. Mech. Anal. **84**, 307–352 (1984)
7. J.T. Beale, T. Nishida, Large-time behaviour of viscous surface waves. Lecuter Notes Num. Appl. Anal. **8**, 1–14 (1985)
8. J. Bergh, J. Löfström, *Interpolation Spaces, An Introduction*. Grundlehren der mathematischen Wissenschaften 223, A Series of Comprehensive Studies in Mathematics (Springer, New York, 1976)
9. M.E. Bogovskiĭ, Solution of the first boundary value problem for the equation of continuity of an incompressible medium. Dokl. Acad. Nauk SSSR. **248**, 1037–1049 (1976); English transl: Soviet Math. Dokl. **20**, 1094–1098 (1976)
10. M.E. Bogovskiĭ, Solution of some vector analysis problems connected with operators div and grad (in Russian), in *Trudy Seminar S. L. Sobolev*, vol. 80 (Akademia Nauk SSR, Sibirskoe Otdelenie Matematik, Nowosibirsk , 1980), pp. 5–40
11. J. Bourgain, Vector-valued singular integrals and the H^1-BMO duality, in *Probability Theory and Harmonic Analysis*, ed. by D. Borkholder (Marcel Dekker, New York, 1997), pp. 1–19
12. I.V. Denisova, A priori estimates for the solution of a linear time-dependent problem connected with the motion of a drop in a fluid medium. Trudy Mat. Inst. Steklov. **188**, 3–21 (1990) (in Russian); English transl.: Proc. Steklov Inst. Math. **188**, 1–24 (1991)
13. I.V. Denisova, Problem of the motion of two viscous incompressible fluids separated by a closed free interface. Acta Appl. Math. **37**, 31–40 (1994)
14. I.V. Denisova, V.A. Solonnikov, Solvability in Hölder spaces of a model initial-boundary value problem generated by a problem on the motion of two fluids. Zap. Nauchn. Sem. Leningrad. Otdel. Mat. Inst. Steklov. (LOMI) **181**, 5–44 (1991) (in Russian); English transl.: J. Math. Sci. **70**, 1717–1746 (1994)
15. I.V. Denisova, V.A. Solonnikov, Classical solvability of the problem on the motion of two viscous incompressible fluids. Algebra i Analiz **7**, 101–142 (1995) (in Russian); English transl.: St.Petersburg Math. J. **7**, 755–786 (1996)
16. R. Denk, R. Schnaubelt, A structurally damped plate equations with Dirichlet-Neumann boundary conditions. J. Differ. Equ. **259**(4), 1323–1353 (2015)
17. R. Denk, M. Hieber, J. Prüß, $\mathcal{R}$-*Boundedness, Fourier multipliers and problems of elliptic and parabolic type*, vol. 166, no. 788 (Memoirs of AMS, Providence, 2003)
18. Y. Enomoto, Y. Shibata, On the $\mathcal{R}$-sectoriality and its application to some mathematical study of the viscous compressible fluids. Funk. Ekvaj. **56**, 441–505 (2013)
19. G.P. Galdi, *An Introduction to the Mathematical Theory of the Navier–Stokes Equations, Steady Problems*. Springer Monographs in Mathematics, 2nd edn (Springer, Berlin, 2011), ISBN 978-0-387-09620-9 (eBook). https://doi.org/10.1007/978-0-387-09620-9. Springer, New York
20. Y. Giga, Sh. Takahashi, On global weak solutions of the nonstationary two-phase Stokes flow. SIAM J. Math. Anal. **25**, 876–893 (1994)
21. G. Grubb, V.A. Solonnikov, Boundary value problems for the nonstationary Navier–Stokes equations treated by pseudo-differential methods. Math. Scand. **69**, 217–290 (1991)
22. E. Hanzawa, Classical solutions of the Stefan problem. Tohoku Math. J. **33**, 297–335 (1981)

23. Y. Hataya, Decaying soluiton of a Navier–Stokes flow without surface tension, J. Math. Kyoto Univ. **49**, 691–717 (2009)
24. Y. Hataya, A remark on Beale-Nishida's paper. Bull. Inst. Math. Acad. Sin. (N.S.) **6**(3), 293–303 (2011)
25. Y. Hataya, S. Kawashima, Decaying solution of the Navier–Stokes flow of infinite volume without surface tension. Nonlinear Anal. **71**(12), 2535–2539 (2009)
26. M. Köhne, J. Prüss, M. Wilke, Qualitative behavior of solutions for the two-phase Navier–Stokes equations with surface tension. Math. Ann. **356**, 737–792 (2013)
27. D. Lynn, G. Sylvester, Large time existence of small viscous surface waves without surface tension. Comm. Part. Differ. Eqns. **15**, 823–903 (1990)
28. I.Sh. Mogilevskiĭ, V.A. Solonnikov, Solvability of a noncoercive initial boundary-value problem for the Stokes system in Hölder classes of functions. Z Anal. Anwend. **8**(4), 329–347 (1989)
29. I.Sh. Mogilevskiĭ, V.A. Solonnikov, On the solvability of an evolution free boundary problem for the Navier–Stokes equations in the Hölder spaces of functions, in *Mathematical Problems Relating to the Navier–Stokes Equations*, ed. by G.P. Galdi. Series on Advances in Mathematics for Applied Sciences, vol. 11 (World Scientific, Singapore, 1992), pp. 105–181
30. P.B. Mucha, W. Zajączkowski, On local existence of solutions of the free boundary problem for an incompressible viscous self-gravitating fluid motion. Applicationes Mathematicae **27**, 319–333 (2000)
31. U. Neri, *Singular Integrals*. Lecture Notes in Mathematics, vol. 200 (Springer, Berlin, 1971)
32. A. Nouri, F. Poupaud, An existence theorem for the multifluid Navier–Stokes problem. J. Differ. Equ. **123**, 71–88 (1995)
33. M. Padula, V.A. Solonnikov, On the global existence of nonsteady motions of a fluid drop and their exponential decay to a uniform rigid rotation. Quad. Mat. **10**, 185–218 (2002)
34. M. Padula, V.A. Solonnikov, On local solvability of the free boundary problem for the Navier–Stokes equations. Problemy Mat. Analiza **50**, 87–112 (2019); English trans. J. Math. Sci., **170**(4), 522–553 (2010)
35. J. Prüss, G. Simonett, On the two-phase Navier–Stokes equations with surface tension. Interfaces Free Bound. **12**, 311–345 (2010)
36. J. Prüss, G. Simonett, Analytic solutions for the two-phase Navier–Stokes equations with surface tension and gravity. Progr. Nonlinear Differ. Equ. Appl. **80**, 507–540 (2011)
37. J. Prüss, G. Simonett, *Moving Interfaces and Quasilinear Parabolic Evolution Equations*. Birkhauser Monographs in Mathematics, (Springer, Berlin, 2016), ISBN: 978-3-319-27698-4
38. H. Saito, Global solvability of the Navier–Stokes equations with a free surface in the maxial $L_p - L_q$ regularity class. J. Differ. Equ. **264**(3), 1475–1520 (2018)
39. H. Saito, Y. Shibata, On decay properties of solutions to the Stokes equations with surface tension and gravity in the half space. J. Math. Soc. Japan **68**(4), 1559–1614 (2016)
40. H. Saito, Y. Shibata, On the global wellposedness of free boundary problem for the Navier Stokes systems with surface tension, Preprint arXiv:1912.10121 [math.AP]
41. B. Schweizer, Free boundary fluid systems in a semigroup approach and oscillatory behavior. SIAM J. Math. Anal. **28**, 1135–1157 (1997)
42. Y. Shibata, Generalized resolvent estimates of the Stokes equations with first order boundary condition in a general domain. J. Math. fluid Mech., **15**(1), 1–40 (2013)
43. Y. Shibata, On the $\mathcal{R}$-boundedness of solution operators for the Stokes equations with free boundary condition. Differ. Int. Eqns. **27**(3–4), 313–368 (2014)
44. Y. Shibata, Local well-posedness of free surface problems for the Navier–Stokes equations in a general domain. Discret. Contin. Dyn. Sys. Series S **9**(1), 315–342 (2016)
45. Y. Shibata, On the $\mathcal{R}$-bounded solution operators in the study of free boundary problem for the Navier–Stokes equations, in Y. Suzuki. *Springer Proceedings in Mathematics & Statistics*, ed. by ed. Y. Shibata, vol. 183 (Mathematical Fluid Dynamics, Present and Future, Tokyo, 2016), pp.203–285

46. Y. Shibata, Global wellposedness of a free boundary problem for the Navier–Stokes equations in an exterior domain. Fluid Mech. Res. Int. **1**(2), (2017). https://doi.org/10.15406/fimrij.2017.01.00008

47. Y. Shibata, Global well-posedness of unsteady motion of viscous incompressible capillary liquid bounded by a free surface. Evol. Equ. Control. The. **7**(1), 117–152 (2018). https://doi.org/10.3934/eect.2018007

48. Y. Shibata, Local wellposedness for the free boundary problem of the Navier–Stokes equations in an exterior domain . Commun. Pure Appl. Anal. **17**(4), 1681–1721 (2018). https://doi.org/10.3934/cpaa.2018081

49. Y. Shibata, On L_p–L_q decay estimate for Stokes equations with free boudary condition in an exterior domain. Asymptotic Anal. **107**(1–2), 33–72 (2018). https://doi.org/10.3233/ASY-171449

50. Y. Shibata, S. Shimizu, On a resolvent estimate for the Stokes system with Neumann boundary condition. Differ. Int. Eqns. **16**(4), 385–426 (2003)

51. Y. Shibata, S. Shimizu, On a resolvent estimate of the interface problem for the Stokes system in a bounded domain. J. Differ. Equ. **191**, 408–444 (2003)

52. Y. Shibata, S. Shimizu, On a free boundary problem for the Navier–Stokes equations. Differ. Int. Eqns. **20**, 241–276 (2007)

53. Y. Shibata, S. Shimizu, On the L_p–L_q maximal regularity of the Neumann problem for the Stokes equations in a bounded domain. J. Reine Angew. Math. **615**, 157–209 (2008)

54. Y. Shibata, S. Shimizu, On the maximal L_p–L_q regularity of the Stokes problem with first order boundary condition; model problems. J. Math. Soc. Japan **64**(2), 561–626 (2012)

55. Y. Shibata, Suma'Inna, On the maximal L_p–L_q theory arising in the study of a free boundary problem for the Navier–Stokes equations, FMRIJ-18-eBook-220, 2018

56. S. Shimizu, Maximal regularity and viscous incompressible flows with free interface, Parabolic and Navier–Stokes equations, Banach Center Publ. **81** (2008), 471–480.

57. S. Shimizu, Local solvability of free boundary problems for two-phase Navier–Stokes equations with surface tension in the whole space, in *Parabolic Problems*. Progress in Nonlinear Differential Equations and Their Applications, vol. 80 (Birkhäuser/Springer Basel AG, Basel, 2011), pp. 647–686

58. G. Simonett, M. Wilke, Stability of equilibrium shapes in some free boundary problems involving fluids, in *Handbook of Mathematical Analysis in Mechanics of Viscous Fluids*, ed. by Y. Giga, A. Novtný, chap 25 (Springer International Publishing AG, Berlin, 2018), pp. 1221–1266 http://doi.org/10.1007/978-3-319-13344-7_27

59. V.A. Solonnikov, Solvability of the evolution problem for an isolated mass of a viscous incompressible capillary liquid. Zap. Nauchn. (LOMI) **140**, 179–186 (1984) (in Russian); English transl.: J. Soviet Math. **32**, 223–238 (1986)

60. V.A. Solonnikov, Unsteady motion of a finite mass of fluid, bounded by a free surface. Zap. Nauchn. Sem. (LOMI) **152**, 137–157 (1986) (in Russian); English transl.: J. Soviet Math. **40**, 672–686 (1988)

61. V.A. Solonnikov, On the transient motion of an isolated volume of viscous incompressible fluid. Izv. Acad. Nauk SSSR. **51**, 1065–1087 (1987) (in Russian); English transl.: Math. USSR Izv. **31**, 381–405 (1988)

62. V.A. Solonnikov, On nonstationary motion of a finite isolated mass of self-gravitating fluid. Algebra i Analiz **1**, 207–249 (1989) (in Russian); English transl.: Leningrad Math. J. **1**, 227–276 (1990)

63. V.A. Solonnikov, On an initial-boundary value problem for the Stokes systems arising in the study of a problem with a free boundary. Trudy Mat. Inst. Steklov **188**, 150–188 (1990) (in Russian); English transl.: Proc. Steklov Inst. Math. **3**, 191–239 (1991)

64. V.A. Solonnikov, Solvability of the problem of evolution of a viscous incompressible fluid bounded by a free surface on a finite time interval. Algebra i Analiz **3**, 222–257 (1991) (in Russian); English transl.: St. Petersburg Math. J. **3**, 189–220 (1992)

65. V.A. Solonnikov, in *Lectures on evolution free boundary problems: classical solutions*, ed. by L. Ambrosio, P. Colli, J.F. Rodrigues. Lecture Notes in Mathematics (LNM), vol. 1812 (Springer, Berlin, 2003), pp.123–175

66. V.A. Solonnikov, On the linear problem arising in the study of a free boundary proiblem for the Navier–Stokes equations. St. Petersburg Math. J. **22**, 1023–1049 (2011)
67. V.A. Solonnikov, I.V. Denisova, Classical well-posedness of free boundary problems in viscous incompressible fluid mechanics, in ed. by Y. Giga, A. Novtný. *Handbook of Mathematical Analysis in Mechanics of Viscous Fluids*, chap 24 (Springer International Publishing AG, Berlin, 2018), pp.1135–1220. http://doi.org/10.1007/978-3-319-13344-7_27
68. E.M. Stein, *Singular Integrals and Differentiability Properties of Functions* (Princeton University Press, Princeton, 1970)
69. Sh. Takahashi, On global weak solutions of the nonstationary two-phase Navier–Stokes flow. Adv. Math. Sci. Appl. **5**, 321–342 (1995)
70. N. Tanaka, Two-phase free boundary problem for viscous incompressible thermo-capillary convection. Jpn. J. Math. **21**, 1–42 (1995)
71. A. Tani, Small-time existence for the three-dimensional Navier–Stokes equations for an incompressible fluid with a free surface. Arch. Ration. Mech. Anal. **133**, 299–331 (1996)
72. A. Tani, N. Tanaka, Large-time existence of surface waves in incompressible viscous fluids with or without surface tension. Arch. Ration. Mech. Anal. **130**, 303–314 (1995)
73. H. Tanabe, *Functional Analytic Methods for Partial Differential Equations*. Pure and Applied Mathematics: A Series of Monographs and Text Books. (Dekker, New York, 1997), ISBN 0-8247-9774-4
74. L. Weis, Operator-valued Fourier multiplier theorems and maximal L_p-regularity. Math. Ann. **319**, 735–758 (2001)

LECTURE NOTES IN MATHEMATICS  Springer

Editors in Chief: J.-M. Morel, B. Teissier;

Editorial Policy

1. Lecture Notes aim to report new developments in all areas of mathematics and their applications – quickly, informally and at a high level. Mathematical texts analysing new developments in modelling and numerical simulation are welcome.

   Manuscripts should be reasonably self-contained and rounded off. Thus they may, and often will, present not only results of the author but also related work by other people. They may be based on specialised lecture courses. Furthermore, the manuscripts should provide sufficient motivation, examples and applications. This clearly distinguishes Lecture Notes from journal articles or technical reports which normally are very concise. Articles intended for a journal but too long to be accepted by most journals, usually do not have this "lecture notes" character. For similar reasons it is unusual for doctoral theses to be accepted for the Lecture Notes series, though habilitation theses may be appropriate.

2. Besides monographs, multi-author manuscripts resulting from SUMMER SCHOOLS or similar INTENSIVE COURSES are welcome, provided their objective was held to present an active mathematical topic to an audience at the beginning or intermediate graduate level (a list of participants should be provided).

   The resulting manuscript should not be just a collection of course notes, but should require advance planning and coordination among the main lecturers. The subject matter should dictate the structure of the book. This structure should be motivated and explained in a scientific introduction, and the notation, references, index and formulation of results should be, if possible, unified by the editors. Each contribution should have an abstract and an introduction referring to the other contributions. In other words, more preparatory work must go into a multi-authored volume than simply assembling a disparate collection of papers, communicated at the event.

3. Manuscripts should be submitted either online at www.editorialmanager.com/lnm to Springer's mathematics editorial in Heidelberg, or electronically to one of the series editors. Authors should be aware that incomplete or insufficiently close-to-final manuscripts almost always result in longer refereeing times and nevertheless unclear referees' recommendations, making further refereeing of a final draft necessary. The strict minimum amount of material that will be considered should include a detailed outline describing the planned contents of each chapter, a bibliography and several sample chapters. Parallel submission of a manuscript to another publisher while under consideration for LNM is not acceptable and can lead to rejection.

4. In general, **monographs** will be sent out to at least 2 external referees for evaluation.

   A final decision to publish can be made only on the basis of the complete manuscript, however a refereeing process leading to a preliminary decision can be based on a pre-final or incomplete manuscript.

   Volume Editors of **multi-author works** are expected to arrange for the refereeing, to the usual scientific standards, of the individual contributions. If the resulting reports can be forwarded to the LNM Editorial Board, this is very helpful. If no reports are forwarded or if other questions remain unclear in respect of homogeneity etc, the series editors may wish to consult external referees for an overall evaluation of the volume.

5. Manuscripts should in general be submitted in English. Final manuscripts should contain at least 100 pages of mathematical text and should always include

   - a table of contents;
   - an informative introduction, with adequate motivation and perhaps some historical remarks: it should be accessible to a reader not intimately familiar with the topic treated;
   - a subject index: as a rule this is genuinely helpful for the reader.
   - For evaluation purposes, manuscripts should be submitted as pdf files.

6. Careful preparation of the manuscripts will help keep production time short besides ensuring satisfactory appearance of the finished book in print and online. After acceptance of the manuscript authors will be asked to prepare the final LaTeX source files (see LaTeX templates online: https://www.springer.com/gb/authors-editors/book-authors-editors/manuscriptpreparation/5636) plus the corresponding pdf- or zipped ps-file. The LaTeX source files are essential for producing the full-text online version of the book, see http://link.springer.com/bookseries/304 for the existing online volumes of LNM). The technical production of a Lecture Notes volume takes approximately 12 weeks. Additional instructions, if necessary, are available on request from lnm@springer.com.

7. Authors receive a total of 30 free copies of their volume and free access to their book on SpringerLink, but no royalties. They are entitled to a discount of 33.3 % on the price of Springer books purchased for their personal use, if ordering directly from Springer.

8. Commitment to publish is made by a *Publishing Agreement*; contributing authors of multiauthor books are requested to sign a *Consent to Publish form*. Springer-Verlag registers the copyright for each volume. Authors are free to reuse material contained in their LNM volumes in later publications: a brief written (or e-mail) request for formal permission is sufficient.

Addresses:
Professor Jean-Michel Morel, CMLA, École Normale Supérieure de Cachan, France
E-mail: moreljeanmichel@gmail.com

Professor Bernard Teissier, Equipe Géométrie et Dynamique,
Institut de Mathématiques de Jussieu – Paris Rive Gauche, Paris, France
E-mail: bernard.teissier@imj-prg.fr

Springer: Ute McCrory, Mathematics, Heidelberg, Germany,
E-mail: lnm@springer.com

Printed in the United States
By Bookmasters